D0820626

Geochemical Investigations in Earth and Space Science: A Tribute to Isaac R. Kaplan

SPECIAL PUBLICATION SERIES

No. 1. *Magmatic Processes: Physicochemical Principles.* B. O. Mysen (Editor), 1987.

No. 2. *Fluid-Mineral Interactions: A Tribute to H. P. Eugster.* R. J. Spencer and I-Ming Chou (Editors), 1990.

No. 3. *Stable Isotope Geochemistry: A Tribute to Samuel Epstein.* H. P. Taylor, Jr., J. R. O'Neil and I. R. Kaplan (Editors), 1991.

No. 4. *Victor Moritz Goldschmidt: Father of Modern Geochemistry.* Brian Mason, 1992.

No. 5. *Mineral Spectroscopy: A Tribute to Roger G. Burns.* M. D. Dyar, C. McCammon and M. W. Schaefer (Editors), 1996.

No. 6. *Mantle Petrology: Field Observations and High-Pressure Experimentation, A Tribute to Francis R. (Joe) Boyd.* Y. Frei, C. M. Bertka and B. O. Mysen (Editors), 1999.

No. 7. *Water–Rock Interactions, Ore Deposits, and Environmental Geochemistry: A Tribute to David A. Crerar.* R. Hellman and S. A. Wood (Editors), 2002.

No. 8. SEG Special Volume 10 (also Geochemical Society Special Publication No. 8). *Volcanic, Geothermal and Ore-Forming Fluids: Rulers and Witnesses of Processes within the Earth.* S. F. Simmons and I. Graham (Editors), 2003.

No. 9. *Geochemical Investigations in Earth and Space Science: A Tribute to Isaac R. Kaplan.* R. J. Hill, J. Leventhal, Z. Aizenshtat, M. J. Baedecker, G. Claypool, R. Eganhouse, M. Goldhaber and K. Peters (Editors), 2004.

Geochemical Investigations in Earth and Space Science: A Tribute to Isaac R. Kaplan

Edited by:

Ronald J. Hill

U.S. Geological Survey, Denver, Colorado, USA

Joel Leventhal

U.S. Geological Survey, Denver, Colorado, USA

Zeev Aizenshtat

Department of Organic Chemistry and Casali Institute

The Hebrew University, Jerusalem, Israel

Mary Jo Baedecker

U.S. Geological Survey, Reston, Virginia, USA

George Claypool

Consultant, Lakewood, Colorado, USA

Robert Eganhouse

U.S. Geological Survey, Reston, Virginia, USA

Martin Goldhaber

U.S. Geological Survey, Denver, Colorado, USA

Kenneth Peters

U.S. Geological Survey, Menlo Park, California, USA

Publication No. 9

THE GEOCHEMICAL SOCIETY

QE515
G42.

Published by

ELSEVIER B.V.
Sara Burgerhartstraat 25
P.O. Box 211, 1000 AE Amsterdam
The Netherlands

ELSEVIER Inc.
525 B Street, Suite 1900
San Diego, CA 92101-4495
USA

ELSEVIER Ltd
The Boulevard, Langford Lane
Kidlington, Oxford OX5 1GB
UK

ELSEVIER Ltd
84 Theobalds Road
London WC1X 8RR
UK

Copyright © 2004 by The Geochemical Society

All rights reserved. No part of this publication may
be reproduced, stored in a retrieval system, or
transmitted in any form or by any means, electronic,
mechanical, photocopying, recording, or otherwise,
without written permission or The Geochemical Society

First edition 2004

Library of Congress Cataloguing in Publication Data
A catalog record is available from the Library of Congress.

British Library Cataloguing in Publication Data
Geochemical investigations in earth and space science : a
    tribute to Isaac R. Kaplan. - (The Geochemical Society
    special publications series; v. 9)
    1. Geochemistry
    I. Hill, Ronald J. II. Kaplan, Isaac R.
    551.9

ISBN: 0-444516-47-6

⊗ The paper used in this publication meets the requirements of ANSI/NISO Z39.48-1992 (Permanence of
Paper. Printed in Great Britain.

TABLE OF CONTENTS

Section C. Petroleum Geochemistry

Section D. Geochemistry of Ancient and Recent Sediments

PREFACE

Geochemical Investigations in Earth and Space Science: A Tribute to Isaac R. Kaplan is the written product of a technical session organized for the 2002 Geological Society of America in recognition of Isaac (Ian) Kaplan's many contributions to various fields of geochemistry. This volume is a collection of 29 papers written by former students, post-doctoral researchers, friends and colleagues from countries all over the world (including Sweden, Japan, Taiwan, New Zealand, Australia, Israel and the United States).

To say that Ian's career is defined by diverse, multidisciplinary, integrated studies is an understatement. Since 1953, Ian's scientific curiosity has led him to investigate the stable isotope geochemistry and biogeochemistry of carbon, nitrogen, sulfur, oxygen and hydrogen. Ian and his students, post-docs and colleagues have applied isotope geochemistry to a wide variety of problems including: 1) the isotopic and carbon chemistry of extraterrestrial materials, 2) the inorganic and organic geochemistry of interstitial water and the microbiology marine sediments, 3) the biogeochemistry of sapropels, humic materials, kerogens and petroleum, 4) paleo-dietary reconstruction of humans and other mammals, and 5) environmental and atmospheric chemistry. Ground-breaking research conducted in Ian's laboratory resulted in the widespread use of stable isotopes to investigate the biology of iron and sulfate-reducing bacteria, methanogenic archaebacteria, cyanobacteria, fungi and marsupials. In addition, a series of experiments into the natural and laboratory simulated diagenesis and thermal alteration of organic matter led to advances in our understanding of the formation of petroleum and natural gas. An abiding interest in the application of organic geochemistry to environmental problems culminated in a variety of studies on the origin and fate of anthropogenic pollutants in natural environments. Thus, long before the term 'forensic geochemistry' was coined, Ian was developing and applying the principles that define the science today. As Ian enters his sixth decade of scientific investigation, it is fair to say that his work has touched or influenced innumerable scientists either directly or indirectly. It is certainly the case for the editors of this volume.

The first section of the volume is dedicated to stable isotopic studies of all kinds. Included in this section are investigations of carbon, sulfur, oxygen and boron from terrestrial and extraterrestrial systems. The second section of the volume focuses on geochemical investigations of environmental and atmospheric chemistry problems. The third section of the volume is concerned with various aspects of petroleum geochemistry, and the final section includes investigations of recent and ancient sediments.

We thank the Geochemical Society for providing the publication series necessary to make this volume possible. The Department of Earth and Space Sciences and the Institute of Geophysics and Planetary Physics at the University of California, Los Angeles and Shell International gave valuable financial support. The editors thank all of the referees who carried out reviews of the individual chapters in this volume. Finally, we wish to thank Lynne Newton for the tireless effort she put forth in compiling the history of the Kaplan group, Rosario De Leon for all of her help in compiling the list of invitees for the symposium and celebration and to symposium participant themselves, who traveled from Canada, Sweden, Japan, Taiwan, New Zealand, Australia, Israel and the United States to make the celebration a success.

RONALD HILL
JOEL LEVENTHAL
ZEEV AIZENSHTAT
MARY JO BAEDECKER
GEORGE CLAYPOOL
ROBERT EGANHOUSE
MARTIN GOLDHABER
KENNETH PETERS
October, 2003

A Tribute to ISAAC R. KAPLAN:
Scientist, Mentor and Friend

Ian and Helen Kaplan, 2003.

Isaac R. (Ian) Kaplan was born July 10, 1929, in Baranowicze, Poland, now a part of Belarus. He and his family moved to Christchurch, New Zealand in 1937. Ian attended Canterbury University in New Zealand where he earned a Bachelor's degree in Chemistry in 1952 and a Master's degree in Chemistry in 1953. He subsequently served as research officer in the Commonwealth Scientific and Industrial Research Organization (CSIRO), Division of Fisheries and Oceanography, Cronulla, N.S.W., Australia, from 1953–1957 before attending the University of Southern California (USC) to pursue a Ph.D., which he received in Biogeochemistry in 1961. While a graduate student at USC, Ian was a lecturer in geochemistry at the Geology Department (1959–1961). Following graduation, Ian completed a Postdoctoral Research Fellowship in Geochemistry with Sam Epstein at the California Institute of Technology (1961–1962). In October 1962, he accepted a position as Jacob Ziskind Visiting Scholar and Guest Lecturer, Department of Microbiological Chemistry, Hebrew University, Jerusalem, Israel, where he remained until 1965. Ian then returned to the United States where he took a position as Associate Professor of Geology and Geochemistry at UCLA, retiring in 1993 as Emeritus Professor.

During his career Ian and his colleagues published over 300 papers on environmental, sedimentary and petroleum geochemistry. He has served as associate editor and editor for several professional journals and is a Fellow of several professional societies including the American Association of Petroleum Geologists, American Chemical Society, American Institute of Chemists, American Association for the Advancement of Science, Geological Society of America, Geochemical Society and the European Association for Geochemistry, the Russian Academy of Natural Science. He has served on numerous advisory committees for the State of California and Federal agencies and has received numerous awards, including the 1993 Treibs Medal bestowed by the Geochemical Society for pioneering research in Organic Geochemistry. In 1998, he was elected a Foreign Member of the Russian Academy of Natural Science, and in 2002 he received the American Association of Petroleum

Geologists Division of Environmental Geology President's Award for career achievement in the field of environmental geology. In 1975 Ian founded Global Geochemistry Corporation (GGC), serving as its president until 1999 when GGC was acquired by ZymaX Envirotechnology, Inc., and emerged as ZymaX Forensics, Inc./Global Geochemistry Corporation, San Luis Obispo, California. Since then he has been consulting in the field of environmental forensic geochemistry and working on manuscripts.

Ian started his career as an assistant to Laurens Baas Becking at Commonwealth Scientific and Industrial Research Organization (C.S.I.R.O.) in what proved to be a long and fruitful friendship and scientific collaboration. Together they published classic papers on the sulfuretum, the influence of microbial process in sedimentary diagenesis and Eh and pH relationships in sediments. During this period Ian began collaborating with John Hulston, a high school friend and contributor to this volume, investigating sulfur and carbon isotope systematics in geothermal and volcanic systems. While at C.S.I.R.O., Ian also met someone who turned out to be the most important person in his life, a young woman by the name of Helen Fagot. A year after their marriage in 1957, Ian and Helen departed for southern California, so that Ian could pursue a Ph.D. in biogeochemistry at the University of Southern California with Drs. K. O. Emery and S. C. Rittenberg, pioneers in the field of marine diagenesis. At USC Ian started his groundbreaking work on bacterial sulfur isotope fractionation and the distribution of sulfur phases in recent sediments. This work prompted research into the sulfur cycle by groups around the world and laid the foundation for our understanding of bacterial sulfur isotope fractionation and the contemporary distribution of sulfur phases. In 1961, Ian joined Sam Epstein's group as a post doctoral fellow at the California Institute of Technology and started what would become a life long scientific relationship and friendship with Sam. At CalTech, Ian studied the fractionation of carbon isotopes during photosynthesis and was introduced to the geochemistry of meteorites. He also collaborated with William Holser (then at Chevron Oil Co.) on the measurement of large numbers of globally distributed anhydrites, whose depositional ages ranged from pre-Cambrian to Holocene. This resulted in the early publication (1966) of data providing evidence for changes in the chemical composition of the ocean over time, with implications for concomitant changes in the atmospheric oxygen content during the Phaenerozoic epochs. Upon completion of his post doctoral fellowship with Sam Epstein, Ian was offered the Jacob Ziskind Visiting Scholar and Guest Lecturership in the Department of Microbiological Chemistry at Hebrew University in Jerusalem, Israel, where he conducted research on microbial processes in the supersaline Dead Sea. During this period, Ian came in contact with an outstanding group of students and asked many of them to join him when he accepted a faculty position at UCLA. Thus, Ian's first students included Yeshu Kolodny, Arie Nissenbaum, Sam Ben-Yaakov and Bobby Joe Presley. They were joined by visiting scholars, Robert Brooks, on sabbatical leave from Massey University, New Zealand and Ted Belsky, a Postdoctoral Fellow from University of California, Berkeley. Together, the first incarnation of the 'Kaplan group' vigorously addressed processes that influence the composition of interstitial waters in recent marine sediments. They studied bacterial processes and their effects on isotope fractionation and developed analytical methods for trace element distribution in interstitial waters. Sam Ben-Yaakov, an electronic engineer, developed the first instruments for in situ measurement of Eh, pH and carbonate saturation in the deep ocean. Yeshu Kolodny conducted research on the geochemistry of phosphorite nodules and developed a method for age dating their formation based on uranium disequilibrium. Ted Belsky measured the distribution and concentration of organic gases in meteorites.

By 1968, the research group began to expand. The second wave of students, lab technicians and visiting scholars arrived between 1967 and 1969. This group included Mary Jo Baedecker, George Claypool, Joel Cline, Martin Goldhaber, Emil Kalil, Ronald McCready, Robert Sweeney and Chari Petrowski. Joel Leventhal worked as a research assistant for one summer, designing a system to collect and quantify shipboard measurements of organic gases from Deep Sea Drilling Program. Paul Doose made his first appearance in the lab as an undergraduate lab technician. 1967 also saw the arrival of Lynne Newton as administrative assistant and Ed Ruth as staff research associate and eventual lab manager. The investigation of processes occurring in recent sediments continued. Mary Jo Baedecker was involved in the characterization of organic matter in sediments of Saanich Inlet; her research also provided the earliest evidence of archaebacterial products in Dead Sea sediments. Marty Goldhaber continued Ian's early work on microbial controls on sulfur geochemistry. Joel Cline investigated the microbial processes controlling the distribution of nitrogen compounds in seawater and recent sediments, while George Claypool was exploring methanogenesis in recent sediments and the formation of clathrates (gas hydrates). George and Ian were among the first to study naturally occurring gas hydrates, an area of significant research today. Emil Kalil addressed uranium distribution in recent sediments, following the work of Bobby Joe Presley and Yeshu Kolodny. Chari Petrowski examined the light element geochemistry of meteorites and lunar samples.

The decade of the 1970's was an extremely productive time in the Kaplan lab. The group continued to investigate microbial processes in recent sediments and became more involved in the transformations of organic matter. The investigation of lunar samples and meteorites reached its apex at this time, and the group began to investigate anthropogenic contaminants in recent sediments and the processes controlling petroleum generation. Some of the first laboratory simulations of hydrocarbon generation were performed during this time. Ian also co-founded Global Geochemistry Corporation (GGC) in 1975 and served as president of the company from 1977 through 1999. Ian credits much of what he learned regarding petroleum geochemistry and particularly environmental geochemistry to his investigation of many hundreds of case studies at GGC on the geochemistry of petroleum and its refined products. The application of geochemical principles to molecular characterization, age dating, and kinetics of diagenetic transformations, gelled into the field of study he refers to as "forensic environmental geochemistry."

Graduate students who joined the group in the 1970's include Paul Doose, Brian Rohrback, Mark Sandstrom, Kon-Kee Liu, Ken Peters, Bob Eganhouse, Shan-tan Lu, Henry Halpern, Peter Jenden and Erdem Idiz. Dave Winter worked in the lab for the first time in 1970 as an undergraduate lab assistant and became an integral part of the group upon graduation in 1973. He assumed responsibility for running the stable isotope ratio mass spectrometers and providing laboratory technique and methods expertise for students and faculty. Paul Doose made his second appearance in the lab, working on methanogenesis in recent sediments. Brian Rohrback, Mark Sandstrom and Kon-Kee Liu arrived in 1974 and were followed closely by Ken Peters in 1975. Brian and Ken conducted some of the first pyrolysis experiments simulating hydrocarbon generation. The Kaplan group was among the first to recognize bitumen as an intermediate product in petroleum generation. K.K. Liu continued work on the oceanic nitrogen cycle, measuring concentrations and $^{15}N/^{14}N$ changes in inorganic nitrogen compounds in the Santa Barbara Basin and off the coast of Peru. In 1976, Mark Sandstrom and Indira Venkatesen performed some of the first environmental organic geochemistry studies on eastern Bering Sea and Gulf of Alaska sediments. Bob Eganhouse joined the group 1976. As a doctoral student and post-doctoral research associate he carried out studies of the organic matter in urban runoff and municipal wastes and their impact on the coastal ecosystem off southern California. Shan-tan Lu, Henry Halpern, Peter Jenden and Erdem Idiz entered in 1978 and 1979 and brought the investigation of petroleum geochemistry and petroleum generation processes to the forefront of lab activities as the 1980's approached. Shan-tan explored the generation of petroleum from coals using pyrolysis techniques and quantitatively described the hydrocarbons generated by heating coal. Henry Halpern evaluated the role of minerals in hydrocarbon generation, initiating what would become a significant research effort in the lab during the 1980's. Peter Jenden documented changes in shale diagenesis and sandstone cementation and later, as an employee at GGC and Chevron, carried out studies on the origin of natural gas. Erdem examined interactions between organic matter, trace metals and sulfur in recent and ancient sediments, documenting the importance of sulfur in petroleum generation. Robert Haddad, who went on to earn a Ph.D. with Chris Martens at University of North Carolina, was an undergraduate researcher in the lab in 1978 and 1979, and later became an employee at GGC where he worked on the Colorado uranium deposits.

The Kaplan lab also hosted numerous visiting scientists and postdoctoral researchers in the late 1960's and 1970's. It was at this time that Ian began collaborating with five other scientists who would become life-long friends, John Smith, Bill Schopf, Zeev Aizenshtat, Raphael Ikan and Keith Kvenvolden. John Smith joined the group as a visiting scholar in 1968 from C.S.I.R.O. in Australia and worked for two years on the geochemistry of meteorites and lunar samples. Raphael (Rafi) Ikan, who entered the scene in 1972 as a visiting professor from The Hebrew University, worked on humic acids in sediments, analyzed the lipid fraction of Dead Sea sediments and described the diagenetic pathway leading to pristane and phytane. Zeev Aizenshtat made his first journey to UCLA in 1970 as a postdoctoral researcher and for two years worked on naturally occurring lipids and perylene in marine sediments. Zeev and Rafi would make many other visits to the lab over the years. Ian and Bill collaborated on the organic geochemistry of Precambrian sediments and Ian also was part of Bill's Precambrian Paleobiology Research Group (PPRG) that examined the origin and evolution of life on earth. Ian and Keith shared an interest in the geochemistry of petroleum seeps, lunar samples and environmental geochemisty that resulted in many collaborative projects over the years.

Other visiting scholars in the 1970's included Hitoshi Sakai who traveled to UCLA in 1971 as a visiting professor from Okayama University in Japan. Hitoshi analyzed lunar samples and investigated anhydrite sulfur and oxygen isotope fractionation with Israel Zak, who arrived at about the same time as a visiting professor from the Hebrew University. Charles Curtis, from Shefield, U.K., was a visiting lecturer in the Earth and Space Sciences department in 1970–1971 who studied trace metals in sedimentary rocks. Bob Sweeney, whose work on pyrite formation and habit had a major impact in this field and is one of the most highly cited of all of Kaplan's papers, graduated in 1972 and continued work on carbon and sulfur in recent sediments as a post doctoral fellow until 1975 when he joined GGC. Bohdan Bubela was a visiting scientist from the Baas-Becking Geobiology Institute in Canberra, Australia in 1974 and explored the use of chemostats for growing anaerobic bacteria. Ryoshi Ishiwatari was a visiting professor from Tokyo Metropolitan University, Japan from 1974–1975 and studied humic substances in recent marine sediments. Ryoshi also was involved in some of the first artificial maturation experiments with Brian Rohrback and Mariko Ishiwatari (Ryoshi's wife) designed to simulate hydrocarbon generation. Dan Stuermer joined as a postdoctoral researcher in 1975 and began exploring the diagenesis of organic matter, particularly the conversion of humic acids to kerogen as a intermediate step in the formation of petroleum. John Kerridge started at UCLA in 1973 and remained part of Ian's lab until 1994. John became the primary cosmochemistry researcher in the lab and continued work on organic matter in meteorites and temporal solar compositional changes. David Ward (1975–1977) and Steve Zinder (1977–1979), postdoctoral fellows in microbiology, performed studies on the symbiotic relationships involved in methanogenesis during acetate fermentation in collaboration with Bob Mah of the Public Health School. Berndt Simoneit considered the diagenesis and transformation of organic matter in recent sediments and also characterized organic matter in natural petroleum seeps as a research associate between 1976 and 1979. Jaime Bonilla-Ruiz visited from Venezuela from 1976–1978 and worked on oceanographic studies of organic-rich sediments. Shmuel Brenner came from Hebrew University at the same time and studied the organic geochemistry of recent sediments in Alaska, the Guaymas Basin and the Atlantic. David Toth was also a postdoctoral researcher at this time working on kinetics of organic diagenesis. Indira Venkatesan, a member of the group since 1976, remains a fixture in the Kaplan lab today, studying the flux of carbon compounds in recent marine sediments. Phillip Crisp was a visiting scientist from Australia from 1977–1978 and performed some of the initial work on the analysis of petroleum hydrocarbons in marine sediments. Amati Katz landed in 1979 from Hebrew University as a visiting professor and studied the distribution of heavy metals in marine sediments along the California coast. Min-Sung Lee, a professor of Economic Geology, visited Ian's lab from Korea in 1979–1980 and worked on meteorite mineral geochemistry. Greg Rau, a postdoctoral fellow from 1979–1981, worked on biogeochemistry and the ecology of coastal, pelagic and hydrothermal vent communities. He also was involved in marine pollution studies and measurements of stable isotope composition of human bones. The studies performed in the lab in the decade of the 1970's, spawned the field of Forensic Environmental Geochemistry.

The 1980's in the Kaplan lab were equally productive. Research centered on petroleum genesis and environmental topics with a new emphasis on atmospheric chemistry. Petroleum geochemistry research was focused on factors affecting petroleum generation and oil composition including the influence of minerals, brines, nuclear radiation and pressure. Environmental geochemistry studies into atmospheric chemistry and specifically the study of organic acid deposition were initiated. Bruce Barraclough, Jose Sarto, Jeremy Dahl, Brad Huizinga, Thomas Dorsey, Henry Ajie, Julie Bartley and Ronald Hill were graduate students in Ian's lab in the 1980's. Jose Sarto continued researching the factors affecting petroleum generation by performing some of the first experiments designed to address the role of brines in hydrocarbon generation. In collaboration with Eli Tannenbaum, Brad Huizinga further explored the role of minerals in hydrocarbon generation. Jeremy Dahl addressed the effect of nuclear radiation from uranium-enriched shale on petroleum composition in his study of the Alum Shale, Sweden. Thomas Dorsey studied controls on atmospheric carbon dioxide and factors influencing carbon transport to the deep sea. Henry Ajie compared carbon and nitrogen isotopes of bone collagen and osteocalcin protein for reconstruction of mammalian paleodiets and for assessing the reliability of ^{14}C age measurements. Julie Bartley worked with Ian and Bill Schopf on variations in the carbon isotopes of Precambrian environments. Ron Hill was the last graduate student to pass through Ian's lab. He evaluated the role of pressure in hydrocarbon generation and oil cracking and performed some of the early experiments in this area, quantifying chemical changes resulting from elevated pressures, emulating maturation resulting from burial.

In the 1980's, the Kaplan lab had many postdoctoral fellows conducting research on various topics. Kimitaka Kawamura was in the lab from 1981–1984 and worked on methods for the analysis of polar organic substances in acid rain. He showed that a significant amount of the acidity in Los Angeles originated from organic acids either directly released from automobile exhaust or from atmospheric oxidation processes. During this same time period, Masao Minagawa, a postdoctoral fellow from Japan, examined the behavior of carbon, hydrogen and nitrogen in biochemical systems and nitrogen isotopes in sedimentary organic matter. Likewise, Spencer Steinberg was in the lab at this time working on high performance liquid chromatography applications in atmospheric chemistry and lignin geochemistry. Eli Tannenbaum researched the role of minerals in petroleum formation; this resulted in a series of classic papers on the subject. In the mid to late 1980's, Hiroshi Sakugawa continued research on atmospheric oxidation processes resulting in organic and inorganic acid deposition in Los Angeles. Hiroshi also used radioactive and stable carbon isotope measurements of atmospheric carbon monoxide to determine its mode of formation. Paul Mankiewicz worked at GGC after receiving a Ph.D. in Civil Engineering from UCLA and later joined Exxon Production Research as a petroleum geochemist. In 1984, Rosario De Leon, an undergraduate student, joined the group as secretary and administrative assistant. She continued to work at UCLA until she graduated with a Master's degree in Public Health (1994) at which time she then joined GGC as a lab technician and later as office manager at ZymaX Forensics. In 2002 she accepted a job in the Contract Audit office at UCLA.

By 1994, Ian had retired to emeritus status at UCLA, Ian's last student had completed his degree, John Kerridge had moved to the University of California, San Diego, Ed Ruth had moved to Civil and Environmental Engineering Department at UCLA and Dave Winter was soon headed to the Department of Geology at the University of California, Davis. In all, 22 Ph.D. students and four Master's student completed degrees under Ian's tutelage. Indira Venkatesan remains at UCLA as the last group member and now has the pleasure of interacting with Ian in his emeritus role. In 1999, Global Geochemistry Corporation merged with ZymaX Envirotechnology to become ZymaX Forensics Inc./Global Geochemistry Corporation. Ian completed a three-year commitment to the new company in 2002 and transitioned to semi-retirement. We organized the symposium held at the *Geologic Society of America* meeting in 2002 to commemorate Ian's move to the next stage of his career – grandfather, scientist, mentor and friend. Some things never change. Ian remains active at UCLA, still works on projects for ZymaX, and continues to publish results of his work. Friends, colleagues, former students and postdoctoral fellows have carried on the Kaplan legacy in the greater scientific community. We dedicate this volume to you, Ian, with utmost sincerity and love. You are a great scientist, mentor and most of all a great friend.

RONALD HILL
JOEL LEVENTHAL
ZEEV AIZENSHTAT
MARY JO BAEDECKER
GEORGE CLAYPOOL
ROBERT EGANHOUSE
MARTIN GOLDHABER
KENNETH PETERS
October, 2003

Geochemical Investigations in Earth and Space Science: A Tribute to Isaac R. Kaplan
© The Geochemical Society, Publication No. 9, 2004
Editors: R.J. Hill, J. Leventhal, Z. Aizenshtat, M.J. Baedecker, G. Claypool,
R. Eganhouse, M. Goldhaber and K. Peters

Boron isotopes in DSDP cherts: fractionation and diagenesis

YEHOSHUA KOLODNY[1] and MARC CHAUSSIDON[2]

[1]Institute of Earth Sciences, Edmond Safra Campus, Givat Ram, The Hebrew University,
Jerusalem 91904, Israel
[2]CRPG-CNRS, Vandoeuvre-les-Nancy, France

Abstract—Twenty samples of cherts from the Central Pacific (18 of them from DSDP, Leg17, Hole 167) were analyzed for B content and $\delta^{11}B$ by ion-probe. In all samples $\delta^{18}O$, $^{87}Sr/^{86}Sr$ and Ge have been previously measured. $\delta^{11}B$ of DSDP cherts varies between -9.3 and $+8‰$, indicating a "bulk" fractionation of $32-45‰$ with respect to present day seawater B ($39.5‰$), the highest fractionation of any marine phase. This fractionation is similar to the previously observed values in opal-A radiolarians (having $\delta^{11}B$ of $+2$ to $+4.5‰$, *Earth Planet. Sci. Lett.* **117**, 1993, 567–580). Siliceous sediments are a significant sink of boron in the ocean, an even more significant sink of ^{10}B. Assuming no secular variation in $\delta^{11}B_{sw}$ our results can be the reflection of a three-stage precipitation–dissolution process of DSDP cherts: the first leading to opal-A and involving a large fractionation of B isotopes between tetrahedral boron in solution and opal-A, the second and third resulting in quartz via opal-CT. The dissolution–reprecipitation occurred at variable temperatures and water/rock ratios. The deepest cherts underwent diagenesis at higher temperature (low $\delta^{18}O$ values) and in closed system (low $\delta^{11}B$ values). Boron isotopic composition of marine cherts cannot be a clue to the paleochemistry (specifically paleo-pH) of seawater but indicates that some mineral phases might fractionate B isotopes differently than carbonates and thus be used in conjunction with carbonates to reconstruct paleo-pH and $\delta^{11}B$ values of seawater.

INTRODUCTION

THE FIRST MEASUREMENT of boron in siliceous sediments can be traced back to the founding fathers of geochemistry (GOLDSCHMIDT and PETERS, 1932). RANKAMA and SAHAMA (1950) in their classic text noted that (p. 488) in some structures [BO₄] tetrahedra replace silica tetrahedra. "Accordingly, boron replaces silicon in oxygen tetrahedra". Most of the research on the geochemistry of boron in the past 70 years focused on investigating one aspect of the tetrahedral substitution of boron for silicon—in clays. The basic assumption here was that since clayey sediments are strongly enriched in B, they are also the best monitors of changes both in the concentration and in the isotopic composition of boron in the hydrosphere.

GOLDSCHMIDT and PETERS (1932) were also the first to note the high concentration of boron in seawater (4.5 ppm) as opposed to its low content in fresh waters (10 ppb in average river, LEMARCHAND et al., 2002). This contrast served as a basis for the numerous attempts to use boron content as a paleosalinity proxy (see LERMAN, 1966; WALKER, 1968), again primarily utilizing clays as the indicator phase.

Several reports on boron concentration in cherts appeared in the literature (HARDER, 1974; KOLODNY et al., 1980; TRUSCOTT and SHAW, 1984). Retreating to the classics again, the boron–silica substitution should not be surprising in view of GOLDSCHMIDT (1958, p. 280) mentioning that both BPO₄ and BAsO₄ are closely structurally related to cristobalite. A close association between boron and high SiO₂ in deep-sea (DSDP) cherts from the Central Pacific ocean was noted by HEIN et al. (1981) who also observed a positive correlation between boron content and $\delta^{18}O$ in those cherts. KOLODNY et al. (1980) combined oxygen isotopic analysis with α-track mapping of boron in cherts of the Campanian Mishash Formation in southern Israel, to demonstrate the schizohaline origin of these rocks.

With the realization that the isotopic geochemistry of boron may yield interesting information on such parameters as paleo-pH (SCHWARTZ et al., 1969; SPIVACK et al., 1987, 1993; PALMER et al., 1998; PEARSON and PALMER, 2000; LÉCUYER et al., 2002) and the interaction between the

Y. Kolodny and M. Chaussidon

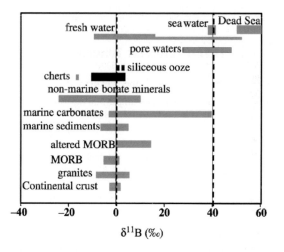

FIG. 1. Ranges of δ^{11}B values in different reservoirs on Earth (after schemes of VENGOSH et al. (1992), PALMER and SWIHART (1996) and LEEMAN and SISSON (1996)). The data on siliceous oozes and two Triassic cherts are from ISHIKAWA and NAKAMURA (1993). All other chert data (marked in black) were added in this study.

oceanic crust and seawater (SPIVACK and EDMOND, 1987) the range of subjects for B-isotope geochemical studies has been extended to carbonates (VENGOSH et al., 1991; HEMMING and HANSON, 1992) as well as to fluid phases such as interstitial waters (BRUMSACK and ZULEGER, 1992; SPIVACK and YOU, 1997; KOPF et al., 2000).

Isotopic analyses of boron in cherts are still very few. The most quoted values are the three measurements of two siliceous oozes and one Triassic chert by ISHIKAWA and NAKAMURA (1993) (see later) as part of their survey of the boron isotopic composition of marine sediments and sedimentary rocks.

The purpose of the present study is to measure and understand the isotopic composition of boron in purely marine cherts, thus expanding the database of δ^{11}B in sediments and sedimentary rocks (see Fig. 1). For this we use a sequence of cherts from the Deep Sea Drilling Project (DSDP). This suite of samples offers the possibility to determine the post-deposition changes in δ^{11}B that accompany silica transformations during diagenesis. These transformations pose strong limits on the use of B isotopes in marine cherts as a potential paleo-tracer.

METHODS

Deep sea chert samples

Twenty samples of cherts from the Central Pacific were selected for analysis. Of those, 18 are in stratigraphic sequence from DSDP Hole 167 (Leg 17). Hole 167 was drilled atop the Magellan Rise in the Central Pacific. It penetrated about 1200 m of nanofossil chalk, limestones and cherts, Tithonian to Pleistocene in age. Starting with core 33, at a depth of 605 m below sea floor (mbsf), cherts were recovered in the sequence. The cherts occur either as part of the carbonate sequence or, very often as broken fragments in the core catcher (marked "cc" in Table 1).

Between the Tithonian and the Late Cenomanian (1185–850 m, cores 61–94) sedimentation was continuous. The sedimentation rate was about 3–5 m/m.y. (SCHLANGER et al., 1973). Above the Late Cenomanian, a sequence of hiatuses or stratigraphic gaps appear, associated with angular unconformities, representing probably the upper Cenomanian, Santonian and the longest one in all of the Paleocene to the middle Eocene (DOUGLAS et al., 1973). Starting in the Middle Eocene and throughout the Tertiary, sedimentation at Site 167 was continuous and sedimentation rates high, varying between 9 and 18 m/m.y.

Table 1. Sample location, depth, age, boron concentration (by ion-probe and ICP-AES), $\delta^{11}B$, standard deviation from the mean of $\delta^{11}B$ and $\delta^{18}O$ for DSDP samples

Sample	Depth (mbsf)	Age (Myr)	B conc. (ppm) (ICP-AES)	B conc. (ppm) (I-probe)	1/B (ICP-AES)	$\delta^{11}B$ (‰)	$1\sigma(\delta^{11}B)$ (‰)	$\delta^{18}O$ (‰)
167-33R-1W (127–129)*	605	42.5	67.8		0.0147	1.1	0.73	36.7
167-33R-1W (147–149)	605	43	75.8	74.5	0.0132	−0.1	0.05	34.6
167-34-cc*	620	43.5	69.5	95.9	0.0144	0.6	0.04	34.8
167-36-cc*	640	55	48.9		0.0205	4.4	1.3	34.4
167-38-cc	660	59.3	73.5	60.3	0.0136	0.8	1.26	34.8
167-40-cc	675	64	60.8	50.0	0.0165	−0.1	1.19	34.7
167-42-1 (130–131)	690	68.8	86.1	80.7	0.0116	2.7	0.27	35.7
167-44-1 (113–114)	710	71.1	85.0	83.5	0.0118	4.0	0.11	35.8
167-50-cc	770	74.3	97.3	104.3	0.0103	7.7	1.19	35.3
167-60-2 (81–82)	850	94.3	69.6	70.3	0.0144	1.4	0.55	35.1
167-61-2 (0–3)	860	97	85.9		0.0116	−5.2	1.99	34.1
167-65-1 (64–66)	890	101	70.9	68.2	0.0141	3.0	1.22	33.3
167-68-2 (100–101)	920	111.5	74.3	56.0	0.0135	−0.7	2.26	34.6
167-70-3	940	111.5	56.3	71.0	0.0178	−9.3	1.63	34.0
167-71-2 (37–38)	950	119	66.0	68.8	0.0152	−5.4	0.33	33.3
167-71-2 (42–44)	950	119	54.5		0.0183	−6.9	1.86	33.3
167-73-2 (90–92)	960	125.7	64.5		0.0155	−4.2	0.68	32.4
167-92-1 (64–67)	1160	141	53.3		0.0188	4.1	1.76	31.2
169-10-cc (0–1)	220	105	53.7	43.7	0.0186	1.0	0.99	31.9
195-B-3-1 (87–88)	390	130	52.8	75.5	0.0189	−0.6	0.07	32.9

*Samples containing a small (undetermined) amount of opal-CT.

(WINTERER, 1973; SCHLANGER et al., 1973), but no chert younger than the Middle Eocene was recorded at this site. Two samples (one from each) are from Holes 169 and 195. All the samples reported on here were previously analyzed for $\delta^{18}O$ (KOLODNY and EPSTEIN, 1976), $^{87}Sr/^{86}Sr$ in the adjacent carbonate (KOEPNICK et al., 1985), and Ge content (KOLODNY and HALICZ, 1988). All the relevant results are tabulated in KOLODNY and HALICZ (1988). The thermal history of these cherts was analyzed by PISCIOTTO (1981). In most cases the original samples have been exhausted prior to our study; we hence resampled the cores at the designated or closest depths. Obviously, extrapolating results from one sampling site to the world ocean is risky. On the other hand there is something to be said in favor of analyzing a sequence that has been studied by several other tools.

We made a maximum effort to ascertain that the boron in our analyses is indeed related to a siliceous phase. All samples were analyzed by X-ray diffraction, specifically searching for opal-CT peaks. All, but three samples are practically pure microcrystalline quartz. In three samples (marked by " * " in Table 1), small opal-CT peaks were observed. The results for these samples should be viewed as possibly representing an admixture of opal-CT to quartz. All samples were examined under a scanning electron microscope both on a scale of 200 μ (field width) and 100 μ. On both scales, and at spot-counts, only Si and O peaks were observed (in some samples an Al peak slightly above background appeared). No cathodoluminescence is seen on all examined cherts in the visible wavelength region.

The solutions that were prepared by reaction of the cherts with HF, HNO_3 for boron analysis (see later) were also analyzed for other elements (Table 2). Aluminium (as reflecting clay) concentration averaged 0.5% and did not exceed 1%; the total of non-Si and Al cations did not exceed 0.5% in all but one case (1%), and averaged 0.4%. Most importantly B content is not correlative with the concentration of any of the measured elements.

Organic matter content was estimated by combustion of the samples on an elemental analyzer (EA-IRMS) coupled to a DeltaPlus-XL Finnigan mass-spectrometer. None of the samples contained more than 0.2% organic carbon.

To sum up our special tests, there is good evidence that all the analyzed samples consist practically of pure silica. In all but three, quartz is by far the predominant phase.

Y. Kolodny and M. Chaussidon

Table 2. Chemical composition of 18 DSDP cherts

Sample	B (ppm)	Na (ppm)	K (ppm)	Mg (ppm)	Ca (ppm)	Sr (ppm)	SO$_4$ (ppm)	Li (ppm)	Ba (ppm)	Al (ppm)	C (org) (%)
167-33R-1W (127–129)	67.8	927	19	206	3870	22.3	609	2.6	99.5	1190	
167-33R-1W (147–149)	75.8	605	30	81.5	1920	9.5	428	1.4	28.3	6480	
167-34-cc	69.5	1110	357	685	4040	18.8	772	1.6	111	1200	0.18
167-36-cc	48.9	1190	509	227	1690	10.6	813	4.3	17.7	4630	
167-38-cc	73.5	1180	61	160	715	8.9	225	6.7	2.0	8390	
167-40-cc	60.8	1220	631	139	1380	11.7	243	8.1	4.4	3740	0.04
167-42-1 (130–131)	86.1	1070	611	163	767	6.2	151	3.8	3.0	7850	
167-44-1 (113–114)	85.0	1330	729	169	412	5.9	88.0	3.5	3.2	5290	
167-50-cc	97.3	1090	576	172	368	4.8	135	3.3	2.7	8290	
167-61-2 (0–3)	85.9	931	810	197	708	10.9	135	5.6	28.4	7120	
167-65-1 (64–66)	70.9	1170	938	319	1690	4.4	165	7.9	3.1	4050	
167-68-2 (100–101)	74.3	2100	1110	268	1280	22.9	284	19.9	8.9	6860	0.05
167-70-3	56.3	1840	869	265	845	12.2	213	24.5	4.1	6410	
167-71-2 (37–38)	66.0	1460	680	184	1720	10.5	284	15.0	3.5	7280	0.08
167-71-2 (42–44)	54.5	1150	494	158	1580	3.8	270	13.6	2.8	6690	
167-73-2 (90–92)	64.5	1750	522	154	1330	14.5	339	12.6	11.5	3880	
167-92-1 (64–67)	53.3	44	186	168	1060	2.6	765	25.8	6.5	4050	0.1
169-10-cc (0–1)	53.7	1510	1708	502	2890	7.4	344	61.6	6.7	3090	
Blank	1.8	0.0	0.6	5.0	106	1.2	152	0.0	0.2	50	

Specifically, there is no indication of the presence of a significant amount of either clay minerals or organic matter. Thus all our conclusions obtained on bulk samples apparently apply to chert as a pure, monomineralic quartz rock.

Determination of boron concentration and isotopic compositions by ion microprobe

The B contents and isotopic composition were determined by ion microprobe using a Cameca ims 3f instrument at CRPG-CNRS (Nancy). The analytical technique used was previously described for silicates and glasses (CHAUSSIDON and JAMBON, 1994; CHAUSSIDON et al., 1997) and details can be found there. Because of their generally high B concentrations (50–100 ppm range), the problems of surface contamination and reduced precision due to low counting rates are not encountered for the analysis of B in cherts. The instrumental mass fractionation was determined using the GB4 glass standard (CHAUSSIDON et al., 1997). It was reproducible on the GB4 glass to better than ±0.8‰ (2 sigmas) during the analytical sessions. The B isotopic composition is given in $\delta^{11}B$ notation relative to NBS 951, which has a $^{11}B/^{10}B$ ratio of 4.04558. Because of their heterogeneous appearance, i.e. color variations, several analytical points (between 3 and 6) were systematically measured at different locations on each polished sample of DSDP chert. The $\delta^{11}B$ values given in Table 1 correspond to the average of 3–6 measurements made on each chert sample. A general feature in these analyses is that a significant range of $\delta^{11}B$ values, of up to 4‰ for a few samples, is present in nearly all samples. This range is indicated in Table 1 by the 1 sigma on the mean of the different analyses made on a given sample.

The B concentrations were also measured by ion microprobe on a few chert samples, using a set of glass standards having B concentrations between 0.2 and 1000 ppm as previously described (CHAUSSIDON et al., 1997).

Determination of boron concentration measurements by ICP-AES

As a double-check, the concentration of boron in all DSDP samples was redetermined by ICP atomic emission spectroscopy (ICP-AES). The chert samples were ground in a corundum mortar (Diamonite®) to pass a 200 mesh sieve, and were brought into solution following a modified version of the procedure of NAKAMURA et al. (1992). About 200 mg of sample powder was reacted with HF, HNO$_3$ and mannitol in sealed PFA Teflon®

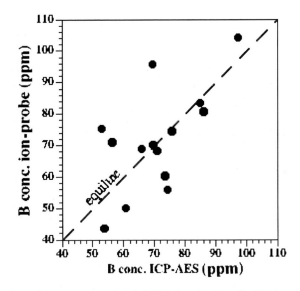

FIG. 2. Comparison of boron concentrations in DSDP cherts as determined by ion-probe (counting on discrete points of the polished rock) with an ICP-AES determination of bulk rock specimens. A line of perfect agreement (unity slope) is marked.

Advanced Composite Vessels (ACVs) and heated in a CEM microwave sample preparation system (Model MDS-2000). The peak pressure reached in the vessel was about 80 psi; temperature was not recorded, but apparently did not exceed about 70°C; total reaction time was 40 min. The vessels were left to cool overnight, and the fluoride was evaporated to dryness at 40°C. The dry residue was then dissolved in water and analyzed on a Perkin-Elmer Optima-3000 system. The 249.773 nm boron line was used. Samples were run in triplicate with a precision of 1–2% (RSD). The results of the comparison between wet chemical and ion probe analyses are shown in Fig. 2. Considering the fact that chert samples are certainly not perfectly homogeneous on the sampled scale (200 mg in wet vs. about 500 μm^3) the observed agreement seems excellent. Such agreement between boron *concentrations* determined by the two methods, suggests that the average boron *isotopic compositions*, as determined by ion probe, might also represent well the bulk isotopic composition of a sample.

RESULTS

B concentrations and isotopic compositions in DSDP cherts

Boron concentration in the analyzed DSDP cherts varies between 44 and 104 ppm and $\delta^{11}B$ value between -9.3 ± 1.6‰ and $+7.7 \pm 1.2$‰ (Table 1, Fig. 3). The range of $\delta^{11}B$ values reported here overlaps and includes those observed in opal-A radiolarians and diatoms ($+2.2$ to $+4.5$‰, ISHIKAWA and NAKAMURA, 1993). Some of the DSDP cherts analyzed by us have very low $\delta^{11}B$ values lending further support to the general claim of LEMARCHAND et al. (2002) that boron sinks from the ocean are ^{10}B enriched. The two chert samples from Holes 169 and 195 have $\delta^{11}B$ values within the range defined by cherts from Hole 167 but tend to have low B concentrations and low $\delta^{18}O$ values.

Three of the four DSDP chert samples located at the top of core 167 contain traces of opal-CT (Table 1), implying that they have not been totally transformed into quartzose chert. These three samples have high $\delta^{18}O$ values compatible with isotopic equilibrium with seawater (KOLODNY and EPSTEIN, 1976) and have $\delta^{11}B$ values between -0.1 and $+4.4$‰ similar to those observed for opal-A radiolarians deposited on the seafloor (ISHIKAWA and NAKAMURA, 1993).

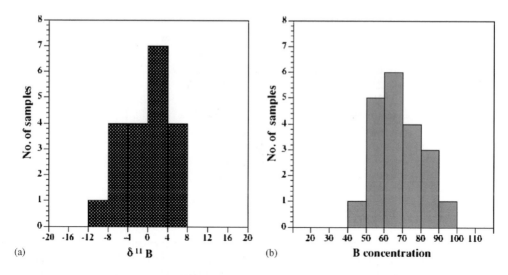

FIG. 3. Histograms of δ^{11}B (a) and boron concentrations (b) in 20 DSDP cherts.

Correlations between B concentrations, δ^{11}B values, δ^{18}O values and depth

A rather systematic variation of B concentration and δ^{11}B values is observed with depth for samples from DSDP Hole 167 (Fig. 4). B concentrations increase regularly from ≈ 70 ppm at 605 mbsf to a maximum of ≈ 100 ppm at 770 mbsf (sample 167-50-cc) and decrease then to reach $\approx 40-50$ ppm at 1200 mbsf. The δ^{11}B values also show a pattern of change with depth: δ^{11}B values increase from $\approx 0\%o$ at 605 mbsf to $+7.7\%o$ at 770 mbsf and then decrease to negative values at greater depth. A similar profile is observed for δ^{18}O values (Fig. 4): δ^{18}O tends to slightly increase from 600 mbsf down to $\approx 700-800$ mbsf where it reaches a value close to $+36\%o$ and then decreases continuously with depth down to a value of $+31\%o$ at ≈ 1200 mbsf.

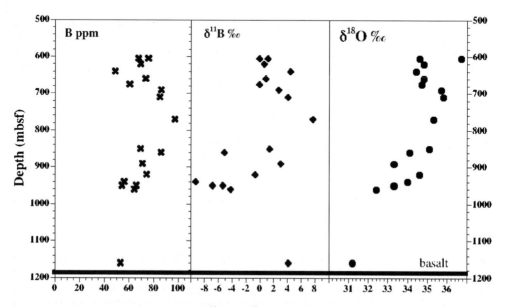

FIG. 4. Variation of B concentration, δ^{11}B and δ^{18}O with depth below seafloor (mbsf) in cherts from Hole 167, on Magellan Rise. δ^{18}O data from KOLODNY and EPSTEIN (1976), see KOLODNY and HALICZ (1988). Note parallel behavior of the three variables, which reach a maximum at around 800 mbsf. Basalt is marked at bottom of hole.

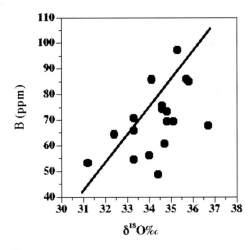

FIG. 5. Variation of $\delta^{18}O$ with B concentration in cherts of Hole 167. The present DSDP cherts scatter around a correlation line previously suggested by HEIN *et al.* (1981) for cherts from Leg DSDP 62 in the mid-Pacific. As a first order observation, B is lost by cherts during diagenesis with increasing temperature.

These profiles of B concentrations and $\delta^{11}B$ values are reminiscent of those previously observed for pore waters, sediments and carbonates in other DSDP holes (YOU *et al.*, 1993; SPIVACK and YOU, 1997). The deepest sample of the core, sample 167-92-1 (64–67), has a quite high $\delta^{11}B$ value of $\approx +4‰$ compared to other deep samples. The proximity of a basaltic layer, within 25 m distance, to this sample was previously noted (KOLODNY and EPSTEIN, 1976) and was invoked as a possible source of perturbation of the $\delta^{18}O$ value of this chert. The same might be true for its $\delta^{11}B$ value though this can only happen through a complex process of leaching and reprecipitation since oceanic basalts have in general slightly negative $\delta^{11}B$ values (CHAUSSIDON and JAMBON, 1994).

Despite these similarities in variations with depth, correlations between B concentrations, $\delta^{11}B$ and $\delta^{18}O$ values are not obvious at first glance. There is a general tendency of B concentrations to decrease with $\delta^{18}O$ values (Fig. 5) in agreement with previous data for cherts from DSDP Leg 62 in the mid-Pacific (HEIN *et al.*, 1981; HEIN and YEH, 1981): high B concentrations close to ≈ 100 ppm are associated with high $\delta^{18}O$ values close to $+36‰$, while low B concentrations (down to ≈ 10 ppm) are associated with low $\delta^{18}O$ values of $\approx +28‰$. However, a large range of ≈ 30–40 ppm in B concentrations is associated with cherts having a given $\delta^{18}O$ value.

A clear linear correlation is observed between $\delta^{11}B$ values and 1/B concentrations (Fig. 6) for most of the samples from Hole 167 (15 samples out of 18). The three samples which do not fit this linear correlation are sample 167-92-1 (64–67) which is the deepest of the section and is close to a basaltic dike (see earlier and KOLODNY and EPSTEIN, 1976) and samples 167-36-cc[*] and 167-61-2 (0–3). These two latter samples do not show any significant mineralogical or petrographic difference from all the other cherts. The heterogeneity of B concentration must, however, be noted: a 10–30% relative variation is shown in Fig. 2 by the comparison between bulk and *in situ* B concentration analyses. Such a heterogeneity might explain the noise around the $\delta^{11}B$ vs. 1/B correlation. The regression line calculated in Fig. 6 intersects the $\delta^{11}B$ axis (1/B = 0) at $+24.6‰$.

Finally a rough positive trend can be observed between $\delta^{11}B$ values and $\delta^{18}O$ values (Fig. 7). The correlation seems worse than between $\delta^{11}B$ and 1/B but once again it is tempting to ascribe part of this scatter to an heterogeneity of the samples since contrary to B concentrations and isotopic compositions which were measured on the same piece of chert (but at different scales, bulk or micrometer scale), the $\delta^{11}B$ values were measured on pieces of cherts resampled recently in the DSDP section as close as possible to the samples analyzed 25 years ago for $\delta^{18}O$ values (see sample description).

Y. Kolodny and M. Chaussidon

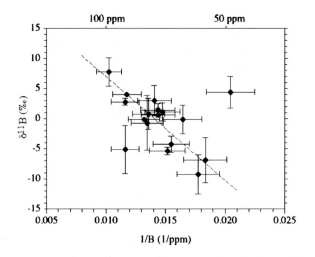

FIG. 6. Variation of $\delta^{11}B$ with 1/[B] in Hole 167 cherts (errors are two sigmas on $\delta^{11}B$ values). Sample 167-92-1 (64–67) directly overlying basalt was excluded from the plot. Fifteen of the 18 samples studied define a correlation line which intercepts the Y-axis at $\delta^{11}B = +24.6‰$. This value is close to that of the $\delta^{11}B$ value of $B(OH)_4^-$ dissolved in seawater (see text) and incorporated by silica during diagenesis. This plot suggests two extreme conditions of diagenesis: (i) diagenesis at low temperature in a pore fluid dominated by seawater corresponding to the cherts having high $\delta^{11}B$ (and high $\delta^{18}O$) and (ii) diagenesis at higher temperature in a closed sytem where the $\delta^{11}B$ of the pore fluid is dominated by B coming from the dissolution of silica.

DISCUSSION

The $\delta^{11}B$ values of recent siliceous sediments as well as of deep-sea cherts presented in this study raise two inter-related questions: (1) what is the B isotopic fractionation associated with B removal from seawater by silica precipitation? and (2) what are the processes that can explain the B isotopic variability, especially the considerably lower $\delta^{11}B$ values of ancient cherts? This variability can be viewed as reflecting one of two "end-member" alternative processes.

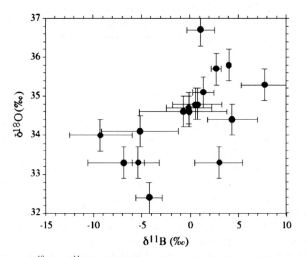

FIG. 7. Variation of $\delta^{18}O$ vs. $\delta^{11}B$ for DSDP cherts. Despite the scatter, a positive trend is present. This trend is consistent with diagenesis under variable conditions (see text): high T and closed system for deep samples having low $\delta^{18}O$ and low $\delta^{11}B$ values and low T and seawater dominated pore fluid for samples above 700 mbsf having high $\delta^{18}O$ and high $\delta^{11}B$ values.

Changes in $\delta^{11}B$ of siliceous sediment could reflect primarily secular variation in $\delta^{11}B$ of seawater. LEMARCHAND et al. (2002) basically adopted such an approach for the simpler case of carbonates, in which diagenesis could be minimal. They suggested a model for $\delta^{11}B$ evolution of seawater that shows a maximum range of about 6% in $\delta^{11}B_{sw}$ (for the equivalent age-range of 120–49 Myr). Alternatively, assuming that secular variation of $\delta^{11}B_{sw}$ is negligible (i.e. $\delta^{11}B_{sw}$ is more or less constant in time) the observed variation could be a reflection of the diagenetic history of DSDP cherts. Obviously reality can also reflect some mixture of the two extreme cases mentioned here. These questions are discussed.

B isotopic fractionation during silica precipitation from seawater

The distribution of the two boron coordination species in solution is strongly pH dependent and the fractionation factor between $B(OH)_3$ and $B(OH)_4^-$ was estimated as being about 20% at 20°C (KAKIHANA et al., 1977). Assuming that only tetrahedral boron in solution enters a given mineral lattice, the boron isotopic composition of this mineral can be approximated as that of tetrahedral boron, if the major boron isotopic fractionation step is indeed the isotopic equilibrium in solution between $B(OH)_3$ and $B(OH)_4^-$. Whereas this simplistic view describes quite well the observations on carbonates, it seems a rather crude approximation in the case of silicates, as demonstrated by the $\delta^{11}B$ values of the cherts in the present study.

It has been shown that B incorporation into carbonates occurs predominantly via $B(OH)_4^-$ adsorption on the growing mineral (HEMMING et al., 1995) and results in an isotopic fractionation between carbonate and seawater which is thus grossly that between dissolved $B(OH)_4^-$ and seawater. A few permil variations of this isotopic fractionation have been observed among biogenic carbonates such as foraminifera, brachiopods and corals (SANYAL et al., 1996; LÉCUYER et al., 2002; ROLLION-BARD et al., 2003): these variations are not well understood but could result, in the case of corals for instance, from some small scale pH variations in the calcifying fluid (ROLLION-BARD et al., 2003).

B isotopic fractionation between the dissolved B species and silicates seems to follow a slightly different scheme. PALMER et al. (1987) determined experimentally the B isotopic fractionations between dissolved boron and adsorbed boron on clay minerals: they found values of $-31\pm2\permil$ for $B(OH)_3$ and $-8\pm2\permil$ for $B(OH)_4^-$. The theoretical reasons for this fractionation of 8‰ between dissolved $B(OH)_4^-$ and $B(OH)_4^-$ adsorbed on clay have not been investigated, but it must be suggested that some additional isotopic fractionation might occur between adsorbed $B(OH)_4^-$ and structurally incorporated $B(OH)_4^-$. Though this question has not yet been clarified experimentally, it is possible, using the relationships determined for clays to discuss the $\delta^{11}B$ values of silica precipitated from seawater. For seawater at a pH of 8.2 having a $\delta^{11}B$ value of $+39.5\permil$, the 20‰ isotopic fractionation between dissolved $B(OH)_3$ and $B(OH)_4^-$ implies that these species have $\delta^{11}B$ values of $\approx +42$ and $\approx +22\permil$, respectively. Thus, if we consider as a first approximation that only dissolved $B(OH)_4^-$ is incorporated into silica, this silica should have a $\delta^{11}B$ value of $\approx +14\permil$, if the B isotopic fractionation during $B(OH)_4^-$ incorporation in silica is similar to that determined for clays. This prediction is clearly too high when compared to the present set of marine cherts. The three cherts which have undergone the least diagenesis (i.e. that contain traces of opal-CT) have $\delta^{11}B$ values ranging from -0.1 to $+4.4\permil$ (Table 1), values similar to those previously determined for opal-A radiolarians deposited on the seafloor ($\delta^{11}B$ ranging from $+2.1$ to $+4.5\permil$, ISHIKAWA and NAKAMURA, 1993). There is no mineralogical or theoretical reason to suggest that siliceous sediments containing opal-A are not directly precipitated from seawater. If so, their low $\delta^{11}B$ values must be taken at face value as representing silica precipitation from seawater, thus exhibiting an extremely large boron isotopic fractionation of approx. -35 to $-40\permil$. This fractionation is similar to that of about 40‰ observed by LEMARCHAND et al. (2003) for boron adsorption onto goethite and humic acid at acid pH. Finally, this fractionation between seawater and silica would imply an isotopic fractionation between dissolved $B(OH)_4^-$ and $B(OH)_4^-$ incorporated in silica of approx. -18 to $-22\permil$.

This large B isotopic fractionation (approx. -35 to $-40\permil$) observed during B removal from seawater by silica precipitation has an important consequence: silica precipitation must be considered as a significant flux in the control of ^{11}B-rich seawater boron. Assuming a rough figure of 4×10^{14} g yr^{-1} as the amount of biogenic silica removed from the ocean (DEMASTER, 1981) and an average B content of precipitated silica of 75 ppm, 3×10^{10} g yr^{-1} of boron are removed annually from the ocean by siliceous sediments. This is about half the amount removed by carbonates (LEMARCHAND et al., 2002). However, because the B isotopic fractionation between seawater and silica is about twice that between seawater and carbonate, silica precipitation results in a ^{10}B sink similar to that due to carbonate precipitation.

Can chert $\delta^{11}B$ values record seawater paleo-pH?

One of the major motives for the study of B isotopic compositions of marine carbonates is to attempt to determine seawater paleo-pH (e.g. SPIVACK et al., 1987, 1993; PALMER et al., 1998; PEARSON and PALMER, 2000; LÉCUYER et al., 2002). This approach suffers from the fact that paleo-pH is calculated from carbonate $\delta^{11}B$ values assuming that the $\delta^{11}B$ value of coeval seawater is the same as that of the present ocean. Despite the very convincing approach developed for foraminifera (PEARSON and PALMER, 2000), this assumption cannot be tested rigorously and can be questioned on the basis of models describing the evolution of seawater $\delta^{11}B$ value in response to changes in variables such as continental weathering or the rate of generation of oceanic crust (LEMARCHAND et al., 2000, 2002). The large difference between B isotopic fractionation during carbonate precipitation ($\approx -23\permil$) and silica precipitation (approx. -35 to $-40\permil$) suggests that it should be theoretically possible to determine seawater paleo-pHs and paleo-$\delta^{11}B$ values from the B isotopic analysis of associated carbonate and silica in marine sediments. Unfortunately, several arguments indicate that variations in paleo-pH and/ or seawater $\delta^{11}B$ values are not the first order control on chert $\delta^{11}B$ values.

The range of $\delta^{11}B$ variation observed in the cherts analyzed in this study is about 17‰ (Figs. 3, 4 and 6), much larger than even the most extreme secular change suggested for the Phanerozoic ($\approx \pm 2\permil$, LEMARCHAND et al., 2002). Thus it is obvious that $\delta^{11}B$ variations in cherts are not a simple reflection of seawater $\delta^{11}B$ values. As previously stressed (see Section 4.1), silica precipitation must be considered as an important ^{10}B sink from seawater. In this respect, though it is rather difficult to discern any time-related pattern of chert $\delta^{11}B$ values (Fig. 4), a set of low $\delta^{11}B$ values are observed at about 1000 m depth, which is equivalent to 120–100 Myr BP (see KOLODNY and HALICZ, 1988). Interestingly, this is a time span that shows a $\delta^{11}B$ maximum on the secular curve of LEMARCHAND et al. (2002). It could thus be conceivable that an increased rate of ^{10}B removal from seawater (through a complex process of diagenesis, see Section 4.3) could by mass balance increase the $\delta^{11}B$ value of seawater. This apparent qualitative consistency is probably coincidental since it has been shown that cherts undergo extremely slow diagenesis, so that chert formation occurs up to about 50 Myr after the deposition of the original siliceous sediment. In that case the $\delta^{11}B$ curve would be shifted by some 40–50 Myr towards the present with respect to the curve proposed by LEMARCHAND et al. (2002) and would no longer be consistent with it. Such a shift was suggested for $\delta^{18}O$ in DSDP cherts by KOLODNY and EPSTEIN (1976), and accepted later by several investigators (e.g. HEIN and YEH, 1981).

There is thus no clear indication from the DSDP analyzed chert $\delta^{11}B$ record that ancient cherts can be used to reconstruct seawater paleo-pH or paleo-$\delta^{11}B$ values, in agreement with previous statements on the importance of diagenesis in the control of chert $\delta^{11}B$ values (ISHIKAWA and NAKAMURA, 1993). However, it must be stressed that this is probably not the case for silica having undergone limited diagenesis such as in opal-A or opal-CT.

Chert diagenetic history

The complex history of chert diagenesis must be considered in detail in order to try and elucidate its B isotopic evolution during diagenesis. Basically, cherts are the result of a three-stage diagenetic process: the first stage is the formation of opal-A at equilibrium with seawater, followed by its dissolution and crystallization to opal-CT, and then by a redissolution and crystallization to microcrystalline quartz. The transition of amorphous siliceous sediment to microcrystalline quartz (mcq, chert) has been the subject of numerous studies and different approaches (see summaries in KASTNER, 1981; KNAUTH, 1994). The consensus opinion is probably that favoring the "maturation sequence" of opal-A → opal-CT → mcq (HEATH (1973), depicted by KNAUTH (1994) as path "A" on his diagram). Each stage in this sequence is viewed as a dissolution–reprecipitation step. The factors determining these transitions are believed to be temperature, time, and host rock facies (CALVERT, 1974; KASTNER, 1981). Thus whereas all three forms of siliceous rocks are represented in the Miocene Monterey Formation of California, the youngest abundant quartzose cherts were drilled by DSDP in rocks of Eocene age, and opal-CT (mostly as porcellanite) abundance increases sharply in deep sea drillings only in rocks older than the Miocene. The importance of temperature in the opal-A → opal-CT and the opal-CT → quartz transitions has been repeatedly stressed and argued over. Furthermore, the opal-A to opal-CT transition has been monitored in most of the later DSDP and ODP drill-holes. The depth of this transition varies within a broad range. It appears (HEIN et al., 1981) that temperature may compensate for time in the transition and that once silicification begins it may progress rather rapidly to completion. The arguments were based both on field occurrence and on oxygen isotope studies. Whereas HEIN et al. (1978) concluded that opal-A transforms into opal-CT between the temperatures of 35 and 50°C, MATHENEY and KNAUTH (1993) and BEHL (1992) lowered the required range to 17–21°C and 2–33°C, respectively. BOHRMANN et al. (1994) presented evidence that purity of the diatomaceous sediment enhances formation of opal-CT to short times and very low temperatures. Thus, alternative paths of diagenesis have been suggested, amounting to different degrees of shortcutting of the full diagenetic sequence (see KNAUTH, 1994): both opal-CT and mcq-chert may be very early siliceous deposits. The possibility of such a complex diagenetic history must obviously result in rather complex relationships involving boron introduction into the silica mineral and modifications of its $\delta^{11}B$ values.

Boron isotopic changes during chert diagenesis

Although diagenesis is expected to act continuously in the sedimentary pile, increasing with depth as temperature increases, the present profiles with depth of [B], $\delta^{11}B$ and $\delta^{18}O$ values (Fig. 4) show a maximum at 700–800 mbsf. Such a profile can *a priori* be explained by the combination of (i) an addition of B through fluid percolation at 700–800 mbsf and its progressive incorporation during diagenesis in chert above and below this level with (ii) a change in the conditions of diagenesis (*e.g.* temperature, pH, fluid/rock ratio) with depth resulting in variations of the B and O isotopic fractionations between recrystallizing cherts and pore fluids.

In a simple model of dissolution and recrystallization of silica at equilibrium with pore water having the composition of seawater, it can be anticipated that later generations of silica will be depleted in both ^{11}B and ^{18}O. This is true for oxygen because as temperature increases and recrystallization proceeds, the isotopic fractionation between silica and water decreases. For B, this comes from the large isotopic fractionation between silica and B dissolved in water (see Section 4.2). The magnitude of this fractionation is not well constrained but it is likely large because at each dissolution step dissolved B coming from silica is partitioned in the fluid between $B(OH)_3$ and $B(OH)_4^-$ (with still a 20‰ isotopic fractionation between $B(OH)_3$ and $B(OH)_4^-$), while only $B(OH)_4^-$ will participate again in silica precipitation. Experimental studies of the B isotopic fractionation during the illitization of smectite, if extrapolated to 25°C, show that when Si–O bonds are broken and reformed, B incorporated in illite is depleted in ^{11}B by ≈ 31‰ relative to the fluid (WILLIAMS et al., 2001). It is thus clear that if a parallel is drawn for B between the process of illitization of smectite and the transformations of opal to microcrystalline quartz, the B isotopic fractionation associated with this transformations will depend critically on the $\delta^{11}B$ of pore water in which diagenesis takes place. This $\delta^{11}B$ depends in turn on the relative amounts in pore fluids of B of seawater origin and of B released by the dissolution of the first generation of opal. If seawater B is dominant in the pore water fluid (*i.e.* if the pore fluid has a $\delta^{11}B$ value of $+40$‰) it can be anticipated that latter generation of microcrystalline quartz will have $\delta^{11}B$ values at around $+10$‰. Alternatively, if B released from the dissolution of the first generation chert is the dominant source of B in the fluid (*i.e.* δ^{11}B value of the pore fluid close to 0‰), then extremely low δ^{11}B values of -30‰ can be expected for the microcrystalline quartz. Of course pH variations in the pore fluid can result in additional isotopic effects on the recrystallizing cherts. Finally note that O and B isotopes differ strongly regarding the evolution of the isotopic composition of the pore water, since the water/rock ratios will be drastically different for the two elements: it is conceivable to have a pore fluid keeping a δ^{18}O value close to that of seawater while its δ^{11}B value is significantly decreased from the seawater value by silicate dissolution.

The compositions of DSDP cherts are in agreement with this first order view of B isotopic fractionation during diagenesis in the sedimentary pile. In fact, the linear trend observed between the δ^{11}B values and l/[B] (Fig. 5) can be interpreted as a mixing line between two extreme chert compositions, one having [B] $\approx$ 100 ppm and δ^{11}B $\approx +8$‰ and the other having [B] $\approx$ 50 ppm and δ^{11}B ≈ -10‰. It is important to notice that neither of these two end-members correspond to silica at equilibrium with seawater (δ^{11}B $\approx +2 \pm 2$‰ for cherts having traces of opal-CT in the top of the sequence). Thus Fig. 5 demonstrates that δ^{11}B values can either increase or decrease during diagenesis. The only possible cause for that is a variation of the δ^{11}B of the pore fluid. Though it might be coincidental, the Y intercept of the correlation line between chert δ^{11}B values and l/[B] is $+24$‰, close to the expected δ^{11}B value of B(OH)$_4^-$ dissolved in seawater as well as to that of marine carbonates (see Sections 4.1 and 4.2 on B isotopic fractionations) so that the range of [B] and δ^{11}B values observed in cherts could well be explained by an addition of seawater or seawater-derived (from carbonate dissolution) B to the cherts during their diagenesis. The high δ^{18}O values ($\approx +36$‰) of the cherts having high δ^{11}B values argue in favor of an interaction at low temperature between silica and a fluid dominated by seawater, rather than a process involving carbonate dissolution. In fact the δ^{18}O values of carbonate precipitated at around 10°C are of $\approx \pm 32$‰ (O'NEIL *et al.*, 1969), which is too low to account for the trend observed in Fig. 7. On the other hand, silica equilibrated with seawater at low temperature is characterized by higher δ^{18}O values (*e.g.* $+36$‰ at 23°C and $+39$‰ at 10°C; LABEYRIE, 1974).

It is also clear from the positive correlation between [B] and δ^{18}O and between δ^{11}B and δ^{18}O (Figs. 6 and 7) that B and ^{10}B are generally lost by the cherts with increasing diagenesis, this effect being the most prominent for the deepest samples (Fig. 4). In fact the low δ^{18}O of cherts are believed to result at first order from a progressive O isotopic reequilibration of silica with pore waters at temperatures increasing with burial depth (KOLODNY and EPSTEIN, 1976). Thus it is likely that the end-member having the lowest [B] and δ^{11}B value represents the "typical" composition of chert after diagenesis in a more or less closed system. Following this line of reasoning, the chert end-member having high [B] and δ^{11}B values, even higher than biogenic silica deposited on the sea floor, could result from a recrystallization during diagenesis in the presence of a B- and ^{11}B-rich fluid, dominated by seawater. The reason for this enhanced interaction with seawater after deposition on the seafloor for cherts at around 800 mbsf and a few hundred meters on both sides (Fig. 4), can tentatively be found in the deposition history of the sedimentary pile at site DSDP 167 (though it is not an unusual history for any oceanic sediment column). In fact at the end of the Cenomanian (sample 167-60-2, depth 850 mbsf), about 100 million years ago, the cherts of Cenomanian to Tithonian age were buried under less than 300 m of sediment (1160–850 mbsf). The younger cherts, which formed in the next 70 million years were exposed for long periods of their history to very slow or zero sedimentation rates, thus to rather complex diagenesis at shallow depth, involving dissolution, bioturbation, predation (SCHLANGER *et al.*, 1973).

CONCLUSIONS

The present results demonstrate that the major control on the B isotopic composition of cherts is diagenetic. As a first approximation it can be suggested that cherts lose B and ^{11}B during diagenesis so that old cherts are characterized by associated low B contents and low δ^{11}B and δ^{18}O values. However the present suite of DSDP cherts also shows that cherts can be enriched in ^{11}B during diagenesis if seawater B is dominant in the pore fluid. Thus, marine cherts cannot be used to reconstruct seawater paleo-pH. However biogenic silica is precipitated with a B isotopic fractionation relative to seawater of -35 to -40‰, which is much larger than that of marine carbonates ($-23\%_0$). This suggests that phases may exist in marine sediments that fractionate B isotopes differently from carbonates and that may be used in combination with carbonates to determine both paleo-pH and paleo-δ^{11}B values of seawater. In addition, because of this large B isotopic fractionation, biogenic silica precipitation must be considered as an important flux in the control of seawater δ^{11}B values since the removal flux of ^{10}B is of same order of magnitude than that due to carbonate precipitation.

Acknowledgements—We gratefully acknowledge the assistance of the National Science Foundation (US) in obtaining the samples for this study from the DSDP depository. A. Katz carried out the ICP-AES boron analyses, and E. Barkan the organic matter analysis. A. Matthews, M. Kastner and an anonymous reviewer critically read the manuscript and made several useful suggestions. We are most grateful to all of them.

REFERENCES

BEHL R. J. (1992) Chertification in the Monterey Formation of California and deep-sea sediments of the West Pacific. Ph.D. Thesis, University of California, SantaCruz.

BOHRMANN G., ABELMANN A., GERSONDE R., HUBBERTEN H. and KUHN G. (1994) Pure siliceous ooze, a diagenetic environment for early chert formation. *Geology* **22**, 207–210.

BRUMSACK H. J. and ZULEGER E. (1992) Boron and boron isotopes in pore waters from ODP-Leg 127, Sea of Japan. *Earth Planet. Sci. Lett.* **113**, 427–433.

CALVERT S. E. (1974) Deposition and diagenesis of silica in marine sediments. In *Pelagic Sediments: On Land and Under the Sea* (eds. K. J. HSÜ and H. C. JENKYNS) Vol. 1, pp. 273–299. Spec. Publ. Int. Assoc. Sediment.

CHAUSSIDON M. and JAMBON A. (1994) Boron content and isotopic composition of oceanic basalts: geochemical and cosmochemical implications. *Earth Planet. Sci. Lett.* **121**, 277–291.

CHAUSSIDON M., ROBERT F., MANGIN D., HANON P. and ROSE E. (1997) Analytical procedures for the measurement of boron isotope compositions by ion microprobe in meteorites and mantle rocks. *Geostand. Newslett.* **21**, 7–17.

DEMASTER D. J. (1981) The supply and accumulation of silica in the marine environment. *Geochim. Cosmochim. Acta* **45**, 1715–1732.

DOUGLAS R. G., ROTH P. H. and MOORE T. C. (1973) Biostratigraphic synthesis, Deep-Sea Drilling Project Leg-17. *Initial Reports of the Deep Sea Drilling Project* **17**, 905–909.

GOLDSCHMIDT V. M. (1958) *Geochemistry*. Oxford University Press, New York, 730 pp.

GOLDSCHMIDT V. M. and PETERS C. (1932) Zur geochemie des bors. *Nachr. Ges. Wiss. Güttingen. Math-Phys. Kl*, 528–545.

HARDER H. (1974) Boron. In *Handbook of Geochemistry* (ed. K. H. WEDEPOHL), pp. 5-I-1–5-K-13. Springer, Berlin.

HEATH G. R. (1973) Cherts from the eastern Pacific, Leg 16, Deep-Sea Drilling Project. *Initial Reports of the Deep Sea Drilling Project* **7**(Part 2), 991–1008.

HEIN J. R. and YEH H. W. (1981) Oxygen-isotope composition of chert from the Mid-Pacific Mountains and Hess Rise, Deep-Sea Drilling Project Leg-62. *Initial Reports of the Deep Sea Drilling Project* **62**, 749–758.

HEIN J. R., SANCETTA C. and MORGENSON L. A. (1981) Petrology and geochemistry of silicified Upper Miocene chalk, Costa Rica Rift, Deep-Sea Drilling Project Leg-69. *Initial Reports of the Deep Sea Drilling Project* **69**, 749–758.

HEIN J. R., SCHOLL D. W., BARRON J. A., JONES M. G. and MILLER J. (1978) Diagenesis of Late Cenozoic Diatomaceous deposits and formation of bottom simulating reflector in Southern Bering Sea. *Sedimentology* **25**, 155–181.

HEMMING N. G. and HANSON G. N. (1992) Boron Isotopic composition and concentration in modern marine carbonates. *Geochim. Cosmochim. Acta* **56**, 537–543.

HEMMING N. G., REEDER R. J. and HANSON G. N. (1995) Mineral-fluid partitioning and isotopic fractionation of boron in synthetic calcium carbonate. *Geochim. Cosmochim. Acta* **59**, 371–379.

ISHIKAWA T. and NAKAMURA E. (1993) Boron isotope systematics of marine-sediments. *Earth Planet. Sci. Lett.* **117**, 567–580.

KASTNER M. (1981) Authigenic silicates in deep-sea sediments: formation and diagenesis. In *The Sea, The Oceanic Lithosphere*, (ed. C. EMILIANI), Vol. 7, pp. 915–980. Wiley, New York.

KAKIHANA H., KOTAKA M., SATOH S., NOMURA M. and OKAMOTO M. (1977) Fundamental studies on the ion-exchange separation of boron isotopes. *Bull. Chem. Soc. Jpn* **50**, 158–163.

KNAUTH L. P. (1994) Petrogenesis of Chert. In *Silica: Physical Behavior, Geochemistry and Materials Applications*. (ed. P. J. HEANEY, C. T. PREWITT and G. V. GIBBS), Vol.29, pp. 233–258. Min. Soc. Am., Rev. Mineral.

KOEPNICK R. B., BURKE W. H., DENISON R. E., HETHERINGTON E. A., NELSON H. F., OTTO J. B. and WAITE L. E. (1985) Construction of the seawater Sr-87/Sr-86 curve for the Cenozoic and Cretaceous-supporting data. *Chem. Geol.* **58**, 55–81.

KOLODNY Y. and EPSTEIN E. (1976) Stable isotope geochemistry of deep sea cherts. *Geochim. Cosmochim. Acta* **40**, 1195–1209.

KOLODNY Y. and HALICZ L. (1988) The geochemistry of germanium in deep-sea cherts. *Geochim. Cosmochim. Acta* **52**, 2333–2336.

KOLODNY Y., TARABLOUS A. and FRIESLANDER U. (1980) Participation of fresh water in chert diagenesis—evidence from stable isotopes and boron α-track mapping. *Sedimentology* **27**, 305–316.

KOPF A., DEYHLE A. and ZULEGER E. (2000) Evidence for deep fluid circulation and gas hydrate dissociation using boron and boron isotopes of pore fluids in forearc sediments from Costa Rica (ODP Leg 170). *Mar. Geol.* **167**, 1–28.

LABEYRIE L. (1974) New approach to surface seawater paleotemperatures using O^{18}/O^{16} ratios in silica of diatom frustules. *Nature* **248**, 40–42.

LÉCUYER C., GRANDJEAN P., REYNARD B., ALBARÈDE F. and TELOUK P. (2002) $^{11}B/^{10}B$ analysis of geological materials by ICP-MS Plasma 54: application to the boron fractionation between brachiopod calcite and seawater. *Chem. Geol.* **186**, 45–56.

LEEMAN W. P. and SISSON V. B. (1996) Geochemistry of boron and its implication for crustal and mantle processes. In *Boron: Mineralogy, Petrology and Geochemistry*, 33(ed. E. S. CREW and L. M. ANOVITZ), Min. Soc. Am., Rev. Mineral., pp. 645–707.

LEMARCHAND D., GAILLARDET J., LEWIN E. and ALLEGRE C. J. (2000) The influence of rivers on marine boron isotopes and implications for reconstructing past ocean pH. *Nature* **408**, 951–954.

LEMARCHAND D., GAILLARDET J., LEWIN E. and ALLEGRE C. J. (2002) Boron isotope systematics in large rivers: implications for the marine boron budget and paleo-pH reconstruction over the Cenozoic. *Chem. Geol.* **190**, 123–140.

LEMARCHAND E., GAILLARDET J. and SCHOTT J. (2003) Boron sorption and isotopic fractionation on humic acid and goethite. *European Geophysical Society Meeting, Nice; Geophys. Res. Abst.* **5**, 10065.

LERMAN A. (1966) Boron in clays and estimation of paleosalinities. *Sedimentology* **6**, 267–286.

MATHENEY R. K. and KNAUTH L. P. (1993) New isotopic temperature estimates for early silica diagenesis in bedded cherts. *Geology* **21**, 519–522.

NAKAMURA E., ISHIKAWA T., BIRCK J. L. and ALLEGRE C. J. (1992) Precise boron isotopic analysis of natural rock samples using a boron mannitol complex. *Chem. Geol.* **94**, 193–204.

O'NEIL J. R., CLAYTON R. N. and MAYEDA T. K. (1969) Oxygen isotope fractionation in divalent metal carbonates. *J. Chem. Phys.* **51**, 5547–5558.

PALMER M. R. and SWIHART G. R. (1996) Boron geochemistry: an overview. In *Boron: Mineralogy, Petrology and Geochemistry* (eds. E. S. CREW and L. M. ANOVITZ), Vol. 33, pp. 709–744. Min. Soc. Am., Rev. Mineral.

PALMER M. R., SPIVACK A. J. and EDMOND J. M. (1987) Temperature and pH controls over isotopic fractionation during adsorption of boron on marine clay. *Geochim. Cosmochim. Acta* **51**(9), 2319–2323.

PALMER M. R., PEARSON P. N. and COBB S. J. (1998) Reconstructing past ocean pH–depth profiles. *Science* **282**, 1468–1471.

PEARSON P. N. and PALMER M. R. (2000) Atmospheric carbon dioxide concentrations over the past 60 million years. *Nature* **406**, 695–699.

PISCIOTTO K. A. (1981) Distribution, thermal histories, isotopic compositions, and reflection characteristics of siliceous rocks recovered by the Deep Sea Drilling Project. In *The Deep sea Drilling Project: A Decade of Progress*, SEPM Special Publ., Vol. 32, 129–147.

RANKAMA K. and SAHAMA T. G. (1950) *Geochemistry*. The University of Chicago Press, Chicago.

ROLLION-BARD C., CHAUSSIDON M. and FRANCE-LANORD C. (2003) pH control on oxygen isotopic composition of symbiotic corals. *Earth Planet. Sci. Lett.* **215**, 275–288.

SANYAL A., HEMMING N. G., BROECKER W. S., LEA D. W., SPERO H. S. and HANSON G. N. (1996) Oceanic pH control on the boron isotopic composition of foraminifera: evidence from culture experiments. *Paleooceanography* **11**, 513–517.

SCHWARTZ H. P., AGYEI E. K. and MCMULLEN C. C. (1969) Boron isotopic fractionation during adsorption from seawater. *Earth Planet. Sci. Lett.* **6**, 1–5.

SCHLANGER S. O., DOUGLAS R. G., LANCELOT Y., MOORE T. C. and ROTH P. H. (1973) Fossil preservation and diagenesis of pelagic carbonates from the Magellan Rise, central North Pacific Ocean. *Initial Reports of the Deep Sea Drilling Project* **17**, 407–427.

SPIVACK A. J. and EDMOND J. M. (1987) Boron isotope exchange between seawater and the oceanic-crust. *Geochim. Cosmochim. Acta* **51**, 1033–1043.

SPIVACK A. J. and YOU C. F. (1997) Boron isotopic geochemistry of carbonates and pore waters, Ocean Drilling Program Site 85 I. *Earth Planet. Sci. Lett.* **152**, 113–122.

SPIVACK A. J., PALMER M. R. and EDMOND J. M. (1987) The sedimentary cycle of the boron isotopes. *Geochim. Cosmochim. Acta* **51**, 1939–1949.

SPIVACK A. J., YOU C. F. and SMITH J. (1993) Foraminiferal boron isotopic ratios as a proxy for surface ocean pH over the past 21 Myr. *Nature* **363**, 149–151.

TRUSCOTT M. G. and SHAW D. M. (1984) Boron in chert and Precambrian siliceous iron formations. *Geochim. Cosmochim. Acta* **48**, 2313–2320.

VENGOSH A., KOLODNY Y., STARINSKY A., CHIVAS A. R. and MCCULLOCH M. T. (1991) Coprecipitation and isotopic fractionation of boron in modern biogenic carbonates. *Geochim. Cosmochim. Acta* **55**, 2901–2910.

VENGOSH A., STARINSKY A., KOLODNY Y., CHIVAS A. R. and RAAB M. (1992) Boron-isotope variations during fractional evaporation of sea water: new constraints on the marine vs. nonmarine debate. *Geology* **20**, 799–802.

WALKER L. T. (1968) Evaluation of boron as paleosalinity indicator and its application to offshore prospects. *Bull. Am. Assoc. Petrol. Geol.* **52**, 751–766.

WILLIAMS L. B., HERVIG R. L., HOLLOWAY J. R. and HUTCHEON I. (2001) Boron isotope geochemistry during diagenesis. Part I. Experimental determination of fractionation during illitization of smectite. *Geochim. Cosmochim. Acta* **65**, 1769–1782.

WINTERER E. L. (1973) Regional problems, Deep-Sea Drilling Project Leg-17. *Initial Reports of the Deep Sea Drilling Project* **17**, 911–922.

YOU C. F., SPIVACK A. J., SMITH H. J. and GIESKES J. M. (1993) Mobilization of boron in convergent margins: implications for the boron geochemical cycle. *Geology* **21**, 207–210.

Geochemical Investigations in Earth and Space Science: A Tribute to Isaac R. Kaplan
© The Geochemical Society, Publication No. 9, 2004
Editors: R.J. Hill, J. Leventhal, Z. Aizenshtat, M.J. Baedecker, G. Claypool,
R. Eganhouse, M. Goldhaber and K. Peters

Significance of δ^{34}S and evaluation of its imprint on sedimentary organic matter: I. The role of reduced sulfur species in the diagenetic stage: A conceptual review

ZEEV AIZENSHTAT and ALON AMRANI

Department of Organic Chemistry, Casali Institute, The Hebrew University, Jerusalem 91904, Israel

Abstract—Both carbon and sulfur cycles in the geosphere are biogenically and chemically interwoven. Sulfate is the main source for sulfur in marine sediments. The incorporation of sulfur into biogenic organic matter (OM) via the assimilatory process has very little isotopic discrimination. The pioneering work by Kaplan and Rittenberg showed that sulfate-reducing bacteria (SRB) oxidize organic carbon to CO_2 while producing H_2S depleted in the ^{34}S isotope. The use of sulfate as an electron acceptor during bacterial dissimilatory processes produces H_2S that can be up to 72‰ depleted in ^{34}S relative to the sulfate. Carbon source, SRB species and hence rate of sulfate reduction may influence the overall isotopic fractionation $\Delta SO_4^{2-} \rightarrow S^{2-}$. In addition, the supply of sulfate (open versus closed system) is important for determining isotopic fractionation. The H_2S formed by the SRB quickly reacts with available iron to form pyrite (FeS_2) via the precursor FeS. Sulfide-oxidizing bacteria may form elemental sulfur, that at pH $\sim$7–9, reacts with sulfide to form polysulfides. Polysulfides were found to be chemically the most reactive species of sulfur with OM. Isotopically, polysulfides carry the dissimilatory δ^{34}S value. Hence, if the secondary sulfur enrichment in sedimentary organic matter (SOM) is by chemical reaction with the polysulfides, then the δ^{34}S values for the OM, rich in sulfur, will gradually be imprinted by the dissimilatory process.

Most of the information on δ^{34}S values of the reduced sulfur in sediments derives from both acid "volatile" sulfide (FeS) and pyrite (FeS_2). In a few cases, both sulfate and other sulfur species were isotopically compared. Due to analytical difficulties, organic sulfur and elemental sulfur isotope (δ^{34}S) ratios were studied only in cases where the secondary enrichment led to OM rich in sulfur. This secondary enrichment forms type II-S kerogens.

Three such natural cases of secondary enrichment of OM will be discussed:

(a) Solar lake—young cyanobacterial mat (Sinai, Egypt);
(b) Dead Sea—immature asphalts and bituminous rocks (Senonian Ghareb Formation, Israel);
(c) Monterey Formation selected samples (Miocene Formation, California, USA).

These three case studies are typified by low–medium maturity. The Solar Lake mats are at the early stages of diagenesis, the Monterey and the Ghareb Formations have already formed type II-S kerogens. In most cases, the pyrite records the most ^{34}S-depleted sulfur in the sediments and sedimentary rocks, whereas sulfate is the most ^{34}S enriched. The organically bonded sulfur has a wider range of isotopic compositions probably due to its dependence on timing and multiple step reactions discussed in this review. It is our intention in this review to offer a feasible mechanistic approach to connect δ^{34}S ratios recorded with depositional environment and diagenetic processes.

INTRODUCTION

THE MOST quantitatively significant sink for reduced sulfur is pyrite that is considered to be the end product of sulfur diagenesis in anoxic marine sediments (BERNER, 1970; GOLDHABER and KAPLAN, 1974; BERNER and RAISWELL, 1983; KUMP and GARRELS, 1986). However, another important sink for reduced sulfur in marine sediments and sedimentary rocks is the organic sulfur (OS). In the last two decades, a significant progress has occurred in the understanding of the OS formation in marine sediments (for reviews see SINNINGHE DAMSTÉ and DE LEEUW, 1990; KREIN, 1993). Sulfur stable isotope data prove that incorporation of inorganic reduced sulfur species into organic matter (OM) during early diagenesis is the most important source of OS (DINUR et al., 1980; AIZENSHTAT et al., 1983; FRANCOIS, 1987; MOSSMANN et al., 1991; ZABACK and PRATT, 1992; ANDERSON and PRATT, 1995; CANFIELD et al., 1998; PASSIER et al., 1999; WERNE et al., 2003). The main source for inorganic reduced sulfur species in marine sediments is the dissimilatory sulfate reduction by sulfate-reducing bacteria (SRB) (KAPLAN et al., 1963). The diagenetic sulfur enrichment of OM-rich sediments typically shows the following: (a) pyrite which is always lighter (^{34}S depleted) relative to all other species; (b) sulfate which is (i) open sea water $\delta^{34}S_{sulfate}$ or (ii) heavier (^{34}S enriched) gypsum or anhydrite if the system was closed. Elemental sulfur is generally isotopically similar to the kerogen δ^{34}S values and is always ^{34}S enriched relative to the pyrite by an average of 10‰ (ANDERSON and PRATT, 1995). In all sediments of this early stage of maturity, the extracted bitumens show slight enrichment in ^{34}S (IDIZ et al., 1990). These observations are still under intense investigation and yet not satisfactorily addressed.

This paper focuses on the formation of type II-S kerogens and their ^{34}S isotopic imprint. These kerogens are defined as containing high organically bonded sulfur (S/C 0.04–0.8 atomic ratio). The high content of sulfur cannot be attributed to the selective preservation of the assimilatory bio-incorporated sulfur. Hence, it is suggested that type II-S kerogens indicate sulfur-rich euxinic depositional environments with very active SRB activity.

It is our intention in this review to offer a feasible mechanistic approach to connect δ^{34}S ratios of these kerogens with depositional environment and diagenetic processes, by reviewing and evaluating the δ^{34}S data of three case studies integrated with theoretical and laboratory simulation experiments data. The δ^{34}S of the later and higher temperatures catagenetic stage (i.e. kerogens compared with bitumens and oils) that are controlled by thermal mechanisms will be discussed in Part II of this review (this volume). Furthermore, our research indicates that dissolved polysulfide species (S_x^{2-}) are key reactants intermediate in sulfurization of OM; hence, polysulfides geochemistry will also be stressed.

CHEMICAL SULFATE REDUCTION

Since the earth's atmosphere became oxygen rich, the sedimentary supply of sulfur is marine sulfate dominated (HOLSER et al., 1988; BERNER, 1989). The δ^{34}S values recorded for ocean sulfate fluctuate with geological time, averaging $+20 \pm 10$‰. The geological record of ocean sulfate is documented by the $CaSO_4$ minerals (gypsum, anhydrite). The crystallization of these minerals from seawater does not have a significant isotopic fractionation. Some seawater sulfate is incorporated into carbonate minerals; however, STAUDT and SCHOONEN (1995) claim that most of this sulfate is released upon diagenetic recrystallization. The major pathways for the removal of sulfate from the marine environment into the sediments are: (i) precipitation of evaporites; (ii) reduction to H_2S and subsequent formation of mineral sulfides (mostly pyrite); (iii) reduction to H_2S (to form other reduced sulfur species) and subsequent incorporation into OM. Although the first two pathways were recognized during early studies of the sulfur cycle (e.g. SAKAI, 1957; KAPLAN et al., 1963; BERNER, 1970; GOLDHABER and KAPLAN, 1974), the organic geochemistry of sulfur has become a research focus more recently (VAIRAVAMURTHY et al., 1995).

The reduction of sulfate (S^{6+}) to sulfide (S^{2-}) is an 8-electron reaction (see Fig. 1a), which requires a very high-energy investment, and if conducted without a catalyst has a very high activation energy. Although the chemical reduction of sulfate to H_2S will not occur at low temperatures, it has been calculated to have an isotope kinetics effect of 25‰ (^{34}S depletion) for H_2S produced (BIGELEISEN, 1949; TUDGE and THODE, 1950; HARRISON and THODE, 1957). This theoretical value for the production of ^{34}S-depleted H_2S is temperature dependent as well as controlled catalytically. The presence of elemental sulfur was found by WILLIAMSON and RIMSTIDT (1990) to reduce the activation energy via multi-step addition of elemental sulfur to form catanated sulfur compounds such as polysulfides and polythionates (Fig. 1b). Figure 1c describes the difference between the multi-step enzymatic bio-catalytic and a one-step chemically controlled reduction. The change of the formal valence of sulfur from $+6$ to -2 is then energetically feasible even at moderate temperatures.

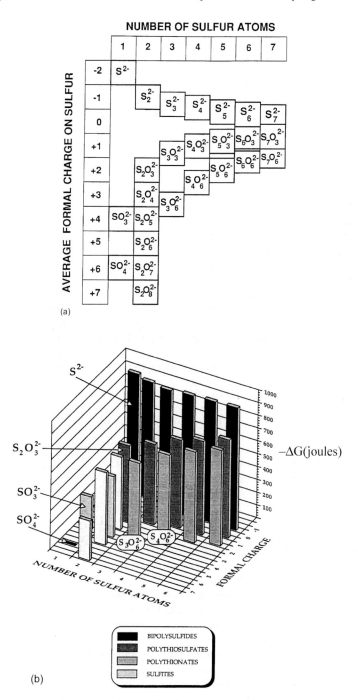

Fig. 1. (a) Schematic description of formal charge on sulfur atom versus number of sulfur atoms. All possible steps from S^{7+} to the S^{2-} by addition of S_x^0, $x = 1-7(S_8^0)$. (b) The thermodynamic stability of sulfur species relative to sulfate, 3D diagram adapted from WILLIAMSON and RIMSTIDT (1990), showing the formal charge changes via addition of sulfur atoms to form polythionates and polysulfates to polysulfides. (c) Conceptual reaction profile and activation energy comparison of catalyzed multi-step reaction to chemical one 8-electron mechanism. The overall energy required for the reduction of the sulfate (S^{6+}) to the sulfide (S^{2-}) is marked. Addition of elemental sulfur to form the intermediate valences will reduce the incremental activation energies thus facilitating the reaction under lower temperatures (see discussion in text and references cited).

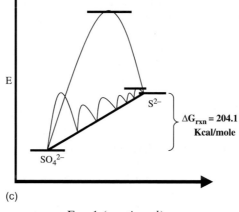

(c)

FIG. 1 (*continued*)

ASSIMILATORY AND DISSIMILATORY SULFATE REDUCTION

The biogenic reduction of sulfate follows two biosynthetic pathways: (1) assimilatory incorporation of sulfur into the amino acids (*e.g.* cysteine) with very small sulfur isotope fractionation; (2) dissimilatory production of sulfide coupled to oxidation of OM to CO_2 (see Scheme 1). If we take, for example, methane as the electron donor, the reaction is

$$CH_4 + SO_4 + 2H^+ \rightarrow H_2S + CO_2 + 2H_2O, \qquad \Delta G_r^0 = -26.3 \text{ kcal/mol.} \qquad (1)$$

This dissimilatory sulfate reduction by bacteria to H_2S has a marked effect on the $\delta^{34}S$, up to $-72‰$, though the average is about $-45 \pm 5‰$ (*e.g.* KAPLAN *et al.*, 1963; KAPLAN and RITTENBERG, 1964; GOLDHABER and KAPLAN, 1974; FRY, 1988; HABICHT *et al.*, 1998; PETERSON, 1999; DETMERS *et al.*, 2001; WORTMANN *et al.*, 2001). Since the pioneering work by Kaplan (KAPLAN *et al.*, 1963; KAPLAN and RITTENBERG, 1964) and subsequent studies of the Kaplan group relating to the sulfur isotope fractionation during microbial reduction, many studies have been reported on cultures of SRB. The diversity of the dissimilatory isotope fractionation of some 32 prokaryotes (SRB) in pure cultures under optimal growth conditions

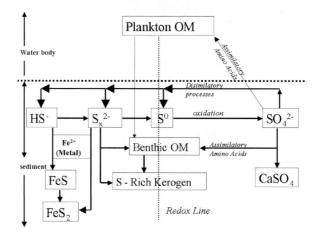

SCH.1. Depositional environment schematic sulfur cycle in marine water body and upper water–sediment interface. The redox line (gray dashed perpendicular) is placed at S^0 and relates only to the redox state of the sulfur, not the oxic or euxinic conditions. The black dashed line separates between sediment and water bodies. For the present paper, we would like to emphasize the major $\delta^{34}S$ ratio changes caused by the dissimilatory reduction of sulfate. For more detailed explanations on the formation of sulfur rich OM, see AIZENSHTAT *et al.* (1983, 1995, 2004) and KREIN and AIZENSHTAT (1993, 1994).

showed much lower isotope fractionation (-2 to $-40‰$) than that observed for natural environments (DETMERS et al., 2001). Since these experiments were conducted with single organic molecules as carbon source (benzoate, lactate, pyruvate, etc.), it is difficult to compare these results to those from marine depositional environments. In addition, it is important to note that in most cases studied, the sulfur isotope fractionation was found to be independent of the sulfate concentration or its reduction rate.

REDUCED SULFUR SPECIES IN MARINE SEDIMENTS

There are three principal species of reduced sulfur: (i) H_2S that is produced mainly by SRB as earlier discussed. At pH above 7 the sulfide anion (HS^-) is quantitatively the dominant species (SCHWARZENBACH and FISHER, 1960; BOULEGUE and MICHARD, 1978; BOULEGUE, 1978). In order to simplify, we will use the term sulfide for both species; (ii) elemental sulfur (S^0) produced from bacterial or chemical oxidation of sulfide (JORGENSEN, 1988); (iii) polysulfides (S_x^{2-}) produced from bacterial or chemical oxidation of H_2S (FOSSING and JORGENSEN, 1990) or by reaction between H_2S or HS^- and elemental sulfur according to Eq. (2):

$$2HS^- + S_8^0 \rightleftharpoons 2\ ^-S-S-S-S-S^-\ (S_x^{2-}, x = 2-6) \tag{2}$$

At pH above 7 there is a minor concentration of the hydropolysulfide HS_x^- and the dominant species is the polysulfide S_x^{2-} (GIGGENBACH, 1972; LICHT et al., 1986; LICHT and DAVIS, 1997). Thus, at pH 7.5–9 that prevails in marine sediments, polysulfides are produced chemically and stabilized. The number of sulfur atoms catanated in polysulfides form is influenced by pH. As the pH increases, the number of sulfur atoms in a polysulfide molecule decreases. At pH 7.5–9.0 the most abundant species of polysulfides in aquatic media are S_4^{2-}, S_5^{2-} and S_6^{2-} (SCHWARZENBACH and FISHER, 1960; BOULEGUE and MICHARD, 1978; BOULEGUE, 1978). The reactivity of these reduced sulfur species towards OM is a key characteristic of their behavior and will be discussed in detail later.

The abundance of reduced sulfur species in marine sediments

There is uncertainty as to the reported concentrations of polysulfides and elemental sulfur in recent sediments since the chemical treatment (mainly acidification) prior to the extraction of the elemental sulfur decompose the polysulfides (MOSSMANN et al., 1991; ROZAN et al., 2000; NERETIN et al., 2003) according to Eq. (3):

$$S_x^{2-} + 2H^+ \longrightarrow H_2S + (x-1)S^0 \tag{3}$$

Moreover, chemical decomposition of sulfur-rich OM (i.e. polysulfide cross-linked polymer) can also release elemental sulfur (MOSSMANN et al., 1991). Therefore, the concentration of reduced sulfur species and especially polysulfides is a difficult analytical problem. With this caveat, the concentration of dissolved sulfide in anoxic marine sediments is generally the largest followed by elemental sulfur and polysulfides. In most cases elemental sulfur dominates at the water sediment interface (oxic–anoxic) and polysulfides are detected at intermediate depth. At greater depth, sulfide is dominant and small concentration of elemental sulfur (including polysulfide) is present (MOSSMANN et al., 1991; ROZAN et al., 2000; LUTHER et al., 2001, NERETIN et al., 2003). Polysulfide concentrations in microbial mat from a Great Marsh, Delaware was up to 40 μmol/l (LUTHER et al., 2001). In extreme cases such as Solar Lake, Sinai, Egypt, the polysulfides concentration in microbial mats can reach 150 μmol/l (JORGENSEN et al., 1979).

δ^{34}S OF REDUCED SULFUR IN PYRITE AND OTHER IRON SULFIDES IN SEDIMENTS

The early studies of biogenically reduced sulfur isotope SRB signature were performed on the chemically stable pyrite (GOLDHABER and KAPLAN, 1974). Since the formation of iron sulfides in the sedimentary environment is kinetically very fast and the fractionation between dissolved sulfide and pyrite associated with it is negligible (~1‰, PRICE and SHIEH, 1979; WILKIN and BARNES, 1996; BOTTCHER et al., 1998), these data will record the dissimilatory isotope imprint. It is therefore common that the pyrite is often the most [34]S-depleted species in the sediments (see following discussion and references cited). In very young as well as in more mature sediments deposited under reducing conditions, acid volatile sulfides (AVS), mostly iron sulfides, are also recorded. Some investigators claim that these AVS are precursors of pyrite, but if polysulfides are formed, direct pyrite precipitation may occur (LUTHER, 1991; WANG and MORSE, 1995). The chemistry of iron sulfides in sediments as well as the formation of pyrite and its crystalline forms have been studied and reviewed extensively (RICKARD et al., 1995; FURUKAWA and BARNES, 1995; WANG and MORSE, 1995; GOLDHABER, 2004). If the supply of sulfate (~+20‰) is unlimited (i.e. an open system), then on average the δ^{34}S values recorded for the pyrite are $-30 \pm 5‰$. However, in cases where the supply of sulfate is restricted (a partially closed system), the pyrite will gradually become isotopically heavier. In this paper, we review only the diagenetic stage, the formation of secondary pyrite under higher temperature conditions ($T > 100°C$) will be discussed in Part II (this volume).

ISOTOPE EFFECTS DURING MIXING OF SULFUR SPECIES IN MARINE ENVIRONMENTS

In order to understand the significance of the sulfur isotope composition of OM, we need to understand factors influencing the sulfur isotope composition of polysulfides.

Schematically, the formation of polysulfides may be conceptualized through the following simplified reactions:

$$3S^{2-} + S^{6+} \rightleftharpoons 4S^0 \tag{4}$$

$$S^0 + S^{2-} \rightleftharpoons S_2^{2-} \tag{5}$$

If the aqueous sulfide (S^{2-}) is about $-20‰$ and SO_4^{2-} (S^{6+}) $+20‰$, the S^0 formed under strictly stoichiometric calculation (Eq. (4)) will be $-10‰$. The coupling of sulfide and sulfate up to 200°C (autoclave conditions) will not yield 1:1 mixing. These sulfur isotope-mixing reactions have been reviewed as part of our study of thermal sulfate reduction (TSR) in GOLDSTEIN and AIZENSHTAT (1994). The conclusion of that review is that the isotope mixing between H_2S and SO_4^{2-} is very slow at pH 7 up to even 200°C (90% isotope mixing in 10,000 years, 200°C). However, if elemental sulfur is present (formed by bacterial sulfide oxidation), then the isotope mixing may occur via the formation of polysulfides (Eq. (5)). The incremental changes in the formal charge on the sulfur atoms allow lower temperature mixing since the energetic steps (Fig. 1b) between the sulfur species are lower (WILLIAMSON and RIMSTIDT, 1990).

FOSSING and JORGENSEN (1990) reported that no isotope mixing was observed between radioactive ^{35}S sulfate tracer and reduced sulfur compounds including polysulfides. However, there are no laboratory experimental data on δ^{34}S chemical isotope mixing of polysulfides and sulfate at low temperatures. From field data, we can assume that such mixing is not favorable under diagenetic marine sediments, since in most cases, sulfate is significantly [34]S enriched compared to the coexisting reduced sulfur species (for examples see Fig. 2 and AIZENSHTAT et al., 1983; MOSSMANN et al., 1991; BRUCHERT and PRATT, 1996; CANFIELD et al., 1998; WERNE et al., 2003). Laboratory experiments with ^{35}S-labeled sulfur species showed rapid isotope mixing between H_2S, S, S_x^{2-}, FeS at pH 7.6 and 20°C (FOSSING and JORGENSEN, 1990). Recently, we demonstrated a total isotope exchange of stable sulfur isotopes according to Eq. (5) at pH 8.5 and 25°C (AMRANI and AIZENSHTAT, 2003). Polysulfides serve as mediator for this isotope mixing. Thus, we can consider the reduced sulfur species pool as isotopically homogeneous.

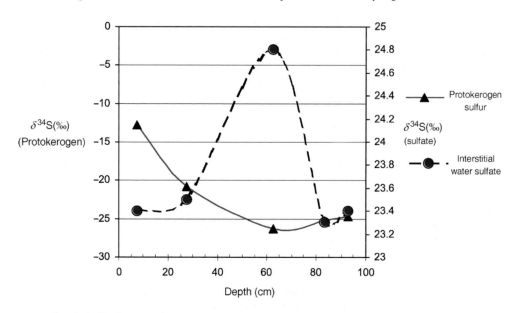

FIG. 2. Sulfur isotopes of protokerogens and interstitial waters sulfate Solar Lake (Sinai, Egypt). The sediment core of 1 m was extensively studied and reported in AIZENSHTAT et al. (1983) and STOLER (1981). $\delta^{34}S_{org}$ of protokerogens and $\delta^{34}S_{sulfate}$ from interstitial waters are plotted versus depth.

DEPOSITIONAL CONDITIONS FOR SULFUR-RICH OM

The formation of marine sediments rich in OM requires a rather unique set of environmental conditions (see review by AIZENSHTAT et al. (1999)). The present discussion concentrates primarily on depositional environments that support secondary sulfur enrichment of OM (DINUR et al., 1980; FRANCOIS, 1987). SRB consume large amounts of the OM during the dissimilatory sulfate reduction (see Eq. (1)) and therefore a very high OM productivity is required to lead to a sulfur-rich OM sediment (AIZENSHTAT et al., 1999). It has been shown recently that these environments do not have to be strictly anaerobic (obligatory) but could accommodate both prolific photosynthetic CO_2-fixing bacteria as well as SRB (KLEIN et al., 2001; SIGALEVICH et al., 2000; SIGALEVICH and COHEN, 2000). This can explain the occurrence of polysulfides even in an oxygen-rich aquatic system (GUN et al., 2000). The preservation of high concentrations of OM and reduced species of sulfur commonly occurs in hypersaline conditions (AIZENSHTAT et al., 1999). These saline environments support formation and accumulation of benthic matter such as communal bacterial mats (AIZENSHTAT et al., 1983; PETERSON, 1999). Benthic bacterial mats are highly prolific for CO_2 fixation and formation of sulfide and polysulfides in the interstitial water (see Scheme 1) (AIZENSHTAT et al., 1983, 1984, 1999). Another very important factor for the formation of marine sediments rich in sulfurized OM is the rate of sulfide production by the SRB that must exceed the rate at which it can be removed by iron minerals (JORGENSEN, 1982; CHANTON et al., 1987). In that case, dissolved sulfide can accumulate in the sediment and undergo further reactions such as sulfide oxidation to form elemental sulfur and polysulfides that react with OM.

THE SECONDARY SULFUR ENRICHMENT OF OM

The incorporation of inorganic species of reduced sulfur into OM during the early diagenesis stage is widely recognized (VALISOLALAO et al., 1984; BRASSELL et al., 1986; FRANCOIS, 1987; SINNINGHE DAMSTÉ et al., 1989; KOHNEN et al., 1990a; WAKEHAM et al., 1995; LÜCKGE et al., 2002). Moreover, the depletion of the sedimentary organic matter (SOM) in ^{34}S compared to the coexisting marine sulfate is generally interpreted as an evidence for this secondary enrichment of OM by dissimilatory reduced sulfur species. If only assimilatory sulfur is the source for the sedimentary S in the OM, it will carry the isotope signature of sulfate in the depositional environment (*e.g.* KAPLAN *et al.*, 1963; FRY, 1988). OM with this isotopically heavy signature ($\sim 20‰$) is extremely rare in benthic accumulated OM (AIZENSHTAT *et al.*, 1983; MOSSMANN *et al.*, 1991; ANDERSON and PRATT, 1995; CANFIELD *et al.*, 1998; PASSIER *et al.*, 1999; WERNE *et al.*, 2003). Thus, in most cases, dissimilatory reduced sulfur species are the most important fraction of the OM sulfur in marine sediments.

Mechanisms of reduced sulfur species incorporation into OM

The bisulfide anion (HS^-) is the dominant species of reduced sulfur species at pH values prevailing in marine sediments in chemical equilibrium with other reduced sulfur species (SCHWARZENBACH and FISHER, 1960; BOULEGUE and MICHARD, 1978; BOULEGUE, 1978). Elemental sulfur and polysulfides concentration are lower (see previous discussion). Reactivity of these various sulfur species towards OM in aquatic media is different. Because elemental sulfur solubility in water is very low, its reactivity at diagenetic temperatures is negligible. The temperature required to cleave the $S-S$ bond of S_8 to initiate free radical reaction is at about 130°C (LALONDE *et al.*, 1987; KREIN, 1993). Such temperatures are not reached in most diagenetic depositional environments. Moreover, free radicals of sulfur are not stable in aquatic environments. Bisulfide anion (HS^-) and polysulfides (S_x^{2-}) are the main sulfurizing agents in slightly basic environments (*i.e.* pH 7.5–9). However, polysulfides are much more reactive species as was demonstrated theoretically and experimentally (LALONDE *et al.*, 1987; VAIRAVAMURTHY and MOPPER, 1989; LOCH *et al.*, 2002). The nucleophilicity of the polysulfides increases as the chain length of the polysulfide increases hence leading to high activity (LALONDE *et al.*, 1987). At pH 7.5–9 that prevails in marine sediments, the dominant forms of polysulfide species are the more catanated and thus the reactivity of polysulfide anions are even higher. Therefore, polysulfides are likely the most important species involved in diagenetic secondary sulfur enrichment (AIZENSHTAT *et al.*, 1983; FRANCOIS, 1987; MOSSMANN *et al.*, 1991; VAIRAVAMURTHY *et al.*, 1992). Laboratory sulfurization experimental simulations with model organic compounds and polysulfide anions in aquatic environments were performed at relatively low temperatures (up to 50°C) with different reaction media (KREIN and AIZENSHTAT, 1993, 1994; SCHOUTEN *et al.*, 1994; AIZENSHTAT *et al.*, 1995; AMRANI and AIZENSHTAT, 2004). The main conclusions from these experiments were that electrophilic model compounds such as aldehydes, ketones and activated double bonds react rapidly (hours to days) with polysulfide anions and yield polysulfide cross-linked oligomer–polymers similar to those found in marine sediments (see review by KREIN (1993) on the chemical structure of such sedimentary polymers). A growing number of studies suggest the formation of polysulfide cross-linked oligomer–polymer is not only formed from functionalized lipids reaction with polysulfides, but also from carbohydrates as well (VAN KAAM-PETERS *et al.*, 1998; SINNINGHE DAMSTÉ *et al.*, 1998; KOK *et al.*, 2000; AYCARD *et al.*, 2003; VAN DONGEN *et al.*, 2003) despite the fact that carbohydrates are generally thought to be poorly preserved during early diagenesis (ARNOSTI, 1995).

Scheme 2 describes the diagenetic chemically controlled suggested pathways for the incorporation of polysulfides into the SOM. The catagenic thermally controlled section is discussed in Part II.

THE CONTROLLING FACTORS FOR THE $\delta^{34}S$ OM RICH IN SULFUR

OS in marine sediments is enriched in ^{34}S by up to 30‰ with an average of about 10‰ relative to co-existing pyrite in most modern marine sediments and sedimentary rocks (ANDERSON and PRATT, 1995). This phenomenon of heavy sulfur enrichment can be explained by four major constraints either singly or in combination:

   (a) timing of the sulfur incorporation—sulfate open versus closed systems;
   (b) assimilatory compared with dissimilatory sulfur incorporation into OM;
   (c) sulfur species that are incorporated into OM;
   (d) fractionation during sulfur incorporation into OM.

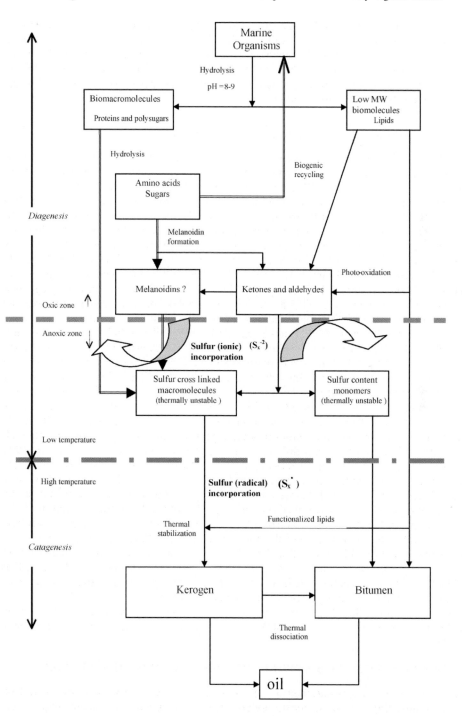

SCH. 2. General scheme of both the diagenetic changes and catagenetic transformations in euxinic depositional environments. For the present discussion we emphasize the role of polysulfides in the enrichment of sedimentary organic matter (SOM) during diagenesis. Mechanisms controlled by the nucleophilicity of S_x^{2-} are detailed in AIZENSHTAT et al. (1995). The catagenesis mechanisms leading to the production of petroleum are controlled by free radicals and are discussed in detail in Part II (this book). This scheme is based on many studies cited in this manuscript as well as in Part II (adapted from AMRANI and AIZENSHTAT, 2003).

Timing of the sulfur incorporation—sulfate open versus closed systems

The best marker for the "early" $\delta^{34}S$ value of biogenically produced sulfide is pyrite. That is because the available iron reacts rapidly and with negligible isotope discrimination with reduced aqueous sulfide. The isotopic value of S in pyrite and consequently in the residual sulfide and polysulfides, which reacts more slowly to form the organically bonded S, is controlled by several mechanisms. The most important of these mechanisms is the rate of sulfate dissimilatory reduction and proportion of sulfate reservoir reduced. If the supply of sulfate is unlimited (*i.e.* an open system) and the SRB obtain available OM and nutrients (N, P), the rate of sulfate reduction will produce ^{34}S depleted, $\Delta\delta^{34}S_{sulfate-sulfide} \sim -45\%_o$. However, if the system is partially closed to sulfate re-supply, the early-formed sulfide will be much lighter than the reduced sulfur formed under closed conditions. Hence, in environments where the reservoir of sulfate is limited and the amount of available iron (Fe^{2+} or Fe^{3+}) is small, the first pyrite to be formed will be much more ^{34}S depleted than the remaining reservoir of reduced sulfur. This means that with the increase of the extent of sulfate reduction in closed systems such as the Monterey Formation (ZABACK and PRATT, 1992; ANDERSON and PRATT, 1995), the isotope fractionation between the starting sulfate and the reduced sulfur species becomes smaller. In extreme cases, complete consumption of all sulfates would result in a pool of reduced sulfur of $\sim 20 \pm 5\%_o$ (ORR, 1986). In very early diagenesis, the very intensive production of H_2S can lead to release of the light isotope gas. Thus, if we apply the Rayleigh distillation, then the residual reduced sulfur species become very heavy (^{34}S enriched) (IDIZ *et al.*, 1990).

These processes of isotope evolution explain the wide diversity of both $\delta^{34}S$ values for the organic-S and the pyrite, as well as the fact that the pyritic $\delta^{34}S$ values are the most ^{34}S depleted. Because of the complexity of the various mechanisms controlling the formation of iron sulfide $\rightarrow$ pyrite (see review by GOLDHABER (2004)) and the slower reaction of the polysulfide reaction with the SOM, the difference of $\delta^{34}S$ values for pyrite compared to the organic-S varies. Those variations $\Delta\delta^{34}S_{polysulfides-pyrite}$ will later control the $\Delta\delta^{34}S_{organic-pyrite}$.

Comparison of assimilatory with dissimilatory processes as source of sulfur incorporated into OM

Prior to the diagenetic processes, the marine OM generally exhibits low sulfur content ($\sim$ S/C 0.01–0.02) as expected if not "secondarily" enriched by dissimilatory reduced sulfur species. The $\delta^{34}S$ for the assimilatory sulfur reflects the associated sulfate $\delta^{34}S$ (*i.e.* $\sim 20\%_o$). If a significant amount of the assimilatory sulfur fraction is preserved in the SOM, then the $\delta^{34}S$ of the OM will be ^{34}S enriched. However, the assimilatory sulfur fraction (mainly peptides) is relatively labile, it is unlikely that it will be directly incorporated into the SOM. It has been suggested that reactive sulfur species formed during the biomolecules degradation may incorporate into the SOM (ANDERSON and PRATT, 1995; BRUCHERT and PRATT, 1996; CANFIELD *et al.*, 1998; PASSIER *et al.*, 1999). We tend to agree that if this enrichment is only moderate (S/C 0.01–0.03) such is the case. However, it is hard to accept the suggested explanation for protokerogens (very young sedimentary insoluble OM) or kerogens of type II-S (S/C 0.04–0.16). In these type II-S geo-polymers, the enrichment via reduced sulfur species is more than 400% compared to original assimilatory sulfur. Moreover, in intense SRB activity environments, reduced sulfur species can reach very high concentration. In the Solar Lake (Sinai, Egypt), 150 μmol/l sulfide in equilibrium with elemental sulfur forms a steady-state pool of polysulfides (JORGENSEN *et al.*, 1979). In this dissolved reduced sulfur-rich water, the biomolecules degraded sulfur will be diluted. Thus, in such cases of intense dissimilatory sulfur incorporation, the OM will carry the $\delta^{34}S$ of the dissimilatory reduced sulfur species (*i.e.* polysulfides and sulfide in water).

Sulfur species incorporated into OM

In most modern marine sediments, the $\delta^{34}S$ of the coexisting OM and elemental sulfur are similar and are ^{34}S enriched relative to the coexisting pyrite by about $10\%_o$ (MOSSMANN *et al.*, 1991; ANDERSON and PRATT, 1995). Because bacterial sulfide oxidation produces elemental sulfur up to only 2‰ ^{34}S enriched compared to the precursor sulfide (FRY et al., 1984; FRY, 1988), oxidation cannot explain this phenomenon. Moreover, elemental sulfur is not a reactive species with OM under diagenetic marine sediment conditions (see previous discussion). MOSSMANN et al. (1991) suggested that polysulfides (and elemental sulfur) are isotopically heavier than pyrite because they are formed by the oxidation of sulfide diffusing across the sediment–water interface whereas pyrite precipitated from pore water sulfide within the sediment. Sulfide diffusing across the sediment–water interface is isotopically heavier than pore water sulfides within surface sediment (CHANTON et al., 1987). Nevertheless, if we consider the rapid chemical isotope mixing of reduced sulfur species discussed before, this mechanism has to assume that the rate of the polysulfides diffusion must exceed the rate of polysulfides incorporation into OM. Otherwise, OS and the inorganic reduced sulfur species will carry the same isotopic imprint.

Fractionation during sulfur incorporation into OM

The incorporation of reduced sulfur species into OM is generally assumed to have negligible δ^{34}S fractionation (MOSSMANN et al., 1991; ANDERSON and PRATT, 1995; PASSIER et al., 1999; WERNE et al., 2003). This assumption, as far as we know, has never been confirmed by experimental data.

Recently, we have demonstrated that ^{34}S enrichment up to 6‰ can occur during simulated laboratory experiments of sulfur incorporation into OM (AMRANI and AIZENSHTAT, 2003). This finding can explain part of the difference between pyrite δ^{34}S and OM δ^{34}S to a value that can be explained solely by this fractionation or in combination with chemical isotope mixing discussed previously. The mechanisms controlling this discrimination are under investigation and will be published elsewhere.

SELECTED CASE STUDIES OF TYPE II-S KEROGEN FORMATION: SULFUR ENRICHED (S/C > 0.04)

The following discussion will focus on studies of the sulfur isotopic imprint as a function of depositional environments.

Solar Lake—young cyanobacterial mat (Sinai, Egypt)

The sulfur cycle in the Solar Lake is very complex due to the many redox changes of the sea sulfate. The lake is a marine, hypersaline, stratified (in winter), heliothermal body. The sediment core discussed here was taken from the western sandbank (details are given in AIZENSHTAT et al. (1983)). The biological–chemical sulfur cycle in the Solar Lake is described in Scheme 1. The benthic communal cyanobacterial mat is a good example of a recent sulfuritum (euxinic controlled environment) exhibiting high rates of microbial mediated metabolism of carbon and sulfur. Because of the very high photosynthetic productivity, the cyanobacteria grow under CO_2 limitation. This limitation leads to very heavy δ^{13}C values recorded for the biomass of -6.5 to -8.6‰ (AIZENSHTAT et al., 1984). Due to the extensive microbial sulfate reduction and limitation of available iron, the lower part of the stratified water column and sediment's pore waters are rich in reduced sulfur species. JORGENSEN et al. (1979) measured 150 μmol/l dissolved sulfide in equilibrium with elemental sulfur that forms a steady-state pool of polysulfides. Pyrite is absent in the sediments; however, minute traces of cubic iron sulfide are detected. The presence of high concentration of polysulfides in the interstitial waters with only minute traces of iron sulfides is due to the very low iron supply (see pyrite section discussion and references cited).

The examination of the sulfur isotope variation in the different species and oxidative states reveals very interesting behavior (Fig. 2). This figure should be examined in relation to Fig. 3 that presents the δ^{34}S values versus S/C atomic ratio. Following the interstitial water sulfate isotope profile with depth into the core shows that the supply of sea water is open for the lake (top) and through the sand barrier seepage (bottom). This keeps the values close to open sea

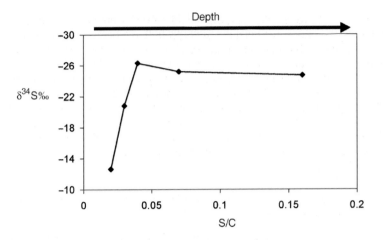

FIG. 3. $\delta^{34}S_{org}$ of protokerogens values from Fig. 2 are plotted versus atomic S/C ratios.

$+20 \pm 2\%o$. Only the mid-section of the core at 45–80 cm indicates some longer residence time for the sulfate ($\sim +25\%o$). Based on these results, we can assume that the SRB function under an essentially open sulfate system. The upper 5 cm of the core is biologically very active and therefore very difficult to isolate insoluble OM (protokerogen). This is due to the chemical treatment to remove minerals that causes the hydrolysis and dissolving of the OM. However, the extractable bitumen from the top 5 cm and from the bottom of the core has a $\delta^{34}S$ value of -12 to $-14\%o$. The $\delta^{34}S$ of elemental sulfur from the sediment core was $\sim -22\%o$ (the separation of this sulfur was carried out by sublimation under high vacuum). This light value indicates the introduction of dissimilatory sulfur since the assimilatory sulfur is expected to be close to the sulfate $\delta^{34}S$ value (*i.e.* $\sim 23\%o$, see Fig. 2). The instability of assimilatory sulfur bound to amino acids is illustrated in Fig. 4a and b. The rapid loss of nitrogen and the increase of the atomic ratio for S/N are marked with depth (Fig. 4a). Figure 4b emphasizes the decrease of the N/C and increase of S/C for the isolated protokerogens with depth. The sulfur incorporation into the protokerogen increases the S/C from 0.01 to 0.16 (see Figs. 3 and 4b). With the enrichment of the protokerogen in sulfur, it becomes ^{34}S depleted with $\delta^{34}S$ values as low as $-26.3\%o$

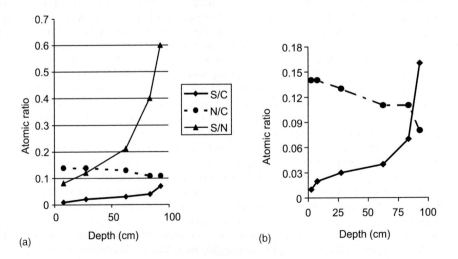

FIG. 4. Solar Lake protokerogen atomic ratios of selected sections same core as for samples given in Fig. 2. (a) S/C, N/C and S/N ratios versus depth (cm), (b) S/C and N/C ratios versus depth zoom (a), see references and Fig. 2.

(see Figs. 2 and 3). However, this organic-bound sulfur is chemically labile. Experimental exposure of the polysulfide cross-linked polymer (protokerogen) to 175°C for 24 h releases large quantities of elemental sulfur and some H_2S. The ratio of elemental sulfur to H_2S is dependent on the availability of organic hydrogen (AIZENSHTAT et al., 1983).

The study of the sulfur cycle and stable isotopes, in a geochemically young sedimentary basin (2400 years, see COHEN et al., 1977), renders a rare insight into the ongoing biogeochemical processes. The very high polysulfide concentration in the pore waters governs the dissimilatory sulfur incorporation into OM at the very early diagenesis stage. The lack of humic and fulvic acids and the formation of protokerogen rich in sulfur support the mechanism of secondary sulfur enrichment proposed. The resulting sulfur isotopic signature is $\delta^{34}S$ depleted and reflects generally the $\delta^{34}S$ of the polysulfides.

Dead Sea area (Jordan Valley Rift)—kerogens, bitumens, immature asphalts and bituminous rocks (Senonian Formation, Israel)

The most intensive organic geochemical studies in Israel and, to a lesser extent, in Jordan have been conducted on the organic-rich bituminous marls of the Upper Cretaceous Ghareb Formation. The bituminous rocks of the Upper Cretaceous sequence in Israel and in Jordan (KOHNEN et al., 1990b) are Senonian in age. They include carbonates as well as phosphorites and siliceous rocks deposited along the southern margins of the Tethys Ocean. The depositional environments are still debated. Candidates include an upwelling model (ALMOGI-LABIN et al., 1993) and hypersaline depositional model of separated sub-basins (SPIRO and AIZENSHTAT, 1977; DINUR et al., 1980; SPIRO, 1980; SPIRO et al., 1983a,b; AIZENSHTAT et al., 1999). However, all researchers agree that the formation of sediments with up to 25% TOC and kerogens of type II-S required a highly reducing environment. Very active and prolific OM production associated with SRB intensive activity is required to supply the large amounts of reduced sulfur that are observed.

Pyrite in the bituminous rocks appears as both framboids and single crystals (<50 μm) that are undetected by X-ray diffraction analysis because they are coated with OM (SPIRO, 1980; SPIRO et al., 1983a,b). Very small amounts of humic matter occur in these samples, but more than 95% of the OM in the bituminous rocks is high sulfur kerogen (8–10% w/w). The richest samples in both TOC and organically bonded sulfur occur in the Nabi Musa (Israel) area (S/C atomic ratio 0.05–0.065). Various maturity parameters indicate that all of the kerogens are immature with an H/C ratio of 1.3–1.6.

The quantitative comparison between pyritic sulfur and organically bonded sulfur reveals almost a 1:1 ratio with very low amount of sulfate, mostly as gypsum (SPIRO, 1980). The gypsum recorded is only filling fossils and its isotopic sulfur ratio yet to be determined. In all samples studied from the Zefa Efe location (Mishor Rotem, Negev, Israel), the pyrite is isotopically the lightest for these samples. The $\delta^{34}S$ for pyrite ranges from -19.4 to $-30.1‰$ ($n = 15$). Kerogens are, on average, heavier by 10‰ (-4 to $-11‰$; $n = 10$) and the bitumen even heavier at $\delta^{34}S \sim -4‰$ ($n = 7$). The Nabi Musa area organic-rich samples show a narrower range with less depleted ^{34}S isotope values for all the reduced sulfur: pyrites $+1.2$ to $-6‰$ ($n = 10$), kerogens $+6$ to $+7‰$ ($n = 15$) and the bitumens $+8$ to $+9‰$ ($n = 10$). This narrow range and ^{34}S enrichment indicates a potential closed system (restricted sulfate supply) in Nabi Musa depositional environment (SPIRO, 1980; DINUR et al., 1980).

There is no doubt that the asphalt seeps, which are found along the Jordan Rift Valley, originate in the Senonian Ghareb Formation (SPIRO et al., 1983b; TANNENBAUM, 1983; TANNENBAUM and AIZENSHTAT, 1984, 1985). All of these immature asphalts are sulfur rich (see Table 1). These asphalts show some similarity to the heavy immature oils of the Monterey Formation (see following discussion) (AIZENSHTAT et al., 1979; NISSENBAUM and GOLDBERG, 1980). Although carbon isotopes of some of these asphalts have been reported (TANNENBAUM, 1983), very few measurements of $\delta^{34}S$ have been reported. The present discussion on $\delta^{34}S$ values of the asphalts and oils is based on unpublished data of Tannenbaum and Coleman. Figure 5 demonstrates the ^{34}S enrichment versus ^{13}C depletion with maturation of the OM rich in sulfur. We examined both carbon and sulfur isotopes of specific organic fractions: hydrocarbon (saturated), aromatics, NSO, more polar resins and asphaltenes. The results for the carbon

Table 1. Samples of the Dead Sea area S/C list for Fig. 5 (Part I) and Fig. 6 (Part II)

Sample	S/C
Amiaz kerogen	0.045
Amiaz asphalt	0.046
Zefa Efe	0.056
Nabi Musa	0.052
Nahal Heimar	0.048
Wadi Heimar	0.049
Float	0.048
Sdom SH5	0.046
IPRG 1	0.061
Lot	0.031
Kidod 3	0.035
Zohar	0.035
Massada	0.020

Locations and geological background: TANNENBAUM (1983) and TANNENBAUM *et al.* (1987).

isotopes showed ^{13}C depletion with increasing maturation and with reduction of polarity (TANNENBAUM, 1983). However, the sulfur isotopes showed a general trend of ^{34}S enrichment with maturation and loss of cross-linked polysulfides. All kerogens were, in general, lighter isotopically than the asphalts derived from them. The data relating to the oils from the same area will be discussed in Part II.

In all samples studied, $\delta^{34}S$ of pyrite is the most ^{34}S depleted. The OM sulfur is on average 10‰ heavier a very small amount of the sulfate minerals gypsum and anhydrite for the Nabi Musa location indicates evaporative conditions and sulfate closed system. Despite the same Senonian Ghareb Formation sedimentary rock for all of the Dead Sea area samples studied, the $\delta^{34}S$ values recorded for different sites show wide variability. This indicates that small differences in sub-basin conditions have a significant imprint on sulfur isotope distribution.

Monterey Formation (Miocene)—kerogens, bitumens, asphalts formation (California, USA)

The Miocene Monterey Formation (California, USA) was suggested by ARNOLD and ANDERSON (1907) to be the principal source for the widespread California asphalts and heavy sulfur-rich oils (see discussion in EMERY (1960)). Stratigraphy of the Monterey Formation is

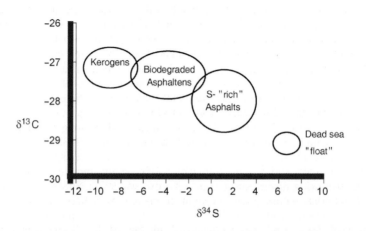

FIG. 5. Carbon and sulfur isotopes for kerogens and asphalts of samples from the Dead Sea area. Detailed geographical description and geological setting are given in DINUR *et al.* (1980), SPIRO *et al.* (1983b) and TANNENBAUM and AIZENSHTAT (1984). Some of the data for the $\delta^{34}S$ reported here were not published earlier and were obtained from collaborative work with Prof. M. Coleman (see detailed report, Part II).

reviewed by ISAACS (1984) and the latest organic geochemistry studies by SCHOUTEN et al. (2001). The age of deposition of the three major members: lower calcareous, middle phosphatic and upper siliceous is from 22 to 5 Ma. The organically rich carbonate member contains signs of benthic fauna whereas the phosphate-rich member indicates evaporitic conditions as well as an influx of cooler waters. The laminated siliceous member records a drop in sea level (ZABACK and PRATT, 1992). All kerogens and asphalt of low maturity exhibit high (4%) to very high (8–9%) sulfur content. The amount of pyrite in the Monterey sediments fluctuates between 5 and 50% of the total sulfur (ZABACK and PRATT, 1992). Good correlation between amount of OM and reduced sulfur in the Monterey was derived from the TOC/total reduced sulfur ratios that were found to be invariant. For the same samples, pyrite S decreases with increase in kerogen-bonded S (ZABACK et al., 1993). In most Monterey samples the residual sulfate is in small concentration compared to other sulfur species, and thus ^{34}S enriched in comparison to the open sea water sulfate (GOLDHABER and KAPLAN, 1974; ORR, 1986; ZABACK and PRATT, 1992; ZABACK et al., 1993).

ORR (1984a,b) studied pyrite, kerogen, asphalt and oil mostly of the Monterey Formation in the Santa Maria basin in the quest for the potential source rocks for the offshore oils of southern California. In 1986, Orr compiled the sulfur isotope data and reported that in all cases, the pyritic S was the lightest and that the associated kerogens were always ^{34}S enriched in comparison. The same general observation with some modification of the observed range is reported by ZABACK and PRATT (1992) and ANDERSON and PRATT (1995). ORR (1986) was the first to name the organically rich kerogens with >4% S as type II-S and agreed with our suggested mechanism of secondary polysulfides enrichment (DINUR et al., 1980).

From the literature data, we can assume that the Monterey sedimentary environments were changing with time and that for cases where both pyrite and organically bound sulfur became less ^{34}S depleted the supply of sulfate was restricted. This in turn made the $\Delta\delta^{34}$S$_{\text{organic-pyrite}}$ smaller. An additional point of interest is the fact that all bitumens and asphalts were ^{34}S enriched in comparison to the kerogens. The Monterey Formation is geographically widespread in comparison with the Dead Sea area and hence samples from different locations and different mineral matrix record different sedimentary conditions. The δ^{34}S changes for heavy oils and more mature petroleum assumed to be generated from the Monterey Formation PSR is discussed in Part II.

SUMMARY AND REMARKS: δ^{34}S$_{\text{ORG}}$ VALUES AS RELATED TO DEPOSITIONAL ENVIRONMENTS AND MECHANISMS OF SULFUR INTRODUCTION INTO THE SOM

The formation of type II-S kerogens is characterized by the sulfur isotope (δ^{34}S$_{\text{org}}$) signature of reduced sulfur species formed subsequent to sulfurization of available iron in the early diagenetic stage. This pyrite formation stage is kinetically controlled, hence, it yields δ^{34}S$_{\text{py}}$ that marks the earliest generated sulfides. Thus, the SRB isotopic dissimilatory fractionation is best recorded by the pyrite δ^{34}S$_{\text{py}}$ value. On the one hand, the SRB activity consumes large quantities of the SOM, whereas, on the other hand, the polysulfide polymer formed protects the residual OM by the cross-linking and formation of water-insoluble protokerogen. The span of isotope values recorded for the sulfur in kerogens depends on the following: (1) open or closed system for the supply of sulfate and timing of SRB activity; (2) the proportion between assimilatory sulfur retained and re-reacted with the SOM; (3) diffusion effects and chemical isotopic mixing of sulfur species; (4) the isotopic fractionation associated with the sulfur incorporation into the SOM.

We tend to suggest that enrichment of more than 400‰ in sulfur (S/C > 0.04) is more likely to happen under benthic hypersaline prolific shallow water depositional environments. The reported investigations of case studies brought in this review support this conclusion.

Acknowledgements—Zeev Aizenshtat wishes to thank Ian Kaplan for his introduction into the interdisciplinary scientific world of organic geochemistry. Many of the works cited and discussed in these Part I and Part II conceptual reviews, although not authored by Kaplan, have been inspired by his early pioneering works and his continued introduction of novel ideas into the field. The second generation represented by Alon Amrani, Baruch Spiro, Eli Tannenbaum, Achikam Stoler, Eitan Krein and many others is now carrying the torch. Special thanks to Dr Max Coleman for some of the data on the Dead Sea samples. Last but not the least, our warm thanks to Dr Martin Goldhaber and anonymous referee for intense editing and suggestions that improved this manuscript.

REFERENCES

AIZENSHTAT Z., DINUR D. and NISSENBAUM A. (1979) Tetrapyrroles and associated compounds in the Dead Sea asphalts. *Chem. Geol.* **24**, 161–174.

AIZENSHTAT Z., STOLER A., COHEN Y. and NIELSEN H. (1983) The geochemical sulphur enrichment of recent organic matter by polysulfides in the Solar Lake. In *Advances in Organic Geochemistry 1981* (ed. M. BJOROY), pp. 279–288. Wiley, New York.

AIZENSHTAT Z., LIPINER G. and COHEN Y. (1984) Biogeochemistry of carbon and sulfur cycle in the microbial mats of the Solar Lake (Sinai). In *Microbial Mats: Stromatolies* (eds. Y. COHEN, D. CASTENHOLTZ and H. HALVOSOM), pp. 281–312.

AIZENSHTAT Z., KREIN E. B., VAIRAVAMURTHY M. A. A. and GOLDSTEIN T. P. (1995) Role of sulfur in the transformations of sedimentary organic matter: a mechanistic overview. In *Geochemical Transformation of Sedimentary Sulfur* (eds. M. A. A. VAIRAVAMURTHY and A. SCHOONEN), pp. 16–39 American Chemical Society Symposium Series 612.

AIZENSHTAT Z., MILOSLAVSKI I. and ASHENGRAU D. (1999) Hypersaline depositional environments and their relation to oil generation. In *Microbiology and Biogeochemistry of Hypersaline Environments* (ed. A. OREN), pp. 89–105. CRC Press, Boca Raton, FL.

AMRANI A. and AIZENSHTAT Z. (2004) Photosensitized oxidation of natural occurring isoprenoid allyl-alcohols as possible pathway for their transformation to thiophenes in sulfur rich depositional environments. *Org. Geochem.* **35**, 693–712.

ALMOGI-LABIN A., BEIN A. and SASS E. (1993) Late Cretaceous upwelling system along the southern Tethys margin (Israel): interrelationship between productivity, bottom water environment, and organic matter preservation. *Paleoceanography* **8**, 671–690.

AMRANI A. and AIZENSHTAT Z. (2003) *Mechanisms of sulfur introduction chemically controlled: $\delta^{34}S$ imprint. 21st IMOG Meeting.* Extended Abstracts Book (Krakow, Poland), pp. 39–40.

ANDERSON T. F. and PRATT L. M. (1995) Isotopic evidence for the origin of organic sulfur and elemental sulfur in marine sediments. In *Geochemical Transformation of Sedimentary Sulfur* (eds. M. A. VAIRAVAMURTHY and M. A. A. SCHOONEN), pp. 378–396. ACS Symposium Series 612.

ARNOLD R. and ANDERSON R. (1907) Geology and resources of the Santa Maria oil district, Santa Barbara County, California. *US Geol. Surv. Bull.* **322**, 161–165.

ARNOSTI C. (1995) Measurement of depth- and site-related differences in polysaccharide hydrolysis rates in marine sediments. *Geochim. Cosmochim. Acta* **59**(20), 4247–4257.

AYCARD M., DERENNE S., LARGEAU C., MONGENOT T., TRIBOVILLARD N. and BAUDIN F. (2003) Formation pathways of proto-kerogens in Holocene sediments of the upwelling influenced Cariaco Trench, Venezuela. *Org. Geochem.* **34**, 701–718.

BERNER R. A. (1970) Sedimentary pyrite formation. *Am. J. Sci.* **268**, 1–23.

BERNER R. A. (1989) Biogeochemical cycles of carbon and sulfur and their effect on atmospheric oxygen over Phanerozoic time. In *The Changing Chemistry of the Oceans* (eds. D. DYRSSEN and D. JAGNER), pp. 347–361. Nobel Symposium 20.

BERNER R. A. and RAISWELL R. (1983) Burial of organic carbon and pyrite sulfur in sediments over Phanerozoic time: a new theory. *Geochim. Cosmochim. Acta* **47**, 855–862.

BIGELEISEN J. (1949) The relative velocities of isotopic molecule. *J. Chem. Phys.* **17**, 675.

BOTTCHER M. E., SMOCK A. and CYPOINKA H. (1998) Sulfur isotopes fractionation during experimental precipitation of iron (II) sulfide and manganese (II) sulfide at room temperature. *Chem. Geol.* **146**, 127–134.

BOULEGUE J. (1978) Solubility of elemental sulfur in water at 298 K. *Phophorous, Sulfur Related Elem.* **5**, 127–128.

BOULEGUE J. and MICHARD G. (1978) Constantes de formation des ions polysulfides S_6^{2-}, S_5^{2-} et S_4^{2-} en phase aqueuese. *Fr. Hydrologie* **9**, 27–33 (in French).

BRASSELL S. C., LEWIS C. A., DE LEEUW J. W., DE LANGE F. and SINNINGHE DAMSTÉ J. S. (1986) Isoprenoid thiophenes: novel products of sediment diagenesis. *Nature* **320**, 160–162.

BRUCHERT V. and PRATT L. M. (1996) Contemporaneous early diagenetic formation of organic and inorganic sulfur in estuarine sediments from St. Andrew Bay, Florida, USA. *Geochim. Cosmochim. Acta* **60**(13), 2325–2332.

CANFIELD D. E., BOUDREAU B. P., MUCCI A. and GUNDERSEN J. K. (1998) The early diagenetic formation of organic sulfur in the sediments of Mangrove Lake, Bermuda. *Geochim. Cosmochim. Acta* **62**(5), 767–781.

CHANTON J. P., MARTENS C. S. and GOLDHABER M. B. (1987) Biogeochemical cycling in an organic-rich coastal marine basin: 8. A sulfur isotopic budget balanced by differential diffusion across the sediment–water interface. *Geochim. Cosmochim. Acta* **51**, 1201–1208.

COHEN Y., GOLDBERG M., KRUMBEIN W. E. and SHILO M. (1977) Solar Lake (Sinai 1). Physical and chemical limnology. *Limnol. Oceanogr.* **22**, 297–307.

DETMERS J., BRUECHERT V., HABICHT K. S. and KUEVER J. (2001) Diversity of sulfur isotope fractionations by sulfate-reducing prokaryotes. *Appl. Environ. Microbiol.* **67**, 888–894.

DINUR D., SPIRO B. and AIZENSHTAT Z. (1980) The distribution and isotopic composition of sulfur in organic rich sedimentary rocks. *Chem. Geol.* **31**, 37–51.

EMERY K. O. (1960) *The Sea of Southern California.* Wiley, New York, 366 pp.

FRANCOIS R. (1987) A study of sulphur enrichment in humic fraction of marine sediments during early diagenesis. *Geochim. Cosmochim. Acta* **51**, 17–27.

FRY B. (1988) Food web structure on Georges Bank from stable C, N, S isotopic compositions. *Limnol. Oceanogr.* **33**, 1182–1190.

FRY B., GEST H. and HAYES J. M. (1984) Isotope effect associated with the anaerobic oxidation of sulfide by the purple photosynthetic bacterium *Chromatium vinosum. Microbiol. Lett.* **22**, 283–287.

FOSSING H. and JORGENSEN B. B. (1990) Isotope exchange reactions with radiolabeled sulfur compounds in anoxic seawater. *Biogeochemistry* **9**, 223–245.

FURUKAWA Y. and BARNES H. L. (1995) Reactions forming pyrite from precipitated amorphous ferrous sulfide. In *Geochemical Transformation of Sedimentary Sulfur* (eds. M. A. VAIRAVAMURTHY and M. A. A. SCHOONEN), pp. 194–205. ACS Symposium, Washington, DC, Series 612.

GIGGENBACH W. (1972) Optical spectra and equilibrium distribution of polysulfide ions in aqueous solution at 20°C. *Inorg. Chem.* **11**, 1201–1207.

GOLDHABER M. B. (2004) Sulfur rich sediments. In *Treatise of Geochemistry* (eds. HOLLAND and TUREKIAN), Chap. 7.17 (in press).

GOLDHABER M. B. and KAPLAN I. R. (1974) The sulfur cycle. In *The sea* (ed. E. D. GOLDBERG), Vol. 5, pp. 569–655. Wiley, New York.

GOLDSTEIN T. P. and AIZENSHTAT Z. (1994) Thermochemical sulfate reduction: a review. *J. Therm. Anal.* **42**, 241–290.

GUN J., GOIFMAN A., SHKROB I., KAMYSHNY A., GINZBURG B., HADAS O., DOR I., MODESTOV A. D. and LEV O. (2000) Formation of polysulfides in an oxygen rich freshwater lake and their role in the production of volatile sulfur compounds in aquatic systems. *Environ. Sci. Technol.* **34**(22), 4741–4746.

HABICHT K. S., CANFIELD D. E. and RETHMEIER J. (1998) Sulfur isotope fractionation during bacterial reduction and disproportionation of thiosulfate and sulfite. *Geochim. Cosmochim. Acta* **62**(15), 2585–2595.

HARRISON A. G. and THODE H. G. (1957) The kinetic isotope effect in the chemical reduction of sulphate. *Trans. Faraday Soc.* **53**, 1–4.

HOLSER W. T., SCHIDLOVSKI M., MACKENZIE F. T. and MAYNARD J. B. (1988) Geochemical cycles of carbon and sulfur. In *Chemical Cycles in the Evolution of Earth* (eds. C. B. GREGOR, R. M. GARRELS, F. T. MACKENZIE and J. B. MAYNARD), pp. 105–173. Wiley, New York.

IDIZ E. F., TANNENBAUM E. and KAPLAN I. R. (1990) Pyrolysis of high-sulfur Monterey kerogens—stable isotopes of sulfur, carbon, and hydrogen. In *Geochemistry of Sulfur in Fossil Fuels* (eds. W. L. ORR and C. M. WHITE), pp. 575–591. ACS Symposium Series 429.

ISAACS C. M. (1984) The Monterey—key to offshore California boom. *Oil Gas J.* **82**, 75–81.

JORGENSEN B. B., REVSBECH N. P., COHEN Y. (1979) Diural cycle of oxygen and sulfide microgradients and microbial photosynthesis in cyanobacterial mat sediments. *Appl. Environ. Microbiol.* **38**, 46–48.

JORGENSEN B. B. (1982) Mineralization of organic matter in the sea bed—the role of sulphate reduction. *Nature* **296**, 643–645.

JORGENSEN B. B. (1988) Ecology of the sulfur cycle: oxidative pathways in sediments. In *The Nitrogen and Sulfur Cycles* (eds. J. A. COLE and S. J. FERGUSON) *Soc. Gen. Microbiol. Symp.* **42**, 31–63.

KAPLAN I. R. and RITTENBERG S. C. (1964) Microbial fractionation of sulfur isotopes. *J. Genet. Microbiol.* **34**, 195–212.

KAPLAN I. R., EMERY K. O. and RITTENBERG S. C. (1963) The distribution and isotopic abundance of sulfur in recent marine sediments of southern California. *Geochim. Cosmochim. Acta* **27**, 297–331.

KLEIN M., FRIEDRICH M., ROGER A. J., HUGENHOLTZ P., FISHBAIN S., ABICHT H., BLACKALL L. L., STAHL D. A. and WAGNER M. (2001) Multiple lateral transfers of dissimilatory sulfite reductase genes between major lineages of sulfate reducing prokaryotes. *J. Bacteriol.* **183**, 6028–6035.

KOHNEN M. E. L., SINNINGHE DAMSTÉ J. S., KOCK-VAN DALEN A. C., TEN HAVEN H. L., RULKOTTER J. and DE LEEUW J. W. (1990a) Origin and diagenetic transformation of C_{25} and C_{30} highly branched isoprenoid sulphur compounds: further evidence for the formation of organically bound sulfur during early diagenesis. *Geochim. Cosmochim. Acta* **54**, 3053–3063.

KOHNEN M. E. L., SINNINGHE DAMSTÉ J. S., RIJPSTRA W. I. C. and DE Leeuw J. W. (1990b) Alkylthiophenes as sensitive indicators of palaeoenvironmental changes. In *Geochemistry of Sulfur in Fossil Fuels* (eds. W. L. ORR and C. M. WHITE), pp. 444–485. American Chemical Society, Washington, DC.

KOK M. D., SCHOUTEN S. and SINNINGHE DAMSTÉ J. S. (2000) Formation of insoluble, nonhydrolyzable, sulfur-rich macromolecules via incorporation of inorganic sulfur species into algal carbohydrates. *Geochim. Cosmochim. Acta* **64**(15), 2689–2699.

KREIN E. B. (1993) Organic sulfur in the geosphere: analysis, structures and chemical processes. In *The Chemistry of the Sulphur-Containing Functional Groups* (eds. S. PATAI and Z. RAPPOPORT), Supplement S, pp. 975–1032. Wiley, New York.

KREIN E. B. and AIZENSHTAT Z. (1993) The phase-transfer catalyzed reactions between polysulfide anions and α,β-unsaturated carbonyl compounds. *J. Org. Chem.* **58**, 6103–6108.

KREIN E. B. and AIZENSHTAT Z. (1994) The formation of isoprenoid sulfur compounds during diagenesis: simulated sulfur incorporation and thermal transformation. *Org. Geochem.* **21**, 1015–1025.

KUMP L. R. and GARRELS R. M. (1986) Modeling atmospheric O_2 in the global sedimentary redox cycle. *Am. J. Sci.* **286**, 337–360.

LaLonde R. T., Ferrara L. M. and Hayes M. P. (1987) Low-temperature, polysulfide reactions of conjugated ene carbonyls: a reaction model for the geologic origin of S-heterocycles. *Org. Geochem.* **11**, 563–571.

Licht S. and Davis J. (1997) Disproportionation of aqueous sulfur and sulfide: kinetics of polysulfide decomposition. *J. Phys. Chem.* **101**, 2450–2545.

Licht S., Fodes G. and Manassen J. (1986) Numerical analysis of aqueous polysulfide solutions and its application to cadmium chalcogenide/polysulfide photoelectrochemical solar cells. *Inorg. Chem.* **25**, 2486–2489.

Loch A. R., Lippa D. L., Carlson D. L., Chin Y. P., Trina S. J. and Roberts A. L. (2002) Nucleophilic aliphatic substitution reactions of propachlor, alachlor, and metolachlor with bisulfide (HS$^-$) and polysulfides (S$_n^{2-}$). *Environ. Sci. Technol.* **36**, 4065–4073.

Lückge A., Horsfield B., Littke R. and Scheeder G. (2002) Organic matter preservation and sulfur uptake in sediments from the continental margin off Pakistan. *Org. Geochem.* **33**, 477–488.

Luther G. W. III (1991) Pyrite formation in an anoxic chemostatic reaction system. *Geochim. Cosmochim. Acta* **55**, 2839–2849.

Luther G. W. III, Glaze B. T., Hohmann L., Popp J. I., Taillefert M., Rozan T. F., Brendel P. J., Theberge S. M. and Nuzzio D. B. (2001) Sulfur speciation monitored in situ with solid-state gold amalgam voltammetric microelectrodes: polysulfides as a special case. *J. Environ. Monit.* **3**, 61–66.

Mossmann J. R., Aplin A. C., Curtis C. D. and Coleman M. L. (1991) Geochemistry of inorganic and organic sulphur in organic-rich sediments from the Peru Margin. *Geochim. Cosmochim. Acta* **55**, 3581–3595.

Neretin L. N., Bottcher M. E. and Grinenko V. A. (2003) Sulfur isotopes geochemistry of the Black Sea water column. *Chem. Geol.* (in press).

Nissenbaum A. and Goldberg M. (1980) Asphalts, heavy oils, ozocorite and gases in the Dead Sea Basin. *Org. Geochem.* **2**, 167–180.

Orr W. L. (1984a) *Sulfur and sulfur isotope ratios in Monterey oils of the Santa Maria basin and Santa Barbara channel, area offshore.* Technical Program Abstracts, 1984 SEPM Annual Meeting, San Jose, California, 10–13, 62 pp.

Orr W. L. (1984b) *Geochemistry of asphaltic Monterey oils from Santa Maria basin and Santa Barbara Channel area offshore.* ACS Division of Geochemistry Abstracts, 187th ACS National Meeting, April 8–13.

Orr W. L. (1986) Kerogen/asphaltene/sulfur relationships in sulfur-rich Monterey oils. *Adv. Org. Geochem.* **10**, 499–516.

Passier H. F., Böttcher M. E. and DeLange G. J. (1999) Sulphur enrichment in organic matter of Eastern Mediterranean sapropel: a study of sulphur isotope partition. *Aquat. Geochem.* **5**, 99–118.

Peterson B. J. (1999) Stable isotopes as tracers of organic matter input and transfer in benthic food web: a review. *Acta Ocol.* **20**(4), 479–487.

Price F. T. and Shieh Y. N. (1979) Fractionation of sulfur isotopes during laboratory synthesis of pyrite at low temperatures. *Chem. Geol.* **27**, 245–253.

Rickard D., Schoonen M. A. A. and Luther G. W. (1995) Chemistry of iron sulfides in sedimentary environments. In *Geochemical Transformation of Sedimentary Sulfur* (eds. M. A. Vairavamurthy and M. A. A. Schoonen), pp. 168–193. ACS Symposium Series 612.

Rozan T. F., Theberge S. M. and Luther G. III (2000) Quantifying elemental sulfur (S0), bisulfide (HS$^-$) and polysulfides (S$_x^{2-}$) using a voltametric method. *Anal. Chim. Acta* **415**(1–2), 175–184.

Sakai H. (1957) Fractionation of sulfur isotopes in nature. *Geochim. Cosmochim. Acta* **12**, 150–169.

Schouten S., de Graaf W., Sinninghe Damsté J. S., van Driel B. G. and de Leeuw J. W. (1994) Laboratory simulation of natural sulphurization: II. Reaction of multi-functionalized lipids with inorganic polysulphides at low temperatures. *Org. Geochem.* **22**, 825–834.

Schouten S., Schoell M., Sinninghe Damsté J. S., Summons R. E. and de Leeuw J. W. (2001) Molecular biogeochemistry of Monterey sediments, Naples Beach, California: II. Stable carbon isotopic composition of free and sulfur bound carbon skeleton. In *The Monterey Formation, from Rock to Molecules* (eds. C. M. Isaacs and J. Rullkoetter), pp. 175–188. Columbia University Press, New York.

Schwarzenbach G. and Fisher A. (1960) Aciditat der sulfane und die zusammensetzung wasseriger polysulfidlosungen. *Helv. Chim. Acta* **43**, 1365–1390 (in German).

Sigalevich P. and Cohen Y. (2000) Oxygen-dependant and possible aerobic metabolism of sulfate reducing bacterium *Desulfovibrio oxylinae* in a chemostat coculture with *Marinobacter* sp. Strain MB in an aerated sulfate depleted chemostat. *Appl. Environ. Microbiol.* **66**, 5019–5023.

Sigalevich P., Baev M. V., Teska A. and Cohen Y. (2000) Sulfate reduction and possible aerobic metabolism of sulfate reducing bacterium *Desulfovibrio oxylinae* in a chemostat coculture with *Marinobacter* sp. Strain MB under exposure to increasing oxygen concentrations. *Appl. Environ. Microbiol.* **66**, 5013–5018.

Sinninghe Damsté J. S. and de Leeuw J. W. (1990) Analysis, structure and geochemical significance of organically bound sulphur in the geosphere: state of the art and future research. *Org. Geochem.* **16**, 1077–1101.

Sinninghe Damsté J. S., Eglinton T. I., de Leeuw J. W. and Schenck P. A. (1989) Organic sulfur in macromolecular sedimentary organic-matter: 1. Structure and origin of sulfur-containing moieties in kerogen, asphaltenes and coal as revealed by flash pyrolysis. *Geochim. Cosmochim. Acta* **53**, 873–889.

Sinninghe Damsté J. S., Kok M. D., Koster J. and Schouten S. (1998) Sulfurized carbohydrate: an important sedimentary sink for organic carbon? *Earth Planet. Sci. Lett.* **164**, 7–13.

Spiro B. (1980) Geochemistry and mineralogy of bituminous rocks in Israel. Ph.D. Thesis, The Hebrew University of Jerusalem (in Hebrew abstract in English).

Spiro B. and Aizenshtat Z. (1977) Bacterial sulphate reduction and calcite precipitation in hypersaline deposition of bituminous shales. *Nature* **269**, 235–237.

SPIRO B., DINUR D. and AIZENSHTAT Z. (1983a) Evaluation of source, environment of deposition and diagenesis of some Israeli "oil shales" n-alkanes, fatty acids, tetrapyrroles and kerogen. *Chem. Geol.* **39**, 189–214.

SPIRO B., WELTE D. H., RULLKOETER J. and SCHAEFER R. G. (1983b) Asphalts, oils and bituminous rocks from the Dead Sea area—a geochemical correlation study. *Am. Assoc. Petrol. Geol. Bull.* **67**, 1163–1175.

STAUDT W. J. and SCHOONEN M. A. A. (1995) Sulfate incorporation into sedimentary carbonate. In *Geochemical Transformation of Sedimentary Sulfur* (eds. M. A. VAIRAVAMURTHY and M. A. A. SCHOONEN), pp. 332–345. ACS Symposium Series 612.

TANNENBAUM E. (1983) Researches in the geochemistry of oils and asphalts in the Dead Sea area, Israel. Ph.D. Thesis, The Hebrew University, Jerusalem, 117 pp. (in Hebrew with English abstract).

TANNENBAUM E. and AIZENSHTAT Z. (1984) Formation of immature asphalt from organic-rich carbonate rocks: 2. Correlation of maturation indicators. In *Advances in Organic Geochemistry 1983* (eds. P. A. SCHENCK, J. W. DE LEEUW and G. W. M. Lijmback) *Org. Geochem.* **6**, 503–511.

TANNENBAUM E. and AIZENSHTAT Z. (1985) Formation of immature asphalt from organic-rich carbonate rocks: 1. Geochemical correlation. *Org. Geochem.* **8**, 181–192.

TANNENBAUM E., STARINSKY A. and AIZENSHTAT Z. (1987) Light oils transformation to heavy oils and asphalts. In *Exploration for Heavy Crude Oils and Natural Bitumen* (ed. R. F. MEYER), pp. 221–231. Studies in Geology American Association of Petroleum Geologists, Series 25, Tulsa.

TUDGE A. P. and THODE H. G. (1950) Thermodynamic properties of isotopic compounds of sulfur. *Can. J. Resour.* **28b**, 567–578.

VAIRAVAMURTHY M. A. and MOPPER K. (1989) Mechanistic studies of organo-sulphur (thiol) formation in coastal marine sediments. In *Biogenic Sulfur in the Environment* (eds. E. S. SALTZMAN and W. J. COOPER), pp. 231–242. ACS, Washington, DC.

VAIRAVAMURTHY M. A., MOPPER K. and TAYLOR B. F. (1992) Occurrence of particle-bonded polysulphides and significance of their reaction with organic matter in marine sediments. *Geophys. Res. Lett.* **19**, 2043–2046.

VAIRAVAMURTHY M. A., ORR W. L. and MANOWITZ B. (1995) Geochemical transformations of sedimentary sulfur: an introduction. In *Geochemical Transformation of Sedimentary Sulfur* (eds. M. A. VAIRAVAMURTHY and M. A. A. SCHOONEN), pp. 1–14. ACS Symposium Series 612.

VALISOLALAO J., PERAKIS N., CHAPP B. and ALBRECHT P. (1984) A novel sulfur containing C$_{35}$ hopanoid in sediments. *Tetrahedr. Lett.* **25**, 1183–1186.

VAN DONGEN B. E., SCHOUTEN S., BAAS M., GEENEVASEN J. A. J. and SINNINGHE DAMSTÉ J. S. (2003) An experimental study of the low-temperature sulfurization of carbohydrates. *Org. Geochem.* **34**(8), 1129–1144.

VAN KAAM-PETERS H. M. E., SCHOUTEN S., KÖSTER J. and SINNINGHE DAMSTÉ J. S. (1998) Controls on the molecular and carbon isotopic composition of organic matter deposited in a Kimmeridgian euxinic shelf sea: evidence for preservation of carbohydrates through sulfurisation. *Geochim. Cosmochim. Acta* **62**, 3259–3283.

WAKEHAM S. G., SINNINGHE DAMSTÉ J. S., KOHNEN M. E. L. and DE LEEUW J. W. (1995) Organic sulfur compounds formed during early diagenesis in the Black Sea. *Geochim. Cosmochim. Acta* **59**, 521–533.

WANG Q. and MORSE J. W. (1995) Laboratory simulation of pyrite formation in anoxic sediments. In *Geochemical Transformation of Sedimentary Sulfur* (eds. M. A. VAIRAVAMURTHY and M. A. A. SCHOONEN), pp. 206–223. ACS Symposium Series 612.

WERNE J. P., LYONS T. W., HOLLANDER D. J., FORMOLO M. J. and SINNINGHE DAMSTÉ J. S. (2003) Reduced sulfur in euxinic sediments of the Cariaco Basin: sulfur isotope constrains on organic sulfur formation. *Chem. Geol.* **27**, 245–253.

WILKIN R. T. and BARNES H. L. (1996) Pyrite formation by reactions of iron monosulfides with dissolved inorganic and organic species. *Geochim. Cosmochim. Acta* **60**, 4167–4179.

WILLIAMSON M. A. and RIMSTIDT J. D. (1990) Thermodynamic and kinetic controls on aqueous oxidation of sulfide minerals. In *Conference for Advancement of Geochemistry. Goldschmid Int.*, pp. 91–103.

WORTMANN U. G., BERNASCONI S. M. and BOTTCHER M. E. (2001) Hyper-sulfidic deep biosphere indicates extreme sulfur isotope fractionation during single-step microbial sulfate reduction. *Geology* **29**(7), 647–650.

ZABACK D. A. and PRATT L. M. (1992) Isotopic composition and speciation of sulfur in Miocene Monterey Formation: re-evaluation of sulfur reactions during early diagenesis in marine environments. *Geochim. Cosmochim. Acta* **56**, 763–774.

ZABACK D. A., PRATT L. M. and HAYES J. M. (1993) Transport and reduction of sulfate and immobilization of sulfide in marine black shales. *Geology* **21**, 141–144.

Geochemical Investigations in Earth and Space Science: A Tribute to Isaac R. Kaplan
© The Geochemical Society, Publication No. 9, 2004
Editors: R.J. Hill, J. Leventhal, Z. Aizenshtat, M.J. Baedecker, G. Claypool,
R. Eganhouse, M. Goldhaber and K. Peters

Significance of $\delta^{34}S$ and evaluation of its imprint on sedimentary sulfur rich organic matter II: Thermal changes of kerogens type II-S catagenetic stage controlled mechanisms. A study and conceptual overview

ZEEV AIZENSHTAT and ALON AMRANI

Department of Organic Chemistry, Casali Institute, The Hebrew University,
Jerusalem 91904, Israel

Abstract—Kerogens of type II-S are rich in sulfur, containing up to 10–12% organically bound sulfur. Most of this sulfur is thermally unstable due to the presence of catanated poly-S-S linkages. The $\delta^{34}S$ values for these kerogens carry the imprint of the pore water polysulfides introduced into the organic matter at the diagenetic stage as described in the previous review (Part I). The catagenetic stage, covered in Part II, is mostly driven by the increase of temperature, leading to rearrangement of both carbon and sulfur bonds that are reformed thermally to stabilized alicyclic and aromatic sulfur-containing structures. The controlling factors for $\delta^{13}C$ changes during these modifications are the release of CO_2 and C_1-C_5 hydrocarbons, mostly CH_4. The sulfur stabilization releases H_2S and S^0 during the forming of the $C-S-C$ moieties and their aromatization. The present report and review examines the influence of the above geochemical changes and the mechanisms controlling them on the stable isotope distribution of the thermally derived products. The understanding of these changes can lead to a better correlation between potential source rocks (PSR) and petroleum generated from them. The released carbon-containing molecules, *i.e.* CO_2 and CH_4 are chemically stable and not reactive (non-reversible reactions); in contrast at elevated temperature the sulfur released (H_2S, S^0) can re-react with the organic matter if not removed. It is therefore very important to examine the sulfur functionality changes during catagenesis that are thermally controlled through the mechanisms leading to sulfur isotope ratios variation. In addition, whether the system is open or closed influences the free radical restructuring of the organic matter and hence will influence the isotopic distribution of sulfur. The thermal cleavage and restructuring of kerogen to produce oil has an impact on the $\delta^{13}C$ of the asphalts and petroleum of 2‰. Moreover, the various fractions such as gas (CH_4), saturates, aromatics, resins and asphaltenes show different $\delta^{13}C$ ranges. The most depleted in ^{12}C is methane. Despite the recognized impact of maturity on carbon isotopes ratios, it has been previously suggested that the decrease in concentration of sulfur from kerogens of type II-S during maturation to generate oil does not cause $\delta^{34}S$ changes. While many hydrous pyrolysis and "dry" pyrolysis thermal simulation experiments were carried out and the thermal behavior of kerogens (type II-S) was studied, very few of these experiments were monitored for $\delta^{34}S$ changes. However, in the last 15 years some studies showed that the loss of sulfur and associated thermal stabilization is reflected in ^{32}S enrichment in H_2S and concurrently, ^{34}S enrichment of the petroleum produced. Based on these experiments we will offer mechanisms for the observed trend. Some new laboratory experiments performed by us in both closed and open systems are reported. Only very rough examination of the various organic sulfur-containing fractions was carried out in these studies. Some natural (geological) sites such as the Monterey Formation (Miocene, California, USA) and Senonian Formation (Dead Sea Area, Israel) are presented for comparison. In the general scheme, the $\delta^{34}S$ signature recorded in the kerogen changes during catagenesis to form petroleum depleted in sulfur and isotopically heavier. This enrichment in ^{34}S could amount to $+4$ to $+8‰$ relative to the kerogen of the PSR in thermally controlled experiments. In a field-based source rock to oil generated comparison, the isotope discrimination could be even higher, leading also to secondary metal sulfides (including relatively heavy pyrite). The chemically controlled thermal sulfate reduction ($\geqq 200°C$) is discussed only briefly.

INTRODUCTION

THIS PAPER deals with the chemical and isotopic evolution of type II-S kerogen during its thermal maturation.

Nucleophilic attack by polysulfides during diagenesis produces sulfur-containing macromolecules of kerogen of the II-S type (see Discussion, Part I). This hypothesis was first proposed based on studies of humic and fulvic acids reacted with polysulfides to form protokerogen (STOLER, 1981; AIZENSHTAT et al., 1983) and kerogen enrichment in organically bound sulfur (DINUR et al., 1980). Sulfur isotopic ratio investigation of bituminous rocks in Israel supported the mechanisms proposed (SPIRO et al., 1983a; TANNENBAUM et al., 1987) but the relevance of the conditions prevailing during diagenesis to the formation of kerogen type II-S has been debated ever since. These sulfur cross-linked polymers proposed to dominate the sulfur functionality of the kerogens were investigated by various analytical methods (PHILP et al., 1988; SINNINGHE DAMSTÉ et al., 1989, 1990, 1992; VORONKOV and DERYAGINA, 1990; ADAM et al., 1991; EGLINTON et al., 1992; KREIN, 1993; LUCKGE et al., 2002). It has been suggested that the high S content of II-S kerogen (S/C ratio of 0.07–0.09, 7–13% S) causes its anomalous thermal instability in comparison to that of "normal" kerogens of type II (LEWAN, 1998). This difference is manifested in the lowering of the activation energy required for the generation of "heavy oils" from type II-S kerogens compared with low sulfur-containing kerogens (LEWAN, 1998).

ORR (1986) suggested that the low thermal stability of the type II-S kerogen is mainly caused by the S–S bonds, which cross-link the polymers. More recent studies (KREIN, 1993; KREIN and AIZENSHTAT, 1995), support this general hypothesis. LEWAN (1998) postulates that the chemical susceptibility to thermal activation of the C–S bond is the reason for "early" heavy oil formation from kerogen of type II-S. The rates of these reactions are controlled by the formation of sulfur radicals generated during the initial stages of the thermal degradation. Available evidence shows that "early formation" of radicals from the polysulfide cross-linked macromolecules (PCLM) rather than the cleavage of the carbon–sulfur bond is responsible for the generation of smaller molecular weight organic sulfur compounds as well as release of inorganic sulfur (NELSON et al., 1995). This depolymerization leads to heavy oils still relatively rich in sulfur and with increase of maturation depletion of organically bound sulfur yielding less sulfur-rich petroleum (see details later). ORR (1986) concludes that sulfur is incorporated in oil without isotopic fractionation relative to its parent kerogen. In fact, most petroleum explorers seek to establish a correlation between oils and kerogens based on equal $\delta^{34}S$ values (ORR, 1974, 1986).

Analytical pyrolysis is one of the most informative methods to study the thermal behavior of kerogens (PHILP et al., 1988; SINNINGHE DAMSTÉ et al., 1989, 1990, 1992; VORONKOV and DERYAGINA, 1990; ADAM et al., 1991; EGLINTON et al., 1992). The objectives include structure determination of the macromolecules and kinetic profiling of the thermal degradation. The sulfur-containing pyroproducts cannot be related directly to the building blocks of the kerogen due to thermal modifications and rearrangements of the fragments. The chemical degradation of both, synthetic PCLM and very young sulfur-rich kerogens (protokerogens), reveals no aromatic rings of thiophenes (see Part I). Most of the sulfur is in S–S type bonds. The pyrolysis of high sulfur type II-S kerogens results in the decomposition of the macromolecular structure into the aromatic thermally stable moieties. The difference in products stems from the different mechanisms controlling the chemical degradation compared with the pyrolysis. The sulfur-containing volatile products produced by high temperature are dominated by light alkylated thiophenes (C_4–C_{10}) that are formed due to the cleavage of longer chains (SINNINGHE DAMSTÉ et al., 1989, 1992; VORONKOV and DERYAGINA, 1990; ADAM et al., 1991; EGLINTON et al., 1992). Hence, it is difficult to derive concrete pathways of sulfur stabilization if very young protokerogens are exposed to thermal treatment (AIZENSHTAT et al., 1995; LUCKGE et al., 2002). However, both types of analytical information are needed for the modeling of oil generation from potential source rocks (PSR). These very young sulfur-rich PCLM release part of the bridging sulfur as S^0. The rest of the sulfur that is in the gas phase is dominated by H_2S with minor quantities of COS and SO_2.

With increase of temperature, the chemical behavior of sulfur changes, and it becomes thermally active in hydrogen extraction, cyclization, reduction and aromatization (KREIN and AIZENSHTAT, 1995; see also Sch. 2 in Part I, Catagenesis section). These reactions are controlled by free radicals. Kerogens of type II-S can be regarded as PCLM. Their thermal maturation during the catagenesis stage causes rearrangement of the sulfur functionality. Specifically, the ratio between elemental sulfur and H$_2$S release is mainly dependant on the reactivity of labile sulfur moieties and availability of hydrogen during maturation (LUCKGE et al., 2002). The possible thermal stabilization of sulfur as thiophenes or thiolanes can be related to the original position of the polysulfide linkage. This has been demonstrated in the comparison of straight-chain and isoprenoid PCLM pyroproducts (unpublished data). These changes in the functionality of sulfur by release from the PCLM and its subsequent reactions to form the thermally stable aromatic sulfur compounds are a major pathway in the maturation of the sulfur-rich kerogens. The number of attacking positions of the C–S bond and C–S$_x$ polysulfides in the PCLM determines the yield and structure of heterocyclic compounds produced during catagenesis. This mechanistic approach is discussed, based on the results of pyrolytic experiments in AIZENSHTAT et al. (1995) and NELSON et al. (1995). Unlike the released small carbon-containing fragments (CO$_2$, CH$_4$, etc.) and light hydrocarbons, the S$_x$ and H$_2$S cleaved from the PCLM are reactive organically as well as inorganically at elevated temperatures. Moreover, as the redox conditions change, the sulfur radicals, e.g. HS$^{\cdot-}$, S$_x^{\cdot}$, RS$^{\cdot}$ may be involved in a variety of possible reactions with both available metals and the organic matter. The sulfur radicals could act as hydrogen extractors resulting in aromatization, or in the presence of water, initiate oxidation (KREIN, 1993; AIZENSHTAT et al., 1995; LEWAN, 1998). In contrast, H$_2$S, at elevated temperatures, is a reducing agent. The differences are also dependant on the nature of the system, either closed or open with respect to sulfur species released (Schs. 1 and 2 discussed in the Section 2). These conditions may differ to a greater or lesser extent from the conditions prevailing during catagenesis.

Thermally driven processes that generate oil from type II-S kerogens cause a decrease in sulfur concentration in the residual heavy petroleum–asphalt formed (ORR, 1986; BASKIN and PETERS, 1992). The relation between API gravity and sulfur content was followed both in petroleum exploration and in various laboratory studies (BASKIN and PETERS, 1992). The early generation of sulfur-rich asphaltic matter was suggested by GRANCH and POSTUMA (1974) for the Venezuela La Luna area and was also suggested for the Dead Sea asphalts by TANNENBAUM and AIZENSHTAT (1984, 1985). No systematic measurements were carried out on the changes

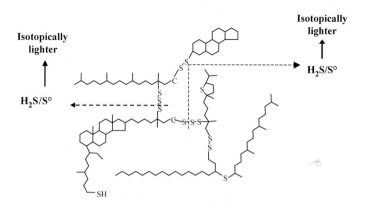

SCH. 1. Schematic thermal cleavage of S$_x$ bonds and release of H$_2$S/S^0 under open system conditions. A simplified PCLM representing immature kerogen of type II-S is given with early cleavage of the S–S bonds. The model is adapted from previous papers by Aizenshtat et al. (1995, 2003). The laboratory controlled experiments show the cleavage of the S–S bonds (marked by arrows) to initiate at 130°C; at higher temperatures other bonds such as C–C and C–S become susceptible to cleavage and rearrangements. The formation of H$_2$S requires hydrogen extraction that leads to aromatization.

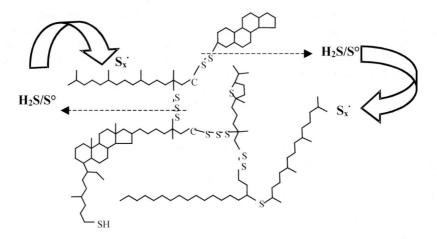

SCH. 2. Schematic thermal cleavage of S_x bonds and release of H_2S/S^0 under closed system conditions (see details in Sch. I). Arrows indicate back reactions of both hydrogen elimination and re-introduction of sulfur to the monomers formed (see also Aizenshtat et al. (1995, 2003)).

in $\delta^{34}S$ that occur during maturation of source rocks and their related oils (TANNENBAUM and AIZENSHTAT, 1984, 1985; ORR, 1986; BASKIN and PETERS, 1992).

Whereas Part I reviews the diagenetic imprint on the $\delta^{34}S$ through the whole sequence of the biogenic dissimilatory reduction of sulfate and the secondary sulfur enrichment, Part II is aimed at the catagenetic changes from immature kerogens of type II-S through thermal maturation to the generation of petroleum. Some new data are presented in the paper for results on pyrolysis experiments on type II-S kerogens exposed to different experimental conditions. In addition, geological samples that were examined for sulfur isotopic compositions are discussed. These new results are discussed in context of previous experiments and analytical data on oils and their related kerogens. An attempt is made to rationalize the sulfur isotope changes in both experimental and field samples based on the mechanisms proposed.

The complex chemical structure and diversity of kerogens drove us to study synthetic PCLM (AIZENSHTAT et al., 1995; KREIN and AIZENSHTAT, 1995). The thermal behavior of these oligo-polymers and their thermal products were much easier to characterize and hence assign mechanisms. During the maturation of sedimentary organic matter, both carbon and sulfur atoms are rearranged and bonds are cleaved. These transformations must have an influence on both carbon and sulfur stable isotopes ratios. For comparison and contrast, a short description of $\delta^{13}C$ changes during the catagenetic processes is given. The focus of the present chapter, however, is the presentation of a mechanistic hypothesis for sulfur-related transformations that is supported by the sulfur isotope composition accompanying the thermally induced changes.

PYROLYSIS EXPERIMENTS AND CHANGES IN $\delta^{34}S$ FOR ORGANIC AND INORGANIC SULFUR SPECIES

Experimental thermal alteration (pyrolysis) provides the best opportunity to characterize changes in type II-S kerogen under thermal stress. In the present section, we follow the changes in sulfur isotope distribution as a result of the different temperatures and pyrolysis conditions.

Open and closed pyrolysis systems

Open systems consist of vessels from which the volatile pyro-products can leave the heated reactor thus avoiding secondary reactions. Closed systems are sealed reactors that confine all reactants and products to a restricted volume. The removal of gaseous products from the open system reactor can be enhanced by purging with inert or chemically active carrier gas (STOLER et al., 2003). In most cases if a small sample is introduced into a very large evacuated and sealed vessel the early production of volatiles can be considered to be produced under "open" conditions. These definitions are relevant to geochemical simulation experiments as the mobility of the products from the hot zone can effect major changes in comparison to the in situ water, petroleum and gas impermeable trap.

The geological setting imposes an even more complicated environment due to the catalytic and chemical influence of mineral surfaces on the thermal process, e.g. formation of metal complexes with the organic matter that catalyze reactions or lower reactivity of species by inorganic removal of sulfur species.

During the discussion on pyrolysis experiments we will refer to the Rock Eval® thermal method, commercially used for the evaluation of PSR and maturation of sedimentary organic matter. This method employs an open system programmed pyrolyser purged with He carrier gas into various detectors for monitoring hydrocarbons, CO_2 and a modified version with a sulfur detector. The use of this method for the evaluation of rocks containing type II-S kerogens is well documented (see SPIRO, 1991; BASKIN and PETERS, 1992).

In the planning of simulation experiments on sulfur-rich organic matter, it is imperative to understand the influence of the choice between open and closed systems. Sch. 1 describes the open system where the lower temperature released $H_2S\backslash S^0$ is thermally removed and the residual polymeric matter undergoes thermal stabilization. Under geological burial conditions the availability of iron and other metals in the sediments could remove the released sulfur, e.g. formation of secondary pyrite. Sch. 2 shows a closed system scenario for the same model type II-S kerogen as in Sch. 1. In Sch. 2 the $H_2S\backslash S^0$ thermally formed is retained and at elevated temperatures forms chemically very active $S_x^{\cdot}$ free radicals. These free radicals can reintroduce sulfur or extract more hydrogen to form aromatics and H_2S that can react in other reactions such as reduction of double bonds. For the present discussion, it is important to emphasize the marked difference in isotope signature: a closed system will minimize the isotope effect because sulfur is conserved whereas an open system will enhance the isotopic discrimination governed by the release of sulfur.

Pyrolysis of type II-S kerogens

In this section, we discuss the pyrolysis of type II-S kerogens. The key feature for the changes in δ^{34}S of the products relates to the type of system applied, closed or open. Whole rock samples as well as isolated kerogens (type II-S) have been exposed to pyrolysis conditions by numerous researchers. Most of these thermally controlled experiments were designed to establish kinetic parameters that could be used for Basin Evaluation modeling (ATTAR, 1978; LEWAN, 1985; STOLER, 1990; EGLINTON et al., 1992; KOOPMANS et al., 1998). Few of these studies monitored the fate of the isotopic distribution of sulfur during the various stages (IDIZ et al., 1990; EVERLIEN et al., 1997). Despite the meager experimental data, relevant review of both carbon and sulfur isotope discrimination related to thermal influence of carbon oxidation and sulfate reduction is given by GOLDSTEIN and AIZENSHTAT (1994). The yield of $H_2S\backslash S$ thermally released from kerogens and asphalts peaks at 400–450°C as was monitored by Rock Eval® type open system reactor. At the same temperature interval the total liquefaction is maximized.

We have obtained new data on the open system pyrolysis of purified kerogen with δ^{34}S$_{kerogen}$ = −4‰ value, derived from Zefa Efe' bituminous rock (Mishor Rotem, Senonian age, Israel). These kerogens evolved H_2S that initially is −3.5‰ and at peak H_2S generation becomes −7.2‰ (detailed description by STOLER et al. (2003)). The δ^{34}S of liquid pyrolysate isolated from these experiments is +4.6‰. This δ^{34}S$_{kerogen-oil\ (liquid\ pyrolysates)}$ of up to +8‰ is much larger than the value recorded for the hydrous pyrolysis experiments reported by IDIZ et al. (1990), which were carried out on samples of the Monterey Formation in closed vessels. However, it is in agreement with the data reported by EVERLIEN et al. (1997)) (Figs. 1 and 2) for the Monterey Formation samples and other rocks with type II-S kerogens. Actually, EVERLIEN et al. (1997) used a closed system for each interval and sampled the cumulative production for each temperature. Closed vessel "dry" experiments on Monterey Formation and other sulfur-rich kerogens and asphalts were isotopically characterized by

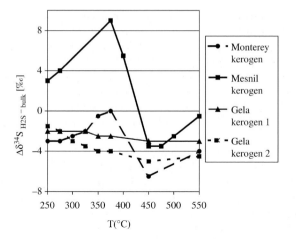

FIG. 1. Sulfur isotope fractionation during closed vessel pyrolysis of several immature type II-S kerogens. This figure is after EVERLIEN et al. (1997). The $\delta^{34}S$ ratios reported as difference between values recorded for the kerogen and H_2S produced are plotted versus temperature.

EVERLIEN et al. (1997)) during thermogenic experiments from 250–550°C parallel to the Rock Eval® (normal heating rate of 25°C/min) profile to determine maturity and potential for oil production. All samples studied were selected to have low maturity of $R_0 \sim 0.4$ (vitrinite equivalent). Asphaltenes showed 50°C offset to higher temperature for the release of volatile sulfur, mostly H_2S. The generation of H_2S in all samples started at 250–300°C with the highest production at 425–500°C (ATTAR, 1978). All H_2S samples showed ^{34}S depletion relative to the starting kerogen or asphaltenes, from 4 to 8‰. The fact that they did not remove pyrite from the samples caused some high temperature pyrite contribution to both the H_2S and sulfur in the bitumen (EVERLIEN et al., 1997).

Differences are to be expected between the hydrous (KOOPMANS et al., 1998), dry, open and closed systems. For example, the 300–350°C hydrous pyrolysis experiments produce heavy asphalts with even more sulfur than in the kerogen. IDIZ et al. (1990) found up to 14% S in the liquid pyrolysate from kerogen with 10% S. In contrast, in open system experiments the liquid pyrolysate that had the highest yield between 425 and 475°C contains less than 50% of the original organically bound kerogen sulfur (STOLER et al., 2003). The hydrous pyrolysis experiments of IDIZ et al. (1990) were conducted in closed vessels, whereas, the experiments reported here for the pyrolysis of samples from the Dead Sea Area (Senonian age, Israel described later) were conducted under controlled open system. Two experiments were conducted under high

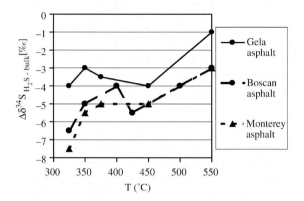

FIG. 2. Sulfur isotopes differentiation for several sulfur-rich asphaltenes during closed vessel pyrolysis. This figure is after EVERLIEN et al. (1997). The $\delta^{34}S$ reported as difference between values recorded for the asphaltenes and H_2S produced are plotted versus temperature.

pressure gas flow of 1600 psi N_2 and 1600 psi H_2. The experiments under hydrogen atmosphere showed almost full conversion of the kerogen, yielding liquid pyrolysates with less than 2% sulfur. This is only 20% of the kerogen's initial organically bound sulfur. Large amounts of H_2S were released (STOLER et al., 2003).

In closed systems H_2S and S^0 thermally produced can remix and re-react at elevated temperatures. These back reactions depend on the minerals present and the dilution factor of the gas pressurized in the autoclave. It is, therefore, much more complex to decipher the mechanisms controlling the $\delta^{34}S$ values of all fractions analyzed specially for closed systems (NELSON et al., 1995).

In previous articles we hypothesized the existence of a high temperature "back" reaction of "stabilized" organic products with thermally activated sulfur such as S˙ radical) (AIZENSHTAT et al., 1995; KREIN and AIZENSHTAT, 1995). To experimentally evaluate this possibility we reacted, under autoclave inert conditions at 300°C, a mixture of hydrocarbons with elemental sulfur. The products and their GC/MS identification were presented in STOLER (1990). A similar experiment was performed without detailed analysis of the products distribution by BAKER and REED (1929). Most of the material polymerizes. Only 15–20% could be extracted by organic solvents (e.g. hexane or methylene chloride). The sulfur isotopes data for all products show that no isotopic discrimination occurred in this experiment (STOLER et al., 2003). All sulfur-containing isolated and identified compounds produced are thiophenes with the structure:

$$R_y \diagdown \overset{\diagup\diagup}{\underset{S}{\diagdown}} \diagup R_x$$

$R_{x+y} + 4C = nC_{11,12,14,17,18(:1)}$

As expected, the unsaturated hydrocarbon $C_{18:1}$ (unsaturation position at carbon 1) reacted much more than the n-saturated hydrocarbons in the mixture. This is due to the higher reactivity of double bonds to sulfur-free radical attack.

In an open system some isotopic discrimination between the rubber-like polymer (PCLM) and the H_2S formed is expected. This is due to two stages: (1) attack to form the C–S_x bonds and (2) the extraction of hydrogens to form the H_2S leading to the aromatic stable thiophenes structure. This discrimination for the open systems was lately demonstrated by us also for the reaction of elemental sulfur with unsaturated hydrocarbons (unpublished results). Nevertheless, since the system described above is totally closed (hydrocarbons and elemental sulfur in ampoule), remixing of the sulfur reservoir and its isotope ratio occurs. The above results support the suggested high temperature re-reaction of previously released reduced sulfur with no isotopic discrimination under closed vessel reactions.

IDIZ et al. (1990) summarize their experimental results on two Monterey Formation (Santa Maria Valley) samples of isolated kerogen pyrolysed in closed ampoules, by the statement: the range of variation of $\delta^{34}S$ value between the pyrolysates and the source kerogen is about 2‰. Moreover, the literature up to 1990 considers this range typical for the difference $\delta^{34}S_{kerogen-oil}$ between source rocks and petroleum (e.g. ORR (1986)). This isotopic change is hypothesized to result by the production of H_2S with ^{32}S enrichment of up to 3‰ compared to the kerogen. The loss of organic sulfur was recorded in kerogen (two sets of experiments 0.056–0.039 S/C and 0.054–0.037 S/C kerogen–kerogen after thermal experiment, respectively), but the percent S in the bitumen (pyrolysates) is very high, 12–14% (IDIZ et al., 1990). Since the experiments were conducted for 10–100 h at 300°C in closed vessels, it is possible that under higher thermal stress and open system more H_2S would have been released. Hence, the isotopic differentiation between the original kerogen and the pyrolysates might have been higher.

The hydrous pyrolysis experiments conducted by BASKIN and PETERS (1992) on Monterey Formation whole sediments and isolated kerogens showed a decrease of sulfur content in the liquid pyrolysate, from 8.8% at 300°C to 5.5% (organic bound S wt%) at 360°C. The total yield of retained original kerogen sulfur decreased to ~25% of the original content. Most of the released sulfur was in the H_2S form with minor elemental sulfur amount.

Based on the pyrolysis experiments listed above we can see a much higher isotope effect in open systems than for closed systems. The ^{32}S isotope enrichment for both H_2S and S released leads to ^{34}S enrichment in the residual S-organic products providing no back reactions occurred to enrich the pyrolysate in the range of 2–8‰.

CASE STUDIES, GEOLOGICAL MATURATION

In this section, we examine correlations of sulfur isotopes with maturation in a stratigraphic sequence. In this section of the overview, we need to examine in parallel information on maturation of the samples chosen and the natural variability of geological samples. Very small changes in depositional conditions even for samples of the same formation and geographical location can cause a wide range (− 15 to + 15‰) of initial δ^{34}S values(see Part I).

Monterey Formation, bitumen, heavy oils and asphalts, petroleum (California, USA)

The Monterey Formation of Miocene age is an important source rock for petroleum in California, USA (ORR, 1986; IDIZ et al., 1990; AIZENSHTAT et al., 1999; SCHOUTEN et al., 2001). In this Formation, extensive sulfate reduction was controlled by insufficient sulfate supply. ZABACK and PRATT (1992) and ZABACK et al. (1993) argue that the observed diversity in pyrite sulfur isotopes from −18 to +15‰ may relate to the timing of its formation, in the diagenesis sequence. Thus, the range of δ^{34}S values for pyrite in this formation is somewhat heavy. The relation between timing and sulfur isotope imprint of conditions of sedimentation is discussed in depth in Part I.

BASKIN and PETERS (1992) have followed the ratio of the sulfur wt% versus API gravity for asphalts and oil shows, proposed to originate from Monterey PSR, and found a good correlation: with lowering of sulfur content the petroleum API increases (lighter oil). Moreover, the maturation trend for oils of the Santa Barbara Channel show good correlation with API gravity. Most geochemists accept that the source rocks for these asphalts and petroleum are Monterey Formation and that the migration of the heavy petroleum is over short distances. The proposed source rocks and some of these oils (as marked in Fig. 3)

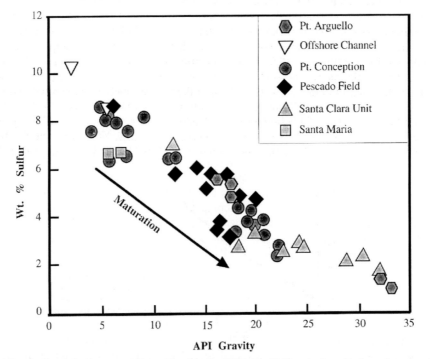

FIG. 3. Correlation between sulfur wt% and gravity (API) for oils from the Santa Barbara channel and offshore Santa Maria basin (Monterey Formation, California, USA). The authors (BASKIN and PETERS, 1992) suggest the line for maturity assuming no biodegradation. See also the detailed description of kerogens isolated and analyzed from the Monterey Formation.

Table 1. Monterey Miocene Formation samples, shown in Figs. 4 and 5: (a) sulfur percent, S/C ratio and Rock Eval® for kerogens and oils, see also description in Figs. 4 and 5; (b) the same data on core samples at depth 1155–2490 m

	S%			S/C		
	Kerogen	Oil		Kerogen	Oil	T_{max} (°C) (RE)[a]
(a) Samples						
(A) Pt. Conception	8.5	2.5 (light) (25–30 API)		0.049	0.012	415
(B) Pt. Argello	7.9	1.5–2.0		0.046	0.013	410
(C) Santa Maria	8.2	6.2 (heavy) (10 API)		0.047	0.029	410
(D) Offshore Channel	10.2	8.0 (heavy) (10 API)		0.059	0.038	404

(b) Bardly Lands samples depth of borehole (feet)	S% bitumen	S/C	T_{max} for kerogens (°C) (RE)[a]
3810	10.1	0.057	400
4620	10.8	0.062	415
5340	9.0	0.050	419
6780	8.4	0.048	425
7140	8.0	0.046	429
8220	5.2	0.030	431

[a] T_{max} for all kerogens of type II-S show also a T_{mid} range at 350–370°C. Vitrinite reflectance for all rock samples $\%R_0$ 0.3–0.35.

Very heavy oils have similar sulfur concentration to the kerogens. Isolation of kerogens is as described by TANNENBAUM and AIZENSHTAT (1984); oil analysis from BASKIN and PETERS (1992).

were analyzed for $\delta^{34}S$ (Table 1 and Fig. 4). The results show that all PSR kerogens are isotopically lighter than the oils. Figure 4 includes also for some of the PSR, the pyrite $\delta^{34}S$ that is much lighter than the associated kerogens.

Figure 5a and Table 1a contain data for a set of samples taken from a $\sim$1500 m thick section of the Monterey shales. The kerogens were isolated (HCl, HCl/HF removal of carbonates and silicates and dilute nitric acid to remove the pyrite) for some of the core samples, but the total rock was used for the Rock Eval® T_{max} determination to estimate maturation levels. The $\delta^{34}S$ was measured for the extracted bitumen of each sample. The increase in T_{max} with depth is correlated with maturation indicating that burial of the samples increased their maturation. Two trends are indicated by the $\delta^{34}S$ values: one starts at $+15.6$ and increases to $+17.7‰$, and the second from $+19.0$ to $+21.3‰$. The data indicate that the bitumen produced at higher level of maturation is enriched with the heavier isotope. The elemental analysis of the bitumen shows that the shallowest (1400 m) sample has $\sim$11% S, whereas the deepest at 2490 m

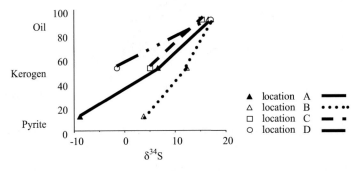

FIG. 4. Sulfur isotopes in sediments and oils of the Monterey Formation location as noted in BASKIN and PETERS (1992). $\delta^{34}S$ are plotted for pyrite, kerogen and oils; the connecting lines are suggested to follow the generation trends from the PSR from the same locality to the oils migrated. All kerogens are of type II-S and oils are typical for the offshore California.

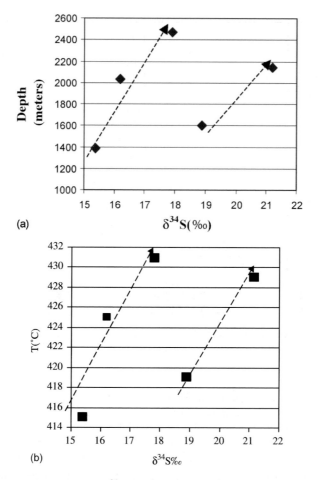

FIG. 5. Monterey Formation cores δ^{34}S: (a) bitumen extracted from the source samples versus depth and (b) bitumen profile of δ^{34}S versus T_{max} from Rock Eval® on whole sample. Samples taken from a drilled well of the same location as reported in Table 1 and BASKIN and PETERS (1992).

contains ~5% S. The gradual loss of sulfur in the course of maturation reflects the loss of the less stable sulfur.

It is noteworthy that the bitumens, kerogens and oils of the Monterey Formation (Miocene section) have δ^{13}C values of −21 to −22.5‰. These values are 5–6‰ heavier than other similar formations, even those rich in organically bound sulfur such as the Dead Sea bituminous rocks (see Section 3.2). These δ^{13}C values are typical for the Miocene era and hence the petroleum derived from these Formations is heavy globally (AIZENSHTAT et al., 1999). The data presented earlier show that during catagenesis of Monterey Formation II-S-kerogens, the thermal reforming of more stable compounds (see introduction) and decreasing of the content of total S in both bitumen and the kerogen, lead to ^{34}S-depleted H_2S and ^{34}S-enriched petroleum.

Senonian age Ghareb Formation, bituminous rocks, asphalts and oil shows; Dead Sea Area Israel

Studies of organic matter in both bituminous rocks of the Ghareb Formation as well as asphalts and petroleum shows of the Northern Negev and the Dead Sea area in Israel have been previously published (SPIRO et al., 1983a,b; TANNENBAUM, 1983; TANNENBAUM et al., 1987; ALMOGY-LABIN et al., 1993). The diagenesis of these sulfur-rich deposits is discussed in

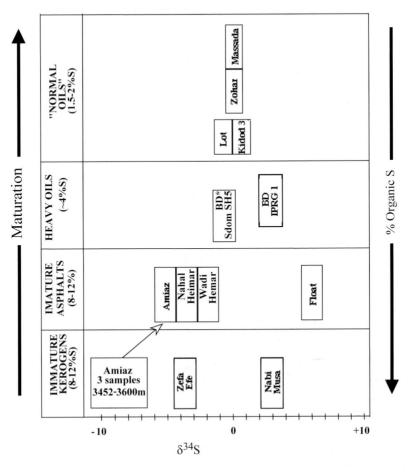

FIG. 6. Kerogens, asphalts, biodegraded asphalts (oil) and oil shows of the Dead Sea area, Israel are plotted on $\delta^{34}S$ versus maturation. See also Fig. 4 in Part I. We assume that the kerogens that are isolated from the Ghareb-Senonian age bituminous rocks (type II-S), represent the deeper, more mature members, which generated the asphalts and the oils. Scale of maturation is arbitrary. See also Table 1 in Part I.

Part I. Despite the geographical proximity and the similar depositional environments proposed for the various Ghareb formation studied sites, we find that the $\delta^{34}S$ values for the kerogens span in the range from $-10‰$ for Amiaz to $+4‰$ at the Nabi Musa deposit (see Table 1 in Part I). Fig. 6 shows the data obtained by our group over the last 20 years for $\delta^{34}S$ of kerogens, asphalts and oil shows. Some previously unpublished isotope results were provided by Drs E. TANNENBAUM and M. COLEMAN (written communication 1989). An attempt to use polarity fractionation of the oils and asphalts (hydrocarbons, aromatics, resins, NSO containing and asphaltenes) to explain $\delta^{34}S$ of the different fractions was not successful. However, it was obvious that the lowest $\delta^{34}S$ values were recorded from the apolar fractions, whereas the aromatic fractions had the highest $\delta^{34}S$ values. This trend supports the hypothesis that with the stabilization of the C–S bonds during maturation, such as the formation of the aromatic thiophenic structures, the sulfur is more enriched in ^{34}S (see Section 4).

CATAGENESIS—THERMAL CHANGES AND THEIR INFLUENCE ON THE SULFUR ISOTOPE DISTRIBUTION

In order to interpret the changes in $\delta^{34}S$ during catagenesis thermal processes we have to understand the stabilization of sulfur during these increased temperature transformations. During the early diagenetic introduction of polysulfides into the organic matter, most of the sulfur is catenated as S–S bonds (AIZENSHTAT *et al.*, 1983; KOHNEN *et al.*, 1991). The immature kerogens of type II-S and to some extent even sulfur-rich coaly matter are thus dominated by cross-linking of S–S bonds. This sulfur carries the isotope signature of the original pore water polysulfides that reacted with the organic matter (see Part I). We have argued that this isotopic imprint reflects the biochemical–geochemical history of the depositional environment (AIZENSHTAT *et al.*, 1999). The bond strength for the C–S and S–S bonds shows that the longer the catenation, the lower the energy needed for the cleavage of the S–S_x bonds (AIZENSHTAT *et al.*, 1995). At advanced stages of catagenesis, when C–S bonds start to break the activation energy required is high. Theoretically, the cleavage of the polysulfide bonds should impact an isotopic effect because of the energy difference between $^{32}S-\ ^{32}S$ and $^{34}S-\ ^{32}S$ bond strengths. However, the release of H_2S requires the extraction of two hydrogen atoms and this may add to the isotopic discrimination. The availability of hydrogen is therefore very important as reflected in the experimental studies (see Section 3.2). Obviously, based on bond strength the carbon–sulfur bond cleavage will yield a larger isotope effect than the S–S_x bond cleavage, particularly if $x > 2$. Hence, if the cleavage of such bonds leads only to rearrangement, one should not expect a large isotopic effect. On the other hand, if this cleavage results in release of volatile molecules such as hydrogen sulfide, this release *will cause an isotopic difference*. The same effect is recognized at still higher degree of maturity (thermally produced gas) for the hydrocarbon-gases (see Sch. 3); the largest effect is of course recorded for methane (GALIMOV, 1973; STAHL, 1977, 1978). In order to manufacture methane, multiple C–C bond cleavage is required, whereas higher ethane, propane and the other wet gases are isotopically closer to the bulk of the hydrocarbons of the petroleum.

Scheme 3 shows the overall direction for the thermal changes from kerogen to the petroleum (various fractions) and gases. The loss of the ^{13}C richer CO_2 leads to the Galimov–Stahl diagram (GALIMOV, 1973; STAHL, 1977, 1978) whereas the formation of the gases (C_{1-4}) relate to the C–C bond rupture during gasification (thermally produced gases) that yields an opposite carbon isotope effect.

The basic difference between the formation of sulfur and the carbon gases and its influence on the isotopic ratios of kerogen, bitumen, oil and gas lies in the chemical activity of the thermally derived fragments. Both elemental sulfur and hydrogen sulfide formed at elevated temperatures could re-react with all fractions (STOLER *et al.*, 2003). Research on changes in both carbon and

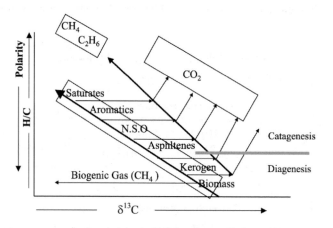

SCH. 3. Schematic carbon stable isotopes ratio from the diagenetic stage—biomass to kerogen through the catagenetic stage to petroleum formation. The petroleum fractions are enrichment in ^{12}C with decrease in polarity (adapted from GALIMOV (1973) and STAHL (1977, 1978)). Biomass range of carbon isotope ratio is determined by the origin and depositional conditions. Carbon dioxide release is always heavier than the residual organic matter (quantitatively important as the transformation, biomass → kerogen and kerogen → petroleum). The largest thermal isotopic effect is the production of methane. The bio-genically derived methane is lightest ^{12}C enriched. Thermally produced methane is −30 to 40‰ lighter than petroleum, whereas the biogenic methane is −50 to 80‰ lighter than the organic matter from which it is derived.

sulfur isotope signatures during the catagenesis has been motivated by the exploration for oil and gas. The basic concept was that the information embedded into the kerogen of the potential source rock could yield good correlation with the generated oil and gas.

GOLDSTEIN and AIZENSHTAT (1994) dedicated two sections of the review to follow thermal sulfate reduction (TSR) influence on carbon and sulfur isotopes. Based on the works cited and in particular the studies by WILLIAMSON and RIMSTIDT (1990), GOLDSTEIN and AIZENSHTAT (1994) conclude: (a) lower activation energy pathway is enabled by the multi-step mechanism provided by polysulfidic catanation (see also Fig. 1a–c in Part I); (b) organic matter oxidation by sulfate is facilitated at temperature > 200°C. The H_2S yielded in these processes is isotopically enriched by ^{32}S in the range of -10 to $-25‰$, depending on the type of organic matter oxidized and in some cases kinetically controlled. The imprint of TSR on the carbon isotopes is also discussed but it is out of the scope of the present review.

CONCLUDING REMARKS

There is no doubt that during maturation of kerogens of the II-S type to produce gas, liquids and a solid residue, there is sulfur isotope fractionation. Scheme 4 gives the general trend for δ^{34}S change from the diagenetic stage (discussed in Part I) through the thermal maturation (catagenesis). The direction of these changes seems to be inversed to the direction of the carbon isotopes. We propose that the high reactivity of the sulfur at the catagenetic–methagenetic stages (i.e. $T > 130$°C) might cause remixing of the released sulfur species erasing any isotopic effects and thermally leading to the re-equilibration of H_2S, S^- and the petroleum sulfur. The various changes depend on the type of conditions prevailing during the maturation, e.g. open or closed.

In the past, researchers used to implicate the pyritic sulfur for being involved in the differences of isotopic changes during catagenesis; we claim that such is *not* the case.

However, we can see some systematic relationships, which emerge from the sulfur isotope profiles from simulated pyrolysis experiments:

(1) The evolved H_2S from the kerogen is always ^{32}S-enriched. Very little, if any, pyritic light sulfur is contributed. On average, the H_2S released is 5‰ lighter than the original δ^{34}S of the organic matter.

(2) All pyro-bitumens are $+5$ to $+8‰$ ^{34}S heavier than the original kerogen.

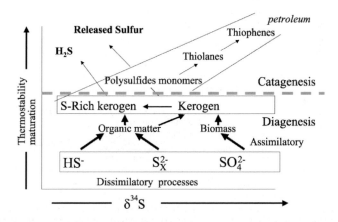

SCH. 4. Schematic sulfur isotope ratio changes for both diagenesis and catagenesis. In order to scale the δ^{34}S value: sea water sulfate $+20 \pm 2‰$ and sulfide in pyrite, if recorded for an open system for sulfate supply is $-20 \pm 5‰$. S-rich kerogen type II-S and stage of enrichment in cross-linked polysulfides marked by arrows could be some 5‰ heavier than the water-dissolved polysulfides. The present review suggests that the H_2S released from kerogens and asphalts is ^{32}S richer than the substrate from which it is derived. The isotopic changes at the aromatization stages are speculative due to lack of direct data. Note: the scale for the δ^{34}S shows only direction and not actual discrimination for each step.

(3) The thermally produced CaS (formed only at high temperatures and dry conditions) is the lightest marker of ^{32}S-enriched H_2S released; no change was recorded in pyrite $\delta^{34}S$ up to 500°C.

(4) The residual kerogen is isotopically heavier although the changes are relatively small.

The general conceptual diagrammatic proposal of the sulfur isotope changes from diagenesis to catagenesis is presented in this review. Future studies and better control on the kinetics as well as thermodynamics of the recorded geochemical changes will enable better understanding of mechanisms involved. These will provide an important tool for basin evaluation and determination of PSR in relation to generation of oil.

The real test for the adaptability of mechanisms proposed to control the catagenesis of type II-S kerogens during their maturation to the formation of asphalt and petroleum is the degree to which they explain the natural cases studied. The geological conditions are much more complex due to the multiplicity of the mineral matrix catalyzing and controlling the transformations. The present review concentrates on the sulfur isotopes as an added marker for both degree of maturation as well as identification of PSR.

(1) With increase of maturation and production of bitumens, asphalts and heavy oils the %S decreases and the residual stabilized sulfur becomes isotopically heavier.

(2) In most cases studied, the petroleum sulfur is ^{34}S enriched.

(3) Most of the sulfur in immature kerogens of type II-S is in polysulfide (S–S) functionality (PCLM) whereas in petroleum most of the sulfur is identified in the aromatic thiophene functionality.

Although the database for the thermal changes in laboratory controlled experiments supplies some indication as to the mechanisms involved, we are still far from the detailed pathways for the geological maturation of sulfur-rich kerogens. Despite the advent of analytical methodology, we still do not know if the use of elevated temperatures to compensate for the long geological times is completely valid.

Acknowledgements—Zeev Aizenshtat wishes to thank Ian Kaplan for his introduction into the interdisciplinary scientific world of organic geochemistry. Many of the works cited and discussed in Part I and Part II conceptual reviews, although not authored by Kaplan, have been inspired by his early pioneering works and his continued introduction of novel ideas into the field. The second generation represented by Alon Amrani, Baruch Spiro, Eli Tannenbaum, Achikam Stoler, Eitan Krein and many others is now carrying the torch. Special thanks to Max Coleman for some of the data on the Dead Sea samples. Last but not least our warm thanks to Dr Martin Goldhaber for extensive editorial help and Dr Murthy Vairamurthy for critical review of this manuscript.

REFERENCES

ADAM P., SCHMIDT J. C., ALBRECHT P. and CONNAN J. (1991) 2-α and 3-β steroid thiols from reductive cleavage of macromolecular petroleum fraction. *Tetrahedron Lett.* **32**, 2955–2957.

AIZENSHTAT Z., STOLER A., COHEN Y. and NIELSEN H. (1983) The Geochemical sulphur enrichment of recent organic matter by polysulfides. In *The Solar Lake. Advances in Organic Geochemistry 1981* (ed. M. BJOROY), pp. 279–288. Wiley, Chichester.

AIZENSHTAT Z., KREIN E. B., VAIRAVAMURTHY M. A. and GOLDSTEIN T. P. (1995) Role of sulfur in the transformation of sedimentary organic matter: a mechanistic overview. In *Geochemical Transformation of Sedimentary Sulfur*, (eds. M. A. VAIRAVAMURTHY and M. A. SCHOONEN), Vol. 612, pp. 16–39. American Chemical Society Series, ACS, Washington, DC.

AIZENSHTAT Z., MILOSLAVSKI I. and ASHENGRAU D. (1999) Hypersaline depositional environments and their relation to oil generation. In *Microbiology and Biogeochemistry of Hypersaline Environments* (ed. A. OREN), pp. 89–105. CRC Press, Boca Raton, FL.

AIZENSHTAT Z., AMRANI A. and KREIN E. B. (2003) Simulated sulfur incorporation and thermal transformation: C_{10}, C_{15}, and C_{20} isoprenoids naturally occurring as model compounds. *Org. Geochem.*, submitted for publication.

ALMOGI-LABIN A., BEIN A. and SASS E. (1993) Late Cretaceous upwelling system along the southern Tethys margin (Israel): interrelationship between productivity, bottom water environment, and organic matter preservation. *Paleoceanography* **8**, 671–690.

ATTAR A. (1978) Chemistry, thermodynamics and kinetics of reactions of sulfur in coal-gas reactions: a review. *Fuel* **57**, 201–211.

BAKER R. B. and REED E. E. (1929) The action of sulfur on normal heptane and normal butene. *J. Am. Chem. Soc.* **51**(Suppl.), 1568–1573.

BASKIN D. K. and PETERS K. E. (1992) Early generation characteristics of a sulfur-rich Monterey kerogen. *Am. Assoc. Petrol. Geol.* **76**, 1–13.

DINUR D., SPIRO B. and AIZENSHTAT Z. (1980) The distribution and isotopic composition of sulfur in organic rich sedimentary rocks. *Chem. Geol.* **31**, 37–51.

EGLINTON T. I., SINNINGHEDAMSTÉ J. S., POOL W., DE LEEUW J. W., EIJKEL G. and BOON J. J. (1992) Alkylpyrroles in a kerogen pyrolysate—evidence for abundant tetrapyrrole pigments. *Geochim. Cosmochim. Acta* **56**, 1545 p.

EVERLIEN G., LEBLOND C. and PRINZHOFER A. (1997) Isotopic characterization of thermogenic hydrogen sulphide from sulfur-rich organic matter. *Extended Abstract of the 18th International Meeting on Organic Geochemistry: Part II*, pp. 803–804. Maastricht, The Netherlands.

GALIMOV E. M. (1973) *Carbon Isotopes in Oil–Gas Geology*. Nedra Press, Moscow.

GOLDSTEIN T. P. and AIZENSHTAT Z. (1994) Thermochemical sulfate reduction: a review. *J. Therm. Anal.* **42**, 241–290.

GRANCH J. A. and POSTUMA J. (1974) On the origin of sulfur in petroleums. In *Advances in Organic Geochemistry* (eds. B. TISSOT and F. BIENNER), pp. 727–739. Technip, Paris.

IDIZ E. F., TANNENBAUM E. and KAPLAN I. R. (1990) Pyrolysis of high-sulfur Monterey kerogens—stable isotopes of sulfur, carbon, and hydrogen. In *Geochemistry of Sulfur in Fossil Fuels* (eds. W. L. ORR and C. M. WHITE), Vol. 429, pp. 575–591. American Chemical Society Series, ACS, Washington, DC.

KOHNEN M. E. L., SINNINGHE DAMSTÈ J. S., KOCK-VAN DALEN A. C. and DE LEEUW J. W. (1991) Di- or polysulphide-bound biomarkers in sulphur-rich geomacromolecules as revealed by selective chemolysis. *Geochim. Cosmochim. Acta* **55**, 1375–1394.

KOOPMANS M. P., RIJPSTRA W. I. C., DE LEEUW J. W., LEWAN M. D. and SINNINGHE DAMSTÈ J. S. (1998) Artificial maturation of immature sulfur- and organic matter-rich limestone from the Ghareb Formation, Jordan. *Org. Geochem.* **28**, 503–521.

KREIN E. B. (1993) Organic sulfur in the geosphere: analysis, structure and chemical process. In *Supplement S: The Chemistry of Sulfur-Containing Functional Groups* (eds. S. PATAI and Z. RAPPOPORT), pp. 975–1032. Wiley, Chichester.

KREIN E. B. and AIZENSHTAT Z. (1995) Proposed thermal pathways for sulfur transformation in organic simulation macromolecules: laboratory simulation experiments. In *Geochemical Transformation of Sedimentary Sulfur*, (eds. M. A. VAIRAVAMURTHY and M. A. A. SCHOONEN), Vol. 612, pp. 110–137. American Chemical Society Symposium Series, ACS, Washington, DC.

LEWAN M. D. (1985) Evaluation of petroleum by hydrous pyrolysis. *Phil. Trans. R. Soc. Lond. A* **315**, 123–134.

LEWAN M. D. (1998) Sulfur-radical control on petroleum formation rates. *Nature* **391**, 164–166.

LUCKGE A., HORSFIELD B., LITTKE R. and SCHEEDER G. (2002) Organic matter preservation and sulfur uptake in sediments from the continental margin off Pakistan. *Org. Geochem.* **33**, 477–488.

NELSON B. C., EGLINTON T. I., SEEWALD J. F., VAIRAVAMURTHY M. A. and MIKNIS F. P. (1995) Transformations in organic sulfur speciation during maturation of Monterey shale: constraints from laboratory experiments. In *Geochemical Transformation of Sedimentary Sulfur* (eds. M. A. VAIRAVAMURTHY and M. A. A. SCHOONEN), Vol. 612, pp. 138–166. American Chemical Society Symposium Series, ACS, Washington, DC.

ORR W. L. (1974) Changes in sulfur content and isotopic ratios during petroleum maturation—study of Big Horn basis Paleozoic oils. *Am. Assoc. Petrol. Geol.* **50**, 2295–2318.

ORR W. L. (1986) Kerogen/asphaltene/sulfur relationships in sulfur-rich Monterey oils. *Advances in Organic Geochemistry, Org. Geochem.* **10**, 499–516.

PHILP R. P., BAKEL A., GALVEZ-SINBALDI A. and LIN L. H. (1988) A comparison of organosulfur compounds produced by pyrolysis of asphaltenes and those present in related crude oils and tar sands. *Org. Geochem.* **13**, 915–926.

SCHOUTEN S., SCHOELL M., SINNINGHE DEMSTÉ J. S., SUMMONS R. E. and DE LEEUW J. W. (2001) Molecular biogeochemistry of Monterey sediments, Napales Beach, California, II: stable carbon isotopic composition of free and sulfur bound carbon skeleton. In *The Monterey Formation, from Rock to Molecules* (eds. M. ISSAC and J. RULLKOETTER), pp. 175–188. Columbia University press, New York.

SINNINGHE DAMSTÉ J. S., EGLINTON T. I., DE LEEUW J. W. and SCHENCK P. A. (1989) Organic sulfur in macromolecular sedimentary organic-matter. 1. Structure and origin of sulfur-containing moieties in kerogen, asphaltenes and coal as revealed by flash pyrolysis. *Geochim. Cosmochim. Acta* **53**, 873–889.

SINNINGHE DAMSTÉ J. S., RIJPSTRA W. I. C. and DE LEEUW J. W. (1990) Characterization of organically bound sulfur in high-molecular-weight, sedimentary organic-matter using flash pyrolysis and Raney Ni desulfurization. In *Geochemistry of Sulfur in Fossil Fuels*, (eds. W. L. ORR and C. M. WHITE), Vol. 429, pp. 486–528. American Chemical Society Series, ACS, Washington, DC.

SINNINGHE DAMSTÉ J. S., DE LAS HERAS X. C. and DE LEEUW J. W. (1992) Molecular analysis of sulfur-rich brown coals by flash pyrolysis–gas chromatography mass–spectrometry: the type-II-S kerogen. *J. Chromatogr.* **607**, 361–376.

SPIRO B. (1991) Effects of minerals on Rock Eval pyrolysis of kerogen. *J. Therm. Anal.* **37**, 1513–1522.

SPIRO B., DINUR D. and AIZENSHTAT Z. (1983a) Evaluation of source, environments of deposition and diagenesis of some Israeli "oil shales": *n*-alkanes, fatty acids, tetrapyrroles and kerogen. *Chem. Geol.* **39**, 189–214.

SPIRO B., WELTE D. H., RULLKOTTER J. and SCHAEFFER R. G. (1983b) Asphalts, oils and bituminous rocks from the Dead Sea area: a geochemical correlation study. *Am. Assoc. Petrol. Geol.* **67**, 1163–1175.

STAHL W. J. (1977) Carbon and nitrogen isotopes in hydrocarbon research and exploration. *Chem. Geol.* **20**, 121–149.

STAHL W. J. (1978) Source rock–crude oil correlation by isotopic type-curves. *Geochem. Cosmochem. Acta* **42**, 1573–1577.

STOLER A. (1981) Solar Lake sulfur species. M.Sc. thesis, submitted to the Hebrew University of Jerusalem (English abstract—most of relevant information published in AIZENSHTAT *et al.*, 1983).

STOLER A. (1990) Studies on sulfur in Israeli oil shales and the thermal behavior of these; catalytical desulfurization. Ph.D. thesis, Hebrew University of Jerusalem, Israel.

STOLER A., SPIRO B., AMRANI A. and AIZENSHTAT Z. (2003) Evaluation of δ^{34}S changes during stepwise pyrolysis of bituminous rocks and type II-S kerogen. *Org. Geochem.* 103–104 (21st IMOG Meeting, extended abstracts).

TANNENBAUM E. (1983) Researches in the geochemistry of oils and asphalts in the Dead Sea area, Israel. Ph.D. thesis, The Hebrew University, Jerusalem, 117 pp. (in Hebrew with English abstract).

TANNENBAUM E. and AIZENSHTAT Z. (1984) Formation of immature asphalt from organic-rich carbonate rocks—2: correlation of maturation indicators. In *Advances in Organic Geochemistry 1983*, (eds. P. A. SCHENCK, J. W. de Leeuw and G. W. M. LIJMBACK), Vol. 6, pp. 503–511. Org. Geochem.

TANNENBAUM E. and AIZENSHTAT Z. (1985) Formation of immature asphalt from organic-rich carbonate rocks—I. Geochemical correlation. *Org. Geochem.* **8**, 181–192.

TANNENBAUM E., STARINSKY A. and AIZENSHTAT Z. (1987) Light oils transformation to heavy oils and asphalts. In *Exploration for Heavy Crude Oils and Natural Bitumen. Studies in Geology* (ed. R. F. MEYER), pp. 221–231. American Association of Petroleum Geologists Series 25, American Association of Petroleum Geologists, Tulsa.

WILLIAMSON M. A. and RIMSTIDT J. D. (1990) Thermodynamic and kinetic controls on e aqueous oxidation of sulfide minerals. In *Goldschmidt International Conference for Advancement of Geochemistry*, pp. 91–103.

VORONKOV M. G. and DERYAGINA E. N. (1990) Thermal reactions and high temperature syntheses of organosulfur compounds. In *Chemistry of Organosulfur Compounds—General Problems* (ed. L. I. BELENKII), pp. 48–60. Ellis Horwood, Chichester.

ZABACK D. A. and PRATT L. M. (1992) Isotopic composition and speciation of sulfur in Miocene Monterey Formation: reevaluation of sulfur reactions during early diagenesis in marine environments. *Geochim. Cosmochim. Acta* **56**, 763–774.

ZABACK D. A., PRATT L. M. and HAYES J. M. (1993) Transport and reduction of sulfate and immobilization of sulfide in marine black shales. *Geology* **21**, 141–144.

Geochemical Investigations in Earth and Space Science: A Tribute to Isaac R. Kaplan
© The Geochemical Society, Publication No. 9, 2004
Editors: R.J. Hill, J. Leventhal, Z. Aizenshtat, M.J. Baedecker, G. Claypool,
R. Eganhouse, M. Goldhaber and K. Peters

The distribution and isotopic composition of sulfur in solid bitumens from Papua New Guinea

M. Ahmed[1], S. A. Barclay[1], S. C. George[1], B. McDonald[2] and J. W. Smith[1,3]

[1]CSIRO Petroleum, PO Box 136, North Ryde, NSW 1670, Australia
[2]CSIRO Exploration and Mining, PO Box 136, North Ryde, NSW 1670, Australia
[3]CSIRO Energy Technology, PO Box 136, North Ryde, NSW 1670, Australia

Abstract—Previous geochemical and petrographic studies have only partly revealed the history and origins of solid bitumens from the East Papuan Basin, Papua New Guinea. In particular, the alteration phenomena, the distribution of hydrocarbon components and the increasing yields of pyritic sulfur indicate major effects of biodegradation and reduction of sulfate to sulfide. Sulfur isotope analysis of the three main sulfur fractions (elemental, sulfate and pyritic) showed these to cluster closely around a $\delta^{34}S$ value near $-25‰$. An unlimited source of non-marine sulfate of constant isotopic composition is visualized. However, one clear trend in the data is the increase in the ^{34}S content of pyrite as abundance increases. Such an isotopic relationship might well occur where, on reduction of a pool of sulfate, an early loss or reduction of ^{32}S species would result in an increase in the ^{34}S contents of the residual sulfate and product pyrite. Where the oxidation of pyrite to sulfate and sulfur occurs, the relatively greater ^{34}S content of the elemental sulfur is consistent with previous data.

INTRODUCTION

Preliminary geochemical and petrographic studies (Barclay et al., 2003) suggest that solid bitumens in the Late Cretaceous Pale and Subu sandstones from the Subu-1 and Subu-2 wells, Aure Scarp, Papua New Guinea (Fig. 1), were derived by biodegradation of migrating crude oil. The crude oil possibly migrated up faults and was altered to varying degrees to form the reservoir bitumens when it reached a fluctuating water table. Two separate oil charges from marine source rocks have been identified based on the molecular geochemistry of the solid bitumens. One (family A) is from a strongly terrestrially influenced marine source rock that may well be Jurassic. A second oil charge (family B) is from a calcareous-influenced marine source rock with an abundance of prokaryotic organic matter input and a less oxic depositional environment than for the family A solid bitumens. Since changes observed in the hydrocarbon distribution with increasing degree of biodegradation correspond with changes in the distribution of sulfur forms, it was expected that the sulfur isotopic data might reveal general reaction pathways.

SAMPLES AND GEOCHEMICAL CHARACTERISTICS

Three sandstone samples containing solid bitumens were selected for analysis (CN409, CN457 and CN250). Gross compositional data together with some selected biomarker parameters for these samples are provided in Table 1. The first two samples are family B solid bitumens and are characterized by significant amounts of 28,30-bisnorhopane (BNH), 29,30-BNH and 2α-methylhopanes, consistent with sourcing from a more reducing marine rock with a calcareous influence (Mello et al., 1988; Subroto et al., 1991; Table 1). These two samples also have a higher maturity, as indicated by methylphenanthrene indices (MPI, Radke and Welte, 1983) consistent with the peak of the oil window. On the other hand the third sample, CN250, is a family A member. This sample does not contain BNHs and has a lower content of 2α-methylhopanes, but it contains high amounts of diterpanes that suggest coniferous organic matter input, in common with other family A solid bitumens. Sample CN250 has a lower MPI (Table 1), indicating a low maturity overprint from indigenous organic matter in the Pale Sandstone.

M. Ahmed *et al.*

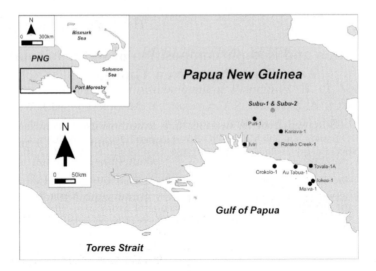

FIG. 1. Location map of the Subu-1 and Subu-2 well sites in Papua New Guinea.

These three samples exhibit different degrees of biodegradation, as shown by the total ion chromatograms (TIC) of their aliphatic and aromatic hydrocarbon fractions (Figs. 2–4). Aliphatic hydrocarbons of CN409 are characterized by a unimodal distribution of C_{11}–C_{34} *n*-alkanes (Fig. 2a). Alkylphenanthrenes are the most abundant class of compounds in the aromatic hydrocarbons of this sample, with substantial amounts of alkylnaphthalenes, alkylfluorenes and polycyclic aromatic hydrocarbons (Fig. 2b). Such a composition may be considered as characteristic of unaltered hydrocarbons. Aliphatic hydrocarbons of CN457 are characterized by a bimodal distribution of C_8–C_{32} *n*-alkanes, which exhibit signs of biodegradation. There has been a depletion of *n*-alkanes over the entire range, leaving two prominent humps of aliphatic unresolved complex mixtures (UCM) (Fig. 3a). Similarly, the aromatic hydrocarbons are also characterized by a UCM hump with few substituted and unsubstituted naphthalenes and phenanthrenes, and dominant polycyclic aromatic hydrocarbons (Fig. 3b). These signatures indicate that CN457 has been partly altered by biodegradation (level 4 of VOLKMAN *et al.* (1984) and TROLIO *et al.* (1999)). The TIC of the aliphatic hydrocarbons of CN250 shows the near complete removal of *n*-alkanes, two humps of UCM and large amounts of relatively resistant steranes (Fig. 4a). Such a molecular composition indicates that the aliphatic hydrocarbons of

Table 1. Gross compositional and some selected geochemical parameters for three sandstone samples containing solid bitumens from the Subu boreholes, East Papuan Basin

	Unaltered (Fig. 2) Subu-1, 142.95 m Pale Sandstone CN409	Biodegraded (level 4) (Fig. 3) Subu-2, 257.27 m Subu Sandstone CN457	Biodegraded (level 5) (Fig. 4) Subu-1, 88.5 m Pale Sandstone CN250
EOM (mg)/kg of rock	747	141	1299
Aliphatic hydrocarbons (EOM%)	3.3	18.1	5.3
Aromatic hydrocarbons (EOM%)	36.6	27.1	36.3
Polars + asphaltenes (EOM%)	60.1	54.8	58.3
19NIP/$C_{30}\alpha\beta$	0.25	0.06	0.34
$C_{31}2\alpha Me/(C_{31}2\alpha Me + C_{30}\alpha\beta$ hopane)	0.58	0.75	0.28
28,30-BNH/$C_{30}\alpha\beta$ hopane	0.04	0.03	<0.01
29,30-BNH/$C_{30}\alpha\beta$ hopane	0.13	0.11	<0.01
MPI	1.17	1.13	0.86

EOM, extractable organic matter; 19NIP, 4β(H)-19-isopimarane; BNH, bisnorhopane; MPI, methylphenanthrene index ($1.5 \times [3\text{-MP} + 2\text{-MP}]/[P + 9\text{-MP} + 1\text{-MP}]$).

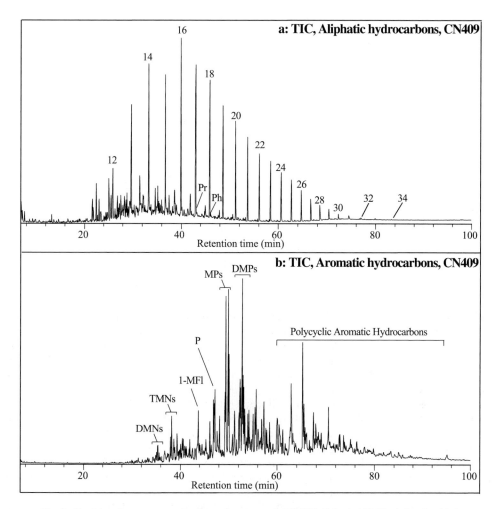

FIG. 2. Total ion chromatograms for the sandstone sample (CN409, Subu-1, 142.95 m) showing (a) the distribution of aliphatic hydrocarbons and (b) the distribution of aromatic hydrocarbons. Numbers refer to n-alkane chain length. Pr, pristane; Ph, phytane; DMNs, dimethylnaphthalenes; TMNs, trimethylnaphthalenes; MFl, methylfluorene; P, phenanthrene; MPs, methylphenanthrenes; and DMPs, dimethylphenanthrenes.

CN250 have been altered by biodegradation (level 5). Relatively higher abundances of polycyclic aromatic hydrocarbons compared to alkyl-substituted hydrocarbons in this sample provide consistent evidence for its alteration (Fig. 4b).

EXPERIMENTAL

The analytical scheme for determination of sulfur distribution and isotopic compositions, adapted from SMITH and BATTS (1974), is shown in Fig. 5.

Sample splits (50–100 g) of the three crushed rock samples (Table 2) were Soxhlet extracted under reflux with dichloromethane and methanol (93:7) for 72 h. The total extract, which consisted of extractable organic matter (EOM) and elemental sulfur, was separated by reflux with bright Cu turnings. More Cu turnings, activated in dilute HCl and pre-cleaned with methanol, were added to the sample flask containing the total extract and refluxed, until the brightness persisted on further additions of Cu turnings. The CuS resulting from direct elemental

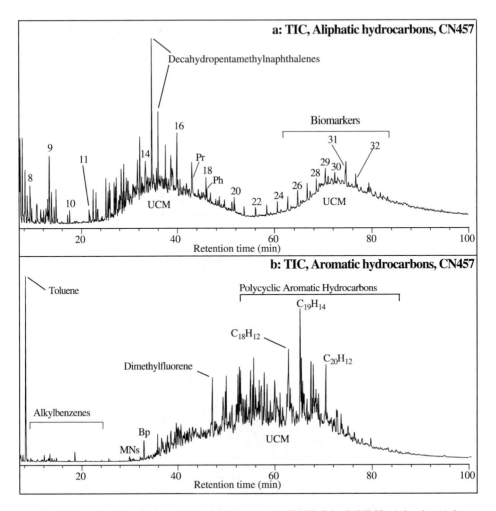

FIG. 3. Total ion chromatograms for the sandstone sample (CN457, Subu-2, 257.27 m) showing (a) the distribution of aliphatic hydrocarbons and (b) the distribution of aromatic hydrocarbons. Abbreviations are defined in Fig. 2. UCM, undifferentiated complex mixture; MNs, methylnaphthalenes; Bp, biphenyl.

interaction, together with excess Cu, was collected by filtration and combined with the insoluble residue from the initial solvent extraction (Fig. 5).

The mix of insoluble carbonaceous and mineral matter and CuS was transferred to a flask purged with N_2 and reacted with cold 10% HCl in a stream of nitrogen. The H_2S generated was carried in the N_2 stream through a water scrubber to remove traces of HCl and then into a 0.1 M $AgNO_3$ solution, where it was collected by precipitation as Ag_2S. Extraction of dissolved sulfide continued until the $AgNO_3$ solution cleared and precipitation was complete. The precipitate was weighed and saved for isotopic analysis of acid volatile sulfide (fraction #5; Fig. 5).

Next, the HCl mixture remaining from the initial separation was heated to boiling and the H_2S generated was collected as Ag_2S as before. The precipitate was weighed and saved for isotopic analysis of elemental sulfur (fraction #1; Fig. 5).

Finally, the hot HCl mixture was filtered. Original soluble sulfate contained in the filtrate was recovered by the addition of $BaCl_2$ and precipitated as $BaSO_4$. The precipitate was dried and weighed, and saved for isotopic analysis of soluble sulfate (fraction #3; Fig. 5).

Material insoluble in the hot 10% HCl was regarded as being comprised of pyrite and carbonaceous remnants (including insoluble bitumen). On reaction with 10% HNO_3 the pyrite was directly oxidized to sulfate in solution and recovered as $BaSO_4$ as above. The precipitate

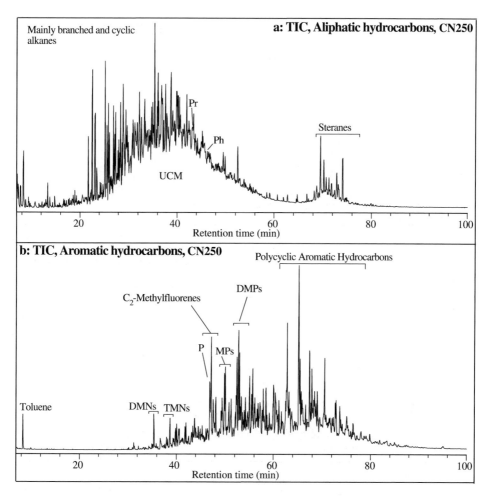

FIG. 4. Total ion chromatograms for the sandstone sample (CN250, Subu-1, 88.5–88.6 m) showing (a) the distribution of aliphatic hydrocarbons and (b) the distribution of aromatic hydrocarbons. Abbreviations are defined in Figs. 2 and 3.

was dried and weighed, and saved for isotopic analysis of pyritic sulfate (fraction #2; Fig. 5). No attempt was made to differentiate isotopically between sulfur derived from pyrite and that from carbonaceous material. These were co-precipitated. The analysis of the iron and sulfur contents in aliquots of the total HNO_3 solubles and differences in the Fe/S ratios of these from the theoretical ratios of Fe/S in pyrite allow an estimation to be made of the possible contribution of sulfur from carbonaceous material. Calculations suggest that in the highly degraded bitumens these contribution may be 20% and in the least degraded bitumen it is < 10%.

Insoluble carbonaceous material remaining at this stage was converted to sulfate by fusion with Eschka mixture ($MgO + Na_2CO_3$, 2:1) in a crucible at 800°C for 2 h. The filtrate was acidified, heated and sulfate was precipitated from the solution by the addition of $BaCl_2$ to form $BaSO_4$. The precipitate was dried and weighed, and saved for isotopic analysis of sulfur in insoluble bitumen + carbonaceous residue (fraction #4; Fig. 5).

The EOM remaining in the elemental-sulfur-free solvent extract solution was recovered by careful evaporation of the solvent in a nitrogen stream. The total EOM was combusted with oxygen in a Parr Bomb and the sulfur retained as $BaSO_4$. The yields of EOM by solvent extraction were too small to support combustion in the Parr Bomb, so 1 ml of hexane was added to promote combustion. Even with such additions, gravimetric yields of $BaSO_4$ were less than 1 mg and no isotopic measurements were achieved on this organic sulfur (EOM) fraction (fraction #6; Fig. 5).

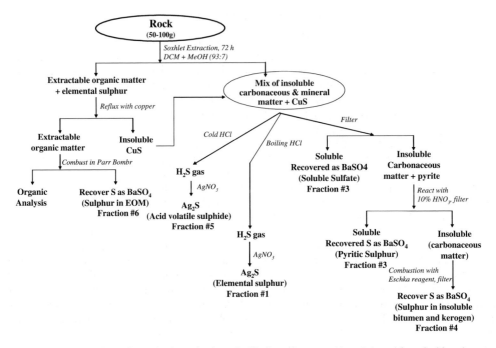

FIG. 5. Analytical scheme for determination of sulfur isotopic compositions (adapted from Smith and Batts, 1974).

Solid sulfur compounds (Ag_2S and $BaSO_4$) were converted to SO_2 for isotopic analysis by modified versions of the methods of KAPLAN *et al.* (1970) and BAILEY and SMITH (1972), respectively. Ag_2S was oxidized to SO_2 by reaction with Cu_2O at 1050°C. $BaSO_4$ was directly decomposed to SO_2 by heating at > 1700°C in a quartz tube (BAILEY and SMITH, 1972). Isotopic determinations were carried out on a Finnigan MAT 252 IRMS, calibrated by a series of IAEA standards. Results are presented as $\delta^{34}S$ relative to Canyon Diablo Troilite, with a precision of $\pm 0.2‰$.

The yields and isotopic composition of all fractions are included in Table 2. The mass of sulfur in each fraction expressed as their percentage of the original rock mass is directly calculated from the mass of $BaSO_4$ or Ag_2S.

Table 2. Sulfur contents and isotopic compositions for three sandstone samples containing solid bitumens from the Subu boreholes, East Papuan Basin

	Unaltered (Fig. 2) Subu-1, 142.95 m Pale Sandstone CN409		Biodegraded (level 4) (Fig. 3) Subu-2, 257.27 m Subu Sandstone CN457		Biodegraded (level 5) (Fig. 4) Subu-1, 88.5 m Pale Sandstone CN250	
	Sulfur (wt/wt%)	$\delta^{34}S^a$ (‰)	Sulfur (wt/wt%)	$\delta^{34}S$ (‰)	Sulfur (wt/wt%)	$\delta^{34}S$ (‰)
Elemental	0.04	−24.7	0.06	−20.1	0.31	−20.9
Pyritic	0.28	−30.6	1.45	−26.5	3.8	−19.2
SO_4	0.25	−28.6	0.42	−23.8	0.37	−26.2
Eschka[b]	0.13	−21.5	0.04	−17.2	0.38	−14.7
Acid volatile	n.d.	n.d.	n.d.	n.d.	n.d.	−42.5
Soluble bitumen	<0.1	n.d.	<0.1	n.d.	<0.1	n.d.

[a] $\delta^{34}S$ values reported relative to Canyon Diablo Troilite (CDT).
[b] Represent insoluble bitumen and carbonaceous residues.
n.d. = not determined.

RESULTS AND SUMMARY

The pyritic sulfur, elemental sulfur and soluble sulfate have a relatively narrow spread of values (-19.2 to $-30.6‰$). The residual sulfur in the organic matter (both in insoluble bitumen and any carbonaceous residue present) is isotopically heavier (Table 2), whilst the one measurement on acid volatile sulfide is isotopically very light. No isotopic measurements on sulfur in the EOM could be made, due to the small amount of sulfur (<0.1 wt%) recovered (Table 2). There is a trend for pyritic sulfur and elemental sulfur to be more abundant and isotopically heavier with greater biodegradation (Table 2).

The average δ^{34}S values approximating $-25‰$ for all the sulfur forms (weighted averages are -27.9, -25.5 and $-19.5‰$ for CN409, CN457 and CN250, respectively) suggest a single non-marine sulfur source in an open system. However, the isotopic data give very little indication of which sulfur form, pyritic or sulfate sulfur, is the sulfur source, and which represents a final product. In this respect the increase in the ^{34}S content (represented by the increasingly positive δ^{34}S values) and yield of pyrite with increasing degrees of biodegradation is consistent with an early reduction of the more active ^{32}S-enriched sulfate to ^{32}S-enriched sulfide (KAPLAN and RITTENBERG, 1964). The continuation of this reaction leads to pyrite with an increased ^{34}S content. The high concentration of elemental sulfur associated with the pyrite in the most biodegraded core section is best explained as a co-product of the oxidation of pyrite to ferrous sulfate (CHOU, 1990; STOCK and WOLNY, 1990). The data reported here may represent progression of the reaction to an unusually marked degree. This same isotopic enrichment of elemental sulfur in ^{34}S relative to sulfate sulfur (Table 2), as observed earlier and demonstrated by SMITH and BATTS (1974) in their study of the forms of sulfur in oxidized high sulfur Australian coals, is found again.

The origin of the light acid volatile sulfide was not pursued as it was present in only one sample, in low abundance, and was isotopically far removed from the bulk of the data.

The source of the sulfur is uncertain. It may have originated from deeply buried evaporites that have not yet been penetrated by any well intersection, or it may have originated from the oxidation of pyritic sulfur external to the Subu sequence. Sulfate-rich porewaters migrated into the Subu sequence where crude oil already present in the Pale and Subu sandstones provided a nutrient source for sulfate-reducing bacteria to form solid bitumen and pyrite. Evidence from textural and petrographic observations also suggests that the pyrite precipitated as a by-product of biodegradation (BARCLAY et al., 2003). The more abundant and isotopically heavier pyritic and elemental sulfur in the more biodegraded samples is consistent with pyrite formation during biodegradation. Once pyrite had been formed, it underwent some diagenetic alteration (oxidation) to form $FeSO_4$ and elemental sulfur.

Acknowledgements—The expertise of Ken Riley of CSIRO Energy Technology in completing all the Parr Bomb experiments is gratefully acknowledged. We thank InterOil for providing the samples and permission to publish these results. We are grateful for the comments of journal reviewers, Baruch Spiro and Zeev Aizenshtat, which helped to improve the manuscript.

REFERENCES

BAILEY S. A. and SMITH J. W. (1972) Improved method for the preparation of sulfur dioxide from barium sulfate for isotope ratio studies. *Anal. Chem.* **44**, 1542–1543.

BARCLAY S. A., LIU K. and HOLLAND D. (2003) Reservoir quality, diagenesis and sedimentology of the Pale and Subu sandstones: revisiting the Eastern Papuan Basin, Papua New Guinea. *Aust. Petr. Prod. Expl. Assoc. J.* **43**(1), 515–535.

CHOU C.-L. (1990) The geochemistry of sulfur in coal. In *Geochemistry of Sulfur in Fossil Fuel* (eds. W. L. ORR and C. M. WHITE), Vol. 429, pp. 30–52. ACS Symposium.

KAPLAN I. R. and RITTENBERG S. C. (1964) Microbiological fractionation of sulfur isotopes. *J. Gen. Microbiol.* **34**, 195–212.

KAPLAN I. R., SMITH J. W. and RUTH E. (1970) Carbon and sulfur concentration and isotopic composition in Apollo II Lunar samples. *Proc. Apollo II Lunar Sci. Conf.* **2**, 1317–1329.

MELLO M. R., GAGLIANONE P. C., BRASSELL S. C. and MAXWELL J. R. (1988) Geochemical and biological marker assessment of depositional environments using Brazilian off-shore oils. *Mar. Petr. Geol.* **5**, 205–223.

RADKE M. and WELTE D. H. (1983) The methylphenanthrene index (MPI): a maturity parameter based on aromatic hydrocarbons. In *Advances in Organic Geochemistry 1981* (ed. M. BJORØY et al.), pp. 504–512. Wiley, Chichester.

SMITH J. W. and BATTS B. D. (1974) The distribution and isotopic composition of sulfur in coal. *Geochim. Cosmochim. Acta* **38**, 121–133.

STOCK L. M. and WOLNY R. (1990) Elementary sulfur in bituminous coal. In *Geochemistry of Sulfur in Fossil Fuel*, (eds. W. L. ORR and C. M. WHITE), Vol. 429, pp. 241–248, ACS Symposium.

SUBROTO E. A., ALEXANDER R. and KAGI R. I. (1991) 30-Norhopanes: their occurrence in sediments and crude oils. *Chem. Geol.* **93**, 179–192.

TROLIO R., GRICE K., FISHER S. J., ALEXANDER R. and KAGI R. I. (1999) Alkylbiphenyls and alkyldiphenylmethanes as indicators of petroleum biodegradation. *Org. Geochem.* **30**, 1241–1253.

VOLKMAN J. K., ALEXANDER R., KAGI R. I., ROWLAND S. J. and SHEPPARD P. N. (1984) Biodegradation of aromatic hydrocarbons in crude oils from the Barrow sub-basin of Western Australia. *Org. Geochem.* **6**, 619–632.

Geochemical Investigations in Earth and Space Science: A Tribute to Isaac R. Kaplan
© The Geochemical Society, Publication No. 9, 2004
Editors: R.J. Hill, J. Leventhal, Z. Aizenshtat, M.J. Baedecker, G. Claypool,
R. Eganhouse, M. Goldhaber and K. Peters

Ventilation of marine sediments indicated by depth profiles of pore water sulfate and δ^{34}S

GEORGE E. CLAYPOOL

8910 West Second Avenue, Lakewood, CO 80226, USA

Abstract—Anoxic marine sediments are usually characterized by systematic decrease in concentration of dissolved sulfate in the pore waters of marine sediments. This depletion in dissolved sulfate can take place over depth scales ranging from a few centimeters up to a few hundred meters. Accompanying the decrease in sulfate concentration is an increase in δ^{34}S, due to preferential utilization of the ^{32}S-isotope during bacterial sulfate reduction. The degree of ^{34}S-enrichment for a given degree of dissolved sulfate depletion is also highly variable, and reflects the extent to which the sediment column is relatively open or closed with respect to diffusive replenishment of sulfate during bacterial sulfate reduction.

Gradients in stable sulfur isotope ratios (^{34}S/^{32}S) and concentration of dissolved sulfate in pore waters of marine sediments with negligible advective sulfate supply have been compiled from 42 sites of the Deep Sea Drilling Project and the Ocean Drilling Program, Legs 4 through 188.The linear sedimentation rates for the sites range from about 5 to 1500 m per million years. The apparent ^{34}S-enrichment factors (from Rayleigh equation) vary from 4 to 72‰ ($\alpha = 1.004 - 1.072$), and are directly proportional to a system "openness" parameter (*i.e.* the diffusive sulfate flux divided by the combined diffusive plus burial fluxes), which ranges from 0.06 to 0.87.

Under conditions represented by the DSDP/ODP cores in this study, microbial sulfate reduction occurs at very slow rates with a relatively large ^{34}S-isotopic enrichment (75‰). This large fractionation is diluted by variable degrees of diffusive replenishment with relatively unfractionated seawater sulfate.

INTRODUCTION

MICROBIALLY MEDIATED OXIDATION of organic matter buried in marine sediments is initially coupled with reduction of dissolved O_2, NO_3^-, and Fe^{3+} (CLAYPOOL and KAPLAN, 1974; FROELICH *et al.*, 1979). These compounds are normally in low concentration in seawater and are rapidly depleted in pore waters, leaving dissolved SO_4^{2-} as the most important electron acceptor for continued remineralization of organic matter. Sediments undergoing organic matter oxidation coupled with sulfate reduction generally exhibit characteristic profiles of decreasing dissolved sulfate with increasing depth of burial (BERNER, 1964).

Depletion of dissolved SO_4^{2-} in marine sediments is displayed over widely varying depth ranges, depending mainly on the interplay of rates of microbial metabolism, sediment accumulation and diffusive sulfate replenishment. Sulfate concentration gradients can be used with sedimentation rates, diffusion coefficients and model assumptions to estimate rates or rate constants for microbial sulfate reduction in sediments (TOTH and LERMAN, 1977; BERNER, 1978; CANFIELD, 1991). There is a general proportionality between rates of sediment accumulation and sulfate reduction, but this relation is only weakly predictive for specific sites because of the influence of other factors on sulfate reduction rates (*e.g.* organic matter composition, microbial population density, etc.).

In addition to the gradient in sulfate concentration observed in marine sediments, there is a parallel gradient in the relative abundance of the two common stable isotopes of sulfur (^{32}S and ^{34}S) in the residual pore water sulfate, such that there is an increasing ^{34}S-enrichment (expressed as δ^{34}S) with decreasing SO_4^{2-} concentration (KAPLAN *et al.*, 1963; HARTMANN and NIELSEN, 1969). However, the degree of δ^{34}S increase for a given concentration decrease is variable, suggesting that isotopic fractionation during microbial sulfate reduction is variable. Variable sulfur isotopic fractionation in marine sediments has been attributed to differences in reaction rates (GOLDHABER and KAPLAN, 1975) and to repeated cycles of sulfide oxidation and reduction (JØRGENSEN, 1990; CANFIELD and THAMDRUP, 1994). This report reviews the pore water sulfur isotope geochemistry in a variety of marine depositional environments cored by the Deep Sea Drilling Project and the Ocean Drilling Program (DSDP/ODP). The apparent variability in ^{34}S-fractionation observed in these environments is related to the "openness" of the sediment column to diffusive replenishment with isotopically light sulfur (JØRGENSEN, 1979) while the underlying fractionation factor appears to be large ($k_{32}/k_{34} = 1.075$) and relatively constant.

The case for a large fractionation factor in bacterial sulfate reduction was previously shown based on detailed diagenetic modeling (GOLDHABER and KAPLAN, 1980; RUDNICKI et al., 2001; Wortmann et al., 2001). This paper shows how the apparent (closed-system) fractionation factor is quantitatively related to system openness or the relative importance of diffusion.

DSDP/ODP PORE WATER SAMPLES AND ANALYTICAL METHODS

Various marine sedimentary environments have been sampled by coring during the 35 year history of DSDP/ODP, from ocean ridges to continental margins and shelf sediments to deep ocean trenches, with sedimentation rates ranging from 5 to 1500 m ma^{-1}. Interstitial water samples obtained by squeezing of sediment cores (MANHEIM, 1966) have been routinely collected and analyzed, initially in shorebased (MANHEIM et al., 1969; PRESLEY, 1971; PRESLEY and CLAYPOOL, 1971) and later in shipboard laboratories (GIESKES et al., 1991). Concentration of major ions, including sulfate, are routinely reported in the Initial Reports and Proceedings volumes of DSDP/ODP. Sulfur isotope analysis is performed in shorebased labs, usually on SO$_2$ gas prepared from BaSO$_4$ precipitates obtained from pore water samples. Various techniques and instruments for isotopic analysis have been employed over the years, and are generally described in cited references. The isotopic composition of sulfur in the residual pore water sulfate is reported as the permil deviation of the ^{34}S/^{32}S isotope ratio (R) relative to that of the meteoritic standard, Canyon Diablo Troilite (CDT):

$$\delta^{34}S = ((R_{sample} - R_{CDT})/R_{CDT})1000‰ \tag{1}$$

RESULTS AND DISCUSSION

A summary of results from 42 DSDP/ODP sites compiled for this study is given in Table 1. The basic results summarized in Table 1 are interpreted sulfate concentration gradients and sulfur isotope ratio gradients, expressed in a normalized logarithmic form. The concentration gradient is combined with a diffusion coefficient and the reported sedimentation rate to provide an estimate of the diffusive proportion of sulfate flux into the sediment, or an index of system openness. The sulfate concentration gradient is also combined with the sulfur isotope ratio gradient in the Rayleigh distillation equation to provide an estimate of the apparent (closed system) fractionation factor. The sources for the δ^{34}S measurements used in the compilation are indicated in the footnotes to Table 1.

Dissolved sulfate normally is added to the sediment column by two mechanisms: (1) burial of sulfate-containing seawater with the sediment and (2) diffusion of dissolved sulfate from overlying seawater or sediments into sulfate-depleted pore waters (here, ignoring addition of sulfate by advection or bioturbation). The flux of sulfate due to burial of seawater is (BERNER, 1980)

$$J_{burial} = \omega\phi(C_0) \tag{2}$$

where ω is the sedimentation rate, ϕ the porosity and C_0 the concentration of dissolved sulfate in seawater. The flux of sulfate into the sediments by diffusion is (BERNER, 1980)

$$J_{diffusion} = -D_s\phi(\partial C/\partial x)_{x=0} \tag{3}$$

where D_s is the whole sediment diffusion coefficient and $(\partial C/\partial x)_{x=0}$ is the sulfate concentration gradient at the sediment–water interface. The proportion of sulfate in the sediment column due

Table 1. Summary of DSDP/ODP holes for which δ^{34}S data have been measured

Location	Site	Sed. rate (m/my)	$\dfrac{-\mathrm{d}\ln(C/C_0)}{\mathrm{d}x}$	$\dfrac{-\mathrm{d}\ln(R/R_0)}{\mathrm{d}x}$	α_a	$J_d/(J_d + J_b)$	Reference
Cent. Atlantic	27	10	2.70×10^{-3}	7.02×10^{-5}	1.027	0.4859	CLAYPOOL (1974)
N. Pacific	34	18	3.30×10^{-3}	1.29×10^{-4}	1.041	0.3909	CLAYPOOL (1974)
E. Caroline basin	63	7	1.18×10^{-3}	6.13×10^{-5}	1.055	0.3711	CLAYPOOL (unpublished)
Ontong Java	64	23	7.92×10^{-3}	1.47×10^{-4}	1.019	0.5466	CLAYPOOL (unpublished)
Cariaco Basin	147	505	3.80×10^{-1}	6.46×10^{-3}	1.017	0.7248	CLAYPOOL (1974)
Caribbean	148	50	1.35×10^{-2}	4.32×10^{-4}	1.033	0.4859	CLAYPOOL (1974)
Astoria Fan	174	230	1.20×10^{-2}	6.60×10^{-4}	1.058	0.1544	CLAYPOOL (1974)
Oregon Slope	175	198	1.50×10^{-1}	3.30×10^{-3}	1.022	0.7261	CLAYPOOL (1974)
Oregon Shelf	176	152	2.10×10^{-1}	1.47×10^{-3}	1.007	0.8286	CLAYPOOL (1974)
N. Pacific	178	151	1.00×10^{-1}	5.30×10^{-4}	1.056	0.1882	CLAYPOOL (1974)
Aleutian Trench	180	1515	2.70×10^{-2}	1.81×10^{-3}	1.072	0.0587	CLAYPOOL (1974)
Aleutian Trench	181	300	4.40×10^{-2}	1.36×10^{-3}	1.032	0.3392	CLAYPOOL (1974)
Aleutian Trench	182	101	1.20×10^{-1}	1.20×10^{-3}	1.010	0.8061	CLAYPOOL (1974)
N. Pacific	467	75	5.40×10^{-2}	9.18×10^{-4}	1.017	0.7159	CLAYPOOL (unpublished)
N. Pacific	471	20	8.00×10^{-3}	2.24×10^{-4}	1.029	0.5833	CLAYPOOL (unpublished)
Blake Ridge	533	85	1.60×10^{-1}	1.81×10^{-3}	1.011	0.8682	CLAYPOOL and THRELKELD (1983)
Peru Slope	682	26	3.54×10^{-2}	3.54×10^{-4}	1.010	0.8263	CLAYPOOL (unpublished)
Peru Trench	685	130	9.43×10^{-2}	2.90×10^{-3}	1.032	0.7174	CLAYPOOL (unpublished)
Peru Slope	688	100	3.05×10^{-1}	3.40×10^{-3}	1.011	0.9143	CLAYPOOL (unpublished)
Japan Sea	794	35	1.74×10^{-2}	5.77×10^{-4}	1.034	0.6350	BRUMSACK et al. (1992)
Japan Sea	795	65	3.20×10^{-2}	8.59×10^{-4}	1.028	0.6328	BRUMSACK et al. (1992)
Japan Sea	797	50	1.43×10^{-2}	5.60×10^{-4}	1.041	0.4995	BRUMSACK et al. (1992)
E. Mediterranean	963	111	4.54×10^{-2}	2.36×10^{-3}	1.055	0.5887	BÖTTCHER et al. (1998)
E. Mediterranean	964	34	4.44×10^{-3}	2.40×10^{-4}	1.057	0.3136	BÖTTCHER et al. (1998)
E. Mediterranean	965	6	1.61×10^{-3}	9.64×10^{-5}	1.064	0.4842	BÖTTCHER et al. (1998)
W. Mediterranean	976	154	9.35×10^{-2}	2.31×10^{-3}	1.025	0.6800	BÖTTCHER et al. (1999)
W. Mediterranean	977	250	5.47×10^{-2}	2.03×10^{-3}	1.039	0.4337	BÖTTCHER et al. (1999)
W. Mediterranean	979	260	3.44×10^{-2}	2.29×10^{-3}	1.071	0.3165	BÖTTCHER et al. (1999)
Blake Ridge	994	68	7.52×10^{-2}	1.65×10^{-3}	1.022	0.7947	BOROWSKI et al. (2000)
J. de Fuca Ridge	1023	488	6.18×10^{-2}	2.75×10^{-3}	1.047	0.3071	MOTTL et al. (2000)
J. de Fuca Ridge	1025	340	7.97×10^{-2}	3.77×10^{-3}	1.050	0.4507	MOTTL et al. (2000)
J. de Fuca Ridge	1026	305	6.12×10^{-2}	2.78×10^{-3}	1.048	0.4126	MOTTL et al. (2000)
J. de Fuca Ridge	1028	344	6.83×10^{-2}	2.90×10^{-3}	1.044	0.4100	MOTTL et al. (2000)
Blake Ridge	1054	45	1.87×10^{-2}	4.68×10^{-4}	1.026	0.5927	BOROWSKI (personal communication)
Blake Ridge	1055	160	1.70×10^{-2}	1.00×10^{-3}	1.062	0.2715	BOROWSKI (personal communication)
Blake Ridge	1059	181	9.59×10^{-2}	1.13×10^{-3}	1.012	0.6497	BOROWSKI (personal communication)
Blake Ridge	1060	238	9.19×10^{-2}	1.06×10^{-3}	1.012	0.5748	BOROWSKI (personal communication)
Blake Ridge	1061	204	8.16×10^{-2}	1.13×10^{-3}	1.014	0.5832	BOROWSKI (personal communication)
Bahama Ridge	1062	126	4.44×10^{-2}	1.61×10^{-3}	1.038	0.5522	BOROWSKI (personal communication)
Bermuda Rise	1063	206	5.17×10^{-2}	1.95×10^{-3}	1.039	0.4676	BOROWSKI (personal communication)
N. Jersey Slope	1073	405	1.96×10^{-1}	2.78×10^{-3}	1.014	0.6288	CLAYPOOL (unpublished)
E. Antarctic Rise	1165	5	8.10×10^{-3}	2.90×10^{-5}	1.004	0.8501	CLAYPOOL et al. (2003)

Sulfate concentration and δ^{34}S values are generalized to logarithmic normalized gradients.

to diffusion is:

$$J_d/(J_d + J_b) = -D_s\phi(\partial C/\partial x)_{x=0}/(-D_s\phi(\partial C/\partial x)_{x=0} + \omega\phi(C_0)) \qquad (4)$$

Canceling the porosity terms, dividing numerator and denominator of the right-hand side of Eq. (4) by C_0 and substituting

$$(\partial C/\partial x)_{x=0}/C_0 = \mathrm{d}(\ln C/C_0)/\mathrm{d}x \qquad (5)$$

gives:

$$J_d/(J_d + J_b) = [-D_s \; d(\ln C/C_0)/dx]/[-D_s \; d(\ln C/C_0)/dx + \omega] \qquad (6)$$

The proportion of dissolved sulfate supplied to the sediment column by diffusion is equal to the diffusion coefficient times the logarithmic normalized concentration gradient divided by the sum of the diffusion coefficient times the logarithmic normalized concentration gradient plus the sedimentation rate. $J_d/(J_d + J_b)$ is the degree to which the sediment column approaches an open system. The sediment column is a completely closed system when $J_d/(J_d + J_b) = 0$, and a completely open system when $J_d/(J_d + J_b) = 1$.

The $\delta^{34}S$ versus depth profile is expressed as the normalized logarithmic gradient $d(\ln R/R_0)/dx$, where $R/R_0 = (\delta + 1000)/(\delta_0 + 1000)$, and δ_0 is the $\delta^{34}S$ of seawater sulfate. The ratio of the sulfate $\delta^{34}S$ gradient to the concentration gradient is equivalent to a term in the Rayleigh equation

$$(\ln R/R_0)/(\ln C/C_0) = (k_{34}/k_{32}) - 1 \qquad (7)$$

where k_{34}/k_{32} is the fractionation factor or ratio of the rate constants for reduction of the $^{34}SO_4$ and $^{32}SO_4$ species. Taking into account the previous notation and the negative slope of the concentration gradient, and defining $\alpha = k_{32}/k_{34}$, the above equation can be rewritten:

$$(d \ln(R/R_0))/(-d \ln(C/C_0)) = 1 - (1/\alpha_a) \qquad (8)$$

This form of the Rayleigh equation assumes a closed system, and the α_a value calculated by Eq. (8) will deviate from the true values for the k_{32}/k_{34} rate constant ratio to the extent that the closed system assumption is not fulfilled. Accordingly, α_a values calculated from Eq. (8) are referred to as apparent fractionation factors.

Table 1 summarizes most of the DSDP and ODP sites where $\delta^{34}S$ of pore water sulfate have been measured. Tabulated for each site are estimated sedimentation rates (ω) from biostratigraphy, logarithmic normalized sulfate concentration and $\delta^{34}S$ gradients, apparent fractionation factors (α_a), and the degree to which the sediment column approaches an open system or proportion of sulfate flux from diffusion ($J_d/(J_d + J_b)$). Gradients are determined for depth intervals just beneath the seafloor, generally extending to depths of 5–20 mbsf depending on sample spacing. Some variability in initial sulfate concentration and $\delta^{34}S$ value was accounted for (e.g. the Mediterranean), but initial values of 28 mM and 20.1‰ typically were assumed.

Figure 1 is a plot of system openness ($J_d/(J_d + J_b)$) against the apparent fractionation factor (α_a), assuming a constant whole sediment diffusion coefficient for sulfate of 3500 m^2 ma^{-1} (1.1×10^{-6} cm^2 sec^{-1}). This value for the average diffusion coefficient minimized the scatter relative to the diagonal line in Fig. 1. The diagonal line is consistent with an ideal relationship

$$(J_d/J_d + J_b)1.000 + (J_b/J_d + J_b)1.075 = \alpha_a \qquad (9)$$

in which the proportion of sulfate flux from diffusion is reduced with no fractionation, while the proportion of the sulfate flux from burial is reduced with a fractionation factor of 1.075. Accordingly, DSDP/ODP sites in which diffusion is the dominant means of supplying sulfate to the sediment column have smaller apparent isotopic fractionation factors ($\alpha_a = 1.00$–1.02). Sites in which the sulfate is mainly that buried with sediment pore waters have larger apparent isotopic fractionation ($\alpha_a = 1.05$–1.07).

Scatter in the data shown in Fig. 1 could be due to several factors, such as variable diffusion coefficients, sample intervals too widely spaced to accurately indicate gradients, poorly constrained sedimentation rates, failure to recover seafloor sediments in cores and consequent depth inaccuracies, an oxidizing (bioturbated) zone near the seafloor, advective supply of sulfate at depth and intense anaerobic methane oxidation.

For example, shallow sediments on the Blake Outer Ridge have been cored on numerous DSDP/ODP Legs represented in Fig. 2 by Sites 533, 994, 1059, 1060 and 1061. Sedimentation rates reported on Leg 172 for Sites 1059-61 were 1.5–3 times greater than were reported for sites previously cored on Legs 11, 76, and 164 in the same general region. Scatter in Fig. 1 would be

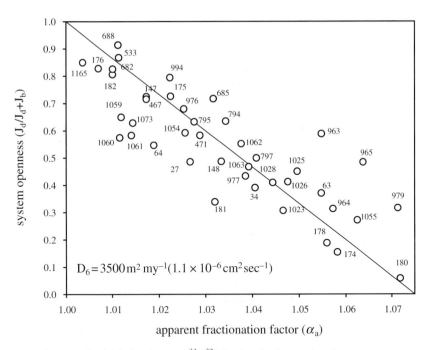

FIG. 1. Apparent (Rayleigh closed system) ^{34}S/^{32}S fractionation factors plotted versus system openness (diffusive sulfate flux as a proportion of diffusive plus burial sulfate flux) for DSDP/ODP sites listed in Table 1. Symbols are indexed by DSDP/ODP site numbers.

somewhat decreased if sedimentation rates for Leg 172 Sites were similar to those interpreted on previous Legs.

   In some areas, there is replenishment of sulfate by advection at depth due to buried evaporites (Mediterranean) or seawater circulation adjacent to mid-ocean ridges (Juan de Fuca). Addition of sulfate at depth will decrease the gradient and diminish both the calculated openness and apparent fractionation properties. Sites or depth intervals in which there were obvious contribution of sulfate other than that from burial or diffusion were excluded from the compilation in Table 1.

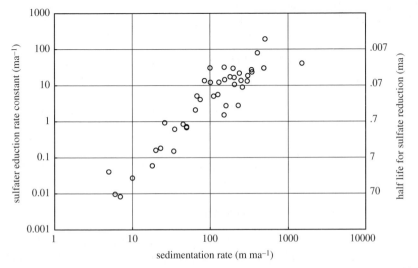

FIG. 2. Log–log plot of rate constant for sulfate reduction versus sedimentation rate for DSDP/ODP sites listed in Table 1.

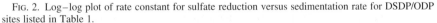

Sulfate reduction proceeds to completion in most of the environments represented by the sites listed in Table 1, leading to the onset of methanogenesis and upward diffusion of methane. Anaerobic methane oxidation linked with sulfate reduction at the base of the sulfate reduction zone can also influence the sulfate concentration gradient. However, with the exception of a few extreme cases, effects of anaerobic methane oxidation are confined to the lower part of the sulfate reduction zone and convert the typical exponential sulfate gradient to a linear gradient (BOROWSKI et al., 1996). The sulfate gradient at the sediment water interface is generally not influenced by methane oxidation at depth, as shown by the fact that carbon isotope ratios of dissolved inorganic carbon in pore water samples from shallow depths do not show effects of addition of methane-derived carbon (CLAYPOOL and THRELKELD, 1983).

The observed sulfate concentration gradients represented by the DSDP/ODP sites listed in Table 1 have other implications for microbiological activity in marine sediments. The product of the concentration gradient and the sedimentation rate provides an estimate of the rate constant for sulfate reduction (TOTH and LERMAN, 1977; BERNER, 1978). A plot of the rate constant for sulfate reduction versus the sedimentation rate for the DSDP/ODP sites listed in Table 1 is shown in Fig. 2. Rate constants vary by more than four orders of magnitude, with the slowest rates requiring about 70 million years for half of the sulfate to be reduced. The most rapid rates are equivalent to a half-life of less than 7000 years for sulfate reduction. Volumetric sulfate reduction rates range from 2 to 5000 mM ma^{-1} (nmol cm^{-3} a^{-1}). The sediment temperatures represented by DSDP/ODP sites in Table 1 range from 0 to 60°C. Neither sulfate reduction rate nor temperature appears to be a significant cause of variation in the isotopic fractionation observed in the marine environments represented in Table 1.

SUMMARY AND CONCLUSIONS

Although there is a wide range of apparent (closed system) fractionation factors for bacterial sulfate reduction in pore waters of marine sediments cored in DSDP/ODP sites, the apparent degree of ^{34}S-fractionation is consistently related to the relative importance of diffusion, primarily controlled by sedimentation rate. Several studies (JØRGENSEN, 1979; GOLDHABER and KAPLAN, 1980; RUDNICKI et al., 2001; WORTMANN et al., 2001) have used diagenetic (transport-reaction) models to show that closed system models underestimate the degree of sulfur isotopic fractionation associated with bacterial sulfate reduction. Moreover, fractionation factors calculated from open-system models are relatively large (1.06–1.08), and consistent with observations of large ^{34}S-depletions in marine sulfides (SWEENEY and KAPLAN, 1980).

A survey of the δ^{34}S of pore water sulfate from a variety of DSDP/ODP sites suggests that a large instantaneous ^{34}S/^{32}S fractionation factor ($\alpha = 1.075$) is characteristic of most deep water marine sediments. However, this tendency for ^{34}S-enrichment in residual sulfate with depth is counteracted by partial diffusive replenishment with isotopically lighter sulfate from overlying sediments or seawater. The result is that the apparent (closed-system) fractionation factor (α_a) is smaller than the actual fractionation factor (α), but directly proportional to the relative importance of diffusion.

Acknowledgements—This work is an updated version of a chapter from an unpublished Ph.D. thesis (CLAYPOOL, 1974) produced at UCLA with advice and direction from Professors I. R. Kaplan and B. J. Presley. Samples and data were provided by DSDP/ODP which has been sponsored by the US National Science Foundation (NSF) and participating countries, now under management of Joint Oceanographic Institutions, Inc. (JOI).

REFERENCES

BERNER R. A. (1964) An idealized model of dissolved sulfate distribution in recent sediments. *Geochim. Cosmochim. Acta* **28**, 1497–1503.
BERNER R. A. (1978) Sulfate reduction and the rate of deposition in marine sediments. *Earth Planet. Sci. Lett.* **37**, 492–498.
BERNER R. A. (1980) *Early Diagenesis*. Princeton University Press, Princeton, NJ.

BÖTTCHER M. E., BRUMSACK H.-J. and DE LANGE G. J. (1998) Sulfate reduction and related stable isotope (^{34}S, ^{18}O) Variations in interstitial waters from the Eastern Mediterranean. In *Proceedings of ODP, Scientific Results* (eds. A. H. F. ROBERTSON, K.-C. EMEIS, C. RICHTER, A. CAMERLENGHI, and SHIPBOARD SCIENTIFIC PARTY), Vol. 160, pp. 365–373.

BÖTTCHER M. E., BERNASCONI H.-J. and BRUMSACK H.-J. (1999) Carbon, Sulfur, and Oxygen Isotope Geochemistry of Interstitial Waters from the Western Mediterranean. In *Proceedings of ODP, Scientific Results* (eds. R. ZAHN, M. C. COMAS, A. KLAUS and SHIPBOARD SCIENTIFIC PARTY), Vol. 161, pp. 413–421.

BOROWSKI W. S., PAULL C. K. and USSLER W. III. (1996) Marine pore water sulfate profiles indicate *in situ* methane flux from underlying gas hydrate. *Geology* **24**, 655–658.

BOROWSKI W. S., HOEHLER T. M., ALPERIN M. J., RODRIGUEZ N. M. and PAULL C. K. (2000) Significance of Anaerobic Methane Oxidation in Methane-Rich Sediments Overlying the Blake Ridge Gas Hydrates. In *Proceedings of ODP, Scientific Results* (eds. C. K. PAULL, R. MATSUMOTO, P. J. WALLACE and W. P. DILLON), Vol. 164, pp. 87–99.

BRUMSACK H.-J., ZUELEGER E., GOHN E. and MURRAY R. (1992) Stable and Radiogenic Isotopes in Pore Waters from Leg 127, Japan Sea. In *Proceedings of ODP, Scientific Results, Pt. 1* (eds. K. A. PISCIOTTO, J. C. INGLE, JR., M. T. VON BREVMANN, J. BARRON and SHIPBOARD SCIENTIFIC PARTY), Vols. 127/128, pp. 635–650.

CANFIELD D. E. (1991) Sulfate reduction in deep-sea sediments. *Am. J. Sci.* **291**, 177–188.

CANFIELD D. E. and THAMDRUP B. (1994) The production of ^{34}S-depleted sulfide during bacterial disproportionation of elemental sulfur. *Science* **266**, 1973–1975.

CLAYPOOL G. E. (1974) Anoxic diagenesis and bacterial methane production in deep sea sediments. Ph.D. thesis, University of California, Los Angeles.

CLAYPOOL G. E. and KAPLAN I. R. (1974) The origin and distribution of methane in marine sediments. In *Natural Gases in Marine Sediments* (ed. I. R. KAPLAN), pp. 99–139. Plenum Press, New York.

CLAYPOOL G. E. and THRELKELD C. N. (1983) Anoxic Diagenesis and Methane Generation in Sediments of the Blake Outer Ridge, Deep Sea Drilling Project Site 533, Leg 76. In *Initial Reports of DSDP* (eds. R. E. SHERIDAN, F. M. GRADSTEIN, and SHIPBOARD SCIENTIFIC PARTY), Vol. LXXVI, pp. 391–402.

CLAYPOOL G. E., LORENSON T. D. and JOHNSON C. A. (2003) Authigenic Carbonates, Methane Generation and Oxidation in Continental Rise and shelf sediments, ODP Leg 188 Sites 1165 and 1166, Offshore Antarctica (Prydz Bay). In *Proceedings of ODP, Scientific Results* (eds. P. E. O'BRIEN, A. K. COOPER, C. RICHTER and SHIPBOARD SCIENTIFIC PARTY), Vol. 188, pp. 1–15. Available online from World Wide Web: E-mail: http://www-odp.tamu.edu/publications/188_SR/VOLUME/CHAPTERS/004.PDF [cited 2003-02-10].

FROELICH P. N., KLINKHAMMER G. P., BENDER M. L., LUEDTKE N. A., HEATH G. R., CULLEN D., DAUPHIN P., HAMMOND D., HARTMAN B. and MAYNARD V. (1979) Early oxidation of organic matter in pelagic sediments of the eastern equatorial Atlantic: suboxic diagenesis. *Geochim. Cosmochim. Acta* **43**, 207–213.

GIESKES J. M., GAMO T. and BRUMSACK H. (1991) Chemical methods for interstitial water analysis onboard. *JOIDES Resolut.* (ODP technical note 15).

GOLDHABER M. B. and KAPLAN I. R. (1975) Controls and consequences of sulfate reduction rates in recent marine sediments. *Soil Sci.* **119**, 42–45.

GOLDHABER M. B. and KAPLAN I. R. (1980) Mechanisms of sulfur incorporation and isotopic fractionation during early diagenesis in the sediments of the Gulf of California. *Mar. Chem.* **9**, 95–145.

HARTMANN U. M. and NIELSEN H. (1969) δS^{34} werte in rezenten meeressedimenten und ihre deutung am beispiel einiger sedimentprofile aus der west lichen ostsee. *Geol. Rundschau* **58**, 621–655.

JØRGENSEN B. B. (1979) A theoretical model of the stable sulfur isotope distribution in marine sediments. *Geochim. Cosmochim. Acta* **43**, 363–374.

JØRGENSEN B. B. (1990) A thiosulfate shunt in the sulfur cycle of marine sediments. *Science* **249**, 152–154.

KAPLAN I. R., EMERY K. O. and RITTENBERG S. C. (1963) The distribution and isotopic abundance of sulfur in recent marine sediments off Southern California. *Geochim. Cosmochim. Acta* **27**, 297–331.

MANHEIM F. T. (1966) A hydraulic squeezer for obtaining interstitial water from consolidated and unconsolidated sediments. US Geol. Survey Prof. Paper 550-C, 256.

MANHEIM F. T., SAYLES F. L. and FRIEDMAN I. (1969) Interstitial Water Studies on Small Core Samples, Deep Sea Drilling Project, Leg 1. In *Initial Reports of DSDP* (ed. M. EWING, J. L. WORZEL, A. O. BEALL, W. A. BERGGREN, D. BURKY, C. A. BURK A. G. FISCHER and E. A. PESSAGNO, JR.), Vol. I, pp. 403–410.

MOTTL M. J., WHEAT C. G., MONNIN C. and ELDERFIELD H. (2000) Data Report: Trace Elements and Isotopes in Pore Water from Sites 1023 Through 1032, Eastern Flank of the Juan de Fuca Ridge In *Proceedings of ODP, Scientific Results* (eds. A. T. FISHER, E. E. DAVIS and C. ESCUTIA), Vol. 168, pp. 105–115.

PRESLEY B. J. (1971) In *Techniques for Analyzing Interstitial Water Samples. Part I: Determination of Selected Minor and Major Inorganic Constituents* (ed. E. L. WINTERER), Vol. VII, pp. 1749–1755. Initial Reports of DSDP, Pt. 2.

PRESLEY B. J. and CLAYPOOL G. E. (1971) In *Techniques for Analyzing Interstitial Water Samples. Part II: Determination of Total Dissolved Carbonate and Carbon Isotope Ratios* (ed. E. L. WINTERER), Vol. VII, pp. 1756–1757. Initial Reports of DSDP, Pt. 2.

RUDNICKI M. D., ELDERFIELD E. and SPIRO B. (2001) Fractionation of sulfur isotopes during bacterial sulfate reduction in deep ocean sediments at elevated temperatures. *Geochim. Cosmochim. Acta* **65**, 777–789.

SWEENEY R. E. and KAPLAN I. R. (1980) Diagenetic sulfate reduction in marine sediments. *Mar. Chem.* **21**, 165–174.

TOTH D. J. and LERMAN A. (1977) Organic matter reactivity and sedimentation rates in the ocean. *Am. J. Sci.* **277**, 465–485.

WORTMANN U. G., BERNASCONI S. M. and BÖTTCHER M. E. (2001) Hypersulfidic deep biosphere indicates extreme sulfur isotope fractionation during single-step microbial sulfate reduction. *Geology* **29**, 647–650.

Geochemical Investigations in Earth and Space Science: A Tribute to Isaac R. Kaplan
© The Geochemical Society, Publication No. 9, 2004
Editors: R.J. Hill, J. Leventhal, Z. Aizenshtat, M.J. Baedecker, G. Claypool,
R. Eganhouse, M. Goldhaber and K. Peters

Factors controlling the carbon isotopic composition of methane and carbon dioxide in New Zealand geothermal and natural gases

JOHN R. HULSTON[*]

Institute of Geological and Nuclear Sciences, P.O. Box 31-312, Lower Hutt, New Zealand

Abstract—Gases from Taupo Volcanic Zone (TVZ) and Taranaki natural gas fields of New Zealand show both similarities and differences. The mantle helium signal is strong in both fields. In Taranaki, CH_4 generally predominates while in TVZ, CO_2 predominates and CO_2, CH_4, H_2 and N_2 concentrations relative to mantle-derived ^3He are correlated with B/Cl ratios. In the high B/Cl areas of TVZ $\delta^{13}C(CH_4)$ values are mainly in the -25 to $-28‰$ range, possibly from high temperature decomposition of kerogen, whereas in the lower B/Cl areas other processes result in $\delta^{13}C(CH_4)$ values of -14 to $-37‰$.

The relatively high heat flow in parts of the Taranaki Basin appears to have accelerated the natural gas generation process in which $\delta^{13}C(CH_4)$ is initially much more negative than the source organics (-27 to $-29‰$) but becomes more positive with increasing maturity of the source rocks. This results in a range of $\delta^{13}C(CH_4)$ values from -32 to $-48‰$.

Isotopic equilibrium between $^{13}CO_2$ and $^{13}CH_4$ does not appear likely in Taranaki. In TVZ apparent isotopic temperatures are higher than reservoir temperatures but show fair correlation. This may relate both to a hydrocarbon maturity effect and to a source effect but this does not fully explain the $\delta^{13}C(CH_4)$ and $\delta^{13}C(CO_2)$ distribution within some fields.

INTRODUCTION

THE GEOTHERMAL AREAS of Taupo Volcanic Zone (TVZ) and the natural gas areas of the Taranaki Basin to the west are both influenced by the heat flow from the Pacific plate subducting beneath the North Island of New Zealand (Fig. 1). Both have helium-3 signals from the mantle and have areas in which the $\delta^{13}C(CH_4)$ values are more positive than are encountered in many other similar areas (CRAIG, 1953; POREDA et al., 1992; Schoell, 1983). It is therefore useful to present a geologic overview of these two regions together to consider the factors controlling the composition of methane and carbon dioxide, those factors that are in common for these systems and those that relate solely to natural gas systems formed by burial of organic matter.

In early studies of $\delta^{13}C$ on coexisting CO_2 and CH_4 in geothermal gases (CRAIG, 1953; HULSTON and McCABE, 1962) close agreement between underground temperatures and isotopic temperatures, evaluated on the basis of theoretical fractionation factors calculated by UREY (1947) and CRAIG (1953), was obtained. These results suggested that the CO_2 and CH_4 were in isotopic and chemical equilibrium. With the advent of more refined calculations by BOTTINGA (1969) and RICHET et al. (1977), these isotopic temperatures were increased by approximately 110°C. HULSTON (1976) suggested that the discrepancy could be explained by a freezing-in of a higher temperature equilibrium composition within the rising fluid as the corresponding reaction rates decrease, preventing the re-equilibration of the gases. SACKETT and CHUNG (1979) and GIGGENBACH (1982), however, suggested that the rates of equilibrium were too slow for this to be a viable explanation. Despite perturbation of the gaseous composition caused by steam–water separation, GIGGENBACH (1980) considered that the composition of fluids discharged from geothermal areas in New Zealand reflected close to complete attainment of chemical equilibrium within the system

$$H_2O, \ CO_2, \ H_2S, \ NH_3, \ H_2, \ N_2 \ \text{and} \ CH_4. \tag{1}$$

This included the reaction $CH_4 + 2H_2O = CO_2 + 4H_2$ that had been suggested as the

---

[*] *Present address*: Isotope Consulting Limited, 7 Earlston Grove, Lower Hutt, NZ 6009.

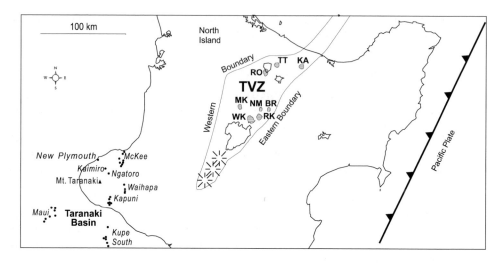

FIG. 1. Location of the Taranaki natural gas fields, Taupo Volcanic Zone (TVZ) and the subducting Pacific plate. WK, Wairakei; RK, Rotokawa; MK, Mokai; NM, Ngatamariki; BR, Ohaaki-Broadlands; RO, Rotorua; TT, Tikitere; KA, Kawerau.

mechanism for the carbon isotope exchange reaction

$$^{13}CH_4 + {}^{12}CO_2 = {}^{13}CO_2 + {}^{12}CH_4. \tag{2}$$

Initially it was thought that the attainment of chemical equilibrium might be sufficient to indicate the attainment of isotopic equilibrium but GIGGENBACH (1982) argued that the attainment of isotopic equilibrium involving CO_2 and CH_4 can be expected to proceed about a hundred times slower than the chemical equilibrium.

Recently, HORITA (2001) succeeded in establishing carbon isotopic equilibrium between CO_2 and CH_4 within a maximum of 2000 h, at temperatures from 200 to 600°C, by reversal from opposite isotopic directions using a commercial Ni catalyst. The equilibrium values obtained were some 0.9‰ greater than those calculated by RICHET et al. (1977), slightly increasing the discrepancy between geothermal field measurements and isotopic equilibrium. HORITA (2001) fitted the results to the following equation, valid ($1\sigma = 0.14‰$) between $T = 473.15$ and 873.15 K (200–600°C):

$$10^3 \ln \alpha(CO_2-CH_4) = 26.70 - 49.137(10^3/T) + 40.828(10^6/T^2) - 7.512(10^9/T^3) \tag{3}$$

Re-adjustment to Horita's results, of the expression used by LYON and HULSTON (1984) to calculate isotope equilibrium temperatures from measured isotopic values gives, for the above temperature range (T in °C, $1\sigma = 3.5°C$) gives:

$$T(°C) = -91.5 + 11,200/(10^3 \ln \alpha(CO_2-CH_4) + 3.6). \tag{4}$$

With this new equilibrium data it seems appropriate to include a revisit of the carbon isotopic data from CO_2 and CH_4 in the TVZ and Taranaki systems to look at the factors controlling this data and the relevance of isotopic equilibrium temperatures calculated for the $\Delta^{13}C(CO_2-CH_4)$ system.

NEW ZEALAND NATURAL GAS AND GEOTHERMAL REGIONS

Geological setting

The Pacific Plate subducts beneath the North Island of New Zealand (Fig. 1). This gives rise to a series of geothermal fields in TVZ (subduction zone depth 80–120 km) and to a high heat flow zone beneath the Taranaki natural gas fields (subduction zone depth 150–200 km). An extinct 2518 m andesitic volcano, Mt. Taranaki is located adjacent to these natural gas fields.

Taranaki natural gas fields

Taranaki Basin is an oil, gas and condensate province. Geological and geochemical evidence (KING and THRASHER, 1996) indicates that the source rocks for both oil and gas are primarily Cretaceous and Eocene terrestrial coal measures, although biomarkers sometimes reflect kerogen with a marine influence. Thermal modeling and geohistory case studies (*e.g.* ARMSTRONG *et al.*, 1996) indicate that the thickest sequences of the late Cretaceous source rocks may have started to generate and expel hydrocarbons in the Paleocene, and continued over differing geographical areas up to the present time (Fig. 2). Locally Eocene Kapuni Group source rocks have also generated and expelled hydrocarbons during the Mid-Miocene.

Surface heat flow estimates in the Taranaki Basin (FUNNELL *et al.*, 1996) range from 47 mW m^{-2} in the SE to a maximum of 74 mW m^{-2} near New Plymouth (where the subduction zone has a depth of *ca.* 200 km) before decreasing again further to the north-west. The heat flow estimate for the Maui field is 64 mW m^{-2}.

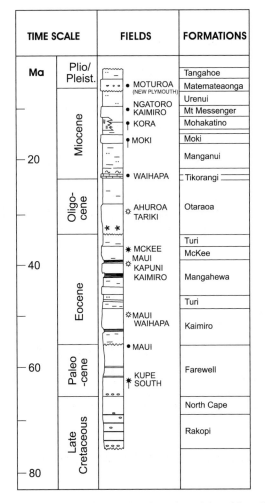

FIG. 2. Stratigraphy of Taranaki Basin hydrocarbon formations. Adapted from KING and THRASHER (1996).

The chemical and isotopic compositions of gases from hydrocarbon systems of the Taranaki Basin reported by HULSTON et al. (2001) show wide variation. The most striking difference between the south-eastern and western groups of gases is the helium content and its isotopic ratio. In the south-east, the Kupe South and most Kapuni natural gases have low total helium concentrations of $10-19$ μM M^{-1} and a minor mantle helium input of $0.03-0.32R_A$ (where R_A = the air ^{3}He/^{4}He ratio of 1.4×10^{-6}). In the west, adjacent to the still active Cape Egmont Fault, the Maui and New Plymouth gases are over an order of magnitude higher in helium concentration (up to 190 μM M^{-1}). Their ^{3}He/^{4}He ratios of $2.6-4.9R_A$ are half that of upper mantle helium issuing from volcanic vents of TVZ. The ^{3}He/C ratio (where C represents the total carbon in the gas phase) is generally high at locations where the surface heat flow is high, often associated with areas of volcanism.

A further striking difference is in the methane carbon isotopic composition δ^{13}C(CH$_4$). In the west, the most positive Maui δ^{13}C(CH$_4$) value is $-32.4‰$ compared to ca. $-42‰$ for most wells in the east with the Kaimiro field reaching $-49‰$.

Taupo volcanic zone geothermal fields

The TVZ (Fig. 1) is roughly 250 by 50 km and extends from the active volcanoes of Tongariro Volcanic Centre in the south-west to beyond White Island to the north-east (GAMBLE et al., 1993). The volcanic activity is less than 2 Ma old and is related to subduction of the Pacific Plate beneath the Australian Plate. COLE and LEWIS (1981) proposed that the TVZ is the southern continuation of the Tonga–Kermadec subduction system that evolved through a gradual rotation of accretionary elements from a NW trend in the Kermadecs to the present NNE trend in TVZ. At the intersection of these two trends about 15,000 km^3 of rhyolitic material, mainly ignimbrites, has been erupted, whereas andesitic volcanism dominates at the SW and NE extremities. Basement rocks are Mesozoic greywackes of the Torlesse and Waipapa terranes. A number of geothermal areas occur within the TVZ, and over the last 30 years most have been surveyed to determine their potential for electricity generation. Many were drilled to depths of around 1 km and some to depths of 2 km. These investigations led to the establishment of the

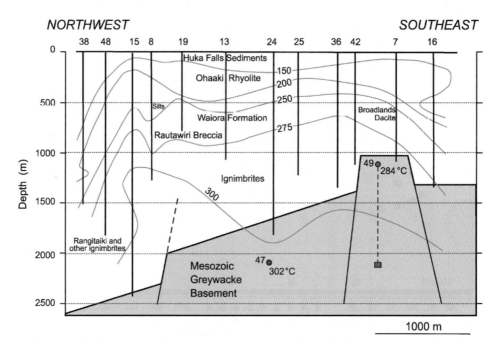

FIG. 3. Cross-section showing the geology of the Ohaaki-Broadlands geothermal field, TVZ. Adapted from HEDENQUIST (1990) and CHRISTENSON et al. (2002).

Wairakei and Ohaaki geothermal power stations. The pumice pyroclastics constitute aquifers in the geothermal areas, with inter-layering sandstones and mudstones often forming impermeable strata. The different geothermal areas are closely associated with active NE trending faults resulting from the subsidence of major structural depressions. These faults may provide ascent channels for hot waters, especially where they are transected by minor NW trending faults. The fairly regularly spaced main geothermal areas of the TVZ are shown in Fig. 1. Some of these systems have been established for at least 0.5 Ma. The present study is concentrated in the central area of the TVZ. Figure 3 shows the stratigraphy of a (NW–SE trending) cross-section of the Ohaaki-Broadlands geothermal field. The stratigraphy of the south-eastern section is typical of the fields on the boundary of TVZ, while the north-western end of the cross-section indicates the likely stratigraphy of the fields in the center of TVZ.

LYON and HULSTON (1984) summarized the data available from three fields in the TVZ namely Wairakei, Broadlands (now referred to as Ohaaki) and Tikitere and obtained $\Delta^{13}C(CH_4–CO_2)$ equilibrium temperatures in the 300–400°C range. These temperatures were some 100°C higher than the observed two-phase discharge temperatures. A possible explanation is that isotopic temperatures represent a "frozen-in" temperature from a deeper reservoir where isotopic equilibrium is being established.

RESULTS AND DISCUSSION

Taranaki natural gas fields

The data for this paper originates from HULSTON et al. (2001) which included measurements by the authors plus data from LYON et al. (1996). The gas isotopic data from two Maui condensate samples listed in LYON et al. (1996) that were excluded from HULSTON et al. (2001) have been included in this study. A summary of the data is shown in Table 1.

The relationship between $\delta^{13}C(CO_2)$ and $\delta^{13}C(CH_4)$ for the Taranaki natural gas fields is shown in Fig. 4a together with the equilibrium temperatures calculated from the calibration of HORITA (2001) discussed above. The production gas wells have isotopic temperatures higher than either the *ca.* 100°C reservoir temperature or the expected maturation temperature of *ca.* 180°C. The shallow exploration holes at New Plymouth, however, have equilibrium temperatures of *ca.* 150°C similar to that expected at depth. In addition, the Maui gases show a near linear relationship (with the exception of the two condensate samples), crossing the equilibrium temperature lines from 220 to 350°C, suggesting a mixing process across the Maui A platform. The plots of $\delta^{13}C(CO_2)$ and $\delta^{13}C(CH_4)$ against the helium isotope ratios shown in Figs. 4b,c indicate that this mixing process also involves the $^3He/^4He$ ratio through either the mantle (^3He) or the radiogenic (^4He) helium components (as discussed in more detail in HULSTON et al. (2001)).

The helium isotopes also provide an indication of the origin of the CH_4 and CO_2. In Fig. 5a, the radiogenic component of the helium (He_{rad}), as defined by HULSTON et al. (2001), divided by the total carbon (C) in the CH_4, C_2H_6, C_3H_8 and CO_2 components of the gas phase is plotted against $\delta^{13}C(CH_4)$. It is observed that for Maui in particular, the gas phase samples with the highest He_{rad}/C ratios have the most positive $\delta^{13}C(CH_4)$ values while the samples with the lowest He_{rad}/C ratio have $\delta^{13}C(CH_4)$ values in the −50 to −40‰ range normally observed in petroleum field gases. HULSTON et al. (2001) suggested that this pattern indicated that the most positive $\delta^{13}C(CH_4)$ values have been in contact with radiogenic host rock for a longer period of time and this time has allowed the gases to reach a greater stage of maturity. Alternatively this gas may have encountered basaltic rocks of greater U + Th content that occur in this area (R. FUNNELL, personal communication).

The $^3He/CO_2$ ratio (and its inverse) may be used as an indicator of the proportion of mantle carbon in these gases. MARTY and ZIMMERMANN (1999) reported $^3He/CO_2$ ratios of $(0.7 \pm 0.2) \times 10^{-9}$ (corrected for the effects of air-contamination and potential fractional degassing) for N-MORB samples. NISHIO et al. (1999) reported $^3He/CO_2$ ratios of $(3.5 \pm 1) \times 10^{-9}$ for Rodrigues Triple Junction Indian Ocean MORB and suggested that this higher value reflected "pristine mantle" whereas lower ratios for MORB indicate carbon

Table 1. Summary of heat flow and $\delta^{13}C$ results for Taranaki hydrocarbon CO_2 and CH_4 gases

Field/Platform/ Region	Heat flow (mW m^{-2})	Average depth (m AH)	Maximum temperature (°C)	Samples CO$_2$(CH$_4$)	Number of outliers	$\delta^{13}C(CH_4)$ (‰) Mean	1σ	$\delta^{13}C(CO_2)$ (‰) Mean	1σ	Isotopic temperature (°C)	Slope	Regression coefficient (r)	Data source
Maui A–C sand	64	3100	95	5 (5)	1	−36.5	0.6	−10.8	1.4	290	−0.45	0.97	a, b
Maui A–D sand	64	3470	104	2 (2)	0	−37.5	0.6	−7.5	0.3	235	−0.54	–	a, b
Maui B–C sand	64	–	93	1 (1)	–	−36.0	–	−6.5	–	240	–	–	b
New Plymouth	74	400	34	4 (5)	0	−41.4	1.2	−0.7	0.5	145	−2.5	–	b
Kaimiro	70	1300	63	0 (3)	–	−47.0	1.8	–	–	–	–	–	b
Ngatoro	67	1550	70	0 (2)	–	−38.9	2.3	–	–	–	–	–	b
McKee region	60	2300	82	1 (1)	–	−41.3	–	−16.1	–	290	–	–	b
				0 (6)	–	−42.5	1.6	–	–				
Waihapa	57	2990	84	0 (1)	–	−43.0	–	–	–	–	–	–	b
Kapuni	53	4750	118	4 (5)	0	−42.1	0.5	−14.4	0.1	260	−3.6	−0.90	a, b
Kupe South	50	3100	99	0 (2)	–	−42.6	0.5	–	–	–	–	–	a

Carbon isotope data sources: (a) HULSTON et al. (2001); (b) LYON et al. (1996).

Surface heat flow, depth and temperature data: FUNNELL et al. (1996); R. FUNNELL (personal communication).

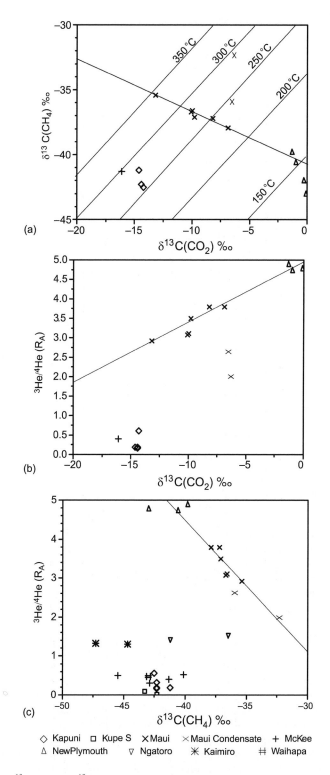

FIG. 4. (a) $\delta^{13}C(CH_4)$ vs. $\delta^{13}C(CO_2)$ relationship for Taranaki natural gases; helium isotopic ratios (relative to air) vs. (b) $\delta^{13}C(CO_2)$ and (c) $\delta^{13}C(CH_4)$. The line fits shown exclude the two Maui condensate phase samples (shown as lighter ×). Isotherms are in °C.

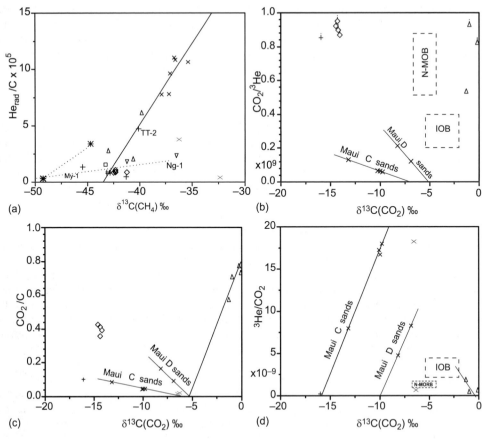

◇ Kapuni □ Kupe S ✕ Maui ⨯ Maui Condensate ✛ McKee △ New Plymouth ▽ Ngatoro ✳ Kaimiro ⊞ Waihapa

FIG. 5. (a) $\delta^{13}C(CH_4)$ vs. the radiogenic helium/total carbon ratio (b–d) $\delta^{13}C(CO_2)$ plots to estimate isotopic endpoints for high and low CO_2 components of the natural gases.

recycling to the mantle. The results shown in Fig. 5b indicate that the Maui gases contain less CO_2 than either of these values (suggesting some loss of CO_2) while all other Taranaki gases, with the exception of one sample from New Plymouth contain considerably more. Extrapolation of the Maui data to zero $CO_2/^3He$ indicates $\delta^{13}C(CO_2)$ values between −5 and −7‰, within the range detected for the MORBs. Extrapolation to zero CO_2/C (Fig. 5c) also indicates a mantle $\delta^{13}C(CO_2)$ end-point for the New Plymouth samples. The other (non-mantle) end-point for samples with higher CO_2 content is indicated as ^3He tends to zero in the $^3He/CO_2$ plot shown in Fig. 5d. This non-mantle component of between −1 and 0‰ for New Plymouth indicates a carbonate source while −16.6‰ for Maui C sands, similar to the −14.5‰ value for Kapuni, suggests an organically influenced source. The Maui D sand samples, however, extrapolate to ca. −10‰ suggesting a mixture of organic and carbonate.

It would thus appear from the evidence above that, particularly in the Maui and New Plymouth fields that CH_4 and CO_2 have different origins. GIGGENBACH (1997) investigated the relationship between CH_4/CO_2 temperature under various "rock-buffers" and concluded that the CH_4/CO_2 ratios in Taranaki and the NZ geothermal systems were slightly lower than equilibrium values calculated for a liquid dominated system controlled by magnetite–hematite. He suggested that the deviations might be assumed to be due to attainment of equilibrium with rock at somewhat higher-than-measured temperatures. If so, then the equilibrium would be moving to increase CH_4 through the reduction of CO_2. This could explain why the $CO_2/^3He$ ratio is lower than mantle values in the Maui fields. SHERWOOD LOLLAR et al. (1997) found similar sub-mantle values in the Pannonian and Vienna basins of Hungary and Austria, respectively. They considered the loss of CO_2 to be due to possible sinks, including precipitation as carbonate and reduction to graphite. However, if this loss of CO_2 occurred through re-equilibration processes, a change of $\delta^{13}C(CO_2)$ in the residual would be expected. This would then alter the interpretation of Fig. 5b,c but at this stage it is not possible to define this process with certainty.

Taupo volcanic zone geothermal fields

The sources and summary of the geothermal $\delta^{13}C$ data used for this paper are listed in Table 2. Figure 6 shows the $\delta^{13}C(CO_2)$ vs. $\delta^{13}C(CH_4)$ relationship for these samples except for the recent Ohaaki data of CHRISTENSON et al. (2002) which has to be omitted (a) to avoid further overcrowding the plots and (b) to avoid possible effects due to the recent run-down of this field. The Wairakei, Ohaaki and Tikitere areas indicate $\Delta^{13}C(CH_4-CO_2)$ equilibrium temperatures in the 300–400°C range. These temperatures are some 100°C higher than the observed two-phase discharge temperatures. The Mokai (north) field (HULSTON, 1986; HULSTON and LYON, 1999), however, shows isotopic temperatures up to 660°C, despite indicated chemical equilibrium temperatures between CH_4 and CO_2, after allowing for steam–water separation process underground, of ca. 300°C (HENLEY and MIDDENDORF, 1985).

Some insight into this relationship can be gained by the examination of the CO_2, CH_4, $\delta^{13}C(CO_2)$ and $\delta^{13}C(CH_4)$ data individually. In particular, their relationship to ^3He released from the mantle through the subduction zone is useful as the ^3He signal appears to be associated with heat flow from the mantle. POREDA et al. (1992) found, in their study of Icelandic geothermal systems, a relationship between $CH_4/^3He$ and $\delta^{13}C(CH_4)$ with the most positive $\delta^{13}C(CH_4)$ value of $-17.8‰$ being close to the lowest $CH_4/^3He$ ratios. In a study of $^3He/^4He$ ratios in the TVZ, HULSTON and LUPTON (1996) found a relationship between $^3He/^4He$ ratios in the gas phase and the B/Cl ratio of the fluid phase as shown in Fig. 7a. They suggested that the boron source is releasing radiogenic ^4He resulting in a decrease of the $^3He/^4He$ ratio in high B regions. HULSTON (1998) extended this by demonstrating that the $CH_4/^3He$, $H_2/^3He$, $N_{2C}/^3He$ and $CO_2/^3He$ ratios (where N_{2C} represents the non-atmospheric N_2 component, calculated using the atmospheric Ar content of the sample) tended to increase with increasing B/Cl. In these latter plots the scatter, caused by the differing solubility of these gases that occur during steam–water separation, was reduced by selecting data where the Ar/water ratio was close to the air saturated water value. This technique was not used in the plots shown in Fig. 7b–d as this would reduce the dataset available. As a result more scatter is apparent in Fig. 7b–d than in HULSTON (1998).

Figure 7b–d suggests a positive linear correlation between CO_2, CH_4, H_2 and B/Cl, particularly in the production wells. If chemical equilibrium existed in these wells and they were of similar temperature it would be expected that

$$K_C = P_{CO_2}P_{H_2}^4/P_{CH_4}P_{H_2O}^2 \tag{5}$$

would remain constant. Thus, if the B source also introduced CO_2, CH_4 and H_2 then since P_{H_2} appears to the fourth power in the equation for the equilibrium, $H_2/^3He$ would need to be non-linear if chemical equilibrium were being re-established.

$\delta^{13}C(CH_4)$ and $\delta^{13}C(CO_2)$ vs. B/Cl (Fig. 8a,b) indicate relatively narrow ranges of isotopic values for the high B/Cl region superimposed on a wider range of values for the regions with lower B/Cl. As seen in Table 2, the eastern land based fields (Tikitere and Kawerau) have the most negative $\delta^{13}C(CH_4)$ values and Mokai the most positive. The high B/Cl samples have a narrow range of $\delta^{13}C(CH_4)$ values, -28 to ca. $-26‰$. The $\delta^{13}C(CO_2)$ values follow a slightly different trend with Rotokawa being the most negative and Kawerau mid-range. HULSTON (1998) plotted $\delta^{13}C(CO_2)$ against the inverse ratio (Cl/B) and found that the high B/Cl samples from Ohaaki, Rotokawa, etc. appeared to have an end point (as Cl/B tends to zero) of $\delta^{13}C(CO_2)$ ca. $-9‰$. The low B/Cl samples from Mokai and Wairakei indicate $\delta^{13}C(CO_2)$ values ca. $-4‰$. The ca. $-9‰$ value is similar to that found in calcite from a Torlesse greywacke sample (S. WOLDEMICHAEL, personal communication) suggesting that the meta sedimentary greywacke basement rocks are the most likely source of the B, CO_2, CH_4, H_2 and non-atmospheric N_2 in

Table 2. Summary of $\delta^{13}C$ results for geothermal CO_2 and CH_4 gases in Taupo Volcanic Zone

Field	Number of samples	Number of outliers	Maximum temperature (°C)	$\delta^{13}C(CH_4)$ (‰) Mean	1σ	$\delta^{13}C(CO_2)$ (‰) Mean	1σ	Isotopic temperature (°C)	Slope	Regression coefficient (r)	Data source
Tikitere	12	1	260	−28.0	0.4	−7.0	0.4	360	0.77	0.43	b
Kawerau	5	0	280	−28.0	0.2	−6.9	0.8	360			c
Ohaaki	30	2	300	−26.2	0.3	−7.7	0.4	410	0.84	0.40	b, c
Ohaaki (1997–1999)	19	1	280	−26.3	0.5	−7.9	0.8	410	0.55	0.28	f
Rotokawa	5	0	295	−25.9	0.3	−7.8	0.3	420			c
Wairakei	37	3	280	−26.0	0.7	−4.5	0.9	350	−0.76	0.43	a, b
Ngatamariki	3	0	270	−25.2	0.2	−6.8	0.2	410			
Mokai south	7	1	260	−24.5	0.5	−4.6	0.5	380			d, e
Mokai north	11	0	310	−17.3	2.6	−5.8	0.8	660	3.4	0.74	d, e

Carbon isotope data sources: (a) HULSTON and MCCABE (1962), (b) LYON and HULSTON (1984), (c) GIGGENBACH (1995), (d) HULSTON (1986), (e) HULSTON and LYON (1999), (f) CHRISTENSON et al. (2002).

Temperature data: C. BROMLEY (personal communication).

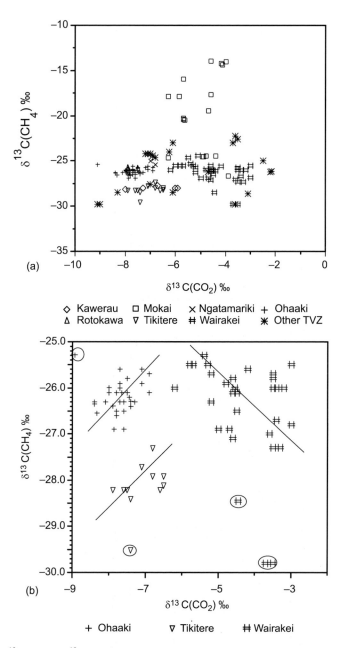

FIG. 6. $\delta^{13}C(CH_4)$ vs. $\delta^{13}C(CO_2)$ plots (a) for TVZ geothermal gases. (b) expansion of (a) for Ohaaki, Tikitere and Wairakei samples only. The regression lines shown exclude the circled points.

TVZ geothermal fluids. $\delta^{13}C(CH_4)$ vs. $CH_4/^3He$ (Fig. 9a) and $\delta^{13}C(CO_2)$ vs. $CO_2/^3He$ (Fig. 9b) confirm the association of the high $CH_4/^3He$ and $CO_2/^3He$ components with $\delta^{13}C(CH_4)$ values of *ca.* $-26‰$ and $\delta^{13}C(CO_2)$ values of $-8‰$.

The above relationships, together with the difference between the isotopic and measured temperatures, would indicate that the CO_2 and CH_4 in the TVZ are not in isotopic equilibrium. However, this does not completely explain all the observed results, particularly at Mokai.

Closer examination of the $\delta^{13}C(CO_2)$ vs. $\delta^{13}C(CH_4)$ relationship in Fig. 6a indicates a trend for some areas (particularly Ohaaki) along the isotopic equilibrium line. Regression coefficients

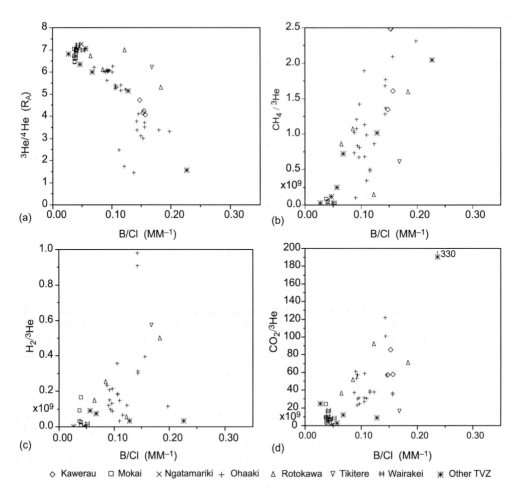

◇ Kawerau   □ Mokai   × Ngatamariki + Ohaaki   △ Rotokawa ▽ Tikitere ⊪ Wairakei   ✱ Other TVZ

FIG. 7. B/Cl in the fluid phase vs. (a) $^3He/^4He$ (b) $CH_4/^3He$ (c) $H_2/^3He$ and (d) $CO_2/^3He$.

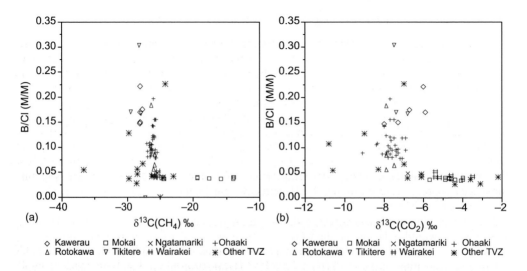

◇ Kawerau   □ Mokai   × Ngatamariki +Ohaaki   ◇ Kawerau   □ Mokai   × Ngatamariki + Ohaaki
△ Rotokawa ▽ Tikitere ⊪ Wairakei   ✱ Other TVZ   △ Rotokawa ▽ Tikitere ⊪ Wairakei   ✱ Other TVZ

FIG. 8. (a) $\delta^{13}C(CH_4)$ and (b) $\delta^{13}C(CO_2)$ vs. B/Cl for TVZ geothermal gases.

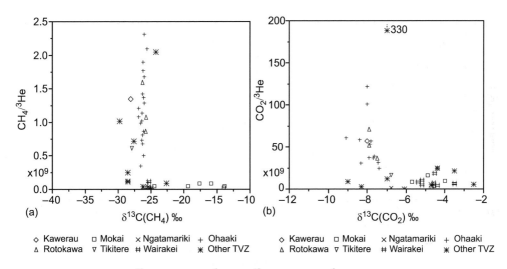

FIG. 9. (a) $\delta^{13}C(CH_4)$ vs. $CH_4/^3He$ (b) $\delta^{13}C(CO_2)$ vs. $CO_2/^3He$ for geothermal gases.

and slopes for the relevant areas have been calculated and are presented in Table 2. The regressions have been calculated assuming equal errors in the $\delta^{13}C(CO_2)$ and $\delta^{13}C(CH_4)$ measurements and significant outliers have been excluded. Several trends are significant, suggesting that the $\delta^{13}C(CO_2)$ and $\delta^{13}C(CH_4)$ are not independent. Since CO_2 predominates this would suggest that some process is operating to change $\delta^{13}C(CH_4)$ when $\delta^{13}C(CO_2)$ changes. Figure 6b shows the correlations for the Ohaaki, Tikitere and Wairakei data alone. Positive correlations are found for Ohaaki and Tikitere—both in the high B/Cl zone. There are sufficient data from Ohaaki to make this regression significant but the limited number of samples from Tikitere reduces the significance level to between 70 and 80%. The slope of the regression for the Wairakei data is negative as found for the Maui samples. LYON and HULSTON (1984) suggested that the $\delta^{13}C(CO_2)$ values in low enthalpy–low CO_2 wells at Wairakei may have been affected by the release of CO_2 from the dissolved HCO_3 phase as the relatively fast CO_2–HCO_3 equilibrium system attempts to re-equilibrate after the loss of CO_2 in the steam phase. The recent data from Ohaaki (CHRISTENSON et al., 2002) has a greater $\delta^{13}C(CO_2)$ spread and lower regression slope (Table 2) than the earlier data, possibly due to beginning of a similar process.

The relationship between $\delta^{13}C(CO_2)$, $\delta^{13}C(CH_4)$ and observed temperature

Table 2 lists the maximum measured temperatures for each of the areas listed. There is a loose correspondence between these temperatures and the measured $\delta^{13}C(CH_4)$ values. The most negative $\delta^{13}C(CH_4)$ values corresponding to the lowest temperatures and the more positive to higher temperatures. As found previously, temperatures calculated assuming isotopic equilibrium between CH_4 and CO_2 are somewhat higher than the observed temperatures. The spread of results in Table 2 is too great to determine if this temperature correspondence is related to the measured $\delta^{13}C(CH_4)$ value alone or to the difference between the $\delta^{13}C(CH_4)$ and $\delta^{13}C(CO_2)$ values. If it were related to the latter then this would be an indication of some type of equilibrium process. If it were related solely to the $\delta^{13}C(CH_4)$ value then presumably this is the result of increasing maturation of organic source of the methane with increasing temperature (or time) as appears to occur in the Taranaki natural gas fields.

Are abiogenic sources of CH_4 indicated in TVZ or Taranaki Basin?

Relatively positive $\delta^{13}C(CH_4)$ values observed in some TVZ and Taranaki Basin samples might suggest abiogenic generation of CH_4 by serpentinization of ultrabasic rocks at depth (BERNDT et al., 1996). In this process relatively high H_2 concentrations are required to reduce

CO_2 or calcite to CH_4. However, H_2 is barely detectable in the Taranaki gas wells and in TVZ the $H_2/^3He$ ratios for Mokai (Fig. 7c) are much lower than at Ohaaki and the Mokai wells with the most positive $\delta^{13}C(CH_4)$ values have the lowest $H_2/^3He$ ratios. This suggests that this process is not occurring to any detectable extent.

SHERWOOD LOLLAR et al. (2002) have used both the relative $\delta^{13}C$ values of higher hydrocarbons relative to CH_4 and the slope of the $\delta^{13}C(CH_4)$ vs. $\delta^2H(CH_4)$ plot as indicators of abiogenic CH_4. No higher hydrocarbon $\delta^{13}C$ data is available for TVZ but HULSTON et al. (2001) report for several Taranaki wells $\delta^{13}C(C_2)$ and $\delta^{13}C(C_3)$ values more positive than $\delta^{13}C(CH_4)$. This indicates a thermogenic hydrocarbon pattern (viz. produced from the thermal decomposition of high molecular weight organic matter).

Another indicator of abiogenic hydrocarbons is that the relative concentrations of C_1-C_4 n-alkanes would follow the Schulz–Flory distribution—indicative of hydrocarbons produced by the polymerization of lower homologs, in which:

$$\alpha_{SF} = C_2/C_1 = C_3/C_2 = C_4/C_3 = \cdots \tag{6}$$

Figure 11 of GIGGENBACH (1997) is a ternary plot of C_1, C_2 and C_3 for Taranaki and TVZ data, indicating that relative to methane (C_1) and propane (C_3) the TVZ samples contained only ca. 10% of the ethane (C_2) required for a Schulz–Flory distribution. The Taranaki production well samples, however, contained ca. 50% of the C_2 required while the shallower New Plymouth drillhole samples contained ca. 25%. These values are closer to the TVZ but with a α_{SF} $\{(C_3/C_1)^{0.5}\}$ value of 0.1 compared to 0.01–0.03 for the TVZ samples and 0.1–0.3 for the Taranaki production wells. In particular, it should be noted that the Maui wells where $\delta^{13}C(CH_4)$ is the most positive have Schulz–Flory characteristics similar to the other Taranaki production wells. These results suggest that abiogenic hydrocarbons are not measurable in either the Taranaki Basin or in the TVZ.

SHERWOOD LOLLAR et al. (2002) showed a negative trending slope for $\delta^{13}C(CH_4)$ vs. $\delta^2H(CH_4)$ of abiogenic gas from the Kidd Creek Mine in the Canadian Abitibi greenstone belt. $\delta^{13}C(CH_4)$ vs. $\delta^2H(CH_4)$ as shown in Fig. 10 has a positive slope for most fields and

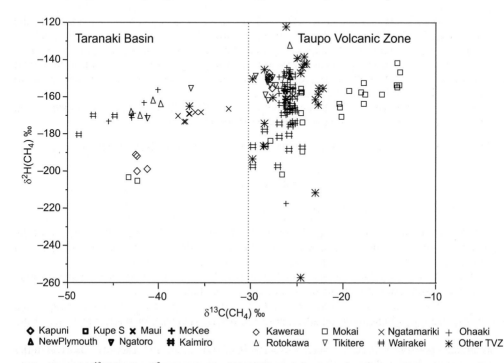

FIG. 10. $\delta^{13}C(CH_4)$ vs. $\delta^2H(CH_4)$ for Taranaki and TVZ. Apart from one TVZ sample the two groups fall either side of the $\delta^{13}C(CH_4) = -30‰$ line but have similar trends.

particularly for the Mokai field where the most positive $\delta^{13}C(CH_4)$ results occur. This is similar to that obtained for hydrocarbon natural gases (SCHOELL, 1983), providing additional evidence that the Mokai methane and other TVZ and Taranaki Basin methanes are not of abiogenic origin.

Are the TVZ and Taranaki basin source rocks related?

Geological and geochemical evidence (see "Taranaki natural gas fields" section) indicates that the Taranaki source rocks are primarily Cretaceous and Eocene terrestrial coal measures, although biomarkers sometimes reflect kerogen with a marine influence. Although petroleum seeps of terrigenous origin have been found in Pleistocene sediments in TVZ (CZOCHANSKA et al., 1986), these appear to be of relatively shallow origin and the most likely source rocks for the geothermal CH_4 and CO_2 are kerogen and carbonates from the Mesozoic greywackes of the Torlesse terrane (see "Taupo Volcanic Zone geothermal fields" section). The only carbon isotopic data available for these appears to be a measurement (HIRNER and LYON, 1989) from the Moawhanga tunnel south of Lake Taupo, which contained 0.8% kerogen with a $\delta^{13}C$ isotopic value of $-25.1‰$. WOLDEMICHAEL (1998) reported $\delta^{13}C = -5.0$ to $-15‰$ for calcite in greywackes from the Manaia Hill terrane, north-west of TVZ. A more recent measurement from the Torlesse terrane close to TVZ (S. WOLDEMICHAEL, personal communication) yielded $\delta^{13}C = -8.6‰$.

The most significant geochemical difference between the Taranaki and TVZ gases is that, with the exception of some New Plymouth gases, CH_4 predominates in Taranaki natural gases while CO_2 predominates in the geothermal gases of TVZ.

The unanswered questions

Despite the arguments presented in this paper against carbon isotopic equilibrium between CO_2 and CH_4, chemical equilibrium and an abiogenic methane source in TVZ and Taranaki, several questions still remain:

(1) Why does $\delta^{13}C(CH_4)$ (or $\Delta^{13}C(CO_2-CH_4)$) appear to be related to temperature in the geothermal wells of TVZ. Is it maturation of hydrocarbons or partial isotopic equilibrium?
(2) Why is $\delta^{13}C(CH_4)$ in the Mokai wells much more positive than in other parts of TVZ?
(3) Given the perturbations caused by steam–water separation in existing geothermal production wells, what is the best technique for accurately determining just how close to chemical equilibrium these geothermal wells are?
(4) Why do the majority of the Taranaki natural gas wells follow a hydrocarbon distribution line parallel to, but not on the Schulz–Flory distribution?

The answers to these questions are left to the next generation of isotope geochemists.

CONCLUSIONS

The main findings of this paper are as follows:

(1) Both the geothermal fields of the TVZ and the natural gas fields in the Western part of the Taranaki Basin show evidence of mantle fluid influence, particularly ^3He and CO_2 arising from the Pacific Plate subducting from the east.
(2) Within both these regions some areas have $\delta^{13}C(CH_4)$ isotope ratios that are more positive than normally encountered in geothermal and natural gas fields.
(3) In the natural gases from Maui, the wells with the most positive $\delta^{13}C(CH_4)$ values have lower $^3He/^4He$ ratios indicating the addition of radiogenic ^4He with increasing maturity of the methane.
(4) In the geothermal areas the more positive $\delta^{13}C(CH_4)$ values are generally observed in the highest temperature wells but the temperatures calculated on the assumption of isotopic equilibrium between CH_4 and CO_2 are much higher than the measured temperatures.

(5) In the Maui and New Plymouth fields, $CO_2/^3He$ values are close to mantle values and the $\delta^{13}C(CO_2)$ values indicate a mixture of organically sourced CO_2, marine carbonate CO_2 and mantle CO_2.

(6) In the TVZ, $CO_2/^3He$ exceeds the mantle ratio by factors of $10-100$, with the highest ratios corresponding to the areas (mainly close to the SE boundary of the TVZ) with the highest B/Cl ratios. $\delta^{13}C(CO_2)$ values of -7 to $-8\permil$ (possibly from calcite in the greywacke basement rocks) predominate in these high B/Cl areas, while in the lower B/Cl areas $\delta^{13}C(CO_2)$ values between -6 and $-3\permil$ predominate suggesting a mixture of mantle and marine carbonate sources.

(7) No evidence of abiogenic CH_4 was found in either Taranaki natural gases or in the geothermal areas of TVZ.

(8) Despite the above, a number of unanswered questions still remain.

Acknowledgements—I would like to pay tribute to two New Zealanders who have contributed in various ways to this paper. Firstly to my friend Ian Kaplan (a New Zealander from an early age) who has encouraged me throughout my scientific career and who was responsible for the extension of my geothermal research into the high helium-3 natural gas fields of Taranaki. Secondly to the late Werner Giggenbach (who came to New Zealand in the late 1960s) whose innovative thought on geothermal and natural gas fields provided stimulus for further investigations to test his interesting hypotheses. Advice and assistance from my colleagues at the Institute of Geological and Nuclear Sciences is gratefully acknowledged. Constructive reviews by Ron Hill, George Claypool and Ian Graham led to significant improvements in this manuscript and are greatly appreciated. I would also like to thank my wife Patricia for her support and encouragement over 42 years of isotopic research. This paper was funded in part by contract C05X0201 from the NZ Foundation for Research, Science and Technology.

REFERENCES

ARMSTRONG P. A., CHAPMAN D. S., FUNNELL R. H., ALLIS R. G. and KAMP P. J. J. (1996) Thermal modelling and hydrocarbon generation in an active-margin basin: Taranaki Basin, New Zealand. *AAPG Bull.* **80**, 1216–1241.

BERNDT M. E., ALLEN D. E. and SEYFRIED W. E. J. (1996) Reduction of CO_2 during serpentinization of olivine at 300°C and 500 bar. *Geology* **24**, 351–354.

BOTTINGA Y. (1969) Calculated fractionation factors for carbon and hydrogen isotope exchange in the system calcite–carbon dioxide–graphite–methane–hydrogen–water vapour. *Geochim. Cosmochim. Acta* **33**, 49–64.

CHRISTENSON B. W., MROCZEK E. K., KENNEDY B. M., VAN SOEST M. C., STEWART M. K. and LYON G. (2002) Ohaaki reservoir chemistry: characteristics of an arc-type hydrothermal system in the Taupo Volcanic Zone, New Zealand. *J. Volcan. Geotherm. Res.* **115**, 53–82.

COLE J. W. and LEWIS K. B. (1981) Evolution of the Taupo-Hikurangi subduction system. *Tectonophysics* **72**, 1–21.

CRAIG H. (1953) The geochemistry of the stable carbon isotopes. *Geochim. Cosmochim. Acta* **3**, 53–92.

CZOCHANSKA Z., SHEPPARD C. M., WESTON R. J., WOOLHOUSE A. D. and COOK R. A. (1986) Organic geochemistry of sediments in New Zealand. Part I. A biomarker study of the petroleum seepage at the geothermal region of Waiotapu. *Geochim. Cosmochim. Acta* **50**, 507–515.

FUNNELL R., CHAPMAN D., ALLIS R. and ARMSTRONG P. (1996) Thermal state of the Taranaki Basin, New Zealand. *J. Geophys. Res.* **101**(B11), 25197–25215.

GAMBLE J. A., WRIGHT I. C. and BAKER J. A. (1993) Seafloor geology and petrology in the oceanic to continental transition zone of the Kermadec–Havre–Taupo Volcanic Zone arc system, New Zealand. *NZ J. Geol. Geophys.* **36**, 417–435.

GIGGENBACH W. F. (1980) Geothermal gas equilibria. *Geochim. Cosmochim. Acta* **44**, 2021–2023.

GIGGENBACH W. F. (1982) Carbon-13 exchange between CO_2 and CH_4 under geothermal conditions. *Geochim. Cosmochim. Acta* **46**, 159–165.

GIGGENBACH W. F. (1995) Variations in the chemical and isotopic composition of fluids discharged from the Taupo volcanic zone, New Zealand. *J. Volcan. Geotherm. Res.* **68**, 89–116.

GIGGENBACH W. F. (1997) Relative importance of thermodynamic and kinetic processes in governing the chemical and isotopic composition of carbon gases in high-heat flow sedimentary basins. *Geochim. Cosmochim. Acta* **61**, 3763–3785.

HEDENQUIST J. W. (1990) The thermal and geochemical structure of the Broadlands-Ohaaki geothermal system, New Zealand. *Geothermics* **19**, 151–185.

HENLEY R. W. and MIDDENDORF K. I. (1985) Geothermometry in the recent exploration of Mokai and Rotokawa geothermal fields, New Zealand. *Trans. Geotherm. Res. Coun.* **9**(Part 1), 317–324.

HIRNER A. V. and LYON G. L. (1989) Stable isotope geochemistry of crude oils and possible source rocks from New Zealand: 1. Carbon. *Appl. Geochem.* **4**, 109–120.

HORITA J. (2001) Carbon isotope exchange in the system CO_2–CH_4 at elevated temperatures. *Geochim. Cosmochim. Acta* **65**, 1907–1919.

HULSTON J. R. (1976) Isotope work applied to geothermal systems at the Institute of Nuclear Sciences, New Zealand. *Geothermics* **5**, 89–96.

HULSTON J. R. (1986) Further isotope evidence of the origin of methane in geothermal area. *Proc. 5th Int. Symp. Water–Rock Interaction, Iceland*, 270–273.

HULSTON J. R. (1998) Correlations between B/Cl ratios and other chemical and isotopic components of Taupo Volcanic Zone, NZ geothermal fluids—evidence for water–rock interaction as a major source of boron and gas. In *Proc. 9th Int. Symp. Water–Rock Interaction* (eds. G. B. AREHART and J. R. HULSTON), pp. 629–632. Balkema, Rotterdam.

HULSTON J. R. and LUPTON J. E. (1996) Helium isotope studies of geothermal fields in the Taupo Volcanic Zone, New Zealand. *J. Volcan. Geotherm. Res.* **74**, 297–321.

HULSTON J. R. and LYON G. L. (1999) Isotopic evidence for the origin of methane from New Zealand Geothermal Systems. *Geologisches Jahrbuch. Reihe D: Mineralogie, Petrographie, Geochemie, Lagerstaettenkunde* **107**, 115–125.

HULSTON J. R. and McCABE W. J. (1962) Mass spectrometer measurements in the thermal areas of New Zealand. Part I—Carbon dioxide and residual gas analyses. *Geochim. Cosmochim. Acta* **26**, 383–398.

HULSTON J. R., HILTON D. R. and KAPLAN I. R. (2001) Helium and carbon isotope systematics of natural gases from Taranaki Basin, New Zealand. *Appl. Geochem.* **16**, 419–436.

KING P. R. and THRASHER G. P. (1996) *Cretaceous–Cenozoic Geology and Petroleum Systems of the Taranaki Basin, New Zealand*, Institute of Geological & Nuclear Sciences Monograph 13, 243 p., 6 enclosures. Lower Hutt, New Zealand.

LYON G. L. and HULSTON J. R. (1984) Carbon and hydrogen isotopic compositions of New Zealand geothermal gases. *Geochim. Cosmochim. Acta* **48**, 1161–1171.

LYON G. L., GIGGENBACH W. F. and SANO Y. (1996) Variations in the chemical and isotopic composition of Taranaki gases and their possible causes. *NZ Petrol. Conf.* **1**, 171–174.

MARTY B. and ZIMMERMANN L. (1999) Volatiles (He, C, N, Ar) in mid-ocean ridge basalts: assessment of shallow-level fractionation and characterization of source composition. *Geochim. Cosmochim. Acta* **63**, 3619–3633.

NISHIO Y., ISHII T., GAMO T. and SANO Y. (1999) Volatile element isotopic systematics of Rodrigues Triple Junction Indian Ocean MORB: implications for mantle heterogeneity. *Earth. Planet. Sci. Lett.* **170**, 241–253.

POREDA R. J., CRAIG H., ARNORSSON S. and WELHAN J. A. (1992) Helium isotopes in Icelandic geothermal systems: I. ^{3}He, gas chemistry, and ^{13}C relations. *Geochim. Cosmochim. Acta* **56**, 4221–4228.

RICHET P., BOTTINGA Y. and JAVOY M. (1977) A review of hydrogen, carbon, nitrogen, oxygen, sulphur and chlorine stable isotope fractionation among gaseous molecules. *Annu. Rev. Earth Planet. Sci.* **5**, 65–110.

SACKETT W. M. and CHUNG H. M. (1979) Experimental confirmation of the lack of carbon isotope exchange between methane and carbon oxides at high temperatures. *Geochim. Cosmochim. Acta* **43**, 273–276.

SCHOELL M. (1983) Genetic characteristics of natural gases. *AAPG Bull.* **67**, 2225–2238.

SHERWOOD LOLLAR B., BALLENTINE C. J. and O'NIONS R. K. (1997) The fate of mantle-derived carbon in a continental sedimentary basin: integration of C/He relationships and stable isotope signatures. *Geochim. Cosmochim. Acta* **61**, 2295–2307.

SHERWOOD LOLLAR B., WESTGATE T. G., WARD J. A., SLATER G. D. and LACRAMPE-COULOUME G. (2002) Abiogenic formation of alkanes in the Earth's crust as a minor source for global hydrocarbon reservoirs. *Nature* **416**, 522–524.

UREY H. C. (1947) The thermodynamic properties of isotopic substances. *J. Am. Chem. Soc.* **57**, 562–581.

WOLDEMICHAEL S. (1998) Stable isotope studies of calcite from very-low grade metamorphic Greywacke Terranes of the North Island, New Zealand. In *Proc. 9th Int. Symp. Water–Rock Interaction* (eds. G. B. AREHART and J. R. HULSTON), pp. 429–432. Balkema, Rotterdam.

Geochemical Investigations in Earth and Space Science: A Tribute to Isaac R. Kaplan
© The Geochemical Society, Publication No. 9, 2004
Editors: R.J. Hill, J. Leventhal, Z. Aizenshtat, M.J. Baedecker, G. Claypool,
R. Eganhouse, M. Goldhaber and K. Peters

Insights on the origin of pristane and phytane in sediments and oils from laboratory heating experiments

RYOSHI ISHIWATARI[1,*] and MARIKO ISHIWATARI[2,*]

[1]Department of Chemistry, School of Science, Graduate School of Tokyo Metropolitan University,
Hachioji, Tokyo 192-0397, Japan
[2]School of Engineering, Engineering Research Institute, Graduate School of Tokyo University, Yayoi,
Bunkyo-ku, Tokyo 113-8656, Japan

Abstract—We investigated a formation pathway of pristane (Pr) and phytane (Ph) in crude oils and ancient sediments from the phytyl side chain of chlorophyll and degradation of Pr and Ph during thermal maturation by laboratory heating experiments of pure dihydrophytol (DHP), Pr and Ph, followed by measurement of carbon isotopic compositions. In the first heating experiment, dihydrophytol (DHP) was pyrolyzed in a closed system at 320–350°C for 1–12 h. DHP yields large amounts of Pr and prist-1-ene with lesser amounts of Ph and phytenes. The Pr/Ph ratio which was obtained after hydrogenation of prist-1-ene and phytenes increases from 0.7 to 21 with increasing thermal degradation of DHP. We conclude that DHP originating from chlorophyll pigments can be an important precursor of Pr under geological conditions. Pr and Ph produced from DHP are slightly depleted in ^{13}C compared to the original DHP and $\delta^{13}C$ of Pr is similar to that of Ph. This result supports the idea of a common source for both Pr and Ph, primarily chlorophyll pigments.

In the second heating experiment which was intended to evaluate the role of thermal stress on the fate of Pr and Ph in petroleum, Pr and Ph were heated at 380 or 420°C under vacuum for 5–800 h, and the shift in their carbon isotopic composition was studied as a function of thermal decomposition. The decomposition of Pr and Ph follows apparently first-order kinetics. Pr/Ph ratio does not change significantly during thermal decomposition of 90% of the initial amount of these isoprenoids. The fractionation factors for both Pr and Ph are practically identical: $\alpha = 0.9994 \pm 0.0002$ ($n = 23$). The effect of temperature on the fractionation factor is negligible. Although Pr and Ph are enriched in ^{13}C with increasing thermal decomposition, the magnitude of the enrichment is small and 90% of Pr and Ph must be thermally decomposed to cause a $1.4 \pm 0.4‰$ increase in $\delta^{13}C$ values. The results also indicate that Pr/Ph ratio does not change significantly during thermal decomposition. Therefore, we conclude that small changes in Pr/Ph ratios and in $\delta^{13}C$ of Pr and Ph with increasing thermal maturity may be a notable feature of occurrence of thermocracking of Pr and Ph in crude oils.

INTRODUCTION

PRISTANE (Pr: 2,6,10,14-tetramethylpentadecane) and phytane (Ph: 2,6,10,14-tetramethyl-hexadecane) are common biomarkers present in most petroleum and geological samples. The primary source of these isoprenoids is considered to be the phytyl side chain of chlorophyll pigments in photosynthetic organisms (PETERS and MOLDOWAN, 1993; LI et al., 1995; HUNT, 1996), although other sources (e.g. Archaebacteria) have been proposed (CHAPPE et al., 1980; ILLICH, 1983; GOOSSENS et al., 1984; ROWLAND, 1990). Possible pathways for the formation of these isoprenoids from chlorophylls were compiled by DIDYK et al. (1978).

The origin of Pr in petroleum, however, has been controversial for a long time. LARTER et al. (1983) suggested that Pr is bound to kerogens through non-hydrolyzable C–C and/or C–O bonds. GOOSSENS et al. (1984) claimed that tocopherols can be precursors of Pr in petroleum

---

Present address: Geotec Inc., 3-16-11 Takaido-nishi, Suginami-ku, Tokyo 168-0071, Japan.

because α-tocopherol produced abundant prist-1-ene, a probable precursor of Pr, as a major product during flash pyrolysis or thermal degradation.

It is well known that chlorophylls and phytol yield phytenes and phytadienes as major products, but pristenes are minor products generated by flash pyrolysis or other heating (VAN DE MEENT et al., 1980; ISHIWATARI et al., 1991, 1997, 1999). The extremely low yield of prist-1-ene from chlorophylls and phytol was raised by GOOSSENS et al. (1984) as one of the reasons for the unlikely origin of Pr from these materials. On the other hand, catagenetic decomposition of chlorophylls condensed with phenolic compounds yields abundant prist-1-ene (LI et al., 1995). This work indicates that polymeric structures containing chlorophyll pigments can be important precursors for Pr. However, the pathways for Pr formation from chlorophylls are still poorly understood.

In previous studies, we examined a pathway for Pr formation from chlorophylls (ISHIWATARI et al., 1991). Chlorophyll a was heated at 250°C over times of 3 min to 24 h, and then pyrolyzed at 470°C. The results showed that the ratio of prist-1-ene to the sum of phytadienes and phytenes increases from 0.01 to 0.1 with increasing time of preheating, although the ratio was extremely low. This experimental result suggested a transformation of chlorophyll into a prist-1-ene producing material upon preheating. DHP is another possible precursor for Pr. Although DHP is an important biological breakdown product of chlorophylls, it has not been considered as a direct precursor of Pr. For example, DHP was thought to change into Ph via phytene(s) and Pr via phytanic acid during diagenesis (DIDYK et al., 1978).

HAYES et al. (1990) studied carbon isotopic compositions of Pr and Ph in Greenhorn Formation sediments (Cenomanian–Turonian). They reported that the $\delta^{13}C$ values of Pr and Ph in the sediment samples are similar, and the 4.5‰ depletion of $\delta^{13}C$ of Pr and Ph relative to porphyrins within the range for living organisms. These results suggest that Pr and Ph in these sediments originate from a common biological precursor, chlorophyll (HAYES et al., 1990). LI et al. (1995) compiled data for a large number of crude oils worldwide. Although Pr and Ph showed a wide range of $\delta^{13}C$ (−34 to −24‰), the $\delta^{13}C$ values for Pr and Ph in each oil were similar. Similar relationships are recognized for oils from northern Japan (ISHIWATARI et al., 2001, unpublished result). $\delta^{13}C$ values for both Pr and Ph in each Japanese oils are similar, but the $\delta^{13}C$ values of all samples scatter in a wide range from −25.4 to −19.4‰. These observations suggest a common source for both Pr and Ph, primarily chlorophyll pigments.

The first objective of this work was to study the formation of Pr and Ph with laboratory pyrolysis of DHP. In a previous paper, we reported pyrolytic formation of C_{19} isoprenoid hydrocarbons from DHP (ISHIWATARI et al., 2000). This work is a continuation of our previous study. We determined carbon isotopic compositions of Pr and Ph generated from DHP in order to gain understanding of the pathway for generation of these isoprenoids from chlorophyll. As a second objective, we examined changes in $\delta^{13}C$ of Pr and Ph as a function of their thermal decomposition by laboratory heating experiment using authentic standards.

Carbon isotopic compositions of Pr and Ph in petroleum are usually controlled by two factors: (1) $\delta^{13}C$ values of primary producers, and (2) post-depositional modification (probably largely due to thermal decomposition). It is important to establish a method for differentiating these two factors. One approach is to estimate quantitatively the shift in $\delta^{13}C$ of Pr and Ph with the degree of thermal decomposition. BJORØY et al. (1992) conducted hydrous pyrolysis experiments of different types of source rocks to examine the effect of thermal decomposition on carbon isotopic compositions of various types of hydrocarbons including Pr and Ph (Paris Basin shale with type II kerogen, Green River shale with type I kerogen, Jurassic coal with type III kerogen and a marine limestone with type II kerogen). Except for the limestone sample, the Pr and Ph become enriched in ^{13}C (1.4–3.4‰) at high pyrolysis temperatures (330–360°C). They claimed that the positive shift in $\delta^{13}C$ for Pr and Ph is due to preferred cleavage of $^{12}C-^{12}C$ bonds at elevated temperatures. However, no enrichment in ^{13}C is observed for the limestone sample, rather Pr and Ph become depleted in ^{13}C by 1.1–1.3‰. Therefore, the decrease in $\delta^{13}C$ of Pr and Ph for the limestone sample cannot be explained by thermal decomposition, and other factors, such as addition of Pr and Ph released from their precursors of which $\delta^{13}C$ values are different at the time of biosynthesis (cf. KOOPMANS et al., 1999), must be considered.

EXPERIMENTAL

Heating of dihydrophytol

DHP was prepared by hydrogenation of standard phytol (Aldrich Chem.). Phytol in hexane was hydrogenated by bubbling H_2 gas with PtO_2 for 10 min. The hexane solution was evaporated, the residue dissolved in benzene/methanol (3:1), and then, PtO_2 removed by filtration. The benzene/methanol was evaporated from the filtrate and the remaining residue, dissolved in hexane, which was passed through a silica-gel column with hexane and methanol, successively. DHP was obtained from the methanol fraction and its identification confirmed by GC–MS analysis. For the heating experiment, about 2–10 mg DHP was sealed under vacuum in a Pyrex tube (4 mm i.d. × 100 mm in length), and heated at 320–350°C for 1–12 h.

Heating of Pr and Ph

Pr (98%) and Ph (98.9%) were obtained from Aldrich Chem. and Chem. Service, respectively, without further purification. For the heating experiment, about 1–100 mg Pr, Ph, or a mixture of Pr and Ph (1 mg, each) was sealed under vacuum in a Pyrex tube (9 mm i.d. × 120 mm in length), and heated at 380–420°C for times ranging from 1 to 800 h.

Instrumentation

A Shimadzu GC 9A equipped with a flame ionization detection (FID) and a 0.25 μm DB-5 coated fused silica capillary column (30 m × 0.25 mm i.d., J&W Scientific) was used to identify and quantify the products. The GC oven temperature was programmed from 100 to 250°C at 5°C/min. The C_{21} n-alkane was used as an internal standard.

The $\delta^{13}C$ values of individual compounds were measured on an HP Model 5890 gas chromatograph equipped with a DB-5 coated fused silica capillary column (30 m × 0.32 mm i.d., 0.25 μm film thickness, J&W Scientific) connected to a Finnigan MAT Delta-S mass spectrometer via combustion interface (CuO, NiO and Pt wires) maintained at 940°C. The carrier gas was helium and column injection was applied. The GC oven temperature was programmed as follows: injection at 50°C, 30°C/min to 120°C, 5°C/min to 310°C, isothermal for 15 min. Co-injected deuterated n-alkanes $C_{16}D_{34}$ and $C_{24}D_{50}$ were employed as isotopic standards. Analyses were generally done in duplicate or triplicate with standard deviations (± 1 s) better than 0.4‰. Carbon isotopic ratios are expressed as $\delta^{13}C$ relative to the PDB standard.

RESULTS AND DISCUSSION

Formation of Pr and Ph by pyrolysis of dihydrophytol

When heated at 320–350°C for 1–12 h, DHP decomposes to produce C_{19} isoprenoids (Pr and prist-1-ene) as major products, and C_{20} isoprenoids (Ph and phytenes) as minor products. As shown in Fig. 1A, the yield for C_{19} isoprenoids (expressed as % of the initial DHP on mole to mole bases) is generally higher than that for C_{20} isoprenoids, and it increases with progressive decomposition of DHP. The yield of C_{20} isoprenoids is <2% of the initial DHP for all heating experiments. The maximum yield of C_{19} isoprenoids (27% of the initial DHP) was observed when 82% of the original DHP was degraded. The C_{19} isoprenoid/C_{20} isoprenoid ratio ranges from 0.7 to 21 and increases with progressive degradation of DHP (Fig. 1B). This result indicates that DHP can be an important precursor of Pr. To our knowledge, the high ratios of C_{19} isoprenoids to C_{20} isoprenoids from DHP have not been reported previously.

The DHP-pyrolysis products (C_{19} isoprenoids and C_{20} isoprenoids) were hydrogenated to form Pr and Ph by bubbling with H_2 gas and PtO_2 in hexane. The resulting Pr and Ph were measured for their carbon isotopic composition. Figure 2 shows selected gas chromatograms of the generated Pr and Ph with DHP. Pr and Ph produced from DHP are slightly depleted in ^{13}C compared to the original DHP. $\delta^{13}C$ values of Pr obtained by reduction of C_{19} isoprenoids are

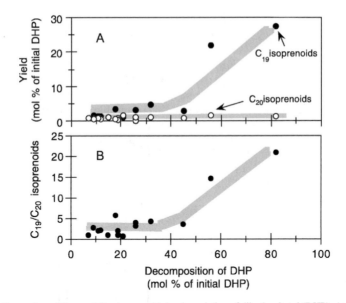

FIG. 1. Generation of C_{19}- and C_{20}-isoprenoids by degradation of dihydrophytol (DHP). A: Changes in yield of C_{19}- and C_{20}-isoprenoids; B: Changes in the ratio of C_{19}-isoprenoids to C_{20}-isoprenoids. DHP was heated at 320–330°C for 1–12 h.

almost identical to those of Ph from C_{20} isoprenoids (Fig. 2 and Table 1). This result agrees with the observations reported for many crude oils and ancient sediments, where $\delta^{13}C$ values of Pr and Ph are similar (HAYES et al., 1990; LI et al., 1995). This result supports an idea of a common source for both Pr and Ph, primarily chlorophyll pigments.

Geochemical implications–We hypothesize that during burial heating phytol forms Pr via DHP. This pathway results in the formation of isoprenoids with Pr/Ph ratios larger than one, which are often observed in petroleum. The following steps are proposed: (1) formation of DHP from chlorophyll-derived phytol by biological reactions under oxic/anoxic conditions, (2) formation of Pr/pristenes and phytenes/phytadienes by thermal decomposition of DHP or DHP-derived moieties in kerogen and (3) formation of Pr and Ph by reduction under geological conditions.

The following fact is favor of the step 1 of the above hypothesis. A significant amount of DHP occurs in common sediments. PRAHL et al. (1984) showed conversion of phytol to DHP during passage of green alga through the guts of copepods. SUN et al. (1998) reported that benthic organisms grazing on planktonic detritus produce significant quantities of DHP under oxic conditions. Formation of DHP under oxygenated conditions deserves attention in relation to the use of the Pr/Ph ratio as a paleoenvironmental indicator. A high Pr/Ph ratio is generally associated with oxic sedimentary conditions of paleoenvironments (e.g. DIDYK et al., 1978; PETERS and MOLDOWAN, 1993). This fact could be reconciled with our hypothesis that DHP or DHP-derived materials are a precursor of Pr. The formation of DHP by microbial reduction of phytol in young sediments also has been suggested (e.g. BROOKS and MAXWELL, 1974; VAN FLEET and QUINN, 1979). However, GROSSI et al. (1998) did not observe the formation of DHP in an incubation experiment of free phytol in anoxic sediment slurries. These facts indicate that the microbial formation of DHP from phytol in sediments may take place under certain conditions. Further study is needed on factors regulating DHP formation under oxic/anoxic depositional conditions.

It is interesting to note that a considerable fraction of total DHP and phytol is present in bound form (SUN et al., 1998). This fraction becomes extractable only after saponification with KOH/MeOH in sediments. Since there is a possibility that this fraction could be incorporated into kerogen during diagenesis, DHP in bound form may be an important intermediate to isoprenoid-derived moieties in kerogen.

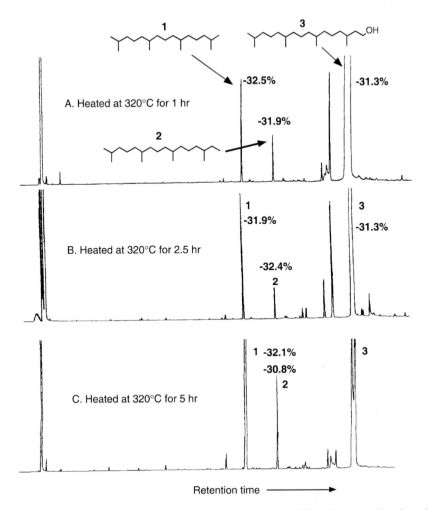

FIG. 2. Gas chromatograms of pyrolyzates (after reduction) of DHP with their carbon isotopic compositions. 1: pristane, 2: phytane and 3: DHP.

The results of this study provide an alternative explanation for the formation of Pr and Ph during hydrous pyrolysis of rock samples reported by KOOPMANS *et al.* (1999). According to KOOPMANS *et al.* (1999), both the yield of these isoprenoids and the Pr/Ph ratio increase with increasing temperature of heating during hydrous pyrolysis of rock samples from the Gessoso-solfifera, Ghareb, and Green River Formations. They explained that the increase in the Pr/Ph ratio is caused by the presence of two types of isoprenoid precursors, *i.e.* Pr precursor and Ph precursor, and the former is more abundant than the latter in kerogen, rather than different timing of Pr and Ph generation. However, their result can also be explained by the decomposition of a single precursor, such as DHP-derived moieties, in rock samples, and does not necessarily need two types of precursors. The curve obtained by KOOPMANS *et al.* (1999) showing the increase of Pr/Ph ratios with increasing heating temperature closely resembles the curve that we obtained for the C_{19} isoprenoids/C_{20} isoprenoids and the extent of DHP degradation (Fig. 1).

Changes in the isotopic composition of Pr and Ph during thermal decomposition

Decomposition of Pr and Ph during heating–During thermal decomposition of Pr at 380–420°C, low-molecular-weight isoprenoids ($C_{18:1}$, $C_{16:1}$, $C_{15:1}$, $C_{15:0}$, $C_{14:1}$, $C_{14:0}$, $C_{13:1}$ and

Table 1. Composition and isotopic data for C_{19} and C_{20} isoprenoids generated from thermal degradation of dihydrophytol (DHP)

| Sample No. | Heating | | DHP remaining[a] | Isoprenoids produced[b] | | | $\delta^{13}C$ of isoprenoid after reduction (‰) | | |
	Temperature (°C)	Time (h)		C_{19} (mol/mol)	C_{20} (mol/mol)	C_{19}/C_{20}	Pr	Ph	DHP
Initial DHP (unheated)									
9202	320	1	0.91	1.7	0.6	2.8	-32.5 ± 0.2	-31.9 ± 0.0	-31.2
8D25	320	2.5	0.55	2.9	0.8	3.6	-31.9 ± 0.2	-32.1 ± 0.9	-31.3
9203	320	5	0.18	27.3	1.3	21.0	-32.1 ± 0.1	-30.8 ± 0.0	$-$

[a] Fraction of the initial amount of DHP.
[b] Ratio of isoprenoids produced to the initial amount of DHP.

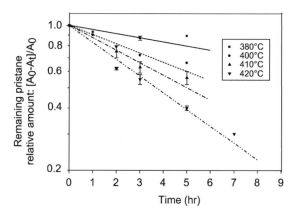

FIG. 3. Decomposition of pristane as a function of time of heating at various temperatures. Vertical bars = standard error ($\pm\sigma$).

$C_{13:0}$; $C_{n:m}$: n indicates carbon number in a molecule and m indicates number of carbon–carbon double bond) were produced. The abundance of these low-molecular-weight isoprenoids was very low. These results are similar to those reported by KISSIN (1993), where the thermal degradation of Pr produces C_{10}–C_{16} isoprenoids. Therefore, the mechanism of thermocracking of Pr and Ph may be radical chain reactions in similar to that proposed by KISSIN (1993), *i.e.* (1) formation of Pr or Ph radial, (2) fission of the radial into an olefin molecule and a smaller alkyl radicals and (3) the chain transfer reactions.

The decomposition reaction of Pr approximates first-order kinetics. Figure 3 is a plot of $\log[(A_0 - A_t)/A_0]$ versus heating time of early 7 h at 380–420°C. A_0 represents the amount of unheated Pr, and $(A_0 - A_t)$ is the amount of Pr at heating time t. A good straight-line fit was obtained from the linear regression analysis. Figure 4 shows the Arrhenius plot of rate constants for Pr decomposition. The slope of the line yields the Pr decomposition activation energy: $E_a = 127$ kJ/mol and $A = 2.0 \times 10^5$ s^{-1}.

Changes in the relative remaining for Pr and Ph during their decomposition were examined for the experiments where both Pr and Ph were heated in the same glass tube or under the same heating condition (temperature and heating period). The results indicate that a Pr/Ph ratio changes little during their decomposition, as shown in Table 2. These results imply that thermal decomposition of Pr and Ph once generated from their precursor does not play an important role for changes in the Pr/Ph ratios of petroleum during thermal maturation, which are generally observed (*e.g.* quoted in PETERS and MOLDOWAN (1993)).

Isotopic behavior during thermal decomposition–Pr and Ph become isotopically heavier with progressive thermal decomposition at 380 and 420°C. Variations of $\delta^{13}C$ for Pr and Ph during thermal decomposition in a closed system are assumed to follow a Raleigh-type fractionation,

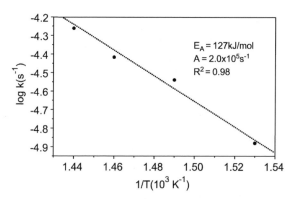

FIG. 4. Arrhenius plot of rate constants of pristane decomposition.

Table 2. Changes in the remaining amount of Pr relative to that of Ph during their thermal decomposition

| Run No. | Heating | | Remaining isoprenoid $(f)^a$ | | |
	Temperature (°C)	Time (h)	Pr $(f_{Pr})^a$	Ph $(f_{Ph})^a$	$f_{Pr}{}^a/f_{Ph}{}^a$
1	380	360	0.20	0.21	1.1 ± 0.2
2	380	360	0.27	–	
3	380	804	0.05	0.06	1.3 ± 0.4
4	380	804	0.11	–	
5	420	12	0.40	0.45	0.9
6	420	24	0.30	0.32	
7	420	24	0.34	–	1.0 ± 0.1
8	420	24	–	0.33	
9	420	48	0.06	0.07	
10	420	48	0.11	–	1.1 ± 0.3
11	420	48	–	0.09	

^{a} Fraction of the initial amount.

and are expressed by the following equation:

$$\delta^{13}C_R - \delta^{13}C_{R0} = 1000(\alpha - 1)\ln f = \Delta \ln f, \qquad \alpha = 1 + \Delta/1000$$

where $\delta^{13}C_R$ and $\delta^{13}C_{R0}$ represent $\delta^{13}C$ values of the substrate (Pr or Ph) remaining after heating and those of the unheated substrate, respectively; α is the isotopic fractionation factor for Pr or Ph; and f represents the ratio of the weight of the substrate (Pr and Ph) remaining after heating to the initial weight of the substrate.

Figure 5 shows a good correlation ($R^2 = 0.95$) between the magnitude of the isotopic shift and the fractions of remaining Pr and Ph after heating at 380 and 420°C. The results in Table 3 indicate that the fractionation factors for both Pr and Ph are practically identical. No temperature effect on the fractionation factor was observed within the range of heating temperatures (380–420°C). Therefore, average Δ and α values from a total of 23 data points for Pr and Ph are calculated to be $-0.61 \pm 0.16‰$, and $\alpha = 0.9994 \pm 0.0002$.

Geochemical implications–The results of this study predict that 90, 99 and 99.9% of the initial amount of Pr or Ph in the geological samples must be thermally decomposed to obtain positive shifts in $\delta^{13}C$ of 1.4 (± 0.4), 2.8 and 4.2‰, respectively. As already described, BJORØY *et al.* (1992) observed maximum positive carbon isotope shifts of 1.4–3.4‰ for Pr and Ph in their laboratory hydrous pyrolysis of sedimentary rocks. We calculated the extent of thermal

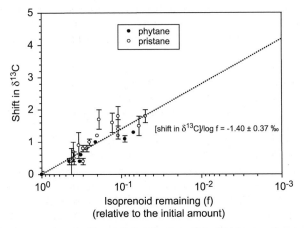

FIG. 5. Relation between isoprenoids remaining (fraction of the initial amount) and positive shift in $\delta^{13}C$ of isoprenoids. Open and closed circles indicate pristane and phytane, respectively. Vertical bars = standard error ($\pm \sigma$).

Table 3. Carbon isotope fractionation during thermal decomposition of Pr or Ph

Sample No.	Heating Temperature (°C)	Time (h)	Remaining isoprenoid $(f)^a$	$\delta^{13}C$ (‰ ± S.D.)	Δ^b
Phytane	–	0	1.00	− 28.9 ± 0.1	
Ph-4-1	380	360	0.21	− 27.9 ± 0.0	− 0.64
Ph-4-2	380	804	0.06	− 27.4 ± 0.0	− 0.53
				For 380°C Ph	− 0.59 ± 0.04
Ph-3-1	420	12	0.45	− 28.5 ± 0.0	− 0.50
Ph-3-5	420	24	0.33	− 28.5 ± 0.0	− 0.36
Ph-3-4	420	24	0.32	− 28.3 ± 0.0	− 0.53
Ph-3-8	420	48	0.09	− 27.8 ± 0.1	− 0.46
Ph-3-7	420	48	0.07	− 27.6 ± 0.0	− 0.49
				For 420°C Ph	− 0.47 ± 0.06
				For Ph total	− 0.50 ± 0.08
Pristane	–	0	1.00	− 26.8 ± 0.3	
Ph-4-1	380	360	0.20	− 25.6 ± 0.0	− 0.75
Ph-4-5	380	360	0.27	− 26.0 ± 0.1	− 0.61
Ph-4-2	380	804	0.05	− 25.0 ± 0.2	− 0.60
Ph-4-6	380	804	0.11	− 25.4 ± 0.3	− 0.63
				For 380°C Pr	− 0.65 ± 0.07
Pr-25	420	5	0.39	− 26.3 ± 0.2	− 0.53
Pr-29	420	7	0.30	− 26.4 ± 0.1	− 0.33
Pr-46	420	9	0.19	− 25.1 ± 0.3	− 1.02
Pr-47	420	11	0.13	− 25.2 ± 0.3	− 0.78
Ph-3-1	420	12	0.40	− 26.1 ± —	− 0.76
Pr-48	420	13	0.11	− 25.0 ± 0.3	− 0.82
Ph-3-4	420	24	0.30	− 26.0 ± 0.1	− 0.66
Ph-3-6	420	24	0.34	− 25.9 ± 0.4	− 0.83
Ph-3-7	420	48	0.06	− 25.3 ± 0.3	− 0.53
Ph-3-9	420	48	0.11	− 25.6 ± 0.0	− 0.54
				For 420°C Pr	− 0.68 ± 0.20
				For Pr total	− 0.67 ± 0.17
				For total isoprenoids	− 0.62 ± 0.17

^{a} Fraction of initial amount.
$^b \Delta = (\delta^{13}C_R - \delta^{13}C_{R0})/\ln f$.

decomposition of isoprenoids from their $\delta^{13}C$ shift assuming that the enrichment in ^{13}C for Pr and Ph is caused only by thermocracking of isoprenoids. The calculation indicates that the positive shift of 1.4–3.4‰ corresponds to thermal decomposition of 90–99.6% of Pr and Ph in the unheated samples. If such an extensive thermal decomposition of Pr and Ph is unlikely to occur, $\delta^{13}C$ of Pr and Ph must have been controlled by a mechanism other than thermocracking: e.g. $\delta^{13}C$ values of Pr and Ph reflect those of their precursors having different $\delta^{13}C$ at the time of biosynthesis.

Another case deserving of discussion is an apparent large increase in $\delta^{13}C$ for Pr or Ph with progressive oil maturity for a group of the genetically related oils from the Akita–Yamagata and the Niigata Basins in the northeast Japan (ISHIWATARI et al., 2001). These oils are considered to be genetically related. The oils from both the basins are derived from Middle Miocene marine source rocks and these oils give similar $\delta^{13}C$ values, i.e. − 22.3 ± 0.5‰ ($n = 52$) for the Akita–Yamagata Basin vs. − 22.2 ± 0.7 ($n = 53$) for the Niigata Basin, indicating that $\delta^{13}C$ of source kerogens are similar (WASEDA, 1992). However, the source materials of both the oils are not completely identical because biomarker studies indicate that oils from the Akita–Yamagata Basin show lower terrestrial input and generally lower maturity than those from the Niigata Basin (SAKATA et al., 1990; CHAKHMAKHCHEV et al., 1996). WASEDA (1993) studied carbon and hydrogen isotopic composition of total oils including condensates, saturate and aromatic fractions from both the basins. He found that isotopic compositions of total oils and saturate hydrocarbons become heavier with increasing maturity and concluded that these results can be explained by kinetic isotope effects of thermocracking of oils. We analyzed carbon isotopic

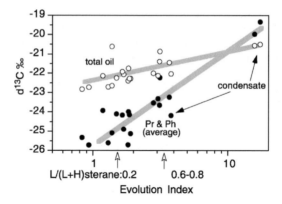

FIG. 6. Carbon isotope compositions of total oils, Pr and Ph versus thermal maturity (Evolution Index) for oils from northeast Japan (ISHIWATARI et al., 2001). Data for total oils are quoted from WASEDA (1993). Data for biodegraded oils are omitted. Definition of Evolution Index is a ratio of n-heptane to trans-1,3-dimethyl-cyclopentane (HIRATSUKA, 1976). A thermal maturity index of $L/(H+L)$-sterane (definition is given in the text) is quoted from SAKATA et al. (1990) for comparison. Open circles: total oils; closed circles: Pr and Ph (average values).

composition of individual n-alkanes and isoprenoids from the same oil samples that WASEDA (1993) used and reported the results for n-alkanes (ISHIWATARI et al., 1999). Oil maturity was expressed by Evolution Index (EI), which is defined as a ratio of n-heptane to trans-1,3-dimethyl cyclopentane (HIRATSUKA, 1976). Although the EI is not commonly used in organic geochemistry, it is confirmed by us to roughly correlate with a steroid maturity index of $-L/(L+H)$, which is defined as a ratio of low-molecular-weight (e.g. $C_{21} + C_{22}$) steroid hydrocarbons (e.g. steranes, triaromatic steroids) to low-plus-high-molecular weight (e.g. C_{21}, C_{22}, C_{27}, C_{28} and C_{29}) steroid hydrocarbons (LEWAN et al., 1986; SAKATA et al., 1990; WINGERT and POMERANTZ, 1986). The steroid $L/(L+H)$ ratio was proposed as a maturity index for highly mature oils (SUZUKI et al., 1987).

As shown in Fig. 6, $\delta^{13}C$ for Pr and Ph from most oils with EI of <4 fall within a range of approximately 3‰ (-26 to -23‰), and $\delta^{13}C$ for Pr and Ph appears to increase by 1–2‰ with progressive oil maturity. Furthermore, Pr and Ph from two condensates are isotopically by 3–4‰ heavier than those in oils with EI of 3–4. It seems unlikely that positive $\delta^{13}C$ shifts by totally 5–6‰ for these oils and condensates could be caused only by thermocracking of free Pr and Ph, because 99.99–99.999% of the initial isoprenoids should have been thermally decomposed. Indeed, there is evidence for considerable thermal decomposition of isoprenoids for these oils and condensates. Ph/n-C_{18} alkane ratio decreases with progressive maturity from 1.0 for the oils of lower maturity (EI = 1) to 0.1 for condensates (EI = 17) as shown in Fig. 7. This decreasing trend in Ph/n-C_{18} alkane ratio is in accordance with a well-known fact that ratios of Pr/n-C_{17}

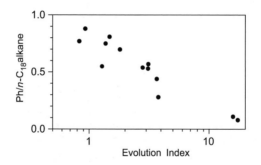

FIG. 7. Decrease in Ph/n-C_{18} alkane ratio with thermal maturity (Evolution Index) for oils from northeast Japan (ISHIWATARI et al., 2001). Definition of Evolution Index is the same as in Fig. 6.

alkane or Ph/n-C$_{18}$ alkane decrease with thermal stress, since isoprenoids are thermally less stable than n-alkanes (TISSOT et al., 1971; ISHIWATARI et al., 1977). The decrease in Ph/n-C$_{18}$ alkane ratio from 1.0 to 0.1 suggests that at least 90% of Ph in the original oils was decomposed during transition from oils to condensates. Our thermal decomposition experiment indicates that the 90% decomposition of Pr and Ph causes only a 1.4‰ positive shift in their δ^{13}C, as described already. Therefore, the 5–6‰ positive shift in δ^{13}C of Pr and Ph for the Japanese oils and condensates can not be explained only by thermocracking.

As a possible mechanism other than thermocracking, the following scenario might be presented, although an effect of δ^{13}C at the time of photosynthesis on Pr and Ph to some extent might not be completely ruled out: Pr and Ph precursors are present as high-molecular-weight fractions (probably as asphaltene) in oils. They release Pr and Ph during transition from oils to condensates under thermal stress. Pr and Ph with heavy carbon isotope compositions must be produced from their precursors during last stages of transition to condensates.

CONCLUSIONS

(1) In laboratory heating experiments, DHP yields substantial amounts of C$_{19}$ isoprenoids (Pr and prist-1-ene) with lesser amounts of C$_{20}$ isoprenoids (Ph and phytenes). The Pr/Ph ratio, which was obtained after hydrogenation of prist-1-ene and phytenes, increases from 0.7 to 21 with increasing thermal degradation of DHP. This result implies that DHP is an important precursor of Pr under geological conditions. Pr and Ph produced from DHP are slightly depleted in ^{13}C compared to the original DHP. There is no difference in δ^{13}C between Pr and Ph. We conclude that DHP originating from chlorophyll can be a precursor for Pr in geological samples based on our experimental results, their occurrence in zooplankton and sediments, and possible incorporation into kerogen during diagenesis.

(2) The decomposition of Pr and Ph follows apparently first-order kinetics. Activation energy for Pr decomposition: E_a = 127 kJ/mol and A = 2.0 × 10^5 s^{-1} (R^2 = 0.98). Pr/Ph ratio does not change significantly during decomposition of 90% of the initial amount of these isoprenoids. The carbon isotope fractionation factors for Pr and Ph are nearly identical: α = 0.9994 ± 0.0002 (n = 23). The temperature effect on the fractionation factor is negligible. Pr and Ph are enriched in ^{13}C with increasing thermal decomposition, indicating that 90% of Pr and Ph must be thermally decomposed to cause a 1.4 ± 0.4‰ increase in δ^{13}C values. The heating experiments of Pr and Ph revealed little effect on δ^{13}C of Pr and Ph until the last stages of decomposition. Therefore, we conclude that small changes in Pr/Ph ratios and in δ^{13}C of Pr and Ph with increasing thermal maturity may be a notable feature of occurrence of thermocracking of Pr and Ph in crude oils.

Acknowledgements—Drs Keita Yamada, David Brincat and Hiroshi Naraoka are gratefully acknowledged for their support during this research and for measuring carbon isotopic compositions of isoprenoids using GC-IRMS. Prof. Hitoshi Sakai is acknowledged for his comments for improving the manuscript. We also appreciate Drs Ken Peters and John Guthrie for reviewing the paper and for their critical and constructive comments which have substantially improved the initial manuscript. This work was partly supported by the Japan National Oil Corporation (JNOC).

REFERENCES

BJORØY M., HALL P. B., HUSTAD E. and WILLIAMS J. A. (1992) Variation in stable carbon isotope ratios of individual hydrocarbons as a function of artificial maturity. *Org. Geochem.* **19**(1–3), 89–105.

BROOKS P. W. and MAXWELL J. R. (1974) Early stage fate of phytol in a recently-deposited lacustrine sediment. In *Advances in Organic Geochemistry 1973* (eds. B. TISSOT and F. BIENNER), pp. 978–991. Editions Technip, Paris.

CHAKHMAKHCHEV A., SUZUKI N., SUZUKI M. and TAKAYAMA K. (1996) Biomarker distributions in oils from the Akita and Niigata Basins, Japan. *Chem. Geol.* **133**, 1–14.

CHAPPE B., MICHAELIS W. and ALBRECHT P. (1980) Molecular fossils of Archaebacteria as selective degradation products of kerogen. In *Advances in Organic Geochemistry 1979* (eds. A. G. DOUGLAS and J. R. MAXWELL), pp. 265–274. Pergamon Press, Oxford.

DIDYK B. M., SIMONEIT B. R. T., BRASSELL S. C. and EGLINTON G. (1978) Organic geochemical indicators of palaeoenvironmental conditions of sedimentation. *Nature* **272**, 216–222.

GOOSSENS H., DE LEEUW J. W., SCHENCK P. A. and BRASSELL S. C. (1984) Tocopherols as likely precursors of pristane in ancient sediments and crude oils. *Nature* **312**, 440–442.

GROSSI V., HIRSCHLER A., RAPHEL D., RONTANI J.-F., DE LEEUW J. W. and BERTRAND J.-C. (1998) Biotransformation pathways of phytol in recent anoxic sediments. *Org. Geochem.* **29**(4), 845–861.

HAYES J. M., FREEMAN K. H., POPP B. N. and HOHAM C. H. (1990) Compound-specific isotopic analyses: a novel tool for reconstruction of ancient biogeochemical processes. *Org. Geochem.* **16**(4–6), 1115–1128.

HIRATSUKA R. (1976) Geochemical consideration on generation and evolution of petroleum. Evolution and alteration of petroleum. *J. Jpn. Assoc. Petrol. Technol.* **41**(6), 338–350 (in Japanese).

HUNT J. M. (1996) *Petroleum Geochemistry and Geology*, 2nd Ed., 743 p. W.H. Freeman and Company, New York.

ILLICH H. A. (1983) Pristane, phytane, and lower molecular weight isoprenoid distributions in oils. *Am. Assoc. Petrol. Geol. Bull.* **67**(3), 385–393.

ISHIWATARI R., ISHIWATARI M., ROHRBACK G. B. and KAPLAN I. R. (1977) Thermal alteration experiments on organic matter from recent marine sediments in relation to petroleum genesis. *Geochim. Cosmochim. Acta* **41**, 815–828.

ISHIWATARI M., ISHIWATARI R., SAKASHITA H., TATSUMI T. and TOMINAGA H. (1991) Pyrolysis of chlorophyll a after preliminary heating at a moderate temperature: implications for the origin of prist-1-ene on kerogen pyrolysis. *J. Anal. Appl. Pyrol.* **18**, 207–218.

ISHIWATARI M., ISHIWATARI R. and YAMADA K. (1997) Two possible pathways of formation of Pr and phytane from chlorophylls on diagenesis: carbon isotopic results from laboratory heating of chlorophyll-a. *Chem. Lett.*, 355–356.

ISHIWATARI M., ISHIWATARI R. and YAMADA K. (1999) Compositional and carbon isotopic behavior of C_{19} and C_{20} isoprenoid hydrocarbons produced by laboratory heating of phytol: a study of formation of Pr and phytane from chlorophylls on diagenesis. *Chem. Lett.*, 43–44.

ISHIWATARI M., YAMADA K. and ISHIWATARI R. (2000) Pyrolytic formation of C_{19} isoprenoid hydrocarbons from dihydrophytol: in relation to the genesis of Pr in petroleum. *Chem. Lett.*, 206–207.

ISHIWATARI R., ISHIWATARI M., YAMADA K., NARAOKA H., OHYAMA K., OKUI A. and KISAMORI S. (2001) *Carbon isotope composition of phytane and pristane in Japanese crude oils and their implications*, 20th International Meeting on Organic Geochemistry (Abstracts Vol. 1), pp. 415–416.

KISSIN Y. V. (1993) Catagenesis of light acyclic isoprenoids in petroleum. *Org. Geochem.* **20**(7), 1077–1090.

KOOPMANS M. P., RIJPSTRA W. I. C., KLAPWIJK M. M., DE LEEUW J. W., LEWAN M. D. and SINNINGHE DAMSTE J. S. (1999) A thermal and chemical degradation approach to decipher pristane and phytane precursors in sedimentary organic matter. *Org. Geochem.* **30**, 1089–1104.

LARTER S. R., SOLLI H. and DOUGLAS A. G. (1983) Phytol-containing melanoidins and their bearing on the fate of isoprenoid structures in sediments. *Advances in Organic Geochemistry 1981*, pp. 513–523, Wiley, New York.

LEWAN M. D., BJORØY M. and DOLCATER D. L. (1986) Effects of thermal maturation on steroid hydrocarbons as determined by hydrous pyrolysis of Phosphoria Retort Shale. *Geochim. Cosmochim. Acta* **50**, 1977–1987.

LI M., LARTER S. R., TAYLOR P., JONES D. M., BOWLER B. and BJOROY M. (1995) Biomarkers or not biomarker? A new hypothesis for the origin of pristane involving derivation from methyltrimethyltridecylchromans (MTTCs) formed during diagenesis from chlorophyll and alkylphenols. *Org. Geochem.* **23**(2), 159–167.

PETERS K. E. and MOLDOWAN J. M. (1993) *The Biomarker Guide*, 363 p. Prentice-Hall, Englewood Cliffs, NJ.

PRAHL F. G., EGLINTON G., CORNER E. D. S. and O'HARA S. C. M. (1984) Copepod fecal pellets as a source of dihydrophytol in marine sediments. *Science* **224**, 1235–1237.

ROWLAND S. J. (1990) Production of acyclic isoprenoid hydrocarbons by laboratory maturation of methanogenic bacteria. *Org. Geochem.* **15**(1), 9–16.

SAKATA M., KANEKO N. and SUZUKI N. (1990) Biomarkers in petroleum from sedimentary basins in Northeast Japan. *J. Jpn. Assoc. Petrol. Technol.* **55**(1), 48–53 (in Japanese).

SUN M.-Y., WAKEHAM S. G., ALLER R. C. and LEE C. (1998) Impact of seasonal hypoxia on diagenesis of phytol and its derivatives in Long Island Sound. *Mar. Chem.* **62**, 157–173.

SUZUKI N., SAKATA S. and KANEKO N. (1987) Biomarker maturation levels and primary migration stage of Neogene Tertiary crude oils and condensates in the Niigata sedimentary basin, Japan. *J. Jpn. Assoc. Petrol. Technol.* **52**(6), 499–510 (in Japanese).

TISSOT B., CALIFET-DEBYSER Y., DEROO G. and OUDIN J. L. (1971) Origin and evolution of hydrocarbons in early Toarcian shales, Paris Basin, France. *Am. Assoc. Petrol. Geol. Bull.* **55**(12), 2177–2193.

VAN DE MEENT D., DE LEEUW J. W. and SCHENCK P. A. (1980) Origin of unsaturated isoprenoid hydrocarbons in pyrolysates of suspended matter and surface sediments. In *Adv. in Organic Geochemistry 1979* (eds. A. G. DOUGLAS and J. R. MAXWELL), pp. 469–474. Pergamon Press, Oxford.

VAN FLEET E. S. and QUINN J. G. (1979) Early diagenesis of fatty acids and isoprenoid alcohols in estuarine and coastal sediments. *Geochim. Cosmochim. Acta* **43**, 289–303.

WASEDA A. (1992) Carbon and hydrogen isotopes of crude oils in Japan. *Res. Org. Geochem.* **8**, 29–33 (in Japanese).

WASEDA A. (1993) Effect of maturity on carbon and hydrogen isotopes of crude oils in Northeast Japan. *J. Jpn. Assoc. Petrol. Technol.* **58**(3), 199–208 (in Japanese).

WINGERT W. S. and POMERANTZ M. (1986) Structure and significance of some twenty-one and twenty-two carbon petroleum steranes. *Geochim. Cosmochim. Acta* **50**, 2763–2769.

Geochemical Investigations in Earth and Space Science: A Tribute to Isaac R. Kaplan
© The Geochemical Society, Publication No. 9, 2004
Editors: R.J. Hill, J. Leventhal, Z. Aizenshtat, M.J. Baedecker, G. Claypool,
R. Eganhouse, M. Goldhaber and K. Peters

Light-element geochemistry of the lunar surface

JOHN F. KERRIDGE

Department of Chemistry, University of California, San Diego, CA, USA

Abstract—Measurements of the abundance and isotopic composition of carbon, nitrogen and sulfur in lunar samples, carried out in the laboratory of Ian Kaplan at UCLA, led to several important insights into lunar, and possibly solar, processes. Sulfur isotope systematics in the lunar regolith record both addition of meteoritic sulfur and loss of sulfur from the lunar surface. Regolith nitrogen reveals complex, long-term isotopic variations of large magnitude which may originate in the Sun. Only the carbon isotope systematics defy explanation at this time.

INTRODUCTION

IN THE SPRING of 1973, the grant on which I was employed at UCLA came to rather an abrupt end, leaving me with an uncertain future. Fortunately, my office happened to be next to that of Ian Kaplan, an old friend from meteorite analysis in the early 60s. With characteristic generosity, he suggested that I work in his lab until I could find a more permanent arrangement. The "temporary" arrangement lasted until 1993, though along the way, in yet another act of generosity, Ian gave me part of his lab for my own use. Insofar as I was able to spend those 20 years doing worthwhile science, therefore, it was entirely due to the thoughtful and unselfish behavior of Ian Kaplan.

When I started work with Ian, he made me a fascinating, though somewhat forbidding, offer. His group, consisting of Chari Petrowski, Ed Ruth and Dave Winter, occasionally supplemented by visitors such as John Smith, Hitoshi Sakai, Yehoshua Kolodny, and Sherwood Chang, had been one of the first to receive lunar samples and had been busily analyzing samples from every mission from Apollo 11 on. Consequently, they had amassed a large quantity of data but, in Ian's view, had not had sufficient time to interpret those data in useful depth. My task, besides helping to operate the equipment, was to try to make sense of what appeared to be a substantial, but confusing, set of data.

Given Ian's interests and expertise, it is hardly surprising that the data consisted primarily of measured abundances and stable-isotopic compositions of carbon, nitrogen and sulfur in a wide range of lunar samples, from pristine igneous rocks to heavily processed soils. (The group also derived abundances of hydrogen, though at the time they lacked a suitable mass spectrometer on which to determine its deuterium content, and for some samples they obtained measures of the quantity of metallic iron present and the proportion of carbon that was present as methane. However, those analyses played a rather small role in what followed.)

Prior to Apollo 11, it was thought that interest in carbon, and the other biogenic elements, on the lunar surface would focus primarily on the issue of whether there was, or had ever been, life on the Moon. Given what we knew of lunar surface conditions, it seemed highly unlikely that any life could be present, but the lure of finding extraterrestrial life was sufficient to keep the issue on the front burner, and the fear of lunar pathogens was sufficient to maintain at least a semblance of planetary protection. Fortunately, it was quickly apparent that the lunar samples were sterile, which was just as well considering the frequency with which quarantine was breached following the return of Apollo 11.

However, even though the lunar surface was devoid of bugs, and even of fossils, it still contained measurable amounts of carbon and nitrogen, so the immediate challenge was to figure out what they were telling us. By the time I joined the group, there was general agreement in the science community that the carbon and nitrogen on the lunar surface were not of lunar origin but were derived from one or more extralunar sources, one of which appeared to be the solar wind. It had been recognized, from long before the first samples came back, that the lunar surface is bombarded by the solar wind which, given its high energy, would probably implant solar atoms into the outermost surfaces of lunar mineral grains. In fact, a group from the University of Bern, under Johannes Geiss, had capitalized on this phenomenon, by building a solar-wind collector which the astronauts deployed on the surface during several missions. That prescient experiment provided the community with much-needed ground-truth data when sample analysis revealed substantial quantities of the noble gases in lunar soils: Their relative contents and isotopic compositions matched those from the Bern experiment, proving beyond any doubt the existence of a "solar signature" in the lunar regolith.

However, unlike the solar-wind-derived noble gases, whose isotopic composition varied little from sample to sample, nitrogen, and to a less extent carbon, revealed substantial intersample differences in isotopic composition, suggesting that their story, or stories, were more complicated. Those were the stories which Ian wanted me to help unravel.

In fact, it turned out that each element, including sulfur, involved a different story, so in what follows, I shall consider each element separately. I shall divide each element's "story" into two parts: First, what we concluded in 1973–1974; Second, how those conclusions stand up today.

But first, a word about the samples used. We analyzed a wide range of samples from every Apollo mission but for our interpretive work focused primarily on data from Apollo 16 soils. The main reason for this choice was the essentially uniform major-element chemistry of the soils from that site. At the time, we were concerned that the behavior of the light elements on the lunar surface could be influenced by the composition of the mineral substrates, in which we had good reason to believe those elements were implanted. Consequently, use of samples from a reasonably homogeneous site, though one which still exhibited a wide range in exposure histories, served to eliminate one potential variable from the story.

CARBON, PART 1

In 1973, we knew the following about carbon. Lunar crystalline rocks were essentially devoid of carbon. We tended to obtain abundances of around 1 ppm, but suspected that even those low values probably represented terrestrial contamination. (In fact, it was not until several years later that Dave DesMarais, following his post-doc in Ian's lab, succeeded in beating down analytical blank levels for carbon to a point where he was able to demonstrate abundances of about 0.1 ppm in lunar rocks (DesMarais, 1983).) Lunar soils, however, revealed contents of carbon that ranged from a few ppm to well over 100 ppm. Our values compared reasonably well with those from other labs on splits of the same soils, but it was clear that inter-lab differences were significantly greater than the reproducibility that we obtained from duplicate analyses, which we performed for the great majority of samples. It was not difficult to assign most of those differences to terrestrial contamination. The challenge of analyzing for trace amounts of carbon in a carbon-based biosphere hardly needs stressing. The protocol in Ian's lab was rigorous in its control of contamination, but many other labs were more casual so that much of the early lunar carbon data are not worth the paper they were printed on. Furthermore, few other groups analyzed the lunar carbon isotopically and, since we were observing inter-sample variations in $\delta^{13}C$ of up to 20‰ in soil samples, it was obvious that even good-quality abundance data, on their own, would never yield the full story. Nonetheless, even poor-quality data showed that carbon abundances in the soil tended to increase with maturity of the soil, though use of better-quality data sharpened that correlation dramatically. ("Maturation" is the term given to the sum of processes which convert fresh, impact-derived fragments of lunar crystalline rocks into fine, heavily irradiated soil; "maturity" is the quantitative, though presently uncalibrated, measure of the extent of that conversion, and is generally based on the progressive reduction of indigenous Fe^{++} to ultra-fine-grained superparamagnetic Fe^0 by solar-wind hydrogen (Morris, 1976).) Based on those correlations, as noted earlier, some groups had invoked the solar wind as a major, or even sole, contributor of carbon to the lunar surface.

Armed with the foregoing information, and a sense that $\delta^{13}C$ values tend also to increase with increasing maturity, I embarked on the construction of a computer model intended to quantify all the processes that I thought might be affecting carbon on the lunar surface (viz. implantation by

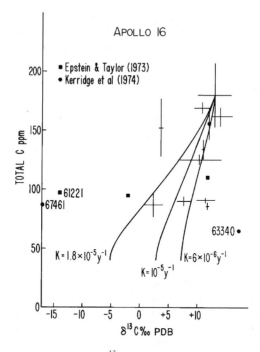

FIG. 1. Carbon abundances as a function of $\delta^{13}C$ values for a suite of 16 Apollo 16 soils, compared with predictions of a model based primarily on solar-wind implantation and proton stripping. The curves represent predictions for three trial values of the model parameter K, which is a rate constant incorporating both production and diffusive loss of methane. The numbered data points are for samples known from other evidence to be anomalous. The remaining data appear to be loosely consistent with the model predictions. From KERRIDGE *et al.* (1974).

solar wind, reaction with solar-wind-derived hydrogen, isotopically selective loss of methane by volatilization, comminution of rock fragments, and incorporation of regolith grains into agglutinates). Given my then-total lack of experience in computer modeling, this was a more ambitious undertaking than it might appear, and I quickly sought the aid of a friend, David Lesley, a math professor at San Diego State. Thanks to David, we got the program up and running and generated sets of model data which, while not reproducing the experimental data precisely, at least seemed to be consistent with them (Fig. 1). We therefore submitted them for publication. We received two reviews. One was from a group with whose (non-quantitative) interpretations we were in general agreement, and whom we had referenced copiously. They recommended rejection. The second was from a group whose work, in their own estimation, we had rendered obsolete. They recommended publication. Suffice it to say that the paper eventually appeared (KERRIDGE *et al.*, 1974).

Irrespective of reviewer prejudices, it was clear that more work was needed. However, with an apparently viable model for carbon under our belts, it seemed like time to tackle nitrogen and sulfur.

NITROGEN, PART 1

In 1973, few groups had analyzed lunar samples for nitrogen, and even fewer had analyzed its isotopes. Despite storage of lunar samples under dry nitrogen, terrestrial contamination was much less of a problem than it was for carbon, though it was necessary to outgas the sample overnight at about 150°C to get rid of most of the NASA-derived nitrogen. Fortunately, that had no effect on nature's nitrogen. Again, as with carbon, lunar rocks were devoid of nitrogen, whereas soils contained amounts that could exceed 100 ppm. However, whereas intersample variations in $\delta^{13}C$ were of the order of 20‰, those for $\delta^{15}N$ exceeded 15 per cent. That's right, per cent. Clearly, whatever was happening with nitrogen, we were looking at a major phenomenon.

At first, the data seemed utterly intractable. Contrasting with the robust correlation between nitrogen abundance and any and every measure of soil maturity, the isotopic compositions of those samples yielded no hint of a relationship with maturity nor, hence, with nitrogen content. I spent the early summer of 1974 searching desperately for something, anything, that would reveal a trend with $\delta^{15}N$. Then, one day I happened to notice in the literature that the soil sample which had provided our lowest $\delta^{15}N$ value (a 5% depletion relative to atmospheric nitrogen), an otherwise breathtakingly ordinary sample known as 65501, also was characterized by an anomalously large ^{21}Ne age. I quickly plotted our $\delta^{15}N$ data against all the ^{21}Ne data for aliquots of the same samples we had analyzed that I could find in the literature. Bingo! A trend, less than perfect, but much more convincing than anything that I had seen in several months (Fig. 2). With admittedly a lot of scatter, as the ^{21}Ne "age" increased, the $\delta^{15}N$ value decreased. The quotes around "age" mean that I have to say something about ^{21}Ne.

Neon-21 is the least abundant of the stable isotopes of neon, and for that reason is the most affected by production of the neon isotopes by cosmic-ray-induced nuclear reactions, termed spallation, on the major rock-forming elements, magnesium, aluminum and silicon. Also, the fact that there are three stable neon isotopes enables the amount of spallogenic ^{21}Ne to be estimated rather precisely. Consequently, ^{21}Ne spallation "ages" had become popular in meteorite research as a means of calculating the length of time that a meteorite had spent in space before impacting the Earth. Similarly, ^{21}Ne "ages" of lunar soils were regarded as a useful means of estimating how long an individual soil had resided on the lunar surface within the production depth of ^{21}Ne by galactic cosmic rays (about 2 m). Given that any Apollo soil consists of billions and billions of individual particles, the precise physical significance of the ^{21}Ne "age" was not immediately apparent, but nonetheless those ages, for a suite of samples, tended to behave in a sufficiently systematic fashion that the "age" could be taken as a useful property of the soil. Though when it came to evaluating the relationship between $\delta^{15}N$ and ^{21}Ne age, the meaning of that age was less than obvious.

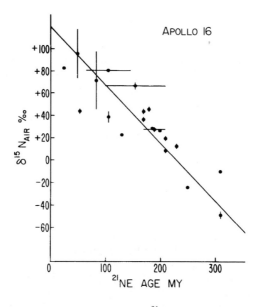

FIG. 2. Nitrogen isotopic compositions as a function of ^{21}Ne spallation ages for a suite of 20 Apollo 16 soils. The spallation "ages" are relative and not absolutely calibrated, but samples exposed longer ago on the lunar surface clearly contain lighter N than those exposed more recently. From KERRIDGE (1975).

The most logical interpretation seemed to be that the ^{21}Ne was behaving as a rather rough measure of how long ago the soil's constituents had first been derived from rocks, a parameter termed "antiquity". This differs from maturity in that a soil can spend substantial periods of time buried within the regolith below the depth at which maturation occurs, typically a few millimeters. So it looked as though the longer ago that a soil began its exposure on the lunar surface, the lower was its δ^{15}N value. Because it looked also as though most, if not all, of the nitrogen in the soil was implanted by the solar wind, and because the lack of any relationship between δ^{15}N and maturity made it unlikely that the nitrogen isotopes were being fractionated on the lunar surface, the implication was that the isotopic composition of solar-wind nitrogen had changed with time. This took us out of geochemistry and into the realm of nuclear astrophysics.

Consequently, I took our data and discussed them with the late Willy Fowler, one of the founders of nuclear astrophysics. I explained to him that I thought that the Sun had begun with nitrogen that was isotopically much lighter than that in the Earth and that nuclear reactions on the most abundant oxygen isotope, ^{16}O, had produced significant amounts of ^{15}N in the solar interior over time, with the result that δ^{15}N in the contemporary solar wind was much higher than the terrestrial value. I had to acknowledge to him that my best attempts to quantify that ^{15}N production fell at least three orders of magnitude short, but that I hoped he might be able to rectify that shortcoming. Not long after our meeting, I received a letter from him stating that "after cogitation of some duration" he had concluded that I must be wrong, but that we should publish anyway. I relayed this to Ian, who encouraged me to submit a Report to *Science*, though for reasons of intellectual unselfishness, or just possibly self-preservation, he declined to be named as co-author.

So the possibility of a "Secular change in the isotopic composition of solar nitrogen" entered the literature, albeit buried, for inscrutable reasons known only to the AAAS, among the pages in *Science* dedicated to biomedical research (KERRIDGE, 1975). It subsequently transpired that, unbeknownst to me, Bob Clayton and Richard Becker had independently arrived at the same conclusion (BECKER and CLAYTON, 1975), and it has been gratifying that at least one of them has consistently affirmed that the UCLA publication took precedence.

It will have been apparent from the above that the identification of the isotopic variability of lunar-surface nitrogen with a long-term change in solar composition rested on less than solid ground, but in early 1975 it still looked like the best bet.

SULFUR, PART 1

The most obvious difference between sulfur and the elements considered so far is that sulfur is by no means absent from lunar crystalline rocks. Anorthosites, and other crystalline rocks from the lunar highlands, contain several hundred ppm of sulfur, while mare basalts can contain more than 1000 ppm. What was curious is that lunar soils nonetheless contain significantly more sulfur than the rocks in their immediate vicinity, and that the regolith sulfur is enriched in ^{34}S by up to 13‰, compared to rock sulfur, which averages close to the standard terrestrial value. It was not difficult to postulate that sulfur, being a volatile element, could be lost from the lunar surface, by impact heating or solar-wind sputtering, with mass-dependent, *i.e.* isotopically selective, fractionation (CLAYTON *et al.*, 1974). However, that would lead to an inverse relationship between δ^{34}S and sulfur content. In practice, both δ^{34}S and sulfur abundance tend to correlate with each other and with sample maturity (Fig. 3). Furthermore, it was clear that, unlike the case for carbon and nitrogen, the solar wind was utterly incapable of supplying enough sulfur to the lunar surface to influence those systematics. Fortunately, by 1974, it was well known that meteoritic input to the lunar surface, though unlikely to be the source of extralunar carbon and nitrogen, nonetheless contributed measurable quantities of volatile trace elements such as bismuth and thallium. Also, it had been established by Harry Thode and Ted Rees that sulfur in the lunar regolith tended to reside on the surfaces of soil particles (REES and THODE, 1974), even though it was distributed throughout the volume of their parental rocks. It was therefore straightforward to conclude that a combination of impact heating of lunar rock material and vaporization of sulfur from impacting meteorites leads to deposition of sulfur-rich material on the surfaces of soil particles, and that both the deposition process itself and subsequent

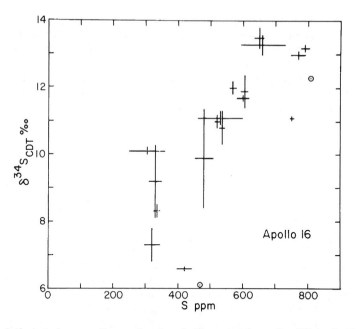

FIG. 3. Sulfur isotopic composition as a function of sulfur content for a suite of 20 Apollo 16 samples. Note that samples with less sulfur are isotopically lighter, the opposite of what would be predicted for loss of sulfur from the lunar surface. From KERRIDGE *et al.* (1975).

impact-heating and/or solar-wind sputtering would lead to preferential loss of the lighter isotope, ^{32}S, from the lunar surface. (Meteoritic sulfur has an average isotopic composition close to that of indigenous lunar sulfur.) A relatively simple model of this scenario provided good agreement with the observational data (Fig. 4), and the resulting estimate of the throughput of meteoritic

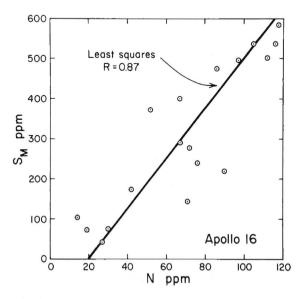

FIG. 4. The quantity of meteorite-derived sulfur, S_M, in the samples illustrated in Fig. 3, as a function of nitrogen content. Values of S_M were calculated from observed sulfur contents using a model based on mixing of indigenous and meteoritic sulfur followed by isotopically competitive loss of sulfur from the lunar surface, see text. Nitrogen content is employed here as a surrogate for soil maturity, see for example, KERRIDGE (1993). The clear correlation of S_M with maturity constitutes evidence for addition of meteoritic sulfur to the lunar regolith. From KERRIDGE *et al.* (1975).

sulfur was consistent with calculations based on, for example, bismuth contents. It is important to appreciate that the amount of sulfur lost from the moon by this combination of processes can amount to about 50% of that present at some time in the soil, and that without the isotopic clue to its former presence, there would be no way to know that it was ever there.

Despite a savage, though poorly informed, attack by Jerry Wasserburg at a Lunar Science Conference on "Origin and Evolution of the Lunar Regolith" in 1974, this interpretation of the sulfur isotope systematics on the lunar surface seemed to be in good accord with the data at the start of 1975 (KERRIDGE *et al.*, 1975).

CARBON, PART 2

Agreement between our model predictions and the actual carbon data was sufficiently tenuous that more data were clearly needed. We therefore requested and obtained a number of additional Apollo 16 soil samples, analyses of which doubled our database. Our immediate reaction to the new data was regret: The larger data set yielded a clearly worse relationship with the model. Although our data on duplicates continued to show excellent precision for most analyses, a handful still showed markedly worse precision and we wondered whether the old demon, terrestrial contamination, might, despite our best efforts, still be perturbing occasional analyses and even possibly skewing the comparison with the model predictions. We therefore devised a test to identify suspect data. This was based on the fact that addition of terrestrial carbon, from virtually any common source, to a lunar sample, regardless of its isotopic composition, would systematically both increase the sample's apparent carbon content while simultaneously lowering its $\delta^{13}C$ value towards that characteristic of terrestrial contamination, namely, $-25‰$. After normalizing each measured carbon content to some measure of maturity, such as nitrogen content, and calculating an average $\delta^{13}C$ value for the data set, it was straightforward to calculate, for each analysis revealing an excess of carbon, what the isotopic composition of that excess carbon would have to be in order to generate the observed $\delta^{13}C$ value, assuming as a first approximation that the true isotopic composition is not far removed from the average value. If the $\delta^{13}C$ value calculated for the excess carbon equaled $-25‰$, within experimental uncertainties, that analysis was deemed contaminated. It is important to note that those contaminated analyses were simply discarded, no attempt being made to correct the apparent contamination. As a result, the carbon database was reduced by some 18%, the corrected data set revealing greatly improved correlations with a number of other parameters but not, alas, with our model predictions, Fig. 5 (KERRIDGE *et al.*, 1978). It was clear that our model had perished from a surfeit of good data.

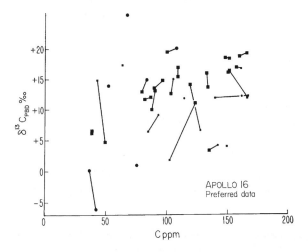

FIG. 5. Carbon isotopic composition as a function of carbon content for a suite of 30 Apollo 16 soils, minus data for a few samples revealing clear evidence for terrestrial contamination, see text. Of those plotted, the larger samples represent the most reliable analyses, with the remainder still possibly affected by contamination. Note the lack of any clear pattern to the data.

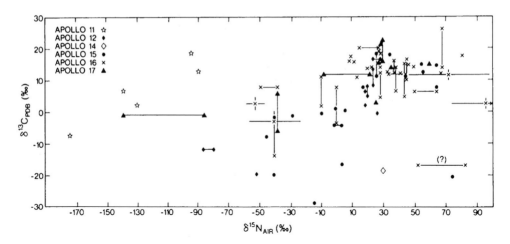

FIG. 6. A comparison of $\delta^{13}C$ and $\delta^{15}N$ values for a large number of Apollo samples from all missions. A hint of a positive trend has been proposed (BECKER, 1980).

Surprisingly, that is almost the end of the story about carbon on the lunar surface. In the past 25 years or so, very few papers have appeared on this topic, and most of those have contributed little or nothing to our understanding. There are, however, three exceptions. First is the heroic effort by Dave DesMarais, alluded to earlier, to determine the true carbon contents of lunar rocks (DESMARAIS, 1983). Second, in 1980, Richard Becker published a meticulous review of all available carbon data for soils and regolith breccias in which he, too, failed to find any statistically meaningful trends in the data (BECKER, 1980). He did suggest, however, that if a handful of samples, characterized by apparent terrestrial contamination or possibly unusual exposure histories, are ignored a broad trend might exist in which very low $\delta^{13}C$ values might be associated with very low $\delta^{15}N$ values, and vice versa (Fig. 6), but I must beg leave to differ with him on this.

Finally, some most intriguing new results were presented at the Lunar and Planetary Conference in March 2002. Hashizume, Chaussidon, Marty and Terada reported the results of an ion-microprobe analysis of carbon and nitrogen in the surfaces of some mineral grains extracted from an ancient lunar regolith breccia. The preliminary carbon data still have very large uncertainties, but appear to reveal carbon isotopic components whose compositions range far outside the envelope observed in bulk analyses. If those data hold up, we may be on the verge of a renaissance in the study of carbon in the lunar regolith.

NITROGEN, PART 2

In contrast to the case for carbon, the intervening years have seen a great deal of attention paid to nitrogen on the moon. It is impossible here to summarize all the studies that have been carried out. In brief, the major experimental developments have been the following.

The range of bulk $\delta^{15}N$ values has doubled, to more than 30%, driven in large part by the discovery of ancient regolith breccias characterized by ultra-low $\delta^{15}N$ values. Within individual samples, substantial isotopic inhomogeneity has been observed, indicative of at least two isotopically distinct components in essentially all regolith samples. Thanks to the pioneering efforts of Otto Eugster at the University of Bern, the use of ^{21}Ne as a measure of antiquity has been supplanted by parentless ^{40}Ar (EUGSTER et al., 1983). This is generated within the moon, by decay of ^{40}K, diffuses to the lunar surface where it escapes into space, and is then ionized by the solar radiation and accelerated by the interplanetary magnetic field back into the lunar surface, with sufficient energy to become efficiently implanted. Because the flux of such ^{40}Ar is assumed to be controlled by the decay rate of ^{40}K, it follows that the amount of implanted ^{40}Ar can serve as a chronometer dating the epoch when a sample was exposed to the solar wind. Unfortunately, although the analyses themselves are relatively straightforward, the all-important calibration of

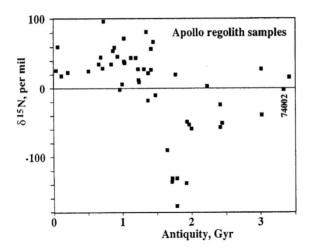

FIG. 7. Nitrogen isotopic compositions as a function of antiquity for a suite of regolith samples from several Apollo missions. Antiquity is calculated from contents of parentless ^{40}Ar, see text. Enlargement of the database has introduced added complexity to the trend observed in Fig. 2, and has substantially increased the magnitude of isotopic variation observed.

the technique has been laborious and technically challenging, though now adequately accomplished. Also, at least in its present form, the approach cannot resolve multiple exposures at different epochs, but nonetheless it has permitted a much more direct understanding of sample antiquity. Aided by this improved measure of antiquity, and a large increase in the number of samples analyzed, the long-term trend shown by δ^{15}N in the regolith has now been extended back to about 3 Gyr, revealing a distinctly more complex form than originally found (Fig. 7).

A major recent technical accomplishment has been the application of the ion microprobe to the analysis of nitrogen in the surface regions of individual lunar grains. This has revealed, among other discoveries, that the isotopic composition of implanted nitrogen is a function of depth within a grain, such that the more deeply implanted component is isotopically lighter than that closest to the surface (HASHIZUME et al., 2000). The same depth dependence has also been revealed by progressive etching of samples (MATHEW et al., 1998).

It is worth noting, also, that our understanding of the distribution of the solar noble gases in lunar regolith samples has been vastly improved since 1975, most spectacularly by means of the stepwise etching approach pioneered by Peter Signer and Rainer Wieler, and their colleagues at ETH, Zurich. Space does not permit me to do justice to all their accomplishments, but the most striking is the identification of two isotopically distinct components within the light noble gases: An isotopically light component, close to the surface of a soil grain, which they identify as solar wind (due to its exact correspondence with the results of the Bern solar-wind collection experiment mentioned earlier); and an isotopically heavy component, located at somewhat greater depth, which they have tentatively identified as due to suprathermal corpuscular radiation from the sun (WIELER et al., 1986). These noble-gas components apparently correlate closely with the nitrogen components described above, though the astute reader will have noticed that the mass dependence is of opposite sign in the two cases (Fig. 8).

Parallel with the production of novel and provocative data has been a proliferation of theories designed to explain the nitrogen isotope systematics on the lunar surface. A major landmark was the appearance in 1982 of a paper from the Bern group (GEISS and BOCHSLER, 1982) which reviewed both the then-known data and the theories that had been advanced to explain them. They concluded that no existing theory adequately satisfied the observations, including our original suggestion involving nuclear production of ^{15}N within the sun (their criticisms echoed our own reservations about the process). They noted that nitrogen in the lunar regolith was about an order of magnitude more abundant than it should be based on measured contents of the solar noble gases and the solar abundance proportions. They concluded that this disparity was more

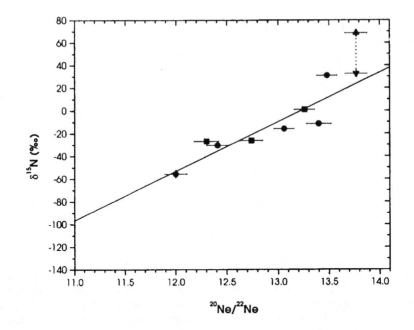

FIG. 8. Isotopic compositions of nitrogen and neon released from a lunar ilmenite separate subjected to stepwise etching by hydrofluoric acid. Close to grain surfaces, on the right of the plot, isotopically heavy nitrogen is associated with neon whose $^{20}Ne/^{22}Ne$ ratio equals that observed directly in the solar wind, whereas at greater depths within the grains, lighter nitrogen is associated with isotopically heavier neon, attributed by WIELER *et al.* (1986) to solar energetic particles. From MATHEW *et al.* (1998).

likely due to an excess of nitrogen than to a partial loss of the noble gases. Thus was established the notion that regolith nitrogen contained, or even largely consisted of, a nonsolar component. Several studies (*e.g.* WIELER *et al.*, 1996, 1999) since then have yielded indirect evidence in support of that idea, so that the concept of a non-solar nitrogen component has tended to dominate the discussion of lunar nitrogen for the past 20 years, alas.

Numerous such hypotheses have been advanced in that time, with the putative nonsolar nitrogen component variously identified as either isotopically heavier or isotopically lighter than the solar wind, and derived from any of a number of possible sources, including the lunar interior, the terrestrial atmosphere, meteorite or micrometeorite impact, interstellar dust, molecular-cloud material. However, none of these hypotheses has ever managed to explain how addition of an isotopically distinct component to the implanted solar wind could fail to produce a systematic relationship between nitrogen abundance and $\delta^{15}N$ value, yet no such relationship can be perceived in the experimental data (KERRIDGE, 1993). Furthermore, the long-term variation in $\delta^{15}N$ is either left unexplained or attributed to some fortuitous, *ad hoc* mixing phenomenon. In short, and the full story is very, very long, we presently have no viable model for the nitrogen isotope systematics in the lunar regolith. My personal conviction is that regolith nitrogen is predominantly solar in origin and that the isotopic variations observed are the result of solar phenomena, but that is, to say the least, a minority view.

SULFUR, PART 2

Nothing published since 1975 calls into question our model invoking influx of meteoritic sulfur and isotopically competitive loss of part of the resulting sulfur population. Curiously, a paper appeared shortly afterwards purporting to show, without benefit of isotopic data, that the lunar regolith contained meteoritic sulfur, but without the isotopic evidence for the substantial amount of "missing" sulfur, it represented an exercise in futility. Our model remains the best available description of the behavior of sulfur on the lunar surface, with the caveat that it is still not clear whether the lost sulfur is expelled by impact vaporization or solar-wind sputtering or, perhaps most likely, a combination of the two.

An interesting post-script to this story appeared in the mid-1990s. One of the earliest findings in the Apollo samples was the existence of a thin rind of amorphous material surrounding many of the mineral grains in the finest fraction of the lunar regolith. These were discovered by Michel Maurette's group in Orsay and studied intensively by them over the subsequent decade or so (*e.g.* BORG *et al.*, 1980). Their conclusion that the amorphous rims were produced by solar-wind-induced radiation damage was never seriously challenged and became an article of faith within the lunar community. With hindsight, it is somewhat surprising that this hegemony persisted in view of clear evidence for volatile elements, such as sulfur, being concentrated on grain surfaces, as argued by Rees and Thode, for example. However, that was how the amorphous-rim story rested throughout the 1970s and 1980s.

But in 1993, a paper appeared by Lindsey Keller and Dave McKay (he of the martian "worm" (MCKAY *et al.* 1996)) claiming that the amorphous rims consisted of vapor-deposited material, presumably resulting from micrometeorite impacts into the lunar surface (and possibly from solar-wind sputtering of nearby grain surfaces), a conclusion that is clearly compatible with the sulfur data (KELLER and MCKAY, 1993). A furious, though mercifully short-lived, fire-fight ensued as Maurette, and his supporters in Bob Walker's group at Washington University, attacked Keller and McKay's findings (BERNATOWICZ *et al.*, 1994; KELLER and MCKAY, 1994). The issue was never definitively resolved, at least in public, though the strength of at least some of the arguments on each side suggested that both processes were probably operative. It is certainly true that the original Orsay interpretation of the rims was overly simplistic.

AFTERWORD

So, we did not find life on the Moon but, thanks to Ian's foresight in setting up a lab capable of analyzing trace quantities of the biogenic elements, we learned a great deal about a variety of processes on the lunar surface, and even possibly in the interior of the Sun. And, thanks also to Ian, I was able to make the transition from crystallographer to isotope geochemist, thereby considerably extending my scientific career.

Acknowledgements—I thank Kurt Marti for advice and support, Richard Becker and Tim Swindle for constructive reviews, NASA for nearly 30 years of financial support, and, of course, Ian Kaplan, without whom I would not have been in a position to write this paper.

REFERENCES

BECKER R. H. (1980) Evidence for a secular variation in the $^{13}C/^{12}C$ ratio of carbon implanted in lunar soils. *Earth Planet. Sci. Lett.* **50**, 189–196.

BECKER R. H. and CLAYTON R. N. (1975) Nitrogen abundances and isotopic compositions in lunar samples. *Proc. Lunar Sci. Conf.* **6**, 2131–2149.

BERNATOWICZ T. J., NICHOLS R. H., HOHENBERG C. M. and MAURETTE M. (1994) Vapor deposits in the lunar regolith. *Science* **264**, 1779–1780.

BORG J., CHAUMONT J., JOURET C., LANGEVIN Y. and MAURETTE M. (1980) Solar wind radiation damage in lunar dust grains and the characteristics of the ancient solar wind. In *The Ancient Sun*, pp. 431–461, Pergamon, New York.

CLAYTON R. N., MAYEDA T. K. and HURD J. M. (1974) *Proc. Lunar Sci. Conf.* **5**, 1801–1808.

DESMARAIS D. J. (1983) Light element geochemistry and spallogenesis in lunar rocks. *Geochim. Cosmochim. Acta* **47**, 1769–1781.

EUGSTER O., GEISS J. and GRÖGLER N. (1983) Dating of early regolith exposure and the evolution of trapped $^{40}Ar/^{36}Ar$ with time. *Lunar Planet. Sci.* **XIV**, 177–178.

GEISS J. and BOCHSLER P. (1982) Nitrogen isotopes in the solar system. *Geochim. Cosmochim. Acta* **46**, 529–548.

HASHIZUME K., CHAUSSIDON M., MARTY B. and ROBERT F. (2000) Solar wind record on the moon: deciphering presolar from planetary nitrogen. *Science* **290**, 1142–1145.

KELLER L. P. and MCKAY D. S. (1993) Discovery of vapor deposits in the lunar regolith. *Science* **261**, 1305–1307.

KELLER L. P. and MCKAY D. S. (1994) Vapor deposits in the lunar regolith. *Science* **264**, 1780 p.

KERRIDGE J. F. (1975) Solar nitrogen: evidence for a secular change in the ratio of nitrogen-15 to nitrogen-14. *Science* **188**, 162–164.

KERRIDGE J. F. (1993) Long-term compositional variation in solar corpuscular radiation: evidence from nitrogen isotopes in the lunar regolith. *Rev. Geophys.* **31**, 423–437.

KERRIDGE J. F., KAPLAN I. R. and LESLEY F. D. (1974) Accumulation and isotopic evolution of carbon on the lunar surface. *Proc. Lunar Sci. Conf.* **5**, 1855–1868.

KERRIDGE J. F., KAPLAN I. R. and PETROWSKI C. (1975) Evidence for meteoritic sulfur in the lunar regolith. *Proc. Lunar Sci. Conf.* **6**, 2151–2162.

KERRIDGE J. F., KAPLAN I. R. and PETROWSKI C. (1978) Carbon isotope systematics in the Apollo 16 regolith. *Lunar Planet. Sci.* **IX**, 618–620.

MATHEW K. J., KERRIDGE J. F. and MARTI K. (1998) Nitrogen in solar energetic particles: isotopically distinct from solar wind. *Geophys. Res. Lett.* **25**, 4293–4296.

MCKAY D. S., GIBSON E. K., THOMAS-KEPRTA K. L., VALI H., ROMANEK C. S., CLEMETT S. J., CHILLER X. D. F., MAECHLING C. R. and ZARE R. N. (1996) Search for past life on Mars: possible biogenic activity in martian meteorite ALH84001. *Science* **273**, 924–930.

MORRIS R. V. (1976) Surface exposure indices of lunar soils: a comparative FMR study. *Proc. Lunar Sci. Conf. 8th*, 315–335.

REES C. E. and THODE H. G. (1974) Sulphur concentrations and isotope ratios in Apollo 16 and 17 samples. *Proc. Lunar Sci. Conf. 5th*, 1963–1973.

WIELER R., BAUR H. and SIGNER P. (1986) Noble gases from solar energetic particles revealed by closed-system stepwise etching of lunar soil minerals. *Geochim. Cosmochim. Acta* **50**, 1997–2017.

WIELER R., KEHM K., MESHIK A. P. and HOHENBERG C. M. (1996) Secular changes in the xenon and krypton abundances in the solar wind recorded in single lunar grains. *Nature* **384**, 46–49.

WIELER R., HUMBERT F. and MARTY B. (1999) Evidence for a predominantly non-solar origin of nitrogen in the lunar regolith revealed by single grain analyses. *Earth Planet. Sci. Lett.* **167**, 47–60.

Geochemical Investigations in Earth and Space Science: A Tribute to Isaac R. Kaplan
© The Geochemical Society, Publication No. 9, 2004
Editors: R.J. Hill, J. Leventhal, Z. Aizenshtat, M.J. Baedecker, G. Claypool,
R. Eganhouse, M. Goldhaber and K. Peters

Stable and radiocarbon isotopes and carbon cycling in coastal sediments

M. I. Venkatesan[1], I. R. Kaplan[1,2] and J. Southon[3]

[1]Institute of Geophysics and Planetary Physics, University of California, Los Angeles, CA 90095-1567,
U.S.A.
[2]Department of Earth and Space Sciences, University of California, Los Angeles, CA 90095-1567, U.S.A.
[3]Earth System Science, University of California, Irvine, CA 92696-3100, U.S.A.

Abstract—Carbon cycling in coastal sediments was investigated using stable isotopes of carbon and nitrogen and radiocarbon isotope of various compound class fractions from a subsurface sediment section (30–50 cm) of Santa Monica Basin (SMB) and a composite of six surface grab sediments from McMurdo Sound, Antarctica. The range of stable isotope ratios reflects a mixture of terrestrial and marine inputs in the SMB sediments. By contrast, the uniformity of the carbon isotopic signature in the Antarctic sediments is characteristic of a dominantly planktonic contribution. Additional information on sources, particle processing and organic matter preservation was obtained by integrating the stable isotope data with radiocarbon measurements and molecular composition of the same fractions.

Inputs of carbon from submarine seeps in coastal sediment are obvious from the relatively older age observed in the aliphatic and aromatic hydrocarbon fractions from the sediments of SMB. On the other hand, eroded sedimentary sequences containing coal could have imparted a non-recent age to the aliphatic and aromatic fractions in the Antarctic sediments. A major biogenic origin is evident from the younger age of most of the Antarctic sediment organic fractions compared to those from SMB. The humic and fulvic acid fractions from both the regions of investigation exhibit relatively young age in the range of 760–1700 BP. In contrast, protokerogen is much older (5470 and 8770 BP, respectively) than the humic and fulvic acids in both the sample locations. This suggests that protokerogen is more likely derived from the selective accumulation and preservation of refractory residues of biological organic components rather than being generated by *in situ* humification. These results have important implications to the carbon cycling in coastal sediments and in understanding the formation of kerogen.

INTRODUCTION

THE MECHANISM of production, transformation and accumulation of marine organic matter is an important component of the global carbon cycle. The information gathered over the past several years regarding the composition and fate of organic compounds in suspended oceanic particles and sediments (*i.e.* CRISP *et al.*, 1979; WAKEHAM *et al.*, 1980; HEDGES *et al.*, 1986; VENKATESAN *et al.*, 1987 among numerous others) has helped in understanding the biogeochemical cycling of carbon in the ocean. Because of the natural cosmogenic and bomb-released components, ^{14}C has also been used as a tracer for both long-term and short-term oceanic processes. Marine organic matter, impacted from introduction of recently synthesized terrestrial organic components of vegetation is expected to reflect a high ^{14}C post-nuclear bomb activity of atmospheric carbon dioxide with $\Delta^{14}C > 150\%o$ relative to pre-1950 background values. In contrast, geopolymers such as oil seeps, eroded kerogen or coal fragments, anthropogenic petroleum products and spills exhibit undetectable amounts of $\Delta^{14}C$. The radiocarbon ages of the fractions can, therefore, be correlated to their origin (MANNING *et al.*, 1990) which, in turn, helps understand the processes controlling input and the biogeochemical fate of carbon in coastal ecosystems.

The radiocarbon age of surface sediments averages around 2400 yr in the basins of southern California and around 2900 yr in the Atlantic Ocean (EMERY and BRAY, 1962; BERGER *et al.*, 1966; BENOIT *et al.*, 1979). Obviously, the measured ^{14}C age of organic matter in the surface sediment of the California basins is not unique to this environment, nor does it reflect a lack of deposition of planktonic debris during the past 2400 yr, as the surface waters of the Southern California Bight are highly productive. For example, the ^{14}C abundance of pore-water DIC in a core taken from 16 m water depth in the Santa Barbara Channel showed a modern origin of CO_2, while the carbon in the surrounding sediment averaged >2500 yr BP (BAUER et al., 1990). The measured ^{14}C value of $-416\%_0$ in the Santa Monica Basin (SMB) by WILLIAMS et al. (1991) implying unexpectedly "old" ages for the near surface carbon, can only be partly explained by a probable redeposition and mixing of ancient with modern organic carbon. The residence time of the majority of organic matter in the euphotic zone is only of the order of few months and not 2000 yr. The apparent age of the near-surface sediment is either linked to the nature of the organic matter settling onto the seafloor or its subsequent history in the sediment column. Thus, the bulk radiocarbon data are paradoxical. Apparently, the organic matter being decomposed on the seafloor represents recently deposited plankton, whereas the bulk of the organic matter buried in the sediment is recycled and refractory carbon, supporting the conclusions reached by SWEENEY and KAPLAN (1980) using $\delta^{15}N$ as a tracer and by TRUMBORE et al. (1991) from C-14 measurements of bulk organic carbon. Several questions still remain to be answered from the available bulk radiocarbon measurements on sedimentary organic matter in attempting to better understand the oceanic carbon cycle. (i) Why is organic carbon in coastal areas around 2000 yr old in undisturbed surface sediment? (ii) What fraction of organic matter is modern, recycled or derived from shale, coal or petroleum? and (iii) what fraction dominates and determines the mean age of TOC?

Pathways of carbon flow in natural environments have also been reconstructed using bulk stable carbon and nitrogen isotopes as well as compound-specific isotopic analysis of individual biomarker lipid component (HAYES et al., 1990) based on the fractionations involved during primary (photosynthetic) and secondary (heterotrophic) processes. Recently, compound-specific radiocarbon analysis of individual biomarker lipid has been shown to be a valuable technique to determine the source of marine organic matter (EGLINTON et al., 1997).

The biomarker concept is a powerful tool for understanding the biogeochemistry when compounds have unique biological origin, but when multiple sources are identified for a given compound, interpretation can become more complex. Likewise stable isotope signals can be complicated because of mixed sources. However, radiocarbon measurement provides an additional important parameter to help evaluate source/age of the components (DRUFFEL and WILLIAMS, 1990; DRUFFEL et al., 1992). The combined biomarker/stable isotope/radiocarbon isotope approach should, therefore, offer a more comprehensive approach for resolving the complex marine carbon cycle.

Our preliminary isotopic study indicated that the hydrocarbon fraction in the sediments from the SMB from southern California Borderland and McMurdo Sound, Antarctica, contain "old" carbon indicative of fossil carbon input in both the study locations (VENKATESAN and KAPLAN, 1994). In this report, we present $^{13}C/^{12}C$ and $^{15}N/^{14}N$ data and the ^{14}C age of several individual organic component fractions isolated from SMB sediment core section and compare the results with those of a more pristine and less complex area of the McMurdo Sound. Isotopic data for lipid classes such as aliphatic and aromatic hydrocarbons, fatty acids, sterols, alcohols, etc. as well as for fulvic and humic acid and protokerogen are discussed. We have also integrated the isotopic data with our past molecular studies in the same regions (i.e. VENKATESAN et al., 1980; VENKATESAN and KAPLAN, 1992; VENKATESAN, 1988) to obtain new information on the origin, age and processing/preservation of depositing and accumulating sedimentary organic matter.

METHODS

Study area and samples

A core from a single site in the center of the SMB was collected in June of 1992 (University of Southern California ship, Sea Watch) from 904 m water depth ($33°41.20'N$, $118°56.11'W$). Bottom sections from 30 to 50 cm were composited for this study. SMB has a high sedimentation rate (1 mm/yr) and bioturbation is minimal especially in the deeper basin due to oxygen-depleted waters below the sill depth of 737 m and should preserve the organic matter from rapid decomposition (HUH *et al.*, 1987). The composited sediment used in this study had 2.56% of total organic carbon and 0.44% of nitrogen with an atomic C/N ratio of 6.79. The significantly higher than 1% total organic carbon reflects most probably reduced condition, if not total anoxia, of the sediment interval investigated here. This sediment horizon predates the onset of anoxic depositional conditions in the basin at around 25 cm (MASIELLO and DRUFFEL, 2003). This sub-bottom horizon was specifically chosen to represent pre-anthropogenic and "pre-bomb" background levels for ^{14}C. Age-dating by Pb-210 of a 30 cm core (CROSS-I, BC 89) collected in 1987 from a close by location (33°41.50'N, 118°56.30'W, 908 m) also near the center of SMB implies, by analogy, that sediment from sub-bottom depth below 10 cm in the core analyzed here was deposited prior to 1900 (VENKATESAN and KAPLAN, 1990). Another core collected in 1991 from a nearby station (core, NOAA II, from 33°41.00'N, 118°56.22'W, 906 m) had a Pb-210 age of 1872 AD at the 10–12 cm horizon (HUH, 1998; VENKATESAN, 1998).

Several surface grab samples (~top 6–8 cm) from McMurdo Sound, Antarctica were collected using a gravity corer during the austral summer of 1989–1990 as part of the NSF/Polar program in the cruise designated as DF90. Sediment samples were composited from six locations, three each from Granite Harbor and New Harbor. The current samples were taken close to the Granite Harbor sediment locations which were found to be pristine, enriched in biogenic components (VENKATESAN, 1988) and sufficiently removed from the McMurdo Research Station to prevent contribution of anthropogenic inputs (VENKATESAN and MIRSADEGHI, 1992).

The sediment samples from individual locations had % TOC content in the range 0.4–0.92 typical of Ross Ice Shelf sediments (VENKATESAN, 1988) and the composite sample had an organic carbon content of 0.89%. Total organic nitrogen was not analyzed in these samples. Sedimentation rate in the region is about 2.7 mm/yr based on Pb-210 chronology and surface mixed layer is limited to the top 7 cm of the bed (MACPHERSON, 1987). The main sources of the seafloor sediment in the Sound are wind-blown sand, glacially eroded soil, rock outcrops, diatomaceous and supraglacial debris from the McMurdo Ice Shelf. Sediments comprise as much as 25% of biogenic silica of the total sample (BARRETT *et al.*, 1984; LEVENTER and DUNBAR, 1987).

Compound class fractionation

About 500 g of wet sediment was extracted with methanol first and then with methylene chloride and processed by established procedures (VENKATESAN *et al.*, 1987) to isolate several classes of compounds as follows: The lipid extract was treated with acid-cleaned copper granules to remove elemental sulfur, saponified, methylated and then eluted as hydrocarbon, fatty acid methyl esters, ketones, alcohols, sterols and non-elutable polar (NEP) fractions from a preparative thin layer plate (Analtech). The hydrocarbon was partitioned by silica gel chromatography to resolve it into an aliphatic and aromatic fraction. All the fractions except NEP compounds were analyzed by gas chromatography and/or gas chromatography/mass spectrometry to check for the integrity of separation and identification of compounds and for quantitation. Fatty acid methyl ester fraction was once more saponified to split off the ester again to remove the methyl carbon added during methylation to avoid the degree of uncertainty in isotopic composition from the methylation procedure.

The residual sediment after organic solvent extraction was extracted repeatedly with 0.2N NaOH to isolate the fulvic/humic acid fraction. Humic acid was then precipitated with hydrochloric acid to separate it from fulvic acid in the aqueous phase. Fulvic acid was purified by adsorption and subsequent elution from a column of Amberlite resin. Humic acid was purified by redissolving in NaOH and reprecipitation with hydrochloric acid. The residual sediment was then treated sequentially with hydrofluoric acid of increasing concentration to remove the silicates, washed several times with water and then dried to recover protokerogen (STUERMER *et al.*, 1978). Procedure blank for each compound class fraction was also combusted for carbon dioxide measurement.

Isotope determination

The lipid fractions were dried under purified nitrogen and combusted after evacuation in closed quartz tubes with CuO in the presence of Ag metal at 850°C. The carbon dioxide was then split into two aliquots for ^{14}C and ^{13}C analyses. The smaller aliquot of carbon dioxide and N_2 gas were purified and made available for δ^{13}C and δ^{15}N measurement on a Finnigan MAT 250 dual collector mass spectrometer (CLINE and KAPLAN, 1975; MINAGAWA *et al.*, 1984). The reference standards used for δ^{13}C and δ^{15}N were Chicago Pee Dee Belemnite and atmospheric nitrogen, respectively. Some nitrogen samples were either too small to measure or, when measured, provided questionable values which are, therefore, not reported. A major fraction of the CO_2 was sent to Lawrence Livermore National Laboratory for Δ^{14}C measurements, where samples containing ≥ 200 μg carbon were graphitized to manufacture targets for AMS analysis.

Control experiments/quality assurance

The carbon yield efficiency and reliability of the solvent extraction/separation procedures were tested with *n*-alkane standard (tetracosane which is of petrochemical origin). A known weight of the pure standard was equilibrated with methylene chloride overnight in a closed flask, rotoevaporated and subjected to silica column chromatography using the procedure described above to isolate the aliphatic fraction containing tetracosane. 50 g of Green River Shale from the Mahogany Zone (GRMZ) was extracted with solvent and after asphaltene was precipitated, the hexane phase was subjected to silica column chromatography described above to isolate the aliphatic fraction. The aliphatic fraction from the solvent extracted tetracosane and GRMZ, the neat tetracosane standard and GRMZ were combusted to measure δ^{13}C and δ^{14}C. These experiments were conducted in duplicate. The carbon yield from the processed tetracosane was calculated. Organic solvent extract of leaves from a living plant was concentrated, converted to carbon dioxide and analyzed for stable and radiocarbon isotopes to check for old carbon contamination from the solvent during sample processing. An old wood sample was extracted with solvent and the humic and fulvic acids were isolated and analyzed to check for Δ^{14}C content. This wood was of Pliocene age collected from the Mojave Desert, California. No ^{14}C had been used at or near the laboratory where the study was conducted. Procedural blanks corresponding to the above treatments were saved for carbon isotope measurement.

RESULTS AND DISCUSSION

Control experiments/quality assurance

The results of the control experiments showed that carbon dioxide from neat and tetracosane in the aliphatic fraction were recovered at the level of 93–96%. δ^{13}C and Δ^{14}C values obtained for the standard before and after extraction and column procedure are very similar demonstrating that these treatments did not bias the results significantly (Table 1). Results from the GRMZ were consistent for the duplicates and the aliphatic carbon was about 2‰ lighter than the shale. The Δ^{14}C (-985 and -986‰) of aliphatic fraction from GRMZ showed that no more than 1.5% of any contaminant carbon was of modern origin.

Lipid extract from the leaves had a Δ^{14}C value of $+29$‰, characteristic of modern carbon suggesting that there was no apparent negative Δ^{14}C input from old/dead contaminant carbon during processing of the sample.

The wood is of Pliocene age and should be radiocarbon-dead. However, the finite age measured here for the humic and fulvic acid fractions probably reflects field contamination from microbial degradation and diagenetic breakdown products from the parent wood.

Carbon dioxide from the procedure blanks for all the individual compound class fractions including those from humic and fulvic acid fraction was collected. Background reading of the monometer in the vacuum line was 0.65 μmol which was consistent over the 1 month period during gas collection. None of the procedure blank had carbon dioxide more than $1-1.2$ μmol (inclusive of the background reading) which was too low to give any stable or radiocarbon

Table 1. Santa Monica Basin sediment: isotopic data

Sample	$\delta^{13}C$ (‰)	$\delta^{15}N$ (‰)	$\Delta^{14}C$ (‰)	^{14}C Age[a] (BP)
Tetracosane (neat)	−31.74		−998	
Tetracosane (neat)	−31.76		−997	
Tetracosane (aliphatic)	−31.77		−983	
Tetracosane (aliphatic)	−31.77		−985	
GRMZ shale (not ext)	−27.59		nd	
GRMZ shale (not ext)	−27.49		nd	
GRMZ shale (aliphatic)	−29.60		−985	
GRMZ shale (aliphatic)	−29.51		−986	
Leaf (lipid extract)	−36.95		+29	Modern
Wood humic acid	−25[b]		−979	31,180 ± 620[c]
Wood fulvic acid	−25[b]		−858	15,670 ± 110[c]
SM sediment (not ext)	−22.13	+10.31	−298	2850 ± 70
SM sediment (residue)	−22.41	+7.52	−285	2700 ± 80
Protokerogen	−27.82		−494	5470 ± 50
Humic acid	−21.85	+8.02	−185	1650 ± 70
Fulvic acid	−20.14		−90	760 ± 60
Lipid	−24.04	+5.06	−609	7530 ± 70
Aliphatic	−25.75		−943	22,970 ± 210
Aromatic	−24.09		−860	15,790 ± 100
Ketone	−27.39		−537	6180 ± 70
Sterol	−24.88		−547	6360 ± 70
Alcohol	−25.02		−433	4560 ± 60
Fatty acid	NR[d]		−917	19,970 ± 120
NEP	−24.80	+8.83	−466	5040 ± 120

[a] Sample preparation (graphitization) backgrounds have been subtracted, based on measurements of samples of ^{14}C-free coal. Backgrounds were scaled relative to sample size.

[b] $\delta^{13}C$ values are the assumed values according to STUIVER and POLACH (1977) when given without decimal places.

[c] Derived from Pliocene-age wood. The finite ages for these fractions reflect *in situ* input of young humic and fulvic acids in the field from the breakdown products of wood, *i.e.* possibly from microbial degradation. See text for more details.

[d] Not reported. See text.

isotope measurements when attempted. No blank correction for laboratory workup was, therefore, done. The insignificant carbon dioxide levels from the blank samples also confirm our concordant results free of contamination from the control experiments described above.

Carbon dioxide collected from the fractions for isotope measurements was substantially higher than minimum requirements and ranged from 40 to 500 μmol (480–6000 μg C) from the SM Basin and from 20 to 290 μmol (240–3480 μg C) from the Antarctic sediment. Protokerogen from the Antarctic sediment was the smallest sample size used in this study yielding 20 μmol of carbon dioxide. However, nitrogen values were uniformly low ranging from 2 to 4 μmol from the SM Basin and from 2 to 10 μmol in the Antarctic sediment.

Santa Monica Basin

Stable carbon isotopes data—The data from stable isotope and radiocarbon measurements are summarized in Table 1. Analytical precision is ± 0.1‰ for C. Stable carbon isotope ratios of the different fractions span from − 20.14 to − 27.82‰ with the unextracted and extracted sediments exhibiting values, as expected, in the midrange around − 22‰ (Fig. 1). The only exception is the fatty acid fraction which is anomalously depleted in $\delta^{13}C$. Stable carbon isotope numbers of the different compound classes from the SMB samples, in general, suggest a mixing of terrestrial and marine components with humic and fulvic acids implying major contribution from marine sources (SWEENEY and KAPLAN, 1980; STUERMER *et al.*, 1978).

The gas chromatogram of the aliphatic fraction exhibits well-resolved peaks with terrestrial *n*-alkanes from C_{27} to C_{33} dominating the profile with a small unresolved complex envelope above the baseline at carbon numbers between C_{27} and C_{31}. The relative depletion of $\delta^{13}C$ of the aliphatic hydrocarbon and an *n*-alkane profile exhibiting a carbon preference index (CPI) of 5.2

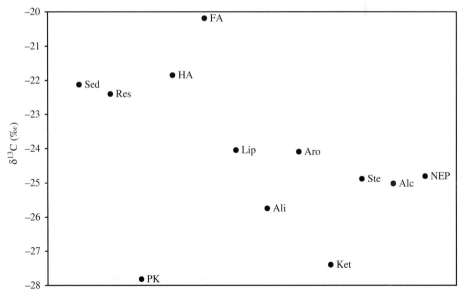

FIG. 1. Stable carbon isotope of organic fractions from the sediments of Santa Monica Basin. Sed: unextracted sediment; Res: residual sediment after organic solvent extraction; PK: protokerogen; HA: humic acid; FA: fulvic acid; Lip: lipid (organic solvent extract); Ali: aliphatic; Aro: aromatic; Ket: ketone; Alc: alcohol; Ste: sterol; NEP: non-elutable polar fraction, near origin from thin layer chromatography.

(calculated from C_{20} to C_{32} after BRAY and EVANS, 1961) are reflective of terrestrial plant components (COLLISTER et al., 1994). The aliphatic fraction in this study is relatively enriched in $\delta^{13}C$ compared to the compound specific ratios of $\delta^{13}C$ of individual n-alkanes from the subsurface sediment (2.5–7.5 cm) from the SMB reported by PEARSON and EGLINTON (2000). Our sample is also from the central SMB but from a different location and from a much deeper horizon compared to that in their study. Aliphatic hydrocarbon in our study, as a compound class, may reflect the average isotopic composition of all the alkanes and alkenes in the SMB sediment. Long-chain alkenes and diatom-derived highly branched isoprenoid alkenes which are of marine origin could also contribute to the observed carbon fractionation of the aliphatic lipid component (BRASSELL et al., 1986; ROWLAND and ROBSON, 1990; EGLINTON et al., 1997). The extent of this isotopic contribution is currently unknown in the sample although relatively lower levels of alkenes were identified in the current sample compared to the surface samples from the basin in our prior study (VENKATESAN et al., 1980). The dominant land plant input is from C_3 plants with $\delta^{13}C$ values from -25 to $-28‰$ and there is no literature evidence of C_4 plants inputs in the natural runoff in the study region.

Because of the high CPI in our sample, we believe that the measured $\delta^{13}C$ value we obtained ($-25.75‰$) reflects a contribution of terrestrial vegetation. In addition, the lower isotope value of the protokerogen, aromatic and the aliphatic hydrocarbons relative to bulk carbon may also imply contribution from Monterey shale and petroleum in addition to that from land plants. SCHOUTEN (1995) found that the average $\delta^{13}C$ for n-alkanes ranging from C_{24} to C_{33} in the Monterey Shale varied from ~-28 to $-32‰$. Similar isotopically light alkanes have also been reported in crude oils (WILLHELMS et al., 1994). Natural petroleum seepage and possibly Monterey Shale have been shown to contribute to hydrocarbons in the sediments from this region based on the presence of C_{28}-bisnorhopane and other petroleum biomarkers in the aliphatic fraction (SIMONEIT and KAPLAN, 1980; VENKATESAN et al., 1980; VENKATESAN and KAPLAN, 1992). These petroleum biomarkers were identified in the aliphatic fraction also from the current sediment suggesting input from fossil carbon, albeit at low levels.

The fatty acid fraction is depleted in ^{13}C significantly more than expected for terrestrially derived acids. For example, stable carbon isotopic composition of the C_{28} fatty acid in the sediment from the western North Pacific was $-31.8‰$, typical of higher plants origin (BALLENTINE et al., 1988; UCHIDA et al., 2001). One possible explanation for anomalously light $\delta^{13}C$ value of fatty acid fraction ($-82.04‰$) in the current sample is that a portion of the fatty acids could have been produced from methane with $\delta^{13}C$ around -60 to $-100‰$ by methane-oxidizing bacteria (HAYES et al., 1987; FREEMAN et al., 1990; SUMMONS et al., 1994; HINRICHS et al., 1999) augmenting the inputs from higher plants. For lack of additional evidence to support this argument, this data is not included in Table 1 and Fig. 1.

The $\delta^{13}C$ value of the sterol fraction ($-24.88‰$, Table 1) and the presence of the C_{29} sterol, β-sitosterol in dominant amounts relative to other sterols, would imply terrestrial contribution to the organic carbon (HUANG and MEINSCHEIN, 1976; VENKATESAN et al., 1987). This value is also within the range of $\delta^{13}C$ determined by PEARSON et al. (2000) for sterols such as cholesterol, $C_{26}\Delta^{5,22}$, $C_{27}\Delta^{5}$ and $C_{28}\Delta^{5,22}$ and many other sterols from the surface sediments of Santa Barbara (SBB, 0–1 cm) and SM basins (0–5.5 cm). They did not find any ^{13}C depletion in the C_{29} sterols as would be expected from a significant terrigenous inputs and concluded that these sediments do not contain a quantitatively significant terrestrial component. Similar conclusion may not be applicable to our sample which comes from a much deeper horizon than their study.

Stable nitrogen isotopes data—Only a few fractions had nitrogen levels sufficiently large enough for isotope measurements (Table 1). Analytical precision is $\pm 0.2‰$ for N. The $\delta^{15}N$ values within the range of $+5.06$ to $+10.31‰$ indicate that the bulk of this element is of marine origin from nitrate fixation by plankton ($\sim +9‰$; PETERS et al., 1978; STUERMER et al., 1978) mixed with smaller proportion of terrigenous nitrogen ($\sim +2‰$; SWEENEY et al., 1978). The relatively lower ^{15}N enrichment of the lipid fraction compared to all other fractions measured here could imply that the nitrogen may have been synthesized by microflora in the sediments, possibly from dissolved ammonia in interstitial water.

Radiocarbon data—The results from the above compound classes from the SMB sediment show large variations in the radiocarbon age although all of them bear signatures of "old" or pre-bomb $\Delta^{14}C$ prior to the 1950s when atmospheric nuclear weapons testing began to be carried out (Table 1, Fig. 2). The bulk phase total organic matter in the unextracted and extracted sediment showed ^{14}C ages 2700–2850 BP. It is highly unlikely that this apparent age is caused by the

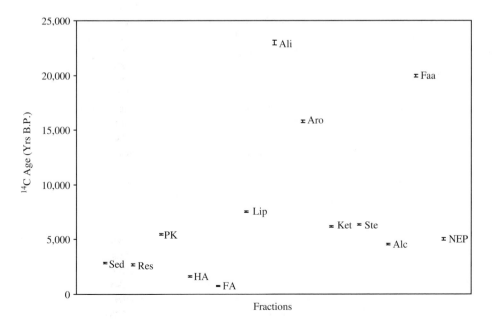

FIG. 2. Radiocarbon age of the organic fractions from the sediments of Santa Monica Basin. For explanation of abbreviations of compound class fractions, refer to Fig. 1. Faa: fatty acid.

mixing of sediments by burrowing organisms because the individual lipid classes of the bulk organic matter exhibit ^{14}C ages from 760 to 22,970 BP. That bioturbation is minimal and the sediment strata are best preserved in the study location is documented by the exponential decrease of excess ^{210}Pb with depth over the entire core length of 30 cm in CROSS-1 BC 89 and the absence of sewage sterol, coprostanol, below the sub-bottom depth of 27.5 cm sediment sections (HUH *et al.*, 1987; VENKATESAN and KAPLAN, 1990). We believe that this large age range represents a mixture of organic matter derived from multiple sources, marine, terrigenous and petroleum. It has to be noted, however, that both deep-ocean heterotrophic and chemoautotrophic microbial processes can potentially take up significant quantities of ambient deep-ocean inorganic carbon (*e.g.* RAU, 1991). Because such DIC is relatively depleted in ^{14}C, the biomass/POC formed by such organisms, while "modern", will also be ^{14}C-depleted. For example, DIC in the SMB can contribute to old carbon (Δ^{14}C) at the level of $-50‰$ (RAU, 1991) to the sediment/lipid fractions. But, as will be discussed below, the Δ^{14}C of the SMB sediment and all other fractions except fulvic acid is considerably older than expected from the DIC contribution alone.

The wide range in the depletion of ^{14}C values of the chemical suites is consistent with the expectations based on the different source of the compound classes investigated. Selective decomposition of the lipid classes may also contribute to the differences in their age. The total lipid fraction before being separated into various compound suites reflects an average radiocarbon age of all the lipid components. As expected the alcohol and NEP fraction are the youngest in contrast to the hydrocarbons. The aliphatic and aromatic fractions are strongly ^{14}C-depleted (-943 and $-860‰$) and appear to be derived from fossil carbon. Since the sediment horizon is pre-anthropogenic, the fossil carbon here is composed of contribution from natural seepage and vegetation in paleosols as well as indigenous marine biota (SIMONEIT and KAPLAN, 1980; VENKATESAN *et al.*, 1980; VENKATESAN and KAPLAN, 1992). A recent model (PEARSON and EGLINTON, 2000) based on compound-specific Δ^{14}C study of *n*-alkanes in sediments from the same general location also corroborates this presence of a mixture of biogenic and fossil carbon in the pre-bomb horizons. It was estimated that $\sim 13\%$ of fossil material (5% from petroleum and 8% from shale) could contribute to the alkanes of the pre-bomb sediments. The non-petroleum higher plant waxes deposited in the SMB are of recent biogenic origin and are the dominant alkane component as evident from the molecular composition as well as the above-discussed model. Similarly, the non-recent age of the aromatic fraction in this pre-anthropogenic horizon derives from petroleum and shale mixed with storm water runoff and atmospheric transport from natural combustion, such as forest fires.

UCHIDA *et al.* (2001) found that high molecular weight fatty acids in the sediments of western North Pacific derived from higher plant waxes tend to have non-recent age with increasing carbon numbers. C_{24} fatty acid in their sample dated 2550 BP whereas the C_{28} acid was 3250 BP. This old age may be due to the long residence time in soil and lake sediments as well as the result of river transport of old terrestrial particulate organic carbon to the ocean (HEDGES *et al.*, 1997; RAYMOND and BAUER, 2001). Fallout of aerosol particles in which higher plant fatty acids could be as old as 5860 BP could also result in the non-recent age of the sediment fatty acids (MATSUMOTO *et al.*, 2001). The anomalously "old" carbon found in the fatty acid fraction in the current sample cannot be entirely attributed to such an input. Nor does it come from petroleum which is known to contain only minor amounts of fatty acids (TISSOT and WELTE, 1984). It is probably the result of fossilized biogenic component which could very well be bacterial cell detritus as is also implied by the strong depletion of ^{13}C in the fatty acid fraction. In addition, bacteria in general, could continue to ingest old dissolved polar carbon compounds in the sediment column which are subsequently incorporated into the sediment pool. This could, indeed, be true of any other lipid component imparting age greater than ~ 3000 BP. A further refinement with compound-specific radiocarbon data may be necessary to identify such a process (EGLINTON *et al.*, 1996). The old age may also suggest that the time frame of incorporation of fatty acids into sediments is rather unusually long and that not all biogenic components are incorporated into the sediment at the same rate. The available limited data does not allow an unequivocal interpretation of this observation.

Sterols and ketones exhibit comparable radiocarbon ages in the SMB sediment (Table 1 and Fig. 2). Recently, PEARSON et al. (2000) found that C_{29} sterols and algal and zooplankton sterols in the SMB surface (0–0.75 cm) sediments showed very similar modern ^{14}C content and concluded that there is no detectable terrestrial component and that all of the major sterols were derived from phytoplankton production from the surface ocean. However, the sterol fraction from the current study (30–50 cm) exhibits much older age than their sample from 4.5 to 5.5 cm. Assuming the sedimentation rate remained the same in the horizon deeper than 10–12 cm in the core (see "Study area and samples" section; HUH, 1998) and applying no correction for sediment compaction, the deposition age of the 30–50 cm horizon should at least have been between 1280 and 1560 AD, thus representing approximately 430–710 yr of deposition at the time of sampling in 1991. This time interval of a few hundred years could probably be attributed to be from recently deposited phytoplankton detritus. However, the old radiocarbon age of 6360 ± 70 for sterols obtained here cannot be explained by such a process alone. Apparently it reflects additional inputs from old detrital organic matter possibly originating from land erosion. This interpretation is supported by the presence of vascular plant sterol such as β-sitosterol, in dominant proportion relative to other sterols in our sample. Radiocarbon ages of alkenones have been reported to be about 4200 yr older on an average than coexisting planktonic foraminifera in the Bermuda rise drift sediments (OHKOUCHI et al., 2002). In the absence of alkenones data in our sample, contribution from such ketones to the total ketones fraction and/or to the radiocarbon age of the fraction is currently unknown.

Alcohols exhibit a slightly younger age than ketones and sterols. However, hydroxy fatty acids present in sediments could retain free hydroxyl group after methylation into esters and could have eluted here along with alcohols. It should be noted that since hydroxyl acids were not isolated in our procedure, their contribution to carbon isotope data of the alcohol fraction can not be excluded.

Based on compound-specific $\Delta^{14}C$ studies, PEARSON et al. (2001) have reported that many of the lipid fractions from SMB sediments in the surface to ~8 cm horizon like alcohols, sterols and fatty acids derive from marine euphotic zone primary production or subsequent heterotrophic consumption of this biomass. The ages of corresponding fractions in our sample from SMB, but from deeper horizon (30–50 cm), are much older to be attributed entirely to such an origin. This reflects multiple sources of organic matter contributions to the sediments (as discussed above) as a function of time.

Protokerogen measurements provided a much older ^{14}C age than the humic and fulvic acids. Humic acids are major components of the organic matter from marine and non-marine sediments (HEDGES et al., 1986). Because of the refractory nature of humic acids, it may be expected that the old ages derived from surface sediment carbon originates from soil humification (i.e. TISSOT and WELTE, 1984 among others) in comparison to other more bioavailable fractions. Despite the above assumption, humic and fulvic acids are, indeed, the youngest of the chemical fractions analyzed in this study. This suggests that protokerogen is more likely derived from the refractory residues of biological organic components rather than being generated by humification. The refractory biological residues may also have been transported into the coastal shelf sediments, possibly from soil profiles. It is conceivable that only minor amounts of protokerogen could have been eroded from immature Pliocene or Miocene source rock outcrops since the radiocarbon age of the protokerogen fraction is of the order of only 5000 yr and not radiocarbon-dead (Table 1).

Radiocarbon ages of the fractions measured here are in variance with the conventional thinking on the evolution of organic matter, but consistent with several recent reports on the selective accumulation of refractory organic carbon. The refractory carbon in protokerogen could have originated from marine biota (planktonic and even crustaceans). For example, algaenans, a major component of microalgae and bacterans from bacteria, are resistant to acid hydrolysis and have been detected in recent and ancient sediments (TEGELAAR et al., 1989a; BLOKKER et al., 1998 among others). Further, LARGEAU and DE LEEUW (1995) showed that a major part of the refractory organic carbon is comprised of cell wall macromolecular material from the similarity in composition of its pyrolytic products with those from algaenan. Based on the above and other findings and also on the comparative study of microscopic examination of recognizable entities in kerogen and their extant counterparts, TEGELAAR et al. (1989b) proposed

Table 2. Antarctic sediment: isotopic data

Sample	$\delta^{13}C$ (‰)	$\delta^{15}N$ (‰)	$\Delta^{14}C$ (‰)	^{14}C Age (BP)
Antarctic sediment (Residue)	−24.24	+2.14	−236	2160 ± 120
Protokerogen	−35.60		−664	8770 ± 120
Humic acid	−26.13	+2.96	−188	1670 ± 70
Fulvic acid	−25.16		−191	1700 ± 70
Aliphatic	−24.30		−650	8430 ± 70
Aromatic	−27.62		−226	2050 ± 80
Ketone	−27.0		−176	1560 ± 80
Sterol	−28.71		−150	1300 ± 60
Alcohol	−29.53		−191	1700 ± 60
Fatty acid	−26.41		−79	660 ± 90
NEP	−30.78		−184	1640 ± 60

a new and simple mechanism of kerogen formation whereby the bulk of the "amorphous" or protokerogen could largely be derived from the selective accumulation and preservation of resistant biomacromolecules of the lipid fraction. Our results on the ^{14}C age dating of humic and fulvic acids and protokerogen do not refute these arguments and thus, have important implications on the carbon cycling in the ocean.

McMurdo Sound, Antarctica

Stable carbon isotopes data—As shown in Table 2 and Fig. 3, the extracted sediment residue exhibits $\delta^{13}C$ value (−24.24‰) which is lower than the SMB sediment (−22.41‰) and Southern Ocean site in the South Pacific sector in the Antarctic Circumpolar Current region (WANG and DRUFFEL, 2001) but falls within the range of values, −23.90 to −26.50‰, obtained from the unextracted sediments earlier from the same region (VENKATESAN, 1988) and from the Ross Ice Shelf (SACKETT, 1986). The lower values are characteristic of mixed inputs from the cold water plankton, especially the diatom-ooze and older recycled kerogen from the shale fragments (SACKETT, 1986). With the exception of ketone and aliphatic fractions, all other lipid fractions have a lower $\delta^{13}C$ value in the

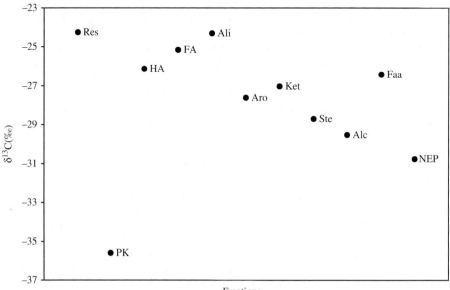

FIG. 3. Stable carbon isotope of organic fractions from the sediments of McMurdo Basin, Antarctica. For explanation of abbreviations of compound class fractions, refer to Figs. 1 and 2.

Antarctic sediments than in SMB. Although $\delta^{13}C$ data are not available for the total lipid fraction from the Antarctic sediment, the uniformly ^{13}C depleted carbon numbers of the lipid classes ranging from -24.30 to $-30.78‰$ are typical of major Antarctic plankton inputs, which are known to be more enriched in ^{12}C over plankton from temperate and tropical regions (RAU et al., 1982; WANG et al., 1996).

Unlike the SMB, where a $\delta^{13}C$ fatty acid anomaly was noted, the fatty acid fraction in the Antarctic sediments exhibits a $\delta^{13}C$ value comparable to other fractions, but lower than the bulk extracted sediment and within the range of C_{16} and C_{18} fatty acids characteristic of marine origin (UCHIDA et al., 2001). This conclusion is also consistent with the carbon number distribution of alkanes, alkenes, fatty acids, alcohols and sterols which reflects the predominance of marine-derived carbon with little recognizable higher plant debris in the sediment. Similar molecular distribution of the lipid classes has previously been reported in sediments from the same general locations in the McMurdo Sound (VENKATESAN, 1988).

The humic and fulvic acids and protokerogen are all uniformly strongly depleted in ^{13}C compared to the SMB sediment with protokerogen exhibiting the lowest $\delta^{13}C$ value (i.e. $-35.6‰$). The stable carbon isotope data in the current humic acid fraction falls within the range of $\delta^{13}C$ values previously reported for acid insoluble fraction from numerous sediments in the Ross Sea. Specifically, values ranging from -23.7 to $-28.9‰$ have been measured in the Western Ross Sea including McMurdo Sound (ANDREWS et al., 1999). The lower ^{13}C values in the Antarctic sediments relative to SMB sediment are most probably related to their origin in the detritus from cold water plankton in addition to indurated shale fragments on the Antarctic land mass.

Stable nitrogen isotopes data—The $\delta^{15}N$ value (Table 2) of the solvent-extracted sediment residue and the humic acid fall within close range ($+2.14$ and $+2.96‰$). The values are closer to that of atmospheric nitrogen ($+0‰$) than the marine end member ($+7$ to $+15‰$) which is relatively enriched in the heavy isotope (SWEENEY and KAPLAN, 1980; STUERMER et al., 1978). The low $\delta^{15}N$ value may in part also reflect nitrogen contribution from the coaly residues eroded from the Antarctic Shelf (STACH et al., 1975) or low values of $\delta^{15}N$ in dissolved marine nitrate in the Antarctic water mass used for planktonic metabolism. The $\delta^{15}N$ of the bulk Antarctic diatomaceous sediment is $+3$ to $+5‰$ and the topmost sediment layer including the fluff layer at the sediment/water interface has $\sim 1‰$ lower $\delta^{15}N$ than the underlying sediment (SIGMAN, 1999). These authors also found that diatom-bound nitrogen, isolated after hydrochloric or perchloric acid treatment of several Antarctic sediments, characteristically contain $\delta^{15}N = +0.5$ to $+2.00‰$ which is lower by $\sim 3‰$ than that for the bulk diatomaceous sediment. This difference is attributed to the diagenesis of the bulk sediment in the ocean in contrast to the preservation of the diatom-bound nitrogen by the microfossil matrix. The $\delta^{15}N$ ($+2.96‰$) of the humic acid in our study after hydrochloric acid treatment is in the range for diatom-bound nitrogen consistent with the composition of the sediments which is rich in diatom debris (LEVENTER and DUNBAR, 1987). Only the sediment residue after solvent extraction was analyzed for nitrogen isotope which had a similar $\delta^{15}N$ ($+2.14‰$) value to that of humic acid. However, a $\delta^{15}N$ value of the bulk sediment before any treatment is not available for further comparison with the results of SIGMAN (1999).

Radiocarbon data—Based on the sedimentation rate of 2.7 mm/yr, the top 8 cm of the grab sample from McMurdo Sound should represent approximately 30 yr of accumulation and presumably include sediment deposited in the "post-bomb" era. Relatively younger age is also expected in the lipids of the Antarctic sediments as humic detritus is not significant in the region. Further, the organic-rich diatomaceous debris which accumulates rapidly in this environment could dilute any reworked or land-derived detritus (DOMACK et al., 1999). With the exception of the aliphatic and protokerogen fractions, all other fractions are close to or younger than 2000 BP (Table 2, Fig. 4) and exhibit younger age than the corresponding organic fractions from the SMB. The $\Delta^{14}C$ values of the acid insoluble and lipid fractions from the Southern Ocean sediment core in the top 9 cm sections studied by WANG and DRUFFEL (2001) range from -461 to $-937‰$ and are comparable to the data presented here for the sediment, protokerogen and lipid fractions (Table 2). Transport of terrestrial and fossil sources and reworked marine carbon were believed to contribute an "old age" to their sediments.

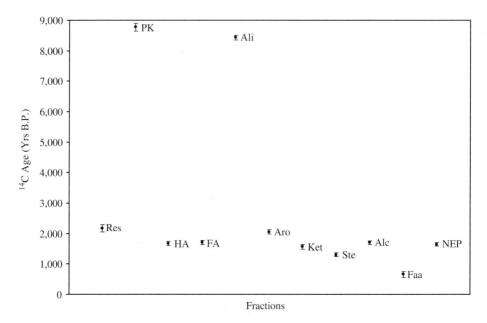

Fig. 4. Radiocarbon age of the organic fractions from the sediments of McMurdo Basin, Antarctica. For explanation of abbreviations of compound class fractions, refer to Figs. 1 and 2.

The older age imparted by the protokerogen (8770 BP), the aliphatic fraction (8430 BP) and to some degree by the aromatic fraction (2050 BP) in the Antarctic sample are consistent with stable carbon ratio described by STACH *et al.* (1975) and may originate from admixture of coal residues introduced into McMurdo Sound. The mono-and/or di-methylated phenanthrenes found in higher amounts relative to the parent homolog and the presence of retene and simonellite in the McMurdo sediment analogous to the previously studied cores may originate from eroded coal outcrops in the seafloor (VENKATESAN, 1988). The average age of the core top sediment in the Ross Sea is 1300–3200 yr older than ages obtained on living marine carbonates around Antarctica (BERKMAN and FORMAN, 1996). This is believed to reflect variable sedimentation rates in the region from glacial land erosion in addition to about 15% input from "old" reworked material in the Western Ross Sea (ANDREWS *et al.*, 1999). This is also supported by the data documenting the distribution of recycled palynomorphs in the core top sediments from the Ross Sea (TRUSWELL and DREWRY, 1984). The discrepancy between the measured and the reservoir-corrected age could also arise from old DIC uptake by the diatoms which is not limited by CO_2 availability (GIBSON *et al.*, 1999; TORTELL *et al.*, 1997). However, the $\Delta^{14}C$ values for the organic fractions measured in the current study demonstrate that the fossil carbon input is much less in proportion to the recent autochthonous carbon in the Antarctic sediments in comparison to the SMB sediment and unlike most other aquatic and oceanic regimes where ancient sedimentary organic carbon has previously been found to be a significant component of the suspended load from rivers and marine sediments (*i.e.* MASIELLO and DRUFFEL, 2001; BLAIR *et al.*, 2003).

Unlike the SMB sediment, fatty acid fraction in the Antarctic sediment is the youngest (660 BP). Fatty acid was the major lipid component (10.5 μg/g dry sediment) from this sediment and was about 80 times greater than that of *n*-alkane fraction. The $\Delta^{14}C$ value of $-79‰$ is in between the values reported (-198 and $-62‰$), respectively, for marine fatty acids, C_{16} and C_{18}, from the western North Pacific sediment (UCHIDA *et al.*, 2001). The young age is also consistent with the stable carbon isotope ratios characteristic of marine origin. Similarly the relative ^{13}C depletion of the ketone, sterol, alcohol and NEP fractions in the sediment and their young age as reflected in the $\Delta^{14}C$ value ranging from -150 to $191‰$, suggest significant inputs from marine organisms. Further, the relatively young age of the fatty acid fraction also suggests that atmospheric transport of old reworked biogenic fatty acid carbon by wind blown aerosols from the northern continents is either absent or trivial in this region, unlike that detected in western North Pacific sediments (UCHIDA et al., 2001). The dissolved inorganic carbon in the northwestern corner of Ross Sea, close to McMurdo Shelf was found to have $\Delta^{14}C$ in the range of -110 to 165‰ (Stations, 772, 775, 777, 779 and 780 in the WOCE cruise S4P in KEY and SCHLOSSER, 1999). This suggests that the sediment and all the compound classes except aliphatic fraction and protokerogen reflect largely the age imparted by the dissolved carbon uptake, more specifically the DIC (RAU, 1991).

Humic and fulvic acids exhibit comparable young age to those of the other lipid fractions, consistent with a marine source, although fulvic acid appears older than that from SMB sediment (1700 vs. 760 BP, respectively). In contrast, protokerogen shows much older age (8770 BP) than all other fractions in the Antarctic sediment suggesting low-level contribution from fossil/coal derived carbon. Antarctic kerogen is also older than that from SMB. However, if the fossil carbon contribution was significant, the Antarctic protokerogen would have exhibited much older age than measured here. The data indicates that the Antarctic protokerogen is also largely and more likely a refractory component of biological organic matter rather than being derived from the humification process and, thus, reinforces arguments related to protokerogen origin in SMB sediment.

CONCLUSIONS

Several significant findings become obvious when radiocarbon results are considered together with stable isotope and biomarkers data. (i) SMB receives inputs from old reworked material from submarine natural petroleum seepage and possibly eroded Monterey Shale fragments in addition to biogenic (marine and terrestrial) sources as reflected in the molecular and stable carbon isotopic composition and the older ^{14}C age of the aliphatic and aromatic fractions. (ii) The non-recent ^{14}C age of the fatty acid fraction suggests possible input from fossil biogenic component, most probably bacterial detritus. (iii) The relatively older age of the ketone and sterol fraction could be from old diagenetically altered land derived organic matter. (iv) $\delta^{15}N$ of the measured fractions suggests recent input from plankton. (v) The young age of the humic and fulvic acid fractions (760–1650 yr) suggests that ^{14}C ages of carbon in excess of ~3000 yr in these sediments probably do not originate from humification. This is contrary to the expectations based on the refractory nature of humic compounds. (vi) Protokerogen is much older than the humic and fulvic acids. This observation, together with the humic and fulvic acids data suggest that protokerogen is more likely a refractory component of biological organic matter rather than being derived from the humification process.

The uniformly lighter $\delta^{13}C$ value of the lipid classes in McMurdo Sound sediment is characteristic of cold water plankton input and DIC. This is also corroborated by the $\delta^{15}N$ value of humic acid which is similar to that reported for diatom-bound nitrogen. As expected from their major origin in biogenic components, most of the fractions in the Antarctic sediment exhibit young radiocarbon age. (i) The older age imparted by the aliphatic fraction and to some degree by the aromatic fraction in the Antarctic sample could be traced to an origin in coal residues being reworked into this region. (ii) The young ^{14}C age suggests significant inputs from marine organisms to the fatty acid, ketone, sterol, alcohol and NEP fractions in the sediment. (iii) Humic and fulvic acids exhibiting young age comparable to the other lipid fractions are consistent with a marine source. (iv) Much older ^{14}C age of protokerogen than all other fractions again indicates that it may not be generated from a humification process, but more likely derives from refractory biomacromolecules.

Most of the fractions in the Antarctic sediment have lower $\delta^{13}C$ and exhibit younger ^{14}C age than the sediment from SMB. The data also document the presence of old organic carbon besides the recent biogenic inputs in sediments from both study locations. Further, the results from humic and fulvic acids and, protokerogen have significant implications on the carbon cycling in the ocean and in understanding the formation of kerogen. The data, thus, provide a test for the source of various compound classes (rather than the bulk carbon) and help understand important benthic processes, such as fluxes, transformations and diagenesis in both water and sedimentary columns. The radiocarbon data add a new dimension to the existing database. The combined database provides considerably more information on multiple sources of organic matter and its preservation than can be obtained by any one technique alone.

Acknowledgements—We thank Mr. Ed Ruth for collecting the core from the Santa Monica Basin and for GC/MS analysis, Dr Tom Dorsey for collecting the samples from McMurdo Sound and Dr Robert Dunbar, his colleagues and Dr Amy Leventer for rendering invaluable help in collecting the McMurdo Sound samples, Rosario P. DeLeon for laboratory assistance, Dr Robert Key for the WOCE data on the Ross Sea and Prof. Sue Trumbore for the wood sample. We also thank the anonymous reviewers for their helpful and critical reviews and, in particular, gratefully acknowledge the detailed comments of one reviewer which helped significantly improve the original version. Funding was provided to M. I. V. by Lawrence Livermore National Laboratory/Center for Accelerator Mass Spectrometry, Task 35 and DPP8816292 (NSF). This is Institute of Geophysics and Planetary Physics, UCLA Contribution No: 5822.

REFERENCES

ANDREWS J. T., DOMACK E. W., CUNNINGHAM W. L., LEVENTER A., LIGHT K. J., JULL A. J. T., DEMASTER D. J. and JENNINGS A. E. (1999) Problems and possible solutions concerning radiocarbon dating of surface marine sediments, Ross Sea, Antarctica. *Q. Res.* **52**, 206–216.

BALLENTINE D. B., MACKO S. A. and TUREKIAN V. C. (1988) Variability of stable carbon isotopic compositions in individual fatty acids from combustion of C_4 and C_3 plants: implications for biomass burning. *Chem. Geol.* **152**, 151–161.

BARRETT P. J., STOFFER P., GLASBY G. P. and PLUGER W. L. (1984) Texture, mineralogy and composition of four sediment cores from Granite Harbour and New Harbour, southern Victoria Land, Antarctica. *NZ J. Geol. Geophys.* **27**, 477–485.

BAUER J. E., SPIES R. B., VOGEL J. S., NELSON D. E. and SOUTHON J. R. (1990) Radiocarbon evidence of fossil-carbon cycling in sediments of a near shore hydrocarbon seep. *Nature* **348**, 230–232.

BENOIT G. J., TUREKIAN K. K. and BENNINGER L. K. (1979) Radiocarbon dating of a core from Long Island Sound. *Estuar. Coast. Mar. Sci.* **9**, 171–180.

BERGER R., TAYLOR R. E. and LIBBY W. F. (1966) Radiocarbon content of marine shells from the California and Mexican west coast. *Science* **153**, 864–866.

BERKMAN P. A. and FORMAN S. L. (1996) Pre-bomb radiocarbon and the reservoir correction for calcareous marine species in the Southern Ocean. *Geophys. Res. Lett.* **23**, 363–366.

BLAIR N. E., LEITHOLD E. L., FORD S. T., PEELER K. A., HOLMES J. C. and PERKEY D. W. (2003) The persistence of memory: the fate of ancient sedimentary organic carbon in a modern sedimentary system. *Geochim. Cosmochim. Acta* **67**, 63–73.

BLOKKER P., SCHOUTEN S., ENDE H. V. D., de LEEUW J. W., HATCHER P. G. and DAMSTE J. S. S. (1998) Chemical structure of algaenans from fresh water algae *Tetrahedron minimum, Scenedesmus communis* and *Pediastrum boryanum. Org. Geochem.* **29**, 1453–1468.

BRASSELL S., EGLINTON G., MARLOWE I. T., PFLAUMANN U. and SARNTHEIN M. (1986) Molecular stratigraphy: a new tool for climatic assessment. *Nature* **320**, 129–133.

BRAY E. E. and EVANS E. D. (1961) Distribution of *n*-paraffins as a clue to the recognition of source beds. *Geochim. Cosmochim. Acta* **22**, 2–15.

CLINE J. D. and KAPLAN I. R. (1975) Isotopic fractionation of dissolved nitrate during denitrification in the eastern tropical North Pacific ocean. *Mar. Chem.* **3**, 271–299.

COLLISTER J. W., RIELEY G., STERN B., EGLINTON G. and FRY B. (1994) Compound-specific $\delta^{13}C$ analyses of leaf lipids from plants with differing carbon dioxide metabolism. *Org. Geochem.* **21**, 619–627.

CRISP P. T., BRENNER S., VENKATESAN M. I., RUTH E. and KAPLAN I. R. (1979) Organic chemical characterization of sediment-trap particulates from San Nicolas, Santa Barbara, Santa Monica and San Pedro Basins, California. *Geochim. Cosmochim. Acta* **43**, 1791–1801.

DOMACK E., HALL B. and HAYES J. M. (1999) Accurate Antarctic dating techniques sought by Quaternary community. *EOS* **80**, 591 p.

DRUFFEL E. R. M. and WILLIAMS P. M. (1990) Identification of a deep marine source of particulate organic carbon using bomb ^{14}C. *Nature* **347**, 172–174.

DRUFFEL E. R. M., WILLIAMS P. M., BAUER J. E. and ERTEL J. R. (1992) Cycling of dissolved and particulate organic matter in the open ocean. *J. Geophys. Res.* **97**, 15639–15659.

EGLINTON T. I., ALUWIHARE L. I., BAUER J. E., DRUFFEL E. R. M. and MCNICHOL A. P. (1996) Gas chromatographic isolation of individual compounds from complex matrices for radiocarbon dating. *Anal. Chem.* **68**, 904–912.

EGLINTON T. I., BENITEZ-NELSON B. C., PEARSON A., MCNICHOL A. P., BAUER J. E. and DRUFFEL E. R. M. (1997) Variability in radiocarbon ages of individual organic compounds from marine sediments. *Science* **277**, 796–799.

EMERY K. O. and BRAY E. E. (1962) Radiocarbon dating of California Basin sediments. *Bull. Am. Assoc. Petrol. Geol.* **48**, 1839–1856.

FREEMAN K. H., HAYES J. M., TRENDEL J.-M. and ALBRECHT P. (1990) Evidence from carbon isotope measurements for diverse origins of sedimentary hydrocarbons. *Nature* **343**, 254–256.

GIBSON J. A. E., TRULL T., NICHOLS P. D., SUMMONS R. E. and MCMINN A. (1999) Sedimentation of ^{13}C-rich organic matter from Antarctic sea-ice algae: a potential indicator of past sea-ice extent. *Geology* **27**, 331–334.

HAYES J. M., TAKIGIKU R., OCAMPO R., CALLOT H. J. and ALBRECHT P. (1987) Isotope compositions and probable origins of organic molecules in the Eocene Messel Shale. *Nature* **329**, 48–51.

HAYES J. M., FREEMAN K. H., POPP B. N. and HOHAM C. H. (1990) Compound-specific isotopic analyses: a novel tool for reconstruction of ancient biogeochemical processes. *Org. Geochem.* **16**, 1115–1128.

HEDGES J. I., ERTEL J. R., QUAY P. D., GROOTES P. M., RICHEY J. E., DEVOL A. H., FARWELL G. W., SCHMIDT F. W. and SALATI E. (1986) Organic carbon-14 in the Amazon River system. *Science* **231**, 1129–1131.

HEDGES J. I., KEIL R. G. and BENNER R. (1997) What happens to terrestrial organic matter in the ocean? *Org. Geochem.* **27**, 195–212.

HINRICHS K. U., HAYES J. M., SYLVA S. P., BREWER P. G. and DELONG E. F. (1999) Methane-consuming archaebacteria in marine sediments. *Nature* **398**, 802–805.

HUANG W. Y. and MEINSCHEIN W. G. (1976) Sterols as source indicators of organic materials in sediments. *Geochim. Cosmochim. Acta* **40**, 323–330.

HUH C. A. (1998) Historical contamination of the Southern California Bight-Metals. Historical Contamination in the Southern California Bight. NOAA Technical Memorandum NOS ORCA 129. US Dept. of Commerce/NOAA, Maryland, 47 pp.

HUH C. A., ZAHNLE D. L., SMALL L. F. and NOSHKIN V. E. (1987) Budgets and behaviors of uranium and thorium series isotopes in Santa Monica Basin sediments. *Geochim. Cosmochim. Acta* **51**, 1743–1754.

KEY R. M. and SCHLOSSER P. (1999) S4P Final report for AMS ^{14}C samples, Ocean Tracer Laboratory. Technical report 99-1. Princeton University Press, Princeton, NJ.

LARGEAU C. and DE LEEUW J. W. (1995) Insoluble, nonhydrolyzable, aliphatic macromolecular constituents of microbial cell walls. In *Advances in Microbial Ecology* (ed. J. G. JONES), Plenum Press, pp. 77–117. New York.

LEVENTER A. and DUNBAR R. B. (1987) Diatom flux in McMurdo Sound, Antarctica. *Mar. Micropaleon.* **12**, 49–64.

MACPHERSON A. J. (1987) The Mackay Glacier/Granite Harbour system (Ross Dependency, Antarctica)—a study in near shore glacial marine sedimentation. Ph.D. Thesis, Victoria University of Wellington, New Zealand, 173 pp.

MANNING M. R., LOWE D. C., MELHUISH W. H., SPARKS R. J., WALLACE G., BRENNINKMEIJER C. A. M. and MCGILL R. C. (1990) The use of radiocarbon measurements in atmospheric studies. *Radiocarbon* **32**, 37–58.

MASIELLO C. A. and DRUFFEL E. R. M. (2001) Carbon geochemistry of the Santa Clara river. *Global Biogeochem. Cycles* **15**, 407–417.

MASIELLO C. A. and DRUFFEL E. R. M. (2003) Organic and black carbon ^{13}C and ^{14}C through the Santa Monica Basin sediment oxic–anoxic transition. *Geophys. Res. Lett.* **40**, 1185–1188.

MATSUMOTO K., KAWAMURA K., UCHIDA M., SHIBATA Y. and YONEDA M. (2001) Compound-specific radiocarbon and δ^{13}C measurements of fatty acids in a continental aerosol sample. *Geophys. Res. Lett.* **28**, 4587–4590.

MINAGAWA M., WINTER D. A. and KAPLAN I. R. (1984) Comparison of Kjeldahl and combustion methods for measurement of nitrogen isotope ratios in organic matter. *Anal. Chem.* **56**, 1859–1861.

OHKOUCHI N., EGLINTON T. I., KEIGWIN L. D. and HAYES J. M. (2002) Spatial and temporal offsets between proxy records in a sediment drift. *Science* **298**, 1224–1227.

PEARSON A. and EGLINTON T. I. (2000) The origin of *n*-alkanes in Santa Monica Basin surface sediment: a model based on compound-specific Δ^{14}C and δ^{13}C data. *Org. Geochem.* **31**, 1103–1116.

PEARSON A., EGLINTON T. I. and MCNICHOL A. P. (2000) An organic tracer for surface ocean radiocarbon. *Paleoceanography* **15**, 541–550.

PEARSON A., MCNICHOL A. P., BENITEZ-NELSON B. C., HAYES J. M. and EGLINTON T. I. (2001) Origins of lipid biomarkers in Santa Monica Basin surface sediment: a case study using compound-specific Δ^{14}C analysis. *Geochim. Cosmochim. Acta* **65**, 3123–3137.

PETERS K. E., SWEENEY R. E. and KAPLAN I. R. (1978) Correlation of carbon and nitrogen stable isotope ratios in sedimentary organic matter. *Limnol. Oceanogr.* **23**, 598–604.

RAU G. H. (1991) Another recipe for bomb ^{14}C dilution. *Nature* **350**, 116 p.

RAU G. H., SWEENEY R. E. and KAPLAN I. R. (1982) Plankton ^{13}C : ^{12}C ratio changes with latitude: differences between northern and southern oceans. *Deep Sea Res.* **29**, 1035–1039.

RAYMOND P. A. and BAUER J. E. (2001) Riverine export of aged terrestrial organic matter to the North Atlantic Ocean. *Nature* **409**, 497–500.

ROWLAND S. J. and ROBSON J. N. (1990) The widespread occurrence of highly branched acyclic C_{20}, C_{25} and C_{30} hydrocarbons in recent sediments and biota—a review. *Mar. Environ. Res.* **30**, 191–216.

SACKETT W. M. (1986) Organic carbon in sediments underlying the Ross Ice Shelf. *Org. Geochem.* **9**, 135–137.

SCHOUTEN S. (1995) Structural and stable carbon isotope studies of lipids in immature sulphur-rich sediments. Ph.D. Thesis. University of Groningen, The Netherlands.

SIGMAN D. M. (1999) The isotopic composition of diatom-bound nitrogen in Southern Ocean sediments. *Paleoceanography* **14**, 118–134.

SIMONEIT B. R. T. and KAPLAN I. R. (1980) Triterpenoids as molecular indicators of paleoseepage in recent sediments of the Southern California Bight. *Mar. Environ. Res.* **3**, 113–128.

STACH E., MACKOWSKY M.-T. H., TEICHMULLER M., TAYLOR G. H., CHANDRA D. and TEICHMULLER R. (1975) *Stach's Textbook of Coal Petrology*, 2nd Ed., Gebruder Borntraeger, Berlin, 428 pp.

STUERMER D. H., PETERS K. E. and KAPLAN I. R. (1978) Source indicators of humic substances and proto-kerogen: stable isotope ratios, elemental compositions and electron spin resonance spectra. *Geochim. Cosmochim. Acta* **42**, 989–997.

STUIVER M. and POLACH H. A. (1977) Discussion: reporting of ^{14}C data. *Radiocarbon* **19**, 355–363.

SUMMONS R. E., JAHNKE L. and ROKSANDIC Z. (1994) Carbon isotopic fractionation in lipids from methanotrophic bacteria: relevance for interpretation of the geochemical record of biomarkers. *Geochim. Cosmochim. Acta* **58**, 2853–2863.

SWEENEY R. E. and KAPLAN I. R. (1980) Natural abundances of ^{15}N as a source indicator for near-shore marine sedimentary and dissolved nitrogen. *Mar. Chem.* **9**, 81–94.

SWEENEY R. E., LIU K. K. and KAPLAN I. R. (1978) Oceanic nitrogen isotopes and their uses in determining the source of sedimentary nitrogen. In *Stable Isotopes in the Earth Sciences* (eds. B. W. ROBINSON, R. E. SWEENEY and K. K. LIU), DSIR Bulletin 220, pp. 9–26. New Zealand Department of Scientific and Industrial Research, Wellington, New Zealand.

TEGELAAR E. W., MATTHEZING R., JANSEN J. B. H., HORSFIELD B. and DE LEEUW J. W. (1989a) Possible origin of *n*-alkanes in high-wax crude oils. *Nature* **342**, 529–531.

TEGELAAR E. W., de Leeuw J. W., DERENE S. and LARGEAU C. (1989b) A reappraisal of kerogen formation. *Geochim. Cosmochim. Acta* **53**, 3103–3106.

TISSOT B. P. and WELTE D. H. (1984) *Petroleum Formation and Occurrence*. Springer, Berlin, 699 pp.

TORTELL P. D., REINFELDER J. R. and MOREL F. M. M. (1997) Active uptake of bicarbonate by diatoms. *Nature* **390**, 243–244.

TRUMBORE, S. E., ANDERSON R. F. and BISCAYE P. (1991) Carbon and C-14 budget for upper slope sediments of the Middle Atlantic Bight. Presented at *Fourth Chemical Congress of North America*, August 25–30, New York.

TRUSWELL E. M. and DREWRY D. J. (1984) Distribution and provenance of recycled palynomorphs in surficial sediments of the Ross Sea. *Mar. Geol.* **59**, 187–214.

UCHIDA M., SHIBATA Y., KAWAMURA K., KUMAMOTO Y., YONEDA M., OHKUSHI K., HARADA N., HIROTA M., MUKAI H., TANAKA A., KUSAKABE M. and MORITA M. (2001) Compound-specific radiocarbon ages of fatty acids in marine sediments from the western North Pacific. *Radiocarbon* **43**, 949–956.

VENKATESAN M. I. (1988) Organic geochemistry of marine sediments in Antarctic region: marine lipids in McMurdo Sound. *Org. Geochem.* **12**, 13–27.

VENKATESAN M. I. (1998) Part II. Historical trends in the deposition of organic pollutants in the Southern California Bight. Historical Contamination in the Southern California Bight. NOAA Technical Memorandum NOS ORCA 129. US Department of Commerce/NOAA, Maryland, 35 pp.

VENKATESAN M. I. and KAPLAN I. R. (1990) Sedimentary coprostanol as an index of sewage addition in Santa Monica Basin, Southern California. *Environ. Sci. Technol.* **24**, 208–214.

VENKATESAN M. I. and KAPLAN I. R. (1992) Vertical and lateral transport of organic carbon and the carbon budget in Santa Monica Basin, California. *Prog. Oceanogr.* **30**, 291–312.

VENKATESAN M. I. and KAPLAN I. R. (1994) Sources and sinks of organic carbon in coastal marine environments from C-13 and C-14 measurements. *EOS* **75**(44), 326 p.

VENKATESAN M. I. and MIRSADEGHI F. H. (1992) Coprostanol as sewage tracer in McMurdo Sound, Antarctica. *Mar. Pollut. Bull.* **25**, 328–333.

VENKATESAN M. I., BRENNER S., RUTH E., BONILLA J. and KAPLAN I. R. (1980) Hydrocarbons in age-dated sediment cores from two basins in the southern California Bight. *Geochim. Cosmochim. Acta* **44**, 789–802.

VENKATESAN M. I., RUTH E., STEINBERG S. and KAPLAN I. R. (1987) Organic geochemistry of sediments from the continental margin off southern New England, USA—Part II. Lipids. *Mar. Chem.* **21**, 267–299.

WAKEHAM S. G., FARRINGTON J. W., GAGOSIAN R. B., LEE C., DEBAAR H., NIGRELLI G. E., TRIPP B. W., SMITH S. O. and FREW N. M. (1980) Organic matter fluxes from sediment traps in the equatorial Atlantic Ocean. *Nature* **286**, 798–800.

WANG X. C. and DRUFFEL E. R. M. (2001) Radiocarbon and stable carbon isotope compositions of organic compound classes in sediments from the NE Pacific and Southern Oceans. *Mar. Chem.* **73**, 65–81.

WANG X. C., DRUFFEL E. R. M. and LEE C. (1996) Radiocarbon in organic compound classes in particulate organic matter and sediment in the deep northeast Pacific Ocean. *Geophys. Res. Lett.* **23**, 3583–3586.

WILLHELMS A., LARTER S. R. and HALL K. (1994) A comparative study of the stable carbon isotopic composition of crude oil alkanes and associated crude oil asphaltene pyrolysate alkanes. *Org. Geochem.* **21**, 751–759.

WILLIAMS P. M., ROBERTSON K. J., SOUTAR A., GRIFFIN S. M. and DRUFFEL E. R. M. (1991) Isotopic signatures (^{14}C, ^{13}C, ^{15}N) as tracers of sources and cycling of soluble and particulate organic matter in the Santa Monica Basin, California. *Prog. Oceanogr.* **30**, 253–290.

Geochemical Investigations in Earth and Space Science: A Tribute to Isaac R. Kaplan
© The Geochemical Society, Publication No. 9, 2004
Editors: R.J. Hill, J. Leventhal, Z. Aizenshtat, M.J. Baedecker, G. Claypool,
R. Eganhouse, M. Goldhaber and K. Peters

Holocene chronostratigraphic beachrocks and their geologic climatic significance

GERALD M. FRIEDMAN[*]

Department of Geology, Brooklyn College and Graduate Center of the
City University of New York, Brooklyn, NY, USA

Abstract—Rocks that form by cementation in the intertidal parts of beaches are known as beachrock. Nearly all modern beachrocks are restricted to warm climatic belts between latitudes 35°N and 35°S. This suggests that in addition to seawater supersaturated for $CaCO_3$, high temperatures within the beach sediments are necessary for cementation. The particles composing beachrocks generally are well-sorted beach sands cemented by carbonate.

In this study gently seaward-dipping prograding successions of beachrock have been compared for three different geographic settings: (1) an island in the southern Caribbean offshore from the mainland of Venezuela, part of the Caribbean Plate, (2) the north shore of Puerto Rico, and (3) the shoreline of the rift valley of the Gulf of Aqaba, a steep-sided tectonic valley that forms the northern segment of the Red Sea.

The purpose of this study is to determine variables involved in cementing beach sediment into beachrock. Beachrocks are markers of relative sea level. The process of cementation reflects the geochemistry and temperature conditions of the interstitial waters at the time of cementation. Since the seaward-dipping successions reflect progressive stages of cementation, a geologic history of the shoreline may be unraveled.

The four- to five levels of prograding beachrock in the three studied areas represent fluctuations in relative sea-level due to eustacy. Beachrocks of the Gulf of Aqaba cements are of decreasing radioactive ages (~ 7000 yr BP to modern) paralleling increasing strontium isotope ratios (0.708930–0.709248) reflecting chronostratigraphic variation. The strontium isotopic values of these cements are among the highest for Phanerozoic carbonates and point to rapid continental chemical weathering and erosion recording fast unroofing. Superimposed on this orogeny the high values for carbon and strontium isotope ratios may relate to Red Sea sea-floor spreading, related tectonic uplifting, weathering, and erosion. The oxygen isotopic composition of Gulf of Aqaba beachrock, recorded by cement reflects temperature decrease of ambient seawater for approximately half of Holocene time.

The computed temperature excursion of the Red Sea beachrock cement implies temperature decrease between the ages 7070 ± 380 and 2620 ± 230 yr BP, an interval of approximately 4 1/2 thousand years, during which the average Red Sea seawater temperature fell from 33 to 17°C. This discovery is at variance with the climate change debate which involves increasing temperatures.

Changes in $\delta^{13}C$ values and $^{87}Sr/^{86}Sr$ ratios of the beachrock cement reflect tectonic activity recording uplift and attendant weathering.

The levels of prograding beachrock in the three studied areas represent sea-level fluctuations and/or tectonic activity. Earthquakes are indicated for the North American Plate, where at least three levels of beachrock were generated during the last few hundred years.

The $^{87}Sr/^{86}Sr$ ratios of the beachrock cements of the southern Caribbean offshore islands of Venezuela (Los Roques Island) are above that of modern seawater. By contrast the $^{87}Sr/^{86}Sr$ ratios of Puerto Rico samples overlap those of modern seawater, and the Red Sea values are below that of modern seawater. The increase in the $^{87}Sr/^{86}Sr$ ratios of the beachrock cement with decreasing age suggests that weathering ratios were high or that the weathered rocks were exceptionally radiogenic, or reflect changes in the global isotope mass balance.

---

[*] *Present address*: Northeastern Science Foundation, affiliated with Brooklyn College, Rensselaer Center of Applied Geology, 15 Third Street, P.O. Box 746, Troy, NY 12181, USA.

INTRODUCTION

THIS STUDY uses field relations, petrography and Sr, O, and C isotopic analyses to examine nearshore marine processes that convert Holocene unconsolidated beach sediment into consolidated beachrock.

Lithification may be the result of inorganic diagenesis or biomineralization. Precipitation of lime mud (micrite) as a cement may be the first stage of cementation followed by deposition of prismatic cement (TAYLOR and ILLING, 1969; MEYERS, 1987; HOLAIL and RASHED, 1992; NEUMEIER, 1999). Various hypothesis have been proposed to explain early diagenesis in the beachrock environment (SCOFFIN and STODDART, 1983): (1) Precipitation from evaporating seawater, (2) CO_2-degassing from carbonate-concentrated continental mixing or brackish waters at the vadose–phreatic contact (HANOR, 1978), and (3) biomineralization created by microbial precipitation (KRUMBEIN, 1979).

In this study gently seaward-dipping successions of beachrock are compared for a Caribbean island off the coast of South America (Figs. 1 and 2), the north shore of Puerto Rico (Figs. 1 and 3), and the northern Red Sea (Figs. 4 and 5). These successions represent fluctuating relative sea-level change due to eustacy and/or tectonic activity, including major earthquakes.

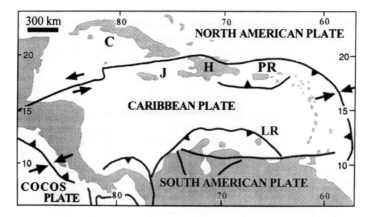

FIG. 1. Geologic provinces of the Caribbean are shown. Islands of the Greater Antilles: C—Cuba, H—Hispaniola, J—Jamaica, PR—Puerto Rico, and LR—Los Roques Islands. Beachrocks on Los Roques Islands (South American Plate) and northeastern Puerto Rico (Caribbean Plate) are described. (*Source*: R. N. ABBOT, Jr., GRANVILLE DRAPER, SHANTANO KESHAV, 2001).

FIG. 2. Photograph of prograding beachrock sequence on Atlantic shoreline of Francisqui Cayo, Los Roques Island, Venezuela. Note low angle of dip to upper right. Arrow points to baseball cap which serves as scale. (Photo: EMMA C. KUMMEROW, Lagoven, Venezuela).

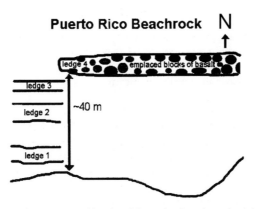

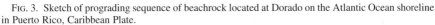

FIG. 3. Sketch of prograding sequence of beachrock located at Dorado on the Atlantic Ocean shoreline in Puerto Rico, Caribbean Plate.

What is beachrock?

Intertidal cementation of beach sediment forms beachrock. Oxygen isotopes in beachrock cement may provide a clue to anthropogenic and pre-anthropogenic climatic and tectonic history to allow the reconstruction of such changes over several thousand years.

Beachrocks are thin-layered (centimeter-thick) deposits inclined seaward at low angles with a dip of 5–15° (Figs. 2 and 5).

Beachrock typically extends vertically from low tide, to a short distance above high tide, during which the waves break over the beachrock surface. Beach sand that has migrated into deeper water, or has been blown into higher inland elevations normally remains unlithified (PURSER, 1980). Because the zone of sedimentation is strongly restricted, beachrocks are indicators of relative sea level (*e.g.* HOPLEY, 1986). Cementation in nearly all modern beachrocks is restricted to warm climates between latitudes 35°N and 35°S, although sporadic beachrocks have been reported from high-latitude coasts (KNEALE and VILES, 2000). Warm temperature is required to achieve calcium-carbonate supersaturation in seawater, and to degas CO_2 from beach interstitial groundwater.

A typical beachrock with several inclined beds (Figs. 2 and 5) corresponds always to a seaward prograding sequence with a temporal succession from the land to the sea. Generally the beachrock is assumed to be formed below a sediment cover in the beach; in that case the included beds reflect the beach stratifications and the cementation can be contemporaneous over a large distance.

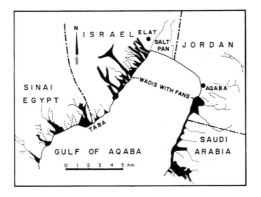

FIG. 4. Index map of northern Gulf of Aqaba showing location of area between Elat and Taba, source of studied beachrock.

Fig. 5. Beachrock with abundant included terrigenous pebbles paving intertidal zone and extending into supratidal zone on west side of Gulf of Aqaba.

The chief beachrock cements are high-magnesian calcite, low-magnesian calcite, and aragonite (Table 1). Superimposed beachrock levels can result from shoreline progradation if enough sediment is available (SCOFFIN and STODDART, 1983). Aragonite-cemented sandstone from the outer continental Atlantic shelf resembles beachrock, but its cementation appears to have been controlled by methane (ALLEN et al., 1969; FRIEDMAN et al., 1971).

Purpose of research and historical background

One purpose of this study is to use radiocarbon, strontium isotopes, stable carbon and oxygen isotopes, and petrography with the aim to determine timing and rates of formation of the carbonate cements between the sand particles, to link Holocene geochemical and paleoceanographic changes to geologic events. These various techniques determine residence time of each beachrock level, and date the timing of the sea-level changes and/or tectonic activity, such as earthquakes, and the changing of temperature and composition of the seawater at each event of cementation. $\delta^{18}O$ values provide a tell-tale sign of climate change (but also the composition of seawater).

Important reviews on beachrock formation are included in GINSBURG (1953), NESTEROFF (1954), RUSSELL (1960), STODDART and CANN (1965), FRIEDMAN (1968), SCOFFIN (1970), FRIEDMAN and GAVISH (1971), SCHMALZ (1971), DAVIES and KINSEY (1973), SIESSER (1974), FRIEDMAN (1975), GAVISH (1975), HANOR (1978), KRUMBEIN (1979), SCOFFIN and STODDART (1983), GAVISH et al. (1985), HOPLEY and DAVIES (1986), MEYERS (1987), EL-SAYED (1988),

Table 1. X-ray diffraction mineralogical analysis. Beachrock, northwestern shore of Gulf of Aqaba, Red Sea (prograding sequence from Elat #1-99 (youngest) to Elat #4-99 (oldest))

Mineral constituents	Relative abundance (%)			
Sample identification	Elat #1-99	Elat #2-99	Elat #3-99	Elat #4-99
Quartz	28	33	37	41
Plagioclase feldspar	40	27	20	22
K-feldspar	10	9	10	12
Low-magnesian calcite	1	7	16	7
High-magnesian calcite	5	9	9	8
Aragonite	5	8	2	3
Halite	trc	trc	trc	trc
Kaolinite	1	1	1	2
Chlorite	3	2	2	2
Illite/Mica	6	2	2	2
Mixed-layer Illite/Smectite	1	2	1	1
(% Illite layers in M.L. Illite/Smectite)	(30–40)	(30–50)	(40–50)	(30–50)
Total	100	100	100	100

STRASSER *et al.* (1989), GUO and FRIEDMAN (1990), HOLAIL and RASHED (1992), KENDALL *et al.* (1994), GISCHLER and LOMANDO (1997), CAMOIN (1999), NEUMEIER (1999), WEBB *et al.* (1999), and KNEALE and VILES (2000).

Rapid conversion of beach sand to beachrock

Loose particles can be cemented into beachrocks very rapidly. Pottery fragments in eastern Mediterranean beachrocks, Coca Cola and rum bottles in Caribbean beachrocks, and debris from World War II in Pacific beachrocks confirm that cementation is occurring today. Loose particles are stabilized below the beach, where microbial binding and cementation can take place. Once consolidated, erosion will expose the beachrock, and cementation will continue to incorporate Coca Cola bottles and other material caught in fissures. CULLIS (1904) described rapid cementation and mineralogical changes in the beachrock borings of the atoll of Funafuti. Captain MORESBY (1835) reports annual beachrock harvesting by natives in Indian Ocean islands, which also exemplify how fast cementation is in this kind of setting. EASTON (1974) reports how a coconut is incorporated in beachrock. FRANKEL (1968) described the wreckage of an aircraft in the northern islands of the Great Barrier Reef in which parts have become cemented into the surrounding beach material and well lithified. EMERY *et al.* (1954) recorded cartridge cans, glass bottles, and shrapnel fragments cemented into beachrock.

DALY (1924) attempted an *in situ* experiment to determine the cementation rate. The experiment was not a success; however, a rapid rate was proved, as Daly explained vividly: "In May, 1919, with the help of the staff of the Marine Laboratory at Tortugas (Florida), a large cask was filled with shelf detritus, dug up at a point offshore from the laboratory. The sides and ends of the cask were perforated, to permit the easy flow of seawater through the mass, and the whole was buried at the proper depth in the beach. It was hoped that, after a year or two, the cask might be opened, with the object of seeing whether its contents had become in any degree cemented together. Unfortunately, the hurricane of September 1919 tore up the cask and carried it several hundred meters along the key. There left exposed to the air, this material cannot be used as a test. The experiment should be duplicated. On the other hand, this hurricane itself improved upon an artificial experiment. Like that of 1910, it threw up on the beach a large quantity of shelf sand, which a year afterwards, was found by Dr A. G. Mayor to have been hardened into beach-rock."

My own experience in Joulter's Cay, Bahamas, reflects the rapidity of marine carbonate cementation in a beachrock setting (FRIEDMAN, 1998). Over a period of 10 yr, together with students and professional geologists I visited Joulter's Cay once a year. A year after one of our visits we found a sardine can from a previous visit. The cementation of the particles (ooids) was surprising. A bivalve shell had become cemented to ooids which had lithified on the underside of the can. 382 g of ooids and skeletal material had filled the can and lithified therein or had become cemented to the outside of the can (FRIEDMAN, 1998). This process is remarkable for 1 year's residence and contrasts with textbook statements that "time is needed—cementation probably requires substantial amounts of time" (TUCKER and WRIGHT, 1990, p. 325).

A similar experience relates to beachrock of the Mediterranean shore of Israel. A sample of tin-can beachrock consists of carbonate cement (75%) and insoluble residue (25%). The carbonate consists of 53% aragonite, 23% low-magnesian calcite, and 24% insoluble residue. X-ray diffraction identified the insoluble residue (25%) as quartz. Once again a rapid rate of cementation created this beachrock.

DARWIN'S EARLY ACCOUNT OF BEACHROCK

One of the first geological accounts of beachrock is that of CHARLES DARWIN (1896) in his book "Geological Observations on the Volcanic Island and Parts of South America Visited during the Voyage of HMS Beagle", in which he described his observation of 1832. Under the heading of "Recent Conglomerate" he noted: "I found fragments of brick, bolts of iron, pebbles, and large fragments of basalt, united by a scanty base of impure calcareous matter into a firm conglomerate".

METHODS

In the field, beachrock sections were located and sampled. Subsequently the rocks were thin-sectioned. Traditional petrographic microscopy studies of the samples included the texture of the rock, composition of particles, and cement, and pore types.

Special attention was given to the amount and kinds of carbonate minerals present. The presence of non-cement carbonate particles, such as fossil fragments, carbonate lithoclasts, or micrite-size allochthons would influence the results of this study. The carbonate bomb was used to determine the concentration of $CaCO_3$ (MÜLLER and GASTNER, 1971). Staining (FRIEDMAN, 1959) differentiated aragonite, high Mg-calcite, and low Mg-calcite, and X-ray diffraction confirmed mineralogical composition (MÜLLER, 1967; GAVISH and FRIEDMAN, 1973; MILLIMAN, 1974).

For $^{87}Sr/^{86}Sr$, radiocarbon, and stable isotope analysis the carbonate rocks were cleaned thoroughly in an ultrasonic cleaner. The samples were then hydrolyzed with HCl, under vacuum, and the carbon dioxide was recovered for analysis. ^{13}C analysis was made from a small portion of the same evolved gas.

For strontium-isotopic ratio error information, the 1σ error is approximately ± 0.000010. The errors often quoted in strontium literature are for 2σ errors.

For O and C isotope ratios, the 1σ error is approximately ± 0.2.

For ^{14}C ages, the errors are 1σ errors.

"^{14}C years BP" identifies the given ages as "uncalibrated" (*i.e.*, they are not calibrated to be "calendar years"). Dates are based upon the Libby half life (5570 yr) for ^{14}C. The error is ± 1 s as judged by the analytical data alone. Our modern standard is 95% of the activity of NBS Oxalic Acid. The age is referenced to the year 1950 AD.

"^{13}C corrected" identifies the given ages as "conventional" in that they have been corrected for natural isotope fractionation.

BEACHROCK IN STUDY AREAS

Sites studied in the present paper include Los Roques Islands in the southern Caribbean, located on the Caribbean Plate (Figs. 1 and 2), north shore of Puerto Rico (Figs. 1 and 3), and the Gulf of Aqaba (Figs. 4–6). This study resulted from a search along the shorelines for prograding beachrock levels that reflect sea-level fluctuations, tectonic activity, and shoreline migration, including major earthquakes.

Caribbean Plate: Los Roques Islands, Venezuela

The islands consist of a complex basement assemblage of igneous and metamorphic rocks on which carbonate sediments have accumulated and modern reefs are well developed. Beachrocks are scattered, but are especially prominent at Francisqui Cayo in the uppermost northeastern location of the island.

Puerto Rico

Formerly flat-lying, shallow-water carbonate facies have been tilted downwards by strong recent plate tectonic motion in the Caribbean Plate on Puerto Rico. The North American Plate extends beneath the northern Caribbean plate (DILLON et al., 1996; NEALON et al., 2000).

In Puerto Rico beachrock lines much of the north shore (MERRIN, 1955). A succession of four levels of beachrock, which progrades seaward, was sampled at Dorado, along the northeastern shore of Puerto Rico (Fig. 3).

Gulf of Aqaba, Red Sea

The Red Sea was opened by sea-floor spreading. The northern section was opened to a width of 120–140 km as of 25–30 Ma (GIRDLER and STYLES, 1974; COCHRAN, 1983;

(a)

(b)

FIG. 6. (a) and (b). Beachrock with abundant included terrigenous pebbles paving intertidal zone and extending into supratidal zone on west side of Gulf of Aqaba.

JOFFE and GARFUNKEL, 1987). Part of the motion of the Arabian Plate was turned into the Gulf of Aqaba along the Dead Sea transform (BEN AVRAHAM *et al.*, 1979a,b). About 70 km of slip occurred only 4–5 Ma ago. The average relative motion during the past 5 Ma was about 1.5 cm/yr (FREUND *et al.*, 1970; GARFUNKEL, 1981, 1988).

The Gulf of Aqaba, a slightly hypersaline body of partially enclosed marine water, occupies a narrow fault-bounded tectonic valley, about 10–26 km wide with precipitous shores and submarine slopes. Threshold of the gulf is narrow (about 100 m), except where alluvial fans are present (FRIEDMAN, 1968, 1985, 1988). Sediment is terrigenous sand or gravel or skeletal fragments, especially from reefs. Reefs are usually present. They may be dead in places, drowned on seaward side, emergent on intertidal side.

At the northern-most tip of the west shore of the Gulf of Aqaba (Figs. 4 and 5), a prograding succession of beachrock levels is comparable to the shoreline setting in the Caribbean Plate at Puerto Rico (Figs. 1 and 3).

At one site beachrock progrades across a coral reef which has been radiocarbon-dated at 4770 ± 140 yr BP (FRIEDMAN, 1965) and signifies beachrock migration into the sea in mid-Holocene times.

PETROGRAPHY AND MINERALOGY OF BEACHROCK

The studied beachrock at the three sites is composed of (1) terrigenous debris derived from nearby bedrock, (2) carbonate particles of local origin, and (3) carbonate cement (FRIEDMAN, 1988). Only petrographic and isotopic information from the carbonate cement is employed here, and it includes: (1) the radiocarbon date of lithification, (2) stable isotopes and interpreted paleoclimatic setting, and (3) strontium-isotopic ratios.

The most common beachrock lithology is terrigenous calcipebblestone (Fig. 6a,b) and calcisandstone in which skeletal debris is minor or absent, terrigenous rock fragments are

Table 2. Petrographic analysis of cement in Los Roques, Venezuela (Caribbean Plate) beachrock (see Fig. 2; Table 3)

Cement	Sample
Cryptocrystalline, equant high-Mg calcite	(Oldest bed) B51
First, cryptocrystalline; second, acicular aragonite	B55
Similar to previous, but lengths of acicular crystals may be more than 30 μm. Crystals extend to center of pores	(Youngest bed) B52

dominant, and carbonate cement is abundant. Terrigenous debris does not affect radiocarbon dating nor carbon or strontium isotopic measurements, because in the samples used the debris does not generally contain carbonates. The samples selected for study contain fossils only sporadically or at best fossils are entirely absent. Where present, K-feldspar, clay minerals, or mica-chlorite which could affect $^{87}Sr/^{86}Sr$, are minor.

Particles

Terrigenous particles in the studied beachrocks consist of igneous and metamorphic rock fragments, quartz, K-feldspar, plagioclase, mica-chlorite, and/or mixed-layer clay minerals (Table 1). The rock fragments include basalt, andesite, trachite, granite, granite gneiss, and quartz–mica schist. Many quartz particles exhibit wavy extinction suggesting a metamorphic origin. Minor well-sorted, rounded, and coated calcareous skeletons include foraminifera, ostracodes, bryozoans, corals, molluscs, coralline algae, echinoid spines and plates, and biogenic intraclasts.

Cement

Beachrock displays a seaward to landward prograding sequence with a temporal succession from land to sea. In the Caribbean Plate of Los Roques Island the cement style changes, from an early high-magnesian calcite cryptocrystalline cement, to a two-generation cement consisting of an equant cryptocrystalline cement covered by acicular aragonite rim cement (Table 2). This observation agrees with that of others who state that the micritic cement is necessary for the development of the subsequent prismatic cement, which probably results from an abiotic precipitation during higher fluid circulation (NEUMEIER, 1999). This significant change of style is developed in a relatively short span of time (Table 3)

Three kinds of cement are present in the Red Sea samples (Table 1). Two- or three-generation aragonite cement consists of dark, cryptocrystalline aragonite rimming pores. Second-generation aragonite cement consists of transparent acicular crystals that grew towards the centers of pores (FRIEDMAN and GAVISH, 1971). The third kind of cement is

Table 3. Strontium isotope ($^{87}Sr/^{86}Sr$) values, radiocarbon ages, and stable isotopes ($\delta^{13}C$, $\delta^{18}O$) of prograding beachrock ledges, Francisqui Cayo, Los Roques Island, Venezuela (see Fig. 2)

Sample	$^{87}Sr/^{86}Sr$	Radiocarbon ages[a] (^{14}C yr BP–^{13}C corrected)	$\delta^{18}O_{PDB}$ ‰	$\delta^{13}C_{PDB}$ ‰
B51	0.709180	1135 ± 70	− 1.2 ± 0.2	+3.6 ± 0.2
B55	0.709227	910 ± 70	− 1.1 ± 0.2	+3.6 ± 0.2
B52	0.709249	750 ± 70	− 1.1 ± 0.2	+3.6 ± 0.2

Note: These dates are based upon the Libby half life (5570 yr) for ^{14}C. The error is ± 1 s as judged by the analytical data alone. Our modern standard is 95% of the activity of NBS Oxalic Acid. The age is referenced to the year 1950 AD.
[a] ^{13}C analysis was made from a small portion of the same evolved gas.

peloidal, finely crystalline or cryptocrystalline, tan to semi-opaque high-magnesian calcite (13 mol% $MgCO_3$).

The origin of the mineral cements in beachrock has been a subject of great interest for many years. Physical–chemical mechanisms of beachrock formation may be a consequence of the rise in ionic concentration resulting from the evaporation of seawater as it draws through the beach or by variation in the chemistry of marine vadose waters. CO_2 degassing is a chemical process generating cement, but cyanobacteria and other microbes may also be involved, not only in stabilizing the beach sediment, but also in raising the pH by photosynthesis to a level at which calcium carbonate is precipitated. According to DAVIES and KINSEY (1973) pH goes up during the day to 9.5 in beachrock, but precipitation of $CaCO_3$ occurs at night after the waters of high pH have drained into the sediment, where precipitation occurs.

Biologically controlled $CaCO_3$ precipitation (principally of aragonite) by pH changes has been demonstrated in laboratory experiments using a variety of bacterial strains. Organically mediated cementation in beachrock, probably occurs extracellularly (as it does in laboratory experiments involving calcium-carbonate precipitation by bacteria), as a result of fluctuations in the concentration of CO_2 and HCO_3^-, that are controlled organically. Films of non-living organic matter coating mineral particles may also be involved in chemical- or biochemical processes that lead to cementation (GINSBURG, 1953; NESTEROFF, 1954; HANOR, 1978; KENDALL et al., 1994; KRUMBEIN, 1979; MEYERS, 1987; CAMOIN, 1999; NEUMEIER, 1999; RUSSELL, 1960).

STRONTIUM ISOTOPE (^{87}SR/^{86}SR) VALUES, RADIOCARBON AGES, AND STABLE ISOTOPES OF BEACHROCK CEMENT

Los Roques, Venezuela: Table 3 lists $^{87}Sr/^{86}Sr$, radiocarbon ages, and stable fisotopes ($\delta^{13}C$ and $\delta^{18}O$) of the prograding ledges from Francisqui Cayo, Los Roques Island (Fig. 2). The ages of the beachrock are young, only one sample age exceeds 1000 radiocarbon years. The strontium isotope values for the most part overlap that of modern seawater or are higher (Fig. 7). The oldest beachrock has slightly lower $^{87}Sr/^{86}Sr$ values than modern seawater.

Puerto Rico: Table 4 provides the same geochemical data for the Puerto Rico site (Fig. 7). Except for one sample the beachrock ages are modern ($<$100 yr old). The exceptional older sample dates back 1700 yr. The strontium isotope values are close to modern seawater, but three samples are lower.

Red Sea: The studied ledges of prograding Red Sea beachrocks of the northern Gulf of Aqaba (Figs. 5 and 7, Table 5) increase in age progressively from 2620 ± 230 to 3570 ± 170, to 6220 ± 280, and to 7070 ± 380 yr BP. Similar ledges of prograding beachrock are present elsewhere along the shoreline of the western gulf, but were not sampled for this study.

Strontium isotopic record ($^{87}Sr/^{86}Sr$): Strontium-isotopic analyses of beachrock cement (Fig. 7; Tables 3, 4 and 6) reflect $^{87}Sr/^{86}Sr$ secular variation in seawater. The seawater Sr-isotope curve is a global ocean record (CAPO and DEPAOLO, 1990; QUINN et al., 1991; FARRELL et al., 1995). When marine carbonate forms, the $^{87}Sr/^{86}Sr$ ratio of ocean water is incorporated into its structure without fractionation (FAURE, 1986). Carbonate cements in marine settings precipitate from waters in strontium isotopic equilibrium with the global ocean (BURKE et al., 1982; KOEPNICK et al., 1985; HODELL et al., 1989; VEIZER, 1989; FARRELL et al., 1995). Plots of radiocarbon age dates against $^{87}Sr/^{86}Sr$ ratios reveal past geochemical changes and cycles of strontium isotopes in the global ocean (MCARTHUR et al., 2001; VEIZER et al., 1999).

The values for the modern-day seawater vary depending on whose lab runs them (from 0.70917 to 0.70921). In this study the seawater values are reported as 0.70918. The modern ocean has a present-day $^{87}Sr/^{86}Sr$ ratio of 0.709180 ± 0.000020 (HESS et al., 1986; HODELL et al., 1989, 1990, 1991), or a marine signal of 0.70919–0.70920 (RUTBERG et al., 2000). The actual global average $^{87}Sr/^{86}Sr$ of riverine water is unlikely to be any lower than about 0.7114 (JONES and JENKYNS, 2001), but GROSS et al. (2001) observed values of 0.7103 below the Grand Canyon. PALMER and EDMOND (1989) estimated a global river $^{87}Sr/^{86}Sr$ ratio of 0.7119. Fig. 7 is a plot of $^{87}Sr/^{86}Sr$ vs. radiocarbon ages for the cements of beachrock from the Caribbean Plate, Puerto Rico Microplate, and Gulf of Aqaba, Red Sea collected in this study. The modern ages

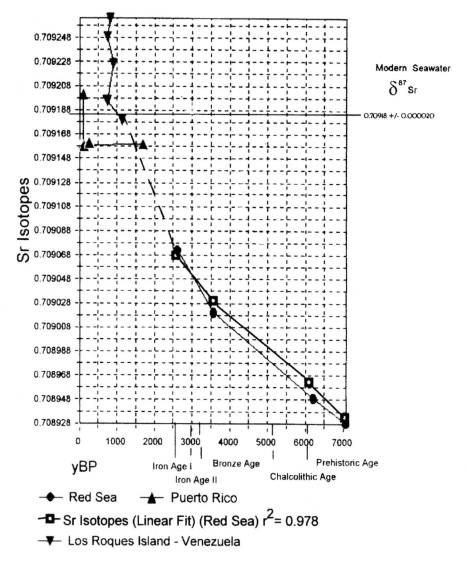

FIG. 7. Plots of strontium isotopic ratios against radiocarbon ages for carbonate cement of prograding beachrock sequences of Caribbean Plate (Los Roques Island, Venezuela), Puerto Rico, and Gulf of Aqaba, Red Sea. For the Red Sea prograding beachrock sequence one graph shows the data points and a second the computed linear fit ($r^2 = 0.978$). The horizontal line at 0.70918 gives the present-day equilibrium $^{87}Sr/^{86}Sr$ ratio of the modern ocean. Dashed line between Red Sea and Caribbean Plate (Los Roques Island, Venezuela) data suggests a chronostratigraphic curve for beachrock cement.

(younger than iron age) for beachrock, yielded $^{87}Sr/^{86}Sr$ ratios close to that of modern seawater. The plot of age against $^{87}Sr/^{86}Sr$ ratios for the prograding sequence of Red Sea beachrocks follows a linear pattern ($r^2 = 0.978$) in which decreasing age generates carbonate cements of progressively higher strontium-isotope ratio.

Strontium isotopic ratios of the beachrock cements are among the highest for Phanerozoic carbonates and probably for all geologic time. They are close to those of the Late Cambrian carbonates which are deposits previously considered to have the highest strontium isotopic ratios (MONTAÑEZ et al., 1996; Friedman, 1996, 2002). The anomalously high $^{87}Sr/^{86}Sr$ values reported here may record rapid unroofing. As a result of extensive chemical weathering, erosion and high uplift rates, seawater $^{87}Sr/^{86}Sr$ records peak values. MONTAÑEZ et al. (1996) related high strontium isotopic values in the Cambrian to the isostatic rebound and associated unroofing

Table 4. Strontium isotope (^{87}Sr/^{86}Sr) values, radiocarbon ages, and stable isotopes (δ^{13}C, δ^{18}O) of prograding beachrock ledges, Dorado, Puerto Rico (see Fig. 5)

Sample	^{87}Sr/^{86}Sr	Radiocarbon ages (^{14}C yr BP–^{13}C corrected)	δ^{18}O$_{PDB}$	δ^{13}C$_{PDB}$
99-1	0.709201	Modern, <100 yr (99.23 ± 0.76)% of the modern (1950) ^{14}C activity	−0.2	+1.6
99-2	0.709159	Modern, <100 yr (103.17 ± 1.20)% of the modern (1950) ^{14}C activity	−1.8, −1.8	+4.7, +4.9
99-3	0.709160	1700 ± 110 ^{14}C yr BP	−0.8	−1.3
99-4	0.709159	Modern <100 yr (99.24 ± 1.20)% of the modern (1950) ^{14}C activity	−0.8	+1.4

The carbonate rocks were cleaned thoroughly in an ultrasonic cleaner. The samples were then hydrolyzed with HCl, under vacuum, and the carbon dioxide was recovered for analysis.

and weathering of ^{87}Sr-rich high-grade metamorphic zones (MILLER, 1983). Chemical weathering of ^{87}Sr-rich phyllosilicates expedited transport of dissolved radiogenic strontium to the ocean (MONTAÑEZ et al., 1996). The episodic variation of the seawater ^{87}Sr/^{86}Sr has been attributed to either variation in the Sr flux or the Sr-isotopic composition of the riverine-dissolved load derived from the weathering of the continental crust (BICKLE et al., 2001, p. 737).

The high average rate of increase of ^{87}Sr/^{86}Sr in the Gulf of Aqaba beachrock cement indicates that continental weathering rates were exceptionally high or that the weathered rocks were exceptionally radiogenic. ARMSTRONG (1971) suggested that peaks in seawater strontium-isotope ratio resulted from enhanced glacial erosion of old shields with elevated ^{87}Sr contents. HODELL et al. (1990) countered that glaciation is a subordinate mechanism for increased radiogenic strontium supply. BICKLE et al. (2001, p. 737) concluded that metamorphism of carbonates may cause elevation of their ^{87}Sr/^{86}Sr ratios and that uplift of metamorphosed carbonates may be a consequence of collisional orogens, which contributes to the elevation of seawater ^{87}Sr/^{86}Sr ratios.

The increase in the ^{87}Sr/^{86}Sr ratio of the beachrock cement with decreasing age is not only related to regional tectonics. The evolution of the ratio of strontium-87 to strontium-86 in beachrock cements reflects changes in the global strontium isotope mass balance.

The Venezuela samples are higher in ^{87}Sr/^{86}Sr than present-day seawater probably reflecting tectonic pulses. Fig. 7 connecting the prograding Red Sea strontium isotope values and those of the Caribbean Plate (Los Roques Island, Venezuela) (note dashed line in Fig. 7) generates an approximate strontium isotope stratigraphy for much to most of the Holocene.

Table 5. Radiocarbon age determinations and δ^{13}C$_{PDB}$ of beachrock samples from four levels of a prograding sequence, northernmost west shore of Gulf of Aqaba, Red Sea

	Radiocarbon ages ^{14}C yr BP (^{13}C corrected)			
	Elat #1-99	Elat #2-99	Elat #3-99	Elat #4-99
δ^{13}C$_{PDB}$	2620 ± 230 ^{14}C yr BP +3.2‰	3570 ± 170 ^{14}C yr BP +2.3‰	6220 ± 280 ^{14}C yr BP +2.2‰	7707 ± 380 ^{14}C yr BP +1.9‰

Table 6. Strontium isotope (^{87}Sr/^{86}Sr) values and stable isotopes (δ^{13}C, δ^{18}O) of prograding samples from four levels of a prograding sequence, northernmost west shore of Gulf of Aqaba, Red Sea

Sample identification	^{87}Sr/^{86}Sr	δ^{18}O$_{PDB}$	δ^{13}C$_{PDB}$
Elat# 1-99	0.709072	−0.1,[a] −0.2	+3.4,[a] +3.2
Elat# 2-99	0.709020	−0.7	+2.8
Elat# 3-99	0.708949	−1.9	+2.3
Elat# 4-99	0.708928	−3.5	+1.1

[a] Duplicate analyses on separate aliquots of the original sample.

OXYGEN ISOTOPES

Table 6 shows the oxygen-isotopic record for the Red Sea prograding beachrock ledges, and Table 7 lists oxygen-isotope values for nearshore and shallow-water non-lithified carbonate sediment of the Gulf of Aqaba, Red Sea.

In contrast to marine organisms the oxygen isotope ratios of beachrock cements were precipitated in equilibrium with the surrounding seawater. For this reason the oxygen isotopic composition of beachrock may be used in climate reconstructions. Depending on the climatic setting, beachrock cement $\delta^{18}O$ primarily reflects sea-surface temperature (SST), but salinity may also affect it. Where absolute temperature reconstruction may be uncertain, relative temperature variations can be determined. Sea-surface salinity affects the temperature signal, but relative climatic changes may be resolved. One of the purposes of this study has been to compute the paleotemperature at the time of precipitation of the interstitial carbonate cement. The $\delta^{18}O$ value of $CaCO_3$ is related to the $\delta^{18}O$ of seawater (δ_ω) and the temperature (t) in °C by the equation

$$t(°C) = 16.9 - 4.2(\delta_c - \delta_\omega) + 0.13(\delta_c - \delta_\omega)^2,$$

where δ_c is the $\delta^{18}O$ value of $CaCO_3$ and δ_ω is related to the $\delta^{18}O$ of seawater (ARTHUR et al., 1983; EPSTEIN et al., 1953). The oxygen isotope composition of seawater is variable, depending on the rate of evaporation from the surface, and because of possible mixing with freshwater which rivers or ground water discharge. In the Gulf of Aqaba increased salinity introduces a slight complication. In this study, I have computed the temperatures from the above equation, but not knowing the δ_ω of the original water I have set it at 0°C and used modern water samples from the shallow-water setting to calibrate for the unknown δ_ω. The mean annual SST is 24°C (FRIEDMAN, 1968, p. 900, Fig. 5, quoting data from ASHBELL (1963)).

Following the equation and the "isotopic temperature scale" of EPSTEIN et al. (1953) (ARTHUR et al., 1983) Fig. 8 and Table 8 present the computed mean sea-surface paleotemperatures and the adjusted temperatures obtained by relating modern average SST (24°C) to computed temperature (t°C) for prograding beachrock samples from four levels of a beachrock sequence of the Red Sea. These data show that conditions are warmer in more recent time. For the mid-Holocene centuries between 7070 ± 380 and 2620 ± 230 ^{14}C yr BP, an interval of approximately 4 1/2 thousand years, a decrease of wide-ranging SST is indicated (Fig. 8). 6000–5000 yr BP corresponds to the Flandrian or Atlantic period when climate in many places was warmer and more humid and sea levels in most oceans was +2 m. RITCHIE et al. (1985) suggested a humid tropical climate with annual monsoon rainfall of at least 400 mm during the mid-Holocene based on sediment and pollen evidence from the eastern Sahara. Stable isotopic composition of mid-Holocene fossil Porites spp. corals from the northern Gulf of Aqaba reveals heavier values compared with modern corals, meaning enrichment in $\delta^{18}O$ and cooler temperature (MOUSTAFA et al., 2000). Mesopelagic pteropoda suggest an arid continental climate in the Red Sea region for the past 4000–5000 yr and a humid climate for the preceding 5000 yr (ALMOGI-LABIN et al., 1991). As Fig. 8 indicates the beachrock data of the Red Sea indicate progressive cooling in the Holocene suggesting a major cooling event. Oxygen isotopes confirm that the climate in Britain 5000 yr ago was about 2°C warmer than the area today (EVANS et al., 2001). This conclusion contradicts the general impression gained from reading news reports which project a warming trend for the recent past (100 yr).

The implied temperature variation (16°C warmer at 7700 BP) may indicate dramatic change in climate. BARD (2002) cites this temperature swing at 8200 BP. He claims that unstable models

Table 7. Stable oxygen- and carbon-isotope determinations for spot samples of modern shallow-water non-cemented carbonate sand, Gulf of Aqaba, Red Sea

Location	$\delta^{18}O_{PDB}$	$\delta^{13}C_{PDB}$
Interuniversity Institute, Elat	(1) −0.5 ± 0.2	+2.7 ± 0.2
Interuniversity Institute, Elat	(2) −0.5 ± 0.2	+2.7 ± 0.2
Interuniversity Institute, Elat	(3) −0.5 ± 0.2	+3.1 ± 0.2
Interuniversity Institute, Elat	(4) −0.2 ± 0.2	+2.7 ± 0.2

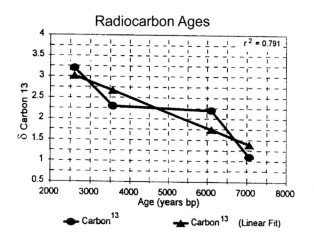

FIG. 8. Plots of carbon-isotope ratios against radiocarbon ages for a prograding beachrock sequence from the Gulf of Aqaba, Red Sea. Plot with elliptical (rounded) pattern provides data points, triangular pattern represents linear fit ($r^2 = 0.791$). This diagram shows increase in the heavy carbon isotope with time through much of the Holocene.

of climate swings are "in fact, very hard to answer". "The climate system is complex because it is made up of several components (such as the atmosphere, oceans, and ice sheets), each of which has its own response times and thermodynamic properties".

The computed SSTs cited in this paper (Table 8; Fig. 9) were determined to indicate a trend of temperature changes, not specific paleotemperatures. The assumption that the present temperature of 17°C was identical at 2620 BP is not necessarily valid; the purpose of the computations is, as previously stated, to imply a direction of change during an approximately 4 1/2 thousand year Holocene interval.

Carbon-isotopic record

As already noted, a plot of age against $^{87}Sr/^{86}Sr$ for the prograding sequence of Red Sea beachrocks illustrates a rise in strontium isotope values with coincident rise in age (Fig. 7). A curve of strontium isotopes against $\delta^{13}C$ documents a rapid rate of increase in values for both $^{87}Sr/^{86}Sr$ and $\delta^{13}C$ (Fig. 10), similar to the pattern observed by MONTAÑEZ et al. (1996, 2000) for a Cambrian marine carbonate sequence. Cambrian carbonate and Holocene beachrock strontium isotope levels are remarkably similar (FRIEDMAN, 2002), possibly reflecting similar conditions of "increased oceanic nutrient levels, primary productivity, and organic carbon burial driven by increased continental flux and associated oceanic sedimentation (MONTAÑEZ et al., 2000, p. 6)".

In Fig. 9 which presents measured data and a linear fit ($r^2 = 0.791$), values of $\delta^{18}O$ increase through much of the Holocene, probably for reasons described above. This shift indicates major paleoceanographic and climatic changes brought on by tectonic forcing and attendant

Table 8. Computed mean sea-surface temperatures for prograding beachrock samples from four levels of a sequence, northernmost west shore of Gulf of Aqaba, Red Sea

Sample	Radiocarbon date (^{14}C yr BP)	Computed temperature ($t°C$)[a]
Elat 1	2620 ± 230	17
Elat 2	3573 ± 170	20
Elat 3	6220 ± 280	25
Elat 4	7070 ± 380	33

[a] Paleotemperature computed according to the equation of EPSTEIN et al. (1953).

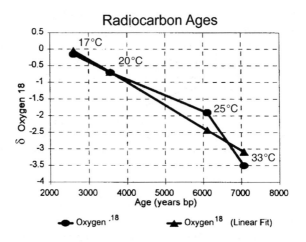

FIG. 9. Plots of oxygen-isotopic ratios against radiocarbon ages of interparticle cement for Gulf of Aqaba, Red Sea prograding sequence. For data and temperature reconstructions, see Tables 7 and 8. Plots with elliptical (rounded) pattern provide data points, triangular pattern represents linear fit ($r^2 = 0.930$). This diagram shows a progressive warming trend in the Red Sea during the past 7000 yr. $t°C$ = paleotemperature computed according to the equation of EPSTEIN et al. (1953).

weathering and records the cumulative effects of erosion, climate and continental weathering (HODELL et al., 1990; RICHTER et al., 1992). MONTAÑEZ et al. (2000) discuss the evolution of strontium and carbon isotope composition of Cambrian oceans.

The Red Sea was opened by sea-floor spreading. Part of the motion of the Arabian Plate was turned into the Gulf of Aqaba along the Dead Sea transform. About 70 km of the slip occurred only 4–5 Ma ago (BEN-AVRAHAM et al., 1979a,b). An age of 20–30 Ma BP is derived for the initial continental breakup of the Gulf of Aqaba. The lands bordering the Gulf in the area of

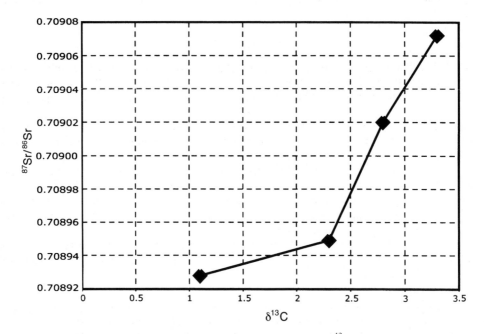

FIG. 10. Plot of strontium isotope ratios for carbonate cement against $\delta^{13}C$ of prograding beachrock sequence of Gulf of Aqaba, Red Sea.

the studied beachrock sequence were uplifted, the uplift reading about 4 km. The uplifted areas were eroded, exposing the pre-Cambrian basement (REISS and HOTTINGER, 1984). "All primary land forms produced by tectonic movements are modified and degraded by mass wasting, erosion, and sedimentation" (REISS and HOTTINGER, 1984). These tectonic phases and the temporal relationships, which radiocarbon and strontium-isotopic values of the beachrocks reflect, link them to oceanic processes and changes in seawater composition. A fossil reef in the intertidal zone of the northern Gulf of Aqaba gave a radiocarbon age of 4770 ± 140 yr BP (FRIEDMAN, 1965). The fossil reef is overlain by beachrock, which is full of igneous and metamorphic pebbles of local derivation and is about one-quarter to one-half meter thick. The intertidal reef and overlying beachrock reflect Holocene tectonic movement of the Gulf's shoreline or lowering of sea level. Another line of evidence for tectonic movement along the same shores has been obtained from radiocarbon dating of sediments that were sampled by cores drilled in an isolated sea-marginal, hypersaline pool named the Solar Lake or Sun Lake, located about 20 km southwest of the sampled prograding sequence of beachrock. The cores drilled within the pool, close to the bar, which separates it from the open gulf, reveal a succession of organic-rich, microbial-mat sediments which are underlain by carbonate mud, rich in *Cerithium* gastropod shells. The microbial-mat sediments typify a very shallow, restricted, hypersaline water environment similar in salinity to the present-day range of salinity of the waters in this pool (70–146‰). On the other hand, the *Cerithium*-rich sediments represent a littoral, open-sea environment. The compilations of the radiocarbon dating of these sediments (FRIEDMAN *et al.*, 1973; NEEV and FRIEDMAN, 1978) indicate that the transition from the open-marine environment (carbonate mud, rich in *Cerithium* gastropod shells) to the restricted, hypersaline lagoonal environment (microbial-mat sediments) occurred sometime between 340 ± 590 yr BP (the youngest age of the *Cerithium*-rich samples analyzed) and 2465 ± 155 yr BP (the oldest age of the microbial-mat sediments analyzed). These changes of the environments (shoaling of facies and the separation of the lagoon from the open gulf) occurred as a result of the tectonic uplifting of the coastal zones. Another phase of tectonic activity occurred at this site sometime after 1910 ± 115 yr BP, *i.e.* during Roman times (KRUMBEIN and COHEN, 1974, Fig. 4). This phase, which involved mainly the subsidence of the pool's trough, is interpreted on the basis of cores that reveal an alternating sequence of microbial mats and gypsiferous sediments. This alternating sequence indicated phases of subsidence of the pool's bottom, where it became too deep for the proliferation of microbial mats.

CONCLUSIONS

Studied beachrock cements from three different geographic settings (an island off the coast of South America, the north shore of Puerto Rico, and the shoreline of the Gulf of Aqaba, Red Sea) consist of prograding beachrock ledges representing sea-level fluctuations and/or tectonic activity.

The beachrock cements of the ledges are of decreasing radioactive ages paralleling increasing strontium isotopic ratios reflecting chronostratigraphic variation (Fig. 7). The strontium isotopic values of the cements are among the highest for Phanerozoic carbonates and point to rapid continental chemical weathering. They reveal patterns of changes in strontium isotopic composition.

The oxygen isotopic composition of beachrock cement reflects temperature decrease of ambient seawater for approximately half of Holocene time. This discovery of cooling is at variance with the climate-change debate.

The changes in $\delta^{13}C$ and strontium isotope values in prograding ledges of beachrock cement (Fig. 10) record a rapid rate of increase in values of both $\delta^{13}C$ and strontium isotopes, and likely record uplift and erosion.

Acknowledgements — In my 1968 paper on the carbonate sediments of the Red Sea I expressed my appreciation to I. R. Kaplan for his interest in this study and for assistance in the field. The present paper is based on the original research, but has been extended more than three decades beyond the original program (FRIEDMAN, 1968).

For reviewing this manuscript grateful thanks are extended to: André Strasser, General Secretary of the International Association of Sedimentologists, Pérolles, Fribourg, Switzerland; Leon E. Long, University of Texas at Austin, USA; Maurice E. Tucker, President of the International Association of Sedimentology, University of Durham, England; Isabel P. Montañez, Associate Editor for the *Journal of Sedimentary Research* and co-editor for *Sedimentology*, University of California, Davis; and Urs Neumeier, University of Geneva, Switzerland.

I am grateful to Richard Reesman of Geochron Laboratories, Cambridge, Massachusetts, for stable isotopic analyses, to Samuel A. Bowring and the TIMS laboratories of the Massachusetts Institute of Technology for strontium isotope analysis, and to Mineralogy of Tulsa, Oklahoma, for mineral analysis.

REFERENCES

ALLEN R. C., GAVISH E., FRIEDMAN G. M. and SANDERS J. E. (1969) Aragonite-cemented sandstone from outer continental shelf off Delaware Bay: submarine lithification mechanism yields product resembling beachrock. *J. Sedim. Petrol.* **39**, 136–149.

ALMOGI-LABIN A., HEMLEBEN C., MEISCHNER D. and ERLENKEUSER H. (1991) Paleoenvironmental events during the last 13,000 years in the central Red Sea as recorded by pteropoda. *Paleoceanography* **6**, 83–98.

ARMSTRONG R. L. (1971) Glacial erosion and the variable isotopic composition of strontium in sea water. *Nat. Phys. Sci.* **230**, 132–133.

ARTHUR M., ANDERSON T. F., KAPLAN I., BEIZER J. and LAND L. S. (1983) Society for Economic Paleontologists and Mineralogists, Short Course #10, 385 p.

ASHBELL D. (1963) Climate conditions of Elath. In *Elath: Israel Exploration Society*, 18th Archeological Convention, Jerusalem, Israel, pp. 242–256.

BARD E. (2002) Climate shock: abrupt changes over millennial time scales. *Phys. Today* **55**, 32–38.

BEN-AVRAHAM Z., ALMAGOR G. and GARFUNKEL Z. (1979a) Sediments and structure of the Gulf of Elat (Aqaba), northern Red Sea. *Sedim. Geol.* **23**, 239–267.

BEN-AVRAHAM Z., GARFUNKEL Z., ALMAGOR G. and HALL J. K. (1979b) Continental breakup by a leaky transform: the Gulf of Elat (Aqaba). *Science* **206**, 214–216.

BICKLE M. J., HARRIS N. B. W., BUNBURG J. M., CHAPMAN H. J., FAIRCHILD I. J. and AHMAD T. (2001) Controls on the $^{87}Sr/^{86}Sr$ ratio of carbonates in the Garhwal Himalaya, headwaters of the Ganges. *J. Geol.* **109**, 737–753.

BURKE W. H., DENISEN R. E., HETHERINGTON E. A., KOEPNICK R. B., NELSON H. F. and OTTO J. B. (1982) Variations of seawater $^{87}Sr/^{86}Sr$ throughout Phanerozoic time. *Geology* **10**, 516–519.

CAMOIN G. F. (1999) Microbial mediation in carbonate diagenesis. *Sedim. Geol.* **126**, 104 p.

CAPO R. C. and DEPAOLO J. D. (1990) Seawater strontium isotopic variations from 2.5 million years ago to the present. *Science* **249**, 51–55.

COCHRAN J. R. (1983) A model for the development of Red Sea. *Am. Assoc. Petrol. Geol. Bull.* **67**, 42–69.

CULLIS C. G. (1904) The mineralogical changes observed in the cores of the Funafuti borings, sec. 14, 392–420. In *The Atoll of Funafuti* (ed. H. E. ARMSTRONG). Royal Society of London, Harrison & Sons, London, 428 p.

DALY R. A. (1924) The geology of American Samoa, 93–144. In *A Memorial to Alfred Goldsborough Mayor*. Vol. 19, no. 340, Carnegie Institute, Washington, 217 p.

DARWIN C. (1896) *The Structure and Distribution of Coral Reefs*, 3rd Ed., D. Appleton & Co., New York, 344 p.

DAVIES P. J. and KINSEY D. W. (1973) Organic and inorganic factors in Recent beachrock formation, Heron Island, Great Barrier Reef. *J. Sedim. Petrol.* **43**, 59–81.

DILLON W. P., EDGAR N. T., SCANLON K. M. and COLEMAN D. F. (1996) A review of the tectonic problems of the strike-slip northern boundary of the Caribbean Plate and examination by Gloria. In *Geology of the United States Sea Floor; the View from Gloria* (eds. J. V. GARDNER, M. E. FIELD and D. C. TWICHELL), pp. 135–164. Cambridge University Press, Cambridge.

EASTON W. H. (1974) An unusual inclusion in beachrock. *J. Sedim. Petrol.* **44**, 693–694.

EL-SAYED M. K. (1988) Beachrock cement in Alexandria, Egypt. *Mar. Geol.* **80**, 29–35.

EMERY K. O., TRACEY J. I. and LADD H. S. (1954) Geology of Bikini and nearby atolls: U.S. Geol. Surv. Prof. Paper, no. 260A, Bikini and nearby atolls: Part 1, Geology, 265 p.

EPSTEIN S., BUCHSBAUM R., LOWENSTAM H. A. and UREY H. C. (1953) Revised carbonate-water isotopic temperature scale. *Bull. Geol. Soc. Am.* **64**, 1315–1326.

EVANS M. J., DERRY L. A., ANDERSON S. P. and FRANCE-LANORD C. (2001) Hydrothermal source of radiogenic Sr to Himalayan rivers. *Geology* **29**, 803–806.

FARRELL J. W., CLEMENS S. C. and GROMET L. P. (1995) Improved chronostratigraphic reference curve of late Neogene seawater $^{87}Sr/^{86}Sr$. *Geology* **23**, 403–406.

FAURE G. (1986) *Principles of Isotope Geology*, 2nd Ed., Wiley, New York, 589 p.

FRANKEL E. (1968) Rate of formation of beachrock. *Earth Planet. Sci. Lett.* **4**, 439–440.

FREUND R., GARFUNKEL Z., ZAK I., GOLDBERG M., WEISSBROD T. and DERIN B. (1970) The shear along the Dead Sea rift. *Philos. Trans. R. Soc. London, Ser. A* **267**, 107–130.

FRIEDMAN G. M. (1959) Identification of carbonate minerals by staining methods. *J. Sedim. Petrol.* **29**, 87–97.

FRIEDMAN G. M. (1965) A fossil shoreline reef in the Gulf of Elat (Aqaba). *Isr. J. Earth Sci.* **14**, 86–90.

FRIEDMAN G. M. (1968) Geology and geochemistry of reefs, carbonate sediments, and waters, Gulf of Aqaba (Elat), Red Sea. *J. Sedim. Petrol.* **38**, 895–919.

FRIEDMAN G. M. (1975) The making and unmaking of limestones or the downs and ups of porosity. *J. Sedim. Petrol.* **45**, 379–398.

FRIEDMAN G. M. (1985) Gulf of Elat (Aqaba). Geological and sedimentological framework, 39–71. In *Hypersaline Ecosystems* (eds. G. M. FRIEDMAN and W. E. KRUMBEIN). 428 p., Springer, Berlin.

FRIEDMAN G. M. (1988) Modern sedimentation in the Gulf of Elat (Aqaba): a review, 933–956. In *Triassic–Jurassic Rifting* (ed. M. WARREN), 998 p. Developments in Geotectonics 2, Part B. Elsevier, Amsterdam.

FRIEDMAN G. M. (1996) Strontium isotopic signatures reflect an origin of dolomite by fresh-water effluent; the Pine Plains Formation (Wappinger Group, Cambrian) of southeastern New York. *Carbon. Evapor.* **11**, 134–140.

FRIEDMAN G. M. (1998) Rapidity of freshwater calcite cementation—implications for carbonate diagenesis and sequence stratigraphy: perspective. *Sedim. Geol.* **119**, 1–4.

FRIEDMAN G. M. (2002) Highest Phanerozoic strontium isotopic ratios of pre-rift late Cambrian passive margin in New York State USA: products of continental weathering and orogenesis. *Sedim. Geol.* **147**, 143–153.

FRIEDMAN G. M. and GAVISH E. (1971) Mediterranean and Red Sea (Gulf of Aqaba) beachrocks, 13–16. In *Carbonate Cements* (ed. P. O. BRICKER). Johns Hopkins University Press, 376 p, Baltimore, MD.

FRIEDMAN G. M., SANDERS J. E. and ALLEN R. C. (1971) Marine lithification mechanism yields rock resembling beachrock, 50–53. In *Carbonate Cements* (ed. O. P. BRICKER). John Hopkins University Press, 376 p, Baltimore, MD.

FRIEDMAN G. M., AMIEL A. J., BRAUN M. and MILLER D. S. (1973) Generation of carbonate particles and laminites in algal mats; example from sea-marginal hypersaline pool, Gulf of Aqaba, Red Sea. *Am. Assoc. Petrol. Geol. Bull.* **57**, 541–557.

GARFUNKEL Z. (1981) Internal structure of the Dead Sea leaky transform (rift) in relation to plate tectonics. *Tectonophysics* **80**, 81–108.

GARFUNKEL Z. (1988) Relation between continental rifting and uplifting: evidence from the Suez rift and northern Red Sea. *Tectonophysics* **150**, 33–49.

GAVISH E. (1975) Recent and Holocene beachrocks along the coasts of Sinai, Gulfs of Elat and Suez. *Rapp. Comm. Mer. Medit.* **23**, 131–132.

GAVISH E. and FRIEDMAN G. M. (1973) Quantitative analysis of calcite and Mg-calcite by X-ray diffraction: effect of grinding on peak height and peak area. *Sedimentology* **20**, 437–444.

GAVISH E., KRUMBEIN W. E. and HALEVY J. (1985) Geomorphology, mineralogy, and groundwater geochemistry as factors of the hydrodynamic system of the Gavish sabkha. In *Hypersaline Ecosystems—The Gavish Sabkha, pp. 186–217* (eds. G. M. FRIEDMAN and W. E. KRUMBEIN). 484 p. Springer, Berlin.

GINSBURG R. N. (1953) Beachrock in south Florida. *J. Sedim. Petrol.* **23**, 85–92.

GIRDLER R. W. and STYLES P. (1974) Two stage Red Sea floor spreading. *Nature* **247**, 7–11.

GISCHLER E. and LOMANDO A. J. (1997) Holocene cemented beach deposits in Belize. *Sedim. Geol.* **110**, 277–297.

GROSS E. L., PATCHETTE P. J., DALLEGGE T. A. and SPENCER J. E. (2001) The Colorado River system and Neogene sedimentary formations along its course: apparent Sr isotopic connections. *J. Geol.* **109**, 449–461.

GUO B. and FRIEDMAN G. M. (1990) Petrophysical characteristics of Holocene beachrock. *Carbon. Evapor.* **5**, 223–243.

HANOR J. S. (1978) Precipitation of beachrock cements: mixing of marine and meteoric waters vs. CO_2-degassing. *J. Sedim. Petrol.* **48**, 489–501.

HESS J., BENDER M. L. and SCHILLING J. G. (1986) Evolution of the ratio of strontium-87 to strontium-86 in seawater from Cretaceous to present. *Science* **231**, 979–984.

HODELL D. A., MUELLER P. A., McKENZIE J. A. and MEAD G. A. (1989) Strontium isotope stratigraphy and geochemistry of the late Neogene ocean. *Earth Planet. Sci. Lett.* **92**, 165–178.

HODELL D. A., MEAD G. A. and MUELLER P. A. (1990) Variation in the strontium isotopic composition of seawater (8 Ma to present): implications for chemical weathering rates and dissolved fluxes to the oceans. *Chem. Geol. (Isotope Geosci. Sect.)* **80**, 291–307.

HODELL D. A., MUELLER P. A. and GARRIDO J. R. (1991) Variations in the strontium isotopic composition of seawater during the Neogene. *Geology* **19**, 24–27.

HOLAIL H. and RASHED M. (1992) Stable isotope composition of carbonate-cemented Recent beachrock along the Mediterranean and the Red Sea coasts of Egypt. *Mar. Geol.* **106**, 141–148.

HOPLEY D. (1986) Beachrock as a sea-level indicator, 157–173. In *Sea-level Research: A Manual for the Collection and Evaluation of Data* (ed. O. van de Plassche). Geo Books, Norwich.

HOPLEY D. and DAVIES P. J. (1986) The evolution of the Great Barrier Reef. *Oceanus* **29**, 7–12.

JOFFE S. and GARFUNKEL Z. (1987) Sedimentary basins within the Dead Sea Rift and other rift zones. *Tectonophysics* **141**, 5–22.

JONES C. E. and JENKYNS H. C. (2001) Seawater strontium isotopes, oceanic anoxic events, and seafloor hydrothermal activity in the Jurassic and Cretaceous. *Am. J. Sci.* **301**, 112–149.

KENDALL C. G. St. C., SADD J. L. and ALSHARHAN A. R. (1994) Holocene marine cement coatings on beach-rocks of the Abu Dhabi coast-line (UAR); analogues for cement fabrics in ancient limestones. *Carbon. Evapor.* **9**, 119–131.

KNEALE D. and VILES H. A. (2000) Beach cement: incipient $CaCO_3$-cemented beachrock development in the upper intertidal zone, North Uist, Scotland. *Sedim. Geol.* **132**, 165–170.

KOEPNICK R. B., BURKE W. H., DENISON R. E., HETHERINGTON E. A., NELSON H. F., OTTO J. B. and WAITE L. E. (1985) Construction of the seawater $^{87}Sr/^{86}Sr$ curve for the Cenozoic and Cretaceous: supporting data. *Chem. Geol.* **58**, 55–81.

KRUMBEIN W. E. (1979) Photolithotropic and chemoorganotrophic activity of bacteria and algae as related to beachrock formation and degradation (Gulf of Aqaba), Sinai. *Geomicrobiol. J.* **1**, 139–209.

KRUMBEIN W. E. and COHEN Y. (1974) Biogeneklastische, und evaporitische Sedimentation in einem mesothemen, monomiktischen ufernahen See (Golf von Aqaba). *Geologische Rundschau* **63**, 1035–1065.

MCARTHUR J. M., HOWARTH R. J. and BAILEY T. R. (2001) Strontium isotope stratigraphy: LOWESS Version 3: Best fit to the marine Sr-isotope curve for 0–509 Ma and accompanying look-up table for deriving numerical age. *J. Geol.* **109**, 155–170.

MERRIN S. (1955) Beachrock in Northeastern Puerto Rico: First Caribbean Geological Congress, Antigua, BWI, 15 p.

MEYERS J. H. (1987) Marine vadose beachrock cementation by cryptocrystalline magnesian calcite—Maui, Hawaii. *J. Sedim. Petrol.* **57**, 558–570.

MILLER R. M. (1983) The Pan-African orogen of southwestern Africa/Namibia. *Geol. Soc. South Africa Spec. Publ.* **11**, 431–515.

MILLIMAN J. D. (1974) *Marine Carbonates.* 375 p. Springer, Berlin.

MONTAÑEZ I. P., BANNER J. C., OSLEGER D. A., BORG L. E. and BOSSERMAN P. J. (1996) Integrated Sr isotope variations and sea-level history of Middle to Upper Cambrian platform carbonates: implications for the evolution of Cambrian sea water $^{87}Sr/^{86}Sr$. *Geology* **24**, 917–920.

MONTAÑEZ I. P., OSLEGER D. A., BANNER J. L., MACH L. E. and MUSGROVE M. L. (2000) Evolution of the strontium and carbon isotope composition of Cambrian Oceans. *Geol. Soc. Am. Today* **10**, 1–7.

MORESBY R. (1835) Report on the Northern Atolls of the Maldives. *Geograph. Soc. J.* **5**, 398–403.

MOUSTAFA Y. A., PÄTZOLD J., LOYA Y. and WEFER G. (2000) Mid-Holocene stable record of corals from the northern Red Sea. *Int. J. Earth Sci.* **88**, 742–751.

MÜLLER G. (1967) *Methods in Sedimentary Petrology.* 283 p. Hafner Publishing Company, New York.

MÜLLER G. and GASTNER M. (1971) The "Karbonat-Bombe", a simple device for the determination of the carbonate content in sediments, soils, and other materials. *Neues Jahrbuch Miner. Mh.* **10**, 446–469.

NEALON J. W., DILLON W. P., TEN BRINK U. S., DANFORTH W. W., ROTH E. G., UCHUPI E., ACOSTA J., MUNOZ A., PALOMO R. and VAN DER HILST R. (2000) Tectonics of the Puerto Rico Virgin Island Microplate, Geological Society of America, Abstracts of Programs, p. A436.

NEEV D. and FRIEDMAN G. M. (1978) Late Holocene tectonic activity along the margins of the Sinai subplate. *Science* **202**, 427–429.

NESTEROFF W. (1954) Sur la formation des gres de plage on "beachrock" en Mer Rouge. *Comptes Rendus des Seances de l'Academie des Sciences* **238**, 2547–2548.

NEUMEIER URS (1999) Experimental modelling of beachrock cementation under microbial influence. *Sedim. Geol.* **126**, 35–46.

PALMER M. R. and EDMOND J. M. (1989) The strontium isotope budget of the modern ocean. *Earth Planet. Sci. Lett.* **92**, 11–26.

PURSER B. H. (1980) Les paleosabkhas du Miocene inferieur dans le sud-est de l' Iran. *Bull. des Centres de Recherches Exploratoin-Production Elf-Aquitaine* **4**, 235–244.

QUINN T. M., LOHMANN K. C. and HALLIDAY A. N. (1991) Sr isotopic variation in shallow water carbonate sequences: stratigraphic, chronostratigraphic, and eustatic implications of the record at Enewetak atoll. *Paleoceanography* **6**, 371–385.

REISS Z. and HOTTINGER L. (1984) *The Gulf of Aqaba: ecological micropaleontology.* 354 p. Springer, Berlin.

RICHTER F. M., ROWLEY D. B. and DEPAOLE D. J. (1992) Sr isotope evolution of seawater: the role of tectonics. *Earth Planet. Sci. Lett.* **109**, 11–23.

RITCHIE J. C., EYLES C. H. and HAYNES C. V. (1985) Sediment and pollen evidence for an early to mid-Holocene humid period in the eastern Sahara. *Nature* **314**, 352–355.

RUSSELL R. J. (1960) *Preliminary Notes on Caribbean Beachrock. Second Caribbean Geological Conference.* University of Puerto Rico, pp. 43–49.

RUTBERG R. L., HEMMING S. R. and GOLDSTEIN S. L. (2000) Reduced North Atlantic deep water flux to the glacial southern ocean inferred from neodymium isotope ratios. *Nature* **405**, 935–938.

SCHMALZ R. F. (1971) Formation of beachrock at Eniwetok Atoll, pp. 17–24. In *Carbonate Cements* (ed. O. P. BRICKER). 376 p. Johns Hopkins University Press, Baltimore, MD.

SCOFFIN T. P. (1970) The trapping and binding of subtidal carbonate sediments by marine vegetation in Bimini Lagoon, Bahamas. *J. Sedim. Petrol.* **40**, 249–273.

SCOFFIN T. P. and STODDART D. R. (1983) In *Chemical Sediments and Geomorphology; Precipitates and Residua in the Near-surface Environment* (eds. A. S. GOUDIE and K. PYE), pp. 401–425. University of Oxford, Sch. Geogr. Academic Press, Oxford, London.

SIESSER W. G. (1974) Relict and Recent beachrock from Southern Africa. *Geol. Soc. Am. Bull.* **85**, 1849–1854.

STODDART D. R. and CANN J. R. (1965) Nature and origin of beachrock. *J. Sedim. Petrol.* **35**, 243–246.

STRASSER A., DAVAUD E. and JEDONI Y. (1989) Carbonate cements in Holocene beachrock; example from Bahiret el Bian, southeastern Tunisea. *Sedim. Geol.* **62**, 89–100.

TAYLOR J. C. M. and ILLING L. V. (1969) Holocene intertidal calcium carbonate cementation, Qatar, Persian Gulf. *Sedimentology* **12**, 69–107.

TUCKER M. E. and WRIGHT V. P. (1990) *Carbonate Sedimentology.* 482 p. Blackwell Scientific Publications, Oxford.

VEIZER J. (1989) Strontium isotopes in seawater through time. *Annu. Rev. Earth Environ. Sci.* **17**, 141–167.

VEIZER J., ALA D., AZMY K., BRUCKSCHEN P., BUHL D., BRUHN F., CARDEN G. A. F., DIENER A., EBNETH S., GODDERIS Y., JASPER T., KORTE C., PAWELLEK F., PODLAHA O. G. and STRAUSS H. (1999) $^{87}Sr/^{86}Sr$, $\delta^{13}C$ and $\delta^{18}O$ evolution of Phanerozoic seawater. *Chem. Geol.* **161**, 59–88.

WEBB G. E., JELL J. S. and BAKER J. C. (1999) Cryptic intertidal microbialites in beachrock, Heron Island, Great Barrier Reef; implications for the origin of microcrystalline beachrock cement. *Sedim. Geol.* **126**, 317–334.

Geochemical Investigations in Earth and Space Science: A Tribute to Isaac R. Kaplan
© The Geochemical Society, Publication No. 9, 2004
Editors: R.J. Hill, J. Leventhal, Z. Aizenshtat, M.J. Baedecker, G. Claypool,
R. Eganhouse, M. Goldhaber and K. Peters

Molecular markers and their use in environmental organic geochemistry

ROBERT P. EGANHOUSE

Water Resources Discipline, US Geological Survey, 12201 Sunrise Valley Drive, MS 432,
Reston, VA 20192-0002, USA

Abstract—Molecular markers are organic substances that carry information about sources of organic matter or contamination. The source/marker relation can be used to indicate the presence of a given source material (qualitative), or, under appropriate conditions, to estimate the amount of a source material (quantitative source apportionment) in the environment. Assemblages of markers can also be used as process probes. In this instance, systematic differences and/or similarities in the physical–chemical properties of markers are coupled with compositional changes in marker composition to infer the operation of natural processes. This paper provides an overview of what molecular markers are, what types of markers are present in the environment, the requirements for the use of markers, and some common applications. To illustrate how molecular markers can answer specific environmental questions, three case studies are presented. The first case study examines the impact of municipal waste on a large urban harbor (Boston Harbor). Linear alkylbenzenes (unreacted residues of linear alkylbenzenesulfonate surfactants) and coprostanol (a fecal indicator) provide information on the sources and likely transport pathways of municipal wastes in a complex hydrologic system. The marker data are also used to estimate the proportion of sewage-derived polychlorinated biphenyls (PCBs) in polluted harbor sediments. The second case study concerns a portion of the continental shelf off southern California (Palos Verdes) where discharge of municipal wastewaters has led to extensive contamination of sediments and biota. Long-chain alkylbenzenes (surfactant residues), PCBs and the pesticide, DDT (dichlorodiphenyltrichloroethane), are used to develop sedimentation rate estimates for several time periods by molecular stratigraphy. These results, when combined with other information, allow conclusions to be drawn about the most likely transport pathway of sediments at the study site and to predict the fate of historically deposited contaminants. Finally, an investigation of a crude-oil spill in Bemidji, MN illustrates how monoaromatic hydrocarbons can be exploited as process probes, providing insights into the relative importance of different attenuation processes in a contaminated aquifer. The results show that natural attenuation of the monoaromatic hydrocarbons is occurring at this site and is dominated, not by physical and/or chemical processes, but by biodegradation.

INTRODUCTION

ORGANIC CHEMICALS generated as a result of human activity enter the environment in vast quantities (on the order of hundreds of tons/yr). These inputs span a large range of temporal and spatial scales, and the media to which the chemicals are introduced (air, water, soil) vary. Some chemicals are applied intentionally over large areas (*e.g.* pesticides) or were used in a wide range of commercial products (*e.g.* polychlorinated biphenyls (PCBs)). Others are undesirable by-products of industrial activity (*e.g.* combustion-derived polycyclic aromatic hydrocarbons (PAHs), polychlorinated dibenzodioxins (PCDDs)) or are man-made chemicals that emanate from a limited number of point sources (surfactants, pharmaceuticals). Many synthetic organic chemicals are individual compounds (*e.g.* lindane, atrazine), but complex mixtures of congeneric substances are not uncommon (*e.g.* PCBs, toxaphene). For these reasons, identification of the sources, transport pathways and fates of anthropogenic organic contaminants represents a significant challenge. Molecular markers, compounds whose structures or isotopic compositions are related to specific sources, offer a means of extracting such information directly from the environment.

The mobility of a chemical depends on its tendency to partition among environmental compartments (atmosphere, hydrosphere, lithosphere, biosphere). Environmental partitioning, in turn, is determined by the physical–chemical properties of a substance and where and how it is released. GOUIN *et al.* (2000) have shown that the behavior of different chemicals can, to a first approximation, be predicted based on where they plot on a two-dimensional property space diagram (log air–water partition coefficient versus log octanol–water partition coefficient). Compounds, such as the low-molecular-weight *n*-alkanes (C_{1-8}), catechol, and DDT, have physical–chemical properties that strongly favor partitioning into air, water, and soil/sediment phases, respectively. Other chemical classes or mixtures, such as the chlorobenzenes and PCBs, have a wide range of properties that favor partitioning among multiple phases. These are multimedia contaminants, and, if sufficiently persistent, they can be transported long distances and become globally distributed (WANIA and MACKAY, 1996).

This paper presents some basic background information about what molecular markers are, requirements for their use, and the types of applications that have been developed over the last 40 yr. Three case studies are offered as a means of illustrating some of the ways in which molecular markers can be used to answer specific environmental questions. These questions probably could not have been readily addressed using conventional environmental chemistry approaches. The objective of this paper is to stimulate interest in the molecular marker approach and to foster its use in environmental organic geochemistry. The interested reader is directed to EGANHOUSE (1997) for a more extensive discussion of molecular markers, their uses and limitations.

BACKGROUND

Molecular markers as information carriers

Molecular markers are carriers of information. This information is contained in their isotopic composition ($\delta^{13}C$, $\Delta^{14}C$, $\delta^{15}N$, δD, $\delta^{37}Cl$, etc.), chemical structure, or the particular mixture of molecules composing an assemblage. If the isotopic composition, structure, or molecular distributions of an individual or an assemblage of chemicals are unique to a given source, these substances can be used as source indicators.

Molecular markers can also be used to investigate the operation of natural processes (*i.e.* process probes). The most common applications involve assemblages of compounds whose properties differ systematically (*i.e. homologs*—an homologous series comprises compounds that differ by the chain length of a single alkyl substituent) or whose properties are similar or identical (*i.e. isomers*). Because chemicals partition among or are degraded within environmental media on a structure-specific basis, changes in marker composition can be used to infer the relative importance of different processes (*e.g.* volatilization, sorption, biodegradation, etc.).

Types of molecular markers

As discussed in EGANHOUSE (1997), molecular markers can be classified as belonging to one of the three groups. *Contemporary biogenic markers* are synthesized by living organisms and can be found in the source organisms themselves (*e.g.* abietane) or, with little or no alteration, in the contemporary environment (*e.g.* dehydroabietic acid). *Fossil biomarkers*, good examples of which are the ubiquitous hopanoids, are biomolecules (*e.g.* bacteriohopanepolyols) that have been transformed through diagenetic and/or catagenetic processes. Even so, they largely retain the stereospecific carbon skeleton of the original biological precursor. The third group comprises *anthropogenic markers*. The presence of these compounds in the environment signals the influence of human activity or the input of wastes (TAKADA and EGANHOUSE, 1998). Anthropogenic markers can be broken into subgroups that include: (1) non-toxic synthetic chemicals (*e.g.* linear alkylbenzenesulfonate surfactants (LAS) used in commercial detergents), (2) non-toxic biogenic chemicals (*e.g.* coprostanol, urobilin), (3) toxic synthetic chemicals (*e.g.* PCBs, toxaphene), and (4) by-product chemicals (*e.g.* PCDDs or PAHs). The latter arise incidentally from human activities such as combustion of organic matter, synthesis of chemicals, or other industrial processes.

Requirements

There are three principal requirements for the *ideal* molecular marker: (1) source-specificity, (2) persistence or predictable behavior, and (3) massive production/usage (TAKADA and EGANHOUSE, 1998).

Ideally, a molecular marker should derive from a single source. However, this condition is rarely met. The Boston Harbor case study describes the use of waste-specific molecular markers (coprostanol, linear alkylbenzenes (LABs)) that serve as particle tracers in a complex urban harbor. In this instance, the markers arise exclusively from municipal wastes. Thus, the requirement of source-specificity is satisfied. Alternatively, when there is a massive local input of what ordinarily are non-source-specific contaminants, such compounds can be used as molecular markers of that local input. This situation is typified by the Palos Verdes Shelf, CA and the Bemidji, MN case studies where DDT (+PCBs) and volatile monoaromatic hydrocarbons (MAHs), respectively, serve as markers of local inputs.

A molecular marker should be persistent or its behavior should be predictable. Ideally, one would desire conservative behavior, but given the diversity of catabolic enzymes, this is a rather demanding requirement for organic substances. In practice, information on the persistence and behavior of markers is often limited. Few studies have been undertaken to systematically determine the susceptibility of any of the commonly used markers to photolysis, chemical reaction, or biodegradation under an appropriate range of environmental conditions (EGANHOUSE, 1997). Even fundamental physical–chemical properties (*e.g.* vapor pressure, aqueous solubility, octanol–water partition coefficient), which are essential for multimedia fate modeling, are rarely known with confidence, and the database, at best, is spotty. This highlights an area where more reliable data are clearly needed. It is important to add that knowledge of the environmental behavior of a marker (*e.g.* LABs) *vis-à-vis* a non-source-specific contaminant (*e.g.* PCBs) is essential for quantitative source apportionment of the latter (HEDGES and PRAHL, 1993; TAKADA *et al.*, 1997).

Finally, a molecular marker should be produced on a massive scale and/or its use should be widespread. This requirement effectively defines the range of applicability of a marker insofar as a substance must be readily detectable in a variety of environments for it to yield useful results. Markers, like other organic contaminants, are subject to the same range of spatial and temporal input functions and they are discharged to various environmental media.

Applications

Table 1 summarizes various ways in which molecular markers have been applied in environmental studies during the last 40 yr.

Source indicators–As source indicators, molecular markers are applied either qualitatively (source identification) or quantitatively (source apportionment). The requirements for quantitative source apportionment are considerably more stringent and the number of examples in the literature is fewer than for source identification (*e.g.* SCHAUER *et al.*, 1996; EGANHOUSE and SHERBLOM, 2001; TAKADA *et al.*, 1997). Specific biogenic markers have been used for chemotaxonomic purposes (KATES, 1997) and to characterize the composition of benthic and pelagic communities present in aquatic ecosystems (FINDLAY and WATLING, 1997). Many fossil biomarkers have served as indicators of organic matter provenance (PETERS and MOLDOWAN, 1993) or for paleoclimate reconstruction (BRASSEL *et al.*, 1986). Biomarkers are also used frequently to identify sources of fossil fuel contamination in the contemporary environment (KAPLAN *et al.*, 1997; VOLKMAN *et al.*, 1997). Finally, there are numerous studies in which molecular markers associated with municipal wastes, urban runoff, or combustion of fossil fuels have been used to infer the effect of various point and non-point sources of contamination (see references in TAKADA and EGANHOUSE, 1998).

Table 1. Applications of molecular markers

Source indicators

Source identification (qualitative)
    Contemporary biogenic markers: chemotaxonomy, biogenic inputs
    Petroleum source rock identification, archaeological geochemistry and paleoclimate reconstruction, forensic geochemistry
    Impacts of major pollutant sources (*e.g.* municipal wastewaters, urban runoff, combustion of fossil fuels)[a]
Source apportionment (quantitative)
    Microbial community composition
    Oil spills
    Emissions to the atmosphere
    Non-specific contaminants[a]

Process probes

Transport
    Pathways[a,b,c]
    Processes
        Dilution
        Resuspension
        Burial[b]
        Mixing[b]
Fate
    Phase transfer processes
        Volatilization
        Vaporization
        Dissolution[c]
        Sorption[c]
    Degradation processes[c]
    Bioaccumulation

Superscripted applications indicate how case studies in this paper reflect the use of molecular markers as source indicators and process probes: [a]Boston Harbor case study; [b]Palos Verdes Shelf, CA case study; [c]Bemidji, MN oil-spill site case study.

Process probes–There are two general ways in which molecular markers are applied as process probes. Both rely upon the ability to relate a prospective marker to a specific source and, in some cases, a known input function. Markers can yield information on the pathways along which particles move (hydrophobic markers) or the direction of ground- and surface-water flow (hydrophilic markers). In addition, molecular markers provide unique approaches with which to understand processes of dilution, resuspension, burial, and mixing. Case studies presented below describe some aspects of these applications (see footnote to Table 1 for a key to these applications).

In terms of the environmental fate of contaminants, three general process types can be distinguished: phase transfer, degradation, and bioaccumulation. All of these process types depend upon knowledge of the physical–chemical properties of the markers, transformation rates and pathways, measurements of the compositions of mixtures of compounds, and an understanding of the physics of the system under investigation.

CASE STUDIES

Boston Harbor

Boston Harbor is a shallow, glacially carved, tidally dominated estuary. Flushing occurs primarily through two channels, President Roads and Nantasket Roads, which connect the harbor with Massachusetts Bay (Fig. 1). For nearly 125 yr (BOTHNER *et al.*, 1998), the wastes of metropolitan Boston and surrounding communities were discharged directly into the harbor with minimal or no treatment. Treated effluent and sludge were released from two plants (Deer Island and Nut Island) through a system of outfalls. When the capacities of the treatment plants were exceeded during wet weather, untreated sewage + runoff was discharged through a relief system consisting of 108 combined sewer overflows (CSOs). As recently as 1990, almost

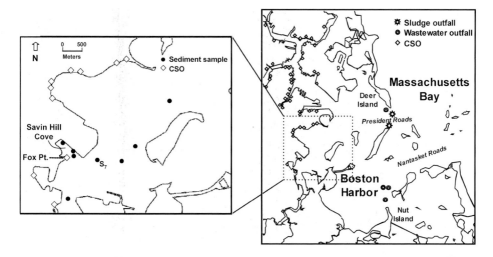

FIG. 1. Map of Boston Harbor showing locations of municipal wastewater and sludge outfalls, major channels to Massachusetts Bay and combined sewer overflows (CSOs). Inset depicts study area adjacent to the Fox Point CSO including stations where surficial sediment samples were collected (modified from EGANHOUSE and SHERBLOM, 2001).

nothing was known about the effects these CSOs were having on water quality in the harbor. For this reason, a study was undertaken to determine the sources of organic contaminants to one of the largest CSOs (at Fox Point; see inset, Fig. 1) and to assess the fate of these materials after discharge to Boston Harbor (EGANHOUSE and SHERBLOM, 2001). In this instance, molecular markers not only served as qualitative source indicators, but also revealed information on transport pathways and allowed for quantitative source apportionment of a class of contaminants of concern, the PCBs.

Samples of Fox Point CSO effluent and harbor receiving water along with sediments near the CSO outfall (Fig. 1) were collected during varying weather conditions. These samples, as well as sludges obtained from Deer Island and Nut Island treatment plants, were analyzed for two molecular markers of municipal wastes, the LABs (surfactant chemicals) and coprostanol (a fecal indicator). PCBs were also measured. LABs and coprostanol were found in the effluent during all sampling periods, indicating that even when weather conditions were dry, sewage entered the CSO system, presumably by leakage or through illegal connections. Moreover, concentrations of the LABs and PCBs in the CSO effluent were highly correlated ($r^2 = 0.87$) suggesting that both of these substances were coming from the same source (namely sewage). Compositional features of the LABs, notably the I/E ratio (the ratio of internal to external isomers; see TAKADA and ISHIWATARI, 1989 for definition), which increases with biodegradation, indicated that sewage entering the CSO had not undergone appreciable alteration. Apparently the CSO was carrying untreated sewage even during dry weather.

The chemical composition of harbor receiving waters was found to resemble that of the CSO effluent under all weather conditions, but more so during periods of rainfall (EGANHOUSE and SHERBLOM, 2001). Thus, the CSO discharge was having a measurable effect on the local aquatic environment. By contrast, sediments deposited near the CSO outfall differed in chemical composition from the CSO effluent and harbor receiving waters. WALLACE et al. (1991) compared mass-emission rates of suspended solids, particulate organic carbon and trace metals from the Fox Point CSO with the rates of accumulation of these materials in nearby sediments (based on ^{210}Pb dating) and showed that less than 4% of the sedimentation near the CSO could be accounted for by the discharge of effluent solids. Thus, the provenance of the sediments, originally expected to be mainly the CSO, appeared to be non-local. Molecular markers were used to elucidate the non-local sources of sediment contamination.

Data on the concentration of coprostanol and the ΣLAB/coprostanol concentration ratios of CSO effluent (E_4), receiving water (W_4), and sediment (S_7) samples, as well as Deer Island and

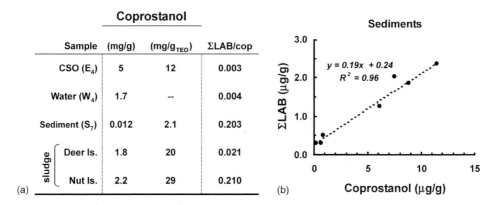

(a)

(b)

FIG. 2. (a) Concentrations of coprostanol and total linear alkylbenzene/coprostanol (ΣLAB/cop) concentration ratios of CSO effluent particles (E₄), municipal sludges, receiving water particles (W₄), and surficial sediments collected near the Fox Point CSO outfall (S₇) and (b) scatter plot of total linear alkylbenzene (ΣLAB) and coprostanol concentrations in surficial sediments from the study site (modified from EGANHOUSE and SHERBLOM, 2001). *Note:* E₄ and W₄ refer to CSO effluent and receiving water samples, respectively, collected during high flow conditions as described in EGANHOUSE and SHERBLOM (2001). TEO, total extractable organic matter. Sediment station (S₇) location is shown in Fig. 1.

Nut Island sludges, are shown in Fig. 2a. Whereas coprostanol concentrations of sludges from the two treatment plants are similar, the ΣLAB/coprostanol ratios differ by a factor of 10. This difference is explained by the fact that at the time of this study LABs in Deer Island waste effluent (and sludge) came principally from residues normally present in linear alkylbenzene-sulfonate surfactant-based detergents (EGANHOUSE et al., 1983). By comparison, Nut Island treatment plant received *additional* inputs from an industrial LAB sulfonating facility. The paired CSO effluent and receiving water samples (E4 and W4), both have low ΣLAB/coprostanol concentration ratios that are nearer the ratio found for Deer Island sludge than Nut Island sludge. This result is consistent with the fact that the Fox Point CSO serves a portion of the same drainage area as the Deer Island treatment plant. As noted above, the Deer Island plant did not receive inputs from the LAB sulfonating facility.

Sediments at station S₇, which have much lower coprostanol concentrations (because of dilution during sedimentation), are characterized by a ΣLAB/coprostanol concentration ratio that is virtually identical to that of the Nut Island sludge. This is illustrated in Fig. 2b, where the slope of the least-squares line (ΣLAB versus coprostanol) for all sediment samples is approximately 0.2. Assuming the ΣLAB/coprostanol marker ratio is conserved, this indicates that the sediment contamination at station S₇ is not primarily the result of wastes released through the CSO (or the Deer Island outfall). Rather, it most likely reflects transport of wastewater and/or sludge particles from the Nut Island treatment plant outfalls (see Fig. 1 for locations of Nut Island wastewater and sludge outfall termini). Possible transport pathways from these outfalls are shown in Fig. 3. As discussed in EGANHOUSE and SHERBLOM (2001), the proposed pathways are consistent with available information on tidal currents in Boston Harbor. Transport of municipal wastes, including sludge, out of the harbor and into Massachusetts Bay was an inefficient process (SIGNELL and BUTMAN, 1992; STOLZENBACH et al., 1993), and a large quantity of this material apparently found its way to quiet embayments along the inner margins of Boston Harbor where rapid sedimentation is favored (KNEBEL et al., 1991).

Although the number of sediment samples in this study was limited, significant correlations between the concentrations of the ∑PCBs and the molecular markers were found (Fig. 4). The similarity of the y-intercepts of the regression lines indicates that the background concentration of *non-sewage derived* PCBs is about 0.1–0.2 μg/g. Thus, as much as 60–80% of the PCBs in the heavily contaminated sediments near the Fox Point CSO are derived from sewage (EGANHOUSE and SHERBLOM, 2001).

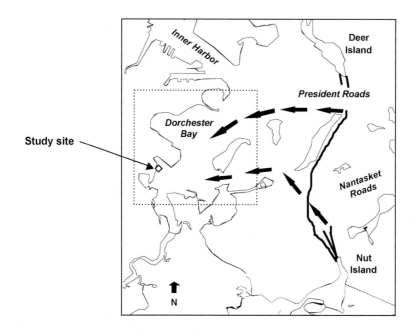

FIG. 3. Map of Boston Harbor showing probable transport routes (arrows) of effluent particles from wastewater and sludge outfalls to study site.

Palos Verdes Shelf, CA

Sediments deposited on the narrow continental shelf off Palos Verdes, CA are heavily contaminated by a variety of inorganic and organic substances (see references in EGANHOUSE *et al.*, 2000). This contamination is the direct result of discharge, through a submarine outfall system, of treated municipal wastewaters (Fig. 5). Among the contaminants of greatest concern in the sediments are DDT (and its metabolites) and PCBs. For example, it has been estimated that 100–250 tons of total DDT presently reside in the shelf and upper slope sediments off Palos Verdes (LEE *et al.*, 2002; MURRAY *et al.*, 2002; MACGREGOR, 1976; MCDERMOTT *et al.*, 1974). The DDT originated from chemical manufacturing wastes that were introduced through sewer lines to the Los Angeles County Sanitation Districts' (LACSD) treatment plant in Carson, CA and subsequently released through the outfall system.

In the early 1990s, a major multidisciplinary research program was initiated to determine the quantity of contaminants in these sediments and to predict the long-term fate of DDT and the PCBs. Central to these investigations was a predictive modeling effort (SHERWOOD *et al.*, 2002)

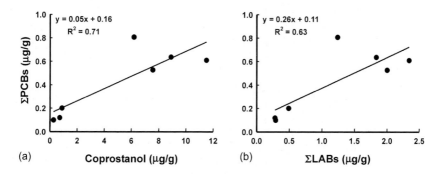

FIG. 4. Scatter plots of total polychlorinated biphenyl (ΣPCB) versus (a) coprostanol and (b) ΣLAB concentrations in surficial sediments from study site. Lines and statistics for linear regression analysis are shown (modified from EGANHOUSE and SHERBLOM, 2001).

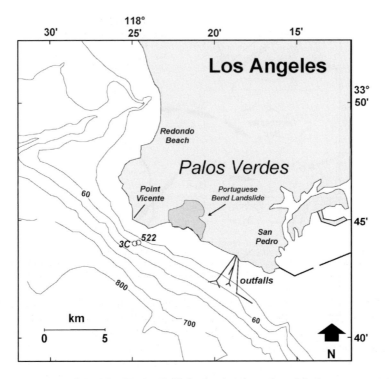

FIG. 5. Map of the Palos Verdes Shelf showing locations of municipal wastewater outfall system operated by the Los Angeles County Sanitation Districts (LACSD), the Portuguese Bend Landslide (PBL), and sediment-coring stations (3C, 522) (modified from EGANHOUSE et al., 2000).

that relied on establishing sedimentation rates at key locations on the shelf. It was recognized that numerous sources (e.g. effluent particles, cliff erosion, landslide, biogenic detritus, river inputs, etc.) contributed sediment to the shelf and slope. During the period when DDT inputs were occurring (1947–1971), however, wastewater solids emanating from the outfall system were believed to represent the largest source.

Correspondence between the vertical distribution of total organic carbon (TOC) at station 3C (approximately 6–8 km downcurrent from the outfall system) and the mass emissions of suspended solids from the outfall system during the period 1946–1981 is illustrated in Fig. 6. Following World War II and up until 1971, the monotonic increase in emissions of suspended solids from the LACSD paralleled the population trend in Los Angeles. Thereafter, solids emissions declined in response to improved source control and advances in waste treatment (STULL et al., 1996). The vertical concentration profile of TOC in the 3C (1981) core records the historical trend in effluent solids emissions and indicates that, for this period, the outfalls dominated sedimentation of organic carbon on the shelf. The decline in emissions of suspended solids from the outfalls after 1971 became a matter of concern because of the potential for remobilization of heavily contaminated sediments that had been laid down in earlier years.

Because of difficulties in applying conventional radioisotopic methods of geochronology (e.g. ^{210}Pb) at the site, a study was undertaken to estimate sedimentation and mass accumulation rates by molecular stratigraphy. Selected persistent organic contaminants (i.e. DDT + metabolites, PCBs) and waste-specific molecular markers (the long-chain alkylbenzenes; EGANHOUSE and PONTOLILLO, 2000) were used to reconstruct the history of waste emissions. Molecular stratigraphy involves correlating known or predicted rates of input of a chemical to a sedimentary system with the vertical distribution of that same chemical in a sediment column (VALETTE-SILVER, 1993). Dates are assigned to inflection points in vertical concentration profiles, and average sedimentation rates (cm/yr) can then be estimated for the corresponding

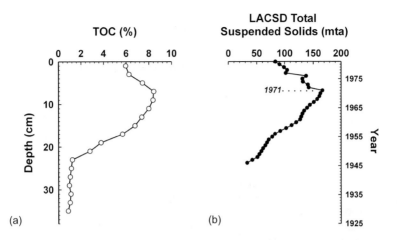

FIG. 6. (a) Vertical concentration profile of total organic carbon (TOC) in sediment core collected from station 3C in 1981 and (b) historical emissions (mta = metric tons yr^{-1}) of suspended solids from the LACSD wastewater outfall system, 1946–1981 (modified from EGANHOUSE and PONTOLILLO, 2000).

depth intervals. Chemical analyses of a gravity core collected by the LACSD in 1981 at station 3C were compared directly with those made on a box core collected by the USGS in 1992 at station 522, slightly inshore of station 3C (Fig. 5; EGANHOUSE and PONTOLILLO, 2000). The objective was to examine the variation in sedimentation (cm/yr) and mass accumulation (g/cm^2 yr) rates over time and to assess the relative importance of wastewater emissions *vis-à-vis* other sediment sources such as the Portuguese Bend Landslide (PBL; see Fig. 5).

The vertical concentration profiles of three molecular markers in the 3C (1981) and 522 (1992) cores are shown in Fig. 7. p,p'-DDE (1,1-dichloro-2,2-bis (p-chlorophenyl) ethylene) is the major persistent DDT metabolite in shelf sediments (60–70% of all DDT compounds; EGANHOUSE *et al.*, 2000), 6-C$_{12}$ (6-phenyldodecane) is the most persistent and abundant

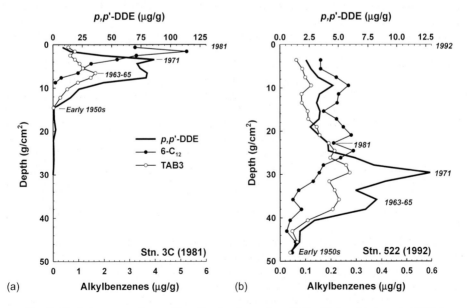

FIG. 7. Vertical concentration profiles of molecular markers (6-C$_{12}$ (6-phenyldodecane), TAB3 (a tetrapropylene-based alkylbenzene) and p,p'-DDE) in sediment cores collected from the Palos Verdes Shelf in (a) 1981 (station 3C) and (b) 1992 (station 522) and age-dates assigned based on historical discharge/usage patterns (modified from EGANHOUSE and PONTOLILLO, 2000). Location of sediment coring stations is shown in Fig. 5.

Table 2. Estimated sedimentation (cm/yr) and mass accumulation (g/cm^2-yr) rates for cores from
stations 3C (1981) and 522 (1992) based on molecular stratigraphy

Station/date	1955–1965	1965–1971	1971–1981	1981–1992
3C/1981				
cm/yr	0.8	1.3	0.7	—
g/cm^2-yr	0.6	0.7	0.3	—
522/1992				
cm/yr	0.9	1.3	0.8	2.1
g/cm^2-yr	1.1	1.4	0.9	2.2

constituent of the LABs (see TAKADA and EGANHOUSE, 1998), and TAB3 is a highly persistent member of the *tetrapropylene-based alkylbenzenes* (TABs: see TAKADA and EGANHOUSE, 1998), a class of surfactant markers related to the LABs. Dates have been assigned according to known and/or estimated historical discharge/usage patterns for these compounds as discussed in EGANHOUSE and PONTOLILLO (2000). Based on the date assignments, average sedimentation (cm/yr) and mass accumulation (g/cm^2 yr) rates were computed by dividing the differences in depth (cm) and mass accumulation (g/cm^2) by the time interval between collection of the two cores (11.2 yr). These results are given in Table 2.

The estimated sedimentation and mass accumulation rates for three time periods (1955–1965, 1965–1971, 1971–1981) at station 3C are reasonably consistent with the historical trends in solids emissions from the outfall system (*cf.* Table 2, Fig. 6). The sedimentation rates for station 522 during the same time periods are similar to those for station 3C. However, mass accumulation rates are systematically higher at station 522 than at station 3C, and there is a marked increase in both sedimentation rate and mass accumulation rate in the 1981–1992 period based on molecular stratigraphy (station 522 only). The latter observation is significant because emissions of wastewater effluent solids from the LACSD outfall system during the post-1981 period continued to decline (see Fig. 5 in EGANHOUSE and PONTOLILLO, 2000). Thus, the increase in sedimentation for the period 1981–1992 near station 522 (and presumably 3C) cannot be ascribed to an overriding influence of the outfall system.

As discussed by EGANHOUSE and PONTOLILLO (2000), the cause of the increased sedimentation in the 1980s and early 1990s was most likely mobilization of coarser sediment from the Portuguese Bend Landslide, which resulted from large winter storms in 1982–1983 and 1988. This hypothesis is supported by the observation of sedimentological and mineralogical indicators of PBL sediment in the vicinity of stations 3C/522 (DRAKE, 1994; WONG, 2002; REYNOLDS, 1987). Prior to this work, it was presumed that transport of PBL debris primarily would be to the southeast via littoral currents (Fig. 8; KAYEN *et al.*, 2002). Offshore transport and subsequent entrainment by poleward-moving currents could be envisioned as a means of bringing PBL debris to the vicinity of stations 3C/522. However, a more plausible explanation for the observed increase in sedimentation near stations 3C/522 in the 1980s and early 1990s and the systematically higher sediment accumulation rates at station 522 than station 3C for all time periods is that PBL debris was mobilized and transported offshore near to its source. Offshore transport would be favored in the vicinity of the PBL because poleward moving subthermocline currents are deflected to the west at Long Point owing to coastal bathymetry (Fig. 8; DRAKE, 1994). Under these circumstances, sediments deposited at station 522 would be expected to receive PBL inputs earlier and to a greater extent than those at station 3C. This case study provides an example of how data from molecular markers combined with sediment data can be used to elucidate sedimentation processes and transport pathways that might not have been identified using other techniques.

Bemidji, MN oil-spill site

In 1979, a high-pressure pipeline carrying a light paraffinic crude oil ruptured near the town of Bemidji, MN. The nearly instantaneous release of approximately 1670 m^3 of oil resulted in its

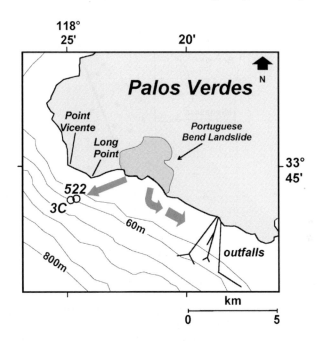

FIG. 8. Map of the Palos Verdes Shelf showing locations of wastewater outfall system, Portuguese Bend Landslide (PBL), sediment coring stations (3C, 522) and possible transport routes (arrows) of PBL debris.

accumulation within topographic lows and spraying of oil to the southwest of the pipeline (Fig. 9). After cleanup operations were completed, approximately $410 \, \text{m}^3$ of oil remained unaccounted for. A portion percolated through the unsaturated zone and formed an irregularly distributed body of residual oil that came to rest on the water table. Soluble constituents of the oil entered ground water (forming a contaminant plume), and degradation of these substances resulted in production of a complex mixture of organic transformation products, as well as alteration of the geochemical conditions in the aquifer (EGANHOUSE et al., 1993a; BAEDECKER et al., 1993; BENNETT et al., 1993). Within a few years after the spill, degradation of the oil led to development of a lens of highly contaminated anoxic ground water beneath and 70–80 m downgradient from the oil body.

Although a complex mixture of volatile aliphatic, aromatic, and alicyclic hydrocarbons was found in ground water near the residual oil body, roughly 85% of the volatile organic compounds was composed of MAHs (EGANHOUSE et al., 1993a). A number of these compounds are known or suspected carcinogens. Therefore, a study was undertaken to understand processes controlling the transport and fate of the MAHs within the anoxic plume. Because the source of these aromatic hydrocarbons was known with certainty, they were considered molecular markers of the crude oil and were used as process probes.

Paired oil and ground-water samples were collected along the axis of the oil body in the general direction of ground-water flow; ground-water samples were also collected upgradient and downgradient from the oil body (Fig. 9). These samples were analyzed for benzene and all alkylbenzenes having 1–4 alkyl carbon substituents (C_{1-4}-benzenes; 36 compounds in all) using purge-and-trap high-resolution gas chromatography/mass spectrometry (EGANHOUSE et al., 1993b). Analysis of the paired oil and ground-water samples revealed that oil from the trailing edge of the oil body was depleted in MAHs compared with oil from the leading edge of the oil body (see EGANHOUSE et al., 1996). The extent of depletion of each MAH was found to be directly related to its aqueous solubility. A sequential batch equilibration simulation showed that most of the depletion could be explained by water washing of the oil as ground water passed beneath and through the residual oil body. At the same time, ground-water samples collected

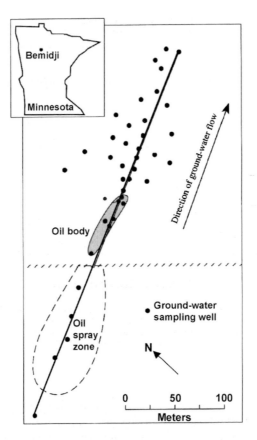

Fig. 9. Map of the Bemidji, MN crude oil-spill site showing approximate location of the residual oil body, the oil spray zone, and ground-water sampling wells along the general direction of ground-water flow. Main sampling transect is depicted as the bold line in general direction of ground-water flow (modified from EGANHOUSE et al., 1996).

along the axis of the oil body in the direction of ground-water flow (Fig. 9) showed increasing concentrations of MAHs. Together, these observations indicated that MAHs were dissolving from the oil into the ground water and that this process was leading to minor, but measurable, changes in composition of the residual oil. In order to assess the causes of observed downgradient changes in MAH composition, it was necessary to determine if ground water had reached saturation with respect to MAHs prior to being transported away from the oil body.

Oil–water equilibration experiments were carried out using several oil samples collected from the field site (EGANHOUSE et al., 1996). The laboratory results were then compared to theoretical calculations based on the aqueous solubilities of individual MAHs. Good agreement was found between the laboratory results and theoretical calculations. This set the stage for computing the degree of saturation of MAHs in ground water within the anoxic plume downgradient from the oil. The 21 MAHs for which data were available fell into two groups. Compounds that clearly were undersaturated (-30 to -90%; e.g. toluene, xylenes) tended to be less persistent within the anoxic plume downgradient from the oil. Compounds that were near saturation ($\pm 30\%$; e.g. benzene, ethylbenzene) were more persistent within the anoxic plume downgradient from the oil. These findings are summarized in Fig. 10, where the minimum apparent disappearance rate (in units of % change in concentration/meter) within the anoxic plume is plotted versus the % difference between measured and equilibrium concentrations. Additional data on the temporal variability of MAH concentrations in ground water immediately downgradient from the oil showed the same type of compound grouping (EGANHOUSE et al., 1996). The more persistent

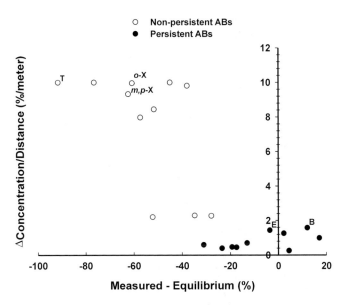

FIG. 10. Scatter plot of minimum disappearance rate (Δconcentration/distance) versus the difference between equilibrium concentrations of alkylbenzenes (ABs) and those measured in ground water immediately downgradient from the oil body (modified from EGANHOUSE *et al.*, 1996). *Note*: B, benzene; T, toluene; E, ethylbenzene; *o*-X, *ortho*-xylene; *m,p*-X, *meta-* + *para*-xylenes.

MAHs (*e.g.* benzene, ethylbenzene) showed little temporal variation (< about 20%), whereas the less persistent MAHs (*e.g.* toluene, xylenes) showed greater temporal variability (20–139%). These findings suggest that compounds that are persistent under the conditions prevailing within the anoxic plume approach or attain saturation and show little variation in concentration with time because dissolution rates exceed removal rates. On the other hand, compounds that are subject to rapid removal under the prevailing environmental conditions are undersaturated because removal rates exceed dissolution rates.

To gain some understanding of the removal processes responsible for attenuation of MAHs at the Bemidji site, a number of homologous series (*e.g.* the *n*-alkylbenzenes) and isomer groups were measured. The physical–chemical properties of homologous series differ systematically. For example, the log octanol–water partition coefficient (log K_{ow}), which is a predictor of the tendency of a compound to sorb to particulate matter, increases from a value of 2.16 for benzene to 3.71 for *n*-propylbenzene (see Fig. 11a; data from SHERBLOM and EGANHOUSE, 1988). If sorption was the principle mechanism responsible for the decreasing concentration of MAHs with distance downgradient from the oil, it is expected that benzene would be transported farthest and *n*-propylbenzene least. As shown in Fig. 11a, this clearly is not the case at the Bemidji oil-spill site. Here, the distributions of benzene + *n*-C_{1-3}-benzenes are plotted against distance. The rapid attenuation of toluene and, by comparison, the persistence of *n*-propylbenzene indicate that sorption does not control the transport and fate of these compounds. Because chemical degradation is not likely to be important, biological degradation is the most plausible explanation for attenuation of these MAHs within the anoxic plume. By the same reasoning, isomers having similar physical–chemical properties would be expected to behave the same if physical processes were dominant. However, three C_3-benzene isomers having near identical log K_{ow} values (SHERBLOM and EGANHOUSE, 1988) show strikingly different distributions (Fig. 11b). This result, again, indicates that structure-specific biological degradation, not sorption, exerts a dominant control on the transport and fate of these compounds. Although ancillary supporting evidence for biodegradation of MAHs at this site exists (BAEDECKER *et al.*, 1993; COZZARELLI *et al.*, 1994), this case study is a good illustration of how molecular markers can serve as process probes.

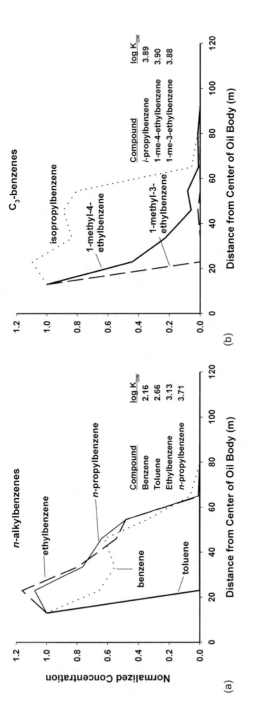

Fig. 11. Distribution of selected monoaromatic hydrocarbons along the main sampling transect at the Bemidji, MN oil-spill site: (a) homologous series (benzene + n-C$_{1-3}$-benzenes); (b) isomeric C$_3$-benzenes (modified from Eganhouse et al., 1996).

CONCLUSIONS

Molecular markers can be valuable tools for environmental investigations. With appropriate selection of marker compounds that are related to specific sources of organic matter and/or contamination, many types of information can be obtained. This information includes a better understanding of sources in complex multi-input systems as well as insights into the transport and fate of markers and associated contaminants. Unfortunately, the lack of data on the behavior of these compounds in the environment limits their utility for quantitative source apportionment. Studies are needed to ascertain the susceptibility of molecular markers to photolytic, chemical, and biological degradation processes. In addition, there is an ongoing need for reliable physical–chemical property data and systematic studies of the behavior of marker compounds *vis-à-vis* non-source-specific organic contaminants of concern. These data are essential for successful application of modeling approaches to predicting the transport and fate of hazardous materials in air, land, and sea. Hydrophilic markers increasingly are being discovered as analytical limits are overcome (*e.g.* pharmaceuticals by electrospray high-performance liquid chromatography/mass spectrometry/mass spectrometry). Such developments and more widespread adoption of the molecular marker approach should advance our understanding of Man's impact on the environment.

Acknowledgements—The author wishes to thank Dr Ian Kaplan, to whom this volume is dedicated, for providing the excellent scientific environment at the University of California, Los Angeles, where the author completed his doctoral and post-doctoral studies. The author also would like to acknowledge former students and research associates who have contributed to the data and interpretations discussed in this paper: P. Sherblom, C. Phinney (Boston Harbor), J. Pontolillo, T. Leiker (Palos Verdes, CA), M. J. Baedecker, I. Cozzarelli, T. Dorsey, C. Phinney, S. Westcott (Bemidji, MN). Finally, the author appreciates the helpful comments and suggestions of M. J. Baedecker, I. Cozzarelli, and Angel Martin who reviewed an early draft of this paper.

REFERENCES

BAEDECKER M. J., COZZARELLI I. M., EGANHOUSE R. P., SIEGEL D. I. and BENNETT P. C. (1993) Crude oil in a shallow sand and gravel aquifer: III. Biogeochemical reactions and mass balance modeling in anoxic groundwater. *Appl. Geochem.* **8**, 569–586.

BENNETT P. C., SIEGEL D. I., BAEDECKER M. J. and HULT M. F. (1993) Crude oil in a shallow sand and gravel aquifer: I. Hydrogeology and inorganic geochemistry. *Appl. Geochem.* **8**, 529–549.

BOTHNER M. H., BUCHHOLTZ TEN BRINK M. and MANHEIM F. T. (1998) Metal concentrations in surface sediments of Boston Harbor—changes with time. *Mar. Environ. Res.* **45**(2), 127–155.

BRASSEL S. C., EGLINTON G., MARLOWE I. T., PFLAUMANN U. and SARNTHEIN M. (1986) Molecular stratigraphy: a new tool for climatic assessment. *Nature* **320**(6058), 129–133.

COZZARELLI I. M., BAEDECKER M. J., EGANHOUSE R. P. and GOERLITZ D. F. (1994) The geochemical evolution of low-molecular-weight organic acids derived from the degradation of petroleum contaminants in groundwater. *Geochim. Cosmochim. Acta* **58**, 863–877.

DRAKE D. E. (1994) The natural recovery of contaminated effluent-affected sediment on the Palos Verdes Margin: background information and results of the USGS natural recovery research. Southern California Damage Assessment Witness Reports, 131 pp.

EGANHOUSE R. P. (1997) Molecular markers and environmental organic geochemistry: an overview. In *Molecular Markers in Environmental Geochemistry* (ed. R. P. EGANHOUSE), pp. 1–20. American Chemical Society, Washington, DC.

EGANHOUSE R. P. and PONTOLILLO J. (2000) Depositional history of organic contaminants on the Palos Verdes Shelf, California. *Mar. Chem.* **70**, 317–338.

EGANHOUSE R. P. and SHERBLOM P. M. (2001) Anthropogenic organic contaminants in the effluent of a combined sewer overflow: impact on Boston Harbor. *Mar. Environ. Res.* **51**, 51–74.

EGANHOUSE R. P., BLUMFIELD D. L. and KAPLAN I. R. (1983) Long-chain alkylbenzenes as molecular tracers of domestic wastes in the marine environment. *Environ. Sci. Technol.* **17**, 523–530.

EGANHOUSE R. P., BAEDECKER M. J., COZZARELLI I. M., AIKEN G. R., THORN K. A. and DORSEY T. F. (1993a) Crude oil in a shallow sand and gravel aquifer: II. Organic geochemistry. *Appl. Geochem.* **8**, 551–567.

EGANHOUSE R. P., DORSEY T. F., PHINNEY C. S. and WESTCOTT A. M. (1993b) Determination of C_6–C_{10} aromatic hydrocarbons in water by purge-and-trap capillary gas chromatography. *J. Chromatogr.* **628**, 81–92.

EGANHOUSE R. P., DORSEY T. F., PHINNEY C. P. and WESTCOTT A. W. (1996) Processes affecting the fate of monoaromatic hydrocarbons in an aquifer contaminated by crude oil. *Environ. Sci. Technol.* **30**, 3304–3312.

EGANHOUSE R. P., PONTOLILLO J. and LEIKER T. J. (2000) Diagenetic fate of organic contaminants on the Palos Verdes Shelf, California. *Mar. Chem.* **70**, 289–315.

FINDLAY R. H. and WATLING L. (1997) Seasonal variation in sedimentary microbial community structure as a backdrop for the detection of anthropogenic stress. In *Molecular Markers in Environmental Geochemistry* (ed. R. P. EGANHOUSE), pp. 49–64. American Chemical Society, Washington, DC.

GOUIN T., MACKAY D., WEBSTER E. and WANIA F. (2000) Screening chemicals for persistence in the environment. *Environ. Sci. Technol.* **34**(5), 881–884.

HEDGES J. I. and PRAHL F. G. (1993) Early diagenesis: consequences for applications of molecular biomarkers. In *Organic Geochemistry: Principles and Applications* (eds. M. H. ENGEL and S. A. MACKO). Plenum Press, New York.

KAPLAN I. R., GALPERIN Y., LU S.-T. and LEE R.-P. (1997) Forensic environmental geochemistry: differentiation of fuel-types, their sources and release time. *Org. Geochem.* **27**(5/6), 289–317.

KATES M. (1997) Diether and tetraether phospholipids and glycolipids as molecular markers for archaebacteria (Archaea). In *Molecular Markers in Environmental Geochemistry* (ed. R. P. EGANHOUSE), pp. 35–48. American Chemical Society, Washington, DC.

KAYEN R. E., LEE H. J. and HEIN J. R. (2002) Influence of the Portuguese Bend landslide on the character of the effluent-affected sediment deposit, Palos Verdes margin, southern California. *Cont. Shelf Res.* **22**, 911–922.

KNEBEL H. J., RENDIGS R. R. and BOTHNER M. H. (1991) Modern sedimentary environments in Boston Harbor, Massachusetts. *J. Sedim. Petrol.* **61**, 791–804.

LEE H. J., SHERWOOD C. R., DRAKE D. E., EDWARDS B. D., WONG F. and HAMER M. (2002) Spatial and temporal distribution of contaminated effluent-affected sediment, Palos Verdes margin, southern California. *Cont. Shelf Res.* **22**, 859–880.

MACGREGOR J. S. (1976) DDT and its metabolites in the sediments off southern California. *Fish. Bull.* **74**, 27–35.

MCDERMOTT D. J., HEESEN T. C. and YOUNG D. R. (1974) DDT in bottom sediments around five southern California outfall systems. Southern California Coastal Water Research Project, 54 pp.

MURRAY C. J., LEE H. J. and HAMPTON M. A. (2002) Geostatistical mapping of effluent-affected sediment distribution on the Palos Verdes shelf. *Cont. Shelf Res.* **22**, 881–897.

PETERS K. E. and MOLDOWAN J. M. (1993) *The Biomarker Guide*. Prentice-Hall, Englewood Cliffs, NJ.

REYNOLDS S. (1987) Sediment dynamics on the Palos Verdes Shelf. Report to the Los Angeles County Sanitation Districts, 59 pp.

SCHAUER J. J., ROGGE W. F., HILDEMANN L. M., MAZUREK M. A., CASS G. R. and SIMONEIT B. R. T. (1996) Source apportionment of airborne particulate matter using organic compounds as tracers. *Atmos. Environ.* **30**(22), 3837–3855.

SHERBLOM P. M. and EGANHOUSE R. P. (1988) Correlations between octanol–water partition coefficients and reversed-phase high-performance liquid chromatography capacity factors. Chlorobiphenyls and alkylbenzenes. *J. Chromatogr.* **454**, 37–50.

SHERWOOD C. R., DRAKE D. E., WIBERG P. L. and WHEATCROFT R. A. (2002) Prediction of the fate of p,p'-DDE in sediment on the Palos Verdes shelf, California, USA. *Cont. Shelf Res.* **22**, 1025–1058.

SIGNELL R. P. and BUTMAN B. (1992) Modeling tidal exchange and dispersion in Boston Harbor. *J. Geophys. Res.* **97**, 15591–15606.

STOLZENBACH K. D., ADAMS E. E., LADD C. C., MADSEN O. S. and WALLACE G. (1993) Boston Harbor study of sources and transport of harbor sediment contamination, Part I: Transport of contaminated sediments in Boston Harbor. MWRA Technical Report, No. 93-12.

STULL J. K., SWIFT D. J. P. and NIEDORODA A. W. (1996) Contaminant dispersal on the Palos Verdes continental margin: I. Sediments and biota near a major California wastewater discharge. *Sci. Total Environ.* **179**, 73–90.

TAKADA H. and EGANHOUSE R. P. (1998) Molecular markers of anthropogenic waste. In *Encyclopedia of Environmental Analysis and Remediation* (ed. R. A. MYERS), pp. 2883–2940. Wiley, New York.

TAKADA H. and ISHIWATARI R. (1989) Biodegradation experiments of linear alkylbenzenes (LABs): isomeric composition of C12 LABs as an indicator of the degree of LAB degradation in the aquatic environment. *Environ. Sci. Technol.* **24**(1), 86–91.

TAKADA H., SATOH F., BOTHNER M. H., TRIPP B. W., JOHNSON C. G. and FARRINGTON J. W. (1997) Anthropogenic molecular markers: tools to identify the sources and transport pathways of pollutants. In *Molecular Markers in Environmental Geochemistry* (ed. R. P. EGANHOUSE), Vol. Symposium Series 671, pp. 178–195. American Chemical Society, Washington, DC.

VALETTE-SILVER N. J. (1993) The use of sediment cores to reconstruct historical trends in contamination of estuarine and coastal sediments. *Estuaries* **16**, 577–588.

VOLKMAN J. K., REVILL A. T. and MURRAY A. P. (1997) Applications of biomarkers for identifying sources of natural and pollutant hydrocarbons in aquatic environments. In *Molecular Markers in Environmental Geochemistry* (ed. R. P. EGANHOUSE), pp. 110–132. American Chemical Society, Washington, DC.

WALLACE G. T., KRAHFORST C. F., PITTS L. C., SHINE J. P., STUDER M. M. and BOLLINGER C. R. (1991) Assessment of the chemical composition of the Fox Point CSO effluent and associated subtidal and intertidal environments: analysis of CSO effluents, receiving water and surface sediments for trace metals prior to CSO modification. Massachusetts Department of Environmental Protection.

WANIA F. and MACKAY D. (1996) Tracking the distribution of persistent organic pollutants. *Environ. Sci. Technol.* **30**(9), 390A–396A.

WONG F. L. (2002) Heavy mineral provinces of the Palos Verdes margin, southern California. *Cont. Shelf Res.* **22**, 899–910.

Geochemical Investigations in Earth and Space Science: A Tribute to Isaac R. Kaplan
© The Geochemical Society, Publication No. 9, 2004
Editors: R.J. Hill, J. Leventhal, Z. Aizenshtat, M.J. Baedecker, G. Claypool,
R. Eganhouse, M. Goldhaber and K. Peters

The geochemical and magnetic record of coal-combustion products in West Virginia reservoir sediments and soils

Martin Goldhaber[1], Ted Callender[2] and Richard Reynolds[1]

[1]USGS, Crustal Imaging Team, Denver Federal Center, MS 973, Denver, CO 80225, USA
[2]USGS, Emeritus, Westerly, RI, USA

Abstract—Western West Virginia lies downwind from numerous coal-fired power plants along the Ohio River Valley. To test whether geochemical and mineralogical impacts from these power plants are detectable on the West Virginia landscape, we obtained reservoir cores, soils, and rocks from two separate sites in West Virginia, one in Harrison County (Hinkel and Deegan Reservoirs) and the other in Roane County (Miletree Run Reservoir). Both have small drainage basins that have the effect of maximizing atmospheric inputs relative to weathering inputs. Sediments from Hinkel Reservoir were dated using the ^{137}Cs method, and by knowledge of the age of the base of the sedimentary section. Major elements in Hinkel Reservoir sediments do not vary systematically over time, suggesting that the depositional history of these sediments has been relatively constant. In contrast, minor elements and sulfur do show dramatic shifts. Zn, Pb, As, S, Cd, Ge, and Hg all peak during the late 1960s and early 1970s. Polyaromatic hydrocarbons associated with combustion processes likewise show this same behavior. Coincident with these maxima is a peak in isothermal remnant magnetization (IRM), a parameter that is proportional to magnetite abundance. We separated the magnetic fraction of the sediments and examined this fraction petrographically. It is dominated by magnetite with a spherical morphology, which is characteristic of magnetite produced by combustion processes. Chemical analyses on local rocks showed that they were not the likely source for the trace element and magnetite enrichments. To compare soils and reservoir sediments, we normalized geochemical data to Ti; the underlying assumption is that Ti will track physical inputs of soil materials into the reservoir sediments. The normalized sediment data for the elements Zn, Pb, As, and S are all higher for sediments deposited during the late 1960s and early 1970s compared to normalized soil data, implying that soils are not the source for the element enrichments. We thus attribute the enrichment to atmospheric inputs. However, a coal-burning zinc smelter within 10 km of the reservoir site that operated until 1971 may have supplied some or all of the anomalous input. There are no nearby major atmospheric pollution sites near Miletree Run Reservoir. A core from this reservoir was dated based on knowledge of the time it was impounded. Like Hinkel Reservoir, there are no systematic shifts in major element contents over its depositional history. Like Hinkel Reservoir, there are also significant shifts in trace elements and IRM that cannot be explained by local soil sources. Magnetic separates from this reservoir are also dominated by spherical combustion-produced magnetite. Covariance over the period 1930–1980 between magnetite and sulfur with US SO_2 production (which comes dominantly from coal combustion) is strongly suggestive of a significant atmospheric input into Miletree Run Reservoir sediments. The nearest upwind power plants are between 50 and 75 km distant. Thus, relatively long distance transmission of particulate matter from the power plants of the Ohio River Valley is likely in this case.

INTRODUCTION

COAL COMBUSTION for heating, as well as steel and power production, has been extensive in the United States, particularly in the north central and northeast portions of the country. Release of sulfur and nitrogen oxides from this process has resulted in a well-characterized impact downwind from the power plants in the northeastern US in the form of acid deposition (HUSAR et al., 1991; SHAW, 1984). Maps of sulfate (sulfuric acid) deposition show high values in parts of Ohio, West Virginia, and Pennsylvania as well as further to the northeast, into the New England states and Canada (EILERS and SELLE, 1991; SHAW, 1984). The impact of acid deposition on lakes and soils is regionally detectible (DRISCOLL et al., 2001; HERLIHY et al., 1993) and is locally substantial (*e.g.* GIBSON *et al.*, 1986). Although acid deposition has been decreasing with time (DRISCOLL *et al.*, 1998, 2001; EPA, 2000), it has not ceased, and the impacts are still apparent (DRISCOLL *et al.*, 2001).

Sulfur and nitrogen oxides are not the only output from coal combustion that may leave a detectible environmental impact. Particulate matter (fly ash) is also released during coal combustion and can be detected near coal-fired power plants (FLANDERS, 1999; GODBEER and SWAIN, 1995; KAPICKA *et al.*, 1999). The focus of the present study was to test whether these particulate emissions have produced a detectible impact on the geochemical landscape downwind from power plants. It is already known that atmospheric inputs can be detected in lake and reservoir sediments (LOCKE and BERTINE, 1986; NORTON, 1986), and we sought to further test whether atmospheric coal fly ash inputs could be identified and quantified.

The deposition and accumulation of fly ash downwind from coal-combustion sites is a concern because it may be significantly enriched in potentially toxic trace elements, including lead (Pb) and arsenic (As), compared to the burned coal (COLES *et al.*, 1979; EARY *et al.*, 1990; HOWER *et al.*, 1999; KAAKINEN *et al.*, 1975). Other elements such as zinc (Zn) and germanium (Ge), of less environmental concern, may also be enriched in fly ash. The relatively high concentrations of As in fly ash reflect partly its presence in pyrite in coal from the Appalachian Basin (GOLDHABER *et al.*, 2002). More importantly, the concentration of these metals and metalloids occurs during the combustion process itself. A suite of elements including As, Ca, Cr, Cu, Ga, Mo, Ni, Pb, Sb, Se, V, and Zn is enriched in the fine fraction of coal fly ash (COLES *et al.*, 1979), because of vaporization in the furnace and subsequent condensation or absorption onto ash particles (KAAKINEN *et al.*, 1975).

Our interest in the issue initially arose from examination of the US Geological Survey national soil geochemical data set (GUSTAVSSON *et al.*, 2001). The Ohio River Valley is the site of a concentration of coal-fired power plants (Fig. 1) and the US Geological Survey

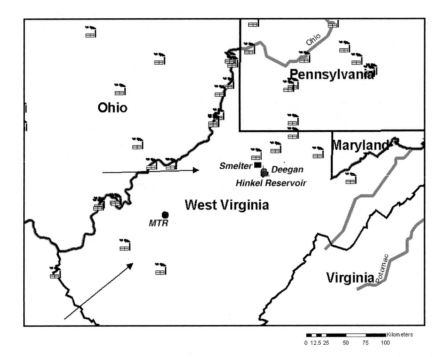

FIG. 1. Location map showing the study sites (filled circles): Hinkel/Deegan Reservoirs and Miletree Run Reservoir. At the scale of the map, Hinkel and Deegan Reservoirs overlie each other. Also shown is the location of a zinc smelter near Hinkel Reservoir (filled square), and the locations of coal-fired power plants (smokestack symbols). The arrows are yearly average wind directions from a NOAA web site (http://ols.nndc.noaa.gov).

data show enrichment in As in soils along the Ohio River Valley. We hypothesized that this correspondence between power plants and trace elements is due to accumulation of power plant combustion products.

This hypothesis was tested by analysis of core samples from reservoirs in West Virginia. Reservoir sampling has been repeatedly shown to be an effective tool for determining temporal trends in element deposition (CALLENDER and RICE, 2000; CALLENDER and VAN METRE, 1997). We chose two sites in West Virginia because they are not adjacent to major urban pollution sources, and they are downwind from the power plants of the Ohio River Valley. Both are among sites with the highest sulfate deposition rates and lowest pH values in stream waters (acid rain impacts) of any in the country (EILERS and SELLE, 1991). The sites were also chosen because they represented a good temporal and depositional history allowing us to track trends for more than half a century. They are also situated in relatively small drainage basins and thus, maximize input from atmospheric sources relative to input from weathering of adjacent rocks.

SITE DESCRIPTIONS

Hinkel (HIN) and Deegan (DGN) reservoirs are located near the town of Bridgeport in Harrison County, West Virginia (Fig. 1; Table 1). Hinkel Reservoir, with a surface area of 15.5 km^2, was formed in 1923 by a dam on Davission Run. Deegan Reservoir, with a surface area of 46.6 km^2, is separated from Hinkel Reservoir by a small wetland, and dates from 1950. Both reservoirs have small drainage areas of 15.8 and 13.2 km^2, respectively. Two push cores and a box core from the deepest portion of Hinkel Reservoir and a push core from the middle of Deegan Reservoir were recovered during November 2000 and analyzed as described below. The study area is within the Elk Creek drainage of the West Fork river watershed. This drainage is dominated by deciduous forest (62.0%), along with mixed conifer and deciduous forest (7.8%). An additional 22.9% of the land is devoted to crops and hay pasture with the remainder devoted to low and high intensity development. The area is underlain by Pennsylvanian-aged sandstone, shale, and coal (RUPPERT and RICE, 2001). Coal mining has been extensive near the study site. The USGS Clarksburg 1:24,000 topographic sheet prepared in 1976 shows coal mining operations within 2.8 km of Hinkel and Deegan Reservoirs and numerous others within a radius of 10 km. These mined areas are now reclaimed. The soil in the area is classified in the Gilpin Series, which are well-drained (strongly leached) moderately deep soils formed in residuum of interbedded shale, siltstone, and sandstone (BERERAGE and YOAKUM, 1980). Red iron oxides are noticeable in the soil profile.

The Miletree Run (MTR) Reservoir, constructed in 1930, is near the town of Spencer in Roane County, West Virginia (Fig. 1; Table 1). It has a surface area of 13 km^2 and a steep-sided drainage area of 1.4 km^2. We obtained push cores, also in November 2000, from behind dam

Table 1. Sample sites and samples

Site	Sample material	Longitude (°)	Latitude (°)	Year completed	Analyses performed[a]
Hinkel Lake				1923	
	Core HIN.1	80.25	39.27		ICP-AES, ICP-MS, Hg, As, S, magnetic data
	Core HIN.2	80.25	39.27		ICP-AES, As
	13 soil samples				ICP-AES, As
	9 rock samples				ICP-AES, As
Deegan Lake	DGN.1	80.25	39.35	1950	ICP-AES, As
Miletree Run		−80.37	38.80	1930	
	Core MTR.1				ICP-AES, As, S, magnetic data
Soils	17 soil samples				ICP-AES, As
	7 rock samples				ICP-AES, As

[a] ICP-AES: 40 elements analyzed by XRAL Laboratories—HCl, HNO_3, $HClO_3$, HF digestion; ICP-MS: 42 elements—HCl, HNO_3, $HClO_3$, HF digestion at USGS; As: hydride generation, AAS at XRAL Laboratories; Hg: atomic fluorescence spectroscopy at USGS Labs; S: LECO induction furnace method at USGS Laboratories; magnetic data: includes both magnetic measurements and magnetic separates (see text).

number 1. The site is underlain by Pennsylvanian and Permian (?) sandstones and shales (RUPPERT and RICE, 2001). Coal mining did not occur near MTR. Soil is similar to that in Harrison County.

METHODS

Field methods

The primary sampling tool was a push corer consisting of a 6.7 cm diameter plastic core barrel attached to aluminum rods and handle. The corer was pushed into the bottom sediment as far as possible and then raised vertically until the bottom of the barrel was barely in the water whereupon a plastic core cap was attached. Push cores were transported in a vertical position to the shore where the sediment was extruded vertically. A secondary sampling tool was a box corer that was lowered slowly into the bottom sediment. This is a clamshell type grab sampler fitted with a 14 cm square plastic box. When the sampler is triggered at the bottom and then hauled up to the surface, relatively undisturbed sediment is retained in the box overlain by bottom water. This sampler was used to compliment the push cores that may exhibit some physical disturbance of the upper few centimeters of surface sediment as the core barrel is pushed down into the bottom sediment. The upper 20 cm of the push cores was sampled at 1 cm intervals and below that at 2 cm intervals using a simple circular mechanical piston attached to a rod. The entire box core (a maximum of 15 cm) was extruded in 1 cm intervals using a square piston attached to a frame with a mechanical worm screw rod. All samples for inorganic analyses were placed in plastic jars or bags and those for organic analyses (HIN) in pre-cleaned glass jars. Rock and soil samples near both HIN/DGN and MTR Reservoirs were collected for comparison with the reservoir sediments. The soil samples were collected around the perimeter of the reservoirs by removing the surface leaf litter to expose the underlying soil, and sampling approximately the upper 5–10 cm of soil into plastic bags. Rock samples were collected from a road cut near the onramp to I-79 from route 50N near Hinkel–Deegan watershed. In order to maximize the abundance of trace elements mobilized during early stages of weathering, rock samples were selected to represent the most weathered material from the outcrop, and included abundant iron staining. An approximately 2 ft (0.6 m) thick coal seam was exposed. The coal, weathered material immediately above the coal, as well as weathered rock samples from a bench approximately 9 m above the coal were collected. In addition, rock samples were collected from highly weathered outcrops immediately surrounding both reservoirs.

Lab methods

The sediment samples were stored on ice and transported to the laboratory where samples for inorganic analyses were frozen, weighed, freeze-dried, and re-weighed to determine water content and porosity. The dry sediment, rock and soil samples were ground for chemical analysis. Samples were analyzed for 40 elements by ICP-AES following acid dissolution (HF, HCl, HNO_3) and for arsenic by a hydride generation, atomic absorption spectroscopy by a contract lab (XRAL Laboratories). Blind duplicates and standards were submitted with the samples, and the results monitored by the USGS sample control group. An additional set of sediment samples from Hinkel Reservoir was submitted to the USGS analytical group for analysis of germanium (Ge) and cadmium (Cd) by acid digestion followed by ICP-MS and for Hg by continuous-flow cold-vapor atomic absorption spectrometry. Total sulfur and total carbon were analyzed at the USGS on a separate aliquot of the MTR samples by LECO® induction furnace instrumentation and total sulfur on the Hinkel samples also using LECO® induction furnace instrumentation. Age dating by the ^{137}Cs method was performed on one core from Hinkel Reservoir. Sediment ^{137}Cs activity was measured on discrete core sections by counting freeze-dried sediment in a fixed-geometry with a high resolution, intrinsic germanium detector gamma-spectrometer (CALLENDER and ROBBINS, 1993). Counting errors were well under 10% and in the region of peak ^{137}Cs activity, the errors were around 2%.

Polyaromatic hydrocarbons (PAHs) and alkyl-substituted PAHs were extracted, isolated, and analyzed from Hinkel Reservoir sediments using a variation of the procedure of FURLONG *et al.* (1995). Variations include the addition of a silica column cleanup step following the gel permeation chromatography cleanup step and the use of selected ion monitoring (SIM) MS to reduce chemical interferences and improve detection limits.

Magnetic methods

Magnetite and related magnetic spinel phases are a common product of coal combustion (LAUF, 1982; LOCKE and BERTINE, 1986) arising from the partial oxidation of pyrite in the low oxygen conditions present during coal combustion (LAUF *et al.*, 1982). In order to take advantage of this potential proxy for the presence of coal-combustion products, we examined magnetic minerals for textures and compositions that are diagnostic of their origins and measured the abundance of magnetic minerals in the reservoir sediment. Spherical magnetite is characteristically produced by combustion processes, and specifically coal combustion, during which the spherical shape may be inherited from pyrite spheres (framboids) present in coal (LAUF *et al.*, 1982). In contrast, angular to rounded (but not spherical) shapes characterize rock-derived magnetic mineral particles in sediment and soil as well as other anthropogenic magnetite-bearing particles, such as rust.

Magnetic minerals were identified and their shapes as well as internal textures were characterized primarily using reflected-light microscopy. In this way, reliable identification of different iron oxide minerals can be made on grains larger than about 3 μm in diameter. The grains were prepared in polished grain mounts after isolation from the bulk sediment in a pumped-slurry magnetic separator (REYNOLDS *et al.*, 2001). Scanning electron microscopy (SEM) of magnetic mineral separates using a JEOL JSM-5800LV instrument enabled examination of the shapes and chemical compositions of particles (GOMES *et al.*, 1999; LOCKE and BERTINE, 1986) larger than about 0.5 μm in diameter. Semi-quantitative analysis of the grains was performed using an Oxford Isis energy-dispersive X-ray analyzer (EDA).

Magnetic measurements (see THOMPSON and OLDFILED, 1986; VEROSUB and ROBERTS, 1995) were made on dried bulk sediment packed into 3.2 cm^3 plastic cubes and normalized for sample mass (typically 2–3 g). The relative abundance of magnetite was determined by measurement of isothermal remanent magnetization (IRM), the magnetization acquired by a sample after exposure to a strong magnetic field, in this study 0.3 T ($IRM_{0.3\ T}$). Remanent magnetization was measured using a 90 Hz spinner magnetometer with a sensitivity of about 10^{-5} A/m. Magnetic susceptibility (MS), a measure of all magnetic materials but mainly magnetite when present, was determined in a 0.1 mT induction at 600 Hz using a susceptometer with a sensitivity better than 4×10^{-7} m^3/kg. A measure of hematite is given by ($IRM_{1.2\ T} - IRM_{0.3\ T}$)/2 (Hard IRM or HIRM) (KING and CHANNEL, 1991). The ratio, $IRM_{0.3}/IRM_{1.2}$, called the S parameter, is a measure of the relative proportion of magnetite to all oxides, including hematite. High S parameter values indicate large amounts of magnetite relative to hematite (a maximum value of 1), and decreasing values indicate increasing amounts of hematite. A measure for magnetic grain size is the ratio of anhysteretic remanent magnetization (ARM) to MS; ARM was imparted in a DC induction of 0.1 mT in the presence of a decaying alternating induction from 100 to 0 mT. The ratio ARM/MS increases as magnetic grain size of magnetite decreases, and it is particularly sensitive to single domain (SD) and small pseudo-single domain (PSD) grain sizes in rocks and sediments in which magnetite is the dominant magnetic mineral.

RESULTS

Hinkel/Deegan Reservoirs

In Hinkel Reservoir Core HIN.2, the peak in ^{137}Cs activity (expressed in dpm/g dry sediment) occurred in the 17–18 cm sampling interval (Fig. 2). The date of deposition for sediment intervals within this core and the other cores from Hinkel and Deegan Reservoirs was based on the location of the pre-reservoir surface, the sampling date, and the peak in ^{137}Cs activity, dated at 1964. The HIN.1 core exhibited depth profiles of sedimentary Pb and As that were identical to

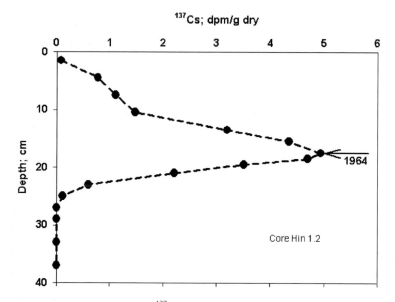

FIG. 2. Plot showing the activity of ^{137}Cs (dpm/g dry weight of sediment) in Hinkel Reservoir sediments. The peak of ^{137}Cs deposition occurred in 1964, and is marked with an arrow.

those from HIN.2 except that the peak concentrations were offset 1 cm deeper in the sediment column (plot not shown). For the Deegan Reservoir core, the bottom of reservoir sediment occurs at 32 cm and is assigned a date of 1950 based on the date of reservoir impoundment. We assumed that the base of the reservoir sediment was reached because leafy material was encountered at the base that differed in character from other sediments recovered.

The content of aluminum (Al), iron (Fe), calcium (Ca), magnesium (Mg), sodium (Na), titanium (Ti), and potassium (K) in core HIN.2 does not vary systematically with sediment depth and thus age (Fig. 3; Table 2). Aluminum has a mean of 9.3 wt% and ranges between 9.0 and 9.9 wt% (one point excluded). Titanium averages 0.47 and ranges between 0.44 and 0.51 wt%. The other element concentrations also fall within relatively narrow ranges. The low Ca content indicates that calcium carbonate if present is a minor phase in the sediments. There are small

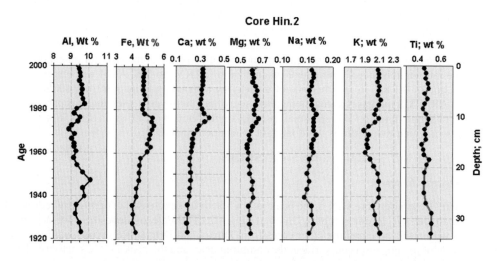

FIG. 3. Depth and age plots of major elements in core HIN.2.

Table 2. Chemical analyses from core HIN.2

Median depth (cm)	Date	Al (wt%)	Ca (wt%)	Fe (wt%)	K (wt%)	Mg (wt%)	Na (wt%)	Ti (wt%)	As (ppm)	Cu (ppm)	Pb (ppm)	Sr (ppm)	Zn (ppm)	PAH[a] (µg/kg)
0.5	2000	9.3	0.33	4.7	2.06	0.65	0.16	0.46	15.0	107	46	139	223	
1.5	1998	9.5	0.32	4.7	2.08	0.60	0.16	0.47	15.2	100	43	140	206	
2.5	1997	9.5	0.32	4.8	2.09	0.61	0.16	0.47	14.5	102	41	141	213	5759
3.5	1995	9.5	0.32	4.7	2.09	0.62	0.16	0.49	16.0	104	45	142	215	
4.5	1993	9.5	0.32	4.8	2.07	0.61	0.16	0.50	14.7	102	41	142	205	
5.5	1991	9.6	0.32	4.7	2.10	0.64	0.16	0.46	16.6	111	45	142	211	
6.5	1989	9.6	0.31	4.7	2.09	0.65	0.15	0.48	17.6	121	45	141	214	
7.5	1987	9.6	0.31	4.8	2.08	0.64	0.15	0.46	15.7	121	43	139	210	6758
8.5	1985	9.7	0.31	4.9	2.12	0.65	0.16	0.45	16.2	120	42	141	214	
9.5	1982	9.8	0.30	4.7	2.09	0.64	0.16	0.47	17.8	114	44	141	203	
10.5	1980	9.3	0.32	4.7	2.05	0.62	0.16	0.50	16.7	118	52	138	211	
11.5	1978	9.2	0.33	4.8	2.03	0.63	0.17	0.48	19.5	133	60	136	235	
12.5	1976	9.5	0.37	5.3	2.09	0.66	0.16	0.46	20.8	85	67	142	261	9123
13.5	1974	9.4	0.34	5.2	2.04	0.64	0.16	0.47	21.1	74	73	138	269	
14.5	1972	9.0	0.29	5.4	1.95	0.61	0.16	0.47	22.6	93	80	130	270	
15.5	1971	8.9	0.27	5.3	1.88	0.60	0.16	0.44	21.9	77	72	127	279	
16.5	1968	9.2	0.25	5.2	1.95	0.58	0.17	0.45	19.1	64	67	130	289	
17.5	1966	9.1	0.24	5.1	1.93	0.58	0.16	0.46	18.8	59	61	128	290	23,620
18.5	1964	9.2	0.24	5.0	1.92	0.56	0.16	0.50	17.6	54	61	129	269	
19.5	1963	9.2	0.23	5.2	1.94	0.56	0.16	0.48	18.1	50	60	130	271	
21	1960	9.3	0.23	5.0	1.90	0.57	0.16	0.46	16.9	51	52	132	263	
23	1957	9.1	0.22	4.6	1.97	0.57	0.15	0.46	16.2	48	41	128	239	14,606
25	1954	9.3	0.22	4.5	2.02	0.58	0.15	0.46	16.3	42	39	131	237	
27	1951	9.7	0.22	4.5	2.08	0.58	0.15	0.47	15.4	37	37	134	234	5194
29	1947	10.1	0.22	4.5	2.09	0.61	0.15	0.52	14.8	37	35	136	240	
31	1943	9.7	0.22	4.3	2.09	0.60	0.15	0.52	15.3	35	30	131	222	
33	1940	9.8	0.21	4.3	2.09	0.61	0.14	0.52	15.5	39	32	130	215	4526
35	1936	9.3	0.19	4.0	2.01	0.57	0.16	0.51	13.0	32	30	123	208	
37	1932	9.3	0.19	4.1	2.04	0.58	0.16	0.50	19.5	32	30	124	210	
39	1927	9.5	0.19	4.0	2.06	0.58	0.16	0.48	14.3	35	29	127	206	
41	1923	9.6	0.19	4.3	2.11	0.59	0.15	0.50	13.5	30	30	130	212	

[a]Combustion PAH = sum of fluoranthene, pyrene, b(a)anthracene, chrysene, b(k)fluoranthene, b(b)fluoranthene, b(e)pyrene, b(a)pyrene, indeno(1,2,3-c,d)pyrene, and b(ghi)perylene. See PRAHL and CARPENTER (1983).

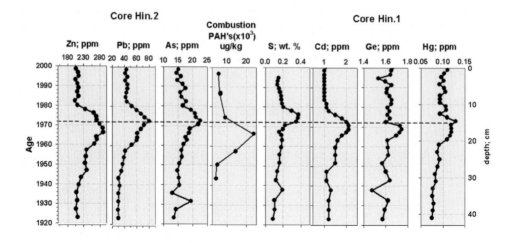

FIG. 4. Depth and age plots of Zn, Pb, As, combustion PAHs, in core HIN.2 and S, Cd, Ge, and Hg in core HIN.1. A dashed reference line is drawn at 1972 corresponding to the peaks in Pb and As.

Table 3. Chemical analyses form core HIN.1

Median depth	Age	Fe (wt%)	Ti (wt%)	As (ppm)	Cd (ppm)	Pb (ppm)	Zn (ppm)	Ge (ppm)	Hg (ppm)	S (wt%)
0.5	2000	4.8	0.45	14.7	1.1	43	214	1.67	0.110	ND[a]
1.5	1998	4.8	0.48	14.7	1.0	41	219	1.65	0.102	ND
2.5	1996	4.8	0.48	14.1	1.0	43	222	1.64	0.096	ND
3.5	1995	4.8	0.48	15.5	1.0	43	217	1.53	0.101	0.151
4.5	1993	4.8	0.45	14.9	1.0	41	216	1.60	0.091	0.139
5.5	1991	4.7	0.45	15.9	1.0	40	213	1.65	0.102	0.136
6.5	1989	4.8	0.47	14.8	1.0	40	217	1.61	0.105	0.154
7.5	1987	5.0	0.48	15.5	1.0	41	209	1.61	0.093	0.167
8.5	1986	4.9	0.48	17.6	1.0	42	220	1.64	0.093	0.179
9.5	1984	4.9	0.47	15.7	1.0	43	205	1.65	0.093	0.182
10.5	1982	5.0	0.46	16.6	1.1	48	224	1.65	0.106	0.189
11.5	1980	4.9	0.45	16.3	1.1	50	221	1.61	0.107	0.205
12.5	1978	4.6	0.46	16.4	1.2	56	219	1.63	0.093	0.296
13.5	1976	4.9	0.44	18.6	1.5	61	246	1.61	0.109	0.373
14.5	1974	5.2	0.44	19.9	1.8	67	254	1.64	0.129	0.367
15.5	1973	5.4	0.45	22.2	2.0	75	280	1.60	0.118	0.346
16.5	1971	5.2	0.48	18.8	2.1	63	283	1.73	0.119	0.224
17.5	1969	5.2	0.45	19.5	2.1	67	283	1.75	0.119	0.200
18.5	1967	4.9	0.44	17.6	2.0	54	267	1.73	0.116	0.154
19.5	1966	5.0	0.47	18.0	1.8	52	267	1.70	0.109	0.182
21	1963	4.7	0.45	15.4	1.6	41	242	1.61	0.091	0.188
23	1959	4.7	0.46	15.2	1.5	39	231	1.62	0.086	0.178
25	1956	4.3	0.46	15.4	1.5	35	230	1.62	0.092	0.160
27	1951	4.4	0.47	14.9	1.5	32	239	1.57	0.087	0.158
29	1947	4.2	0.48	13.5	1.1	32	209	1.60	0.081	0.134
31	1943	4.3	0.51	13.7	1.1	33	219	1.64	0.084	0.120
33	1938	3.6	0.50	12.8	1.3	30	209	1.48	0.077	0.195
35	1933	4.1	0.50	12.8	1.2	29	218	1.62	0.076	0.103
37	1928	4.1	0.49	13.9	1.2	29	205	1.59	0.079	0.104
39	1923	4.1	0.51	12.4	1.2	29	199	1.57	0.077	0.084
41	[b]	4.3	0.51	13.5	1.2	28	218	1.60	0.074	0.078

[a] ND, not determined.
[b] Pre-reservoir leaf layer.

systematic shifts in Al at about 12 cm, and Ca from 15 to 8 cm. Iron gradually increases from concentrations near 4 wt% at the base to values near to 5 wt% at a depth of about 20 cm and fluctuates around this higher concentration to the surface of the core.

The trace elements Zn, Pb, As, together with sulfur (S) from core HIN.2 increase upward to a maximum value at depths between 13.5 and 17.0 cm corresponding to deposition between 1968 and 1974 (Fig. 4; Table 2). The PAH peak is deeper in the core than the other parameters, but the data are limited (see discussion below). The peak Zn encompasses a slightly greater depth interval than does the peak for Pb and As, and occurs at a slightly greater depth. In contrast, S increases to a maximum at a depth of 13.5 cm (shallower than the other elements) and decreases rapidly at lesser depths. The data from core HIN.1 (Table 3) match closely in both concentration and depth. As noted above, Pb and As are offset by 1 cm in core HIN.1 compared to HIN.2, suggesting that the data from the two cores may be interpreted in concert with each other. On this basis, we have also plotted Ge, Cd, and mercury (Hg), as determined on core HIN.1 (Fig. 4). The Ge and Cd more closely match that of Zn than Pb and the Hg peak is close to that of S. Likewise, when plotted as a function of age of deposition (Fig. 4; Table 3), Pb concentration in the Deegan core (data not shown) has the same trend with a maximum at about 1973 as recorded in the Hinkel cores, although the magnitude of the peak is subdued. This correspondence demonstrates that the trends recognized in Hinkel Reservoir also hold in Deegan as well. Because data from the Deegan Reservoir do not add significantly to the longer historical record from Hinkel Reservoir, we will not discuss these data further.

Magnetic minerals vary greatly in core HIN.2 (Fig. 5; Table 4). IRM, indicating magnetite (as confirmed petrographically), increases from 37 cm (1947) to a peak at 15 cm (1973), decreases again to about 12 cm, and remains relatively constant to the top of the core. The maximum in IRM (magnetite) occurs at the same depth as the Pb and As peaks (Fig. 4). In contrast, HIRM (hematite) behaves differently. The HIRM maximum value occurs in the deeper part of the core (40–44 cm), and systematically decreases to values that vary around a mean of 3.8×10^{-4} at depths shallower than 18 cm. The ratio of ARM to MS shows a decrease in magnetite grain size upward in the core to 11.5 cm and lower and constant values above. Finally, the S parameter (range 0.78–0.90) mimics the abundance of magnetite itself.

Miletree Run Reservoir

At Miletree Run Reservoir the base of reservoir sediment occurred at 48 cm. Because it takes about 1 year to completely fill the reservoir, that depth was assigned a date of 1928. We assume the base was reached because the deepest sample was a leafy layer that differed substantially from overlying sediment.

Al, Ca, Fe, and Ti concentrations in the MTR.1 core vary little (Fig. 6; Table 5). Aluminum varies slightly around a mean value of 10 wt%. Calcium is low, implying low calcium carbonate content, and is generally constant except slightly higher values at about 33 cm. Iron increases irregularly from values near 4.0 wt% at the base of the core to about 4.5 wt% at 15 cm depth, and fluctuates around this higher value at shallower depth with a single point maximum at 10.5 cm. Titanium shows no overall trend, but instead, varies narrowly about a mean of 0.48 wt%. Taken together, the data do not indicate any major evolution of overall lake sediment geochemistry since the inception of lacustrine conditions.

Sulfur in core MTR.1 (Fig. 7; Table 5) is generally less than 0.2 wt%, but it increases up core to a sharp maximum at 9 cm depth with two additional subordinate peaks centered at 23 and 37 cm, corresponding to about 1945 and 1970, respectively. The Zn distribution is similar to that of sulfur in that it increases up core, and likewise has two peaks at 23 and 37 cm, although it lacks a distinct peak near 10 cm. Pb content has a peak at 23 cm with another at 37 cm. The Cu profile, on the other hand, is dominated by the peak at 10 cm, but is otherwise nearly constant. Arsenic varies, but has a clear peak at 9 cm and a poorly defined one at 25 cm.

Magnetite content increases irregularly to a sharp maximum at 10.5 cm and varies at shallower depths (Fig. 8; Table 6). There are also subordinate magnetite maxima centered at about 23 and 37 cm, corresponding to those in S, Zn, and Pb. Overall, hematite decreases upward in the core

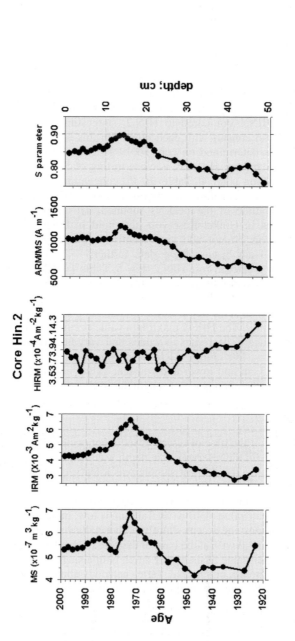

FIG. 5. Depth and age plots of magnetic properties measured in core HIN.2. See text for description of the parameters. Note the maximum in MS and IRM (magnetite content) at about 1972, corresponding to the maxima in Pb and As in Fig. 4.

Table 4. Magnetic data from core HIN.2

Depth (cm)	MS (m³/kg)	IRM (A m²/kg)	ARM/MS (A/m)	HIRM (A m²/kg)	S parameter
1.0	5.32×10^{-7}	4.29×10^{-3}	1046	3.87×10^{-4}	0.85
2.5	5.40×10^{-7}	4.34×10^{-3}	1032	3.79×10^{-4}	0.85
3.5	5.31×10^{-7}	4.27×10^{-3}	1056	3.81×10^{-4}	0.85
4.5	5.35×10^{-7}	4.36×10^{-3}	1066	3.60×10^{-4}	0.86
5.5	5.39×10^{-7}	4.38×10^{-3}	1057	3.89×10^{-4}	0.85
6.5	5.56×10^{-7}	4.48×10^{-3}	1020	3.82×10^{-4}	0.85
7.5	5.68×10^{-7}	4.66×10^{-3}	1032	3.77×10^{-4}	0.86
8.5	5.76×10^{-7}	4.70×10^{-3}	1042	3.67×10^{-4}	0.87
9.5	5.71×10^{-7}	4.69×10^{-3}	1046	3.85×10^{-4}	0.86
10.5	5.30×10^{-7}	5.09×10^{-3}	1133	3.91×10^{-4}	0.87
11.5	5.21×10^{-7}	5.70×10^{-3}	1232	3.75×10^{-4}	0.88
12.5	5.79×10^{-7}	6.08×10^{-3}	1205	3.83×10^{-4}	0.89
13.5	6.27×10^{-7}	6.33×10^{-3}	1141	3.65×10^{-4}	0.90
14.5	6.85×10^{-7}	6.63×10^{-3}	1110	3.74×10^{-4}	0.90
15.5	6.45×10^{-7}	6.15×10^{-3}	1089	3.86×10^{-4}	0.89
16.5	6.11×10^{-7}	5.77×10^{-3}	1066	3.87×10^{-4}	0.88
17.5	5.79×10^{-7}	5.53×10^{-3}	1074	3.79×10^{-4}	0.88
18.5	5.61×10^{-7}	5.34×10^{-3}	1045	3.90×10^{-4}	0.87
19.5	5.59×10^{-7}	5.30×10^{-3}	1021	3.63×10^{-4}	0.88
21.0	5.15×10^{-7}	4.90×10^{-3}	996	3.70×10^{-4}	0.87
22.0	4.78×10^{-7}	4.24×10^{-3}	942	3.59×10^{-4}	0.86
23.0	4.89×10^{-7}	3.93×10^{-3}	819	3.78×10^{-4}	0.84
27.0	4.50×10^{-7}	3.70×10^{-3}	756	3.89×10^{-4}	0.83
29.0	4.20×10^{-7}	3.49×10^{-3}	785	3.81×10^{-4}	0.82
31.0	4.55×10^{-7}	3.32×10^{-3}	730	3.89×10^{-4}	0.81
33.0	4.54×10^{-7}	3.19×10^{-3}	690	3.97×10^{-4}	0.80
35.0	4.58×10^{-7}	3.18×10^{-3}	653	3.95×10^{-4}	0.80
37.0	3.96×10^{-7}	2.79×10^{-3}	719	3.95×10^{-4}	0.78
39.0	4.41×10^{-7}	2.96×10^{-3}	659	4.11×10^{-4}	0.78
41.0	5.49×10^{-7}	3.46×10^{-3}	628	4.27×10^{-4}	0.80

Magnetic separates were prepared from depths of 0–4, 6–9, 10–12, 13–16, 17–20, 30–34, and 40–44 cm.

(Fig. 8; Table 6), slightly increasing to about 10 cm, and decreases at still shallower depths. The S parameter (mostly 0.35–0.62) mimics IRM, with maxima centered at 47, 23, and 9 cm. Magnetic properties in core MTR.1 differ from those in core HIN.2 in several ways. Relative to core HIN.2, core MTR.1 contains less magnetite and more hematite. For these reasons, the S parameter values in MTR.1 are much lower than in HIN.2. The ratio of ARM to MS increases up core (decreasing grain size), with a sharp maxima at 10 and 5 cm.

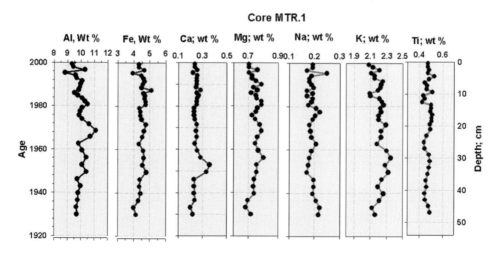

FIG. 6. Age and depth plots of major elements in core MTR.1.

Table 5. Chemical data from core MTR.1

Depth (cm)	Date	Al (wt%)	Ca (wt%)	Fe (wt%)	K (wt%)	Mg (wt%)	Na (wt%)	P (wt%)	Ti (wt%)	As (ppm)	Cu (ppm)	Pb (ppm)	Sr (ppm)	Zn (ppm)	S (wt%)
0.5	2000	9.7	0.26	4.65	2.11	0.735	0.19	0.08	0.473	18.6	45	25	81	119	0.222
1.5	1999	9.3	0.24	4.33	2.08	0.704	0.19	0.07	0.473	14.2	43	27	79	123	0.153
2.5	1998	9.4	0.24	4.33	2.13	0.704	0.19	0.07	0.484	13.1	38	25	80	117	0.124
3.5	1997	10.3	0.26	4.66	2.21	0.761	0.17	0.075	0.473	15.9	53	27	85	132	0.178
4.5	1995	8.8	0.22	3.94	2.11	0.704	0.25	0.065	0.528	12.1	34	28	79	120	0.039
5.5	1994	9.6	0.26	4.47	2.15	0.725	0.19	0.08	0.484	14.3	50	27	82	129	0.183
6.5	1993	9.7	0.25	4.61	2.15	0.725	0.18	0.09	0.446	15.8	70	29	81	128	0.184
7.5	1991	10.1	0.25	4.67	2.25	0.751	0.18	0.08	0.468	14.8	54	31	83	136	0.178
8.5	1990	10.0	0.25	4.56	2.23	0.788	0.19	0.075	0.462	14.6	53	28	84	130	0.154
9.5	1989	9.9	0.25	4.49	2.22	0.735	0.20	0.075	0.512	12.5	49	30	82	129	0.139
10.5	1987	9.8	0.28	5.1	2.19	0.761	0.17	0.105	0.435	20.3	145	29	84	132	0.543
11.5	1986	9.5	0.25	4.61	2.09	0.719	0.19	0.08	0.446	16.5	106	29	80	122	0.288
12.5	1985	9.8	0.27	4.7	2.1	0.751	0.17	0.075	0.424	18.5	101	32	82	125	0.374
13.5	1984	10.1	0.26	4.69	2.21	0.756	0.19	0.07	0.501	14	47	31	84	128	0.206
14.5	1982	10.3	0.25	4.75	2.24	0.788	0.19	0.07	0.501	15.4	38	31	85	134	0.210
15.5	1981	10.5	0.25	4.75	2.26	0.788	0.18	0.07	0.495	15.8	36	33	85	134	0.151
16.5	1979	10.1	0.23	4.36	2.22	0.761	0.21	0.065	0.512	12.3	34	29	84	122	0.073
17.5	1977	10.0	0.23	4.42	2.23	0.751	0.22	0.065	0.506	11.1	34	32	82	124	0.057
18.5	1976	9.9	0.23	4.41	2.2	0.725	0.20	0.06	0.506	10.5	35	28	82	119	0.046
19.5	1974	10.1	0.24	4.49	2.2	0.746	0.19	0.065	0.49	13.4	36	30	82	117	0.091
21	1971	10.6	0.25	4.76	2.29	0.777	0.18	0.07	0.495	14.8	36	33	86	128	0.149
23	1969	11.1	0.25	4.64	2.23	0.788	0.19	0.07	0.462	14.5	38	34	86	130	0.134
25	1966	10.7	0.25	4.53	2.23	0.761	0.19	0.065	0.44	17.6	36	35	84	127	0.121
27	1963	9.8	0.24	4.33	2.18	0.746	0.21	0.065	0.446	10.3	35	32	80	125	0.039
29	1959	10.1	0.26	4.59	2.3	0.767	0.19	0.075	0.484	14	41	35	83	128	0.059
31	1956	10.4	0.28	4.63	2.35	0.803	0.18	0.085	0.49	13.5	41	29	85	124	0.060
33	1953	10.1	0.36	4.55	2.28	0.756	0.18	0.105	0.484	12.5	42	28	87	116	0.040
35	1949	10.4	0.33	4.79	2.32	0.756	0.17	0.09	0.473	17.5	44	34	88	129	0.102
37	1946	9.7	0.23	4.39	2.24	0.74	0.20	0.06	0.457	13.8	35	33	81	137	0.086
39	1943	10.0	0.23	4.35	2.19	0.735	0.20	0.06	0.462	18.1	47	29	81	130	0.094
41	1940	9.8	0.23	4.43	2.26	0.74	0.19	0.05	0.446	13.6	33	26	77	116	0.067
43	1936	9.8	0.24	4.3	2.21	0.698	0.21	0.04	0.451	17.7	30	21	72	100	0.015
45	1933	9.7	0.20	3.99	2.12	0.683	0.22	0.05	0.479	12	65	26	78	115	0.039
47	1930	9.7	0.22	4.15	2.16	0.719	0.22	0.055	0.49	14	33	25	82	118	0.037

Chemical analyses by ICP-AES except S: by LECO® induction furnace.

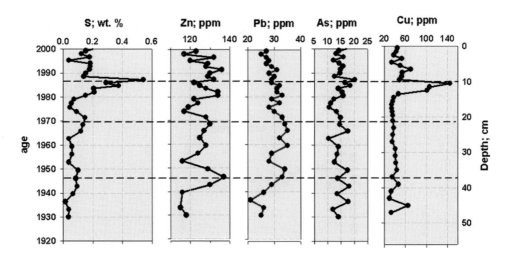

FIG. 7. Depth and age plots of S, Zn, Pb, As, and Cu in core MTR.1. Three reference lines are drawn at about 1946, 1970, and 1986.

Rocks and soils

Generally, contents of trace elements in rock samples (Table 7a) are lower than in reservoir sediments. For example, only one As and one Pb value and no Cu or Zn values from rocks in the vicinity of Hinkel/Deegan Reservoirs are equal to or greater than the data for the same elements from Hinkel Reservoir sediments (compare Table 7a and Fig. 4). The same comparison holds between rocks and reservoir sediments at the Miletree Run site. The comparison between soil (Table 7b) and reservoir sediment chemistry is more complex and is discussed below.

DISCUSSION

Hinkel Reservoir

Initially, we interpreted that the data from Hinkel (and Deegan) reservoirs reflected a significant input from coal-fired power plants. The reasons for reaching that conclusion were: (1) the suite of elements enriched in the sediment matched that expected from coal combustion, (2) the timing of the peak in element concentration coincided with the Clean Air Act of 1970, (3) coal-combustion products could be identified in the sediments and positively correlated with trace element content, (4) other sources of trace metals (specifically Pb from gasoline) could be excluded as element sources, and (5) local rocks are an unlikely source of trace element enrichment. We were aware of a coal-fired power plant in the same county as Hinkel/Deegan Reservoirs (Fig. 1). However, this facility began operations in 1972 (DOE, 2002), and is thus unlikely to have contributed to the buildup in Zn, Pb, As, PAHs, Hg, etc. that began in the late 1950s and peaked before the plant went online. Recently, however, we have become aware of a local source that potentially contributed coal-combustion products and metals to the reservoir sediments, perhaps complicating the interpretation of data. Nonetheless, because the data do reflect an extreme example of coal-related inputs, we will discuss the results in some detail.

Perhaps the most obvious trend in the data from the Hinkel cores is that maxima in elements known to be associated with coal-combustion products (*e.g.* Zn, Pb, As, Ge, and Hg) occur at a depth corresponding to deposition just subsequent to the enactment of the Clean Air Act in 1970 (Fig. 4). Sulfur dioxide emissions from power plants are known to have peaked near 1970 (HUSAR *et al.*, 1991), although S peaks at a slightly shallower depth (younger age) than do the trace elements. The overall suite of elements plotted in Fig. 4 is commonly enriched in coal fly ash as noted above, or in the case of Hg and S, vapor phase output from coal combustion (KAAKINEN *et al.*, 1975). The timing of the decrease in these elements suggested to us decreased

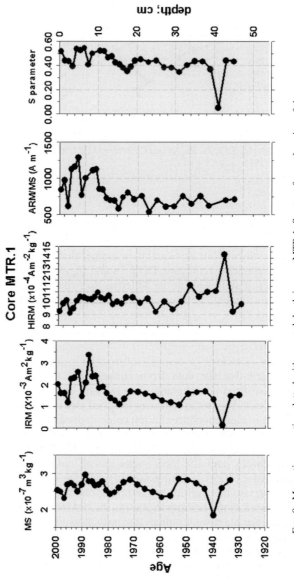

FIG. 8. Magnetic properties plotted with age and depth in core MTR.1. See text for explanation of these parameters.

Table 6. Magnetic data from core MTR.1

Depth (cm)	MS (m³/kg)	IRM (A m²/kg)	ARM/MS (A/m)	HIRM (A m²/kg)	S parameter	IRM.3/MS (A/m)
1	2.57×10^{-7}	2.03×10^{-3}	1050	9.32×10^{-4}	0.52	7891
2.5	2.54×10^{-7}	1.61×10^{-3}	850	1.00×10^{-3}	0.45	6347
3.5	2.50×10^{-7}	1.62×10^{-3}	983	1.03×10^{-3}	0.44	6469
4.5	2.32×10^{-7}	1.21×10^{-3}	622	9.14×10^{-4}	0.40	5218
5.5	2.69×10^{-7}	2.27×10^{-3}	1131	9.56×10^{-4}	0.54	8441
6.5	2.73×10^{-7}	2.33×10^{-3}	1166	1.02×10^{-3}	0.53	8521
7.5	2.67×10^{-7}	2.59×10^{-3}	1296	1.06×10^{-3}	0.55	9692
8.5	2.50×10^{-7}	1.48×10^{-3}	779	1.05×10^{-3}	0.41	5911
9.5	2.70×10^{-7}	2.11×10^{-3}	1009	1.04×10^{-3}	0.50	7817
10.5	2.96×10^{-7}	3.37×10^{-3}	1535	1.04×10^{-3}	0.62	11,388
11.5	2.78×10^{-7}	2.38×10^{-3}	1115	1.06×10^{-3}	0.53	8528
12.5	2.77×10^{-7}	2.42×10^{-3}	1132	1.10×10^{-3}	0.52	8710
13.5	2.67×10^{-7}	1.86×10^{-3}	859	1.05×10^{-3}	0.47	6956
14.5	2.69×10^{-7}	1.92×10^{-3}	851	1.04×10^{-3}	0.48	7127
15.5	2.78×10^{-7}	1.62×10^{-3}	733	1.07×10^{-3}	0.43	5818
16.5	2.55×10^{-7}	1.40×10^{-3}	704	9.94×10^{-4}	0.41	5497
17.5	2.44×10^{-7}	1.28×10^{-3}	700	1.01×10^{-3}	0.39	5263
18.5	2.48×10^{-7}	1.12×10^{-3}	592	9.98×10^{-4}	0.36	4507
19.5	2.61×10^{-7}	1.37×10^{-3}	747	1.05×10^{-3}	0.39	5246
21	2.76×10^{-7}	1.70×10^{-3}	813	1.05×10^{-3}	0.45	6170
23	2.83×10^{-7}	1.67×10^{-3}	716	1.00×10^{-3}	0.45	5910
25	2.70×10^{-7}	1.59×10^{-3}	764	1.04×10^{-3}	0.43	5895
27	2.57×10^{-7}	1.49×10^{-3}	542	9.26×10^{-4}	0.45	5797
29	2.48×10^{-7}	1.29×10^{-3}	704	1.01×10^{-3}	0.39	5177
31	2.35×10^{-7}	1.20×10^{-3}	617	9.48×10^{-4}	0.39	5100
33	2.39×10^{-7}	1.09×10^{-3}	620	1.01×10^{-3}	0.35	4589
35	2.85×10^{-7}	1.60×10^{-3}	761	1.16×10^{-3}	0.41	5601
37	2.82×10^{-7}	1.66×10^{-3}	653	1.06×10^{-3}	0.44	5891
39	2.73×10^{-7}	1.71×10^{-3}	762	1.10×10^{-3}	0.44	6268
41	2.57×10^{-7}	1.33×10^{-3}	627	1.11×10^{-3}	0.38	5190
43	1.84×10^{-7}	1.61×10^{-4}	492	1.43×10^{-3}	0.05	873
45	2.60×10^{-7}	1.49×10^{-3}	704	9.28×10^{-4}	0.45	5742
47	2.82×10^{-7}	1.54×10^{-3}	716	9.92×10^{-4}	0.44	5465

Magnetic separates were prepared from depths of 0–3, 7–8, 10–12, 17–19, 22–26, 30–32, and 38–40 cm.

Table 7a. Rock samples

Location	Description	Al (%)	Ca (%)	Fe (%)	K (%)	Mg (%)	Na (%)	Ti (%)	As (ppm)	Cu (ppm)	Pb (ppm)	Zn (ppm)
Hinkel/Deegan	Weathered shale	6.06	0.79	2.97	1.20	0.56	0.13	0.40	26	6.2	32.3	78.8
	Weathered coal	2.88	13.22	3.43	0.71	5.07	0.14	0.18	5.6	18.6	15.3	138
	Weathered shale	4.92	3.07	3.08	1.16	1.69	0.12	0.32	11.9	27.4	28.5	154
	Shale with iron staining	4.13	4.72	3.29	0.95	1.99	0.37	0.32	4	37.1	17.5	148
	Weathered sandstone	3.49	0.18	2.06	0.62	0.60	0.35	0.37	1.3	<2	12.7	74.5
	Weathered sandstone	2.51	0.14	1.43	0.50	0.45	0.37	0.26	2.4	<2	10.1	48.6
	Misc. rock outcrop	5.50	0.08	2.79	1.14	0.45	0.51	0.34	2.9	14	11	91
	Misc. rock outcrop	2.43	0.01	2.66	0.45	0.08	0.21	0.09	10	6	15	30
	Misc. rock outcrop	3.06	2.16	2.70	0.51	0.29	0.54	0.24	7.1	6	11	57
Miletree Run	Shale	5.72	0.71	3.22	1.45	1.27	0.15	0.33	10.2	104	24.4	132
	Sandstone	3.41	0.20	1.15	0.71	0.66	0.45	0.33	1	<2	11.2	66.8
	Shale outcrop	6.04	0.20	2.36	1.19	0.53	0.93	0.31	2.9	17	11	108
	Shale outcrop	6.71	0.24	3.91	1.53	0.92	0.78	0.37	3.6	14	11	66
	Shale outcrop	5.89	0.22	3.10	1.17	0.76	0.87	0.34	2.9	14	11	91
	Shale outcrop	6.41	0.19	7.04	1.20	1.27	0.92	0.09	10	6	15	30
	Shale outcrop	5.99	0.15	3.00	1.09	0.51	0.84	0.24	7.1	6	11	57

Hg analyses by atomic fluorescence spectroscopy. All other chemical analyses by ICP-AES.

M. Goldhaber, T. Callender and R. Reynolds

Table 7b. Surface soil samples

Site	Sample	Ti (wt%)	As (ppm)	Cu (ppm)	Pb (ppm)	Zn (ppm)	Hg (ppm)	Fe (wt%)	Total S (wt%)	IRM (A m²/kg)
Hinkel/Deegan	Hin-Soil_1	0.451	21.9	27	27	72		3.98	ND[a]	ND
	Hin-Soil_2	0.358	13.9	99	144	322		2.93	ND	ND
	DG_soil	0.424	15.3	59	44	142		4.68	ND	ND
	HDR02B_A	0.505	16.6	21	28	109	0.04	4.00	<0.05	1.87×10^{-2}
	HDR03B_A	0.335	19.3	28	50	112	0.09	3.90	0.08	3.53×10^{-3}
	HDR04B_A	0.305	13.9	43	124	286	0.05	6.21	0.19	3.93×10^{-2}
	HDR05B_A	0.475	26.4	25	30	141	0.05	5.00	<0.05	5.80×10^{-3}
	HDR06B_A	0.55	18.7	27	30	189	0.05	3.85	<0.05	9.87×10^{-3}
	HDR08B_A	0.43	14.9	27	34	190	0.06	3.79	<0.05	1.36×10^{-3}
	HDR09B_A	0.445	13.8	21	30	155	0.06	4.84	<0.05	7.61×10^{-3}
	HDR11B_A	0.46	22.6	23	20	62	0.03	5.61	<0.05	2.49×10^{-3}
	HDR12B_A	0.45	12.6	22	32	151	0.05	3.49	<0.05	2.70×10^{-3}
	HDR13B_A	0.405	14.2	21	51	112	0.1	3.69	0.13	1.79×10^{-2}
Miletree Run	MTR_surf_1	0.385	11.7	25	30	94	0.06	3.57	ND	ND
	MTR_surf_2	0.413	12.4	74	22	101	0.03	4.17	ND	ND
	MTR_surf_3	0.38	6.5	32	30	74	0.07	2.67	ND	ND
	MTR_surf_4	0.292	9.7	26	20	76	0.03	3.1	ND	ND
	MTR_surf_5	0.358	5.4	13	25	74	0.07	2.49	ND	ND
	MTR_surf_6	0.374	6.7	21	28	78	0.07	2.81	ND	ND
	MTR01C_A	0.44	23	35	26	106	<0.02	5.48	<0.05	1.72×10^{-3}
	MTR03C_A	0.475	7.9	30	27	89	0.04	3.95	<0.05	4.38×10^{-3}
	MTR05C_A	0.455	19.1	38	33	105	0.02	5.74	<0.05	3.24×10^{-3}
	MTR06C_A	0.46	11.8	26	27	82	0.04	3.24	<0.05	1.19×10^{-3}
	MTR08C_A	0.445	13.1	15	17	48	<0.02	2.77	<0.05	4.06×10^{-3}
	MTR10B_A	0.415	8.8	26	24	85	0.04	3.45	<0.05	5.67×10^{-4}
	MTR11B_A	0.445	9.3	21	31	89	0.06	2.84	<0.05	1.83×10^{-3}
	MTR12B_A	0.415	4.8	18	15	54	0.02	2.52	<0.05	6.63×10^{-4}
	MTR13B_A	0.355	9.1	22	19	66	0.03	2.94	<0.05	1.04×10^{-4}
	MTR14B_A	0.51	10.3	32	28	80	0.05	3.36	<0.05	1.18×10^{-3}
	MTR15B_A	0.37	11.1	23	25	76	0.03	3.50	<0.05	3.30×10^{-4}

[a] ND, not determined.

atmospheric input resulting from controls at the power plants on particulate matter. Isotope data on Pb from the reservoir sediments is the subject of an ongoing study, but preliminary results exclude gasoline as a source of the Pb anomaly (ROBERT AYUSO, written communication, 2002).

To quantify atmospheric input of these elements to the reservoir sediments, we must recognize any contribution of these elements from local rocks and soils. Rock weathering is a source of natural material, whereas soils may provide both natural and anthropogenic components to the reservoirs. Our approach to tracing these rock and soil inputs is to compare element concentrations normalized to Ti in reservoir sediments to the same ratio in rocks and soils. The underlying assumption is that Ti is insoluble during weathering of sedimentary materials, and thus is conservative and tracks their physical transport (NORTON, 1986).

Mean trace element/titanium ratios in Hinkel sediments are much greater than rock values. This is especially true for Zn and Pb. Thus, element enrichments in Hinkel Reservoir sediments could not have come from local rocks.

Comparisons of depth plots of Ti-normalized values for a series of parameters in HIN.2 to the same ratio in local surface soils (Fig. 9) reveal important information about sources of elements. The soil value is a composite mean of all the soil analyses. The Fe/Ti ratios of the top 10 cm of the core (approximately 1980 onward) are very similar to soil values. Deeper in the core, Fe/Ti values peak at approximately 15 cm (1970) and exceed the ratio in soils. Below this peak, the sediment ratios decrease and fall substantially below values in modern surface soil.

The pattern of normalized IRM values is similar to Fe, including the feature that surface sediments (those deposited after about 1980) are similar to surface soils. The behavior of selected trace elements is also similar in that the maximum values at depths of 15–17 cm (late 1960s to early 1970s) all exceed Ti-normalized modern soil values. Zinc differs from other

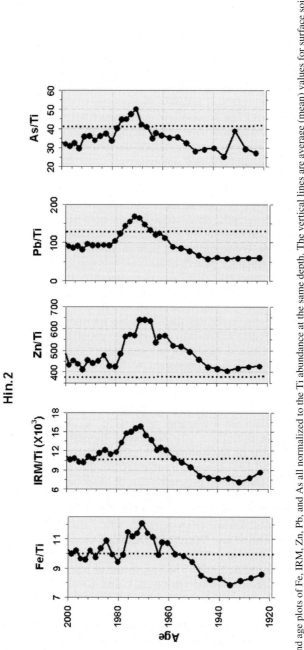

FIG. 9. Depth and age plots of Fe, IRM, Zn, Pb, and As all normalized to the Ti abundance at the same depth. The vertical lines are average (mean) values for surface soil samples in the vicinity of Hinkel/Deegan Reservoirs. Note the similarity between soil and reservoir data in the upper portion of the core (about 1984 and later) for normalized Fe and IRM values. In contrast, reservoir sediments post-1984 are lower than nearby soils and Zn is greater.

parameters in that the normalized values exceed soil ratios at all depths. Pb and As ratios in surface sediments and sediments deposited before 1960 are less than the respective soil values.

Thus, if our assumption is valid that Ti-normalized ratios track natural weathering inputs, the highest Fe, IRM (magnetite), S, Zn, Pb, and As ratios could *not* have been supplied by local soils because reservoir sediment ratios are substantially greater than soil values. In contrast, direct transport of soil particles to the reservoir could explain a substantial portion of the major and trace element inputs of soils to reservoir sediments shallower than 10.5 cm (*i.e.* post-1980). It is likely, however, that these modern soils contain atmospheric input of coal fly ash and associated trace elements. The most direct evidence of fly ash in soils comes from preliminary observations on magnetic separates made from soil samples. Polished sections of magnetic separates from both the surface leaf litter layer and A soil horizon samples (data not shown) taken near Hinkel Reservoir contain many spherical magnetite grains in the 2–30 μm size range. Indirect evidence comes from the patterns of element enrichment and depletion in the soils relative to reservoir sediments. The elements As and Pb that are relatively immobile in soils because of their retention on iron and aluminum oxide surfaces (DRAGUN, 1988) are higher relative to Ti in soils than surface sediments (top 10 cm). Magnetite, whose mobility as a first approximation should be similar to other soil particles, is likewise more concentrated in soils than reservoir sediments deposited prior to 1990. Petrographic observations on the polished sections of magnetite grains in the soil samples revealed no dissolution or evidence of weathering/alteration to other oxides such as goethite and hematite. We would expect such alteration if the magnetite were added from local rocks. This element abundance pattern of As and Pb is consistent with a buildup in soils of the immobile elements and magnetite from an external source. In contrast, elements such as S and Zn that are likely to be relatively mobile in soils (DRAGUN, 1988) are higher in concentration in reservoir sediments than soils. The increase in concentration of reservoir sediments relative to soils for these mobile elements indicates their possible transport into the reservoir from the adjacent drainage basin. It is also noteworthy that prior to about 1960, reservoir sediments have substantially lower Ti-normalized ratios for all parameters (other than Zn) compared with modern soil. This observation is also consistent with buildup in soils over time of Pb, Fe, magnetite, and As.

The sulfur data deserve some additional comment. We might expect that sulfur would correlate with other combustion-related pollutants. However, the peak in sulfur concentration is younger than trace element peaks (Fig. 4). We can think of three likely hypotheses to explain this difference: (1) diagenetic mobility of S within the sediment column, (2) a real difference in the timing of release of gaseous and particulate pollutants from coal combustion, and (3) delayed S delivery to the reservoir due to retention in drainage basin sediments. Diagenetic processes can be discounted because they would tend to cause the sulfur peak to be deeper in the sediment column rather than shallower (MATISOFF and HOLDEN, 1995). Further, the timing of the regional peak on SO$_2$ emissions from coal combustion is well known (HUSAR *et al.*, 1991; see Fig. 15), and we expect it to coincide with particulate emissions. Therefore, the third possibility seems the most likely. The residence time of sulfate in a drainage basin has been reported to be as long as a decade (ROCHELLE and CHURCH, 1987). The key factor in retention is iron oxide abundance because of its ability to adsorb sulfate (ROCHELLE and CHURCH, 1987). The soils of the study area are abundant in iron oxides. We thus conclude that a component of the S flux into the reservoir is related to retardation of S transport on iron oxyhydroxides.

In the absence of rock and soil sources to explain maximum trace-element values, the likely transport vector is the atmosphere. In addition, the timing of the trace-element decrease coinciding with the Clean Air Act of 1970 is suggestive of a relation to coal combustion. However, we did not view this temporal correspondence as a definitive link to inputs from coal combustion, so we sought to develop additional proxies to test for such inputs. The magnetic data provide an excellent one. Concentration of Fe, Pb, and As correlates strongly with magnetite content (IRM, Fig. 10). For Fe, the plot indicates a constant background of non-magnetite iron in the core. The extrapolation to an IRM value of zero yields a background Fe content of about 3.2 wt%, with the additional increment ($\sim$1–2.2 wt%) attributable to variable magnetite content.

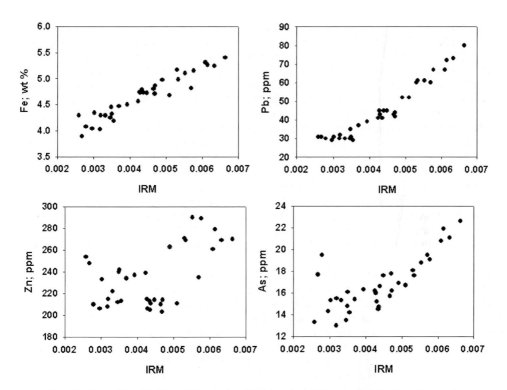

FIG. 10. Plots of Fe, Pb, Zn, and As against IRM (magnetite) for core HIN.2. Note the excellent correlation between Fe, Pb, and As against IRM. The r^2 values for linear regressions of the data are 0.90, 0.93, 0.31, and 0.58 for Fe, Pb, Zn, and As, respectively.

The plots of IRM against As and Pb content likewise show a good linear correlation (Fig. 10). As was the case for the Fe plot, the trace-metal plots indicate that a portion of the Pb and Zn above some background level correlates with magnetite input. For As, the background value is about 9.5 ppm. Pb behaves a little differently in that it extrapolates to a constant background value of about 30 ppm at low but non-zero IRM values. Nonetheless, these plots indicate that either the magnetite itself contains a significant portion of the As and Pb or that an atmospheric sedimentary component with a constant ratio to the magnetite input is rich in these trace elements.

The element–IRM plots are critical because magnetite and related spinel phases are common products of coal combustion. However, there are other natural and anthropogenic sources of magnetite in addition to coal combustion. Observations of shapes and internal textures allow us to distinguish among the possible sources of magnetite. SEM backscatter images of the polished surface of the magnetic fractions enable rapid assessment of morphologies of magnetite particles and the presence of a heavy element such as Fe (Fig. 11). Even at low magnification, many large spherical grains are visible. Several of these are tens of micrometers in diameter and one approaches 100 μm. At higher magnification, the magnetic fraction of this sample (HIN.2, 10–12 cm) is clearly dominated in all size categories by grains with spherical morphology (Fig. 11b and c). Semi-quantitative analysis of the spheres using an energy-dispersive analyzer showed that the major elements present in the spheres are Fe, Al, and Si in variable proportions. In contrast, many of the angular grains were small (typically <20 μm) and dominantly Ti and Fe, in proportions consistent with the minerals ilmenite or titanomagnetite. Between 50 and 90% of the grains in this sample consist of spheres. By mass, the proportion of spherical magnetite is dominant because the larger size grains are typically spherical. Grains with a spherical morphology dominated the magnetic fraction at depths above about 35 cm. Our observations indicate that $>80\%$ the magnetite in the Hinkel core came from combustion-related processes.

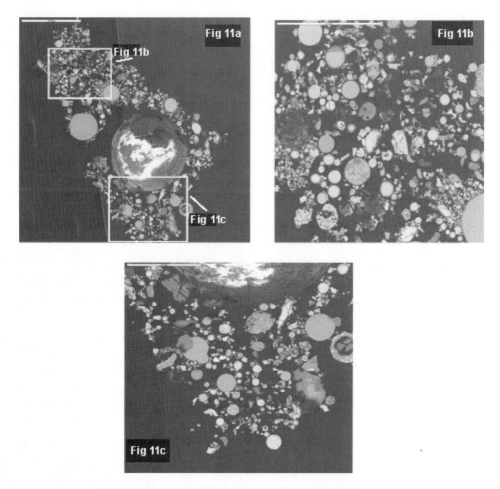

Fig. 11. Scanning electron microscope (SEM) of polished grains from a magnetic separate from core HIN.2 at 10–12 cm (backscatter mode). Fig. 11a is a low magnification image. The white scale bar in the upper left is 200 μm. The large spherical object in the lower center of the image is an air bubble; all other objects are magnetic grains. Note the dominance of spherical grains, including a number that are greater than 50 μm in diameter. Fig. 11b is a close-up of part of the image shown in Fig. 11a (see Fig. 11a for location). The scale bar in the upper left of the figure is 100 μm in length. Note the dominance of spherical grains indicative of a combustion source for the magnetite. Fig. 11c is a close-up of a portion of Fig. 11a (see Fig. 11a for location). The scale bar is 100 μm in length.

Sizes and shapes of the other highly magnetic grains strongly suggest their introduction into the reservoir catchment as mineral dust.

A further test of the presence of coal-combustion products is the presence of combustion-related PAHs (Table 2). Total PAH contents (the sum of a series of combustion-related PAH compounds as defined by PRAHL and CARPENTER (1983)) are highest at a depth of 17.5 cm (Fig. 4). This maximum occurs at a different depth than the trace element and magnetite maxima, but the PAH data are not available at that depth. The PAH results indicate that organic combustion products increase proportionally with the trace elements and magnetite, and these results are thus consistent with the other data showing that coal combustion dominates the trends recognized in the Hinkel Reservoir sediments.

Although the data for Hinkel Reservoir are definitive for a significant contribution from coal combustion, the source may not solely be power plants as we originally hypothesized. Significant contributions may have come from two zinc smelters located adjacent to each other, both operated within about 10 km northwest of the Hinkel Reservoir site (Fig. 1). These facilities used coal-burning retorts to produce spelter (a zinc-based alloy). The Clarksburg Plant operated between 1907 and 1925, and the Meadowbrook Plant from 1910 to 1971 (D. BLEIWAS, USGS Minerals Information Team, written communication, 2002). The area around the plants comprising some 117 acres is now a superfund site (Spelter Smelter site; EPA ID#WV0000634584). The EPA found elevated contents of Zn, Cd, Pb, Cu, and As in tailings piles and soils adjacent to the site (see the EPA web site: http://www.epa.gov/res3hwmd/super/WV/spelter/pad.htm).

Output from the Meadowbrook smelter may have affected Hinkel and Deegan Reservoir sediments. One piece of evidence is that that the suite of trace elements enriched in Hinkel sediments including Zn, Pb, As, Cu, and Cd matches that present at the smelter site. Cadmium in particular is of interest. Cadmium is a significant accessory element in sphalerite ores (VIETS *et al.*, 1992). Based on examination of USGS coal analyses (BRAGG *et al.*, 1997), it is not significantly enriched in Appalachian Basin coals. The fact that the Cd content in reservoir sediments peaks at the same depth/time as the other trace elements suggests a possible link to smelter output. Furthermore, the timing of the peak in the concentration of metals such as Pb and Zn in Hinkel sediments (Fig. 4) is within 4 years of the closure of the smelter facility. We also speculate that the abundance of magnetite recognized in the Hinkel sediments may be related to oxidation of elevated contents of pyrite in the sulfide ore samples that were roasted at the smelters. Also suggestive of a proximal source for constituents is the relatively large size of the magnetic spheres. Many of these spheres are > 10 μm in diameter with many in the $20-100$ μm size range (Fig. 11). Long-distance transport of material of this size and density from power plants over 50 km distant is less likely than from a local source such as the Zn smelters only 10 km to the north.

On the other hand, available production data in USGS archives indicate that the nearby Zn smelting activity does not match temporal evolution of the trace element pattern in the cores. The number of retorts in use at the smelter was relatively steady over the period $1945-1971$ rather than increasing with time as suggested by the peaks in trace elements and magnetite in the late 1960s and early 1970s. The records indicate that the Meadowbrook smelter may have switched from coal to natural gas as a fuel as early as 1960 (DON BLEIWAS, USGS, oral communication, 2002). Furthermore, the peak concentrations in trace elements in Hinkel Reservoir sediments occur at a time when SO_2 emissions in the US were at an all-time high (HUSAR *et al.*, 1991). Finally, our study area is not, on average, downwind from the former smelters; mean wind directions are to the east and northeast (NIEMANN, 1984). Nonetheless, on balance, we cannot discount particulate emissions from the smelter as contributing to the combustion-related inputs that were recognized in the Hinkel sediments. The data also suggest that soils near Hinkel Reservoir contain a component of Zn-smelting input. In any case, the data from Hinkel Reservoir provide a template for what we might expect from coal-combustion inputs into a reservoir and the surrounding soils.

Miletree Run

Because of the recognition that Zn smelters could have influenced the geochemical signatures at the Hinkel/Deegan Reservoir sites, the MTR data assume greater importance for testing the hypothesis of downwind impacts from power plants. In addition, this site is closer to the Ohio River Valley. Taken together, the results indicate inputs from coal combustion.

As was the case for Hinkel Reservoir, the Ti-normalized data help us distinguish weathering from coal-combustion input. When compared to the youngest reservoir sediments, patterns in the Ti-normalized data from MTR are only slightly different from those previously discussed for Hinkel Reservoir (Fig. 12). Zn and S resemble the Hinkel core data in that they are enriched in reservoir sediments relative to soils. Similarly, Pb and magnetite (IRM) are relatively enriched in soil (although there is scatter in the IRM data). However, in contrast to the Hinkel data, Ti-normalized Fe and As in MTR soil samples tend to be depleted compared to reservoir sediments.

Another important similarity with the Hinkel Reservoir plots is that subsurface concentration maxima evident in the non-normalized plots (Fig. 7) are preserved in Ti-normalized ones (Fig. 12). In the MTR data, these maxima occur for S, Zn, Pb, and As

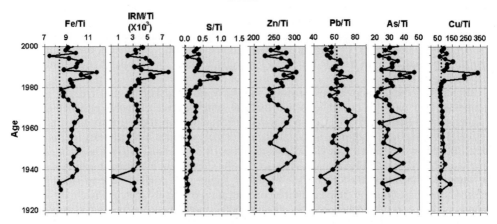

FIG. 12. Age plots of Ti-normalized ratios of Fe, IRM, S, Zn, Pb, As, and Cu, in core MTR.1. The vertical dashed lines are average values for modern soils of the area.

and are centered at 15.5 and 37 cm (1970 and 1946). In each case, the Ti-normalized peak values are higher than modern soil ratios, suggesting an atmospheric source for elements associated with these peaks. An exception is that the IRM/Ti ratio has lower peak values for the 1946 and 1970 peaks than average modern soil.

The Ti-normalized data emphasize the enrichment in Cu at 10.5 cm (1987), a very striking feature of the MTR data. Associated with the copper peak are corresponding major peaks in S and IRM and lesser ones in Fe, Zn, and As (Zn only shows up as a peak in Ti-normalized plots). We believe these maxima at 10.5 cm are unrelated to power plant emissions. Copper sulfate is commonly added to reservoirs to retard algal growth. The manager of the water department in the town of Spencer, Lowell Hardmann, reported to us that copper sulfate was indeed added to the reservoir, but he did not have information on the exact date. Since Cu is not an element associated with power plant emissions and we know of no other significant Cu source nearby, the most likely explanation is that the copper sulfate additions were made in about 1987 coincident with the peak. The dramatic increase in Cu/Ti for reservoir sediments at 10.5 cm compared to nearby soils (Fig. 12) and rocks (Table 7b) is consistent with this conclusion.

We do not yet know why magnetite (IRM values) also spike at the same time as Cu, and why Fe and As contents would peak as well. It has been recognized from other reservoir studies that copper sulfate contains As as an impurity and/or the compound copper arsenate is sometimes added for algal control (RICE et al., 2002). The positive correlation between IRM and Cu may be related to the impact of the copper sulfate addition on the sulfur geochemistry of the lake. This could occur if the additional sulfur directly added to the lake as copper sulfate was converted to the magnetic mineral greigite (Fe_3S_4), the thiospinel of iron. A magnetic sulfide mineral, likely greigite, is present in the sediments, almost exclusively in fine-grained forms within lumens (the interior cell cavity) of organic fragments. Such occurrences, although rare, were observed in each examined sample (depths 0–3, 7–8, 10–12, 17–19, 22–26, 30–32, 38–40 cm). The highest concentrations were found in the sample at 10–12 cm depth, which is characterized by highest IRM, Cu, and S. Nevertheless, the relative and absolute abundances of these fragments, most of which consist of individual particles as ovoidal mineralized cell lumens, appear to be too few to account for variations of 2–3 × in IRM values. A possible, untested explanation for the high IRM values above about 11 cm is the production of ultra-fine-grained (∼0.1 μm) magnetic minerals via bacteriogenic pathways (KONHAUSER, 1998). The higher values of ARM/MS in the upper part of the core (Fig. 8) indicate an overall finer magnetic grain size for magnetic particles than in sediments below. The presence of an iron sulfide phase would be consistent with the peak in Fe and S. For the reason just given, we cannot confirm the trends associated with this copper peak as having any relation to power plant emissions.

Despite the complication arising from the potential addition of copper sulfate to the reservoir, a strong circumstantial case can be made that an atmospheric record of inputs from distal coal-fired power plants is preserved in MTR sediments. This evidence consists of the correlation of sedimentary sulfur and magnetite, a correlation between coal production and elemental and mineralogical trends, and the identification of overwhelmingly abundant fly ash in magnetic separates from the sediments that dominates magnetic properties.

Plots of S, Fe, Zn, and As versus IRM (Fig. 13) provide evidence for atmospheric inputs. Sulfur and IRM are well correlated ($r^2 = 0.73$), but total Fe and IRM are less so ($r^2 = 0.28$). This is just the opposite of the results from Hinkel Reservoir (compare Fig. 10). The close correlation of sulfur and magnetite is unlikely to be related to weathering of local rocks, which are a negligible source of magnetite based on petrographic observations of the magnetic fraction of the sediments. Furthermore, soils are depleted in sulfur compared to reservoir sediments (Fig. 12) and are an insignificant source of this element. What is more likely is that magnetite and sulfur are both supplied by atmospheric transport. As noted earlier, the MTR study site lies within a part of the US significantly impacted by enhanced sulfate deposition from coal combustion. Even if both magnetite and S are supplied from the atmosphere, for the magnetite–S correlation to hold, the residence time of S in the reservoir waters must be short relative to its incorporation into bottom sediments. Increases in lake sediment S content due to atmospheric pollution inputs have been recognized previously (HOLDREN et al., 1984; MATISOFF et al., 1993; NRIAGU and COKER, 1983). The correlation of these two variables is particularly likely given the small drainage area of MTR reservoir, which maximizes atmospheric over weathering additions to the reservoir sediments, and minimizes the potential for retardation of the S input as hypothesized above for the Hinkel Reservoir data. The plots for Zn and As are suggestive of a

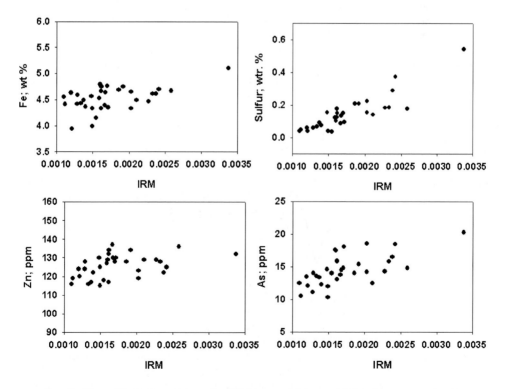

FIG. 13. Plots of Fe, S, Zn, and As against IRM (magnetite) in core MTR.1. In contrast to the plots shown in Fig. 10 (S not shown) for core MTR.2, the correlations with Fe, Zn, and As are less marked. r^2 values for Fe, Zn, and As are 0.28, 0.35, and 0.19, respectively. However, the correlation with S is quite good ($r^2 = 0.73$).

weak correlation with coal-combustion magnetite. Thus, these elements are also likely inputs from atmospheric transport.

Two additional lines of evidence link trends in the MTR data to coal combustion. The first is a comparison of temporal trends in IRM, Zn, and S with SO_2 production in the US. This comparison is shown in Fig. 14. By far the dominant source of SO_2 is fuel combustion (EPA, 2000). For the period from 1930 to about 1980, there is a close correspondence between SO_2 production/combustion and reservoir geochemistry/mineralogy. In particular, SO_2 production maxima during World War II (1945) and the period prior to the Clean Air Act of 1970 are mirrored by peaks in reservoir S and IRM. The correspondence between content of sedimentary trace elements and magnetite related to coal combustion with the source function for these constituents is consistent with a relation between the source and sink.

This linkage is strengthened by the presence of magnetite in fly ash, which dominates the magnetic mineral suite in nearly all samples. Fly ash magnetite varies in relative abundance with respect to detrital Fe–Ti oxides (including titanomagnetite), the other common type of magnetic minerals in the samples. The Fe–Ti oxide grains are typically angular and uniformly small (most < 10 μm), and they represent airborne dust from far beyond the catchment (see REYNOLDS et al., 1999). An estimate of changing relative abundance of fly ash magnetite to mineral-dust magnetic Fe–Ti oxides is given by the curve of IRM/Ti with time (Fig. 12). IRM/Ti represents the ratio of

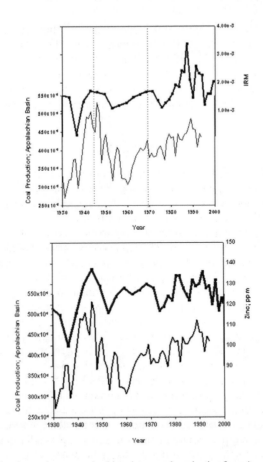

FIG. 14. Upper plot: time series comparing bituminous coal production from Appalachian Basin mines (lower of the two curves) with IRM in core MTR.1. Note the correspondence between peaks in coal production and IRM for the WWII years and 1970. The lower of the two plots is a similar comparison between coal production (lower curve) and Zn content in core MTR.1. The coal production data are from MILICI (1997).

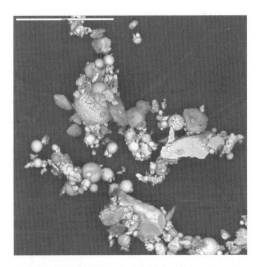

FIG. 15. Scanning electron microscope photo of the magnetic fraction of core MTR.1 from a depth of 36–38 cm. Note the abundance of spheres. The scale bar is 50 μm long.

"all magnetite" to magnetite in the Fe–Ti oxide suite, so that higher values indicate higher relative abundance of fly ash magnetite. Variations in the IRM/Ti curve are consistent with petrographic observations of estimated relative abundances of fly ash magnetite and the Fe–Ti oxide grains. SEM–EDA examination of magnetic separates provides visual evidence for fly ash magnetite (Fig. 15) and its compositions. The spheres exhibit a wide variety of surface and internal metallographic textures and dominantly contain Fe, along with Si, and Al in varying amounts.

SUMMARY

The purpose of this study was to investigate the presence of atmospheric input of particulate matter from coal-fired power plants to reservoir sediments and soils downwind from the plants. Our data show that both the reservoir sediments sampled for this study and adjacent soil samples contain detectable particulate matter from coal combustion.

Hinkel Reservoir sediments contain a diagnostic record of combustion input from atmospheric sources. The evidence of this input comes from examination of the magnetic fraction of the reservoir sediments. Spherical magnetic grains, some with characteristic metallographic textures (Figs. 11 and 15), are characteristic of this source. For Hinkel Reservoir, spherical, combustion-derived magnetite is the dominant morphology of the total magnetic fraction (Fig. 11), and constitutes a significant proportion of the total Fe in the sediment (about 15–35%; Fig. 10). Given the high positive correlation in Hinkel Reservoir sediments between magnetite abundance as reflected in the magnetic (IRM) data and the trace elements Pb and As known to be concentrated in fly ash (Figs. 4 and 10), it is likely that Pb and As were supplied by fly ash as well. Furthermore, peak concentrations of Zn, S, Ge, and Hg, and high values of combustion PAH occur in Hinkel sediments deposited within 5 years of the maxima in IRM, Zn, and Pb. It is likely that a large proportion of the content of this suite of trace elements, S, and organic compounds in Hinkel sediments were supplied via either directly or indirectly (via soil inputs) from atmospheric transport. The data indicate the presence of a component of atmospherically transported particulate material in soil samples adjacent to Hinkel and Deegan Reservoirs. These soil samples contain highly magnetic spherical grains. The soil samples are also enriched relative to local rocks in As and Pb indicating an external source for a portion of these elements as well. What we cannot say with certainty is the source of this atmospherically transported material. A coal-fired power station in Harrison County, WV, located approximately 12 km north of the study area, has three units that went on-stream during the period 1972–1974 and has a large generating capacity of approximately 2000 MW (SCHNAPP, 1999). This plant may have contributed to the post-1970 buildup in the proxies for coal fly ash, but not at all to the buildup in other parameters, which began as early as 1958 (e.g. Fig. 4). A second power plant, slightly further north (in Marion County), has units that went online as early as 1943 but has only modest generating capacity of about 100 MW (SCHNAPP, 1999). These Marion County units are about 25 km from the reservoirs, and may have potentially contributed to the pre-1972 geochemical and mineralogical signals we described. However, neither power plant is, on average, upwind from the study site. A Zn smelter located approximately 10 km northwest of the study site was potentially a significant source of coal-combustion products and metals. Its closure in 1971 coincided approximately with the timing of the maximum abundances of trace elements and PAHs. Because Cd is a common trace contaminant in Zn ore, the Cd enrichment found in reservoir sediments could have come from this source. The presence of a proximal source of atmospheric contamination may also explain the presence of relatively coarse (20 to greater than 100 μm diameter) spherical magnetic grains recognized in the Hinkel Reservoir sediments (Fig. 11). If the smelter was the major contributor of the atmospheric inputs to Hinkel Reservoir that we have identified, we suspect that impacts on soils to its east and northeast (i.e. downwind) would be even more substantial. Given the potentially toxic nature of some of the elements such as Pb, As, Hg, and Cd that we have shown to be enriched in the vicinity of Hinkel Reservoir, these downwind areas should be further geochemically characterized.

The data from Miletree Run Reservoir document the presence of nearly a century of input of atmospherically transmitted coal-combustion products. In contrast to the situation at Hinkel Reservoir, we cannot identify a nearby (<25 km) anthropogenic source for this material. The likely sources are bituminous coal-fired power plants located approximately 50 km to the south and 60–70 km to the west of the site, with the sites to the west being on-average upwind from the study area.

The dominance of fly ash particles in the magnetic fraction of the sediments (Fig. 15) is compelling evidence for atmospheric inputs to these reservoir sediments. Furthermore, magnetite is correlated with sedimentary S and at least moderately correlated with As (Fig. 13). The correlation between magnetite and S is particularly interesting. Given the small drainage area of this reservoir relative to its surface area, it is likely that atmospheric inputs are maximized relative to weathering inputs. Prior to the implementation of the Clean Air Act amendments in 1970, atmospheric SO_2 was at times considerably higher than today, and largely sourced from coal-fired power plants. The study area is among those with the highest atmospheric deposition rates of sulfate (EILERS and SELLE, 1991). Thus, the correspondence between SO_2 production and sedimentary S and magnetite over the period from 1930 to about 1980 (Fig. 14) is consistent with coal-combustion products contributing significantly to these sediments.

As noted in the "Introduction", one of the initial impetuses for this study was a geochemical map showing elevated As in soils along the Ohio River Valley (GUSTAVSSON et al., 2001). The samples for that study were collected during the 1960s and 1970s, a time when controls on outputs from power plants were much less stringent than today. Given the results of this study, in which As is one of the elements introduced by atmospheric transport into reservoir sediments and adjacent soils, it is possible that their sampling did reflect impacts from coal-fired power plants in the area. Although modern outputs of potentially toxic elements from coal-fired power plants are less than they were prior to the Clean Air Act and its amendments (EPA, 1997), the accumulation of potentially toxic substances occurred over a very long period of time prior to the institution of these controls. Interesting questions arise as to the fate in soils of these potentially toxic substances. We are presently pursuing this issue.

Acknowledgements—We are grateful to those who helped make this study possible. Doug Chambers collected many of the soil samples. Jennifer Wilson, Terence Messinger, Katherine Conko, and Carol Skeen provided invaluable help with coring and sample processing in the field. Jean Morrison, Carol Skeen, and Michele Tuttle aided in sample handling and analyses in the lab. Ed Furlong did the PAH analyses, and Jiang Xiao measured magnetic properties and performed magnetic mineral separations. Isabelle Brownfield assisted with the SEM observations. Finally, I would like to express my profound gratitude to Ian Kaplan. He was the best mentor a young scientist could have ever had.

REFERENCES

BERERAGE W. W. and YOAKUM T. D. (1980) *Soil Survey of Harrison and Taylor Counties, West Virginia.* United States Department of Agriculture, Soil Conservation Service.

BRAGG L. J., OMAN J. K., TEWALT S. J., OMAN C. L., REGA N. H., WASHINGTON P. M. and FINKELMAN R. B. (1997) US Geological Survey Coal Quality (COALQUAL) database: version 2.0. US Geological Survey Open File Report OF97-134.

CALLENDER E. and RICE K. C. (2000) The urban environmental gradient: anthropogenic influences on the spatial and temporal distributions of lead and zinc in sediments. *Environ. Sci. Technol.* **34**(2), 232–238.

CALLENDER E. and ROBBINS J. A. (1993) Transport and accumulation of radionuclides and stable elements in a Missouri River reservoir. *Water Resour. Res.* **29**(6), 1787–1804.

CALLENDER E. and VAN METRE P. C. (1997) Reservoir sediment cores show U.S. lead declines. *Environ. Sci. Technol.* **31**(9), 424–428.

COLES D. G., RAGAINI R. C., ONDOV J. M., FISHER G. L., SILBERMAN D. and PRENTICE B. A. (1979) Chemical studies of stack fly ash from a coal-fired power plant. *Environ. Sci. Technol.* **13**, 455–459.

DOE (2002) *Inventory of Electric Utility Power Plants in the United States, 2000.* Vol. DOE/EIA-0095(2000). US DOE, Energy Information Administration, Office of Nuclear, Electric and Alternate Fuels.

DRAGUN J. (1988) *The Soil Chemistry of Hazardous Materials.* Hazardous Materials Control Research Institute.

DRISCOLL C. T., LIKENS G. E. and CHURCH M. R. (1998) Recovery of surface waters in the northeastern U.S. from decreases in atmospheric deposition of sulfur. *Water, Air, Soil Pollut.* **105**, 319–329.

DRISCOLL C. T., LAWRENCE G. B., BULGER A. J., BUTLER T. J., CRONIN C. S., EAGAR C., LAMBERT K. F., LIKENS G. E., STODDARD J. L. and WEATHERS K. C. (2001) Acidic deposition in the northeastern United States: sources and inputs, ecosystems effects, and management strategies. *BioScience* **51**, 180–198.

EARY L. E., RAI D., MATTIGOD S. V. and AINSWORTH C. C. (1990) Geochemical factors controlling the mobilization of inorganic constituents from fossil fuel combustion residues: II. Review of the minor elements. *J. Environ. Qual.* **19**, 202–214.

EILERS J. M. and SELLE A. R. (1991) Geographic overview of the regional case study areas. In *Aciditic Deposition and Aquatic Ecosystems* (ed. D. F. CHARLES), pp. 107–125. Springer, Berlin.

EPA US (1997) *The Benefits and Costs of the Clean Air Act, 1970–1990.* http://www.epa.gov/air/sect812/contsetc.pdf.

EPA US (2000). *National Air Pollutant Emission Trends, 1900–1998.* http://www.epa.gov/ttn/chief/trends/trends98/. US Environmental Protection Agency, Office of Air Quality, EPA-454/r-00-002.

FLANDERS P. J. (1999) Identifying fly ash at a distance from fossil fuel power stations. *Environ. Sci. Technol.* **33**, 528–532.

FURLONG E. T., VAUGHT D. G., MERTEN L. M., FOREMAN W. T. and GATES P. M. (1995) Methods of analysis by the US Geological Survey National Water Quality Laboratory: determination of semivolatile organic compounds in bottom sediment by solvent extraction, gel permeation chromatographic fractionation, and capillary-column gas chromatography/mass spectrometry. US Geological Survey-Open File Report OF 95-0719, 67 pp.

GIBSON J. H., ANDREN A., BRADLEY R. S., CHARLES D. F., HAINES T. A., HUSAR R. B., JOHNSON A. H., KRAMER J. R., MCLAUGHLIN S. B., NORTON S. A., OZHLERT G., STENSLAND G. J., TRIJUNIS J. and WHELPDALE D. M. (1986) *Acid Deposition: Long-Term Trends.* National Academy Press, Washington, DC, p. 506.

GODBEER W. C. and SWAIN D. J. (1995) The deposition of trace elements in the environs of a power station. In *Environmental Aspects of Trace Elements in Coal* (eds. D. J. SWAINE and F. GOODARZI), pp. 178–203. Kluwer Academic Publishers, Dordrecht.

GOLDHABER M., LEE R. C., HATCH J. R., PASHIN J. C. and TREWORGY J. (2002) Role of large scale fluid-flow in subsurface arsenic enrichment. In *Arsenic in Groundwater, Geochemistry and Occurrence* (ed. A. K. S. ALAN WELCH), pp. 127–176. Kluwer Academic Publishers, Dordrecht.

GOMES S., FRANCOIS M., ABDELMOULA M., REFAIT P., PELLESIER C. and EVRAND O. (1999) Characterization of magnetite in silico-aluminous fly ash by SEM, TEM, XRD, magnetic susceptibility, and Mössbauer spectroscopy. *Cem. Concr. Res.* **29**, 1705–1711.

GUSTAVSSON N., BOLVKEN B., SMITH D. B. and SEVERSON R. C. (2001) Geochemical Landscapes of the Conterminous United States: New Map Presentations for 22 Elements. *US Geological Survey Professional Paper P 1648*, p. 35.

HERLIHY A. T., KAUFMANN P. R., CHURCH M. R., WIGINGTON P. J. JR., WEBB J. R. and SALE M. J. (1993) The effects of acidic deposition on streams in the Appalachian Mountain and Piedmont region of the mid-Atlantic United States. *Water Resour. Res.* **29**(8), 2687–2704.

HOLDREN G. R. JR., BRUNELLE T. M., MATISOFF G. and WAHLEN M. (1984) Timing the increase in atmospheric sulphur deposition in the Adirondack Mountains. *Nature* **311**(5983), 245–248.

HOWER J. C., ROBL T. L. and THOMAS G. A. (1999) Changes in the quality of coal combustion by-products produced by Kentucky power plants, 1978 to 1997: consequences of Clean Air Act directives. *Fuel* **78**, 701–712.

HUSAR R. B., SULLIVAN T. J. and CHARLES D. F. (1991) Historical trends in atmospheric sulfur deposition and methods for assessing long-term trends in surface water chemistry. In *Acidic Deposition and Aquatic Ecosystems* (ed. D. F. CHARLES), pp. 65–82. Springer, Berlin.

KAAKINEN J. W., JORDEN R. M., LAWASAINI M. H. and WEST R. W. (1975) Trace element behavior in a coal-fired power plant. *Environ. Sci. Technol.* **9**, 862–869.

KAPICKA A., PETROVSKY E., USTJAK S. and MACHACKOVA K. (1999) Proxy mapping of fly-ash pollution of soils around a coal-burning power plant: a case study in the Czech Republic. In *Geochemical Exploration 1997: Selected Papers from the 18th International Geochemical Exploration Symposium*, (eds. R. BOGOCH and M. SHIRAV), Vol. 66, pp. 291–297, Elsevier, Amsterdam.

KING J. W. and CHANNEL J. E. T. (1991) Sedimentary magnetism, environmental magnetism, and magnetostratigraphy. U.S. National Report to International Union of Geodesy and Geophysics. *Rev. Geophys.* **Suppl.**, 358–370.

KONHAUSER K. O. (1998) Diversity of bacterial iron mineralization. *Earth-Sci. Rev.* **43**(3–4), 91–121.

LAUF R. J. (1982) Microscructures of coal fly ash particles. *Ceram. Bull.* **61**, 487–490.

LAUF R. J., HARRIS L. A. and RAWLSON S. S. (1982) Pyrite framboids as the source of magnetite spheres in fly ash. *Environ. Sci. Technol.* **16**, 218–220.

LOCKE G. and BERTINE K. K. (1986) Magnetite in sediments as an indicator of coal combustion. *Appl. Geochem.* **50**, 345–356.

MATISOFF G. and HOLDEN G. R. JR. (1995) A model for sulfur accumulation in soft water lake sediments. *Water Resour. Res.* **31**(7), 1751–1760.

MATISOFF G., HOLDREN G. JR. and ANONYMOUS (1993) *A model for sulfur accumulation in Adirondack lakes*, Vol. 25. Abstracts with Programs—Geological Society of America. Geological Society of America (GSA), 254 pp.

MILICI R. C. (1997) The coalprod database: historical production data for the major coal-producing regions of the conterminous United States. US Geological Survey Open-File Report OF97-447.

NIEMANN B. L. (1984) Analysis of wind and precipitation data for assessments of transboundary transport and acid deposition between Canada and the United States. In *Meteorological Aspects of Acid Rain* (ed. C. N. BHUMRAKLKAR), pp. 57–92. Butterworths, London.

NORTON S. A. (1986) A review of the chemical record in lake sediment of energy related air pollution and its effects on lakes. In *Water, Air, Soil Pollut.* **30**, 331–345 (ed. H. C. MARTIN). Reidel, Dordrecht.

NRIAGU J. O. and COKER R. D. (1983) Sulphur in sediments chronicles past changes in lake acidification. *Nature* **303**(5919), 692–694.

PRAHL F. G. and CARPENTER R. (1983) Polycyclic aromatic hydrocarbon (PAH)-phase associations in Washington coastal sediment. *Geochim. Cosmochim. Acta* **47**(6), 1013–1023.

REYNOLDS R. L., ROSENBAUM J. G., VAN METRE P., TUTTLE M. L., CALLENDER E. and GOLDEN A. (1999) Greigite (Fe_3S_4) as an indicator of drought—the 1912–1994 sediment magnetic record from White Rock Lake, Dallas Texas, USA. *J. Paleolimnol.* **21**, 193–206.

REYNOLDS R. L., SWEETKIND D. and AXFORD Y. (2001) An inexpensive magnetic mineral separator for fine-grained sediment. US Geological Survey Open File Report OF-01-0281.

RICE K. C., CONKO K. M. and HORNBERGER G. M. (2002) Anthropogenic sources of arsenic and copper to sediments in a suburban lake, northern Virginia. *Environ. Sci. Technol.* **36**, 4962–4967.

ROCHELLE B. P. and CHURCH M. R. (1987) Regional patterns of sulfur retention in watersheds of the eastern US. *Water, Air, Soil, Pollut.* **36**, 61–73.

RUPPERT L. F. and RICE C. L. (2001) Chapter B. Coal resource assessment methodology and geology of the northern and central Appalachian Basin coal regions. In *2000 Resource Assessment of Selected Coal Beds in the Northern and Central Appalachian Basin Coal Regions*, Professional Paper 1625-C. US Geological Survey.

SCHNAPP R. (1999) *Inventory of Electric Utility Power Plants in the United States, 1999*. US Department of Energy, Energy Information Administration, DOEEIA-0095(99).

SHAW R. W. (1984) Atmosphere as delivery vehicle and reaction chamber for acid precipitation. In *Meteorological Aspects of Acid Rain* (ed. C. M. BHUMRALKAR), pp. 33–55. Butterworths, London.

THOMPSON R. and OLDFIELD F. (1986) *Environmental Magnetism*. Allen and Unwin, London.

VEROSUB K. L. and ROBERTS A. P. (1995) Environmental magnetism: past, present, and future. *J. Geophys. Res. B, Solid Earth Planets* **100**(2), 2175–2192.

VIETS J. G., HOPKINS R. T. and MILLER B. M. (1992) Variations in minor and trace metals in sphalerite from Mississippi valley-type deposits of the Ozark region: genetic implications. *Econ. Geol.* **87**(7), 1897–1905.

Geochemical Investigations in Earth and Space Science: A Tribute to Isaac R. Kaplan
© The Geochemical Society, Publication No. 9, 2004
Editors: R.J. Hill, J. Leventhal, Z. Aizenshtat, M.J. Baedecker, G. Claypool,
R. Eganhouse, M. Goldhaber and K. Peters

Projections of fossil fuel use and future atmospheric CO_2 concentrations

PAUL R. DOOSE

Department of Geology, Los Angeles Southwest College, 1600 West Imperial Highway,
Los Angeles, CA 90047-4899, USA

Abstract—Using the published record of atmospheric CO_2 levels from 1743 to 1990 and the published record of world anthropogenic CO_2 emissions from 1860 to 1990 for fossil fuel use, deforestation, biological methane production and cement manufacturing, the Future Atmospheric CO_2 Model (FAC Model) is developed that projects future atmospheric CO_2 levels. Projections of future anthropogenic CO_2 emissions, also developed in this paper, are the input to the FAC Model determining the projections of future atmospheric CO_2 levels. Published data from 1990 to 2001 are used to test the FAC Model. When published anthropogenic CO_2 emissions from 1990 to 1999 are entered into the FAC Model, the projection of atmospheric CO_2 concentrations is parallel to but 2 ppm higher than the instrumental record from Mauna Loa, Hawaii.

World fossil fuel production is modeled using Hubbert's Resource Model (HR Model) and projected into the future. These projections are converted to CO_2 emissions and combined with projections of CO_2 emissions for deforestation, biological methane production and cement manufacturing to form the input to the FAC Model. Proved reserves of world fossil fuel are used to develop a low estimate of future anthropogenic CO_2 emissions and proved reserves plus an estimate of future discoveries of world fossil fuel are used to develop a high estimate of future anthropogenic CO_2 emissions. The two estimates of future anthropogenic CO_2 emissions are put into the FAC Model to generate future atmospheric CO_2 levels. The scenarios presented in this study define a range of anthropogenic impact on atmospheric CO_2 from the low value of 453 ppm to a high value of 632 ppm in the year 2076 and then falling to 450 ppm by the year 2150. The FAC Model predicts that it is unlikely atmospheric CO_2 levels will rise above 650 ppm.

INTRODUCTION

IN THIS PAPER the Future Atmospheric CO_2 Model (FAC Model) is developed that can explore a range of future atmospheric CO_2 levels. The FAC Model is developed from two sets of data: first, the published record of atmospheric CO_2 levels, both from ice cores (KEELING *et al.*, 1989) and direct measurements at Mauna Loa, Hawaii (KEELING and WHORF, 2002, available at http://cdiac.ornl.gov); second, the published record of CO_2 emissions (HOUGHTON, 2002; MARLAND *et al.*, 2002; STERN and KAUFMAN, 1998). The rate of removal of CO_2 from the atmosphere will be modeled. It is assumed that over the short term, a few hundred years, the processes controlling the rate of CO_2 removal will continue to operate as they have for the last few hundred years. Future atmospheric CO_2 levels will be projected by putting estimates of future anthropogenic CO_2 emissions into the model. Estimates of future anthropogenic CO_2 emissions will be made from estimates of future fossil fuel production, deforestation, methane emissions and cement production. Published data from 1743 up to 1990 are used to develop the FAC Model and published data from 1990 to 1999 are used to test the FAC Model.

M. King Hubbert's Resource Model (HUBBERT, 1981), which I will call the HR Model, will be used to project fossil fuel production into the future. Projections of future fossil fuel production will be converted to CO_2 emissions. There is a 3-month lag between fossil fuel production and use, which is negligible compared to the time frame of this study. In the simplest sense, the HR Model for an exhaustible resource shows the production curve will rise to some peak and them come back down, eventually to zero. The area under the production curve is the ultimate cumulative production of the resource. An aggressive rise in the future fossil fuel production curve will be chosen to set an upper limit for future atmospheric CO_2 levels.

WORLD FOSSIL FUEL RESOURCES

For the purpose of developing the long-term projections in this paper, all fossil fuel resource data used will be that available in 1993 and projections shown in the figures start at 1990. Published data from 1990 to 2001 for anthropogenic CO_2 emissions and atmospheric CO_2 levels will be used to test the model. The cumulative production of world fossil fuel (oil, gas and coal) and a projection into the future are shown in Fig. 1. The projection was made using the HR Model (HUBBERT, 1981) and is based on known, proved reserves only. The total area under the curve in Fig. 1 is 47,330 quads (BARABBA, 1989; TAYLOR, 1989; KILGORE, 1993; WEST, 1993). The curve in Fig. 1 will be used to develop the low estimate of fossil fuel production.

Many estimates of future discoveries of world fossil fuel reserves have been made. Works available in 1993, which summarize these estimates (EDMONDS and REILLY, 1985; FRISCH, 1986; HAMMOND et al., 1992), have been used to develop a projection of the ultimate recoverable world fossil fuel resource (Fig. 2). The component parts of this estimate are twice the cumulative production plus proved reserves of world oil or 20,011 quads, twice the cumulative production plus proved reserves of world gas or 11,667 quads and four times the estimated known recoverable world coal or 134,376 quads. The projected production curve is allowed to increase proportional to an aggressive world population growth projection. The population projection increases at 1.87% per year in 1990 and growth changes gradually to a 1% per year increase by the year 2050. The curve in Fig. 2 peaks at the year 2067 and will be used to develop the high estimate of fossil fuel production. For this paper, I wish to explore the impact on atmospheric CO_2 for a range of future fossil fuel production scenarios. Thus, a high-end scenario for population growth is appropriate to explore the high-end impact on the atmosphere.

The projection in Fig. 2 extends well beyond the year 2200 and there are, of course, significant opportunities for deviations: errors in the estimate of fossil fuel resources, errors in population growth rates and errors in *per capita* energy use. These errors could work both ways and some may cancel others. However, the sensitivity of the projection is such that large changes in reserves produce only small changes in the timing of the peak of the resources production curve. Actual future world fossil fuel production will be assumed to fall somewhere between the two limits shown in Figs. 1 and 2.

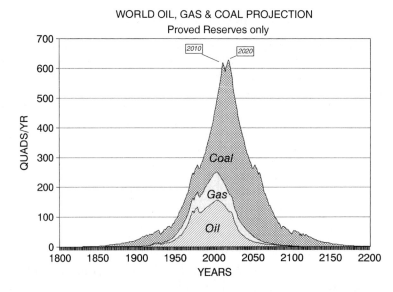

FIG. 1. The combined world fossil fuel use curve shows the sum of the historical worldwide production of oil, gas and coal up to 1990 (BARABBA, 1989; TAYLOR, 1989; KILGORE, 1993; WEST, 1993). The combined world fossil fuel production is projected into the future base on proved reserves of oil and gas, and an estimate of known recoverable coal. The units are in quads (10^{15} BTU).

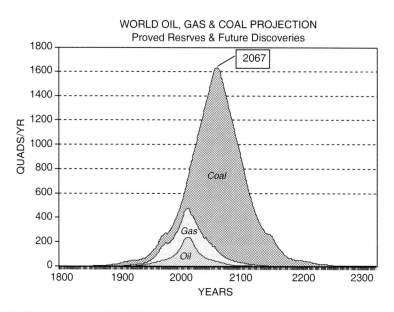

FIG. 2. The combined world fossil fuel production is projected into the future base on proved reserves and estimated future discoveries of oil, gas and coal (BARABBA, 1989; EDMONDS and REILLY, 1985; FRISCH, 1986; KILGORE, 1993; HAMMOND *et al.*, 1992). The units are in quads (10^{15} BTU).

There are fossil fuels which are currently not cost-effective to produce or are not in significant use, such as oil shale. When new technology opens new resources, the production curves can be changed to include them. For this assessment only the fossil fuel sources in Figs. 1 and 2 will be considered.

ATMOSPHERIC CARBON DIOXIDE

Studying published anthropogenic CO_2 emissions as well as trends in atmospheric CO_2 levels allows the development of the FAC Model that can be used to project future atmospheric CO_2 levels. Figure 3 shows ice core CO_2 data back to 1743 (KEELING *et al.*, 1989) and atmospheric CO_2 concentration data from Mauna Loa, Hawaii measured by Keeling and Whorf from 1958 through 2001 (available at http://cdiac.ornl.gov). The Mauna Loa data for only January of each year are used to smooth the curve and see the recent increase.

The source of the increase in atmospheric CO_2 concentration is anthropogenic CO_2 emissions. Figure 4 presents a history of the major anthropogenic sources of world CO_2 emissions from 1860 to 1990 (HOUGHTON, 2002; MARLAND *et al.*, 2002; STERN and KAUFMAN, 1998) and projections of emissions into the future. Figure 4 shows the low estimate of fossil fuel production from Fig. 1; a similar development was done for the high estimate. Clearly, fossil fuel use is the dominant source of world anthropogenic CO_2 emissions with deforestation the second largest contributor. Biogenic methane reacts in the atmosphere to form CO_2. Methane from cattle, sheep and rice patties will grow with world population, as well as CO_2 emissions from the production of Portland cement.

The published world CO_2 emissions from 1860 to 1995 (HOUGHTON, 2002; MARLAND *et al.*, 2002; STERN and KAUFMAN, 1998) are summed in Fig. 5. Figure 5 is divided to show that portion of the anthropogenic CO_2 emissions which stays in the atmosphere to cause the increase and that portion of the anthropogenic CO_2 emissions which have left the atmosphere and gone to various sinks.

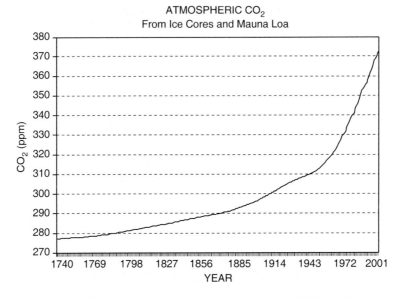

FIG. 3. Atmospheric CO_2 based on ice core data (KEELING *et al.*, 1989) and Mauna Loa data (KEELING and WHORF, 2002, available at http://cdiac.ornl.gov). Only data for the month of January are shown.

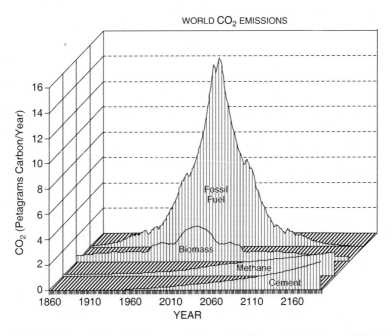

FIG. 4. The history of the major sources of world CO_2 emissions from 1860 to 1993 (HOUGHTON, 2002; MARLAND *et al.*, 2002; STERN and KAUFMAN, 1998). From the largest to the smallest they are: fossil fuel use; biomass, which includes deforestation; methane, which converts to CO_2 in the atmosphere; and cement production. Fossil fuel emissions are projected into the future based on proved reserves shown in Fig. 1. Other curves are projected based on projected population growth.

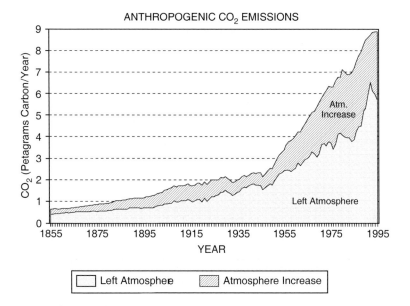

FIG. 5. The world CO_2 emissions from Fig. 4 are summed. The figure is divided to show that portion of the emissions which stays in the atmosphere to cause the increase and the portion of the emissions which are missing and have left the atmosphere for various sinks.

MODELING CO_2 REMOVAL

When developing the FAC Model for the rate of removal of atmospheric CO_2 as a function of the atmospheric CO_2 concentration, there are several approaches that one can take. Looking at the information available, we see in Fig. 3 that by tracing the concentration of CO_2 back in time into the 1700s, it appears to be asymptotically approaching a value of roughly 270 ppm. This, then, would appear to be an equilibrium value maintained over the last few centuries. The magnitude of the CO_2 perturbation (*i.e.* the amount of CO_2 in the atmosphere above 270 ppm) is assumed to drive the rate of removal of CO_2 from the atmosphere. Figure 6 shows the CO_2 removed as a percentage of the CO_2 perturbation.

Ice cores and other evidence indicate that atmospheric CO_2 may have been lower, roughly 200 ppm, in the distant past. However, using 270 ppm as the equilibrium value creates a conservative model. Using lower values would predict a faster rate of atmospheric CO_2 removal. This would predict a lower maximum atmospheric CO_2 concentration and a smaller impact on climate change and conversely for a higher value.

From the data in Fig. 6, it is clear that the CO_2 removed from the atmosphere each year (as a percentage of the amount of atmospheric CO_2 above 270 ppm) increases with increasing anthropogenic emissions. It is appropriate to curve fit the removal curve and develop a model around that relation. The rate of removal of atmospheric CO_2 can be modeled as some function of the CO_2 perturbation (Eq. (1)).

$$R = f(C_{270}) \tag{1}$$

where R is the rate of atmospheric CO_2 removal and C_{270} is the CO_2 perturbation.

$$R = b(C_{270})^x \tag{2}$$

A curve-fit of the data in Fig. 6 gave the function in Eq. (2), where b is 6×10^{-5} and x is 1.5 (both are dimensionless constants). A CO_2 projection model was then developed using the relation in Eq. (2), and is shown in Eqs. (3) and (4).

$$(C_{270i} + E_i) - b(C_{270i} + E_i)^x = C_{270j} \tag{3}$$

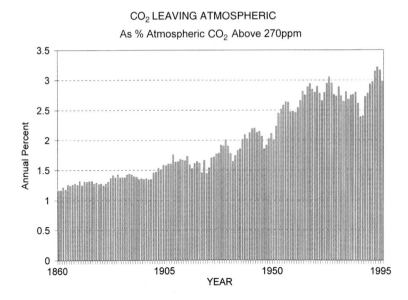

FIG. 6. The CO_2 removed as a percentage of the CO_2 perturbation. The CO_2 perturbation is defined as the total concentration of CO_2 in the atmosphere minus 270 ppm.

$$CO_{2j} = 270 \text{ ppm} + C_{270j} \tag{4}$$

where E is the annual CO_2 emissions, i represents one year and j the subsequent year. CO_{2j} is the total atmospheric CO_2 concentration in the jth year. Figure 4 projects the world anthropogenic CO_2 emissions into the future using the fossil fuel production curve from the HR Model for proved reserves in Fig. 1 and other anthropogenic sources of CO_2. The future anthropogenic CO_2 emissions are summed and run through the FAC Model. The resulting future atmospheric CO_2 concentration projections are shown in Fig. 7. Note that the atmospheric CO_2 concentration, for the low projection based on proved reserves, rises to a peak of 453 ppm in the year 2029 and then decreases. This process was repeated using the high estimate of future fossil fuel resources to generate the high projection in Fig. 7. Here, the CO_2 concentration rises to 632 ppm in the year 2076 and then decreases. The atmospheric concentration of CO_2 is above 600 ppm for 37 years. Let me caution once again that these numbers are calculations based on estimates. Although I will report the actual numbers predicted by the modeling process, the concentrations and dates should be considered estimate ranges of possible values.

The important point to note is that the atmospheric concentration of CO_2 decreases at a significant rate after reaching a peak. Thus, after future anthropogenic CO_2 emissions begin to decrease; atmospheric CO_2 will also decline according to the relationship in Eqs. (3) and (4).

Now we can test the CO_2 projection model. Figure 7 shows the published atmospheric CO_2 concentration through 2001 (KEELING and WHORF, 2002, available at http://cdiac.ornl.gov). The FAC Model projections start with 1991 giving 11 years of comparison. The FAC Model projections are high, predicting a value of roughly 381 ppm for 2001 where the actual value is 371 ppm. Perhaps the use of a very aggressive population growth curve in the HR Model projected fossil fuel emissions too high. Published population growth is lower (LONG, 1999). That the FAC Model is projecting higher than the measured value is a notable result since the high anthropogenic CO_2 emissions curve was intended to set an upper limit on future atmospheric CO_2 levels.

A second test of the FAC Model can be made by using the published record of anthropogenic CO_2 emissions from 1991 through 1999 (HOUGHTON, 2002; MARLAND et al., 2002; STERN and KAUFMAN, 1998). The results are shown in Fig. 8. The projection runs parallel to the measured

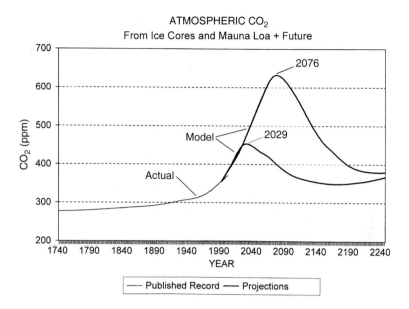

FIG. 7. The future atmospheric CO$_2$ concentration projection if only known, proved reserves of fossil fuel are used (low projection) and the future atmospheric CO$_2$ concentration projection if known, proved reserves of fossil fuel plus estimated future discoveries of fossil fuel are used (high projection). The historical record is shown through 2001 (KEELING and WHORF, 2002, available at http://cdiac.ornl.gov) allowing 11 years of data to be compared to the projections.

record but roughly 2 ppm higher. Thus, if actual anthropogenic CO$_2$ emissions data are entered into the FAC Model, it does give reasonable predictions of the atmospheric concentration of CO$_2$.

The high estimate of anthropogenic CO$_2$ emissions was deliberately constructed at a high rate of use to create an upper limit for anthropogenic CO$_2$ emissions and future atmospheric CO$_2$ levels. The measured record of atmospheric CO$_2$ levels was expected to be lower. That the test using actual anthropogenic CO$_2$ emissions is a little high indicates that the model needs a little

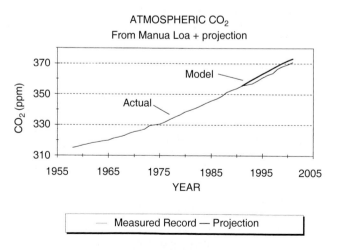

FIG. 8. The future atmospheric CO$_2$ concentration projection using the actual record of CO$_2$ emissions from 1991 through 1999 (HOUGHTON, 2002; MARLAND et al., 2002; STERN and KAUFMAN, 1998). The historical record is shown through 2001 (KEELING and WHORF, 2002, available at http://cdiac.ornl.gov) allowing 11 years of data to be compared to the projections. The projection runs parallel to the historical record but roughly 2 ppm higher.

adjusting. However, the fact that the projections run parallel to the measured record and the lines do not diverge indicates that the FAC Model does a reasonable job of predicting future atmospheric CO_2 levels. Because the projections in Fig. 7 are high, it is possible that atmospheric CO_2 concentrations will never rise above 650 ppm.

DISCUSSION

The FAC Model was developed using only two sets of data, anthropogenic CO_2 emissions data and atmospheric CO_2 concentration data. The rate at which CO_2 leaves the atmosphere was modeled without any assumptions as to where it went. The projection of future atmospheric CO_2 levels is made by inputting anthropogenic CO_2 emissions data into the FAC Model. Both real data and projected future CO_2 emissions data were used.

Over the long term, more than thousands of years, atmospheric CO_2 levels have changed significantly and the cause of these changes is often unknown. For the period 1743–1960, the rate of CO_2 removal from the atmosphere seems to be proportional to the partial pressure of CO_2 in the atmosphere. It is assumed that over the short term, a few hundred years, this relationship will continue.

There are several sinks for atmospheric CO_2. The two most notable being the biosphere and the ocean. It is the author's personal opinion that oceanic deposition of carbonate minerals plays a significant role in the removal of anthropogenic CO_2 from the atmosphere. The potential to remove CO_2 is large. The reservoir of Ca^{++} in the world oceans is 190,000 Pg (GARRELS and MACKENZIE, 1971). If all of the anthropogenic CO_2 emissions projected in the high estimate were deposited as marine calcium carbonate, it would change the concentration of calcium in ocean water by roughly 1% and that assumes no additions. Yet, rivers add 0.5 Pg Ca^{++} to the upper ocean each year (GARRELS and MACKENZIE, 1971) and 10^{17} l of seawater cycle through the hot basalt along ocean ridges each year adding another 0.1 Pg Ca^{++} or more to the deep ocean each year. Carbonate sediments are known to form on ocean ridges and the newly forming shales in Santa Barbara Basin off Southern California contain more than 10% syngenetic carbonate minerals (DOOSE, 1980).

Measurements by QUAY et al. (1992) of the CO_2 flux from the atmosphere to the ocean between the years 1970 and 1990 was 2.1 Pg C/yr on average. Another 1.4 ± 0.1 Pg C/yr of CO_2 were removed by the biosphere (QUAY et al., 1992). Over the 20-year period in Quay's study the average annual anthropogenic CO_2 emissions were 7.0 Pg C/yr of which 3.8 Pg C/yr left the atmosphere. Thus, 55% of the anthropogenic CO_2 emissions leaving the atmosphere entered the ocean during the period 1970–1990.

If the oceanic deposition of carbonate minerals plays a significant role in the removal of anthropogenic CO_2 from the atmosphere, then it is unlikely that the process would saturate. Thus, the relations on which the FAC Model are based would continue over the duration that the FAC Model has been applied and longer. If ocean carbonate deposition is a removal mechanism, it would assure that the FAC Model is valid and its projections reasonable if the projections of future anthropogenic CO_2 emissions are reasonable.

If the projections of the FAC Model are reasonable, then it is unlikely that atmospheric CO_2 levels will ever rise above 650 ppm. This is a significant result. Although this would not preclude global warming from CO_2, it would put limits on the magnitude of the effects. Also, significant is that the CO_2 will not stay long in the atmosphere after anthropogenic CO_2 emissions decline.

Acknowledgements—I thank two anonymous reviewers and Joel Leventhal, who made many helpful and critical suggestions, not all of which were taken. The early part of this work was supported by Southern California Edison Company.

REFERENCES

BARABBA V. P. (1989) *Historical Statistics of the United States Colonial Times to 1970*. US Department of Commerce, Bureau of the Census, Kraus International Publications, New York, 1200 pp.

DOOSE P. R. (1980) The bacterial production of methane in marine sediments. Ph.D. Thesis, University of California, Los Angeles, CA, 240 pp.

EDMONDS J. and REILLY J. M. (1985) *Global Energy, Assessing the Future*. Oxford University Press, New York, 317 pp.

FRISCH J. R. (1986) *Future Stresses for Energy Resources, Energy Abundance: Myth or Reality?* Graham & Trotman, London, UK, 226 pp.

GARRELS R. M. and MACKENZIE F. T. (1971) *Evolution of Sedimentary Rocks*. W. W. Norton and Company, New York, 397 pp.

HAMMOND A. L., PADEN M. E., LIVERNASH R. T., RODENBURG E., HENNINGER N., BRYANT D. and VAN DER WANSEIN M., (1992) *World Resources*. Oxford University Press, New York, NY, 385 pp.

HOUGHTON R. A. (2002) Annual Net Flux of Carbon to the Atmosphere from Land-Use Change: 1850–2000, http://cdiac.ornl.gov.

HUBBERT M. K. (1981) The world's evolving energy system. *Am. J. Phys.* **49**(11), 1029.

KEELING C. D. and WHORF T. P. (2002) Atmospheric CO_2 concentrations (ppmv) derived from in situ as samples collected at Mauna Loa Observatory, Hawaii, http://cdiac.ornl.gov.

KEELING C. D., BACASTOW R. B., CARTER A. F., WHORF T. P., HEIMANN M., MOOK W. G. and ROELOFFZEN H (1989) A three-dimensional model of atmospheric CO_2 transport based on observed winds: 1. Analysis of observational data. In *Aspects of Climate Variability in the Pacific and the Western Americas* (ed. D. H. PETERSON), Geophysical Monograph, 445 pp. American Geophysical Union, Washington, DC, USA.

KILGORE W. C. (ed.) (1993) *Annual Energy Review 1992*. US Government Printing Office, Washington, DC, 350 pp.

LONG J. F. (1999) *World Population Profile: 1998, US Bureau of the Census*. Report WP/98. US Government Printing Office, Washington, DC, 172 pp.

MARLAND G., BODEN T. and ANDERES R. J. (2002) Global CO_2 Emissions from Fossil-Fuel Burning, Cement Manufacture, and Gas Flaring: 1751–1999, http://cdiac.ornl.gov.

QUAY P. D., TILBROOK B. and WONG C. S. (1992) Ocean uptake of fossil fuel CO_2: carbon-13 evidence. *Science* **256**, 74–79.

STERN D. I. and KAUFMAN R. K. (1998) Global Historical Anthropogenic CH_4 Emissions, http://cdiac.ornl.gov.

TAYLOR C. (ed.) (1989) *IGT World Reserves Survey*. Institute of Gas Technology, Chicago, IL, 344 pp.

WEST J. (ed.) (1993) *International Petroleum Encyclopedia*. PennWell Publishing Co., Tulsa, OK, 376 pp.

Geochemical Investigations in Earth and Space Science: A Tribute to Isaac R. Kaplan
© The Geochemical Society, Publication No. 9, 2004
Editors: R.J. Hill, J. Leventhal, Z. Aizenshtat, M.J. Baedecker, G. Claypool,
R. Eganhouse, M. Goldhaber and K. Peters

Geochemistry of coastal tarballs in southern California—A tribute to I. R. Kaplan

KEITH A. KVENVOLDEN and FRANCES D. HOSTETTLER
US Geological Survey, 345 Middlefield Road, MS 999, Menlo Park, CA 94025, USA

Abstract—In the southern offshore California borderland, natural oil seeps occur mainly in the Santa Barbara Channel and Santa Monica Bay. Coastal tar residues (tarballs) from beaches bordering these water bodies were analyzed for six geochemical parameters: stable carbon isotopic compositions ($\delta^{13}C$) and four biomarker ratios (C_{28}/C_{29} hopane, sterane/hopane, refractory index, bisnorhopane index), and the presence or absence of trisnorhopane. The objectives of this study were to group these residues and infer possible sources and transport directions. Three major groups were established. Two groups are likely from natural seeps near the Channel Islands, whereas the third group probably comes from seeps within Santa Monica Bay. Residues from all groups occur on the Channel Islands and on mainland beaches from as far south as San Diego to Point Reyes north of San Francisco.

INTRODUCTION

NATURAL OIL SEEPS are found worldwide (WILSON et al., 1974), including coastal regions both on land and on the sea floor. Whereas oil seeps on land are commonly visible, those in the sea are often invisible because of the overlying water. However, these invisible seeps are commonly revealed because they produce, on the sea surface, oil slicks which can be mapped by various remote-sensing techniques (MACDONALD et al., 1993).

Oil from natural submarine seeps often coagulates as tar residues (collectively called tarballs) that are ubiquitous along the coastline of California, particularly southern California. Tarballs from natural oil seeps can be difficult to distinguish from tarballs resulting from anthropogenic oil spills. This paper focuses on tarballs from the beaches and rocky shorelines of southern California. Because of their distinctive chemical and isotopic signatures, these coastal tarballs are believed to be almost exclusively from natural oil seeps. I. R. Kaplan and associates (see REED and KAPLAN, 1977) pioneered investigations of natural oil and gas seeps in this region.

In the southern California borderland, extending from Point Conception to San Diego (Fig. 1), there are two main areas of natural crude-oil seepage—Santa Barbara Channel and Santa Monica Bay. Prolific oil seepage occurs offshore from Coal Oil Point (Fig. 2) along west–northwest-trending anticlines on the northern margin of the Santa Barbara Channel (HORNAFIUS et al., 1999). In Santa Monica Bay there are at least six active oil seeps along the submarine trace of the northwest-trending Palos Verdes fault (Fig. 3) offshore from Redondo Beach (WILKINSON, 1971). In fact, WILKINSON (1971) estimated that there are probably 50–60 offshore seeps or seep areas between Point Conception in Santa Barbara County and Huntington Beach in Orange County, California. On beaches bordering the Santa Barbara Channel and Santa Monica Bay, tarballs are common (HARTMAN and HAMMOND, 1981).

Previous studies of the natural seeps in the southern California borderland have shown some of the potential and limitations in the application of geochemical methods. For example, REED and KAPLAN (1977) emphasized the use of stable isotopic ratios of sulfur, nitrogen, and carbon in oil and the $\delta^{13}C$ of various chromatographic fractions to distinguish production oil from natural oil seeps. STUERMER et al. (1982) demonstrated that a very heterogeneous distribution of hydrocarbon compositions and concentrations exists in the seep environment. HARTMAN and HAMMOND (1981) measured carbon and sulfur isotopic ratios and total sulfur content to correlate beach tarballs with probable source rock. For our study, we used stable carbon isotopes and various biomarker ratios to group coastal tarballs and to show the geographic distribution of each group along the California mainland coast and on the coasts of two Channel Islands, Santa Cruz and Santa Rosa.

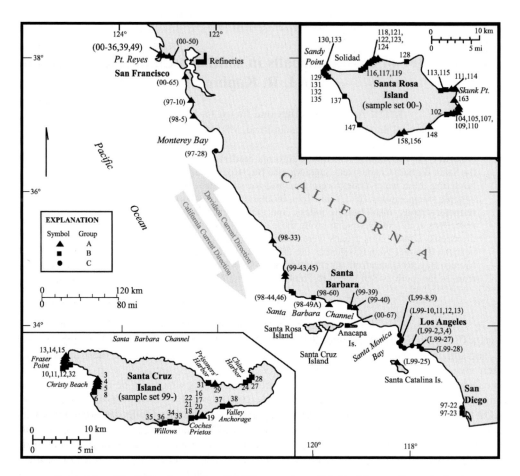

FIG. 1. Map of the California coast from Point Reyes on the north to San Diego on the south, showing the sites where tarballs for this study were collected. Tarballs are differentiated geochemically into three groups: A, B, and C. Sample numbers are designated by year and number (Table 1).

METHODS

Tarballs from 92 coastline locations, mostly in the southern California borderland including Santa Cruz and Santa Rosa Islands, were collected during a 4-year period. In addition, a sample from a known natural oil seep was collected in the Santa Barbara Channel 3 miles north of Anacapa Island. Sample preparation and analysis followed the procedure of KVENVOLDEN et al. (2000): Tarballs were removed from rock and sand with a clean knife and placed in pre-cleaned glass jars for transport to the organic geochemical laboratory in Menlo Park, California. The natural oil-seep sample was collected from the seawater surface and also stored in a pre-cleaned jar. Samples were dissolved in dichloromethane (DCM), filtered through glass wool to remove particulate material, and concentrated by evaporation to remove the DCM solvent. A portion of this extract was removed for determination of the carbon isotopic composition by isotope-ratio mass spectrometry. The results are reported in δ notation in parts-per-thousand (‰) relative to the PeeDee Belemnite (PDB) standard.

A second portion of the extract (~25 mg) was dissolved as completely as possible in 5 ml n-hexane using sonication and mechanical agitation. This solution was loaded onto a liquid–solid chromatography column layered from the bottom to top with activated copper (for sulfur removal), deactivated alumina (5% water), and activated silica gel (Grade 62 and 923). The column was eluted with n-hexane followed by 30% DCM in n-hexane.

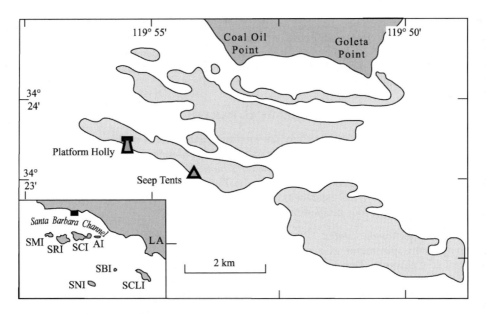

FIG. 2. Map showing areas (in light gray) of natural oil seeps in the Santa Barbara Channel offshore from Coal Oil Point. On the inset map, the Channel Islands (I) are designated: SM, San Miguel; SR, Santa Rosa; SC, Santa Cruz; A, Anacapa; SN, San Nicholas; SB, Santa Barbara; SCL, Santa Catalina. Modified from HORNAFIUS *et al.* (1999).

Two separate hydrocarbon fractions were collected—saturate (*n*-hexane) and aromatic (30% DCM). The saturate and aromatic fractions were analyzed by gas chromatography/mass spectrometry (GC/MS) in the total ion current (TIC) mode and in the selected-ion-monitoring (SIM) mode. Biomarker ratios, listed below, were calculated from GC/MS/SIM mass fragmentograms of mass/charge (*m/z*) 191 (terpanes/hopanes) and 217 (steranes) using

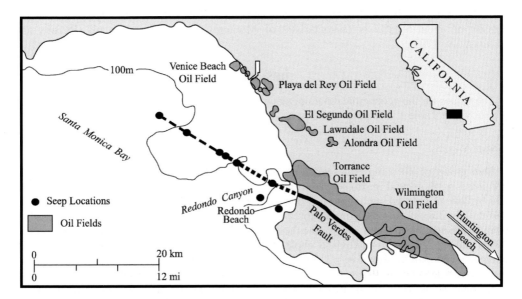

FIG. 3. Map showing known oil seeps along the submarine trace of the Palos Verdes fault (short dash, obscured by water; long dash, inferred) in Santa Monica Bay. Nearby oil fields are indicated. Modified from WILKINSON (1971).

peak heights. The presence or absence of 25,28,30-trisnorhopane was monitored by m/z 177. Peaks on mass fragmentograms of the aromatic fraction for m/z 231 (triaromatic steranes) and m/z 242 (monomethyl chrysenes) were used to determine the refractory index (HOSTETTLER et al., 1999), called here PAH-RI because this parameter is determined from the ratio of two polycyclic aromatic hydrocarbons (PAH).

GEOCHEMICAL PARAMETERS

The parameters chosen for this study are empirical and used as geochemical fingerprints to differentiate the collected tarballs into three major groups. Three biomarker ratios were obtained from peak measurements on mass fragmentograms of the terpanes/hopanes and the steranes, and a fourth ratio (PAH-RI) was determined from the TIC of the aromatic fraction. In addition, the overall pattern of the m/z 217 (steranes) fragmentograms was useful in differentiating groups of tarballs. The following four biomarker ratios were determined:

(1) *BI, bisnorhopane index, 28,30-bisnorhopane/17α,21β(H)-hopane.* CURIALE et al. (1985) used this parameter as a source ratio to characterize Miocene Monterey Formation crude oil of California. The presence of 28,30-bisnorhopane indicated that the oil came from a marine source rock deposited in a highly reducing environment. In addition to BI, the occurrence of another desmethylhopane, 25,28,30-trisnorhopane (T), was determined in the tarballs.

(2) *$C_{28}R/C_{29}R$, sterane ratio, 24-methyl-5α,14α,17α(H)-cholestane (20R)/24-ethyl-5α,14α,17α(H)-cholestane (20R).* This ratio of 20R sterane isomers has been adapted from discussions in GRANTHAM and WAKEFIELD (1988) and WAPLES and MACHIHARA (1991), in which suites of steranes were considered rather than two individual epimers.

(3) *Ster/Hop, sterane–hopane ratio, 5α,14α,17α(H)-cholestane (20R)/17α,21β(H)-hopane.* This ratio, measured directly from SIM fragmentograms, reflects the relative proportion of a ubiquitous and frequently prominent regular sterane to C_{30}-hopane.

(4) *PAH-RI, refractory index, the major peak in each of two suites of highly refractory PAH, the triaromatic steranes, C_{26} to C_{28} (T)/monomethyl chrysenes (C).* This ratio, originally designated RI, was developed by HOSTETTLER et al. (1999) for differentiating tar residues in Prince William Sound and the Gulf of Alaska.

Carbon isotopic compositions of tar residues ($\delta^{13}C$) were determined. The average $\delta^{13}C$ of the tarballs of about $-23‰$ is characteristic of Monterey Formation crude oil (PETERS and MOLDOWAN, 1993).

RESULTS

The values for the geochemical parameters for each sample used to group the coastal tarballs are shown in Table 1. Based on these parameters, three groups are distinguished.

Group A

This group of 40 samples is characterized by high BI ratios where 28,30-bisnorhopane dominates αβ-hopane (Fig. 4a), low concentrations of regular steranes relative to hopanes (Fig. 4a and b), and an unusual sterane profile (Fig. 4b). The sterane profiles (example in Fig. 4b) are particularly diagnostic, consisting of three prominent peaks, the first two tentatively identified from published spectra and retention times as C_{26}-24-nor-5α-cholestane (MOLDOWAN et al., 1991) and C_{27}-27-nor-24-methyl-5α-cholestane (SCHOUTEN et al., 1994), and the third peak being an overlap of the minor m/z 217 fragment ion of 28,30-bisnorhopane which is dominant in the hopane fragmentogram (Fig. 4a). The 24-nor-5α-cholestanes are generally rare, but are known to occur in some Monterey Formation crude oil where they are attributed to widespread late Tertiary deposition of organic-rich diatomaceous sediments (HOLBA et al., 1998). In Group A, regular sterane peaks are minor, if present at all, resulting in questionable parameter values, and leading to very low and uncertain values of the sterane/hopane ratios (0.06 ± 0.03) and to

Table 1. Geochemical parameters for coastal tar residues

Field no.	Description	Location	$\delta^{13}C$	BI	T	$C_{28}R/C_{29}R$	Ster/Hop	PAH-RI
Group A								
97-10	Weathered tar on rock	S. side of Pescadero Creek	−23.2	1.4	+	nc	0.05	47
98-5	Old tar on conglomerate	High tide line, Pigeon Pt.	−23.2	1.5	+	nc	0.14	67
98-33	Fresh tar on rock	Montague State Park	−22.9	1.1	+	nc	0.15	76
98-49A	Fresh floating tar	Offshore Coal Oil Pt.	−23.3	1.5	+	nc	0.03	74
L99-25	Weathered tar on rock	Catalina Island, Twin Harbor	−23.3	1.4	+	nc	0.05	54
99-3	Fresh tar on sand	Santa Cruz I., Christy Beach	−23.6	1.7	+	nc	0.03	54
99-4	Fresh tar on sand	Santa Cruz I., Christy Beach, HTL	−23.4	1.5	+	nc	0.05	49
99-5	Fresh tar on sand	Santa Cruz I., Christy Beach	−23.5	1.9	+	nc	0.05	63
99-8	Fresh tar on sand	Santa Cruz I., Christy Beach	−23.6	1.8	+	nc	0.03	53
99-13	Fresh new tar on mat	Santa Cruz I., Fraser Pt., N	−23.5	1.7	+	nc	0.04	64
99-14	Fresh tar splotch on rock	Santa Cruz I., Fraser Pt., S	−23.4	1.4	+	nc	0.03	45
99-15	Weathered tar splotch on rock	Santa Cruz I., Fraser Pt., S	−23.4	1.7	+	nc	0.04	67
99-19	Weathered tar on rock	Santa Cruz I., Coches Prietos, W	−23.5	1.4	+	0.78	0.08	47
99-29	Fresh tar on sand	Santa Cruz I., Prisoners' Harbor	−23.5	1.4	+	1.1?	0.14	41
99-38	Weathered tar on rock	Santa Cruz Isl., Valley Anchorage	−23.7	1.6	+	nc	0.03	55
99-40	Fresh tar on swash sand	Ventura, CA, Channel Isl. Harbor	−23.4	1.5	+	nc	0.04	53
99-43	Fresh tar on rock	Vandenberg AFB, Minuteman Beach	−23.2	1.6	+	nc	0.06	52
99-45	Fresh tar on rock	Vandenberg AFB, Minuteman Beach	−23.5	1.6	+	nc	0.07	50
00-36	Fresh tar on sand	Pt. Reyes, North Beach, surf zone	−23.3	1.6	+	nc	0.05	53
00-39	Fresh tar, sand, mussel	Pt. Reyes, South Beach, low surf zone	−23.2	1.7	+	0.83?	0.10	48
00-49	Fresh tar on sand	Limintour Beach, mid-tide line	−23.1	1.5	+	nc	0.03	48
00-65	Fresh tar on sand	Fort Funston, surf zone	−23.2	1.9	+	nc	0.05	67
00-104	Large tarball	Santa Rosa I., beach	nd	1.6	+	nc	0.07	46
00-105	Tar on cliff side	Santa Rosa I., pocket beach	nd	1.5	+	nc	0.05	56
00-107	Tar on rock	Santa Rosa I., west of East Pt.	nd	1.7	+	nc	0.03	66
00-109	Tar on cliff side	Santa Rosa I., marine terrace	nd	1.6	+	nc	0.09	57
00-110	Fresh tar on terrace	Santa Rosa I., on terrace	nd	1.5	+	nc	0.03	52
00-111	Fresh floating tar mass	Santa Rosa I., Skunk Pt.	nd	1.5	+	nc	0.03	49
00-114	Fresh tar on rock	Santa Rosa I., Skunk Pt.	nd	1.5	+	nc	0.03	53
00-118	Fresh thick tar on beach	Santa Rosa I., Soledad Pt.	nd	1.7	+	nc	0.07	46
00-121	Tar on rock outcrop	Santa Rosa I., Soledad Pt.	nd	1.7	+	nc	0.03	58
00-122	Tar on cobbles	Santa Rosa I., Soledad Pt.	nd	1.6	+	nc	0.03	52
00-123	Fresh tar on beach	Santa Rosa I., Soledad Pt.	nd	1.8	+	nc	0.05	61
00-124	Fresh floating tar	Santa Rosa I., Soledad Pt.	nd	1.9	+	nc	0.09	49
00-130	Fresh tar on rock	Santa Rosa I., Sandy Pt.	nd	1.5	+	nc	0.03	55
00-133	Fresh tar mat	Santa Rosa I., Sandy Pt.	nd	1.4	+	nc	0.07	27
00-148	Tar on rock	Santa Rosa I., Ford Pt., E	nd	1.4	+	nc	0.03	63

(continued on next page)

Table 1 (continued)

Field no.	Description	Location	δ¹³C	BI	T	$C_{28}R/C_{29}R$	Ster/Hop	PAH-RI
00-156	Tar	Santa Rosa I., South Pt., E	nd	1.6	+	nc	0.06	55
00-158	Tar	Santa Rosa I., Officers' Beach	nd	1.7	+	nc	0.04	55
00-163	Tar on conglomerate	Santa Rosa I., Oat Pt.	nd	1.6	+	nc	0.04	63
Group B								
97-23	Weathered tar on rock	Windansea Beach, San Diego	−23.3	1.7	+	0.94	0.46	78
98-44	Fresh tar on rock	Jalama State Beach	−23.3	1.6	+	0.91	0.31	79
98-46	Fresh tar on rock	Jalama State Beach	−23.3	1.4	+	0.97	0.34	60
98-60	Fresh tar on sand	Refugio State Beach	−23.3	1.1	+	1.1	0.43	37
L99-27	Weathered tar on rock	Huntington Beach #1	−23.3	1.1	+	1.3	0.25	62
99-6	Fresh tar on sand	Santa Cruz I., Christy Beach, HTL	−23.6	1.9	+	1.0	0.35	68
99-10	Fresh tar on rock	Santa Cruz I., Fraser Pt., N	−23.6	1.5	+	1.0	0.48	47
99-11	Weathered old tar	Santa Cruz I., Fraser Pt., N	−23.4	1.1	+	1.0	0.26	30
99-12	Weathered old tar mat	Santa Cruz I., Fraser Pt., N	−23.6	1.4	+	1.0	0.24	40
99-18	Weathered tar on rock	Santa Cruz I., Coches Prietos, W	−23.6	1.6	+	0.96	0.28	60
99-21	Fresh tar on sand	Santa Cruz I., Coches Prietos	−23.7	1.6	+	1.2	0.04	36
99-22	Fresh tar on sand	Santa Cruz I., Coches Prietos	−23.6	1.8	+	0.90	0.17	64
99-24	Fresh tar on rock	Santa Cruz I., China Harbor	−23.3	1.8	+	0.95	0.18	60
99-27	Fresh tar on rock	Santa Cruz I., China Harbor	−23.5	1.8	+	1.1	0.40	75
99-28	Weathered tar on rock	Santa Cruz I., China Harbor	−23.7	1.9	+	1.0	0.47	76
99-31	Weathered tar on rock	Santa Cruz I., Prisoners' Harbor	−23.5	1.0	+	1.1	0.23	20
99-32	Fresh tar on rock	Santa Cruz I., Fraser Pt.	−23.4	1.4	+	1.1	0.31	47
99-33	Weathered tar on rock	Santa Cruz I., Willows	−23.7	1.7	+	1.0	0.28	57
99-34	Weathered tar on rock	Santa Cruz I., Willows	−23.6	1.5	+	1.0	0.31	46
99-36	Fresh tar on sand	Santa Cruz I., Willows	−23.5	1.6	+	0.97	0.18	43
99-37	Weathered tar on rock	Santa Cruz Isl., Valley Anchorage	−23.5	1.7	+	0.92	0.16	56
99-39	Weathered tar on breakwater	Ventura, CA, Channel Isl. Harbor	−23.5	1.7	+	1.0	0.26	58
00-50	Fresh tar on sand	Limintour Beach	−23.2	1.8	+	1.1	0.28	58
00-102	Mixed tar (3 patties)	Santa Rosa I., beach	nd	1.6	+	0.87	0.17	48
00-113	Fresh tar mass on rock terrace	Santa Rosa I., Skunk Pt.	nd	1.2	+	1.0	0.30	31
00-115	Fresh tar on sand	Santa Rosa I., Skunk Pt.	nd	2.0	+	0.79	0.21	59
00-116	Fresh tar on sand	Santa Rosa I., Soledad Pt.	nd	2.1	+	0.94	0.46	84
00-117	Tar on rock outcrop	Santa Rosa I., Soledad Pt.	nd	1.6	+	1.0	0.47	42
00-119	Fresh tar mat, rock outcrop	Santa Rosa I., Soledad Pt.	nd	1.5	+	0.95	0.16	36
00-128	Tar on terrace	Santa Rosa I., Lobo Canyon	nd	1.6	+	0.98	0.35	53
00-129	Fresh massive tar	Santa Rosa I., Sandy Pt., HTL	nd	1.4	+	0.97	0.19	44
00-131	Fresh tar on top of tar mat	Santa Rosa I., Sandy Pt.	nd	2.1	+	0.83	0.23	83
00-132	Tar mat under 00-131	Santa Rosa I., Sandy Pt., intertidal	nd	1.7	+	0.98	0.20	48
00-135	Weathered old tar above HTL	Santa Rosa I., Sandy Pt.	nd	1.5	+	0.88	0.11	43

00-137	Fresh tar on beach	Santa Rosa I., Bee Rock West	nd	1.8	+	0.87	0.23	52
00-147	Tar on beach	Santa Rosa I., Cluster Pt.	nd	1.9	+	1.0	0.58	59
Group C								
97-22	Weathered tar on rock	Windansea Beach, San Diego	−23.2	0.38	—	1.3	0.48	148
97-28	Weathered tar on granite	Monterey Peninsula, 17-mile Drive	−23.1	0.39	—	1.2	0.36	148
L99-2	Weathered tar on sand	Sunset Bch., near offshore platform	−23.4	0.39	—	1.2	0.42	160
L99-3	Weathered tar on sand	Sunset Bch., near offshore platform	−23.4	0.41	—	1.2	0.38	146
L99-4	Weathered tar on sand	Sunset Bch., near offshore platform	−23.3	0.43	—	1.2	0.38	174
L99-8	Fresh tar on sand/cobbles	Redondo Beach, S.	−23.3	0.56	tr	1.2	0.34	163
L99-9	Fresh tar on sand/cobbles	Redondo Beach, S.	−23.2	0.43	—	1.1	0.25	163
L99-10	Fresh tar splotch on rock	Palos Verdes Point	−23.3	0.55	tr	1.2	0.33	163
L99-11	Fresh tar on rock	Palos Verdes, Vicente Lighthouse	−23.5	0.40	—	1.2	0.36	158
L99-12	Fresh tar on outcrop	Palos Verdes, Whites Point	−23.3	0.38	—	1.2	0.41	175
L99-13	Fresh tar on cobbles	San Pedro, Point Fermin	−23.3	0.40	—	1.2	0.40	165
L99-28	Fresh tar on sand	Huntington Beach #2	−23.2	0.42	—	1.3	0.41	137
99-16	Fresh tar on rock	Santa Cruz I., Coches Prietos, E	−23.4	0.40	—	1.3	0.39	144
99-17	Fresh tar on rock	Santa Cruz I., Coches Prietos, E	−23.5	0.45	—	1.2	0.27	149
99-20	Weathered tar on rock	Santa Cruz I., Coches Prietos, W	−23.4	0.41	—	1.2	0.28	152
99-35	Weathered tar on rock	Santa Cruz I., Willows	−23.4	0.62	tr	1.1	0.31	131
Seep								
00-67	Fresh offshore seep	3 miles N. of Anacapa Isl.	−23.4	1.4	+	1.3	0.70	26

nd, not determined; −, not present; tr, trace only; nc, not calculated due to lack of requisite peaks. Column headings are defined in text under "Geochemical Parameters" section.

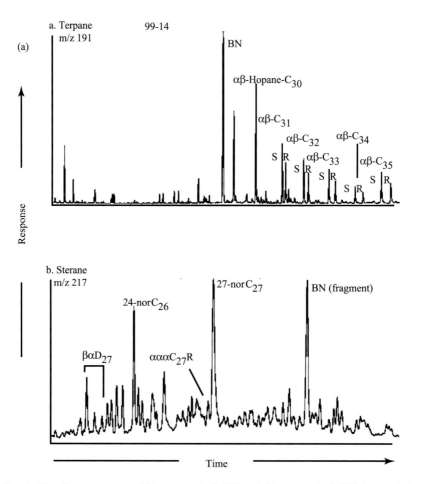

FIG. 4. Mass fragmentograms of (a) terpanes (m/z 191) and (b) steranes (m/z 217) in a typical tar residue from Group A. BN, bisnorhopane.

uncertain values (all <1) for the sterane ratios. The dominance of the 28,30-bisnorhopane is reflected in the high values of BI (1.6 ± 0.2). Trisnorhopane is present in all samples. RI values average 55 ± 9, and the $\delta^{13}C$ average of 22 samples is −23.4 ± 0.2‰ (Table 2).

Group B

Many parameter values for the 36 samples in Group B are similar to those for Group A. For example, trisnorhopane is present in all samples of both groups. Also dominant in Group B,

Table 2. Mean and standard deviation of geochemical parameters of coastal tar residues for Groups A, B, and C and for a natural oil seep

Group	$\delta^{13}C$	BI	T	$C_{28}R/C_{29}R$	Ster/Hop	PAH-RI
A ($n = 40$)	−23.4 ± 0.2 ($n = 22$)	1.6 ± 0.2	+	nc	0.06 ± 0.03	5 ± 9
B ($n = 36$)	−23.5 ± 0.2 ($n = 23$)	1.6 ± 0.3	+	1.0 ± 0.1	0.29 ± 0.12	4 ± 16
C ($n = 24$)	−23.3 ± 0.1 ($n = 16$)	0.44 ± 0.07	− or trace	1.2 ± 0.1	0.36 ± 0.06	153 ± 12
Natural seep ($n = 1$)	−23.4	1.4	+	1.3	0.70	26

Column headings are defined in text under "Geochemical Parameters" section.

as shown in the example terpane fragmentogram (Fig. 5a), is 28,30-bisnorhopane, and BI values of 1.6 ± 0.3 cannot be distinguished from those of Group A (1.6 ± 0.2). Likewise, PAH-RI values of 54 ± 16 for Group B samples overlap those for Group A (55 ± 9). The $\delta^{13}C$ values of −23.5 ± 0.2‰ ($n = 23$) are essentially the same as the values of −23.4 ± 0.2‰ ($n = 22$) for Group A. Differences between the two groups are evident, however, in the sterane ratios and the sterane/hopane ratios. The sterane profiles for Group B (example in Fig. 5b) are quite different from those of Group A. In Group B, the steranes consist mainly of regular C_{27}–C_{29}-ααα- and ββ-sterane stereoisomers. Frequently, the sterane profiles of Group B are dominated by the C_{27}-ααα-cholestane (20R) epimer.

Group C

This group of 16 samples can be distinguished from Group A and B samples by the distinctive values of most geochemical parameters except $\delta^{13}C$ ($− 23.3 ± 0.1‰$), which are essentially the same as in Groups A and B. For Group C samples, BI values (0.44 ± 0.07), sterane ratios (1.2 ± 0.1), sterane–hopane ratios (0.36 ± 0.06), and PAH-RI values (153 ± 12) are all very different from the values in Groups A and B (Table 2). Also Group C samples are characterized by the absence or trace levels of trisnorhopane. Although 28,30-bisnorhopane is present, it fails to dominate the terpane fragmentograms (example in Fig. 6a). Sterane patterns

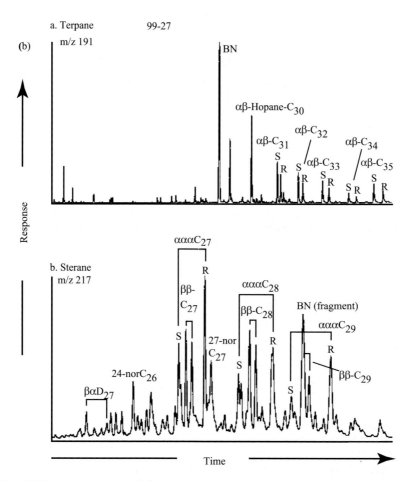

FIG. 5. Mass fragmentograms of (a) terpanes (m/z 191) and (b) steranes (m/z 217) in a typical tar residue from Group B. BN, bisnorhopane.

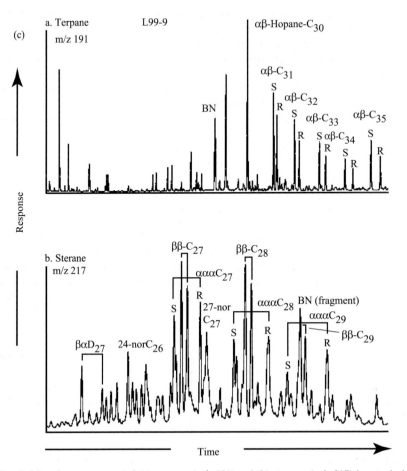

FIG. 6. Mass fragmentograms of (a) terpanes (m/z 191) and (b) steranes (m/z 217) in a typical tar residue from Group C. BN, bisnorhopane.

(example in Fig. 6b) are characterized by relatively high amounts of the C_{27}- and C_{28}-$\beta\beta$-steranes. These steranes are reported to be particularly prominent in Santa Maria basin and Los Angeles basin crude oil (CURIALE *et al.*, 1985).

Natural seep

The single sample (00-67) of a natural crude-oil seep, located offshore from Anacapa Island (Fig. 1), has geochemical parameters that closely match the Group B samples (Table 2). For example, the BI value is 1.4, sterane ratio is 1.3, hopane–sterane ratio is 0.70, and the PAH-RI value is 26. The presence of trisnorhopane in this sample is a characteristic of both Groups A and B. The $\delta^{13}C$ value of $-23.4‰$ matches samples in Groups A, B, and C.

DISCUSSION

Natural crude-oil seeps have been active onshore (HODGSON, 1980) and offshore in the California borderland throughout recorded history (WEAVER, 1969; YERKES *et al.*, 1969) and likely during the Holocene and Pleistocene. In this region modern oil seeps are concentrated in the Santa Barbara Channel (Fig. 2) and Santa Monica Bay (Fig. 3).

Accurate measurements of the rates of submarine oil seepage have been difficult to achieve because direct measurements must be made underwater, and flow rates are usually episodic and ephemeral. In general, measuring techniques have been primitive, and the results typically involved gross extrapolations. Early rate estimates made offshore from Coal Oil Point ranged from 520 to 8300 metric tons (tonnes)/year (ALLEN et al., 1970; WILKINSON, 1971; MIKOLAJ et al., 1972). For the same area, STRAUGHAN and ABBOTT (1971) estimated 4700 tonnes/year (when corrected for a printing error, pointed out in KVENVOLDEN and HARBAUGH (1983)). In Santa Monica Bay, rates of oil seepage range from 100 to 1000 tonnes/year (WILKINSON, 1971; MIKOLAJ et al., 1972). Estimates by FISCHER (1978) for the offshore California borderland, based on mapping of geographical trends of seeps in the Santa Barbara Channel, ranged from 2100 to 35,000 tonnes/year.

Recently, more accurate measurements were made in the Coal Oil Point area by HORNAFIUS et al. (1999). They combined information from two seep tents (huge funnel-like structures placed over natural seep fields), seep-flux buoys that drift across the seep area, and 50 kHz sonar mapping. They concluded that the rate of oil seepage at Coal Oil Point is between 7800 and 8900 tonnes/year. A study by the US Academy of Sciences (NAS, 2003) concluded that the rate of seepage for offshore California is about 20,000 tonnes annually. Although this estimate may seem large, it is still less than the earlier estimate by FISCHER (1978) of 35,000 tonnes/year. This natural crude-oil seepage enters the ocean and commonly rises to the surface of the water because of the low density of the oil and because the oil is often accompanied by natural gas (HORNAFIUS et al., 1999). A portion of the oil evaporates at the sea surface and the remainder commonly emulsifies to form tarballs. It is reasonable to expect that some of these tarballs find their way to the coastline. Thus, one source of coastal tar residues is undoubtedly natural oil seeps.

Geochemical studies of the natural oil seeps in water of the California borderland show the complexity and problems related to determining sources of seep oil and of correlating it with produced crude oil, sediment bitumen, and coastal tar residue or tarballs (REED and KAPLAN, 1977; STUERMER et al., 1982). HARTMAN and HAMMOND (1981) were the first to correlate tarballs with natural oil seeps. They showed that tarballs on beaches of Santa Monica Bay originated from seepage within the Bay but also from seeps as far away as Coal Oil Point (Fig. 1).

We have examined many tarballs from along the California coast and report here the geographic distribution of three geochemically distinct groups of tarballs. A common feature of all of these tarballs is their very similar carbon isotopic composition of about −23‰. The values are well within the range of carbon isotopic compositions of produced crude oil derived from the Miocene Monterey Formation of California (PETERS and MOLDOWAN, 1993). In addition, all of the tarballs contain 28,30-bisnorhopane, a compound typical of Monterey Formation crude oil (CURIALE et al., 1985). Thus, an isotope parameter and a biomarker parameter strongly indicate that the Monterey Formation is the ultimate source of the coastal tarballs. Support for this assertion is the observation that the relative enrichment of $\alpha\beta$-C_{35} homohopane in all tarball groups (see Figs. 4a, 5a, and 6a) is also found in Monterey Formation crude oil (PETERS and MOLDOWAN, 1993, p. 147). Natural oil seeps, so common in the California borderland, are probably the proximate source of much of the coastal tar. Anthropogenic spills (from tankers, pipelines, etc.) of Monterey Formation crude oil could be another proximate source of these tarballs; however, spills of Monterey Formation oil are now rare, although earlier oil spills may have contributed to the coastal tar inventory. There are no means at present to differentiate weathered natural seep oil from weathered spilled oil, having the same ultimate source.

The geographic distribution of the three geochemically distinct groups of tarballs is shown in Fig. 1. Group A tar residues occur as far south as Catalina Island and as far north as Point Reyes and on the shores of both Santa Cruz and Santa Rosa Islands. Group B tarballs have been found as far south as San Diego and occur elsewhere along the mainland coast and on the Channel Islands along with Group A samples. The natural seep sample from 3 miles north of Anacapa Island correlates best, although imperfectly, with Group B tarballs (Tables 1 and 2). Group C tarballs were found in large numbers on the beaches of Santa Monica Bay and to the south on beaches near Palos Verdes and San Diego and on the south side of Santa Cruz Island.

Because Group A and B tar residues seem to be concentrated at or near Frasier Point on Santa Cruz Island and because of the extensive tar mats (extensive coalesced collection of tarballs) on Frasier Point, it is likely that natural oil seeps are present in the nearby offshore. The natural seep offshore from Anacapa Island may be a source of Group B tarballs. Group C tarballs are concentrated on the shore of Santa Monica Bay, and their sources are likely to be known seeps that follow the submarine trace of the Palos Verdes fault (Fig. 3).

Although we have yet to obtain an exact geochemical correlation between tarballs and natural seeps, we believe that the circumstantial evidence points to oil seeps offshore from the Channel Islands and offshore in Santa Monica Bay as the sources of Group A, B, and C tarballs, respectively. If this conclusion is correct, then we have evidence that crude oils from these natural seeps have been transported as tar residues both north and south from their points of origin. HARTMAN and HAMMOND (1981) postulated that the southern California gyre transports seep oil as tar southward during the spring, summer, and fall seasons. During the winter season, the tar is transported northward due to the surfacing of the Davidson Current. Our results support these postulations. Group A, B, and C tarballs are found on mainland beaches both north and south of their regions of origin.

Thus, the early work by Kaplan and his associates (REED and KAPLAN, 1977) has initiated a scientific field which calls attention to the chemistry of marine petroleum seeps and illustrates the potential applications and limitations imposed by geochemical analyses. Our work demonstrates that biomarker chemistry can show subtle differences in coastal tarballs from the California borderland, which have essentially the same stable carbon isotopic compositions.

CONCLUSIONS

(1) The selected set of coastal tarballs from the California coast, particularly from the California borderland, including the beaches of Santa Cruz and Santa Rosa Islands, has very similar carbon isotopic composition (about $-23‰$), which strongly suggests their ultimate origin from the Miocene Monterey Formation.

(2) The occurrence of 28,30-bisnorhopane in all samples supports this conclusion about the ultimate origin of the coastal tarballs.

(3) Three groups of coastal tars are distinguished by their geochemical parameters. Group A and B tars can be separated from Group C tars by the dominance of 28,30-bisnorhopane and the presence of trisnorhopane in Groups A and B, and the lower relative amounts of bisnorhopane and the absence or trace amounts of trisnorhopane in Group C samples. Also, the PAH-RI values of Group C are distinct from those of Group A. Sterane–hopane ratios separate Group A from Groups B and C.

(4) The geographical distribution of the three groups of tarballs suggests that natural oil seeps near the Channel Islands are the proximate sources of Group A and B tar residues, whereas natural seeps in Santa Monica Bay are the proximate source of Group C tarballs.

(5) The location of samples from the three groups of tarballs on the continental coast of California suggests both southward and northward transport from the place of origin. Tar residues from Groups A and B are found as far away as Point Reyes, north of San Francisco.

Acknowledgements—We thank Tom Lorenson and Bob Rosenbauer (USGS) for their efforts to obtain and prepare samples of coastal tarballs and Jeanne Dileo and Bruce Rogers (USGS) for the preparation of the figures. We gratefully acknowledge the guidance on the Channel Islands provided by Dan Richards (National Park Service). We gratefully acknowledge the suggestions from our reviewers, K. E. Peters, L. B. Magoon, and G. E. Claypool.

REFERENCES

ALLEN A. A., SCHLUETER R. S. and MIKOLAJ P. G. (1970) Natural oil seepage at Coal Oil Point, Santa Barbara, California. *Science* **170**, 974–977.
CURIALE J. A., CAMERON D. and DAVIS D. V. (1985) Biological marker distribution and significance in oils and rocks of the Monterey Formation, California. *Geochim. Cosmochim. Acta* **49**, 271–288.
FISCHER P. J. (1978) *Natural gas and oil seeps, Santa Barbara Basin, California.* The State Land Commission 1977. California Gas, Oil, and Tar Seeps, pp. 1–62.
GRANTHAM P. J. and WAKEFIELD L. L. (1988) Variations in the sterane carbon number distribution of marine source derived crude oils through geologic time. *Org. Geochem.* **12**, 61–73.

HARTMAN B. and HAMMOND D. E. (1981) The use of carbon and sulfur isotopes as correlation parameters for the source identification of beach tar in the southern California borderland. *Geochim. Cosmochim. Acta* **45**, 309–318.

HODGSON S. R. (1980) *Onshore Oil & Gas Seeps in California.* Publication No. TR26. California Department of Conservation, Division of Oil and Gas, 97 pp.

HOLBA A. G., TEGELAAR E. W., HUIANINGA B. J., MOLDOWAN J. M., SINGLETARY M. S., MCCAFFREY M. A. and DZOU L. E. P. (1998) 24-Norcholestanes as age sensitive molecular fossils. *Geology* **26**, 783–786.

HORNAFIUS J. S., QUIGLEY D. and LUYENDYK B. P. (1999) The world's most spectacular marine hydrocarbon seeps (Coal Oil Point, Santa Barbara Channel, California): quantification of emissions. *J. Geophys. Res.* **104**(C9), 20703–20711.

HOSTETTLER F. D., ROSENBAUER R. J. and KVENVOLDEN K. A. (1999) PAH refractory index as a source discriminant of hydrocarbon input from crude oil and coal in Prince William Sound, Alaska. *Org. Geochem.* **10**, 873–879.

KVENVOLDEN K. A. and HARBAUGH J. W. (1983) Reassessment of the rates at which oil from natural sources enters the marine environment. *Mar. Environ. Res.* **10**, 223–243.

KVENVOLDEN K. A., ROSENBAUER R. J., HOSTETTLER F. D. and LORENSON T. D. (2000) Application of organic geochemistry to coastal tar residues from central California. *Int. Geol. Rev.* **42**, 1–14.

MACDONALD I. R., GUINASSO N. L. Jr., ACKLESON S. G., AMOS J. F., DUCKWORTH R., SASSEN R. and BROOKS J. M. (1993) Natural oil slicks in the Gulf of Mexico visible from space. *J. Geophys. Res.* **98**(C9), 16351–16364.

MIKOLAJ P. G., ALLEN A. A. and SCHLUETER R. S. (1972) Investigation of the nature, extent, and fate of natural oil seepage off southern California. *4th Offshore Technol. Conf.* **OTC 1549**, I367–I380.

MOLDOWAN J. M., LEE C. Y., WATT D. W., JEGANATHAN A., SLOUGUI N.-E. and GALLEGOS E. J. (1991) Analysis and occurrence of C_{26}-steranes in petroleum and source rocks. *Geochim. Cosmochim. Acta* **55**, 1065–1081.

NAS, US NATIONAL ACADEMY of SCIENCES, (2003) *Oil in the Sea III: Inputs, Fates, and Effects.* National Academy Press, Washington, DC.

PETERS K. E. and MOLDOWAN J. M. (1993) *The Biomarker Guide.* Prentice-Hall, Englewood Cliffs, NJ, 363 pp.

REED W. E. and KAPLAN I. R. (1977) The chemistry of marine petroleum seeps. *J. Geochem. Explor.* **7**, 255–293.

SCHOUTEN S., SINNINGHE-DAMSTE J. S., SCHOELL M. and DE LEEUW J. (1994) A novel sterane, 27-nor-24-methyl-5α-cholestane, in sediments. *Geochim. Cosmochim. Acta* **58**, 3741–3745.

STRAUGHAN D. and ABBOTT B. C. (1971) The Santa Barbara oil spill: ecological changes and natural oil leaks. In *Water Pollution by Oil* (ed. P. HEPPLE), pp. 257–262. Institute of Petroleum, London.

STUERMER D. H., SPIES R. B., DAVIS P. H., NG D. J., MORRIS C. J. and NEAL S. (1982) The hydrocarbons in the Isla Vista marine seep environment. *Mar. Chem.* **11**, 413–426.

WAPLES D. W. and MACHIHARA T. (1991) *Biomarkers for Geologists—A Practical Guide to the Application of Steranes and Triterpanes in Petroleum Geology.* American Association of Petroleum Geologists, Methods in Exploration No. 9, 91 pp.

WEAVER D. W. (1969) *Geology of the Northern Channel Islands.* Pacific Section AAPG and SEPM Special Publications, 200 pp.

WILKINSON E. R. (1971) California offshore oil and gas seeps. *Calif. Oil Fields—Summary Oper.* **57**(1), 5–28.

WILSON R. D., MONAGHAN P. H., OSANIK A., PRICE L. C. and ROGERS M. A. (1974) Natural marine oil seepage. *Science* **184**, 857–865.

YERKES R. F., WAGNER H. C. and YENNE K. A. (1969) *Petroleum development in the region of the Santa Barbara Channel.* Geology, Petroleum Development, and Seismicity of the Santa Barbara Channel Region, California. US Geological Survey Prof. Paper, 679, 13-2.

Geochemical Investigations in Earth and Space Science: A Tribute to Isaac R. Kaplan
© The Geochemical Society, Publication No. 9, 2004
Editors: R.J. Hill, J. Leventhal, Z. Aizenshtat, M.J. Baedecker, G. Claypool,
R. Eganhouse, M. Goldhaber and K. Peters

Behavior of oxy-anions of As, Se, and Mo in full-scale wastewater treatment plants

JEROME O. NRIAGU[1], SHIN JOH KANG[2], JAMES R. MURIN[2] and XIA-QIN WANG[1]

[1]Department of Environmental Health Sciences, School of Public Health, University of Michigan,
109 S. Observatory Street, Ann Arbor, MI 48109, USA
[2]Tetra Tech MPS, 710 Avis Drive, Pittsfield Twp, MI 48108, USA

Abstract—Mass balance calculations and analysis of the redox species in the aqueous and solid phases are used to assess the behavior and removal rates of the oxy-anions of arsenic, selenium, and molybdenum during primary sedimentation, activated sludge, and anaerobic digestion processes of three full-scale wastewater treatment plants (WWTPs). The dominant form of the oxy-anion found in any treatment system was, to a large extent, controlled by slow oxidation rates and interactions at solid–water interfaces. The removal of the oxy-anions of the three elements was limited by the form in which they were loaded into the WWTPs and by internal recycling processes. Under anaerobic conditions, the formation of stable sulfide, organic and carbonate (with arsenic only) complexes resulted in significant release of the elements from the solid phase back into solution. Formation of volatile compounds resulted in the loss of up to 50% of the selenium loading into one facility, suggesting that WWTPs may be a significant source of selenium emitted to the atmosphere.

INTRODUCTION

THE ENVIRONMENTAL BEHAVIOR and fate of heavy metals in sludge-amended soils are mediated and, to a degree, controlled by sewage treatment. Although there have been numerous studies on the principal processes that control the speciation and efficiency of removal of metals in wastewater treatment plants (WWTPs) (BROWN et al., 1973; STERRIT et al., 1981; GOLDSTONE et al., 1990), studies on the behavior of oxy-anions in various treatment systems have been very limited. The oxy-anions of most concern in wastewater treatment programs are those formed by arsenic, molybdenum, selenium, and vanadium. A common feature of the four elements that differentiates them from other metals is that their primary occurrence as oxy-anions means that a change in redox condition can drastically alter the form of the element, and hence its behavior in the treatment system and amended soils.

Oxy-anions are common contaminants in industrial and municipal wastewaters. The ubiquity of arsenic in wastewater, for instance, is related to the large number of uses from which arsenic-rich wastes can be derived: agricultural (as pesticides, insecticides, defoliants, wood preservatives, debarking trees, soil sterilant), livestock (in feed additives, disease prevention, cattle and sheep dips, algaecides), electronics (in solar cells, optoelectronic devices, semiconductors, light emitting diodes as in electronic watches), industrial (in glassware, electrophotography, catalysts, pyrotechnics, antifouling paints, dyes and soaps, ceramics, pharmaceuticals), and metallurgical (alloys, battery plates, locomotive fireboxes, heat exchangers). The chemical behavior of As (and to a limited extent, V) is similar to that of phosphorus, and it has been shown that household washing products (especially phosphate detergents) account for a large fraction of the As (73%) and V in domestic wastewater (JENKINS and RUSSELL, 1994). Drinking water supplies in many parts of the world contain arsenic; one survey, for instance, found that 31% of U.S. groundwater supplies contain over 1.0 μg/l As, 18% above 2 μg/l, 7% above 5 μg/l, and 4% above 10 μg/l (REID, 1994). The recent lowering of As levels in drinking water of the United States (US EPA, 2003) is expected to result in a substantial reduction in loading of this element to WWTPs, and would likely impact the required treatment level for contaminated wastewaters and effluents as well.

The other oxy-anions are also added to industrial and municipal wastewater from a wide variety of sources. Molybdenum concentrations in wastewaters vary widely (from about

1.0 to 2000 μg/l), for instance (ADRIANO, 2001). Because molybdenum is not very toxic to humans, it is increasingly being used to replace chromium in many industrial processes. Molybdenum compounds are used as catalysts, corrosion inhibitors, lubricant additives for the automotive and other industries, and as ingredients in engine oils, grease, dyes, plastic and rubber parts (ADRIANO, 2001). It is used in the production of specific steels, as well as electrical contacts and filaments. In addition, Mo compounds are used in fertilizers for agriculture, in cooling waters and in a number of household products.

Se concentrations in municipal effluents vary widely, from 1.0 to over 1000 μg/l (CAPON, 1991; KAPOOR *et al.*, 1995; ADRIANO, 2001). Extensive use of Se is found in the electronics industries for the manufacture of photocells, rectifiers, photometers, and xerographic equipment. It is used as a pigment in the glass industry, as a sulfur supplement in the rubber industry, as a pigment in plastic and ceramics, and for coating stainless steel and copper. Selenium is an antioxidant found sometimes in mineral and vegetable oils, lubricants and inks and is used in the manufacture of dandruff shampoo (ADRIANO, 2001).

Vanadium has numerous industrial uses such as the processing of steel, chemical production, polymer synthesis, ceramics and electronics production (NRIAGU, 1997). As a catalyst, V is used in the contact process for the production of sulfuric acid and a variety of organic reagents. Vanadium compounds are also used as mordants in dyeing and printing of cotton and silk (NRIAGU, 1997). Because of the multiplicity of sources, the concentrations of vanadium in municipal wastewaters vary widely, from < 1.0 to over 3000 μg/l (BLACKMORE *et al.*, 1996; ADRIANO, 2001).

Limited information and data are available on the concentrations and removal of oxy-anions from wastewater. Most of the available data have been derived from laboratory experiments, pilot-plant studies, and drinking water treatment plants (FRANK and CLIFFORD, 1986; HERING and CHU, 1998; KAPOOR *et al.*, 1995), but the removal from wastewater has not been thoroughly investigated in full-scale treatment plants. More importantly, the effectiveness of various treatment technologies in removing these elements at concentrations of 1.0–15 μg/l (found in many wastewaters) has not been addressed in a critical manner (NURDOGAN *et al.*, 1994; RAO and VIRARAGHAVAN, 1992; PENG and DI, 1994; WAYSAY *et al.*, 1996; YOSHIZAKI and TOMIDA, 2000). An objective of this study is to assess the chemical behavior and transport of oxy-anions in selected full-scale wastewater treatment systems, with emphasis on primary sedimentation, activated sludge, and anaerobic digestion processes. It is currently a common practice to base the removal process on the total metal concentrations without regard to the oxidation state or actual species present (HERING and CHU, 1998). It is not surprising that most wastewater facilities which measure the oxy-anions report highly variable and inadequate removals in their treatment processes (KAPOOR *et al.*, 1995).

The overarching goal of the project on which this paper is based is to improve our understanding of chemical behavior of the oxy-anions of selected metals/metalloids (Se, As, and Mo) in full-scale WWTPs. A significant improvement in the removal efficiency of these redox sensitive pollutants from wastewaters can be expected with a better understanding of their speciation and behavior during various treatment processes. The main focus of this paper is on the speciation, fate and transport of the oxy-anions through primary sedimentation, activated sludge and anaerobic digestion processes in three municipal WWTPs.

Many municipalities in the U.S. are faced with the problem of disposing of increasing amounts of wastewater sludge using environmentally and economically sound methods that carry minimum risk to humans and the ecosystem (LAKE *et al.*, 1984, 1989). Despite the proven fertilizer and soil conditioning value of sludge, the presence of considerable amounts of toxic trace elements has remained a barrier to land application of sludge. Of the trace elements in sludge (CHEN *et al.*, 1974; NEVESSI *et al.*, 1988), redox sensitive ones, notably As, Se, Mo, and V are particularly worrisome because they can be converted into the more toxic and more mobile forms once released into the environment. Although a large volume of information exists on Cd, Cu, Mo, Ni, Zn, and Pb in sludge, little has been published on As, Se, Mo, and V in sludge. Improved understanding of the behavior of oxy-anions in WWTPs should contribute to better assessment of the risks associated with using municipal sludge as a resource, and is the collateral goal of this report.

METHODOLOGY

Sample collection

The WWTPs selected for this study had high concentrations of arsenic (Facility K), selenium (Facility R), and molybdenum (Facility J) and operated a wide range of treatment processes (Table 1).

Sample bottles (500 ml capacity) were decontaminated using the procedure described by NRIAGU *et al.* (1993). The 24-h composite samples were collected from several process locations (including primary influent, primary effluent, primary sludge, secondary effluent, waste activated sludge, anaerobic digester feed, anaerobic digester supernatant, and anaerobic digester sludge) at each of the three participating WWTPs. The same set of samples was collected from each facility on 3–5 separate days during November and December 1998, and three additional sets of samples at the same process locations were obtained between July and October 1999. For mass balance development, composite samples were collected daily for three consecutive days. Equal aliquots of the daily samples from a particular process were combined to get a three-day composite sample.

Aliquots of the samples were filtered immediately after collection through 0.45 μm Millipore polycarbonate membrane filters and both the filtered and unfiltered samples were kept refrigerated and shipped within 24 h after collection in a cooler box to the University of Michigan in Ann Arbor for analysis. No preservative was added to the samples after collection. Each facility maintained its own monitoring program on the treatment processes which provided essential information on the background chemistry.

Speciation of As, Se, and Mo in wastewater samples

The analytical goal of this project was to separate each element into four fractions: total in sludge (after digestion of untreated sample in a microwave system), total dissolved (after filtering through 0.45 μm membrane filter), reduced oxy-anion species (dissolved phase), and oxidized oxy-anion species (dissolved phase). The first two fractions were easily quantified, but the measurement of the Se, Mo, and V redox species presented some challenges as described below.

Total elemental concentrations–Total concentration of each metal in an unfiltered sample was determined after digestion. The unfiltered sample was carefully mixed, and a known amount (approximately 10 g) was weighed into the Teflon bomb of the microwave digestion system. Five milliliters of concentrated ($\sim$ 36N) trace-metal grade nitric acid were added to each sample and the mixture was digested at high temperature and pressure in a microwave digestion unit

Table 1. Summary of participating WWTP processes

Characteristic feature	Facility K	Facility J	Facility R
Significant metals observed	As ($>$10 μg/l)	Mo ($>$40 μg/l)	Se ($>$50 μg/l)
Source of metals	Groundwater and domestic uses	Industrial and domestic uses	Groundwater and domestic uses
Design flow, m^3/day (10^6 gal/day)	287,700 (76)	8330 (2.2)	71,900 (19)
Primary treatment	Clarifiers	Clarifiers	Clarifiers
Secondary treatment process	Activated sludge	Trickling filter– activated sludge	Trickling filter– activated sludge
Tertiary treatment process	–	Pressure sand filtration	–
Phosphorus removal	–	Ferric chloride	–
Nitrification	Biological and denitrification	Biological	–
Sludge conditioning	Anaerobic digestion	Anaerobic digestion	Anaerobic digestion
Sludge dewatering/thickening	Centrifuge	Drying beds	Centrifuge
Sludge disposal	Land application	Land disposal	Land application

(SPEC 7000) following the procedure given in the instruction manual for the instrument. After cooling, the container was opened and the sample was filtered through 0.45 μm Millipore polycarbonate membrane filter. The filtrate was stored in a refrigerator until analyzed.

For the total dissolved metal fraction, the wastewater sample was filtered through a 0.45 μm membrane filter immediately after collection, acidified with trace-metal grade nitric acid to a pH of approximately 2.0 and then stored.

Speciation of the elements in the aqueous phase – For the separation of arsenate from arsenite in solution, the traditional ion-exchange method was used (FRANK and CLIFFORD, 1986; KIM, 1999; KIM and NRIAGU, 2000). A strong anion-exchange resin (AG1-X8, 100–200 mesh, chloride form) was slurry packed into polyethylene columns (20 cm $\times$ 8 mm) and kept moist with Milli-Q (MQ) water. A 50 ml aliquot of the filtered sample was acidified with HCl to a pH of 4–5 and then passed through the column at a rate of 2–3 ml/min (details provided in KIM (1999)). Any As(V) present was retained in the column, and was subsequently eluted using 30 ml of 0.1 M HCl. The uncharged As(III) species is not retained by the resin, and thus remained in the effluent solution. The resins separated out charged species, whether organic or inorganic. Thus, what is designated As(V) in this report may include other negatively charged inorganic species (such as thio-arsenic and carbonato-arsenic complexes) and organo-arsenic compounds (KIM *et al.*, 1999). What is not retained by resin has been called As(III), a fraction that may include neutral and positively charged organic and inorganic As species (KIM *et al.*, 1999). In most cases, inorganic oxy-anions are expected to be the dominant form of As in the samples (see LAKE *et al.*, 1984) and the speciation method used should provide a reasonable estimate of the concentrations of As(III) and As(V) present in the samples.

A modification of the method described in CLESCERI *et al.* (1998) was used to quantify the concentrations of selenite (Se(IV)) in the samples. An aliquot of the filtered sample was reacted with sodium borohydride to distill off any Se(IV) as selenium hydride. The concentration of selenium remaining in solution after the reaction was measured. The difference between original concentration (total dissolved) Se and residual Se in solution is taken as the value for Se(IV). Because of the high concentrations of organo-selenium compounds in the samples, attempts to separate Se(VI) and Se(IV) in the sample by this method (see CLESCERI *et al.*, 1998) were not successful.

Molybdenum forms stable oxy-anions in which it may have oxidation states of 4+ (molybdite) and 6+ (molybdate) (BAES and MESMER, 1976; EARY *et al.*, 1990). It also forms complex polymeric species (including $[Mo_2O_2]^{2-}$; $[Mo_7O_{24}]^{6-}$; $[Mo_8O_{26}]^{4-}$; $[Mo_{12}O_{37}]^{2-}$) as well as stable complexes with organic ligands in wastewater. The speciation of Mo in wastewater clearly requires innovative developmental work that was outside the scope of this study. We limited our effort to measurement of total molybdenum and molybdate ion (MO_4^{2-}) in the samples, with the latter determined using ion chromatography (see instrumental method below).

Partitioning of the elements in solid sludge – An adaptation of the method by TESSIER *et al.* (1979) was used to partition the As, Se, and Mo in sludge samples into the following five fractions: (a) exchangeable, (b) carbonate-bound, (c) oxide, (d) organic-bound, and (e) residual. The fractionation experiments were done using wet samples. Each sample was homogenized and centrifuged for 30 min to separate the solid from the liquid phases. The solid phase was successively processed through the following steps.

(I)   *Exchangeable fraction*: 8 ml 1.0 M Mg(NO$_3$)$_2$ was added to a known weight (10–12 g) of the wet solid and shaken for 2 h at room temperature. The mixture was centrifuged for 30 min and the supernatant was collected and saved for analysis. The residue was washed with 8 ml of MQ-water, centrifuged for 30 min, and the supernatant was discarded.

(II) *Carbonate-associated fraction*: 8 ml of 1.0 M NaOAc (sodium acetate) was added to the residue from Step 1 and shaken for 5 h at room temperature. After centrifuging the mixture for 30 min, the supernatant was collected for analysis. The residue was washed with 8 ml MQ-water, centrifuged for 30 min, and the supernatant was discarded.

(III) *Oxide-bound fraction*: 20 ml 0.04 M NH$_2$OH·HCl (ammonium hydroxylamine) in 25% H$_2$O$_2$ was added to the residue from Step 2 and the mixture was maintained at 85°C for 2 h with occasional shaking. The mixture was centrifuged (30 min), the supernatant was collected and the residue washed with MQ-water (8 ml).

(IV) *Organic-bound fraction*: The residue from Step 3 was reacted with 3 ml 0.02 M HNO_3 and 5 ml 30% H_2O_2 (hydrogen peroxide) at 85°C for 2 h and shaken occasionally. After 2 h, 3 ml of 30% H_2O_2 was added to the mixture which was maintained at 25°C for three more hours with occasional shaking. After cooling, 5 ml of 3.2 M NH_4Ac (ammonium acetate) in 20% HNO_3 was added to the mixture and shaken for 0.5 h at room temperature. The mixture was diluted to 20 ml, centrifuged for 30 min, and the supernatant removed for analysis.

(V) *Residual fraction*: Total concentration of each element was determined after digesting the sample in a microwave oven as described earlier. The residual fraction was obtained as the difference between total concentration and sum of the concentrations of the other four fractions.

Instrumental analysis

Elemental concentrations in various solutions were measured using a Graphite Furnace Atomic Absorption Spectrometer (GFAAS) equipped with L'vov platform (Perkin Elmer, 4100ZL). Electrodeless discharge lamps were used in the analysis of Se and As to enhance the instrumental sensitivity. The concentration of Mo(VI) in wastewater samples (mainly from Facility J) was determined using an ion chromatograph (Alltech Odyssey high performance ion chromatograph equipped with pre-concentrator column and a conductivity detector) equipped with a pre-concentrator column to improve the detection limit for the instrumental method.

The quality assurance program (QAP) for the experiments and analysis of arsenic in the samples followed closely those developed in our lab for the study of metals at ultra-trace levels. Highest quality reagents and chemicals were used throughout the study. Great care was taken to minimize the risk of contamination during all experiments. Although experiments with As(III) were done in a nitrogen atmosphere, the final solutions were analyzed for both As(III) and As(V) to ascertain the degree of oxidation of arsenite during each experiment. A careful blanking scheme was performed to estimate the level of arsenic contamination from every facet of the experiment and during sample analysis. Blanks were also used as diagnostic tools and to generate method detection limits.

The following batch analysis was implemented in the QAP to assess experimental and/or instrumental performance, recovery, precision, and accuracy.

QAP step	Frequency
Instrumental analysis	
Blank levels	Before each experimental batch
Stability	Before each experimental batch
Sensitivity	Before each experimental batch
Resolution	Before each experimental batch
Interference check	Once per week
Blank levels during run	
Calibration blank	Once per experimental batch
Memory check	One per experimental batch
Recovery	
Lab analyte spike, blank matrix	One per experimental batch
Internal standard	One per experimental batch
Precision	
Lab sample replicates (within batch)	One set per experimental batch
Lab sample replicates (different batches) 20%	
Accuracy	
Standard reference material	One per experimental batch

From the QAP implemented for the study (including use of certified reference material), the confidence limit on the reported values is estimated to be $\pm 10\%$ for total element concentrations and $\pm 25\%$ for individual species concentrations.

RESULTS

In a preliminary assessment of the concentrations of oxy-anions of many elements in wastewaters in the U.S. facilities, nine plants (mostly located in Michigan) were asked to submit samples for analysis. Total dissolved concentrations (in filtered samples) of As, Se, V, and Mo in process waters from these WWTPs were determined and the results are given in Table 2. Total dissolved arsenic (TDAs) concentrations varied from <0.5 μg/l (the method detection limit) to 5.9 μg/l and are similar to the values found in municipal water supplies in other parts of the country (ADRIANO, 2001). Selenium concentrations were generally low, with most samples containing less than 1.0 μg/l and the highest concentration observed being 6.6 μg/l. These concentrations are lower than those usually reported in the literature (see ADRIANO, 2001); recent measurements using clean lab procedures show that the results of previous Se measurements by "standard methods" were positively biased. Few previous measurements of V concentrations in wastewater treatment systems in this country have been reported in the literature. The screening level results for the 9 WWTPs ranged from 0.7 to 4.6 μg/l, with no treatment plant revealing unduly high values. Molybdenum concentrations were higher than those of the other three oxy-anions with three of the treatment plants showing concentrations above 20 μg/l. The results revealed the presence of oxy-anions of many elements in most wastewaters but the sample size was too small and too regionalized to form an opinion on the extent of contamination of the wastewater treatment systems in the country with elements that readily form oxy-anions.

After subsequent consultations with the Water Environment Research Foundation (Alexandria, VA) and sampling of several additional facilities, three WWTPs were identified with high arsenic (Facility K), selenium (Facility R), and molybdenum (Facility J) (see Table 1). The rest of this paper deals with results obtained from these three WWTPs.

The physico-chemical characteristics and flow relationships for Facility K are shown in Fig. 1. The mass of total As entering Facility K treatment plant was approximately 2.1 kg/day, while the

Table 2. Concentrations of oxy-anion forming elements in wastewater treatment plants

Facility	Sampling point	Total dissolved concentration (μg/l)			
		As	Mo	Se	V
A	Raw influent	5.9	12	<0.5	1.1
	Final effluent	4.5	10	0.73	1.3
B	Raw influent	0.8	5.9	<0.05	3.9
	Primary effluent	1.6	6.2	<0.5	1.2
	Final effluent	<0.5	7.5	<0.5	1.3
C	Raw influent	<0.5	8.4	<0.5	0.66
	Primary effluent	1.2	8.1	0.62	6
	Final effluent	2.2	8.3	<0.5	4.6
D	Raw influent	1.6	5.9	<0.5	2.4
	Primary effluent	2.3	9.6	1.5	1.4
	Final effluent	<0.5	11	0.74	1.4
E	Raw influent	<0.5	22	6.6	1.3
	Primary effluent	2.5	23	4.4	4.1
	Final effluent	1.6	25	4.9	3.6
F	Raw influent	2.2	12	<0.5	1.1
	Primary effluent	1.1	10	<0.5	1.3
	Final effluent	1.7	13	0.93	1.5
G	Raw influent	1.6	3.8	0.71	1.5
	Primary effluent	2.2	3	2.5	0.92
	Final effluent	1.4	2.8	1.1	2.9
H	Raw influent	2.3	24	1.2	2.8
	Primary effluent	1.6	27	1.1	2.9
	Final effluent	1.3	29	<0.5	1.8
J	Raw influent	1.2	30	1.3	3.4
	Primary effluent	2.2	22	1.6	1.6
	Final effluent	1.7	22	1.6	0.93

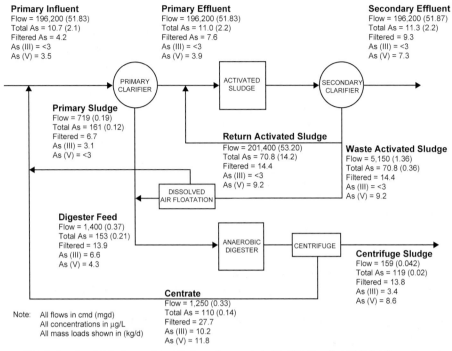

Primary Influent
Flow = 196,200 (51.83)
Total As = 10.7 (2.1)
Filtered As = 4.2
As (III) = <3
As (V) = 3.5

Primary Effluent
Flow = 196,200 (51.83)
Total As = 11.0 (2.2)
Filtered As = 7.6
As (III) = <3
As (V) = 3.9

Secondary Effluent
Flow = 196,200 (51.87)
Total As = 11.3 (2.2)
Filtered = 9.3
As (III) = <3
As (V) = 7.3

PRIMARY CLARIFIER

ACTIVATED SLUDGE

SECONDARY CLARIFIER

Primary Sludge
Flow = 719 (0.19)
Total As = 161 (0.12)
Filtered = 6.7
As (III) = 3.1
As (V) = <3

Return Activated Sludge
Flow = 201,400 (53.20)
Total As = 70.8 (14.2)
Filtered = 14.4
As (III) = <3
As (V) = 9.2

Waste Activated Sludge
Flow = 5,150 (1.36)
Total As = 70.8 (0.36)
Filtered = 14.4
As (III) = <3
As (V) = 9.2

DISSOLVED AIR FLOATATION

Digester Feed
Flow = 1,400 (0.37)
Total As = 153 (0.21)
Filtered = 13.9
As (III) = 6.6
As (V) = 4.3

ANAEROBIC DIGESTER

CENTRIFUGE

Centrifuge Sludge
Flow = 159 (0.042)
Total As = 119 (0.02)
Filtered = 13.8
As (III) = 3.4
As (V) = 8.6

Centrate
Flow = 1,250 (0.33)
Total As = 110 (0.14)
Filtered = 27.7
As (III) = 10.2
As (V) = 11.8

Note: All flows in cmd (mgd)
All concentrations in µg/L
All mass loads shown in (kg/d)

FIG. 1. Species distribution and mass balance for arsenic in Facility K. Effluent/sludge flow rates are in m^3/day (10^6 gal/day in brackets); concentrations shown are in µg/l for dissolved species (filtered solution) and µg/g for sludge; the corresponding mass loading in kg/day is given in parentheses.

outflow from secondary effluent was 2.2 kg/day. The overall mass balance therefore suggested that little or no arsenic was removed during the treatment processes. Mass balances around the primary clarifier, the secondary system, and the anaerobic digester all show a similar lack of significant removal of arsenic during this limited monitoring period (Fig. 1). The flow of arsenic into the anaerobic digester (0.21 kg/day) and the outflow via the centrifuged sludge (0.02 kg/day) suggest that digestion and dewatering served to recycle the arsenic within the system. Over 60% of the As was apparently returned to the influent wastewater via the centrate stream, consistent with the observation that total As concentration in the digester sludge (119 µg/l) was less than in the primary sludge (161 µg/l) and anaerobic digester feed (153 µg/l).

Average TDAs concentrations in Facility K were constant (about 11 µg/l) in primary influent, primary effluent, and secondary effluent (Table 3), further indicating little or no arsenic removal in the system. However, changes in TDAs concentrations in primary sludge (161 µg/g), waste activated sludge (71 µg/g), and anaerobic digester sludge (119 µg/g) suggest that there were significant differences in the behavior of arsenic in the treatment processes. The high concentration in the primary sludge suggests that a large fraction of arsenic being loaded into the plant was in particulate form. The (TDAs)/(total arsenic) ratio increased from about 0.39 in primary influent to 0.69 in primary effluent and 0.82 in the secondary effluent; indicating increasing conversion of influent soluble As to particulate As through the system. The low (TDAs)/(total arsenic) ratio of 0.25 in anaerobic digester supernatant may be related to the fact that the centrifuging system was not particularly effective in removing fine-grained particles from the waste stream during the time of the study. The As(V) concentration was higher in the secondary effluent (7.3 µg/l) compared to the primary effluent (3.5 µg/l), suggesting partial oxidation of the particulate arsenic in the aerated environment.

The physico-chemical characteristics of wastewater samples from Facility R are shown in Table 4 and the flow relationships for selenium are depicted in Fig. 2. The overall mass balance for total Se in the system did not show complete closure during the limited sampling period.

Table 3. Physico-chemical characteristics of wastewater treatment plant (Facility K) with elevated arsenic levels

Parameter	Primary influent[a]	Primary effluent[a]	Primary sludge[a]	Secondary effluent[a]	Waste activated sludge[a]	Anaerobic digester feed[a]	Anaerobic digester sludge	Digester Centrate	Centrifuge sludge
Total As (µg/l)	10.7 (3.7)	11.0 (3.1)	162 (64)	11.3 (0.5)	70.8 (35)	153 (44)	119 (13)	110 (13)	
Total dissolved As (µg/l)	4.2 (3.7)	7.6 (3.4)	6.7 (2.8)	9.3 (1.5)	14.4 (4.3)	13.9 (5.4)	13.8 (7.5)	27.7 (7.6)	
As(III) (µg/l)	<3.0	<3.0	3.1 (2.2)	<3.0	<3.0	6.6 (6.8)	3.4 (1.9)	10.2 (7.7)	
As(V) (µg/l)	3.5 (2.8)	3.9 (1.6)	<3.0	7.3 (1.0)	9.2 (5.0)	4.3 (2.9)	8.6 (9.5)	11.8 (1.8)	
TSS[b] (mg/l)	254 (75)	93.4 (14)	7.4 (1.1)	6.0 (0.5)	0.67 (0.1)	5.0 (0.9)	12.6 (9.7)	0.13 (0.1)	18.9 (1.4)
pH	7.36 (0.5)			7.06 (0.5)					
Flow	203,060 (13,520)	203,060 (13,520)	743 (77)	203,060 (13,520)	4460 (1106)	1386 (170)	141 (30)	2733 (58)	158 (41)
Alkalinity (mg/l CaCO$_3$)	256 (36)	248 (16)	1406 (750)	136 (15)	182 (51)	1499 (825)	3260 (456)		
Temperature (°C)	23.6 (3.4)	23.6 (3.4)		26.7 (2.1)					
Total sulfide (mg/l)	1.5 (0.69)	1.4 (0.87)	46 (32.5)	<1.0	1.3 (1.0)	25 (6.4)	9.0 (7.0)	<2.0	
Dissolved sulfide (mg/l)	<1.0	<1.0	10 (14.7)	<1.0	<1.0	16.6 (29)	7.8 (5.2)	<1.6	<1.0
DOC[b] (mg/l)	36.8 (12.5)	58.6 (17)	465 (203)	6.9 (4.4)	17.2 (3.3)	525 (106)	85 (49)	97 (91)	
Total iron (mg/l)						1060 (69)			
Ferrous iron (mg/l)							8733		8733 (1201)

[a] The standard error (±1 standard deviation) is shown in brackets.
[b] TSS, total suspended solids; DOC, dissolved organic carbon.

Table 4. Physico-chemical characteristics of wastewater treatment plant (Facility R) with elevated selenium levels

Parameter	Primary influent[a]	Primary effluent[a]	Primary sludge[a]	Secondary effluent[a]	Waste activated sludge[a]	Anaerobic digester feed[a]	Centrifuge supernatant	Centrifuge sludge
Total Se (μg/l)	46.5 (9.6)	35.0 (14)	215 (198)	24.5 (7.9)	646 (512)	3179 (4157)	951 (1408)	3697 (4184)
Total dissolved Se (μg/l)	16.8 (7.6)	14.1 (8.2)	11.3 (14)	11.7 (3.6)	32.8 (52)	77.3 (33)	36.9 (15.4)	40.6 (25)
Se(IV) (μg/l)	<3.0	<3.0	5.2 (5.4)	<3.0	14.9 (30)	40.6 (31)	16.2 (10)	15.2 (18)
Total solid (%)					0.66 (0.1)	5.3 (0.6)		2.4 (0.2)
TSS[b] (mg/l)	290 (42)	61.7 (7.3)		7.0 (3.4)	5221 (683)		2388 (2038)	
pH	7.12			7.61		5.95	7.35	7.35
Flow (m^3/day)	51,083 (4288)	50,947 (4204)	2519 (59.6)	50,914 (5558)	597 (110)	132 (5.7)	49.0 (62)	7.1 (8.9)
Alkalinity (mg/l CaCO^3)	305 (44)	280 (29)	383 (58)	240 (14.1)	475 (164)	2514 (986)	4771 (718)	5100 (1020)
Temperature (°C)	-6.7 (1.5)			-6.5 (1.9)				38.4 (1.7)
Eh (mV)			-292 (43)	180 (43)	-225 (28)	-311 (12)	-259 (92)	-334 (44)
Ammonia-N (mg/l)			67.2 (31)		45.6		868 (111)	
Dissolved P (mg/l)			20.0 (2.8)		16		33.0 (5.5)	
Total sulfide (mg/l)	<1.0	<1.0	<1.0	<1.0	<1.0	6.5 (5.0)	23.0 (48)	16.3 (23)
Dissolved sulfide (mg/l)	<1.0	<1.0	3.2 (1.2)	<1.0	2.4 (3.0)	<10	5.4 (5.0)	5.8 (2.0)
DOC[b] (mg/l)	38.5 (2.4)	37.0 (11)	88.3 (28)	14.0 (1.2)	55.0 (44)	620 (339)	168 (121)	245 (78)
Total iron (mg/l)						577 (85)	227 (169)	577 (38)

[a] The standard error (±1 standard deviation) is shown in brackets.
[b] TSS, total suspended solids; DOC, dissolved organic carbon.

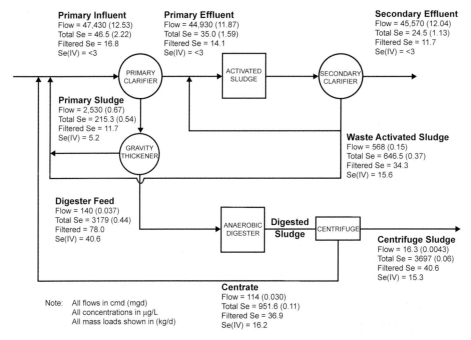

Primary Influent
Flow = 47,430 (12.53)
Total Se = 46.5 (2.22)
Filtered Se = 16.8
Se(IV) = <3

Primary Effluent
Flow = 44,930 (11.87)
Total Se = 35.0 (1.59)
Filtered Se = 14.1
Se(IV) = <3

Secondary Effluent
Flow = 45,570 (12.04)
Total Se = 24.5 (1.13)
Filtered Se = 11.7
Se(IV) = <3

PRIMARY CLARIFIER

ACTIVATED SLUDGE

SECONDARY CLARIFIER

Primary Sludge
Flow = 2,530 (0.67)
Total Se = 215.3 (0.54)
Filtered Se = 11.7
Se(IV) = 5.2

GRAVITY THICKENER

Waste Activated Sludge
Flow = 568 (0.15)
Total Se = 646.5 (0.37)
Filtered Se = 34.3
Se(IV) = 15.6

Digester Feed
Flow = 140 (0.037)
Total Se = 3179 (0.44)
Filtered = 78.0
Se(IV) = 40.6

ANAEROBIC DIGESTER

Digested Sludge

CENTRIFUGE

Centrifuge Sludge
Flow = 16.3 (0.0043)
Total Se = 3697 (0.06)
Filtered Se = 40.6
Se(IV) = 15.3

Centrate
Flow = 114 (0.030)
Total Se = 951.6 (0.11)
Filtered Se = 36.9
Se(IV) = 16.2

Note: All flows in cmd (mgd)
All concentrations in µg/L
All mass loads shown in (kg/d)

FIG. 2. Mass balance and species distribution for selenium in Facility R. Effluent/sludge flow rates are in m^3/day (10^6 gal/day in brackets). Concentrations shown are in µg/l for dissolved species (filtered solution) and µg/g for sludge; the corresponding mass loading in kg/day is given in parentheses.

The Se flow into the plant averaged 1.63 kg/day, while the outflow in the combined secondary effluent and sludge averaged 1.18 kg/day, implying that 28% of all the Se entering the plant was lost during the treatment processes. A loss of 0.09 kg/day was observed in mass of Se entering and exiting the primary clarifier, and a mass discrepancy of approximately 0.09 kg/day was calculated for the activated sludge process. The remaining 0.27 kg/day or about 60% of the total loss was observed around the anaerobic digester system. It is believed that most of the Se lost from the treatment plant was released to the atmosphere (see below).

Average total Se concentration declined from 46.5 µg/l in primary influent and 35 µg/l in primary effluent to 24.5 µg/l in secondary effluent (Fig. 2), indicating some removal of Se by the treatment processes. Total dissolved Se (TDSe) concentration likewise declined from 16.8 µg/l in primary influent and 14.1 µg/l in primary influent to 11.7 µg/l in the secondary effluent. Average total Se concentration was 215 µg/g in primary sludge, 646 µg/g in the activated sludge, and 3697 µg/g in centrifuged anaerobic digester sludge (Fig. 2), consistent with increasing transfer of Se to solid phases with each stage of sludge treatment in this facility. Sludge thickening and dewatering might have contributed to the increase in Se concentration, as evidenced by a 15-fold difference between the primary sludge (215 µg/l) and anaerobic digester feed (3179 µg/l). Average dissolved Se concentration was 78 µg/l in the digester feed compared to 37 µg/l in anaerobic sludge centrifuge effluent, again consistent with immobilization of Se in the sludge.

The movement of Mo through Facility J was most unusual in that average total Mo concentration in primary influent (44 µg/l) was less than that in primary effluent (97 µg/l), secondary effluent (54 µg/l), and anaerobic digester effluent (47 µg/l) (Fig. 3). The loadings of Mo to this particular facility were highly variable over the sampling period, as indicated by the large standard deviations in average Mo levels in sludge and wastewater samples (Table 5). The overall Mo mass-balance for the facility showed no removal through the treatment process; in fact, there seemed to be some memory effect associated with previous high Mo loadings to the system since the influent input (0.33 kg/day) was less than the output from secondary effluent (0.40 kg/day). The mass-balance around individual units serves to highlight the unusual

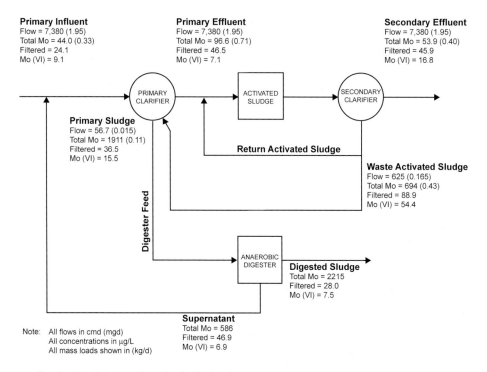

Primary Influent
Flow = 7,380 (1.95)
Total Mo = 44.0 (0.33)
Filtered = 24.1
Mo (VI) = 9.1

Primary Effluent
Flow = 7,380 (1.95)
Total Mo = 96.6 (0.71)
Filtered = 46.5
Mo (VI) = 7.1

Secondary Effluent
Flow = 7,380 (1.95)
Total Mo = 53.9 (0.40)
Filtered = 45.9
Mo (VI) = 16.8

Primary Sludge
Flow = 56.7 (0.015)
Total Mo = 1911 (0.11)
Filtered = 36.5
Mo (VI) = 15.5

Return Activated Sludge

Waste Activated Sludge
Flow = 625 (0.165)
Total Mo = 694 (0.43)
Filtered = 88.9
Mo (VI) = 54.4

PRIMARY CLARIFIER

ACTIVATED SLUDGE

SECONDARY CLARIFIER

ANAEROBIC DIGESTER

Digested Sludge
Total Mo = 2215
Filtered = 28.0
Mo (VI) = 7.5

Digester Feed

Supernatant
Total Mo = 586
Filtered = 46.9
Mo (VI) = 6.9

Note: All flows in cmd (mgd)
All concentrations in µg/L
All mass loads shown in (kg/d)

FIG. 3. Mass balance and species distribution for molybdenum in Facility J. Effluent/sludge flow rates are in m^3/day (10^6 gal/day in brackets). Concentrations shown are in µg/l for dissolved species (filtered solution) and µg/g for sludge; the corresponding mass loading in kg/day is given in parentheses.

behavior of Mo in this facility. The Mo sources $(0.33 + 0.43 = 0.76 \text{ kg/day})$ and sinks $(0.71 + 0.11 = 0.82 \text{ kg/day})$ for the primary clarifier showed a reasonable mass balance, with 87% of the incoming Mo leaving via the primary effluent. For the activated sludge unit, the Mo source (0.71 kg/day) was in reasonable agreement with the sink $(0.40 + 0.43 = 0.83 \text{ kg/day})$ considering the high temporal variability in Mo concentrations. The good mass balance in Mo flow through the primary and secondary treatment units suggests that the anaerobic digester unit was either inoperative or was functioning minimally during the sampling period. It should be noted that since this study, the municipality has identified and eliminated, the principal source of Mo in the city's wastewater collection system and the levels of this metal in Facility J have fallen to $\leq 10 \ \mu g/l$.

DISCUSSION

The concentrations of any of the oxy-anions and their fate in a wastewater treatment systems depend on many factors, including the source and chemical composition of the influents, the mineral phases present, adsorption capacity of the suspended particulates, biological processes in the treatment unit, pH, redox conditions, and reaction time. Data in Figs. 1–3 point to the fact that the oxy-anions have a complex chemistry in sludge ecosystems even where little of a particular element is removed by the wastewater treatment processes. The extent and trajectory of changes varied significantly from metal to metal and process to process as evidenced by the partitioning of these elements in the aqueous phase (Tables 3–5) and various fractions of the sludge (Table 6).

Arsenic

A large number of dissolved species of arsenic can be formed from abiotic and biological processes in a full-scale WWTP. Under aerobic conditions, the principal dissolved As species

Table 5. Physico-chemical characteristics of wastewater treatment plant (Facility J) with elevated molybdenum levels

Parameter	Primary influent[a]	Primary effluent[a]	Primary sludge[a]	Secondary effluent[a]	Waste activated sludge[a]	Anaerobic digester supernatant[a]	Anaerobic digester sludge
Total Mo (µg/l)	44.0 (19)	96.6 (111)	1911 (1189)	53.9 (16)	694 (336)	586 (856)	2215 (1567)
Total dissolved Mo (µg/l)	24.1 (7.8)	46.5 (48)	36.5 (34.7)	45.9 (15)	88.9 (28)	46.9 (49)	28.0 (22)
Mo(VI) (µg/l)	9.1 (7.8)	7.1 (8.1)	15.5 (22)	16.8 (11)	54.4 (27)	6.9 (7.5)	7.5 (11)
TSS[b] (mg/l)	289 (164)	260 (51)		94.6 (110)	6675 (3130)		4.4%
VSS[b] (mg/l)	177 (68)	132 (51)		31.9 (47)	3383 (1545)		
pH	8.04		6.98				6.76
Flow	7391 (1306)	7391 (1306)	5500 (1710)	7391 (1306)			
Alkalinity (mg/l CaCO$_3$)	363 (24)	370 (51)	58.2 (2.1)	263 (17)	626 (325)	1801 (931)	2514 (1144)
Temperature (°C)	18.5 (1.5)	19 (1.8)			380 (8.2)	42.6 (4.0)	42.5 (4.0)
Eh (mV)	+113 (49)	−170 (23)	−204 (26)	130 (56)	31.0 (15)	−203 (72)	−112 (42)
Ammonia-N (mg/l)						168 (125)	125 (39)
Total P (mg/l)						1.3 (1.2)	1.6 (2.2)
Total sulfide (mg/l)	1.6 (2.2)	<1.0	<8.0	<1.0	<1.0	<10	<3.0
Dissolved sulfide (mg/l)	1.1 (1.1)	1.4 (1.8)	5.0 (6.1)	<1.0	1.1 (1.2)	<5.0	<4.0
DOC[b] (mg/l)	18.3 (13)	27.0 (33)	293 (218)	6.9 (1.7)	7.3 (3.4)	24.6 (13)	155 (94)
Total iron (mg/l)			4367 (586)			3867 (115)	3747 (3234)

[a] The standard error (±1 standard deviation) is shown in brackets.
[b] TSS, total suspended solids; VSS, volatile suspended solids; DOC, dissolved organic carbon.

Table 6. Chemical fractions of As, Se, and Mo in solid sludge samples

Fraction	Sample	As[a]	Se[a]	Mo[a]
Exchangeable[b]	Facility K (centrifuge sludge)	1.2 (1.0–1.7)		
	Facility J (activated sludge)			0.8 (0.2–1.6)
	Facility J (digested sludge)			0.5 (0.3–0.6)
	Facility R (activated sludge)		1.7 (0.7–3.3)	
	Facility R (digested sludge)		0.2 (0.1–0.3)	
Carbonate associated[b]	Facility K (centrifuge sludge)	0.7 (0.6–0.7)		
	Facility J (activated sludge)			4.9 (3.3–6.1)
	Facility J (digested sludge)			1.9 (1.0–3.1)
	Facility R (activated sludge)		<0.1	
	Facility R (digested sludge)		<0.1	
Fe/Mn oxide associated[b]	Facility K (centrifuge sludge)	36 (30–51)		
	Facility J (activated sludge)			21 (17–29)
	Facility J (digested sludge)			11 (8.8–16)
	Facility R (activated sludge)		12 (9.2–14)	
	Facility R (digested sludge)		1.9 (1.6–2.1)	
Organic associated[b]	Facility K (centrifuge sludge)	6.7 (4.8–10)		
	Facility J (activated sludge)			24 (23–25)
	Facility J (digested sludge)			30 (26–36)
	Facility R (activated sludge)		18 (12–23)	
	Facility R (digested sludge)		30 (21–38)	
Residual[b]	Facility K (centrifuge sludge)	55 (37–63)		
	Facility J (activated sludge)			49 (39–54)
	Facility J (digested sludge)			52 (42–61)
	Facility R (activated sludge)		68 (61–77)	
	Facility R (digested sludge)		67 (60–77)	

[a] Mean (ranges given in parentheses) percentages relative to total initial concentration of each element. No data shown means that the sample was not analyzed for the particular fraction.
[b] Fractions shown are operationally defined (see "Methodology" section).

are expected to be $H_2AsO_4^-$ (acidic environment) or $HAsO_4^{2-}$ (neutral to alkaline environment). In moderately reducing environments, the dominant species is expected to be H_3AsO_3, while under anaerobic conditions, a number of thioarsenic complexes may be formed, including $AsS(SH)_2^-$, $AsS_2(SH)^{2-}$, and $AsS(OH)(SH)^-$ (HELZ et al., 1995). In treatment units with moderate to high alkalinity, As(III) can form very strong ion pairs (including $As(CO_3)^+$, $As(CO_3)_2^-$, and $As(HCO_3)_n^{(3-n)+}$) with carbonate and bicarbonate ions (KIM et al., 1999). Significant amounts of organoarsenicals can also be formed from arsenate and arsenite through microbiologically-mediated processes or during the decomposition of organic matter; these may include monomethylarsonic acid, MMAA (HCH_3AsO_2OH), dimethylarsinic acid, DMAA ($H(CH_3)_2AsO_2$), oxythioarsenic acid (H_3AsO_3S), methylarsonous acid, MMAA(III) ($CH_3As(OH)_2CH_3AsO$), dimethylarsinous acid, DMAA(III) (($(CH_3)_2AsOH((CH_3)_2As)_2O$) and the so-called "missing" or "refractory" As believed to consist mostly of arsenosugars and arsenobetaine (HASEGAWA et al., 1999). All of these compounds can be formed under the different conditions found in a full-scale WWTP. The behavior of arsenic in Facility K depends on the stability of these soluble species in various treatment units.

Under equilibrium conditions, the reduction of arsenate to arsenite can be described by:

$$HAsO_4^{2-} + 4H^+ + 2e^- \rightarrow H_3AsO_3 + H_2O \tag{1}$$

$$Eh = 0.881 + 0.0296 \log([H_3AsO_3]/([HAsO_4^{2-}][H^+]^4)) \tag{2}$$

In aerobic primary and secondary treatment systems of Facility K with pH $\approx$ 7.3 (Table 3) and Eh > 100 mV, the ratio of As(III)/As(V) is expected to be about 10^{-3}. The concentrations of As(III) in these units were generally below the detection limit in these process streams (Fig. 1), consistent with the predicted behavior in the aerobic systems. Although, according to the speciation scheme used, As(V) was shown to be the dominant dissolved form in most of the primary and secondary treatment streams in this facility (Fig. 1), there were also significant concentrations of other soluble arsenic forms. This is evident from the As(V)/(total dissolved As) ratios of 0.83 in primary influent, 0.51 in primary effluent, and 0.78 in secondary effluent. We suspect that organo-arsenic compounds comprise the rest of the TDAs in these process streams.

The observed speciation of As in the anaerobic digestion system of Facility K was unexpected. The ratios of As(V)/As(III) for average concentrations in the aqueous phase of anaerobic digester feed and centrifuged sludge were 0.67 and 2.6, respectively. According to Eq. (2) above, the expected As(V)/As(III) ratio at equilibrium conditions should be $>10^3$ assuming pH value of 7.3. The disparity between observed and expected ratios points to the presence of other negatively charged species in the system which were misclassified as As(V) by the speciation scheme used in this study. The reduction of As(V) to As(III) leading to the formation of charged thiocomplexes is very possible in the anaerobic digester environment of Facility K:

$$HAsO_4^{2-} + 2HS^- + 3H^+ \rightarrow H_3AsO_3 + H_2O + HS_2^- \tag{3}$$

$$2H_3AsO_3 + 4HS^- \rightarrow 2HAsOS_2^- + 4H_2O + 2e^- \tag{4}$$

The reduction of arsenate to arsenite by sulfide ion (expected in the anaerobic digester) is often accompanied by the formation of elemental sulfur and/or polysulfides represented as HS_2^- in Eq. (3) (HELZ *et al.*, 1995; ROCHETTE *et al.*, 2000). Microbial reduction of arsenate and arsenite under anaerobic conditions is well documented (MACUR *et al.*, 2001) and may result in the formation of arseno-carbonate complexes:

$$2CH_2O + H_3AsO_3 + H_2O \rightarrow As(CO_3)_2^- + 9H^+ + 8e^- \tag{5}$$

$$2CH_2O + HAsO_4^{2-} \rightarrow As(CO_3)_2^- + 5H^+ + 6e^- \tag{6}$$

The high alkalinity of the anaerobic digester environment of Facility K can also favor the formation of stable carbonate and thiocomplexes of arsenic in solution according to the following scheme proposed by KIM *et al.* (1999):

$$As_2S_3 + 2HCO_3^- + H_2O \rightarrow As(CO_3)_2^- + AsO(SH)_2^- + H_2S \tag{7}$$

$$As_2S_3 + HCO_3^- + 2H_2O \rightarrow As(CO_3)(OH)^0 + AsO(SH)_2^- + H_2S \tag{8}$$

It is believed that the presence of high concentrations of sulfide and alkalinity in the anaerobic digester results in aggressive leaching of As out of the sludge into the aqueous phase in the form of carbonate and sulfide complexes of As. Any negatively charged species is likely to be retained by the ion-exchange resin used to partition the species, and apparently would have been misclassified as As(V). This conclusion is supported by the mass-balance data for Facility K which show that most (90%) of the As in the anaerobic digester feed is returned to the primary clarifier.

The behavior of arsenic species in the liquid phases should be closely related to the solid forms of arsenic in the sludge, but little is currently known about the nature of such phases in wastewater treatment systems although stable phosphate analogues have been reported (NRIAGU and MOORE, 1984). Sequential chemical extraction or fractionation schemes provide a useful tool for gaining insight into the chemical forms of an element in sludge and can serve as a useful framework for relating the chemical forms in solution to concentrations in solid sludge. It should be emphasized that the extracted fractions are operationally defined, and need to be used with some caution since no fractionation scheme is totally effective in dissolving each distinct form of a trace metal or oxy-anion in sludge and sediment (KEON *et al.*, 2001).

The fractionation scheme used showed that very little As (1.2%) in the centrifuge (anaerobic digester) sludge from Facility K was in exchangeable form (Table 6). This is somewhat surprising since arsenate is often said to be readily sorbed onto ferric oxyhydroxides and other solids in the sludge (MANCEAU, 1995) and the addition of ferrous chloride to the collection system and headworks of Facility K was intended to enhance this process. Any interaction of arsenate with ferric ions apparently involves much stronger chemical bonding than is generally realized.

It is also conceivable that any As previously adsorbed by the sludge was subsequently converted to other forms during the anaerobic digestion process (see SU and PULS, 2001).

The fact that oxy-anions of As interact non-exchangeably with iron and other metal oxides and hydroxides is well documented (VIRCIKOVA et al., 1995; LIN and WU, 2001), although the nature of co-precipitated phases in the wastewater treatment facilities is unknown. At elevated arsenate concentrations under aerobic conditions (found in some sludge), several compounds can be precipitated including scorodite ($FeAsO_4 \cdot 2H_2O$), parasumplesite ($Fe_2(AsO_4)_3 \cdot 8H_2O$) and rauenthalite ($Ca_3(AsO_4)_2 \cdot 10H_2O$), $Ca_5(AsO_4)_3OH$ (stabite), $Ca_3(AsO_4)_2$, $CaHAsO_4 \cdot 2H_2O$ (pharmacolite), $Fe_2(AsO_4)OH \cdot x\,H_2O$ (olivenite), and $Fe_3(AsO_4)_2 \cdot 8H_2O$ (symplesite) (FOSTER et al., 1998; LEE, 2002). Formation of ferrous and calcium mineral phases would be strongly favored in the environment of an anaerobic digester with high levels of dissolved Ca and Fe(II), and the reactions would serve to reduce the adsorbed As burden in the sludge sample (see ROCHETTE et al., 1998).

Low concentrations (0.6–0.7%) of carbonate-associated fractions for As should be expected; oxy-anions of arsenic rarely react with carbonate ions to form precipitates. The high alkalinity (1499 mg/l $CaCO_3$; Table 3) would furthermore favor the formation of soluble arseno-carbonate complexes. Organic-bound forms accounted for 4.8–10% of the solid phase. This seems reasonable considering that wastewater treatment systems are usually rich in phosphorus, and the ratio of P/As has been shown to be a determining factor in the accumulation of As by aquatic organisms (see NRIAGU, 1994) and implicitly by the microbial population of an anaerobic digester unit.

Oxide phases accounted for 36% (range, 30–51%) of As in sludge samples. Accumulation of As in this phase is somewhat surprising considering the reducing nature of the process and high concentrations of dissolved (ferrous) iron in the digester centrate (297 mg/l) and centrifuge sludge (8733 mg/l) (see KANG et al., 2001). Previous studies using extended X-ray absorption fine structure (EXAFS), energy dispersive analysis of X-rays (EDAX), and infra-red (IR) spectroscopic methods have shown that As(V) forms high-affinity, inner-sphere surface complexes with iron hydroxides (MANNING and GOLDBERG, 1997; MCLAREN et al., 1998) or may be present as iron arsenato oxides or salts (MENG et al., 2001). Under anaerobic conditions and in the presence of high Fe(II) levels, the shuttling of electrons between the ferrous, ferric, and arsenate moieties in the oxide surface may result in the formation of a stable ferrosoferric arsenate solid complex:

$$Fe(OH)_3 + Fe^{2+} + 2AsO_4^{3-} + 3H^+ + 2e^- \rightarrow Fe^{III}Fe^{II}(OH)_2 \cdot AsO_4 + 2H_2O + AsO_3^{3-} \quad (9)$$

We posit that the stabilization of arsenic in such a solid and precipitation of other metal arsenates (especially calcium arsenate) provides a possible explanation for the observed strong association between arsenic and iron in the anaerobic digester environment.

The residual (unidentified) fraction accounted for 55% (range, 37–63%) of the As in the sludge samples analyzed. The nature of the recalcitrant forms of As is unknown (WALTER and GUEVAS, 1999). This fraction can include As held tightly at the internal surfaces of sludge aggregates or bound to humic and fulvic acids (these compounds are not completely destroyed by the acid treatment used to determine organic-bound forms). Under anaerobic conditions, a number of insoluble As sulfides can be precipitated (see CLARKE and HELZ, 2000; TOSSELL, 2000) including CuS (covellite), Cu_2S (digenite), Cu_3AsS_4 (enargite), As_2S_3 (orphiment), As_2S_2 (realgar), and FeAsS (arsenopyrite) by several reaction pathways:

$$2H_3AsO_3 + 3H^+ + 3HS^- \rightarrow As_2S_{3(s)} + 6H_2O; \; \log K = 47.1 \quad (10)$$

$$2H_2AsOS_2^- + H^+ \rightarrow As_2S_3 + HS^- + 2H_2O; \; \log K = 14.9 \quad (11)$$

$$2H_2As_3S_6^- \rightarrow 3As_2S_3 + 3HS^- + H^+ \quad (12)$$

Thermodynamic calculations can be used to show the likelihood for precipitation of arsenic sulfides and formation of arsenic thiocomplexes in the anaerobic sludge environment of Facility K. Substituting average concentrations of As(III) (4.5×10^{-8} M) and dissolved sulfide (2.3×10^{-4} M) in sludge water into the relationships above gives the logarithm of the ion concentration product (log ICP) of 48.5 for Eq. (10) and 20.2 for Eq. (11). The ICP values exceed the equilibrium constants and point to a strong likelihood that the reactions can occur. A recent study of extractants for selective leaching of pure compounds showed that arsenic sulfides can only be recovered after digestion with concentrated nitric acid (KEON *et al.*, 2001), and hence would have been reported in the residual fraction of our scheme. Our results are consistent with those of CARBONELL-BARRACHINA *et al.* (1999) who found insoluble arsenic sulfides to be the dominant form of arsenic present in municipal sewage sludge samples from anaerobic digester systems.

Selenium

Little is currently known about the processes of Se removal in full-scale WWTPs. The limited data available from pilot plant and laboratory studies show that conventional systems are moderately effective for Se(IV) removal but ineffective for Se(VI) removal from water and wastewater. Depending on the pH, iron dosing, and initial Se concentration, the reported removal efficiencies for Se(IV) range from 15 to 20% for alum coagulation, 40–80% for iron coagulation, and 35–45% for lime precipitation (KAPOOR *et al.*, 1995). Removal of selenate was found to be less than 10% for all of these methods (KAPOOR *et al.*, 1995). The reported behavior of selenium in wastewater treatment systems would seem consistent with its environmental chemistry. In general, selenate behaves very much like sulfate in that its sorption by soils, sediments, and geological material is low and its mobility in the environment is high (NEAL and SPOSITO, 1989; ADRIANO, 2001). By contrast, selenite seems to behave like phosphate and is much more readily absorbed by mineral phases, soils, and sediments compared to selenate (ADRIANO, 2001). Potential competition between these ions would imply that high sulfate concentrations can inhibit selenate removal while high phosphate levels will reduce selenite removal from wastewater treatment systems.

The fact that Se(IV) is unstable under aerobic conditions of the primary and secondary treatment units of Facility R is reflected by the low concentrations ($\leq 3\ \mu g/l$) of this species in effluents from these systems (Table 4). There is evidence to suggest that most of the dissolved selenium coming into the treatment plant was either in the form of selenate or complexed with organic compounds (see WSPA, 1994, 1995), which would be consistent with limited removal of Se in the primary and secondary treatment units. Higher concentrations of Se(IV) were observed in the anaerobic digester feed (40.6 $\mu g/l$), centrifuge supernatant (16.2 $\mu g/l$), and centrifuge sludge (15.3 $\mu g/l$) (Table 4). The Se(IV)/(TDSe) ratios were 0.52 in anaerobic digester feed, 0.44 in centrifuge supernatant, and 0.37 in centrifuge sludge, indicating major change in the reduction of Se(VI) and organo-selenium compounds to Se(IV) in the anaerobic digester environment.

The dominant form of Se under anaerobic digester conditions of Facility R (Eh < -200 mV) should be Se^{2-} (EARY *et al.*, 1990). Reduction of selenate to selenite and subsequently to Se^{2-} can be represented simply as

$$SeO_4^{2-} + 2H^+ + 2e^- \rightarrow SeO_3^{2-} + H_2O \tag{13}$$

$$SeO_3^{2-} + 6H^+ + 2e^- \rightarrow Se^{2-} + 3H_2O \tag{14}$$

Possible electron acceptors for this reaction include Fe(III), Mn(III), and organic matter. The available literature suggests that Se^{2-} is very labile and it is thus conceivable that some of the reported Se(IV) in Table 4 may be formed through the oxidation of Se^{2-} during sample collection, storage, processing, and analysis.

Results of the partitioning of Se into various fractions in solid sludge are shown in Table 6. A small fraction of the Se (0.2–4.2%) was present in exchangeable form, implying that a large portion of the sorptively removed Se in the primary and secondary treatment units of Facility R is transformed to more stable phases in the sludge. Previous studies have also found that generally $<5\%$ of the Se in sewage sludge (CAPON, 1991) and soils (ADRIANO, 2001) is in exchangeable form. Low concentrations of carbonate-associated Se were expected, as oxy-anions of Se rarely react with carbonate ions to form precipitates. Only a small fraction (1.9–20%) of the Se was associated with iron oxides, which is consistent with the fact that ferric ion is not effective in removing Se(VI) and organo-selenium compounds from wastewater. The concentration of oxide-bound Se in anaerobically digested sludge (1.9–3.8%) is much lower than the levels in the activated sludge (12–20%), suggesting that some of the Se bound to iron oxyhydroxides is being transformed to other phases during anaerobic treatment.

Organic-bound forms were the second most abundant fraction of Se (12–38%) in the sludge samples analyzed, and most of these compounds in the influent stream were probably derived from human wastes, a major source of this element in municipal effluents (LAWSON and MACY, 1995; ADRIANO, 2001). The higher percentage of organic-bound Se in digested sludge (30% average) compared to activated sludge (18% average) may be attributed to increased sequestration of Se by organic matter in the anaerobic digester environment (WSPA, 1994, 1995). It is, however, more likely that the difference is made up by elemental Se which can be produced via a number of reactions in the anaerobic digester environment (see above).

The dominant form of Se was the residual fraction (60–77%) in both digested and activated sludge. Previous studies of sewage sludge and soils likewise found the recalcitrant fraction to comprise 50–80% of the Se inventory (CAPON, 1991; ADRIANO, 2001). The nature of selenium-bearing compounds in this fraction is unknown. Studies of soils from reducing environments show that the residual fraction was mostly made up of organically bound and elemental Se (WSPA, 1995; LAWSON and MACY, 1995). One would also expect these forms to predominate in the anaerobic sludge environment. The reduction of selenate and selenite in the anaerobic digester environment to elemental selenium:

$$SeO_3^{2-} + 6H^+ + 4e^- \rightarrow Se^0 + 3H_2O \tag{15}$$

can be effected by several agents in the sludge and wastewater (see GARBISU et al., 1996; FOX et al., 2002). The use of ferrous hydroxide to reductively remove Se in water treatment systems as the insoluble elemental form has been proposed by MURPHY (1989), for instance:

$$Na_2SeO_3 + 4Fe(OH)_2 \leftrightarrow Se^0 + 2Fe_2O_3 + 2NaOH + 3H_2O \tag{16}$$

$$Na_2SeO_4 + 6Fe(OH)_2 \leftrightarrow Se^0 + 3Fe_2O_3 + 2NaOH + 5H_2O \tag{17}$$

The efficacy of these reactions in wastewater treatment system remains doubtful, however. The reduction of selenate by ferrous sulfide is deemed to be important in wastewater treatment systems and can result in the formation of highly insoluble ferrohydroxy selenite, the solubility product of which has been estimated to be 10^{-63} (GEERING et al., 1968):

$$SeO_4^{2-} + 2FeS + 6H_2O \rightarrow Fe_2(OH)_4SeO_3 + S_2O_3^{2-} + 8H^+ + 8e^- \tag{18}$$

The use of an organic carbon source to accelerate the reduction of Se(VI) and Se(IV) to elemental Se in soils and sediments has been reported (TOKUNAGA et al., 1996; ARBESTAIN and AROS, 2001). The reduction may be part of the selenium detoxification mechanism (OREMLAND, 1994) or may occur through the following dissimilatory pathway (TOMEI et al., 1992):

$$CH_2O + SeO_3^{2-} + H^+ \rightarrow Se^0 + HCO_3^- + H_2O \tag{19}$$

Oxidation of organic carbon drives most of the redox reactions in wastewater treatment processes, and the high concentration of Se in primary sludge (215 $\mu g/g$) suggests that a large fraction of the Se is delivered to the treatment plant as particulate organic selenium associated with human wastes. CAPON (1991) showed that Se($-$II) and Se(IV) accounted for 70–91% of the selenium in his anaerobically digested samples, an observation that is consistent with the presence of high concentrations of elemental and organic Se. It is conceivable that some of the Se in the residual fraction is bound to iron in a form that could not be extracted by the reagents used to determine the oxide-associated fraction (REFAIT et al., 2000). Formation of insoluble metal selenites and selenides is also feasible and may explain the higher concentration of total selenium in the anaerobic digester sludge compared to the primary sludge (see Table 4).

A wide variety of microorganisms (including those common in wastewater treatment systems) and fungi have the ability to convert SeO_4^{2-}, SeO_3^{2-}, Se(0), and various organo-selenium compounds into the volatile form (DORAN and ALEXANDER, 1977; COOK and BRULAND, 1987;

DUNGAN and FRANKENBERGER, 2001). The principal volatile gas produced by the methylating organisms is dimethylselenide (DMSe)—CH$_3$SeCH$_3$, although other compounds such as dimethyldiselenide (DMDSe)—CH$_3$Se$_2$CH$_3$, methanesalone—(CH$_3$)$_2$SeO$_2$, methaneselenol—CH$_3$SeH, and dimethylselenide sulfide—CH$_3$SeSCH$_3$ may be formed in small quantities (DUNGAN and FRANKENBERGER, 2001). The methylation/demethylation processes resulting in formation of volatile Se gases apparently accounted for the loss of 28% of the Se entering the plant. Losses were observed in all the treatment units, consistent with the fact that methylation of Se can occur under aerobic and anaerobic conditions (COOK and BRULAND, 1987).

Molybdenum

The chemical behavior of Mo should be similar to that of chromium (Cr) in wastewater treatment systems. The two elements can exist in the +3 oxidation state as cations under reducing conditions and tend to form oxides and hydroxides with very low solubilities. In aerated environments, the dominant species are the molybdate, MoO$_4^{2-}$, the oxidized form of [Mo(VI)] and chromate [Cr(VI)], which are anions, in contrast to the cationic forms in reducing environments (MAGYAR *et al.*, 1993; BEAUBIEN *et al.*, 1993). The reversal of charge as a function of oxidation state is expected to change the adsorption characteristics and precipitation reactions of Mo in wastewater systems compared to As and Se.

Processes expected to control the concentration of Mo(VI) oxy-anions in aerobic waters include adsorption and co-precipitation with iron and manganese oxyhydroxides and complexation with organic ligands (ZHENG *et al.*, 2000). The fact that Facility J feeds ferric chloride to the aeration basin effluent (for phosphorus control) may account for the lowering of total Mo concentration in secondary effluent (54 μg/l) compared to the primary effluent (97 μg/l) (Fig. 3). The high residual concentration in the secondary effluent, however, may be related to the fact that MoO$_4^{2-}$ behaves like SO$_4^{2-}$ (both are bi-negatively charged, tetrahedral in structure and in biological systems they tend to be absorbed, transported, and excreted along similar routes), and is weakly adsorbed on mineral surfaces (ADRIANO, 2001). The concentrations of total dissolved Mo were 24 μg/l in primary influent, and about 46 μg/l in both primary and secondary effluents (Table 5), implying some re-mobilization of solid-phase Mo in the treatment units. The Mo(VI)/(total dissolved Mo) ratios of about 0.38 for both primary influent and secondary effluent and only 0.15 for primary effluent suggest (a) that most of the dissolved Mo in waste streams is probably complexed by organic matter, hence the limited effect of iron treatment and (b) the Mo is involved in active precipitation/dissolution processes in the treatment units. The ratios for the influent and effluents are lower than the Mo(VI)/(total dissolved Mo) value of 0.42 for primary sludge and 0.61 for waste activated sludge, which would suggest that sludge was the source of the dissolved Mo(VI) in the primary clarifier and waste activation environment. Increased concentrations of Mo in the aqueous phase of the treatment systems create the potential for continuous recycling of Mo within the wastewater system, hence the limited or no removal of Mo observed.

Under anaerobic conditions, Mo reacts readily with sulfide ion to form a number of stable thiomolybdate complexes (MoO$_n$S$_{4-n}^{2-}$) which have been detected by the oxine chelate method (MAGYAR *et al.*, 1993). A number of insoluble sulfides can be precipitated including MoS$_2$ and MoS$_3$. Very often, the Mo sulfides are co-precipitated with FeS according to the reaction first proposed by BERTINE (1972):

$$FeS(\text{amorphous}) + MoO_2S_2^{2-} + H_2S + 2H^+ \leftrightarrow FeSMoS_3(\text{amorphous}) + 2H_2O \quad (20)$$

In the anaerobic digester environment, one would thus expect sulfide reactions to dominate the Mo chemistry since the oxy-anions of the more reduced forms of this metal are relatively unstable (TUCKER *et al.*, 1997). For Facility J, total dissolved Mo concentration in the anaerobic digester supernatant was high (47 μg/l; Fig. 3) but the Mo(VI)/(total dissolved Mo) ratio was only 0.15, showing substantive formation of soluble species other than molybdate in the system. The dominant form of molybdenum present in the digester supernatant is unknown and is believed to be molybdenum sulfides. Formation of thio-complexes is also suggested by the ratio of Mo(VI)/(total dissolved Mo) in the anaerobic digester sludge (0.27) which was lower than the values of 0.42 in primary sludge and 0.61 for waste activated sludge.

Chemical fractionation of the sludge showed low Mo (0.5–0.8%) concentrations in the exchangeable form in both samples from the activated sludge and digester units of Facility J (Table 6), a feature similar to the other oxy-anions investigated. The carbonate fraction accounted for 1.9–4.9% of the Mo in the sludge samples, suggesting slight formation of Mo–carbonate in the wastewater treatment facility. The percentage of Mo bound to metal oxides ranged from 8.8 to 28. In general, more Mo was retained in metal oxides in activated sludge (21% average) compared to digested sludge (11%) samples. This is consistent with the fact that the ferric ion feed was able to precipitate a substantial fraction of dissolved Mo in wastewater and suggests that some of the Mo so removed was regenerated into the solution during anaerobic treatment. Organic-bound forms were the second most abundant fraction (24–30%). The higher total concentration of Mo in primary sludge (1911 μg/l) compared to waste activated sludge (694 μg/l) suggests that this fraction is associated with high loading of particulate organic Mo into the treatment plant. The dominant form (49–52%) of Mo in the sludge was in the residual fraction. Analysis of a sample of the material used by the factory purported to be the primary source of Mo polluter of the city's wastewater showed that it was not soluble in mineral acids and would thus end up with the residual phase in the sludge. The enrichment of Mo in anaerobic digester sludge (2215 μg/l) relative to the primary sludge (1911 μg/l; Fig. 3) is consistent with immobilization of Mo as insoluble sulfides believed to be an important component of the residual fraction.

Acknowledgements—The research on which this paper is based was funded in part by the United States Environmental Protection Agency (US EPA) through Cooperative Agreement (No. CR825237) with the Water Environment Research Foundation. We thank Myoung-Jin Kim and Janice Lee for laboratory support. Appreciation is extended to the anonymous reviewers for providing comments that resulted in a significant improvement of this paper.

REFERENCES

ADRIANO D. C. (2001) *Trace Metals in the Terrestrial Environment*, 2nd Ed., Springer, New York.
ARBESTAIN M. C. and AROS A. R. (2001) Modeling selenium transfers in straw-amended soils. *Soil Sci.* **166**, 539–547.
BAES C. F. and MESMER R. E. (1976) *The Hydrolysis of Cations*. Wiley, New York.
BEAUBIEN S. E., NRIAGU J. O., LAWSON G. and BLOWES D. W. (1993) Speciation and distribution of chromium in the Great Lakes. *Environ. Sci. Technol.* **26**, 730–736.
BERTINE K. K. (1972) The deposition of molybdenum into anoxic waters. *Mar. Chem.* **1**, 43–53.
BLACKMORE D. P. T., ELLIS J. and RILEY P. J. (1996) Treatment of a vanadium-containing effluent by adsorption/coprecipitation with iron oxyhydroxide. *Water Res.* **30**, 2512–2516.
BROWN H. C., HENSLEY C. P., MCKINNEY G. L. and ROBINSON J. L. (1973) Efficiency of heavy metal removal in municipal sewage treatment plants. *Environ. Lett.* **5**, 103–110.
CAPON C. J. (1991) Sewage sludge as a source of environmental selenium. *Sci. Total Environ.* **100**, 177–205.
CARBONELL-BARRACHINA A. A., JUGSUJINDA A., BURLO F., DELAUNE R. D. and PATRICK W. H. (1999) Arsenic chemistry in municipal sewage sludge as affected by redox conditions and pH. *Water Res.* **34**, 216–224.
CHEN K. Y., YOUNG C. S. and ROHATGI N. (1974) Trace metals in wastewater effluents. *J. Water Pollut. Control Fed.* **46**, 2663–2669.
CLARKE M. B. and HELZ G. R. (2000) Metal–thiometalate transport of biologically active trace elements in sulfidic environments: 1. Experimental evidence for copper thioarsenite complexing. *Environ. Sci. Technol.* **34**, 1477–1482.
CLESCERI L. S., GREENBERG A. E. and EATON A. D. (1998) *Standard Methods for the Examination of Water and Wastewater*, 20th Ed., American Public Health Association, Washington, DC.
COOK T. D. and BRULAND K. W. (1987) Aquatic chemistry of selenium: evidence of methylation. *Environ. Sci. Technol.* **21**, 1214–1219.
DORAN J. W. and ALEXANDER M. (1977) Microbial transformations of selenium. *Appl. Environ. Microbiol.* **33**, 31–37.
DUNGAN R. S. and FRANKENBERGER W. T. (2001) Biotransformations of selenium by *Enterobacter cloacae* SLD1a-1: formation of dimethylselenide. *Biogeochemistry* **55**, 73–86.
EARY L. E., DHANPAT R., MATTIGOD S. V. and AINSOWRTH C. C. (1990) Geochemical factors controlling the mobilization of inorganic constituents from fossil fuel combustion residues: II. Review of the minor elements. *J. Environ. Qual.* **19**, 202–213.
FOSTER A. L., BROWN G. E., TINGLE T. N. and PARKS G. A. (1998) Quantitative arsenic speciation in mine tailings using X-ray absorption spectroscopy. *Am. Mineral.* **83**, 553–568.

FOX P. M., LEDUC D. L., HUSSEIN H., LIN Z. Q. and TERRY N. (2002) Selenium speciation in soils and plants. In *Biogeochemistry of Environmentally Important Elements*, (eds. Y. CAI, and O. C. BRAIDS), Vol. 835, pp. 339–354. Am. Chem. Soc. Symp. Series.

FRANK P. and CLIFFORD D. (1986) Arsenic(III). *Oxidation and Removal from Drinking Water*. Water Engineering Research Lab, U.S. Environmental Protection Agency, Cincinnati, OH.

GARBISU C., ISHII T., LEIGHTON T. and BUCHANAN B. (1996) Bacterial reduction of selenite to elemental selenium. *Chem. Geol.* **132**, 199–204.

GEERING H. R., CAREY E. E., JONES L. H. P. and ALLOWAY W. H. (1968) Solubility and redox criteria for the possible formation of selenium in soils. *Soil Sci. Soc. Am. J.* **32**, 35–40.

GOLDSTONE M. E., KIRK P. W. W. and LESTER J. N. (1990) The behavior of heavy metals during wastewater treatment. *Sci. Total Environ.* **95**, 233–252.

HASEGAWA H., MATSUI M., OKAMURA S., HOJO M., IWASAKI N. and SOHRIN Y. (1999) Arsenic speciation including "hidden" arsenic in natural waters. *Appl. Organometal. Chem.* **13**, 113–119.

HELZ G. R., TOSSELL J. A., CHARNOCK J. M., PATRICK R. A. D., VAUGHAN D. J. and GARNER C. D. (1995) Oligomerization of As(III) sulfide solution: theoretical constraints and spectroscopic evidence. *Geochim. Cosmochim. Acta* **59**, 4591–4604.

HERING J. G. and CHU V. Q. (1998) The chemistry of arsenic: treatment implications of arsenic speciation and occurrence. Presented at the AWWA Inorganic Workshop, San Antonio, TX, February 23–24.

JENKINS D. and RUSSELL L. L. (1994) Heavy metals contribution of household washing products to municipal wastewater. *Water Environ. Res.* **66**, 805–813.

KANG S. J., MURIN J. R. and NRIAGU J. O. (2001) Predicting fate and transformation of oxy-anions in wastewater treatment, Final Report. Water Environment Research Foundation, Alexandria, VA.

KAPOOR A., TANJORE S. and VIRARAGHAVAN T. (1995) Removal of selenium from water and wastewater. *Int. J. Environ. Studies* **49**, 137–147.

KEON N. E., SWARTZ C. H., BRABANDER D. J., HARVEY C. and HEMOND H. F. (2001) Validation of an arsenic sequential extraction method for evaluating mobility in sediments. *Environ. Sci. Technol.* **35**, 2778–2784.

KIM M. J. (1999) Arsenic dissolution and speciation in groundwater of southeast Michigan. Ph.D. Thesis, School of Public Health, University of Michigan.

KIM M. J. and NRIAGU J. O. (2000) Oxidation of arsenic in groundwater using ozone and oxygen. *Sci. Total Environ.* **247**, 71–79.

KIM M. J., NRIAGU J. O. and HAACK S. K. (1999) Carbonate ion and dissolution of arsenic by groundwater. *Environ. Sci. Technol.* **34**, 3094–3100.

LAKE D. L., KIRK P. W. and LESTER J. N. (1984) Fractionation, characterization, and speciation of heavy metals in sewage sludge and sludge amended soils. *J. Environ. Qual.* **13**, 175–183.

LAKE D. L., KIRK P. W. W. and LESTER J. N. (1989) Heavy metal solids association in sewage sludges. *Water Res.* **23**, 285–291.

LAWSON S. and MACY J. M. (1995) Bioremediation of selenite in oil refinery wastewater. *Microbiol. Biotechnol.* **43**, 762–765.

LEE J. S. (2002) Metal arsenates in the environment. Ph.D. Thesis, Department of Environmental Health Sciences, University of Michigan.

LIN T. F. and WU J. K. (2001) Adsorption of arsenite and arsenate within activated alumina grains: equilibrium and kinetics. *Water Res.* **35**, 2049–2057.

MACUR R. E., WHEELER J. T., MCDERMOTT T. R. and INSKEEP W. P. (2001) Microbial populations associated with the reduction and enhanced mobilization of arsenic in mine tailings. *Environ. Sci. Technol.* **35**, 3676–3682.

MAGYAR B., MOOR H. C. and SIGG L. (1993) Vertical distribution and transport of molybdenum in a lake with seasonally anoxic hypolimnion. *Limnol. Oceanogr.* **38**, 521–531.

MANCEAU A. (1995) The mechanism of anion adsorption on iron oxides: evidence for the binding of arsenate tetrahedra on free $Fe(O,OH)_6$ edges. *Geochim. Cosmochim. Acta* **59**, 3647–3653.

MANNING B. A. and GOLDBERG S. (1997) Adsorption and stability of arsenic(III) at the clay mineral–water interface. *Environ. Sci. Technol.* **31**, 2005–2011.

MCLAREN R. G., NAIDO R., SMITH J. and TILLER K. G. (1998) Fractionation and distribution of arsenic in soils contaminated by cattle dip. *J. Environ. Qual.* **27**, 348–354.

MENG X. G., KPRFOATIS G. P., JING C. Y. and CHRISTODOULATOS C. (2001) Redox transformations of arsenic and iron in water treatment sludge during aging and TCLP extraction. *Environ. Sci. Technol.* **35**, 3476–3481.

MURPHY A. P. (1989) A water treatment process for selenium removal. *J. Water Pollut. Control Fed.* **61**, 361–362.

NEAL R. H. and SPOSITO G. (1989) Selenate adsorption on alluvial soils. *Soil Sci. Soc. Am. J.* **53**, 70–74.

NEVESSI A. E., DEWALLE F. B., SUNG S. F. C., MAYER K. and DALSEY R. (1988) Heavy metal variability of different municipal sludges as measured by atomic absorption and inductively coupled plasma emission spectroscopy. *J. Environ. Sci. Health* **A23**(8), 823–841.

NRIAGU J. O. (1994) *Arsenic in the Environment*. Wiley, New York.

NRIAGU J. O. (1997) Occurrence and uses of vanadium. In *Vanadium in the Environment* (ed. J. O. NRIAGU). Wiley, New York.

NRIAGU J. O. and MOORE P. B. (eds) (1984) *Phosphate Minerals*. Springer, Berlin.

NRIAGU J. O., LAWSON G., WONG H. K. T. and AZCUE J. M. (1993) A protocol for minimizing contamination in the analysis of trace metals in the Great Lakes. *J. Great Lakes Res.* **19**, 175–182.

NURDOGAN Y., SCHROEDER R. P. and MEYER C. L. (1994) *Selenium removal from petroleum refinery wastewater, 49th Purdue Industrial Waste Conference Proceedings*. Lewis Publishers, Chelsea, MI.

OREMLAND R. S. (1994) Biogeochemical transformations of selenium in anoxic environments. In *Selenium in the Environment* (eds. W. T. FRANKENBERGER and S. BENTON), pp. 389–420. Marcel Dekker, New York.

PENG F. F. and DI P. (1994) Removal of arsenic from aqueous solution by adsorbing colloid flotation with hydrogen peroxide addition. *Ind. Eng. Chem. Res.* **33**, 922–928.

RAO G. A. K. and VIRARAGHAVAN T. (1992) Removal of heavy metals at a Canadian wastewater treatment plant. *J. Environ. Sci. Health* **A27**, 13–23.

REFAIT P., SIMON L. and GENIN J. M. (2000) Reduction of SeO_4^{2-} and anoxic formation of Iron(II)–Iron(III) hydroxy-selenate green rust. *Environ. Sci. Technol.* **34**, 819–825.

REID J. (1994) Arsenic occurrence: USEPA seeks clear picture. *J. Am. Water Works Assoc.* **86**, 44–51.

ROCHETTE E. A., LI G. C. and FENDORF S. E. (1998) Stability of arsenate minerals in soil under biotically generated reducing conditions. *Soil Sci. Soc. Am. J.* **62**, 1530–1537.

ROCHETTE E. A., BOSTICK B. C., LI G. C. and FENDORF S. E. (2000) Kinetics of arsenate reduction by dissolved sulfide. *Environ. Sci. Technol.* **34**, 4714–4720.

STERRIT R. M., BROWN M. J. and LESTER J. N. (1981) Metal removal by adsorption and precipitation in activated sludge process. *Environ. Pollut.* **24**, 313–327.

SU C. M. and PULS R. W. (2001) Arsenate and arsenite removal by zerovalent iron: kinetics, redox transformation, and implications for in situ groundwater remediation. *Environ. Sci. Technol.* **35**, 1487–1492.

TESSIER A., CAMPBELL P. G. C. and BISSON M. (1979) Sequential extraction procedure for the speciation of particulate trace metals. *Anal. Chem.* **51**, 844–851.

TOKUNAGA T. K., PICKERING I. J. and BROWN G. E. (1996) Selenium transformations in ponded soils. *Soil Sci. Soc. Am. J.* **60**, 781–790.

TOMEI F. A., BARTON L. L., LEMANSKI C. L. and ZOCCO T. G. (1992) Reduction of selenate and selenite to elemental selenium by *Wolinella succinogenes*. *Can. J. Microbiol.* **38**, 1328–1333.

TOSSELL J. A. (2000) Metal–thiometalate transport of biologically active trace elements in sulfidic environments: 2. Theoretical evidence for copper thioarsenite complexing. *Environ. Sci. Technol.* **34**, 1483–1488.

TUCKER M. D., BARTON L. L. and THOMPSON B. M. (1997) Reduction and immobilization of molybdenum by *Desulfovibrio desulfuricans*. *J. Environ. Qual.* **26**, 1146–1152.

US EPA (2003) *Drinking Water Standards*. Office of Ground Water and Drinking Water, U.S. Environmental Protection Agency, Washington, DC, http://www.epa.gov/safewater/standards.html.

VIRCIKOVA E., MOLNAR L., LECH P. and REITZNEROVA E. (1995) Solubilities of amorphous Fe–As precipitates. *Hydrometallurgy* **38**, 111–123.

WALTER I. and GUEVAS G. (1999) Chemical fractionation of heavy metals in a soil amended with repeated sewage sludge application. *Sci. Total Environ.* **226**, 113–119.

WAYSAY S. A., HARON M. D. J., UCHIUMI A. and TOKUNAGA S. (1996) Removal of arsenite and arsenate ions from aqueous solution by basic yttrium carbonate. *Water Res.* **30**, 1143–1148.

WSPA (1994) WSPA Selenium Speciation: Final Report. Western State Petroleum Association, Concord, CA.

WSPA (1995) Selenium Removal Technology Study: Project Final Report. Western State Petroleum Association, Concord, CA.

YOSHIZAKI S. and TOMIDA T. (2000) Principle and process of heavy metal removal from sewage sludge. *Environ. Sci. Technol.* **34**, 1572–1575.

ZHENG Y., ANDERSON R. F., VAN GEEN A. and KUWABARA J. (2000) Authigenic molybdenum formation in marine sediments: a link to pore water sulfide in the Santa Barbara basin. *Geochim. Cosmochim. Acta* **64**, 4165–4178.

Geochemical Investigations in Earth and Space Science: A Tribute to Isaac R. Kaplan
© The Geochemical Society, Publication No. 9, 2004
Editors: R.J. Hill, J. Leventhal, Z. Aizenshtat, M.J. Baedecker, G. Claypool,
R. Eganhouse, M. Goldhaber and K. Peters

An evaluation of hydroxyl radical formation in river water and the potential for photodegradation of bisphenol A

NOBUTAKE NAKATANI, NORICHIKA HASHIMOTO and HIROSHI SAKUGAWA
Graduate School of Biosphere Sciences, Hiroshima University, 1-7-1, Kagamiyama,
Higashi-Hiroshima 739-8521, Japan

Abstract—The influence of the hydroxyl radical ($^{\cdot}$OH) on the photodegradation of the estrogen-like compound, bisphenol A (BPA), was examined in this study. The formation rate of $^{\cdot}$OH, normalized to the vernal equinox solar noon condition of Higashi-Hiroshima (34°N) was in the range $0.70–3.25 \times 10^{-10}$ M s^{-1} in Kurose river water. The total consumption rate constant of $^{\cdot}$OH in river water ranged from 1.66 to 3.89×10^5 s^{-1}. Based on the photochemical formation rate and the total consumption rate constant of $^{\cdot}$OH, steady-state $^{\cdot}$OH concentrations on the order of $3.33–8.35 \times 10^{-16}$ M were determined. The reaction rate constant for $^{\cdot}$OH with BPA determined by competition kinetics was found to be 1.55×10^{10} M^{-1} s^{-1} in water containing nitrate ions that photochemically produced $^{\cdot}$OH. BPA half-lives due to reaction with $^{\cdot}$OH ranged from 0.6 to 1.6 days of midday-May sunlight in the Hiroshima prefecture, similar to previously reported values for biodegradation, suggesting that the reaction of BPA with $^{\cdot}$OH has a significant effect on the fate of BPA in river water.

INTRODUCTION

SOLAR IRRADIATION can cause the degradation and mineralization of organic pollutants and natural organic compounds in surface water via direct and indirect photoreactions. If the substances are active chromophores at wavelengths found in the surface solar spectrum, the direct photoreaction activated by the irradiation of sunlight can occur in natural surface waters. On the other hand, the activation of indirect photoreactions is initiated either through reactions with reactive molecules in ground or excited states that are themselves products of primary photochemistry, or through photosensitized reactions in which the excited state species of some chromophores transfer an electron or energy to the compound (COOPER and HERR, 1987).

The hydroxyl radical ($^{\cdot}$OH) is the one of the transients in photochemical processes and is the most oxidative reactant among free radical species in water (MILL et al., 1980; ZAFIRIOU et al., 1984; BREZONIK, 1994). This transient reacts with organic compounds nonselectively and plays an important role in the transformation of organic compounds (RUSSI et al., 1982; HAAG and HOIGNE, 1985; HAAG and YAO, 1992; BREZONIK and FULKERSON-BREKKEN, 1998). The rate constants for the reaction of $^{\cdot}$OH with organic compounds are within the range of $10^7–10^{10}$ M^{-1} s^{-1} in aqueous solution (BUXTON et al., 1988). Although several mechanisms of $^{\cdot}$OH production in natural water have been presented, the photolysis of nitrate was one of the more important sources of $^{\cdot}$OH (BREZONIK, 1994). RUSSI et al. (1982) concluded that nitrate photolysis was a more important source of $^{\cdot}$OH than dissolved organic compounds. Furthermore, neither direct nor indirect hydrogen peroxide photolysis was a significant source of $^{\cdot}$OH in lake water, whereas nitrate photolysis was a dominant source (HAAG and HOIGNE, 1985). Therefore, nitrate was used as the source of $^{\cdot}$OH to determine photodegradation rate of organic compounds due to reaction with $^{\cdot}$OH in several studies (ZEPP et al., 1987; TORRENTS et al., 1997; BREZONIK and FULKERSON-BREKKEN, 1998; FULKERSON-BREKKEN and BREZONIK, 1998; ANDREOZZI et al., 2003).

Bisphenol A (BPA), which is a commonly used name for 2,2-(4,4-dihydroxydiphenyl)-propane (CAS number: 80-05-7), exhibits extremely weak hormonal activity in test tube assays and in vivo mammalian assays (KRISHNAN et al., 1993; ASHBY and TINWELL, 1998;

MILLIGAN *et al.*, 1998). This compound is industrially produced in large quantities worldwide. In 2000, BPA production and importation into Japan amounted to about 420,000 t (Japan Environmental Agency, 2002). Most of the BPA produced is used as a monomer for the production of polycarbonate and epoxy resin, unsaturated polyester–styrene resin and flame-retardants. The final products of BPA are released to the air, to surface water, or to wastewater treatment plants during use as a coating on cans, as powder paint, as additives in thermal paper, in dental fillings and as antioxidants in plastics (STAPLES *et al.*, 1998). The Henry's constant for BPA (1×10^{-10} atm m^3 mol^{-1}) is much lower than those for low volatility chemicals ($\sim 1 \times 10^{-7}$ atm m^3 mol^{-1}) (HOWARD, 1989), suggesting that BPA is a low volatility compound in the environment. The water solubility and octanol–water partition coefficient values (log K_{ow}) for BPA are 120–300 mg l^{-1} and 2.20–3.82, respectively (STAPLES *et al.*, 1998). Using the water solubility of 120 mg l^{-1} and the log K_{ow} of 3.82, soil–sediment adsorption constant values (K_{oc}) are 314 and 1524, respectively (HOWARD, 1989). These K_{oc} values indicate that soil and sediment are modest sinks for BPA released to the ground or to surface water. Therefore, it is necessary to obtain information on the fate of BPA in aquatic environments as well as monitoring the emission of BPA from anthropogenic sources.

In neutral and acidic methanol solutions, BPA exhibits a slight absorption of UV light wavelengths exceeding 290 nm, while in basic methanol solution BPA exhibits significant absorption of UV exceeding 290 nm (HOWARD, 1989). These results indicate that BPA may well be photolyzed in the environment. In the atmosphere, photodegradation of a small amount of vapor-phase BPA occurs due to direct photolysis or the interaction with ˙OH (STAPLES *et al.*, 1998). In addition to direct photolysis, the photo-oxidation of BPA in surface water may also occur due to the interaction with ˙OH. However, the rate constant for the reaction of ˙OH with BPA is not known.

The objectives of the present study were (1) to determine the steady-state concentration of ˙OH in river water, (2) to determine the rate constants for the reaction between BPA and ˙OH that was produced by nitrate photolysis, and (3) to estimate the lifetime of BPA in the surface river water calculated from the steady-state concentration of ˙OH and the rate constant for the reaction between BPA and ˙OH in river water.

THEORETICAL

In the present study, competition kinetics was used to determine the rate constant for the reaction of ˙OH with BPA. Phenol was produced by the reaction of benzene with ˙OH (RUSSI *et al.*, 1982; ARAKAKI and FAUST, 1998). When benzene as a trapping agent for ˙OH is added to the experimental solution containing nitrate and BPA, the photoformation rate of ˙OH (R_{OH}) and the fraction of ˙OH that reacts with benzene ($F_{benzene,OH}$) are

$$R_{OH} = R_{phenol} / \{(Y_{phenol})(F_{benzene,OH})\} \qquad (1)$$

$$F_{benzene,OH} = (k_{benzene,OH})[\text{benzene}] / \sum(k_{S,OH}[\text{S}]) \qquad (2)$$

where R_{phenol} is the formation rate of phenol in experimental solution (M s^{-1}), Y_{phenol} is the yield of phenol formed per benzene oxidized by ˙OH (mean $\pm$ standard deviation = 0.75 $\pm$ 0.07, in ARAKAKI and FAUST (1998)), $k_{benzene,OH}$ is the reaction rate constant of benzene with ˙OH in experimental solution (7.8×10^9 M^{-1} s^{-1}, in BUXTON *et al.* (1988)), [benzene] is the concentration of benzene in experimental solution (M), [S] is the concentration of a given individual ˙OH scavenger in experimental solution (M), $k_{S,OH}$ is the reaction rate constant of a given individual ˙OH scavenger with ˙OH in experimental solution (M^{-1} s^{-1}), the summation sign (Σ) reflects the numerous ˙OH scavengers present in experimental solution. If the total rate constant for ˙OH consumption excluding BPA and benzene, is significantly lower than that for BPA and benzene, then the following substitution can be made $\sum(k_{S,OH}[\text{S}]) = (k_{benzene,OH}) \times$ [benzene] $+ k_{BPA,OH}[\text{BPA}]$, where the term $k_{BPA,OH}$ is the reaction rate constant of BPA with ˙OH (M^{-1} s^{-1}), and [BPA] is the concentration of BPA in experimental solution (M). Using this substitution, Eq. (2) can be rewritten as

$$F_{benzene,OH} = k_{benzene,OH}[benzene]/\{(k_{benzene,OH})[benzene] + (k_{BPA,OH})[BPA]) \quad (3)$$

Substituting Eq. (3) into Eq. (1) yields the following:

$$R_{OH} = (R_{phenol})\{(k_{BPA,OH})[BPA] + (k_{benzene,OH})[benzene]\}/ \quad (4)$$
$$(Y_{phenol})(k_{benzene,OH})[benzene]$$

Taking the inverse of Eq. (4) and rearranging gives

$$1/R_{phenol} = (k_{BPA,OH})[BPA]/\{(R_{OH})(k_{benzene,OH})[benzene](Y_{phenol})\} \quad (5)$$
$$+ 1/(R_{OH})(Y_{phenol})$$

The formation rate of phenol (R_{phenol}) from the reaction of the ˙OH with added benzene is determined for each BPA concentration of experimental solution containing nitrate which is used as the reactant to produce ˙OH. Eq. (5) shows the relationship between [BPA] and $1/R_{phenol}$ with slope and intercept as shown below:

$$Slope = k_{BPA,OH}/\{(R_{OH})(k_{benzene,OH})[benzene](Y'_{phenol})\} \quad (6)$$

$$Intercept = 1/(R_{OH})(Y_{phenol}) \quad (7)$$

Substituting Eq. (7) into Eq. (6)

$$Slope = (k_{BPA,OH})(Intercept)/\{(k_{benzene,OH})[benzene]\} \quad (8)$$

Eq. (8) can be rewritten to yield an expression describing the rate constant for reaction of ˙OH with BPA in experimental solution, $k_{BPA,OH}$:

$$k_{BPA,OH} = (Slope)(k_{benzene,OH})[benzene]/(Intercept) \quad (9)$$

In this study, the experimental solution containing 1×10^{-3} M of sodium nitrate, 2×10^{-4} M of benzene, and $0-4.77 \times 10^{-4}$ M of BPA was used to determine $k_{BPA,OH}$. As will be seen from the measured 1×10^{-3} M of sodium nitrate solution without BPA (presented later), 2×10^{-4} M of benzene was sufficient to scavenge at least $>98\%$ of ˙OH from nitrate photolysis. Hence, it was assumed that all of ˙OH produced nitrate photolysis is consumed by BPA or benzene in the experimental solution. Thus, $k_{BPA,OH}$ was calculated using Eq. (9) in this study.

The rate constant for reaction of ˙OH with BPA ($k_{BPA,OH}$) can also be expressed as

$$k_{BPA,OH} = -\ln([BPA]_t/[BPA]_0)/\{t[˙OH]_{ss}\} \quad (10)$$

where $[BPA]_t$ is BPA concentration at irradiation time t (M), $[BPA]_0$ is initial BPA concentration (M), t is irradiation time (s), and $[˙OH]_{ss}$ is the steady-state concentration of ˙OH in river water. Eq. (10) can be rewritten as

$$t = -\ln([BPA]_t/[BPA]_0)/\{(k_{BPA,OH})[˙OH]_{ss}\} \quad (11)$$

When t is the half-lives ($t_{1/2}$) of BPA due to reaction with ˙OH (s), then the following substitution can be made $-\ln([BPA]_t/[BPA]_0) = \ln 2$. Using this substitution, Eq. (11) can be rewritten as

$$t_{1/2} = \ln 2/\{(k_{BPA,OH})[˙OH]_{ss}\}. \quad (12)$$

EXPERIMENTAL PROCEDURE

Reagents and Milli-Q water

Chemicals used for this study were highest grade or HPLC grade and were used as received: benzene (99%) and phenol (99%) from Katayama Chemicals; acetonitrile (99%) from Nacalai Tesque Inc.; BPA standard (99.9%) from Kanto Kagaku; 2-nitrobenzaldehyde ($>98\%$) from Tokyo Kasei Kogyo Co., Ltd. Sodium nitrate (99.9%, Katayama Chemicals) was used after

Table 1. Summary of water quality in Kurose river water examples

| Site name | pH | Cl^- | NO_3^- | NO_2^- | SO_4^{2-} | HCO_3^- | CO_3^{2-} | DOC^a |
				$(10^{-6}$ M$)$				$(mg\ l^{-1})$
Namitakiji	7.30	140	<2	0.43	30	93	0.10	1.54
Tokumasakami	7.36	199	53	1.30	132	358	0.46	1.25
Eguma	7.59	324	58	2.64	95	490	1.07	2.22
Ochiai	7.69	711	108	5.70	158	492	1.35	2.33

[a] DOC: dissolved organic carbon.

desiccation at 105°C for 1 h. In this work, all the water used to make up the solutions was ultra pure water (Millipore, >18 MΩ cm, air saturated).

River water samples

River water samples were obtained from four sites in the Kurose River, Hiroshima prefecture, Japan on September 18, 2001. Surface river water samples were collected in polyethylene bags and filtered through pre-cleaned glass fiber filters (0.50 μm pore size) and then stored in pre-cleaned 250 ml brown glass bottles in the dark at 4°C. Concentrations of anions (nitrate, nitrite, sulfate, carbonate and bicarbonate ions) and the pH of the river water samples were determined by ion chromatography (DX-500, Dionex) and a pH meter (HM30S, TOA). Dissolved organic carbon (DOC) in the river water samples was determined by a total organic carbon (TOC) analyzer (TOC-5000A, Shimadzu). A summary of the anion and DOC concentrations is presented in Table 1.

Photochemical instrument and procedure

The photochemical instrument used and the methods for determining the photochemical formation rate, total consumption rate constant and steady-state concentration of ˙OH are described in greater detail in ARAKAKI and FAUST (1998) and ARAKAKI et al. (1999b). In most of this work, the photoproduction of phenol was monitored as a function of time fluorescence (FL) (RF-10XL, Shimadzu) using high performance liquid chromatography (HPLC) (LC-10Ai, Shimadzu) equipped with the C18 reversed phase column (5 μm, 4.6 × 250 mm) (SUPELCO-SIL™ LC-18, SUPELCO). The mobile phase was 50% CH_3CN/50% H_2O (v/v), and the flow rate was 1.0 ml min^{-1}. The daily actinic flux was determined using 2-nitrobenzaldehyde using the same quartz cuvette that was used for the photochemical experiments. All the photochemical data were normalized to the vernal equinox solar noon condition at Higashi-Hiroshima (34°N) where the photolysis rate of 2-nitrobenzaldehyde was 0.00929 s^{-1}.

The rate constant for the reaction of BPA with nitrate-mediated ˙OH was determined using a competitive reaction with benzene. Initial concentrations of reagents used were 1×10^{-3} M of sodium nitrate, 2×10^{-4} M of benzene, and $0–4.77 \times 10^{-4}$ M of BPA. These experimental solutions were irradiated for 0, 15, 30, 45, and 60 min. After irradiation using the solar simulator, photochemical formation rates of phenol in the mixture samples were determined using HPLC–FL and the scavenging constant rate of BPA with ˙OH was calculated.

To investigate the direct photolysis of BPA, the solution containing 4.77×10^{-4} M of BPA without nitrate was irradiated with light of same photochemical instrument. After 1 h irradiation, concentration of BPA was determined by HPLC–FL (INOUE et al., 2000).

RESULTS AND DISCUSSION

Formation rate of ˙OH in authentic river water

ZHOU and MOPPER (1990) reported that there are no differences in ˙OH photoproduction rates between filtered and unfiltered water samples, except for highly turbid sample, and concluded that homogeneous photochemical process dominates ˙OH production in natural water. This result was in agreement with conclusions of some investigators (MILL *et al.*, 1980; HAAG and HOIGNE, 1985; ZEPP *et al.*, 1987). In this study, we also assumed that the photochemical reactions are homogenous reactions. The photochemical formation of ˙OH, resulting from phenol formation, was observed in all of the irradiated river water samples. The formation rates of the ˙OH based on the formation rate of phenol from the reaction of the ˙OH with benzene in river water samples, ranged from 0.70 to 3.25×10^{-10} M s^{-1} (Table 2). In the rain and dew samples, the mean production rates of the hydroxyl radical were 1.0×10^{-10} and 3.5×10^{-10} M s^{-1}, respectively, when normalized to the same actinic flux (ARAKAKI *et al.*, 1999b). HAAG and HOIGNE (1985) reported production rates of the ˙OH in Greifensee lake water were approximately 10^{-11} M s^{-1}. Therefore, the formation rates of ˙OH in Kurose river waters determined in this study are similar to those previously reported for rain, dew and lake water samples.

The formation rate of ˙OH from nitrate and nitrite ion photolysis was estimated based on the photolysis rate constants (nitrate ion: 2.43×10^{-7} s^{-1} in ARAKAKI *et al.* (1998): nitrite ion: 3.2×10^{-5} s^{-1} in ARAKAKI *et al.* (1999a)). The photolysis rate constant of nitrite was two orders of magnitude greater than that of nitrate, whereas concentrations of nitrate ion (<2–108×10^{-6} M) were greater than those of nitrite ion (0.43–5.70×10^{-6} M) in the Kurose river water samples (Table 1). Therefore, ˙OH from nitrite photolysis was greater than that from nitrate photolysis (Table 2). ARAKAKI *et al.* (1999b) reported that almost all of the ˙OH was formed from nitrite ion photolysis in dew whereas the majority of the ˙OH was formed from sources other than nitric acid or nitrous acid photolysis in rainwater. In this study, the concentration of hydrogen peroxide (H$_2$O$_2$) was not determined in Kurose river water. HAAG and HOIGNE (1985) estimated the source of ˙OH photoproduction in lake water, and suggested that neither direct nor indirect H$_2$O$_2$ photolysis was a significant source of ˙OH. If the production rate of ˙OH from direct H$_2$O$_2$ photolysis is greater than that from nitric acid or nitrous acid photolysis, then over 1×10^{-5} M of H$_2$O$_2$ must be present in river water. However, concentrations of H$_2$O$_2$ in river water samples, Hiroshima prefecture, Japan, were observed in the range 0.6–3×10^{-7} M in June (MASUDA, 1999). These results suggest that photoformation of ˙OH associated with direct H$_2$O$_2$ photolysis was not significant in the river water samples. There are other sources of ˙OH such as photolysis of dissolved organic matter (DOM) and Fe(OH)$_2$ (BREZONIK, 1994), and Fenton reaction (ZUO and HOIGNE, 1992; VAUGHAN and BLOUGH, 1998; SOUTHWORTH and VOELKER, 2003) in natural water. Contribution of these additional ˙OH sources remains to be solved by further study.

Total consumption rate constant and steady-state concentration of ˙OH in authentic river water

The total ˙OH consumption rate constant for Kurose river water was in the range 1.66–3.89×10^5 s^{-1} (Table 3). These rate constants are higher than those for rainwater, and similar to those for dew water (ARAKAKI *et al.*, 1999b). Using the reaction rate constant and concentration, the consumption rate constants of ˙OH for each anion (Cl$^-$, NO$_2^-$, NO$_3^-$, CO$_3^{2-}$, HCO$_3^-$, and SO$_4^{2-}$) were estimated (Table 3). In river water, the summation of

Table 2. ˙OH photoformation rates and reactants to produce ˙OH in Kurose river water samples collected in Higashi-Hiroshima city

Site name	˙OH formation rate[a] (10^{-10} M s^{-1})	˙OH formed %[b]		
		NO$_3^-$	NO$_2^-$	Others
Namitakiji	0.70	<1	20	80
Tokumasakami	0.75	17	56	27
Eguma	1.69	6	50	44
Ochiai	3.25	6	56	38

[a] ˙OH formation rates are normalized to clear sky, noon conditions of Higashi-Hiroshima on May 1.
[b] ˙OH formation rate constant for NO$_3^-$: 2.43×10^{-7} s^{-1} (ARAKAKI *et al.*, 1998) and for NO$_2^-$: 3.2×10^{-5} s^{-1} (ARAKAKI *et al.*, 1999a).

Table 3. ˙OH consumption rate constants and reactants to consume ˙OH in Kurose river water samples collected in Higashi-Hiroshima city

Site name	˙OH consumption rate constant (10^5 s^{-1})	˙OH scavenged %[a]						
		HCO_3^-	CO_3^{2-}	NO_2^-	Cl^-	NO_3^-	SO_4^{2-}	Others
Namitakiji	2.10	0.67	0.02	2.05	1.33	0.00	0.02	95.9
Tokumasakami	1.66	3.24	0.12	7.84	2.40	0.02	0.12	86.3
Eguma	2.39	3.08	0.19	11.0	2.71	0.01	0.06	82.9
Ochiai	3.89	1.90	0.15	14.7	3.66	0.01	0.06	79.6

[a] Reaction rate constants for HCO_3^- and CO_3^{2-}: 1.5×10^7 and $4.2 \times 10^8 \text{ M}^{-1} \text{ s}^{-1}$, respectively (WEEKS and RABANI, 1966), for NO_2^-: $1 \times 10^{10} \text{ M}^{-1} \text{ s}^{-1}$ (TREININ and HAYRON, 1970), for Cl^-: $2.0 \times 10^7 \text{ M}^{-1} \text{ s}^{-1}$ (ARAKAKI et al., 1999b) and for NO_3^- and SO_4^{2-}: 5×10^5 and $1.5 \times 10^6 \text{ M}^{-1} \text{ s}^{-1}$, respectively (ARAKAKI and FAUST, 1998).

consumption rate constants of ˙OH for these anions was less than 25% of the total consumption rate constant. It was speculated that the majority of ˙OH was consumed by dissolved organic compounds in the rainwater (ARAKAKI et al., 1999b). The influence of DOM on ˙OH consumption in samples from this study was investigated. The rate constant of DOM (k_{DOM}) has been reported by several workers. ZEPP et al. (1987) obtained a value of $2.5 \times 10^4 \text{ (mg C/l)}^{-1} \text{ s}^{-1}$ for k_{DOM} from Greifensee, Switzerland. HAAG and HOIGNE (1985) reported that k_{DOM} was $2.5–4.0 \times 10^4 \text{ (mg C/l)}^{-1} \text{ s}^{-1}$ for Swiss lake water. For isolates of DOM from surface waters around the United States, WESTERHOFF et al. (1998) reported an average value of $2.9 \times 10^4 \text{ (mg C/l)}^{-1} \text{ s}^{-1}$. For standard Suwannee River fulvic acid and humic acid, k_{DOM} were in the range $3.1–4.4 \times 10^4$ and $1.9–6.8 \times 10^4 \text{ (mg C/l)}^{-1} \text{ s}^{-1}$, respectively (WESTERHOFF et al., 1999; GOLDSTONE et al., 2002; SOUTHWORTH and VOELKER, 2003). Using values of $2–6 \times 10^4 \text{ (mg C/l)}^{-1} \text{ s}^{-1}$ for k_{DOM}, DOM consumed 12–56% of ˙OH in river water.

The lifetime of ˙OH in the river water, based on the reciprocal of the consumption rate constant, was in the range $2.6–6.0 \times 10^{-6}$ s (Table 4). This value is similar to ˙OH lifetimes reported in dew (ARAKAKI et al., 1999b) and cloud waters (ARAKAKI and FAUST, 1998), while those in polluted cloud waters based on a modeling study were in the range $3–66 \times 10^{-6}$ s (JAKOB, 1986). Based on the photochemical formation rate and the total consumption rate constant of ˙OH, steady-state ˙OH concentrations were calculated to be in the range $3.3–8.4 \times 10^{-16}$ M (Table 4), which is similar to reported values in river water (BREZONIK and FULKERSON-BREKKEN, 1998), and in rain and dew water (ARAKAKI et al., 1999b).

Direct BPA photolysis and rate constant for reactions of ˙OH with BPA

BPA exhibits a slight absorption of UV light wavelengths exceeding 290 nm in neutral and acidic methanol solutions (HOWARD, 1989). Therefore, direct BPA photolysis was investigated using the artificial solar light. In a time of less than 1 h of the irradiation with the solar simulator, no significant degradation of BPA was observed in Milli-Q water containing BPA. Hence, we assumed that there is little direct BPA photolysis in surface water.

Table 4. Lifetime and steady-state concentration of ˙OH in Kurose river water samples

Site name	˙OH formation rate[a] $(10^{-10} \text{ M s}^{-1})$	˙OH consumption rate constant (10^5 s^{-1})	˙OH lifetime[b] (10^{-6} s)	$[\text{˙OH}]_{ss}$[c] (10^{-16} M)
Namitakiji	0.70	2.10	4.76	3.33
Tokumasakami	0.75	1.66	6.03	4.51
Eguma	1.69	2.39	4.19	7.09
Ochiai	3.25	3.89	2.57	8.35

[a] ˙OH formation rates are normalized to clear sky, noon conditions of Higashi-Hiroshima on May 1.
[b] ˙OH lifetime = (˙OH consumption rate constant)$^{-1}$.
[c] $[\text{˙OH}]_{ss}$ = (˙OH formation rate)/(˙OH consumption rate constant).

Nitrate also reacts with ˙OH, while it was used as the source of ˙OH to determine rate constants for reactions between ˙OH and BPA. Using 1×10^{-3} M of sodium nitrate solution containing 2×10^{-4} M of benzene without BPA, most of ˙OH (>98%) initiated by nitrate photolysis was scavenged by benzene. Hence, it was assumed that the nitrate-mediated ˙OH is consumed by benzene and BPA in the experiment solutions (1×10^{-3} M of sodium nitrate, 2×10^{-4} M of benzene, and $0–477 \times 10^{-3}$ M of BPA). There was no production of phenol in the reaction for ˙OH with BPA. It suggested that the photoproduction of phenol is only originated from the reaction of benzene with ˙OH.

Our experiment showed that formation rate of phenol from the reaction of ˙OH with benzene was reduced by the addition of BPA. The reciprocal phenol formation rates are linearly correlated to the concentration of BPA (Fig. 1). From the slope and intercept of the regression line, the rate constant for reaction of ˙OH with BPA was calculated to be 1.55×10^{10} M^{-1} s^{-1}. This value is close to diffusion-controlled limits for bimolecular reactions ($\sim 10^{10}$ M^{-1} s^{-1}, in BREZONIK and FULKERSON-BREKKEN (1998)). HAAG and YAO (1992) estimated rate constants for reactions of ˙OH with several drinking water contaminants using the relative rate method. The rate constants for the reactions of ˙OH with PAHs and methoxychlor were found to have values of 1×10^{10} and 2×10^{10} M^{-1} s^{-1}, respectively, close to the value obtained for BPA in this study.

Estimation of BPA degradation

Using the rate constant for the reaction of ˙OH with BPA (1.55×10^{10} M^{-1} s^{-1}) and steady-state ˙OH concentrations in the river water ($3.3–8.4 \times 10^{-16}$ M), half-lives of BPA due to reaction with ˙OH in surface waters were 0.6–1.6 days of midday-May sunlight in the Hiroshima prefecture. Steady-state ˙OH concentrations are changed by light attenuation which affects the photoformation rate of ˙OH. For example, the photoformation rate of ˙OH from nitrate in the surface water in midwinter at latitude 40°N was one-fifth that of midsummer. Furthermore, the photoformation rate of ˙OH decreases with increasing water depth due to the absorption of light in shallow water bodies (ZEPP et al., 1987). Therefore, it is speculated that the steady-state ˙OH concentrations in this study may be near the maximum rate, and that half-lives of BPA in the river water should be larger than those in this study.

Estimated half-lives of BPA in the surface river water due to the reaction of ˙OH were similar to those in biological process. DORN et al. (1987) reported that using three river water samples from near plastics manufacturing facility, more than 90% degradation of BPA within 4 days was observed and the half-lives of BPA were in the range 2.5–4 days. In surface river waters from seven rivers across the United States and Europe, rapid biodegradation of BPA was observed following lag phase ranging of 2–4 days (KLECKA et al., 2001). KANG and

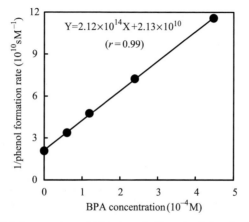

FIG. 1. Relationship between reciprocal of photoproduction rate of phenol and BPA concentration.

KONDO (2002) also found that half-lives for BPA biodegradation in surface waters taken from three rivers were in the range 2–3 days. Therefore, our results suggest that the degradation of BPA due to reaction with ˙OH in surface river water is an important process in controlling the fate of BPA.

Using the rate constant for reaction of ˙OH with BPA found in this study, the degradation of BPA in seawater was also estimated. In natural seawater, ZHOU and MOPPER (1990) reported that the steady-state concentrations of ˙OH were in the range of $0.6–1.4 \times 10^{-17}$ M, which were much lower than those in river water of our study. Using these concentrations and the rate constant for reaction in our study, estimated half-lives of BPA due to reaction with ˙OH in seawater are in the range of 37–86 days. Apparently, it indicates that the degradation rate of BPA reaction with ˙OH is lower in seawater than in river water due to the smaller steady-state concentrations of ˙OH in seawater. In the biodegradation process, rapid degradation of BPA was observed in seawater following the lag phase over 30 days (YING and KOOKANA, 2003). Thus, BPA degradation due to reaction with ˙OH may be not so important in seawater.

CONCLUSION

The photochemical formation rate $(0.70–3.25 \times 10^{-10}$ M s$^{-1})$, the total consumption rate constant $(1.66–3.89 \times 10^5$ s$^{-1})$ and the steady-state concentration of ˙OH $(3.3–8.4 \times 10^{-16}$ M) in river water samples collected in Higashi-Hiroshima, Japan were measured in this study. The measured values were similar to previous values reported for river, rain and dew water samples. In the investigation of production mechanisms of ˙OH, it was found that ˙OH production from nitrite photolysis was greater than that from nitrate photolysis in Kurose river water. The summation of consumption rate constants of ˙OH for major anions occurring in river water was less than 25% of the total consumption rate constant. Based on the reaction rate constant of ˙OH for DOM, it is estimated that DOM accounts for 12–56% of the total consumption rate constant of ˙OH in river water.

Our preliminary study indicated that direct BPA photolysis in river water is very slow and thus an insignificant process. The rate constant for reaction of ˙OH with BPA was measured using competition kinetic between benzene and BPA. The measured rate constant is 1.55×10^{10} M^{-1} s^{-1}, which is close to the diffusion-controlled limit for bimolecular reactions. Using the rate constant for the reaction of ˙OH with BPA and steady-state ˙OH concentrations in the river water, half-lives of BPA due to reaction with ˙OH in surface river waters were estimated to be 0.6–1.6 days. Based on the comparison between half-lives of BPA determined in present study and those in biological process reported by previous investigators, it is concluded that BPA degradation due to reaction with ˙OH is an important process in determining the fate of BPA in surface river water.

Acknowledgements—We thank Dr T. Miyake and Dr K. Takeda for providing many of the laboratory methods and experimental advice. We also are obliged to Dr T. Arakaki for discussions and comments on the manuscript. We also are obliged to Dr J. Kolak for reviewing the manuscript.

REFERENCES

ANDREOZZI R., RAFFAELE M. and NICKLAS P. (2003) Pharmaceuticals in STP effluents and their solar photodegradation in aquatic environment. *Chemosphere* **50**, 1319–1330.

ARAKAKI T. and FAUST B. C. (1998) Sources, sinks, and mechanisms of hydroxyl radical (˙OH) photoproduction and consumption in authentic acidic continental cloud waters from Whiteface Mountain, New York: the role of the Fe(r) (r = II, III) photochemical cycle. *J. Geophys. Res.* **103**, 3487–3504.

ARAKAKI T., MIYAKE T., SHIBATA M. and SAKUGAWA H. (1998) Measurement of photochemically formed hydroxyl radical in rain and dew waters. *Nippon Kagaku Kaishi* No. 9, 619–625 (in Japanese with English abstract).

ARAKAKI T., MIYAKE T., HIRAKAWA T. and SAKUGAWA H. (1999a) pH dependent photoformation of hydroxyl radical and absorbance of aqueous-phase N(III) (HNO$_2$ and NO$_2^-$). *Environ. Sci. Technol.* **33**, 2561–2565.

ARAKAKI T., MIYAKE T., SHIBATA M. and SAKUGAWA H. (1999b) Photochemical formation and scavenging of hydroxyl radical in rain and dew waters. *Nippon Kagaku Kaishi* No. 5, 335–340 (in Japanese with English abstract).

ASHBY J. and TINWELL H. (1998) Uterotrophic activity of bisphenol A in the immature rat. *Environ. Health Perspect.* **106**, 719–720.

BREZONIK P. L. (1994) *Chemical Kinetics and Process Dynamics in Aquatic Systems.* Lewis Publishers/CRC Press, Boca Raton, FL, 754 pp.

BREZONIK P. L. and FULKERSON-BREKKEN J. (1998) Nitrate-induced photolysis in natural waters: controls on concentrations of hydroxyl radical photo-intermediates by natural scavenging agents. *Environ. Sci. Technol.* **32**, 3004–3010.

BUXTON G. V., GREENSTOCK C. L., HELMAN W. P. and ROSS A. B. (1988) Critical review of rate constants for reactions of hydrated electrons, hydrogen atoms and hydroxyl radicals ($^{\cdot}$OH/O^{-}) in aqueous solution. *J. Phys. Chem. Ref. Data* **17**, 513–886.

COOPER W. J. and HERR F. L. (1987) Introduction and overview. In *Photochemistry of Environmental Aquatic Systems* (eds. R. G. ZIKA and W. J. COOPER), pp. 1–9. American Chemical Society, Washington, DC.

DORN P. B., CHOU C.-S. and GENTEMPO J. J. (1987) Degradation of bisphenol A in natural waters. *Chemosphere* **16**, 1501–1507.

FULKERSON-BREKKEN J. and BREZONIK P. L. (1998) Indirect photolysis of acetochlor: rate constant of a nitrate-mediated hydroxyl radical reaction. *Chemosphere* **36**, 2699–2704.

GOLDSTONE J. V., PULLIN M. J., BERTILSSON S. and VOELKER B. M. (2002) Reactions of hydroxyl radical with humic substances: bleaching, mineralization, and production of bioavailable carbon substrates. *Environ. Sci. Technol.* **36**, 364–372.

HAAG W. R. and HOIGNE J. (1985) Photo-sensitized oxidation in natural water via $^{\cdot}$OH radicals. *Chemosphere* **14**, 1659–1671.

HAAG W. R. and YAO C. C. D. (1992) Rate constants for reaction of hydroxyl radicals with several drinking water contaminants. *Environ. Sci. Technol.* **26**, 1005–1013.

HOWARD P. H. (1989) *Handbook of Environmental Fate and Exposure Data for Organic Chemicals, Vol. I.* Lewis Publishers, Chelsea, MI, 574 pp.

INOUE K., KATO K., YOSHIMURA Y., MAKINO T. and NAKAZAWA H. (2000) Determination of bisphenol A in human serum by high-performance liquid chromatography with multi-electrode electrochemical detection. *J. Chromatogr. B* **749**, 17–23.

JAKOB D. J. (1986) Chemistry of OH in remote clouds and its role in the production of formic acid and peroxymonosulfate. *J. Geophys. Res.* **91**, 9807–9826.

Japan Environmental Agency (2002) *Survey of Endocrine Disrupting Chemicals in the Environment (FY2001).* Reports of Ministry of the Environment, Japan, 229 pp. (in Japanese).

KANG J.-H. and KONDO F. (2002) Bisphenol A degradation by bacteria isolated from river water. *Arch. Environ. Contam. Toxicol.* **43**, 265–269.

KLECKA G. M., GONSIOR S. J., WEST R. J., GOODWIN P. A. and MARKHAM D. A. (2001) Biodegradation of bisphenol A in aquatic environments: river die-away. *Environ. Toxicol. Chem.* **20**, 2725–2735.

KRISHNAN A. V., STATHIS P., PERMUTH S. F., TOKES L. and FELDMAN D. (1993) Bisphenol-A: an estrogenic substance is released from polycarbonate flasks during autoclaving. *Endocrinology* **132**, 2279–2286.

MASUDA N (1999) Determination and mechanism of peroxide and OH radical in atmospheric and natural waters. Masters Thesis. Hiroshima University, Higashi-Hiroshima, Japan, 154 pp. (in Japanese).

MILL T., HENDRY D. G. and RICHARDSON H. (1980) Free-radical oxidants in natural waters. *Science* **207**, 886–887.

MILLIGAN S. R., BALASUBRAMANIAN A. V. and KALITA J. C. (1998) Relative potency of xenobiotic estrogens in an acute *in vivo* mammalian assay. *Environ. Health Perspect.* **106**, 23–26.

RUSSI H., KOTZIAS D. and KORTE F. (1982) Photoinduzierte hydroxylierungsreaktionen organischer chemikalien in naturlichen gewassern: nitrate als potentielle OH-radikalquellen. *Chemosphere* **11**, 1041–1048.

SOUTHWORTH B. A. and VOELKER B. M. (2003) Hydroxyl radical production via the photo-fenton reaction in the presence of fulvic acid. *Environ. Sci. Technol.* **37**, 1130–1136.

STAPLES C. A., DORN P. B., KLECKA G. M., O'BLOCK S. T. and HARRIS L. R. (1998) A review of the environmental fate, effects, and exposures of bisphenol A. *Chemosphere* **36**, 2149–2173.

TORRENTS A., ANDERSON B. G., BILBOULIAN S., JOHNSON W. E. and HAPEMAN C. J. (1997) Atrazine photolysis: mechanistic investigations of direct and nitrate-mediated hydroxyl radical processes and the influence of dissolved organic carbon from the Chesapeake bay. *Environ. Sci. Technol.* **31**, 1476–1482.

TREININ A. and HAYRON E. (1970) Absorption spectra and reaction kinetics of NO_2, N_2O_3, and N_2O_4 in aqueous solution. *J. Am. Chem. Soc.* **92**, 5821–5828.

VAUGHAN P. P. and BLOUGH N. V. (1998) Photochemical formation of hydroxyl radical by constituents of natural waters. *Environ. Sci. Technol.* **32**, 2947–2953.

WEEKS J. L. and RABANI J. (1966) The pulse radiolysis of deaerated aqueous carbonate solutions. I. Transient optical spectrum and mechanism. II. pK for OH radicals. *J. Phys. Chem.* **70**, 2100–2106.

WESTERHOFF P., SONG R., AMY G. and MINEAR R. (1998) NOM's role in bromine and bromate formation during ozonation. *J. Am. Water Works Assoc.* **89**, 82–94.

WESTERHOFF P., AIKEN G., AMY G. and DEBROUX J. (1999) Relationships between the structure of natural organic matter and its reactivity towards molecular ozone and hydroxyl radicals. *Water Res.* **33**, 2265–2276.

YING G.-G. and KOOKANA R. S. (2003) Degradation of five selected endocrine-disrupting chemicals in seawater and marine sediment. *Environ. Sci. Technol.* **37**, 1256–1260.

ZAFIRIOU O. C., JOUSSOT-DUBIEN J., ZEPP R. G. and ZIKA R. G. (1984) Photochemistry of natural waters. *Environ. Sci. Technol.* **18**, 356A–371A.

ZEPP R. G., HOIGNE J. and BADER H. (1987) Nitrate-induced photooxidation of trace organic chemicals in water. *Environ. Sci. Technol.* **21**, 443–450.

ZHOU X. and MOPPER K. (1990) Determination of photochemically produced hydroxyl radicals in seawater and freshwater. *Mar. Chem.* **30**, 71–88.

ZUO Y. and HOIGNE J. (1992) Formation of hydrogen peroxide and depletion of oxalic acid in atmospheric water by photolysis of iron(III)–oxalato complexes. *Environ. Sci. Technol.* **26**, 1014–1022.

Geochemical Investigations in Earth and Space Science: A Tribute to Isaac R. Kaplan
© The Geochemical Society, Publication No. 9, 2004
Editors: R.J. Hill, J. Leventhal, Z. Aizenshtat, M.J. Baedecker, G. Claypool,
R. Eganhouse, M. Goldhaber and K. Peters

Organic and inorganic compositions of marine aerosols from East Asia: Seasonal variations of water-soluble dicarboxylic acids, major ions, total carbon and nitrogen, and stable C and N isotopic composition

KIMITAKA KAWAMURA[1], MINORU KOBAYASHI[1], NOBUYUKI TSUBONUMA[1],
MICHIHIRO MOCHIDA[1], TOMOMI WATANABE[1] and MEEHYE LEE[2]

[1]Institute of Low Temperature Science, Hokkaido University, N19 W8, Kita-ku, Sapporo 060-0819, Japan
[2]Department of Earth and Environmental Sciences, Korea University, Seoul 136-701, South Korea

Abstract—Atmospheric particles were collected for 1 year (2001–2002) at a site Gosan on Jeju Island, South Korea. The samples were analyzed for water-soluble dicarboxylic acids and related compounds using a capillary GC and GC/MS. Total carbon and nitrogen contents, as well as their stable isotopic ratios, were also determined using elemental analyzer (EA) and EA/IR/MS, respectively. Elemental and isotopic analyses were also performed after the HCl fume treatment of aerosol samples. The results demonstrate that calcium carbonate of dust origin was not a significant component, except for a few dust event samples. Carbonates most likely reacted with acidic species (such as H_2SO_4, HNO_3 and organic acids) in the atmosphere during a long-range transport, and did not contribute to the total aerosol carbon, except for few strong dust episodes that occurred in spring. This study also demonstrates that nitrate and ammonium largely contribute to aerosol nitrogen, and organic nitrogen is a minor component in the Asian aerosols. Acidic aerosols enriched with sulfate likely adsorb NH_3 gas that is emitted from soils in Asia whereas mineral dust particles containing carbonate may adsorb HNO_3 and volatile organic acids.

Homologous series of dicarboxylic acids (C_2–C_{12}) including unsaturated structures as well as ketocarboxylic acids (C_2–C_9) were detected in the aerosol samples with oxalic (C_2) acid being the most abundant species followed by malonic (C_3) or succinic (C_4) acids. Total concentrations of diacids (130–2070 ng m^{-3}) are one or two orders of magnitude greater than those reported for remote marine aerosols in the North Pacific, but are equivalent to those reported for urban aerosols. Very high concentrations of dusts (up to 880 μg m^{-3}) and water-soluble dicarboxylic acids were often observed in early spring. However, their correlation ($r = 0.27$) is not strong throughout the campaign. Although both mineral dusts and diacids are derived from the Asian continent, they originate from different sources and source regions. Diacids are probably emitted from urban sources in East Asia and produced by secondary photochemical oxidation of their precursors, whereas dusts are derived from the arid interior regions. This study provides evidence for photochemical production of water-soluble organic acids in East Asia and the western North Pacific rim.

INTRODUCTION

ORGANIC ACIDS are ubiquitous in the atmosphere (*e.g.* KAWAMURA and KAPLAN, 1991), natural waters (*e.g.* THURMAN, 1985) and sediments (*e.g.* LEWAN and FISHER, 1994). Due to the carboxyl groups, they play an important role in geochemical processes on the earth surface in several ways. For example, small organic acids such as formic, acetic, and oxalic acids dissolve calcium carbonate and other minerals resulting in a secondary porosity in sedimentary rocks (SURDAM *et al.*, 1984). This process contributes to the migration of oils that are produced from geopolymers such as kerogen (*e.g.* TISSOT and WELTE, 1984). Although the sources and formation mechanisms are still not well understood, microbial oxidation of organic matter and thermal degradation of geopolymers (kerogen and humic acids) are important sources of organic acids (KHRAKA *et al.*, 1983; KAWAMURA *et al.*, 1986; KAWAMURA and KAPLAN, 1987; LEWAN and FISHER, 1994). Small organic acids are abundantly present in oil formation waters (KHRAKA *et al.*, 1983). Organic acids also form organo-metal complexes in the sedimentary and aquatic environments. This process accelerates the dissolution of trace metals in sediments and particles in the natural waters. The formation of the metal complexes has an impact on the geochemical behaviors of metals in river and ocean waters as well as sediments, controlling their distribution and transport on the earth surface.

Organic acids are also present in the atmosphere as gases, adsorbed on particles as well as dissolved in rainwaters and on snow crystals (GROSJEAN *et al.*, 1978; DAWSON *et al.*, 1980; KEENE *et al.*, 1983; KAWAMURA and KAPLAN, 1984; KAWAMURA *et al.*, 1985a,b,c; KAWAMURA and KAPLAN, 1991; NOLTE *et al.*, 1997). Mono- and di-carboxylic acids have been reported in continental aerosols (GROSJEAN *et al.*, 1978; NORTON *et al.*, 1983; KAWAMURA and KAPLAN, 1986; SATSUMABAYASHI *et al.*, 1990; KAWAMURA *et al.*, 2000; WANG *et al.*, 2002), marine aerosols (KAWAMURA and USUKURA, 1993; BARBOUKAS *et al.*, 2000; KAWAMURA and SAKAGUCHI, 1999, MOCHIDA *et al.*, 2003a) and aerosols from polar regions (LI and WINCHESTER, 1993; KAWAMURA *et al.*, 1996a,b; NARUKAWA *et al.*, 2002). Because they are very water soluble, their presence in the aerosols alters the chemical and physical properties of atmospheric aerosols (SAXENA *et al.*, 1995). In general, atmospheric particles act as cloud condensation nuclei (CCN) and contribute to the formation of cloud droplets (*e.g.* YU, 2000). In these processes, organic acids serve as an agent to enhance hygroscopic properties of the particles and play an important role in the cloud processes. This effect is important, as the water-soluble properties of organic aerosols indirectly control solar radiation out at the earth's atmosphere, thus cooling the earth surface (IPCC, 2001). On a regional scale, such as the Asian Pacific region, aerosols produced by anthropogenic activity could reduce the effect of global warming caused by greenhouse gas on a global scale (HUEBERT *et al.*, 2003).

Nitrogen is also an important atmospheric composition and is abundantly present in aerosols and rain (*e.g.* YEATMAN *et al.*, 2001; MACE *et al.*, 2003). For example, total nitrogen (TN) contents in urban aerosols from Tokyo are relatively high comprising 2–15% (av. 5%) of aerosol mass (KAWAMURA *et al.*, 1995). The weight ratios of TN to total carbon (TC) in the aerosols are on average 4.8 (range 2.1–10.9). Occasionally, aerosol nitrogen is about a half the carbon content. Although nitrate and ammonium ions have been most extensively studied in aerosols, their relationship to total aerosol nitrogen has not been studied. Further, stable carbon and nitrogen isotopic ratios of aerosols have not been extensively studied (CHESSELET *et al.*, 1981; CACHIER *et al.*, 1986; NARUKAWA *et al.*, 1999; YEATMAN *et al.*, 2001; MARTINELLI *et al.*, 2002; TUREKIAN *et al.*, 2003). However, they could provide additional information on the source and source region of aerosols and their transformation processes in the atmosphere.

In this study, we conducted an organic geochemical study on the atmospheric particles collected at a site in Gosan on Jeju Island off Korea Peninsula. Particular attention has been paid to atmospheric chemical studies in East Asia because anthropogenic emissions of gas and aerosols in this region are significant on a global scale due to the growing industrial activity (HUEBERT *et al.*, 2003). Further, arid inland areas (such as Gobi and Takla Makan deserts) are important source regions in spring for long-range atmospheric transport of mineral dusts over the Pacific (*e.g.* DUCE *et al.*, 1980; UEMATSU *et al.*, 1983; MORI *et al.*, 2002) and contribute to the sediments in the deep ocean floor (*e.g.* BLANK *et al.*, 1985). The Gosan site has been used as a "super site" of ground stations during an intensive period of the ACE-Asia (Asian Pacific Regional Aerosol Characterization Experiment) campaign (HUEBERT *et al.*, 2003). Jeju Island is located in the pathway of atmospheric transport of the aerosol particles from the Asian continent to the Pacific. Here, we report on 1-year measured data of water-soluble organic compounds (dicarboxylic acids, ketoacids and dicarbonyls). We also report on the chemical characterization of major ions and TC and TN as well as their stable isotopic composition, and discuss the contribution of soil dust to the aerosols and their transformation processes in the atmosphere during a long-range transport.

EXPERIMENTAL

Aerosol sampling (totally 107 samples) was conducted at Gosan site on Jeju Island ($33°29'$N, E$126°16'$E) as part of ACE-Asia campaign from April 2001 to March 2002 on a daily or few days basis. The island (126.08–126.58°E, 33.06–126.58°N, area: 1847 km²) is located at the boundary of the Yellow Sea and the East China Sea, and is surrounded by mainland China, Korea Peninsula, and Kyushu Island, Japan (Fig. 1). The Gosan site is located on a cliff (elevation: 71 m above sea level) on the western edge of the island facing the Asian continent and is isolated from residential areas. The population of the island is 552,000, but major residential areas are located entirely on the northeast side of the island. Dominant winds are westerlies in winter to spring whereas winds from the Pacific (from the south) occur in summer. Aerosol particles were collected on pre-combusted (450°C, 3 h) quartz fiber filters (Pallflex 2500QAT-UP, 20 cm × 25 cm) using a high-volume air sampler (Kimoto AS-810). The sampler was installed on top of a tower (15 m above the ground) during April 2001. After May 2001, the sampler was moved to the rooftop of a trailer, at a height of *ca.* 3 m above ground. Before and after the sampling, filters were stored in a clean glass jar (150 ml) with a Teflon-lined cap. The filter samples were stored at −20°C in a dark room prior to analysis. Several field blanks were taken at the site by attaching the filter to the sampler for few seconds without sucking air. Aerosol mass concentrations were obtained by weighing the quartz filter before and after the aerosol sampling.

Totally, 48 aerosol filter samples were used for the analysis of water-soluble diacids and related compounds using the method described in KAWAMURA and IKUSHIMA (1993) and KAWAMURA (1993) with some modification. Briefly, aliquots of the filters were extracted with pure water (5 ml × 3), followed by the extraction with ethyl acetate (5 ml × 3). The latter solvent was used to extract fatty acids and other lipids. The extracts were combined and concentrated using a rotary evaporator under vacuum and then dried using a nitrogen blow-down

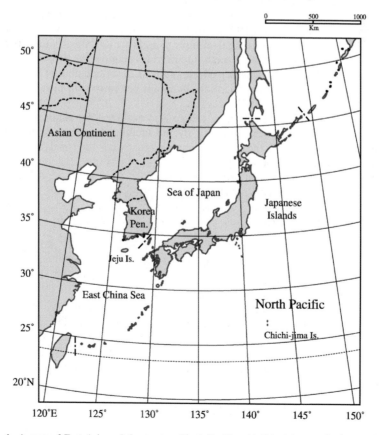

FIG. 1. A map of East Asia and the western North Pacific with Jeju Island (33°29′N, E126°16′E): a sampling location for Asian aerosols.

system under atmospheric pressure. The concentrated extracts were reacted with 14% BF_3/ *n*-butanol in a pear-shaped flask (25 ml) to derivatize carboxyl groups to butyl esters and carbonyl groups to dibutoxy acetals. The derivatives were extracted and then analyzed with a HP 6890 gas chromatograph (GC) installed with a split/splitless injector, fused silica capillary column (HP-5, 0.2 mm × 25 m × 0.52 μm film thickness), and a flame ionization detector. The column oven temperature was programmed from 50°C (2 min) to 120°C at 30°C/min and then to 320°C at 5°C/min.

Identification of the diacid butyl esters was conducted by comparing the GC retention times with those of authentic standards. Mass spectra of the organic compounds were also obtained with a GC/MS (ThermoQuest, Trace MS) using a similar GC condition. Duplicate analyses of several filter samples showed that the analytical errors for major diacids were within 10%. Spiked experiments using authentic standards (oxalic, malonic, succinic and adipic acids) in quartz filter showed that recoveries were 77% for oxalic acid and better than 86% for other diacids. Recoveries of glyoxylic acid, pyruvic acid, and methylglyoxal were 88, 72, and 47, respectively (KAWAMURA and YASUI, in preparation). The procedural blanks showed that contamination levels of the diacids during the analysis were less than 5% of the sample. The data reported here are corrected for the procedural blanks, but not for recoveries.

For the TC and TN analyses, aerosol filter samples ($n = 107$) were cut in small disk (area 3.14 cm^2) and analyzed using an elemental analyzer (EA) (Carlo Erba, NA 1500). Stable carbon and nitrogen isotopic analyses were also conducted using the same EA interfaced to isotope ratio mass spectrometer (IRMS) (ThermoQuest, Delta Plus) by the method described in NARUKAWA *et al.* (1999). The samples were analyzed in duplicate and averaged concentrations and isotopic ratios are reported here after the blank correction. Reproducibility of TC and TN measurements was within 2%. Analytical errors in the carbon and nitrogen isotope ratios are within 0.2 and 0.3‰, respectively. In order to remove carbonate carbon from the aerosol TC, other filter cuts were treated with HCl fume as follows. Each filter cut was placed in a 50 ml glass vial and was exposed to HCl fume overnight in a glass (10 l) desiccator. Excess HCl was removed from the sample with NaOH and P_2O_5 in a desiccator. The HCl-treated filters were analyzed for TC and TN as well as their isotopic ratios as described above. We found a loss of TN content during the HCl treatment due to the evaporative removal of nitrate (as HNO_3) from the filter sample.

Aliquots of filter samples (0.78–3.14 cm^2) were also analyzed for major anions and cations using ion chromatography (Dionex, DX-500) by the method described in NARUKAWA *et al.* (2002). Duplicate analyses of the aerosol samples showed the analytical errors to be within 20%.

RESULTS AND DISCUSSION

Total aerosol mass concentrations

Throughout the 1-year observation of aerosols at Jeju Island, we found enhanced concentrations of total aerosol mass in spring, as shown in Fig. 2a. Total aerosol masses were generally around 100 μg m^{-3} or less, but they significantly increased in spring to more than 400 μg m^{-3}. Especially, we observed the concentration of more than 800 μg m^{-3} in 2002 March. Very high dust concentrations (up to 640 μg m^{-3}) were also recorded in Sapporo, Hokkaido, Japan (KAWAMURA *et al.*, 2002). Higher concentrations observed in spring are associated with strong westerly winds and uplift of mineral dusts from main land China and Mongolia. In fact, very high aerosol mass concentrations (1000–10,000 μg m^{-3}) have been reported in spring near the source regions of the interior Asian continent (MORI *et al.*, 2002). These values obtained in the coastal Southeast China Sea off Korea Peninsula are much higher than those reported in Tokyo in 1988–1989 (54–314 μg m^{-3}, av. 108 μg m^{-3}, KAWAMURA and IKUSHIMA, 1993) and those observed in 1990–1993 at Chichi-jima Island in the western North Pacific (11–292 μg m^{-3}, av. 53 μg m^{-3}, KAWAMURA *et al.*, 2003a). These results suggest that most of the peaks are involved with the outflow of Asian dusts from arid regions in Mongolia and Takla Makan deserts (DUCE *et al.*, 1980; UEMATSU *et al.*, 1983).

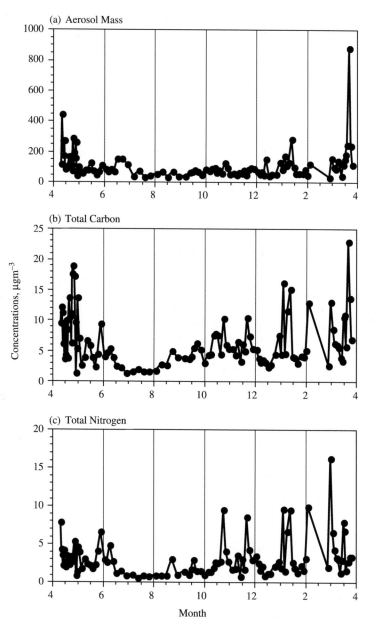

FIG. 2. Seasonal variation in the concentrations of (a) aerosol mass, (b) total carbon, and (c) total nitrogen in the marine aerosols collected in Jeju Island.

Air mass trajectory analyses demonstrated that air masses in winter and spring mostly came from the west (Asia). In contrast, southerly winds often came from the Pacific Ocean in the summer time. We also observed very high concentrations of Ca in aerosol samples with a high mass. The correlation coefficient between aerosol mass and nss-Ca concentration is high ($r = 0.90$) for all samples, as seen in Fig. 3a, although the highest correlations were found to occur at specific times in the year. For example, correlation was very strong for the 2001 spring ($r = 0.997$; April 12–20, see Fig. 3b) and the winter season ($r = 0.98$; October 2002–March 2003) whereas it was very weak ($r = 0.02$) for the summer season (June–September 2002).

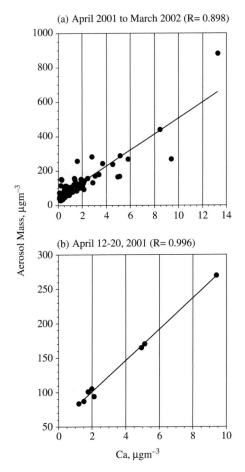

FIG. 3. Relationship between concentrations of total aerosol mass and non-sea salt Ca in the aerosols from Jeju Island.

Dusts are mostly derived from Asian continent in winter to spring and marine contributions are in general not important for the aerosol samples studied over Jeju Island ($r = 0.51$ for both the aerosol concentrations and Na).

TC contents and stable carbon isotopic ratios

Table 1 summarizes the results of TC and TN measurements of the aerosol samples as well as stable carbon and nitrogen isotopic composition and major ions. TC contents ranged from 1.3 to 22.9 μg m^{-3} with an average of 6.5 μg m^{-3}. These values are much lower than those observed in urban areas such as Tokyo (10.2–44.1 μg m^{-3}, av. 22.3 μg m^{-3}, KAWAMURA and IKUSHIMA, 1993) and Beijing (24–85 μg m^{-3}, av. 49 μg m^{-3}, SEKINE *et al.*, 1992). However, they are much higher than those reported for the remote marine atmosphere in the western North Pacific (*e.g.* Chichi-jima, 0.11–1.9 μg m^{-3}, av. 0.63 μg m^{-3}, KAWAMURA *et al.*, 2003a). These results suggest that Jeju Island located in the western Pacific rim is heavily influenced from the Asian continent, where emission of both mineral dusts and organic pollutants are significant. TC/aerosol mass ratios for the Jeju samples (1.2–13.7%, av. 6.6%, see Table 1) are lower than those obtained from year-round observation in Beijing (9.2–23%, av. 17%, SEKINE *et al.*, 1992) and Tokyo (9.2–38%, av. 21%, KAWAMURA and IKUSHIMA, 1993). However, they are much higher

Table 1. Aerosol mass, total carbon (TC) and nitrogen (TN) contents, stable carbon and nitrogen isotonic composition, and major ions in the aerosol samples from East Asia

Components	Minimum	Maximum	Average	Median
Bulk analyses				
Aerosol mass ($\mu g\ m^{-3}$)	31	880	110	83
TC ($\mu g\ m^{-3}$)	1.3	23	6.5	5.3
TOC ($\mu g\ m^{-3}$)	1.1	17	5.9	4.9
Removed C ($\mu g\ m^{-3}$)	–	6.3	0.58	0.20
Removed C (%)	–	31	6.3	4.2
TN ($\mu g\ m^{-3}$)	0.58	16	3.1	2.6
Remained N ($\mu g\ m^{-3}$)	0.32	11	1.9	1.6
C/N weight ratio	0.81	8.3	2.4	2.1
TOC/Remained N weight ratio	1.2	11	4.0	3.6
TC/aerosol mass (%)	1.2	14	6.6	6.2
Isotope analyses (‰)				
$\delta^{13}C$ (TC)	−26.6	−15.5	−23.4	−23.4
$\delta^{13}C$ (TOC)	−26.9	−17.9	−23.5	−23.4
$\delta^{15}N$ (TN)	−3.7	12	4.2	3.8
$\delta^{15}N$ (Remained N)	−5.4	22	6.6	6.8
Ion analyses ($\mu g\ m^{-3}$)				
Na	0.65	13	3.9	3.4
NH_4	0.15	13	2.1	1.5
K	0.11	2.4	0.62	0.48
Mg	0.10	2.0	0.60	0.49
Ca	0.06	13	1.4	0.7
Cl	0.14	24	5.2	3.4
NO_3	0.17	22	5.3	4.0
SO_4	2.4	29	9.0	7.5
MSA	0.00	0.13	0.02	0.00

TOC is defined as total carbon that remained with HCl fume treatment of aerosol sample. Remained N is defined as total nitrogen that remained in aerosols with HCl fume treatment.

than those (0.2–5.4%, av. 1.8%, KAWAMURA et al., 2003a) obtained in Chichi-jima in the western North Pacific (see Fig. 1 for the map).

Figure 4a presents seasonal changes in the TC contents in the aerosol samples collected over Jeju Island without and with HCl treatment. Highest TC values were measured in spring and autumn and less frequently in winter, whereas lowest values were measured in summer. Total organic carbon (TOC) that is defined as TC obtained after HCl treatment of aerosols shows seasonal changes similar to TC (Fig. 4a). The values of TOC are sometimes significantly lower than those of TC. The results indicated that up to 30% of TC were removed from the aerosols by the treatment with HCl fume. However, the highest removal rates were limited to some samples collected in March, May, October and November. Summer and winter samples did not show a significant decline of TC after the HCl treatment except for one sample (KOS43, July 9–16, 2001). The removed carbon may have existed as carbonate and/or volatile organic forms as discussed later.

Figure 4b presents carbon isotope ratios for TC (without HCl treatment) and TOC (with HCl treatment). $\delta^{13}C$ values of TC ranged from −15.5 to −26.6‰ whereas those of TOC ranged from −17.9 to −23.5‰. Interestingly, $\delta^{13}C$ values of TC for the spring samples indicated heavier isotopic ratios more than −20‰ and the rest of the samples showed a range of −23 to −25‰. Such heavier isotopic ratios (> −20‰) have not generally been reported in the previous study on marine aerosol samples (e.g. CHESSELET et al., 1981; CACHIER et al., 1986; TUREKIAN et al., 2003), aerosols affected by forest fires (NARUKAWA et al., 1999) and aerosols from Amazon (MARTINELLI et al., 2002). The heavy isotope ratios may be associated with the presence of carbonate carbon derived from desert dusts, which generally contain calcium carbonate whose $\delta^{13}C$ values are close to zero (CRAIG, 1953). In fact, yellow

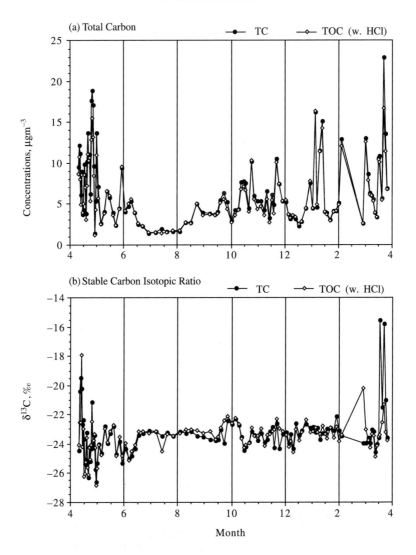

FIG. 4. Seasonal changes in (a) the concentrations of total carbon (TC) and total organic carbon (TOC) and (b) stable carbon isotopic ratios for TC and TOC in aerosol samples. TOC is defined as total carbon that remained after HCl fume treatment.

sand (Kosa) samples collected from the loess layer (1.8–2.5 m from the surface) in Gunsu Province of China (NISHIKAWA *et al.*, 2000) showed that 1.44–1.69% of the dusts were present as inorganic carbon (corresponding to 62–93% of TC, see Table 2).

We analyzed the desert dust samples (yellow sand samples) for carbon and nitrogen isotopic composition with and without HCl treatment (Table 2). Using isotopic mass balance equations as below, carbon isotopic ratios of removed carbon from the Chinese loess samples ($\delta^{13}C_{\text{Removed C}}$) were calculated to be −0.3 to −1.3‰ (Table 2), suggesting that removed carbon was present as carbonate.

$$\text{TC} = \text{TOC} + \text{Removed Carbon (mostly carbonate such as CaCO}_3\text{) by HCl treatment}$$

$$f_{\text{TOC}} + f_{\text{Removed C}} = 1$$

Table 2. Analytical results of standard yellow sand (Kosa) samples collected from Gunsu Province, China

Components	CJ-1 ($<250\ \mu$m)	CJ-2 ($<100\ \mu$m)
Averaged particle size (μm)	38	24
Elemental composition		
Total carbon (TC) (%)	1.76	2.34
Total organic carbon (TOC) (%)	0.11	0.90
Inorganic carbon (IC, CaCO$_3$) (%)	1.69	1.44
Total nitrogen (TN) (%)	0.03	0.08
C/N weight ratio	60	29
Isotopic composition (‰)		
δ^{13}C (TC)	-1.6	-9.8
δ^{13}C (TOC)	-21.5	-23.4
δ^{13}C (IC)	-0.3	-1.3
δ^{15}N (TN)	2.7	2.5

$$f_{TOC} = TOC/TC, \qquad f_{Removed\ C} = (TC - TOC)/TC$$

$$\delta^{13}C_{TC} = f_{TOC} \times \delta^{13}C_{TOC} + f_{Removed\ C} \times \delta^{13}C_{Removed\ C}$$

In the equations above, f_{TOC} and $f_{Removed\ C}$ refer to a fraction of TOC and removed C in TC, respectively. These results suggest that Asian aerosols transported from arid regions should

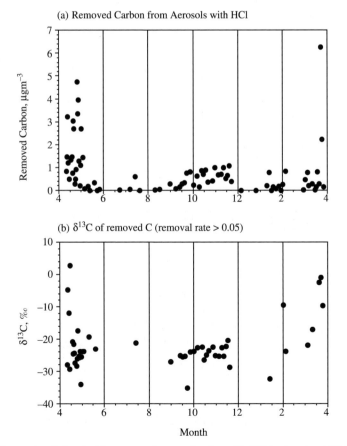

(a) Removed Carbon from Aerosols with HCl

(b) δ^{13}C of removed C (removal rate > 0.05)

Month

FIG. 5. Seasonal variations in (a) amounts of removed carbon with HCl treatment and (b) δ^{13}C of the removed carbon.

contain abundant carbonate carbon when aerosol sampling was conducted during strong dust events.

Using the same mass balance equations, we calculated stable carbon isotopic ratios for the removed carbon from the aerosols by HCl treatment. The results of aerosol samples showed $\delta^{13}C$ ratios for the removed carbon ranging from -35 to $+3‰$ with an average of $-22‰$ and median of $-24‰$. Figure 5 describes the seasonal variation of the amounts of removed carbon from the aerosol samples as well as their $\delta^{13}C$. The $\delta^{13}C$ data are presented for the samples whose removal rates are $>5\%$ because of the low accuracy for the samples with the lower removal rates. These data demonstrate that except for a few samples collected in spring, major portion of the removed carbon by HCl treatment is not carbonate, rather they are largely composed of organic carbon. Low molecular weight organic acids are likely candidates for removed organic carbon. SAKUGAWA and KAPLAN (1995) reported carbon isotopic ratios for formic acid (-28 to $-31‰$, mean: $-30‰$) and acetic acid (-18 to $-21‰$, mean: $-20.5‰$) in Los Angeles atmosphere. More recently, GLASIUS *et al.* (2001) reported $\delta^{13}C$ values of formic plus acetic acids in the atmosphere of Europe to be -23 to $-31‰$.

These volatile organic acids are abundantly present in the urban atmosphere (*e.g.* KAWAMURA *et al.*, 2000), thus it is reasonable to consider that volatile organic acids (present as salts in aerosols) are removed by HCl fume treatment. Although they are mostly present as gases, a small portion of the acids (less than 20%) in the atmosphere may be strongly adsorbed on aerosol particles as well (KAWAMURA *et al.*, 1985c). Thus, formate and acetate that are originally present as particles can be removed from the aerosols by evaporation on exposure to HCl vapor. Some dicarboxylic acids such as oxalic acid could also be removed from aerosols by exposure to HCl. We measured the carbon isotopic ratios of oxalic acid in the selected aerosol samples and found their carbon isotopic values to be in a range from -17.6 to $-21.9‰$ (KAWAMURA *et al.*, 2002, unpublished results). TUREKIAN *et al.* (2003) reported the $\delta^{13}C$ values of oxalate in the Bermuda aerosols to be around $-21‰$.

In contrast, we found very heavy stable carbon isotopic ratios (-5 to $+3‰$) for the removed C on HCl treatment of the aerosol samples whose nss-Ca contents ($3-13\ \mu g\ m^{-3}$) and dust concentrations ($>270\ \mu g\ m^{-3}$) are high. These samples were collected during heavy dust events in March and April (Fig. 5b). This suggests that a major portion of the removed carbon is not organic. They are rather derived from carbonates originated from desert dusts, which may be long-range transported through the atmosphere from arid regions in Asia. Some other samples that were collected in spring showed carbon isotopic ratios between -10 and $-19‰$. Their nss-Ca contents are also high ($3-5\ \mu g\ m^{-3}$) except for one sample. These results indicate that mineral dusts containing carbonate are transported over the western Pacific rim without total dissolution by acidic components such as sulfuric acid. However, such a transport is limited to the spring season when strong dust events generally occur. We believe that carbonates uplifted in the air most likely adsorb acidic gas and particles such as nitric acid, sulfuric acid, and possibly organic acids. Then, they react with carbonates in the aerosols to release CO_2 to the air during a long-range transport over the Asian continent. In fact, we found several aerosol samples collected in spring, in which nss-Ca contents are relatively high ($2-5\ \mu g\ m^{-3}$), but $\delta^{13}C$ ratios of the removed C on HCl treatment are very low (-20 to $-30‰$). These samples indicate that carbonates in the aerosols are seriously reacted with acids. The precursor gases (SO_2, NO_x, and volatile organic compounds) of the acidic species are emitted from industrial locations and converted to inorganic and organic acids in the atmosphere. These acids are generally in sufficient concentration to quantitatively remove aerosol carbonates within the atmosphere of East Asia, except during strong dust events, when the atmospheric titration by the acids is insufficient.

TN contents and nitrogen isotopic ratios

Figure 6a presents a seasonal change in the TN contents and those after the HCl treatment of aerosol samples. By the HCl fume treatment, 7–70% (av. 40%) of nitrogen was removed. Removed nitrogen is present as nitrate. This is supported by the positive correlation ($r = 0.92$) obtained between NO_3^- and removed N with the slope close to unity and intercept of nearly

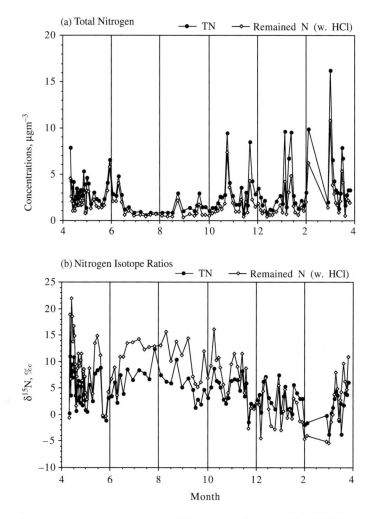

FIG. 6. Seasonal changes in (a) total nitrogen (TN) contents and remained N with HCl fume treatment and (b) nitrogen isotopic ratios of TN and remained N.

zero (see Fig. 7a). Nitrate was found to account on average for 41% (median: 40%, range: 4–90%) of the TN in the Asian aerosols. In contrast, ammonium nitrogen comprises on average 50% of TN (median: 51%, range: 7–90%) in the aerosols. Interestingly, a strong correlation was obtained between NH_4 measured by IC and TN (hereafter defined as remained N) measured by EA after the HCl fume treatment (see Fig. 7b). Further, NO_3 plus NH_4 nitrogen correlate well with TN contents, as seen in Fig. 7c. These results demonstrate that nitrate and ammonium ions account for a significant portion of TN, indicating that organic nitrogen is a less important fraction in the Asian aerosols. Asian aerosols enriched with acidic species may significantly adsorb NH_3 that is released from soils. NH_3 also reacts with acidic gases (HNO_3, H_2SO_4, RCOOH) to result in a small particle, which will be further incorporated to pre-existing aerosols.

Figure 6b presents seasonal changes in the $\delta^{15}N$ for the aerosol TN and remained N (after HCl treatment). The nitrogen isotopic ratios for the bulk aerosols ranged from −3.7 to 12.4‰ with an average of 4.2‰ whereas those after HCl treatment ranged from −5.4 to 21.9‰ with an average of 6.6‰ (Table 1). It was found that $\delta^{15}N$ of aerosols collected in spring and summer generally shift to heavier values on HCl treatment (Fig. 6b) and vice versa for winter samples. The results of spring and summer samples suggest that removed nitrogen

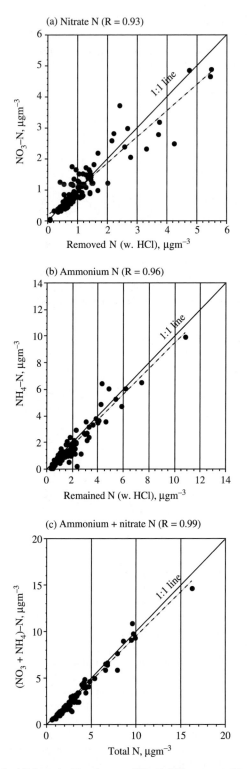

Fɪɢ. 7. Relationship for (a) nitrate (as N) and removed N with HCl treatment, (b) ammonium (as N) and remained N with HCl treatment and (c) (nitrate + ammonium)-N and total N in the aerosol samples from Jeju Island. Nitrate and ammonium were measured by ion chromatography whereas TN and remained N were measured by EA. See the text for the methods.

(mostly as HNO$_3$) should have relatively lighter values. Negative δ^{15}N values (*e.g.*, $-12 \pm 9\permil$) were reported for aerosol NO$_3^-$ in the coastal sites of UK (YEATMAN *et al.*, 2001). Using the similar isotopic mass balance equation for carbon isotopic ratios, δ^{15}N values were calculated for the removed N as follows.

$$\text{TN} = \text{Remained N} + \text{Removed Nitrogen (mostly HNO}_3\text{) by HCl treatment}$$

$$f_{\text{Remained N}} + f_{\text{Removed N}} = 1$$

$$f_{\text{Remained N}} = \text{Remained N}/\text{TN}, \qquad f_{\text{Removed N}} = (\text{TN} - \text{Remained N})/\text{TN}$$

$$\delta^{15}\text{N}_{\text{Remained N}} = f_{\text{Remained N}} \times \delta^{13}\text{C}_{\text{Remained N}} + f_{\text{Removed N}} \times \delta^{15}\text{N}_{\text{Removed N}}$$

Figure 8 presents a seasonal variation of δ^{15}N values for the removed N (mostly as HNO$_3$) together with the amount of removed N. In general the δ^{15}N values are lower in summer ($-1.7\permil$ for the average of May–September samples) than in winter ($+4.6\permil$ for the average of November–February samples), although very light δ^{15}N values were obtained for some spring sample.

The present result shown in Fig. 8 is consistent with the seasonal trend of δ^{15}N of NO$_3^-$ in aerosols (2–$8\permil$) collected in Germany (FREYER, 1991), although our values for removed N are a little lighter. A similar seasonal trend of δ^{15}N values was reported for rainwater NO$_3^-$, although

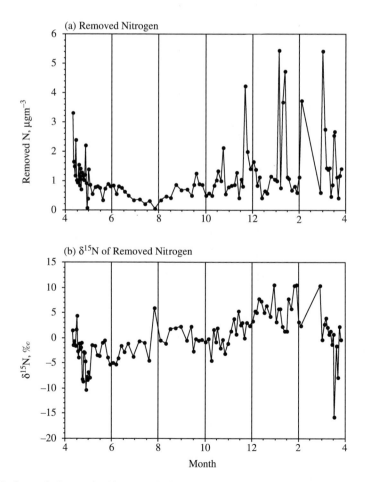

FIG. 8. Seasonal changes in (a) removed nitrogen from aerosol samples by HCl treatment and (b) nitrogen isotopic ratios of removed nitrogen.

the values are much lower (-6 to $+1\permil$) (FREYER, 1991). The heavier isotopic ratios in winter may be due to greater contribution from coal combustion, because NO_x emissions from coal-fired boilers provide heavier $\delta^{15}N$ values ($+6$ to $+13\permil$) than the NO_x emissions (-13 to $-2\permil$) from motor vehicle exhausts (HEATON, 1990). Alternatively, winter samples may have more contribution from soil organic nitrogen because their $\delta^{15}N$ values are heavier (-5 to $+25\permil$) (MOOR, 1974). Our analyses of Chinese loess samples showed $\delta^{15}N$ values of $+2.5$ to $+2.7\permil$ (Table 2). However, high dust samples gave generally lighter isotopic values. For example, $\delta^{15}N$ values of removed N showed a negative correlation with non-sea salt Ca ($r = 0.55$ for January–March, 2002), which can be significantly contributed from desert dusts. Desert dusts contain alkaline salts of Ca, Na and Mg and the aerosols influenced by the dusts may adsorb more gaseous HNO_3. FREYER (1991) reported that $\delta^{15}N$ of gaseous HNO_3 from Germany gives more negative values (-2.2 to $-3.0\permil$) compared to those of aerosol NO_3^- ($+3.0$ to $+6.2\permil$). Heavier nitrogen isotopic ratios were reported for aerosol nitrate in northern ($+15 \pm 3\permil$) and southern UK ($+10 \pm 3\permil$) (YEATMAN *et al.*, 2001). Very light $\delta^{15}N$ values obtained for removed N in spring (Fig. 8b) are not available to explain at this moment.

As seen in Figure 6b, isotopically negative (-5 to $0\permil$) values of $\delta^{15}N$ (remained N) were obtained for the aerosol samples collected in winter and, to a lesser extent, spring seasons. Such isotopically light values are unlikely caused by organic nitrogen whose $\delta^{15}N$ values are in general isotopically more positive (PETERS *et al.*, 1978; MINAGAWA *et al.*, 1984). Rather, they may be affected by the occurrence of ammonium salts in aerosols since ammonium N still remains in the aerosol samples after HCl fume treatment as stated earlier. Our results are consistent with the isotopically light $\delta^{15}N$ values ($-11.8 \pm 1.1\permil$) of ammonium ion, which have been obtained for rainwater samples over the agricultural area of San Joaquin Valley, California (K. K. LIU and I. R. KAPLAN, 2003, personal communication). Ammonia should be abundantly emitted from agricultural areas in Asia and may be trapped by acidic aerosols enriched with sulfate of Asian industrial origin. Further, NH_3 may react with acidic gases (*i.e.* HNO_3, HCOOH, CH_3COOH, HOOC–COOH) in the air to result in the salts (*i.e.* NH_4NO_3, $HCOONH_4$, CH_3COONH_4, $(HN_4)_2C_2O_4$), which should coagulate and then associate with pre-existing particles. Although emission of ammonia in agricultural areas should be greater in summer when soil surface temperatures are elevated, the $\delta^{15}N$ values of the summer aerosols did not show isotopically light values (see Fig. 6b). This is due to the change in wind direction at Gosan site; southerly winds dominate and westerly winds are weakened in summer.

Distributions of water-soluble dicarboxylic acids, ketocarboxylic acids and dicarbonyls

We detected homologous series of both α,ω-dicarboxylic acids (C_2–C_{12}) and ω-oxocarboxylic acids (C_2–C_9) in the aerosol samples. Aromatic (phthalic) diacid and mid-chain keto dicarboxylic acids (kC_3, kC_7) were also detected as well as α-ketoacid (pyruvic acid) and α-dicarbonyls (C_2–C_3). Fig. 9 displays representative chemical structures of dicarboxylic acids, ketoacids and dicarbonyls detected in the aerosol samples. Throughout the 1-year observation, oxalic (C_2) acid was found to be the most abundant water-soluble organic species followed by malonic (C_3) acid and occasionally by succinic (C_4) acid. The predominance of the smallest (C_2) diacid has been reported for the aerosol and rainwater samples from the urban (KAWAMURA and KAPLAN, 1986; KAWAMURA and IKUSHIMA, 1993; WANG *et al.*, 2002) and remote marine regions (SEMPÉRÉ and KAWAMURA, 1996; KAWAMURA and SAKAGUCHI, 1999; MOCHIDA *et al.*, 2003a) as well as the arctic atmosphere (KAWAMURA *et al.*, 1996a), although some Antarctic aerosol samples showed a predominance of succinic acid (KAWAMURA *et al.*, 1996b). Relative abundances of longer-chain diacids generally decrease with an increase in their carbon numbers, except for azelaic acid (C_9) that is more abundant than suberic (C_8) acid. This diacid (C_9) is an oxidation product of biogenic unsaturated fatty acids containing double bond at the C_9-position (KAWAMURA and GAGOSIAN, 1987). It is important to note that phthalic acid (aromatic acid) is often the fourth most abundant diacid in the Gosan aerosol samples collected in winter, following succinic acid. This aromatic acid is directly emitted from combustion sources (KAWAMURA and KAPLAN, 1986) and/or has been considered to generate in the atmosphere by

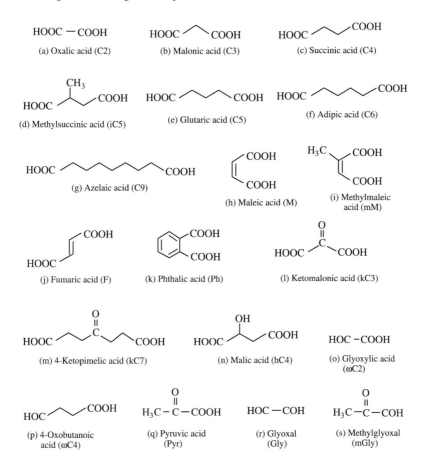

FIG. 9. Representative chemical structures of dicarboxylic acids, ketocarboxylic acids and α-dicarbonyls detected in the aerosol samples collected from Jeju Island.

photochemical oxidation of aromatic hydrocarbons such as naphthalene (KAWAMURA and IKUSHIMA, 1993).

Table 3 presents concentration ranges and average and median concentrations for dicarboxylic acids, ketocarboxylic acids and α-dicarbonyls. Total concentrations of diacids ranged from 130 to 2070 ng m^{-3} with an average of 660 ng m^{-3} (Table 3). These values are higher than those (90–1370 ng m^{-3}, av. 480 ng m^{-3}) reported in urban Tokyo (KAWAMURA and IKUSHIMA, 1993) but are equivalent to the upper range (300–2100 ng m^{-3}) reported in the urban atmosphere from China (WANG et al., 2002). Further, the diacid concentration levels at Gosan site are several times higher than those (6–550 ng m^{-3}, av. 140 ng m^{-3}) reported at Chichi-jima in the western North Pacific (MOCHIDA et al., 2003a), a remote island 1000 km south of Japan (see Fig. 1 for the map). Concentration maxima occur most frequently in spring (April–May), but isolated maxima also occur in October, January, and March (Fig. 10a). These results indicate that the East Asian region over the East China Sea is heavily polluted, because small diacids are directly emitted from combustion processing of fossil fuels and subsequent conversion to carboxylic acids and carbonyls in the atmosphere by the photochemical oxidation of anthropogenic organic precursors (KAWAMURA and KAPLAN, 1986; KAWAMURA and IKUSHIMA, 1993). Ketoacids (2–170 ng m^{-3},- av. 53 ng m^{-3}) and α-dicarbonyls (0.1–84 ng m^{-3}, av. 12 ng m^{-3}) were also detected in the aerosol samples in the Jeju aerosols. Their concentrations are equivalent to those found in Tokyo (SEMPÉRÉ and KAWAMURA, 1994; KAWAMURA and IKUSHIMA, unpublished results).

During the intensive period (April, 2001) of the ACE-Asia campaign, aerosol samples were also collected in the marine boundary layer over the East China Sea, the Japan Sea, and

Table 3. Concentrations of low molecular weight dicarboxylic acids, ketocarboxylic acids and α-dicarbonyls in the aerosol samples from East Asia

Compounds	Concentrations (ng m^{-3})			
	Minimum	Maximum	Average	Median
Dicarboxylic acids				
Saturated straight-chain diacids				
Oxalic, C2	98	1550	473	373
Malonic, C3	10	175	67	57
Succinic, C4	8.3	153	52	41
Glutaric, C5	0.8	34	11	8.8
Adipic, C6	0.6	43	8.4	6.3
Pimeric, C7	0.2	9.3	2.6	2.0
Suberic, C8	0.0	9.6	2.6	2.1
Azelaic, C9	1.2	14	4.7	4.4
Sebacic, C10	0.0	4.7	1.1	0.9
Undecanedioic, C11	0.0	3.9	1.3	1.2
Dodecanedioic, C12	0.0	3.2	0.1	0.0
Branched-chain diacids				
Methylmalonic, iC4	0.0	4.2	1.1	1.0
2-Methylglutaric, iC6	0.0	4.0	1.0	0.8
Methylmaleic, mM	0.4	21	5.3	4.2
Unsaturated diacids				
Maleic, M	0.4	15	4.9	4.1
Fumaric, F	0.5	14	5.2	4.1
Phthalic, Ph	0.5	40	9.8	7.4
Keto or hydroxy diacids				
Ketomalonic, kC3	0.0	9.6	1.9	1.4
4-Ketopimelic, kC7	0.0	13	4.5	3.6
Malic (hydroxysuccinic), hC4	0.0	9.7	3.0	2.6
Total diacids	130	2070	660	540
Ketocarboxylic acids				
α-Ketoacid				
Pyruvic, C3	0.6	54	13	9.1
ω-oxoacids				
Glyoxylic (2-oxoethanoic), ωC2	4.3	133	36	27
3-Oxopropanoic, ωC3	0.0	5.9	1.7	1.2
4-Oxobutanoic, ωC4	0.0	9.5	2.5	1.6
9-Oxononaoic, ωC9	0.0	4.0	1.4	1.3
Total ketoacids	2.1	172	53	45
Dicarbonyls				
Glyoxal, Gly, C2	0.0	3.9	0.5	0.3
Methylglyoxal, mGly, C3	0.7	83	11	8.1
Total dicarbonyls	0.08	84	12	8.5

the western Pacific Ocean near the Japanese Islands using the NOAA research ship *R/V Ron Brown* and NCAR aircraft (C-130). Similar concentrations of dicarboxylic acids were reported in the marine aerosols collected by a ship in the East China Sea (360–1300 ng m^{-3}, av. 760 ng m^{-3}, MOCHIDA *et al.*, 2003b), although much higher concentrations (430–1900 ng m^{-3}, av. 1100 ng m^{-3}) were obtained over the Sea of Japan (MOCHIDA *et al.*, 2003b). The data on diacids (130–1350 ng m^{-3}, av. 672 ng m^{-3}) measured at Gosan in April 2001, are consistent with those obtained on the *R/V Ron Brown*, but are much higher than those obtained in open ocean sampling between Japan and Hawaii (290–660 ng m^{-3}, av. 490 ng m^{-3} for the west of the ordinate 140°E, and 31–620 ng m^{-3}, av. 180 ng m^{-3} for the east of 140°E, MOCHIDA *et al.*, 2003b).

In contrast, aircraft measurements of diacids at the same locations, but at altitudes ranging from 40 to 7000 m demonstrated that the diacid concentrations (44–870 ng m^{-3}, av. 310 ng m^{-3}, KAWAMURA *et al.*, 2003b) are equivalent or several times lower than those obtained at Gosan, Jeju Island. These results demonstrate that the diacid concentrations near

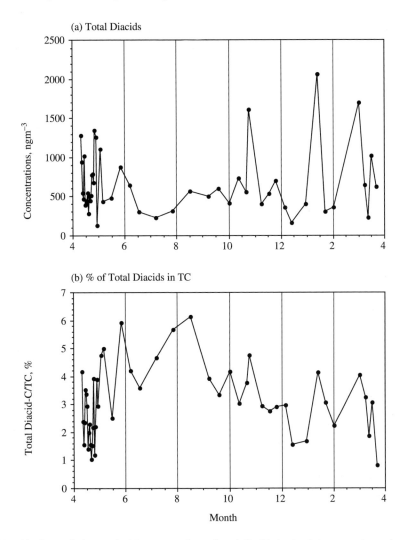

FIG. 10. Seasonal changes in (a) concentrations of total diacids in the Asian aerosol samples and (b) relative abundance of total diacid-carbon in total aerosol carbon.

the ground surface at Gosan are much greater than those of upper troposphere in the Asian Pacific region and those of the western to central North Pacific.

Sources of dicarboxylic acids: correlations with NH₄, NO₃, nss-SO₄ and
other inorganic species

Based on the studies on dicarboxylic acids in both urban and marine aerosols, it is generally accepted that small diacids in the marine aerosols near coastal regions are derived from direct emission through fossil fuel combustion and photochemical oxidation of the anthropogenic and biogenic organic precursors in the atmosphere (KAWAMURA and KAPLAN, 1986; SATSUMABAYASHI et al., 1990; KAWAMURA and IKUSHIMA, 1993; MOCHIDA et al., 2003b). Fig. 11 shows the relationship of concentrations of total diacids with those of ionic species in aerosols. Total diacids positively correlate with NH_4 ($r = 0.83$), NO_3 ($r = 0.86$) and nss-SO_4 ($r = 0.83$), during 1-year observation. Positive correlation ($r = 0.83$) was also found between total diacids and TN (mostly NH_4 and NO_3 ions). These good correlations support that sources

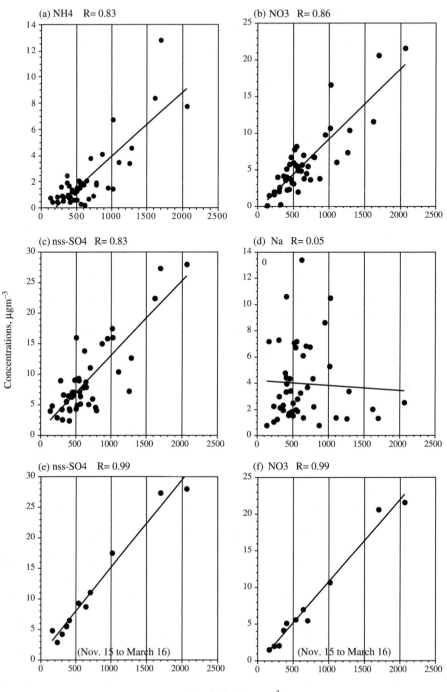

Fig. 11. Relationships between total diacid concentrations and concentrations of (a) NH4, (b) NO3, (c) nss-SO4, (d) Na, (e) nss-SO4 (f) NO3 in the aerosol samples from Jeju Island. Plots (a–d) are representative of all the data (1 year) whereas (e) and (f) give only the data for a limited period (November 15–March 16).

and origin of diacids are involved with combustion sources associated with human activities because nitrate and sulfate are both produced by the atmospheric oxidation of NO_x and SO_2 that are released from combustion sources (FINLAYSON-PITTS and PITTS, 2000).

We found a very good correlation between the diacid concentrations and concentrations of nss-SO_4 and nitrate in the winter aerosol samples collected from mid-November to mid-March. As shown in Fig. 11e and f, the correlation coefficients are very high ($r = 0.99$). This supports that sources of diacids in winter aerosols are strongly connected with combustion sources in Asian countries. In contrast, there is no relationship between the diacids and Na ($r = 0.05$ see Fig. 11d). Similar results were found for Ca ($r = 0.15$), nss-Ca (0.16), Mg (0.05), Cl (0.08) and methanesulfonic acid (MSA) (0.04). Ca, Mg and Cl are important components of seawater whereas MSA is produced in the marine atmosphere by the photochemical oxidation of dimethyl sulfide (DMS) of marine phytoplankton origin (BERRESHEIM et al., 1995). However, diacids positively correlated with nss-K ($r = 0.73$), which is of continental origin. These results suggest that dicarboxylic acids in the aerosols over Jeju Island are derived from continental sources in the Asian industrial areas and also from the photochemical oxidation of anthropogenic organic precursors. We feel that secondary photochemical production of diacids is more significant than the primary emission for the Juju aerosols over the coastal western Pacific.

Photochemical production and destruction of water-soluble dicarboxylic acids

As seen in Fig. 10b, total diacids accounted for 0.8–6.2% of TC (av. 3.1%, median 3.0%) with a maximum in August. These values are consistent with those (1.1–4.9%, av. 3.2%) reported in the western Pacific at 35°N to 40°S (SEMPÉRÉ and KAWAMURA, 2003). However, they are three times higher than those (0.18–1.8%, av. 0.95%) obtained in Tokyo (KAWAMURA and IKUSHIMA, 1993), although a summer time maximum was found in Tokyo. Higher values for Jeju Island indicate that diacid carbon contributes more to the total aerosol carbon possibly due to higher emissions of organic precursor compounds in the source areas than in Tokyo and their subsequent conversion to dicarboxylic acids in the atmosphere.

Alternatively, aerosols collected at Gosan may have experienced atmospheric oxidative reactions more significantly than urban aerosol samples from Tokyo because Jeju Island is located far away from the major pollution sources in Asia and thus atmospheric residence time of the former aerosols is longer than the latter. Diacids may be effectively produced in the air over Asian continent and the East China Sea during a long-range transport. It is of interest to note that higher total diacid-C/TC ratios (1.1–15.8%, av. 8.8%, median 8.4%) have been reported for the marine aerosols collected in the western North to the central Pacific including tropics (KAWAMURA and SAKAGUCHI, 1999).

Throughout the 1-year observation, we found that oxalic (C_2) acid-carbon in TC in the Jeju aerosols increased from spring to summer and decreased toward winter (see Fig. 12a). Similar seasonal trend was obtained for malonic (C_3) acid (see Fig. 12b). These results indicate that small diacids are preferentially produced in the summer time, when solar radiation and ambient temperatures are more suitable for photochemical reaction. However, succinic and other (C_5–C_7) diacids did not show an increase in the diacid-C/TC ratios toward summer. Longer chain (C_4–C_7) diacids may be oxidized to result in smaller diacids (KAWAMURA and SAKAGUCHI, 1999). In contrast, C_8 and C_9 diacids, as well as 4-ketopimelic acid (kC7), do display a maximum in summer (see Fig. 12f and g). Because these long-chain diacids are produced by the oxidation of unsaturated fatty acids, their precursors (biogenic unsaturated fatty acids) are supplied in greater quantity from surface ocean water. This is consistent with the air mass trajectory analyses, which demonstrates that winds originate most often from the Pacific Ocean in summer whereas winds mostly blow from the continents during other seasons of the year. Hydroxysuccinic (malic) acid did not show a seasonal trend (Fig. 12h).

A sharp drop of phthalic acid-C/TC ratios was found in early summer (Fig. 12i). The values decreased by five times from early April (ca. 0.1) to late July (ca. 0.02) and gradually increased toward winter (ca. 0.15). This seasonal variation is in contrast to the trend observed for oxalic and malonic acids. Phthalic acid is an aromatic acid, which has been reported in diesel exhausts as the second most abundant diacid following oxalic acid (KAWAMURA et al., 1986) and been

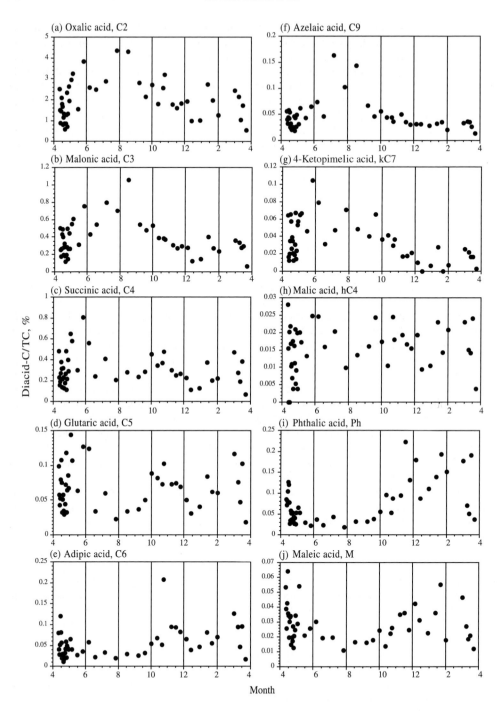

FIG. 12. Seasonal changes in the relative abundances of (a) oxalic acid, (b) malonic acid, (c) succinic acid, (d) glutaric acid, (e) adipic acid, (f) azelaic acid, (g) 4-ketopimelic acid, (h) malic acid, (i) phthalic acid, and (j) maleic (hydroxysuccinic) acid in the total aerosol carbon (TC) in the aerosol samples from Jeju Island.

proposed to generate by photochemical oxidation of aromatic hydrocarbons (KAWAMURA and IKUSHIMA, 1993). Because oceanic air mass reaches Jeju Island primarily in summer, contribution of phthalic acid from anthropogenic sources should be depressed whereas its contribution should be enhanced in winter, especially if polynuclear aromatic hydrocarbon-rich coal is used for heating fuel. However, photochemical degradation of phthalic acid may be responsible for the sharp drop of the diacid-C/TC ratios (Fig. 12i). The atmospheric degradation of phthalic acid should produce oxalic acid and some unsaturated diacids through ring-opening reaction. Interestingly, aliphatic unsaturated diacids (maleic, fumaric and methylmaleic acids) also show lower values of diacid-C/TC ratios in summer and higher values in winter (see Fig. 12j for maleic acid). These results suggest that dicarboxylic acids with unsaturated structure are more susceptible to atmospheric oxidation in summer and may serve as precursors of oxalic acid. Because there is no known anthropogenic source on Jeju Island, these unsaturated diacids are transported from the major cities in Asia to the western Pacific with oxidative production and degradation in the atmosphere.

CONCLUSION

We conducted 1-year observation of atmospheric aerosols at Gosan site, Jeju Island in the East China Sea to better understand the source and behavior of water-soluble dicarboxylic acids. Aerosol samples were also studied for TC and TN contents as well as their isotopic compositions and ionic compositions. Dicarboxylic acids were characterized by a predominance of oxalic acid followed by malonic acid. Concentrations of diacids varied significantly, but they were highest in spring and winter season and lower in summer. Diacids were found to positively correlate with nss-sulfate and nitrate throughout the year. Very strong correlations during the winter season. Low molecular weight diacids accounted for up to 6% of total aerosol carbon. Diacid-C/TC ratios were found to increase in summer mainly due to the photochemical production of small diacids and transformations of organic aerosols in the Asian Pacific region.

Some spring samples collected during strong dust events were found to contain carbonate carbon based on ^{13}C enrichment and high Ca content. However, most aerosol samples were found to contain only negligible amounts of carbonate. We conclude that although mineral dusts are transported over Jeju Island from desert areas of inland Asia, carbonates in the dusts have already reacted with inorganic acids (H_2SO_4 and HNO_3) and possibly with organic acids during their long-range atmospheric transport. Measured stable carbon isotopic data demonstrate that Asian dust aerosols are subjected to serious alteration (titration) in the atmosphere. The amount of organic acids that were removed from the aerosols by the treatment with HCl vapor appears to account for a few percent to 20% of organic carbon in the aerosols. Dicarboxylic acids that are produced by photochemical reactions may be absorbed on alkaline minerals of soil dust origin. Semi-volatile organic acids likely react with NH_3 to result in a formation of new particles, which coagulate and then associate with the pre-existing aerosols.

Acknowledgements—We thank Dr T. Nakatsuka for his help in the isotope analysis. We also acknowledge Drs I.R. Kaplan, K. Ohta and anonymous reviewer for the critical and useful comments. This study was partly supported by the Japanese Ministry of Education, Culture, Sports, Science and Technology through grant-in-aid 01470041 and 14204055. This research is a contribution to the International Global Atmospheric Chemistry (IGAC) Core Project of the International Geosphere Biosphere Program (IGBP) and is part of the IGAC Aerosol Characterization Experiments (ACE).

REFERENCES

BARBOUKAS E. D., KANAKIDOU M. and MIHALOPOULOS N. (2000) Carboxylic acids in gas and particulate phase above the Atlantic Ocean. *J. Geophys. Res.* **105**, 14459–14471.

BERRESHEIM H., WINE P. H. and DAVIS D. D. (1995) Sulfur in the atmosphere. In *Composition, Chemistry and Climate of the Atmosphere* (ed. H. B. SINGH), pp. 251–307. Wiley, New York.

BLANK M., LEINEN M. and PROSPERO J. M. (1985) Major Asian Aeolian inputs indicated by the mineralogy of aerosol and sediments in the western North Pacific. *Nature* **314**, 84–86.

CACHIER H., BUAT-MENARD P., FONTUGNE M. and CHESSELET R. (1986) Long-range transport of continentally-derived particulate carbon in the marine atmosphere: evidence from stable carbon isotope studies. *Tellus* **38B**, 161–177.

CHESSELET R., FONTUGNE M., BUAT-MENARD P., EZAT U. and LAMBERT C. E. (1981) The origin of particulate organic carbon in the marine atmosphere as indicated by its stable carbon isotopic composition. *Geophys. Res. Lett.* **8**, 345–348.

CRAIG H. (1953) The geochemistry of the stable carbon isotopes. *Geochim. Cosmochim. Acta* **3**, 53–92.

DAWSON G. A., FARMER J. C. and MOYERS J. L. (1980) Formic and acetic acids in the atmosphere of the southwest U.S.A. *Geophys. Res. Lett.* **9**, 725–728.

DUCE R. A., UNNI C. K., RAY B. J., PROSPERO J. M. and MERRILL J. T. (1980) Long-range atmospheric transport of soil dust from Asia to the tropical North Pacific: temporal variability. *Science* **209**, 1522–1524.

FINLAYSON-PITTS B. J. and PITTS J. N. Jr. (2000) *Chemistry of the Upper and Lower Atmosphere.* Academic Press, San Diego, pp. 969.

FREYER H. D. (1991) Seasonal variation of $^{15}N/^{14}N$ ratios in atmospheric nitrate species. *Tellus* **43B**, 30–44.

GLASIUS M., BOEL C., BRUUN N., EASA L. M., HORNUNG P., KLAUSEN H. S., KLITGAARD K. C., LINDESKOV C., MOLLER C. K., NISSEN H., PETERSEN A. P. F., KLEEFELD S., BOARETTO E., HANSEN T. S., HEINEMEIER J. and LOHSE C. (2001) Relative contribution of biogenic and anthropogenic sources of formic and acetic acids in the atmospheric boundary layer. *J. Geophys. Res.* **106**, 7415–7426.

GROSJEAN D., VAN CAUWENBERGHE K., SCHMID J. P., KELLEY P. E. and PITTS J. N. Jr. (1978) Identification of C_3–C_{10} aliphatic dicarboxylic acids in airborne particulate matter. *Environ. Sci. Technol.* **12**, 313–317.

HEATON T. H. E. (1990) $^{15}N/^{14}N$ ratios of NO_x from vehicle engines and coal-fired power stations. *Tellus* **42B**, 304–307.

HUEBERT B. J., BATES T., RUSSELL P. B., SHI G., KIM Y. J., KAWAMURA K., CARMICHAEL G. and NAKAJIMA T. (2003) An overview of ACE-Asia: strategies for quantifying the relationships between Asian aerosols and their climatic impacts. *J. Geophys. Res.* **108**(D23), 8633 (doi: 10.1029/2003JD003550).

IPCC (INTERGOVERNMENTAL PANEL FOR CLIMATE CHANGE) (2001) In *The Scientific Basis* (ed. J. T. HOUGHTON), pp. 859. Cambridge University Press, Cambridge.

KAWAMURA K. (1993) Identification of C_2–C_{10} ω-oxocarboxylic acids, pyruvic acid and C_2–C_3 α-dicarbonyls in wet precipitation and aerosol samples by capillary GC and GC–MS. *Anal. Chem.* **65**, 3505–3511.

KAWAMURA K. and GAGOSIAN R. B. (1987) Implications of ω-oxocarboxylic acids in the remote marine atmosphere for photo-oxidation of unsaturated fatty acids. *Nature* **325**, 330–332.

KAWAMURA K. and IKUSHIMA K. (1993) Seasonal changes in the distribution of dicarboxylic acids in the urban atmosphere. *Environ. Sci. Technol.* **27**, 2227–2235.

KAWAMURA K. and KAPLAN I. R. (1984) Capillary gas chromatography determination of volatile organic acids in rain and fog samples. *Anal. Chem.* **56**, 1616–1620.

KAWAMURA K. and KAPLAN I. R. (1986) Motor-exhaust emissions as a primary source for dicarboxylic acids in Los Angeles. *Environ. Sci. Technol.* **21**, 105–110.

KAWAMURA K. and KAPLAN I. R. (1987) Dicarboxylic acids generated from thermal alteration of geopolymers. *Geochim. Cosmochim. Acta* **51**, 3201–3207.

KAWAMURA K. and KAPLAN I. R. (1991) Organic compounds in rainwater. In *Organic Chemistry of the Atmosphere* (eds. L. D. HANSEN and D. J. EATOUGH), pp. 233–284. CRC Press, Boca Raton.

KAWAMURA K. and SAKAGUCHI F. (1999) Molecular distributions of water soluble dicarboxylic acids in marine aerosols over the Pacific Ocean including tropics. *J. Geophys. Res.* **104**, 3501–3509.

KAWAMURA K. and USUKURA K. (1993) Distributions of low molecular weight dicarboxylic acids in the North Pacific aerosol samples. *J. Oceanogr.* **49**, 271–283.

KAWAMURA K., NG L. L. and KAPLAN I. R. (1985a) Determination of organic acids (C_1–C_{10}) in the atmosphere, motor exhausts and engine oils. *Environ. Sci. Technol.* **21**, 105–110.

KAWAMURA K., STEINBERG S. and KAPLAN I. R. (1985b) Capillary GC determination of short-chain dicarboxylic acids in rain, fog and mist. *Int. J. Environ. Anal. Chem.* **19**, 175–188.

KAWAMURA K., NG L. L. and KAPLAN I. R. (1985c) Determination of organic acids (C_1–C_{10}) in the atmosphere, motor-exhausts and engine oils. *Environ. Sci. Technol.* **19**, 1082–1086.

KAWAMURA K., TANNENBAUM E., HUIZINGA B. and KAPLAN I. R. (1986) Volatile organic acids generated from kerogen during laboratory heating. *Geochem. J.* **20**, 51–59.

KAWAMURA K., KOSAKA M. and SEMPÉRÉ R. (1995) Distributions and seasonal changes of hydrocarbons in urban aerosols and rainwaters. *Chikyukagaku (Geochemistry)* **29**, 1–15.

KAWAMURA K., KASUKABE H. and BARRIE L. A. (1996a) Source and reaction pathways of dicarboxylic acids, ketoacids and dicarbonyls in arctic aerosols: one year of observations. *Atmos. Environ.* **30**, 1709–1722.

KAWAMURA K., SEMPÉRÉ R., IMAI Y., HAYASHI M. and FUJII Y. (1996b) Water soluble dicarboxylic acids and related compounds in the Antarctic aerosols. *J. Geophys. Res.* **101**(D13), 18721–18728.

KAWAMURA K., STEINBERG S. and KAPLAN I. R. (2000) Homologous series of C_1–C_{10} monocarboxylic acids and C_1–C_6 carbonyls in Los Angeles air and motor vehicle exhausts. *Atmos. Environ.* **34**, 4175–4191.

KAWAMURA K., KOBAYASHI M., TSUBONUMA N., MOCHIDA M., LEE M. and LEE G. (2002) Water-soluble dicarboxylic acids and total organic carbon in the Asian aerosols: results from Cheju and Hokkaido Islands during the ACE-Asia campaign. *Sixth International Aerosol Conference*, September 8–13, 2002, Taipei, Taiwan, Abstract, pp. 9–10.

KAWAMURA K., ISHIMURA Y. and YAMAZAKI K. (2003a) Four year observation of terrestrial lipid class compounds in marine aerosols from the western North Pacific. *Global Biogeochem. Cycl.* **17**(1), 1003 (doi: 10.1029/2001/GB001810).

KAWAMURA K., UMEMOTO N., MOCHIDA M., BERTRAM T., HOWELL S. and HUEBERT B. (2003b) Water-soluble dicarboxylic acids in the tropospheric aerosols collected by ACE-Asia/C-130 aircraft over East Asia and western North Pacific. *J. Geophys. Res.* **108**(D23), 8639 (doi: 10.1029/2002JD003256).

KEENE W. C., GALLOWAY J. N. and HOLDEN J. D. Jr. (1983) Measurement of weak organic acidity in precipitation from remote areas of the world. *J. Geophys. Res.* **88**, 5122–5130.

KHRAKA Y. K., CAROTHERS W. W. and ROSENBAUER R. J. (1983) Thermal decarboxylation of acetic acid: implications for origin of natural gas. *Geochim. Cosmochim. Acta* **47**, 397–402.

LEWAN M. D. and FISHER J. B. (1994) Organic acids from petroleum source rocks. In *Organic Acids in Geological Processes* (eds. E. D. PITTMAN and M. D. LEWAN), pp. 70–114. Springer, Berlin.

LI S. M. and WINCHESTER J. W. (1993) Water soluble organic constituents in arctic aerosols and snow pack. *Geophys. Res. Lett.* **20**, 45–48.

MACE K. A., DUCE R. A. and TINDALE N. W. (2003) Organic nitrogen in rain and aerosol at Cape Grim, Tasmania, Australia. *J. Geophys. Res.* **108**(D11), 4338 (doi: 10.1029/2002JD003035).

MARTINELLI L. A., CAMARGO P. B., LARA L. B. L. S., VICTORIA R. L. and ARTAXO P. (2002) Stable carbon and nitrogen isotopic composition of bulk aerosol particles in a C4 plant landscape of southeast Brazil. *Atmos. Environ.* **36**, 2427–2432.

MINAGAWA M., WINTER D. A. and KAPLAN I. R. (1984) Comparison of Kjeldahl and combustion methods for measurement of nitrogen isotope ratios in organic matter. *Anal. Chem.* **56**, 1859–1861.

MOCHIDA M., KAWABATA A., KAWAMURA K., HATSUSHIKA H. and YAMAZAKI K. (2003a) The seasonal variations and the origins of dicarboxylic acids in the atmosphere over the western North Pacific. *J. Geophys. Res.* **108**(D6), 4193 (doi: 10.1029/2002JD002355).

MOCHIDA M., KAWAMURA K., UMEMOTO N., KOBAYASHI M., MATSUNAGA S., LIM H., TURPIN B. J., BATES T. S. and SIMONEIT B. R. T. (2003b) Spatial distribution of oxygenated organic compounds (dicarboxylic acids, fatty acids and levoglucosan) in marine aerosols over the western Pacific and off coasts of East Asia: Asian outflow of organic aerosols during the ACE-Asia campaign. *J. Geophys. Res.* **108**(D23), 8638 (doi: 10.1029/2002JD003249).

MOOR H. (1974) Isotopic measurement of atmospheric nitrogen compounds. *Tellus* **26**, 169–174.

MORI I., NISHIKAWA M., QUAN H. and MORITA M. (2002) Estimation of the concentration and chemical composition of kosa aerosols at their origin. *Atmos. Environ.* **36**, 4569–4575.

NARUKAWA M., KAWAMURA K., TAKEUCHI N. and NAKAJIMA T. (1999) Distribution of dicarboxylic acids and carbon isotopic compositions in aerosols from 1997 Indonesian forest fires. *Geophys. Res Lett.* **26**, 3101–3104.

NARUKAWA M., KAWAMURA K., LI S.-M. and BOTTENHEIM J. W. (2002) Dicarboxylic acids in the arctic aerosols and snowpacks collected during ALERT2000. *Atmos. Environ.* **36**, 2491–2499.

NISHIKAWA M., HAO Q. and MORITA M. (2000) Preparation and evaluation of certified reference materials from Asian mineral dust. *Global Environ. Res.* **4**, 103–113.

NOLTE C. G., SOLOMON P. A., FALL T., SALMON L. G. and CASS G. R. (1997) Seasonal and spatial characteristics of formic and acetic acids concentrations in the southern California atmosphere. *Environ. Sci. Technol.* **31**, 2547–2553.

NORTON R. B., ROBERTS J. M. and HUEBERT B. J. (1983) Tropospheric oxalate. *Geophys. Res. Lett.* **10**, 517–520.

PETERS K. E., SWEENEY R. E. and KAPLAN I. R. (1978) Correlation of carbon and nitrogen stable isotope ratios in sedimentary organic matter. *Limnol. Oceanogr.* **23**, 598–904.

SAKUGAWA H. and KAPLAN I. R. (1995) Stable carbon isotope measurements of atmospheric organic acids in Los Angeles, California. *Geophys. Res. Lett.* **22**, 1509–1512.

SATSUMABAYASHI H., KURITA H., YOKOUCHI Y. and UEDA H. (1990) Photochemical formation of particulate dicarboxylic acids under long-range transport in central Japan. *Atmos. Environ.* **24A**, 1443–1450.

SAXENA P., HILDEMANN L. M., MCMURRY P. H. and SEINFELD J. H. (1995) Organics alter hygroscopic behavior of atmospheric particles. *J. Geophys. Res.* **100**, 18755–18770.

SEKINE Y., HASHIMOTO Y., NAKAMURA T., CHEN Z. and MITSUZAWA S. (1992) Characterization of atmospheric aerosols components at Beijing, China. *J. Jpn Soc. Air Pollut.* **27**, 237–245.

SEMPÉRÉ R. and KAWAMURA K. (1994) Comparative distributions of dicarboxylic acids and related polar compounds in snow, rain and aerosols from urban atmosphere. *Atmos. Environ.* **28**, 449–459.

SEMPÉRÉ R. and KAWAMURA K. (1996) Low molecular weight dicarboxylic acids and related polar compounds in the remote marine rain samples collected from western Pacific. *Atmos. Environ.* **30**, 1609–1619.

SEMPÉRÉ R. and KAWAMURA K. (2003) Trans-hemispheric contribution of C_2–C_{10} α,ω-dicarboxylic acids and related polar compounds to water soluble organic carbon in the western Pacific aerosols in relation to photochemical oxidation reactions. *Global Biogeochem. Cycl.* **17**(2), 1069 (doi: 10.1029/2002GB001980).

SURDAM R. C., BOESE S. W. and CROSSEY L. J. (1984) The chemistry of secondary porosity. In *Clastic Diagenesis*, (eds. D. A. MCDONALD and R. C. SURDAM), Vol. 37, pp. 127–149. American Association of Petroleum Geologists Memoirs.

THURMAN E. M. (1985) *Organic Geochemistry of Natural Waters*. Martinus Nijhoff/Dr Junk Publishers, Dordrecht, 497 pp.

TISSOT B. P. and WELTE D. H. (1984) *Petroleum Formation and Occurrence*. Springer, Berlin, 699 pp.

TUREKIAN V. C., MACKO S. A. and KEENE W. C. (2003) Concentrations, isotopic compositions, and sources of size-resolved, particulate organic carbon and oxalate in near-surface marine air at Bermuda during spring. *J. Geophys. Res.* **108**(D5), 4157 (doi: 10.1029/2002JD002053).

UEMATSU M., DUCE R. A., PROSPERO J. M., CHEN L., MERRILL J. T. and MCDONALD R. L. (1983) Transport of mineral aerosol from Asia over the North Pacific Ocean. *J. Geophys. Res.* **88**, 5345–5352.

WANG G. S., LIU N. C. and WANG L. (2002) Identification of dicarboxylic acids and aldehydes of PM10 and PM2.5 aerosols in Nanjing, China. *Atmos. Environ.* **36**, 1941–1950.

YEATMAN S. G., SPOKES L. J., DENNIS P. F. and JICKELLS T. D. (2001) Comparisons of aerosol nitrogen isotopic composition at two polluted coastal sites. *Atmos. Environ.* **35**, 1307–1320.

YU S. (2000) Role of organic acids (formic, acetic, pyruvic and oxalic) in the formation of cloud condensation nuclei (CCN): a review. *Atmos. Res.* **53**, 185–217.

Geochemical Investigations in Earth and Space Science: A Tribute to Isaac R. Kaplan
© The Geochemical Society, Publication No. 9, 2004
Editors: R.J. Hill, J. Leventhal, Z. Aizenshtat, M.J. Baedecker, G. Claypool,
R. Eganhouse, M. Goldhaber and K. Peters

Trace elements in Gulf of Mexico oysters, 1986–1999

BOBBY JOE PRESLEY[1], GARY A. WOLFF[2], ROBERT J. TAYLOR[1] and PAUL N. BOOTHE[1]

[1]Department of Oceanography, Texas A&M University, College Station, TX 77843-3146, USA
[2]Geochemical and Environmental Research Group, Texas A&M University, College Station,
TX 77843-3146, USA

Abstract—As part of the National Oceanic and Atmospheric Administration (NOAA) Status and Trends Mussel Watch Program, oysters (*Crassostria virginica*) have been sampled along the entire US Gulf of Mexico (GOM) coastline once each year since 1986. The same sampling sites were reoccupied each year when possible. As a result, 63 different sites were sampled for at least 7 of the years between 1986 and 1999. Concentrations of Ag, As, Cd, Cu, Hg, Se, Pb and Zn in these oysters are reported here. The data show considerable variation for all metals, with the site-to-site variation generally larger than year-to-year variation at a given site. Metal concentrations at some sites were either much higher or much lower than average GOM concentrations year after year. At other sites metal concentrations decreased or increased for several years in a row and then reversed this trend. Large differences were often found at sites only 10 km or so apart, showing local control of metal concentrations. Only in a few cases can reasons for these differences be suggested, but because several metals commonly followed the same trend at a given site, both natural and human influences are suspected. Mercury concentrations were high near an old chlor-alkali plant, Zn was high near industrial areas and harbors and Pb was high on an Air Force Base. Arsenic concentrations were higher at open water sites than at back bay sites, especially in Florida, possibly due to differences in dissolved phosphate concentrations. Cd and Se concentrations were generally higher in Texas and Louisiana than in Florida but the opposite was true for As and Hg.

INTRODUCTION

MANY COMMON and necessary human activities, including agriculture, mining, manufacturing, transportation and recreation have the potential to increase the input of trace metals to the marine environment. This has led to a concern that the added metals may be harmful to marine organisms or humans. Attempts have been made to document harmful effects directly; however, due to their mostly subtle nature and the great natural variability in indicators of health, this is difficult. Most attempts to assess the health of the marine environment have looked for anomalies in the concentration of various chemicals, thereby implying a potential for effects on organisms. Anomalies can be sought in water and sediment data, but a more direct approach is to use organism data.

Bivalve mollusks are popular organisms for use in monitoring programs due to their ability to concentrate contaminants, their sessile nature, ease of collection and for other reasons (*e.g.* FARRINGTON *et al.*, 1983). The US EPA used bivalves in a "Mussel Watch" program in 1976–1978 (GOLDBERG *et al.*, 1983) and since 1986 National Oceanic and Atmospheric Administration (NOAA) has had such a program as part of their National Status and Trends (NS&T) Program. Metal and organic contaminant data from the NOAA program, which covers the entire US coastline, have been presented in a number of publications (*e.g.* O'CONNOR, 1992, 1994, 1996). O'Connor and others have assessed the current contaminant status of the US coastline and have tried to answer the question "Are things getting better or worse?".

The Trace Element Research Lab (TERL) of the Texas A&M Oceanography Department analyzed all bivalves collected from the US Gulf of Mexico (GOM) coastline for the NOAA NS&T Mussel Watch Program between 1986 and 1999. These samples were, almost without exception, American oysters (*Crassostria virginica*) and they were collected, processed and analyzed in essentially the same way for this entire time period. The data should, thus, be internally consistent and comparable from year to year and from place to place.

METHODS

The methods used in this study are detailed in a NOAA Technical Memorandum (LAUENSTEIN and CANTILLO, 1993) and are outlined in an earlier publication (PRESLEY *et al.*, 1990). Only a brief description is given here. For the first 6 years of the program (1986–1991) three stations 100–500 m apart were sampled at each site along the coastline (Fig. 1), a site being an identifiable oyster reef. After year 6 (1992–1999), only one station was sampled at each site. The distance between sites varied from about 10 to 50 km. Sites were more closely spaced in large bays such as Galveston and Tampa than along open coastlines. Every effort was made to select sites that appeared to be "representative" of an area and to avoid point sources of pollutants. A strong effort was made to reoccupy sites as closely as possible each year and at about the same time of year, always in winter (December–January). As the program progressed, new sites were added, either to fill in gaps in the original pattern or because oysters could not be found at an existing site.

At each station, 20 oysters of a similar size were collected for metals analyses, and additional oysters were collected for organic analyses and other purposes. WRIGHT *et al.* (1985) and JIANN and PRESLEY (1997) have shown that a pooled sample of 20 or more oysters is representative of the population at a site. The oysters for metals were frozen in the field and shipped to the laboratory with no other processing. Details of sampling procedures, exact locations of sites, site descriptions and related information are given in LAUENSTEIN *et al.* (1997). In the laboratory, oyster shells were washed and brushed to remove mud and the soft tissue was removed. The entire soft parts of all 20 oysters were combined to make a pooled sample for each station. The pooled sample was homogenized wet and an aliquot was freeze-dried, or more commonly,

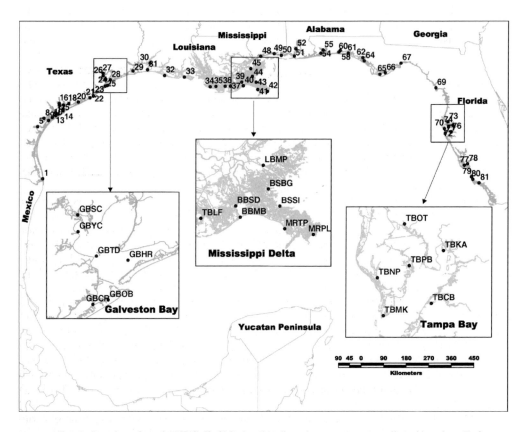

FIG. 1. Locations along the US Gulf of Mexico shoreline where oysters were collected in at least 7 of the years between 1986 and 1999 (see Appendix A for site code).

the entire sample was freeze-dried then homogenized. Only acid-cleaned plasticware and pure alumina were used in sample processing. Analysis of low metal concentration fish tissue showed that the processing did not significantly contaminate the samples with the metals of interest.

An accurately weighed, finely powdered aliquot of about 200 mg of dry tissue was digested in preparation for atomic absorption (AA) analysis. This was done by adding 3 ml of ultra-pure nitric acid to the samples in 60 ml capacity PFA teflon digestion vessels and heating the tightly closed vessels to 130°C in a drying oven. Reference materials, blanks, replicates and spikes were digested with each set of 20 samples. If the concentration obtained for the reference material differed from the certified value by more than 10% or if the spike recoveries were off by more than 10%, the sample set was re-analyzed. Furthermore, the National Research Council of Canada (NRCC) under contract to the NS&T Program Office conducted yearly intercalibration exercises using blind samples. Our laboratory's data were judged by the NRCC to be "Excellent" every year for the 14 years of the program.

Flame AA was used to determine Cu and Zn, cold vapor AA for Hg and graphite furnace AA with Zeeman background correction was used for the other elements. Almost all samples were also analyzed by instrumental neutron activation analysis (INAA) for Se, Zn and Ag, the only elements in high enough concentration to be determined by this method. The AA and INAA results agreed well for Se and Zn but for Ag, INAA gave higher values for some samples. This was traced to a sporadic Ag precipitation in the AA digests, especially at high (>5 ug/g) Ag concentrations. The problem was solved by adding HCl to the digestion but it is believed the INAA Ag data are more reliable so it is used here.

The NS&T Program called for the tissue samples to be analyzed for 15 trace metals (Ag, Al, As, Cd, Cr, Cu, Fe, Hg, Mn, Ni, Pb, Se, Sn, Tl and Zn) but Tl was dropped from the list early in the program. The other 14 metals were determined in all samples even though Sn was often below our detection limit. In this report only data for Ag, As, Cd, Cu, Hg, Pb, Se and Zn will be discussed. Iron, Al and Mn, for which we have reliable tissue data, seem unlikely to be pollutants and while Ni and Cr are potential pollutants, their low concentrations make comparisons between samples difficult.

Dr T. P. O'Connor, a manager of the Mussel Watch Program for NOAA, and his NOAA colleagues have examined all the data supplied to the program by all contractors and have used it to draw conclusions about the status and trends of coastal environmental quality. The result has been a number of publications in journals and books, as well as a number of NOAA Technical Memoranda, all of which list, describe and discuss the data in great detail (contact the NOAA NS&T Program Office at http://nsandt.noaa.gov for the raw data and for a complete list of all publications). No attempt will be made here to critically review these publications but some of their main conclusions will be given. The emphasis here will be on aspects of the GOM data that have received little attention in previous publications.

Recent NS&T publications have discussed both the "Status" and the "Trends" aspect of the Mussel Watch data but the emphasis has been on "Trends". The major underlying purpose of the NS&T Program is to identify human influences on the marine environment and O'CONNOR (1996 and elsewhere) has argued that this can best be done by detecting temporal trends in the data, especially monotonic increases or decreases in concentrations with time. Among other things, decreasing trends in concentration imply that the expensive pollution control devices and practices that have been mandated by the EPA are improving environmental quality.

The NS&T Program has produced data uniquely applicable to trend detection but as O'CONNOR (1996 and elsewhere) has elegantly pointed out, most chemicals at most locations show no statistically significant trend in concentration over time. For example, O'CONNOR (1998) shows for a 10-year time period that out of 2744 combinations of 14 chemicals at 196 sites only 88 increases and 348 decreases are significant at the 95% confidence level. Chance alone predicts 69 increases and 69 decreases, so by this analysis it is quite possible that no real increases occurred in that time period and environmental quality may have improved. This is an important finding.

The O'CONNOR (1998) paper used log-transformation, Spearman rank correlation and non-parametric statistical techniques to avoid skewing trends by unusually high or low data points. This paper will, however, point out some of the unusually high concentrations shown by given metals at given places and times because these might imply human influence at the local level. Local anomalies might have few implications for regional environmental quality but they are nevertheless worthy of attention. Likewise, whereas spatial variations in chemical concentrations along the coastline might have less large-scale or long-term environmental quality significance than variations with time, they are nevertheless deserving of documentation and possible local remedial action.

In a previous paper (PRESLEY *et al.*, 1990), it was pointed out that spatial variation in metal concentrations along the GOM coastline were larger than temporal variations at a given place, based on the first three years of Mussel Watch data. It was further commented, "…with a few exceptions, Man's influence (on metal concentrations) is subtle and very local". It will thus be difficult to answer the question "Are things getting better or worse?". The 14-year data set also shows more obvious spatial variations in metal concentrations than temporal variations but some of each exist, as will be discussed below.

RESULTS AND DISCUSSION

Table 1 gives the mean concentration of each metal in the oysters analyzed during this study. These means are based on one of the largest sets of data on metal concentrations in oysters ever reported. Well over 20,000 oysters from 63 "typical" sites along the GOM shoreline were analyzed over a 14-year period to produce these numbers. We could find no comparable data from elsewhere for comparison. Most oyster data sets are small and often from obviously polluted areas, and most "Mussel Watch" monitoring programs use mussels not oysters. These two organisms differ in average concentration of some metals (*e.g.* O'CONNOR, 1992). Nevertheless, the mean metal concentrations reported here are similar to those reported for Chesapeake Bay oysters (Table 1) and for oysters from other visually pristine places. In other words, there is no evidence of large-scale metal contamination of Gulf Coast oysters. There is also no statistically significant increase or decrease in concentration of any metal over the 14-year time period based on these data. Thus, there is no clear answer to the question "Are things getting better or worse?".

Like the 3-year data set, the 14-year data show large concentration differences from place to place along the Gulf coastline (Appendix A). For example, the 14-year average concentration of Cu and Zn at the highest sites is more than 20 times higher than that at the lowest sites and even if the 10th and 90th percentiles of the concentrations are compared, the differences are more than a factor of 10. Other elements also show large concentration differences between high and low sites. In most cases, reasons for anomalous values are not obvious, but in a few cases they can be inferred from proximity to known or likely sources. An example is the high concentration of Hg at sites in Matagorda Bay, TX. These sites are only a few km from a Hg-contaminated Superfund Site and even higher concentrations have been found in oysters collected closer to the site (PRESLEY, 1994). Oysters at site CBPP on the Eglin Air Force Base in Choctawhatchee Bay, FL that was only sampled for the first

Table 1. Comparison of trace metal concentrations in Gulf of Mexico (GOM) oysters with trace metal concentrations in oysters from elsewhere. All data in ug/g dry weight

Site	Ag	As	Cd	Cu	Hg	Se	Pb	Zn
GOM[a]	3.38	9.24	4.37	178	0.13	2.87	0.68	2901
Chesapeake Bay[b]	5.0	6.6	6.9	266	0.04	2.9	0.34	4900
Taiwan[c]	–	10.8	1.8	229	0.3	–	0.45	783
Australia[d]	1.8	15	5.0	170	–	3.7	0.6	2610

[a] Mean concentrations in *C. virginica* from this study.

[b] Mean concentrations in *C. virginica* from five NS&T sites in Chesapeake Bay sampled yearly, 1986–1997. Data from NOAA NS&T web site.

[c] Median concentrations for *C. gigas* from several coastal areas of Taiwan (JENG *et al.*, 2000).

[d] Calculated mean "background" concentrations for the oyster *S. commercialis* from New South Wales (SCANES and ROACH, 1999).

3 years of the project, were almost 10 times higher than average GOM Pb concentration. This is likely due to some human activity because sediments near the site had Pb concentrations that were five times higher than average but the sediments were not enriched in other metals. Likewise, the higher than average Zn concentrations along the western shoreline of Galveston Bay and in Mobile and Tampa Bays are probably due to human activity, which is much more intense near these sites than it is near most other sites, as is discussed below.

The high metal concentrations noted above, and similar ones elsewhere, attract attention because they suggest pollution and imply potentially harmful biological effects. Identifying pollution from oyster data is, however, not straightforward. As noted above, metal concentrations in oysters vary widely from place to place. Furthermore, this variability occurs over small geographic distances. For example, sites in Tampa Bay, FL, Galveston Bay, TX and other areas show Cu and Zn concentrations that are both much higher and much lower than average GOM values. This is illustrated in Fig. 2 which shows Zn concentrations in oysters from four of the six sites sampled in Galveston Bay. The two high Zn concentration sites (GBYC and GBSC) are along the industrialized western shoreline of upper Galveston Bay (Fig. 1). Almost half of all the wastewater discharge permits for Texas have been issued to the thousands of manufacturing facilities in upper Galveston Bay (*e.g.* NOAA, 1989). The lower Zn concentration sites (GBCR and GBHR) are in visually more pristine lower Galveston Bay where there are far fewer coastal manufacturing facilities and where dissolved trace metal concentrations are lower (WEN *et al.*, 1999). This pattern of higher Zn in oysters near

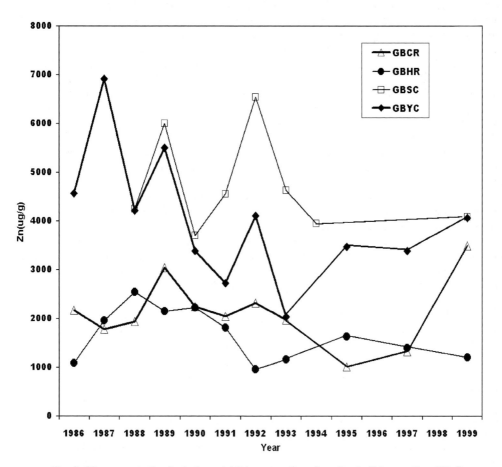

FIG. 2. Zinc concentration (ug/g dry weight) in oysters from four sites in Galveston Bay, TX. Sites GBSC and GBYC are along the industrialized western shoreline. Sites GBHR and GBCR are in the visibly more pristine lower Galveston Bay.

industrialized shorelines was also found in Tampa, Mobile and Corpus Christi Bays. Thus, Zn has a clearer relationship to industrialization than any other metal. There were no other significant correlations between concentrations of metals in oysters and the dissolved metal concentrations reported by WEN *et al.* (1999).

Some of the more toxic metals, such as Ag and Cd, also vary greatly from place to place. Figure 3 shows year-to-year contrasts at three high and three low Cd sites in the Mississippi River Delta area. Note that 5- to 10-fold differences occur year after year. The three low Cd concentration sites are in the Barataria Bay area to the west of the delta. This region is sparsely populated but has been heavily impacted by oil drilling and production. It is an interdistributary estuarine–wetland system that receives Mississippi River water which has crossed the Gulf shelf but has not received direct overflow from the river since the 1930s when levees were built (ORLANDO *et al.*, 1993). The three high Cd concentration sites are closer to the main stream Mississippi River but there was no correlation between salinity and Cd concentrations at these sites. However, a complex of levees and dredged channels complicates fresh water flow in this area (Figs. 1 and 3).

It is tempting to assume that the low Cd, Cu, Zn and other metal concentrations that occur year after year at some sites are "background" and that the higher concentrations observed at other sites result from human influence. SCANES and ROACH (1999) reached essentially this conclusion when they compared metal concentrations in oyster from more and less urbanized estuaries in Australia. However, when O'CONNOR (2002) considered the NS&T metals in bivalves data set for the entire US coastline, only Pb showed a strong correlation to the population density within 20 km of the individual sites. Thus, it may be necessary to consider smaller distance scales and/ or to use factors in addition to population in order to identify sites where human activity has affected metal concentrations in bivalves.

Natural spatial variation is suggested by isolated anomalous concentrations in unlikely places. Examples include the much higher than average Cu concentration in oysters in marshy, rural, isolated south Louisiana (site VBSP), high Cd concentration at a site in rural Texas (site CBCR), high Ag concentration in Galveston's West Bay (site GBCR) and other isolated high concentrations for various metals at various times in apparently pristine places. It may be significant that most of the high concentrations are restricted to small areas of bays or coastlines. It is common to find sites with high concentrations of a given metal within a few km of those with low concentration of that metal. Furthermore, in years when three stations a few hundred meters apart were sampled at a given site, metal concentration at these stations sometimes differed by a factor of five or more. Thus, metal concentrations often seem to be controlled by local activity or conditions and not by regional geology or climate.

Local control is also suggested by the observed large changes in metal concentration in oysters from one year to the next at some sites when nearby sites showed no such changes. Large seasonal changes have been reported by JIANN and PRESLEY (1997) and others, showing that oysters can react to environmental changes in a month or so. At a number of sites, the data show some metals are present at concentrations five or more times higher than the long-term average for only 1 year. There are also many examples of isolated low concentrations for a single year. Human activity varies on small temporal and spatial scales. Consequently, it is tempting to attribute some of these changes in metal concentration in oysters to human activity, even when the specific activity cannot be identified with the data at hand. On the other hand, it is hard to suggest a change in human activity that would result in a large 1-year decrease in concentration at a site when nearby sites show no decrease.

Like isolated high concentrations, long-term temporal trends in metal concentrations suggest human influence, as has been pointed out by O'CONNOR (1996, 1998 and elsewhere). However, at some sites individual metals either increase or decrease in concentration year after year for 5 or more years, only to suddenly reverse that trend. In some cases, several metals show the same temporal changes, with Cu, Zn, Ag and Cd commonly following each other. For example, in 1996 Cd concentrations decreased dramatically at two sites (MRPL, BSSI) on the Mississippi River Delta (Fig. 3), accompanied by drastic decreases in

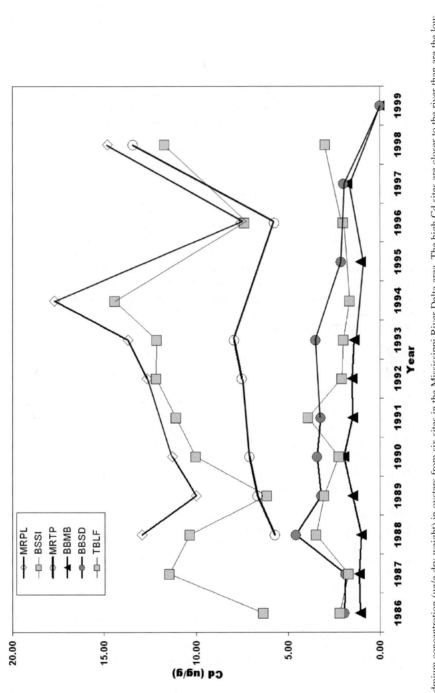

FIG. 3. Cadmium concentration (ug/g dry weight) in oysters from six sites in the Mississippi River Delta area. The high Cd sites are closer to the river than are the low Cd sites.

concentrations of Cu, Zn and Ag and to a lesser extent in Pb and Se. When these sites were next sampled in 1998, Cd concentrations had returned to their former high values (Fig. 3). It seems unlikely that short-term reductions in the amounts of all these metals coming down the Mississippi River or being put into the environment locally caused these changes. However, if changes in salinity, temperature or some other environmental parameter or a slight change in the oyster collection location cause the Cd concentration variation, we have been unable to document it. Stream flow data for the Mississippi River (USGS) were compared with the oyster trace metal concentrations in this area and there was no significant correlation during this time period. There was also no correlation between salinity or temperature at the collection sites and metal concentration, and the sampling crew made no mention of changing the sampling locations.

As suggested above, Cu and Zn concentrations in oysters sometimes correlate well. Figure 4a and b shows all data from all years for two different sites. Note the excellent correlation in each case, but the somewhat different slopes and intercepts. The Cu–Zn correlations for many sites are similar to these, with $r^2 > 0.76$ at 18 of 80 sites. These tight Cu–Zn relationships suggest something other than variations in human input of Cu and Zn to us. On the other hand, the correlation is poor at many sites ($r^2 < 0.3$ at 25 of 80 sites). Certain other metal pairs, such as Cu–As, Cu–Ag and Cu–Cd, also show good correlations at some sites for some time periods but not for others. These observations, once again, suggest that changes in metal concentrations in oysters from time to time and place to place are a complicated function of both natural and human phenomena.

In addition to changes in metal concentrations from site to site and from time to time, it is worth noting a few broad geographic patterns in the 14-year data set. Many of these patterns were noted by PRESLEY *et al.* (1990) based on only 3 years of data. For example, the additional data reinforce the observation of a generally increasing trend for As at sites from Texas and Louisiana to Florida. The 14-year average As concentrations in Florida and non-Florida oysters are 13.47 and 7.92 ug/g, respectively, a significant difference at the 99.9% level of confidence by "*t*" test.

Scrutiny of the data at smaller spatial scales shows that the average concentration of As in Florida oysters is greatly influenced by data from a few sites that are as much as 10 times higher than the others (Fig. 5a). High concentrations are found at sites in Apalachee Bay in north Florida, near Tampa in central Florida and near Naples in south Florida, yet much lower As concentrations are found at some sites in these same locations. The differences are, thus, more local than regional, which is consistent with the spatial concentration differences we have found for other elements.

In examining maps of the collection sites in Florida (Fig. 1), it appears that high As concentration in oysters usually came from areas of open coastline, near open mouths of large bays or mid-bay regions with unrestricted access to the open Gulf. Oysters collected well inside large bays, or in areas with restricted access to the open Gulf generally have lower concentrations of As. For example, two sites near the mouth of Tampa Bay have average As concentrations (ug/g) for the 14-year period of 15 (TBMK) and 56 (TBNP) while the four sites inside Tampa Bay average 3.4 (TBOT), 6.2 (TBPB), 5.4 (TBKA) and 4.5 (TBCB). Differences are also found at the mouth of Charlotte Harbor (23 ug/g) and inside the Harbor (6.6 ug/g) and at other locations.

Outside Florida there is less contrast in As concentration between sites with unrestricted water exchange with the open Gulf and the more restricted (bay) sites, but there appears to be a small difference. The different site concentrations did not correlate with the salinity measured when the oysters were collected or with long-term average salinities as shown in ORLANDO *et al.* (1993). It is suspected that the abundance of bioavailable phosphate or the ratio of phosphate to arsenate in the water column and food of the oysters is more important than salinity. Where phosphate is high, less arsenate is taken up and where phosphate is low more arsenate is taken up. Phosphate data were not obtained as part of NS&T, but BOLER (1995) shows dissolved phosphate to be 3–4 times higher in the inner parts of Tampa Bay than in the outer parts. Furthermore, the Peace river and inner parts of

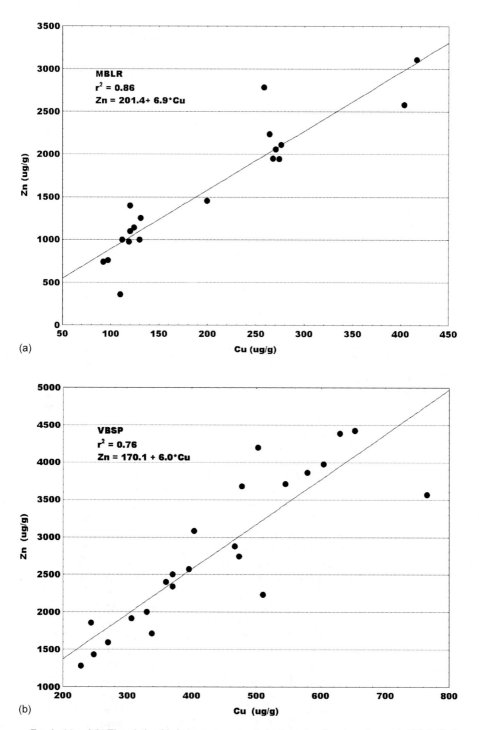

(a)

(b)

FIG. 4. (a) and (b) The relationship between copper and zinc in oysters from two sites on the US Gulf of Mexico shoreline. At 18 of the 80 sites sampled the Cu–Zn r^2 was similar to these (greater than 0.76) but at another 25 sites the r^2 was less than 0.3.

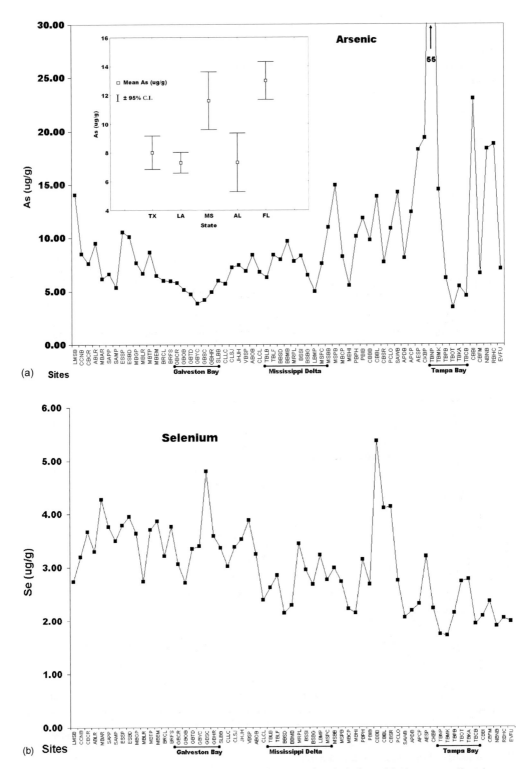

Fig. 5. (a) and (b) Average aresenic and selenium concentrations (ug/g dry weight) in oysters collected at sites along the US Gulf of Mexico shoreline which were sampled for at least 7 of the years between 1986 and 1999. Inset shows mean and 95% confidence intervals for the mean concentration (ug/g dry weight) of arsenic in oysters by state.

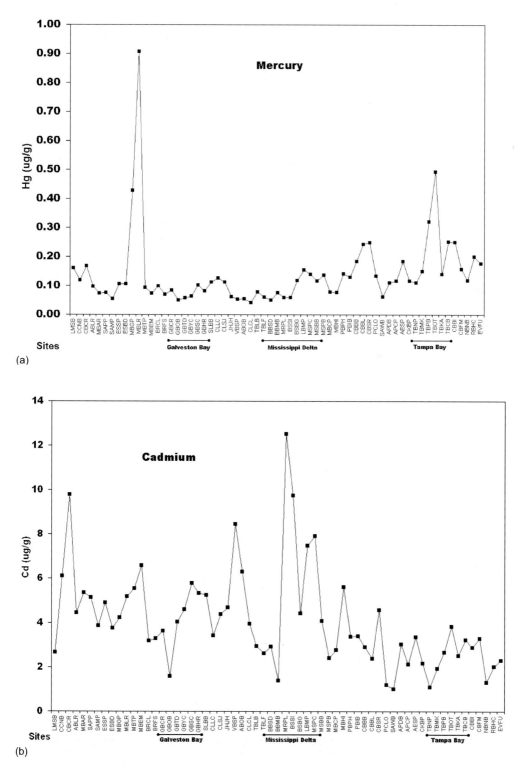

FIG. 6. (a) and (b) Average mercury and cadmium concentrations (ug/g dry weight) in oysters collected at sites along the US Gulf of Mexico shoreline which were sampled for at least 7 of the years between 1986 and 1999.

Charlotte Harbor are much richer in phosphate than is the outer harbor (MILLER and MCPHERSON, 1987).

The distribution of Se, another metalloid, in oysters is quite different from that of As, showing a decrease from Texas to Florida (Fig. 5b). Non-Florida oysters average 3.21 ug/g Se and Florida oysters average 2.32 ug/g (excluding Florida panhandle sites PBPH, PBIB, CBBB, CBBL and CBSR), again a significant difference at the 99.9% level of confidence. Cadmium and Hg concentrations in oysters from Florida and elsewhere in the GOM are also different at the 99.9% level of confidence. Hg, like As, is enriched in Florida oysters, if oysters from Lavaca Bay, TX (MBLR), near a Hg-contaminated Superfund Site, are omitted from the averages (Fig. 6a). Hg averages in Florida and elsewhere are 0.18 and.09 ug/g with the exclusion of the MBLR data point. Cd behaves more like Se, with a decrease from 4.7 ug/g in non-Florida oysters to 2.7 ug/g in Florida (Fig. 6b).

Metal concentration differences in oysters at small spatial scales, even within individual reefs, indicate that there is much variability that cannot be attributed to large-scale phenomena such as temperature, rainfall and salinity. Observations of strong inter-element correlations at many sites, however, are reminiscent of the natural "variation" in metal concentrations commonly seen in sediments. Sediment data can be manipulated to minimize variation due to mineralogy and grain size differences by normalizing (or plotting) the data against a tracer (Al or Fe) of the sediment fraction that bears most of the metal load. Normalization also works well for organic contaminants in tissue but we have seen no published normalization procedure for metals in tissue.

CONCLUSIONS

Average metal concentrations in oysters from sites along the GOM coastline were less than or equal to those reported for oysters from uncontaminated sites elsewhere. Thus, there is no evidence for widespread metal contamination of GOM oysters. Furthermore, long-term trends in metal concentrations in oysters give no clear evidence of either increasing or decreasing metal contamination of the GOM coastal environment. The several-fold variations in concentrations of some metals from site to site and from year to year at a given site seem to be due to both natural and human causes. Isolated high concentrations at a given site or in a given year, high concentrations of metals such as Ag, Cd and Zn at sites near large industrial centers and gradual increases or decreases in concentrations over a several year period all suggest that human activity is affecting metals in oysters. On the other hand, the importance of salinity, temperature and other environmental factors is suggested by low metal concentrations in areas of high human activity, high metals at rural and visibly pristine sites, several metals showing identical trends in concentration with time, abrupt reversals in concentration–time trends, seasonal variability in metal concentrations and regional differences in concentration of some metals. It is, thus, often difficult to determine what causes specific temporal or spatial variations in metal concentrations in oysters, even when the oysters have been analyzed by the same laboratory using the same methods over many years. To attribute causation by comparing data from different laboratories or data from only a few sampling episodes is even riskier. In our opinion, management decisions such as requiring decreases in wastewater inputs to an area should not be based on subtle (< 10-fold) differences in metal concentrations detected by studies of this latter type.

REFERENCES

BOLER R. (1995) *Surface Water Quality 1992–1994: Hillsborough County, FL.* Environmental Protection Commission of Hillsborough County.
FARRINGTON J. W., GOLDBERG E. D., RISEBROUGH R. W., MARTIN J. H. and BOWEN V. T. (1983) U.S. "Mussel Watch" 1976–1978: an overview of the trace metal, DDE, PCB, hydrocarbon and artificial radionuclide data. *Environ. Sci. Technol.* **17**, 490–496.
GOLDBERG E. D., KOIDE M., HODGE V., FLEGAL A. R. and MARTIN J. H. (1983) US Mussel Watch 1976–1978 results on trace metals and radionuclides. *Estuar. Coast. Shelf Sci.* **16**, 69–93.

JENG M.-S., JENG W.-L., HUNG T.-C., YEH C.-Y., TSENG R.-J., MENG P.-J. and HAN B.-C. (2000) Mussel Watch: a review of Cu and other metals in various marine organisms in Taiwan, 1991–98. *Environ. Pollut.* **110**, 207–215.

JIANN K. T. and PRESLEY B. J. (1997) Variations in trace metal concentrations in oysters (*Crassosteria virginica*) collected from Galveston Bay, Texas. *Estuaries* **20**, 710–724.

LAUENSTEIN G. G. and CANTILLO A. Y. (1993) Sampling and analytical methods of the National Status and Trends Program National Benthic Surveillance and Mussel Watch Projects 1984–1992. Vol. III Comprehensive descriptions of elemental analytical methods. *NOAA Technical Memorandum NOS ORCA 71.*

LAUENSTEIN G. G., CANTILLO A. Y., KOKKINAKIS S., FREW S., JOBLING H. J. and FAY R. R. (1997) National Status and Trends Program: Mussel Watch Site Descriptions through 1997. *NOAA Technical Memorandum NOS ORCA 112.*

MILLER R. L. and MCPHERSON B. F. (1987) Concentration and transport of phosphorus and radium-226 in the Peace River and Charlotte Harbor, southwestern, Florida, in 194th National Meeting of the American Chemical Society, New Orleans, LA, 1987. *Preprint of Papers: Am. Chem. Soc.* **27**(2), 389–391.

NOAA, (1989) *Galveston Bay: Issues, Resources, Status and Management.* NOAA Estuary of the Month Series No. 13. NOAA Estuarine Programs Office, Washington, DC.

O'CONNOR T. P. (1992) *Mussel Watch: Recent Trends in Coastal Environmental Quality: Results from the First Five Years of the NOAA Mussel Watch Project.* Special Publication. NOAA/NOS/ORCA, Rockville, MD, 46 p.

O'CONNOR T. P. (1994) The National Oceanic and Atmospheric Administration's (NOAA) National Status and Trends Mussel Watch Program: National Monitoring of Chemical Contamination in the Coastal United States. In *Environmental Statistics, Assessment, and Forecasting* (eds. C. R. COTHERN and N. P. ROSS), pp. 331–349. Lewis Publications, Boca Raton, FL.

O'CONNOR T. P. (1996) Trends in chemical concentrations in mussels and oysters collected along the U.S. coast from 1986 to 1993. *Mar. Environ. Res.* **41**, 183–200.

O'CONNOR T. P. (1998) Mussel Watch results from 1986 to 1996. *Mar. Pollut. Bull.* **37**, 14–19.

O'CONNOR T. P. (2002) National distribution of chemical concentrations in mussels and oysters in the USA. *Mar. Environ. Res.* **53**, 117–143.

ORLANDO S. P. Jr., ROZAS L. P., WARD G. H. and KLEIN C. J. (1993) *Salinity Characteristics of Gulf Coast Estuaries.* NOAA Office of Ocean Resource Conservation and Assessment, Silver Spring, MD, 209 p.

PRESLEY B. J. (1994) *A description of mercury contamination in Lavaca Bay and related biogeochemical factors*, Preworkshop Report for the Lavaca Bay Scientific Workshop, Houston, TX, Feb. 3–5, 1994. Texas General Land Office, Austin, TX.

PRESLEY B. J., TAYLOR R. J. and BOOTHE P. N. (1990) Trace metals in Gulf of Mexico oysters. *Sci. Total Environ.* **97/98**, 551–593.

SCANES P. R. and ROACH A. C. (1999) Determining natural "background" concentrations of trace metals in oysters from New South Wales, Australia. *Environ. Pollut.* **105**, 437–446.

WEN L. S., SHILLER A., SANTSCHI P. H. and GILL G. (1999) Trace metal behavior in Gulf of Mexico estuaries. In *Biogeochemistry of Gulf of Mexico Estuaries* (eds. T. S. BIANCHI, J. R. PENNOCK and R. TWILLEY), pp. 303–346. Wiley, New York.

WRIGHT D. A., MIHURSKY J. A. and PHELPS H. L. (1985) Trace metals in Chesapeake Bay oysters: intra-sample variability and its implications for biomonitoring. *Mar. Environ. Res.* **16**, 181–197.

APPENDIX A

Summary statistics for trace metal concentrations in GOM oysters collected at 63 sites between 1986 and 1999 is given in Tables A1 and A2. All data in ug/g dry weight.

Table A1

Site number	Site	Location	State	Ag Max	Ag Min	Ag Mean	Ag Cv	As Max	As Min	As Mean	As Cv	Cd Max	Cd Min	Cd Mean	Cd Cv	Cu Max	Cu Min	Cu Mean	Cu Cv
1	LMSB	Lower Laguna Madre, South Bay	TX	2.23	0.65	1.33	38	26.0	5.6	14.1	32	4.02	1.49	2.67	30	144	50	105	25
5	CCNB	Corpus Christi, Nueces Bay	TX	2.61	0.77	1.48	37	12.7	6.2	8.51	23	12.4	2.56	6.10	49	223	79	146	26
8	CBCR	Copano Bay, Copano Reef	TX	14.6	1.35	6.52	67	11.3	0.9	7.61	28	15.3	2.93	9.78	35	359	59	216	46
9	ABLR	Aransas Bay, Long Reef	TX	2.93	0.00	1.47	60	14.5	4.3	9.50	33	7.00	0.13	4.44	36	195	20	99	52
10	MBAR	Mesquite Bay, Ayres Reef	TX	8.59	0.68	2.87	78	10.9	3.9	6.13	26	12.9	2.39	5.35	43	201	23	107	42
11	SAPP	San Antonio Bay, Panther Point Reef	TX	7.03	0.99	2.96	64	8.3	4.2	6.60	19	8.34	3.41	5.15	30	291	29	119	67
12	SAMP	San Antonio Bay, Mosquito Point	TX	6.59	0.77	2.33	91	7.9	4.5	5.34	20	5.50	2.42	3.88	26	290	33	117	80
13	ESSP	Espiritu Santo, South Pass Reef	TX	10.4	1.82	4.06	78	24.0	5.5	10.5	43	11.0	3.23	4.90	39	346	53	162	65
14	ESBD	Espiritu Santo, Bill Days Reef	TX	5.09	1.30	2.19	56	12.4	7.0	10.1	16	4.80	3.03	3.76	13	223	80	131	35
15	MBGP	Matagorda Bay, Gallinipper Point	TX	7.06	1.38	2.84	62	11.1	5.1	7.67	19	6.05	0.13	4.24	30	229	73	140	34
16	MBLR	Matagorda Bay, Lavaca River Mout	TX	6.31	1.12	3.08	55	10.7	2.9	6.65	28	8.05	3.43	5.19	26	416	92	197	51
18	MBTP	Matagorda Bay, Tres Palacios Bay	TX	11.2	1.30	4.40	79	15.9	5.1	8.64	36	8.29	2.91	5.54	27	291	58	171	43
20	MBEM	Matagorda Bay, East Matagorda	TX	7.04	1.57	3.65	51	9.0	3.9	6.44	23	13.1	0.31	6.59	40	263	79	187	26
21	BRCL	Brazos River, Cedar Lakes	TX	4.42	1.60	2.84	30	8.5	3.9	5.96	19	4.52	2.39	3.19	20	423	94	191	39
22	BRFS	Brazos River, Freeport Surfside	TX	4.48	1.77	2.80	34	8.2	4.2	5.95	16	4.53	2.22	3.30	21	524	108	240	44
23	GBCR	Galveston Bay, Confederate Reef	TX	26.5	2.21	8.53	78	9.2	4.2	5.76	17	5.61	1.19	3.63	27	295	80	164	35
24	GBOB	Galveston Bay, Offatts Bayou	TX	4.33	0.38	1.32	90	9.2	3.2	5.13	28	5.74	0.66	1.60	68	264	98	154	28
25	GBTD	Galveston Bay, Todds Dump	TX	4.14	1.10	2.31	46	6.7	2.5	4.68	22	5.57	2.96	4.05	19	249	85	162	27
26	GBYC	Galveston Bay, Yacht Club	TX	3.59	0.75	1.95	41	6.7	1.8	3.84	27	6.59	3.48	4.61	17	354	94	168	46
27	GBSC	Galveston Bay, Ship Channel	TX	3.58	1.11	2.24	34	6.3	2.2	4.16	27	8.15	3.59	5.79	23	280	125	168	27
28	GBHR	Galveston Bay, Hanna Reef	TX	4.63	1.34	3.01	43	7.8	3.6	4.90	21	9.24	3.53	5.33	27	262	89	154	29
29	SLBB	Sabine Lake, Blue Buck Point	TX	8.20	4.03	5.80	21	10.3	3.8	5.97	28	8.51	2.00	5.25	26	559	75	264	45
30	CLLC	Calcasieu Lake, Lake Charles	LA	2.64	0.88	1.60	36	20.7	2.6	5.68	70	5.34	1.34	3.41	32	438	68	210	52
31	CLSJ	Calcasieu Lake, St. Johns Island	LA	5.73	1.42	3.21	39	12.0	3.9	7.22	28	6.39	3.10	4.40	23	425	104	216	39

Table A1 (continued)

Site number	Site	Location	State	Ag Max	Ag Min	Ag Mean	Ag Cv	As Max	As Min	As Mean	As Cv	Cd Max	Cd Min	Cd Mean	Cd Cv	Cu Max	Cu Min	Cu Mean	Cu Cv
32	JHJH	Joseph Harbor Bayo, Joseph Harbor Bay	LA	5.16	1.54	3.10	41	12.6	4.1	7.44	27	5.67	2.56	4.69	17	276	100	178	28
33	VBSP	Vermilion Bay, Southwest Pass	LA	13.9	0.68	6.15	56	11.0	4.1	6.89	25	13.0	4.21	8.45	23	765	228	438	32
34	ABOB	Atchafalaya Bay, Oyster Bayou	LA	4.29	1.13	2.70	34	14.0	4.9	8.39	30	11.1	3.60	6.29	29	333	100	198	32
35	CLCL	Caillou Lake, Caillou Lake	LA	2.56	0.91	1.37	34	12.2	4.1	6.79	26	5.59	2.30	3.96	24	206	90	130	22
36	TBLB	Terrebonne Bay, Lake Barre	LA	1.32	0.55	0.86	30	12.3	4.3	6.26	28	4.08	1.25	2.94	27	219	50	102	37
37	TBLF	Terrebonne Bay, Lake Felicity	LA	1.20	0.32	0.73	40	15.0	4.9	8.38	36	5.55	1.13	2.62	41	162	40	92	35
39	BBSD	Barataria Bay, Bayou Saint Denis	LA	2.12	0.33	1.20	48	14.2	4.4	7.95	34	6.99	1.00	2.92	42	326	33	139	58
40	BBMB	Barataria Bay, Middle Bank	LA	1.94	0.26	0.61	82	17.4	5.3	9.68	27	2.44	0.54	1.39	29	171	21	52	56
42	MRPL	Mississippi River, Pass A Loutre	LA	7.68	1.90	4.86	46	12.2	5.5	7.76	28	17.7	7.55	12.5	25	439	169	290	32
43	BSSI	Breton Sound, Sable Island	LA	5.01	1.56	3.10	38	18.8	4.7	8.28	43	14.4	5.50	9.74	27	289	67	176	41
44	BSBG	Breton Sound, Bay Gardene	LA	1.22	0.34	0.76	33	12.4	3.5	6.48	39	6.30	2.39	4.43	22	142	47	79	38
45	LBMP	Lake Borgne, Malheureux Point	LA	4.18	0.73	2.09	45	6.9	2.5	4.94	22	13.3	1.47	7.48	38	599	109	293	43
48	MSPC	Mississippi Sound, Pass Christian	MS	5.77	1.58	3.65	40	14.3	2.9	7.57	40	12.5	1.82	7.92	29	340	69	199	37
49	MSBB	Mississippi Sound, Biloxi Bay	MS	7.45	1.57	3.15	62	27.0	5.5	11.0	43	7.19	1.80	4.09	37	617	100	212	48
50	MSPB	Mississippi Sound, Pascagoula Bay	MS	3.18	1.20	2.39	26	21.0	8.9	14.9	26	3.60	1.27	2.40	24	176	68	121	24
51	MBCP	Mobile Bay, Cedar Point Reef	AL	3.86	1.15	2.28	37	16.0	3.5	8.21	45	4.22	1.12	2.77	29	141	35	82	29
52	MBHI	Mobile Bay, Hollingers Is. Ch	AL	31.7	3.68	9.36	98	8.0	4.1	5.51	19	8.82	3.96	5.62	20	587	170	325	37
54	PBPH	Pensacola Bay, Public Harbor	FL	2.63	1.30	2.21	20	14.5	6.6	10.1	29	4.74	2.41	3.38	21	113	37	80	29
55	PBIB	Pensacola Bay, Indian Bayou	FL	10.3	1.25	3.63	84	18.1	6.3	11.8	28	7.00	2.35	3.40	30	149	30	61	46
58	CBBB	Choctawhatchee Bay, Boggy Bayou	FL	16.7	14.1	15.0	10	11.3	8.5	9.75	14	3.43	2.63	2.91	11	232	145	189	18
60	CBBL	Choctawhatchee Bay, Bens Lake	FL	4.14	3.10	3.66	14	25.1	7.5	13.8	48	2.88	1.97	2.38	17	278	102	203	31
61	CBSR	Choctawhatchee Bay, Off Santa Rosa	FL	7.69	1.05	3.37	52	11.4	3.9	7.70	24	7.70	2.44	4.57	26	119	43	74	28
62	PCLO	Panama City, Little Oyster Bar	FL	1.87	0.40	0.94	54	22.2	5.5	10.8	36	1.50	0.97	1.22	14	40	15	30	26
64	SAWB	St. Andrews Bay, Watson Bayou	FL	3.41	1.59	2.35	27	21.3	9.5	14.2	20	2.24	0.66	1.02	32	450	98	265	38

(continued on next page)

Table A1 (*continued*)

Site number	Site	Location	State	Ag Max	Ag Min	Ag Mean	Ag Cv	As Max	As Min	As Mean	As Cv	Cd Max	Cd Min	Cd Mean	Cd Cv	Cu Max	Cu Min	Cu Mean	Cu Cv
65	APDB	Apalachicola Bay, Dry Bar	FL	2.78	1.01	1.64	36	16.1	4.3	8.09	32	4.03	2.27	3.03	18	73	24	47	29
66	APCP	Apalachicola Bay, Cat Point Bar	FL	2.76	0.72	1.59	42	24.0	6.0	12.4	39	3.01	1.31	2.12	21	232	34	107	55
67	AESP	Apalachee Bay, Spring Creek	FL	1.57	0.41	1.02	38	34.6	6.6	18.2	49	4.31	2.00	3.35	19	47	20	30	28
69	CKBP	Cedar Key, Black Point	FL	1.02	0.11	0.61	41	31.9	7.7	19.3	36	5.54	0.70	2.17	49	94	5	20	88
70	TBNP	Tampa Bay, Navarez Park	FL	3.35	0.00	1.62	75	126	30.8	56.1	65	1.72	0.47	1.10	27	115	48	72	23
71	TBMK	Tampa Bay, Mullet Key Bayou	FL	0.87	0.24	0.47	43	26.0	9.2	14.5	27	3.90	0.96	1.95	39	55	7	24	56
72	TBPB	Tampa Bay, Papys Bayou	FL	2.87	0.57	1.80	41	10.2	4.1	6.21	29	13.2	1.18	2.67	89	328	37	79	72
73	TBOT	Tampa Bay, Old Tampa Bay	FL	3.93	1.62	2.72	28	6.6	1.6	3.42	33	5.44	2.42	3.82	25	203	84	141	29
74	TBKA	Tampa Bay, Peter O. Knight Airport	FL	8.08	2.40	4.05	45	9.7	3.7	5.40	34	7.77	1.01	2.51	64	683	77	247	62
76	TBCB	Tampa Bay, Cockroach Bay	FL	2.53	0.87	1.53	34	9.1	2.4	4.52	32	5.91	1.37	3.22	40	215	35	104	41
77	CBBI	Charlotte Harbor, Bird Island	FL	4.15	1.81	2.77	37	40.0	11.1	23.0	43	4.95	1.80	2.87	33	351	80	132	49
78	CBFM	Charlotte Harbor, Fort Meyers	FL	7.08	1.00	3.43	55	11.9	4.7	6.60	25	5.63	1.97	3.28	34	350	115	203	36
79	NBNB	Naples Bay, Naples Bay	FL	8.91	3.40	6.13	29	30.6	7.8	18.3	40	2.63	0.74	1.32	33	1290	250	468	52
80	RBHC	Rookery Bay, Henderson Creek	FL	3.31	1.28	2.19	30	32.0	9.5	18.7	44	2.93	1.31	2.02	25	195	28	96	44
81	EVFU	Everglades, Faka Union Bay	FL	1.46	0.60	0.97	36	10.0	4.5	7.05	20	4.19	1.20	2.29	31	102	29	52	37
		GOM average		5.84	1.41	2.97	47	17	5.3	9.58	31	6.89	2.1	4.11	30	302	73	156	40

Table A2

Site number	Site	Location	State	Hg Max	Hg Min	Hg Mean	Hg Cv	Pb Max	Pb Min	Pb Mean	Pb Cv	Se Max	Se Min	Se Mean	Se Cv	Zn Max	Zn Min	Zn Mean	Zn Cv
1	LMSB	Lower Laguna Madre, South Bay	TX	0.32	0.05	0.16	42	1.96	0.19	0.63	56	5.26	1.68	2.74	29	6982	297	1815	77
5	CCNB	Corpus Christi, Nueces Bay	TX	0.20	0.00	0.12	45	1.82	0.45	1.62	234	5.78	1.57	3.19	34	6500	2960	4334	22
8	CBCR	Copano Bay, Copano Reef	TX	0.43	0.09	0.17	51	1.20	0.32	0.62	36	7.09	2.00	3.67	38	2341	385	1355	40
9	ABLR	Aransas Bay, Long Reef	TX	0.18	0.00	0.10	46	0.92	0.19	0.52	36	6.00	1.11	3.29	38	5439	179	1285	104
10	MBAR	Mesquite Bay, Ayres Reef	TX	0.16	0.01	0.07	50	1.00	0.19	0.50	46	10.5	2.00	4.28	52	1440	183	694	47
11	SAPP	San Antonio Bay, Panther Point Reef	TX	0.24	0.03	0.08	64	1.00	0.25	0.56	36	4.87	2.97	3.76	13	1372	341	668	45
12	SAMP	San Antonio Bay, Mosquito Point	TX	0.23	0.00	0.06	104	1.20	0.20	0.52	52	5.43	2.26	3.50	25	1800	337	820	61
13	ESSP	Espiritu Santo, South Pass Reef	TX	0.20	0.04	0.11	52	1.00	0.35	0.57	36	8.50	2.92	3.79	35	2259	569	1025	52
14	ESBD	Espiritu Santo, Bill Days Reef	TX	0.14	0.07	0.11	20	0.78	0.29	0.46	26	7.79	2.29	3.96	41	2013	839	1382	34
15	MBGP	Matagorda Bay, Gallinipper Point	TX	0.70	0.09	0.43	34	1.51	0.32	0.74	46	5.98	2.60	3.64	27	2261	913	1568	24
16	MBLR	Matagorda Bay, Lavaca River Mout	TX	1.8	0.00	0.91	66	1.65	0.32	0.69	47	4.20	0.65	2.74	37	3110	360	1600	47
18	MBTP	Matagorda Bay, Tres Palacios Bay	TX	0.26	0.00	0.09	66	1.64	0.24	0.56	57	5.63	0.70	3.71	28	2039	510	1140	37
20	MBEM	Matagorda Bay, East Matagorda	TX	0.10	0.04	0.07	22	0.96	0.24	0.50	34	4.96	2.59	3.87	15	3237	544	1271	60
21	BRCL	Brazos River, Cedar Lakes	TX	0.22	0.05	0.10	41	1.32	0.23	0.75	38	4.38	2.12	3.21	19	3890	1684	2209	27
22	BRFS	Brazos River, Freeport Surfside	TX	0.12	0.00	0.07	37	1.14	0.33	0.80	28	6.82	2.34	3.77	29	6910	1635	4536	32
23	GBCR	Galveston Bay, Confederate Reef	TX	0.14	0.02	0.08	39	1.14	0.24	0.60	39	4.27	1.60	3.06	22	3490	894	2163	31
24	GBOB	Galveston Bay, Offatts Bayou	TX	0.07	0.03	0.05	28	2.00	0.42	1.16	39	4.79	1.74	2.71	26	8180	1648	4849	29
25	GBTD	Galveston Bay, Todds Dump	TX	0.13	0.03	0.06	44	3.16	0.00	0.68	83	4.59	2.42	3.35	15	3675	1550	2157	24
26	GBYC	Galveston Bay, Yacht Club	TX	0.11	0.00	0.06	42	1.20	0.39	0.70	30	5.31	2.53	3.40	17	7222	2036	4301	36
27	GBSC	Galveston Bay, Ship Channel	TX	0.33	0.04	0.10	84	1.80	0.59	1.09	37	6.24	3.60	4.80	15	6664	3527	4668	21
28	GBHR	Galveston Bay, Hanna Reef	TX	0.37	0.03	0.08	109	1.20	0.20	0.54	44	5.30	1.63	3.58	26	3406	950	1811	35
29	SLBB	Sabine Lake, Blue Buck Point	TX	0.19	0.07	0.11	31	0.96	0.23	0.60	35	5.34	1.90	3.37	24	10000	1760	4612	43
30	CLLC	Calcasieu Lake, Lake Charles	LA	0.36	0.06	0.13	58	1.16	0.16	0.52	43	5.40	1.93	3.02	27	5616	555	3250	46

(continued on next page)

Table A2 (*continued*)

Site number	Site	Location	State	Hg Max	Hg Min	Hg Mean	Hg Cv	Pb Max	Pb Min	Pb Mean	Pb Cv	Se Max	Se Min	Se Mean	Se Cv	Zn Max	Zn Min	Zn Mean	Zn Cv
31	CLSJ	Calcasieu Lake, St. Johns Island	LA	0.16	0.04	0.11	25	1.00	0.37	0.57	30	7.86	1.78	3.38	44	3501	1396	2285	28
32	JHJH	Joseph Harbor Bayo, Joseph Harbor Bay	LA	0.11	0.03	0.06	33	2.80	0.19	0.56	93	6.43	1.80	3.53	28	3530	960	1465	36
33	VBSP	Vermilion Bay, Southwest Pass	LA	0.09	0.02	0.05	36	1.31	0.23	0.59	37	6.88	1.90	3.89	29	4424	1282	2779	35
34	ABOB	Atchafalaya Bay, Oyster Bayou	LA	0.08	0.02	0.05	24	1.20	0.22	0.53	41	6.66	1.90	3.25	31	3591	1200	2147	32
35	CLCL	Caillou Lake, Caillou Lake	LA	0.07	0.02	0.04	35	1.03	0.11	0.48	46	3.90	1.30	2.39	21	2605	1200	1838	22
36	TBLB	Terrebonne Bay, Lake Barre	LA	0.13	0.04	0.08	31	1.10	0.27	0.63	32	6.37	1.20	2.61	40	2790	1245	1959	27
37	TBLF	Terrebonne Bay, Lake Felicity	LA	0.17	0.03	0.06	51	1.10	0.19	0.47	51	6.98	1.50	2.85	55	3847	1200	1939	33
39	BBSD	Barataria Bay, Bayou Saint Denis	LA	0.10	0.00	0.05	59	0.95	0.24	0.45	35	4.34	0.80	2.14	40	6085	750	2111	54
40	BBMB	Barataria Bay, Middle Bank	LA	0.11	0.04	0.07	28	1.20	0.19	0.51	43	4.61	0.90	2.29	37	3783	1000	2057	35
42	MRPL	Mississippi River, Pass A Loutre	LA	0.10	0.04	0.06	37	1.38	0.23	0.83	42	4.92	2.76	3.44	20	6204	1791	2958	40
43	BSSI	Breton Sound, Sable Island	LA	0.12	0.02	0.06	46	0.70	0.20	0.49	31	5.55	1.21	2.95	34	4170	722	1830	43
44	BSBG	Breton Sound, Bay Gardene	LA	0.20	0.02	0.12	45	1.40	0.25	0.47	54	5.16	1.90	2.67	31	2870	572	1461	43
45	LBMP	Lake Borgne, Malheureux Point	LA	0.33	0.03	0.15	50	1.56	0.10	0.60	55	4.92	1.70	3.23	26	7252	1171	3651	44
48	MSPC	Mississippi Sound, Pass Christian	MS	0.33	0.08	0.14	41	1.10	0.22	0.55	44	5.15	1.89	2.76	27	6730	1500	3592	41
49	MSBB	Mississippi Sound, Biloxi Bay	MS	0.19	0.08	0.12	28	1.78	0.37	0.84	49	4.60	1.60	2.99	30	8930	2019	4586	32
50	MSPB	Mississippi Sound, Pascagoula Bay	MS	0.21	0.08	0.14	27	1.06	0.20	0.47	43	4.40	1.70	2.73	24	4949	1656	3255	27
51	MBCP	Mobile Bay, Cedar Point Reef	AL	0.16	0.04	0.08	34	0.45	0.16	0.28	28	3.60	1.39	2.21	24	1210	463	868	23
52	MBHI	Mobile Bay, Hollingers Is. Ch	AL	0.12	0.02	0.08	31	0.84	0.20	0.52	38	3.16	1.38	2.14	19	10200	995	6011	39
54	PBPH	Pensacola Bay, Public Harbor	FL	0.25	0.07	0.14	37	0.98	0.00	0.42	49	4.04	1.84	3.15	19	3668	1668	2516	24
55	PBIB	Pensacola Bay, Indian Bayou	FL	0.36	0.07	0.13	55	1.01	0.26	0.43	51	4.12	2.00	2.68	24	3000	438	1112	70
58	CBBB	Choctawhatchee Bay, Boggy Bayou	FL	0.23	0.15	0.18	19	4.27	0.50	1.44	113	6.75	4.54	5.36	16	3194	2634	2963	9
60	CBBL	Choctawhatchee Bay, Bens Lake	FL	0.35	0.16	0.24	33	1.79	0.00	0.79	85	5.63	3.23	4.10	22	3582	2186	2947	20
61	CBSR	Choctawhatchee Bay, Off Santa Rosa	FL	0.45	0.07	0.25	41	0.99	0.08	0.51	48	8.66	2.47	4.13	29	3812	687	1555	49
62	PCLO	Panama City, Little Oyster Bar	FL	0.37	0.07	0.13	60	1.20	0.21	0.45	63	3.90	1.95	2.75	22	4008	1011	2121	49

Table A2 (continued)

Site number	Site	Location	State	Hg Max	Hg Min	Hg Mean	Hg Cv	Pb Max	Pb Min	Pb Mean	Pb Cv	Se Max	Se Min	Se Mean	Se Cv	Zn Max	Zn Min	Zn Mean	Zn Cv
64	SAWB	St. Andrews Bay, Watson Bayou	FL	0.09	0.01	0.06	33	2.38	0.42	1.20	40	3.39	1.00	2.06	27	6281	2590	4213	26
65	APDB	Apalachicola Bay, Dry Bar	FL	0.23	0.04	0.11	36	5.11	0.07	0.52	196	4.33	1.30	2.19	36	2616	72	424	115
66	APCP	Apalachicola Bay, Cat Point Bar	FL	0.21	0.06	0.12	34	0.80	0.00	0.38	48	3.43	1.53	2.32	21	1151	392	728	28
67	AESP	Apalachee Bay, Spring Creek	FL	0.23	0.11	0.18	23	0.76	0.34	0.48	24	4.00	2.31	3.20	17	1189	592	826	25
69	CKBP	Cedar Key, Black Point	FL	0.22	0.04	0.12	37	0.62	0.16	0.33	38	6.31	1.30	2.23	47	1454	160	496	77
70	TBNP	Tampa Bay, Navarez Park	FL	0.16	0.06	0.11	27	1.23	0.40	0.73	27	2.18	1.27	1.74	18	2835	1133	2159	23
71	TBMK	Tampa Bay, Mullet Key Bayou	FL	0.35	0.09	0.15	36	1.06	0.28	0.52	37	2.74	1.10	1.72	26	750	209	383	36
72	TBPB	Tampa Bay, Papys Bayou	FL	0.60	0.10	0.32	36	2.13	0.50	1.22	36	3.50	0.80	2.14	33	8171	1190	2636	53
73	TBOT	Tampa Bay, Old Tampa Bay	FL	0.72	0.30	0.49	25	1.33	0.40	0.81	35	4.59	1.89	2.73	31	8959	5530	7560	13
74	TBKA	Tampa Bay, Peter O. Knight Airport	FL	0.37	0.07	0.14	60	2.46	0.90	1.53	31	4.40	1.89	2.77	28	8630	3013	5066	35
76	TBCB	Tampa Bay, Cockroach Bay	FL	0.51	0.13	0.25	34	1.38	0.08	0.48	58	3.31	1.00	1.93	30	5264	563	1703	62
77	CBBI	Charlotte Harbor, Bird Island	FL	0.45	0.14	0.25	26	0.72	0.12	0.46	37	3.19	1.30	2.08	24	5218	1050	2017	51
78	CBFM	Charlotte Harbor, Fort Meyers	FL	0.25	0.08	0.16	37	1.29	0.19	0.80	39	3.20	1.55	2.36	17	7303	249	3595	47
79	NBNB	Naples Bay, Naples Bay	FL	0.17	0.04	0.12	24	0.73	0.07	0.32	47	2.69	1.10	1.89	24	5670	2000	3209	28
80	RBHC	Rookery Bay, Henderson Creek	FL	0.36	0.10	0.20	31	0.78	0.06	0.38	52	2.80	1.06	2.04	22	1776	450	1059	32
81	EVFU	Everglades, Faka Union Bay	FL	0.31	0.10	0.18	27	0.88	0.06	0.36	65	4.29	1.36	1.98	31	1900	608	1053	34
		GOM average		0.27	0.05	0.14	41	1.39	0.24	0.64	50	5.1	1.8	3.01	28	4396	1178	2391	40

Geochemical Investigations in Earth and Space Science: A Tribute to Isaac R. Kaplan
© The Geochemical Society, Publication No. 9, 2004
Editors: R.J. Hill, J. Leventhal, Z. Aizenshtat, M.J. Baedecker, G. Claypool,
R. Eganhouse, M. Goldhaber and K. Peters

Geochemical differentiation of Silurian from Devonian crude oils in eastern Algeria[*]

KENNETH E. PETERS[1] and STEVE CREANEY[2]

[1]U.S. Geological Survey, 345 Middlefield Road, MS 969, Menlo Park,
CA 94025, USA

[2]ExxonMobil Exploration Company, Houston, TX 77060, USA

Abstract—The ability to distinguish crude oils generated from prolific Silurian and Devonian source rocks provides a means to map their petroleum systems and thereby reduce exploration risk in North Africa. Routine geochemical analyses commonly fail to reliably separate these oils. This study demonstrates that non-routine analyses, such as compound-specific isotopes of light hydrocarbons and gas chromatography–mass spectrometry of diamondoids, can distinguish Silurian from Devonian oils in Algeria.

Oil samples from Zemlet field and the giant Hassi Messaoud field were obtained from the northern part of the study area in Algeria where the Devonian source rock is absent. These oils originated from thermally mature Silurian source rock and migrated updip to fractured Cambro-Ordovician quartzite reservoirs below the Hercynian unconformity. Oil samples from the Assekaifaf, Oued Zenani, Zarzaitine, and Dome fields occur in the southern part of the study area, where both source rocks exist, but migration paths indicate input mainly from Devonian source rock.

Despite higher maturity, the Silurian oil samples have diamondoid isomer concentrations 2–3 times lower than the Devonian samples. Because diamondoids form in source rocks by clay-catalyzed reactions and *increase* relative to other compounds during thermal cracking of oils, the diamondoid concentrations in the oil samples suggest that the Silurian source rock had less clay than the Devonian source rock. Higher dibenzothiophene/2-methylnaphthalene and generally higher sulfur in the Silurian oil samples support a source rock with less clay. For the Devonian oil samples, slightly more iron in clays reacted with sulfides to form pyrite, thus limiting sulfur incorporation into the kerogen and the generated crude oil.

Light hydrocarbons, such as n-hexane, methylcyclopentane, cyclopentane, and methyl-cyclohexane, are depleted in ^{13}C in the Silurian compared to Devonian oil samples ($> 1.1, 0.6$, 1.0, and 0.2‰, respectively). Patterns of isotopic ratios among these compounds also differ (*e.g.* cyclohexane is more depleted in ^{13}C than methylcyclopentane and methylcyclohexane in Silurian, but not Devonian oil samples). The Silurian oil samples have smaller differences in $\delta^{13}C$ between pristane and the C_{17} n-alkane and between phytane and the C_{18} n-alkane than the Devonian samples. Different assemblages of organisms likely contributed to the acyclic isoprenoids versus n-alkanes in the source rocks for these two oil families.

INTRODUCTION

ALGERIA RANKS among the world's top 10 petroleum provinces, with recoverable reserves totaling more than 16,000 MMBO (million barrels of oil), 6000 MMBO of condensate, and 130 TCF (trillion cubic feet) of hydrocarbon gas (MACGREGOR, 1998). Nearly all of the giant fields in the country lie in eastern Algeria (Fig. 1). For example, the giant Hassi Messaoud field had original reserves of 9300 MMBO (~58% of the country's total oil reserves). Late Mesozoic generation and expulsion of petroleum from source rocks in eastern Algeria coincided with the development of large structural traps and giant accumulations of oil and gas in Lower Devonian through Upper Triassic reservoir rocks (DANIELS and EMME, 1995; SONATRACH, 1995). The migration of these hydrocarbons was complex, but excellent regional shale and

---

[*] Presented orally by the principal author at the Symposium to Honor Ian Kaplan (Ron Hill and Ken Peters, Organizers) Annual Geological Society of America Meeting, October 27–30, 2002, Denver, CO.

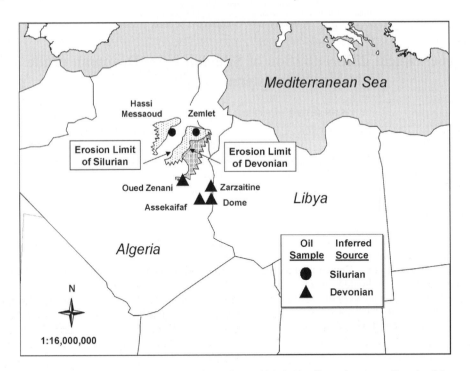

Fig. 1. Map shows locations of fields in Algeria from which the six oil samples were collected and the erosional limits of the Silurian and Devonian subcrops (light and dark stipple, respectively) at the Hercynian unconformity. Complete well names are in Table 1. Although the Dome oil sample could be a mixture, we use the Devonian symbol because it is geochemically more similar to the Devonian than Silurian oil samples.

evaporite seals are likely to have trapped substantial undiscovered petroleum (COCHRAN and PETERSEN, 2001).

Although Lower Silurian and Upper Devonian shales are widely recognized as the main source rocks for petroleum in eastern Algeria (TISSOT et al., 1984), it is generally not possible to geochemically link oil or condensate samples with pods of active source rock (DANIELS and EMME, 1995). Oil-to-source rock relationships in Algeria are inferred from the structural relationships between petroleum accumulations and nearby source rocks. For example, BOOTE et al. (1998) inferred that the regionally extensive Lower Silurian Tannezuft Formation generated 80–90% of the Paleozoic petroleum in the Sahara Desert province of North Africa (Algeria, Libya, Morocco, and Tunisia), while Upper Devonian shales account for the remainder. The truncation of older rocks by the Hercynian unconformity (Carboniferous-Triassic) played a key role in the distribution of source rocks, migration paths between source rocks and reservoir rocks, and subsequent preservation of many oil accumulations in Algeria (ECHIKH, 1998; COCHRAN and PETERSEN, 2001).

The ability to distinguish oil that originated from either Silurian or Devonian source rocks could be the basis to map these petroleum systems and thereby significantly reduce exploration risk in North Africa. Direct geochemical evidence for the origin of accumulated oil could also improve assessment of volumetrics, migration pathways, and entrapment style, which are critical variables that contribute to the productivity of petroleum systems (MAGOON and DOW, 1994). Routine biomarker and isotopic analyses (PETERS and MOLDOWAN, 1993) commonly fail to reliably distinguish the compositionally similar Silurian and Devonian oil samples from Algeria (e.g. DANIELS and EMME, 1995). However, based on such analyses for oil samples from 24 fields in eastern Algeria, DANIELS and EMME (1995) suggested that samples to the north and west of the Devonian shale truncation line (Fig. 1) most likely originated from Silurian source rock, while those to the south and east represent mixtures of Silurian and Devonian oil.

The sampled oils do not correlate with Ordovician, Silurian, Devonian, Carboniferous, or Mesozoic shale extracts that they analyzed. ILLICH *et al.* (1997) classified 65 oil samples from eastern Algeria based on chemometric analysis of biomarker and isotope data. Although their samples have similar biomarker and isotope signatures, the large number of samples in the study may improve the statistical reliability of their classification.

Several factors contribute to the difficulties in geochemically separating Silurian and Devonian oils and relating them to source rock samples: (1) Silurian and Devonian source-rock organic facies are similar because of similar depositional environments and organic matter input (DANIELS and EMME, 1995), (2) most available source rock samples are highly mature (*e.g.* COCHRAN and PETERSEN, 2001), and (3) most Algerian oil samples are light oil (35–45° API gravity) or condensate ($>45°$ API gravity), where useful biomarkers are commonly low or absent due to high thermal maturity.

Silurian source rock

The early Silurian is one of the six most significant time intervals for source rock deposition, accounting for $\sim9\%$ of the world's reserves (KLEMME and ULMISHEK, 1991). Most of this worldwide petroleum resource originated from graptolitic, radioactive (hot), Lower Silurian (Llandovery) black shale deposited during a post-glacial marine transgression across the northern African and Arabian cratons. This radioactive shale is called the Tannezzuft in North Africa (KLETT, 2000a–c) and the Qusaiba Member of the Qalibah Formation in Saudi Arabia (COLE *et al.*, 1994; JONES and STUMP, 1999). Lowermost Silurian shales in North Africa have total organic carbon (TOC) in the range of ~2–17 wt% and originally contained oil-prone type I–II kerogen with estimated hydrogen indices (HI) near 600 mg hydrocarbon/g TOC (DANIELS and EMME, 1995), largely derived from marine prasinophytes (TYSON, 1995). In most Algerian basins, the Silurian shales are now postmature, and consequently have low HI. Although the entire Silurian shale succession is up to 700–1000 m thick, the radioactive shale horizon is usually <30 m thick, suggesting that the anoxic conditions that gave rise to the exceptional source quality may have lasted only a few million years (LÜNING *et al.*, 2000).

Devonian source rock

Upper Devonian source rocks generated $\sim8\%$ of the world's reserves (KLEMME and ULMISHEK, 1991). Much of this oil originated from basal Upper Devonian (Frasnian) black shale, the "Argile Radioactive", which was deposited during a widespread marine transgression across North Africa. These shales contain ~8–14 wt% TOC. Measuring their original generative yields is difficult because most of these rocks are now postmature, but originally they contained type I–II kerogen with inferred HI comparable to the basal Tanezzuft radioactive unit (DAHL *et al.*, 1999).

Purpose

The purpose of this study was to geochemically distinguish Silurian from Devonian oil from Algeria using advanced, non-routine geochemical analyses. Early work on light hydrocarbons from Hassi R'Mel field in Algeria and North Dome field in Qatar indicated that condensate might be correlated to light oil using carbon isotopes of residual heavy hydrocarbon fractions (saturated and aromatic hydrocarbons), light hydrocarbon distributions (*e.g.* C_7 compositions), and compound-specific carbon isotope compositions of the light hydrocarbons (ZUMBERGE *et al.*, 1996). Non-routine analyses used in our study include metastable reaction monitoring-gas chromatography–mass spectrometry (MRM-GCMS) of biomarkers, compound-specific isotope analysis (CSIA; also called isotope-ratio monitoring-gas chromatography–mass spectrometry or IRM-GCMS) of light hydrocarbons, and GCMS of diamondoids.

SAMPLES

We selected six oil samples (Fig. 1, Table 1) where structural relationships and proximity to mature source rock(s) suggested an origin as follows:

Table 1. Location and general geochemical data for oil samples from Algeria

Well	Depth (m)	Test interval	Source rock[a]	Pr/nC_{17}	API gravity (°)	Sulfur (wt%)
Assekaifaf Nord ASN-1	958.6	Lower Devonian	Devonian	0.36	39.7	0.04
Dome à Collenias DCL-1	967.8	Ordovician	Mixed (?)	0.37	43.5	0.04
Hassi Messaoud OMP-17	3350.0	Cambrian	Silurian	0.49	42.6	0.14
Oued Zenani OZN-1	1255.7	Devonian	Devonian	0.43	39.4	0.04
Zarzaitine ZR-313	709.7	Devonian	Devonian	0.46	37.3	0.09
Zemlet El Kalef ZK-1	4026.6	Cambrian	Silurian	0.49	36.7	0.07

[a] Inferred source rock age based on geology. Geochemical analyses discussed below suggest that the Dome oil sample originated mainly from Devonian source rock.

(1) Silurian source rock (Hassi Messaoud and Zemlet oil samples),
(2) Devonian source rock (Assekaifaf, Zarzaitine, and Oued Zenani oil samples), and
(3) both Silurian and Devonian source rock (Dome oil sample).

The Hassi Messaoud and Zemlet oil samples from the northern part of the study area are most likely from Silurian source rock because they occur in fractured Cambro-Ordovican quartzites immediately below the Hercynian unconformity (Fig. 2, top). This unconformity is a migration conduit between effective source rocks and reservoirs throughout much of North Africa (ECHIKH, 1998; COCHRAN and PETERSEN, 2001). The Silurian source rock remains above the top of the reservoir interval at Zemlet, but has been eroded at Hassi Messaoud. Oil migrated updip from thermally mature Silurian source rock into the reservoir in both fields. Devonian source rock is absent in the area and long-distance migration from this source rock to the reservoir rocks

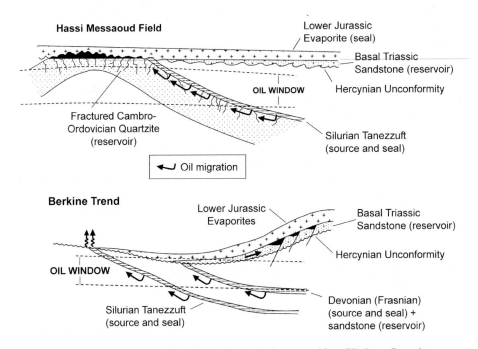

FIG. 2. Schematic secondary migration paths for crude oil generated from Silurian or Devonian source rocks in Algeria. Top—typical of the northern study area, thermally mature Silurian oil migrated into Cambro-Ordovician quartzite at the giant Hassi Messaoud field, where it was trapped below the Hercynian unconformity by overlying Lower Jurassic evaporite seal rock and additional overburden rock. Bottom—typical of the southern study area, migration resulted in loss of Silurian oil along the Berkine Trend, but allowed entrapment of Upper Devonian oil in Devonian sandstone and updip migration (to right in figure) into overlying Triassic sandstone.

in the Hassi Messaoud and Zemlet fields is unlikely. Figure 14 in BOOTE *et al.* (1998) shows a more detailed map and cross-section in the Hassi Messaoud area.

Oil samples from Assekaifaf, Zarzaitine, and Oued Zenani fields occur in Devonian sandstones in the southern part of the study area and most likely originated from adjacent Devonian source rock, even though Silurian source rock is present (Fig. 2, bottom). The Dome oil sample could be a mixture of Silurian and Devonian oils because it occurs in Ordovician reservoir rock. Although not sampled in this study, many oil accumulations from the southern area reside in basal Triassic sandstone reservoir rocks. These oils could originate solely from Devonian source rock due to loss of Silurian oil at the Hercynian unconformity (Fig. 2, bottom). The approach used in this study could be used to identify the source rock for such oil accumulations.

Compositional variations among the Silurian or Devonian oil samples might be caused by differences in source rock organic facies, thermal maturity, or other factors. Our ability to account for these differences is limited by the small number of samples in this study.

METHODS

Gas chromatography of the C_{15+} components in whole oil samples was completed using a Hewlett-Packard 6890 gas chromatograph equipped with a 25-m Ultra-1 column (0.32 mm I.D., 0.52 μm film thickness). The column was programmed from 20 to 300°C at 5°C/min and held at 300°C for 18 min using helium carrier gas. Samples consisted of 1 μl neat injection, where the injector was maintained at 275°C.

Sterane and terpane biomarkers in the saturated hydrocarbon fractions of oil samples were analyzed by MRM-GCMS using a Micromass ProSpec-Q instrument. Compound separation was achieved using a 30-m J&W Scientific DB-5 capillary column (0.32 mm I.D., 0.25 μm film thickness). The column was programmed from 75 to 200°C at 5°C/min using helium carrier gas and 5β-cholane as the internal standard. Further details are in PETERS and MOLDOWAN (1993).

Oil samples (50–60 mg) containing surrogate diamondoid standards were separated using a Hewlett-Packard 6890 gas chromatograph equipped with a 50-m PONA methylsilicone column (0.20 mm I.D., 0.52 μm film thickness) programmed at 35°C for 5 min followed by 4°C/min heating to 310°C. The GC was interfaced with a 5973-MSD for mass spectral analysis.

A Hewlett-Packard 5890 gas chromatograph equipped with a 50-m PONA methylsilicone chromatographic column (0.20 mm I.D., 0.52 μm film thickness) was used to separate C_6–C_{19} hydrocarbons for stable carbon isotope analysis. The column was heated at 35°C for 15 min followed by programmed ramps at 1.5°C/min to 70°C (no hold time), 3°C/min to 130°C (hold 12 min), and 3°C/min to 300°C (hold 20 min) using helium carrier gas. Internal standards included hexene, 2,3-dimethylpentene, *cis*-octene, 1-nonene, and 1-octadecene. Effluent peaks from the column were combusted at 950°C to carbon dioxide and transferred through a Micromass Isochrom 2 interface (350°C) for analysis using a Micromass Prism mass spectrometer. Carbon dioxide was used as the reference standard. C_{15+} hydrocarbons were separated and analyzed using a similar procedure.

DISCUSSION

Gas chromatograms (Fig. 3), high API gravity ($>37°$), and low sulfur (<0.15 wt%, Table 1) indicate little or no biodegradation of the six oil samples. The *n*-alkane distributions appear to be unaltered, maximize at nC_8 or nC_9, and extend beyond nC_{35}. Pristane/nC_{17} ratios are slightly higher for the Hassi Messaoud and Zemlet samples (0.49) than the other oil samples (0.36–0.46, Table 1) due to slight differences in the source rock or secondary processes.

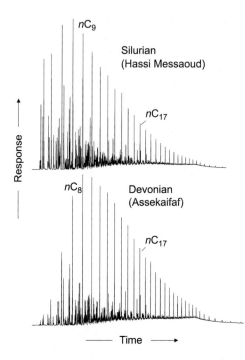

FIG. 3. Representative gas chromatograms for whole oil samples generated from Silurian (Hassi Messaoud, top) and Devonian (Assekaifaf, bottom) source rocks. All six samples are non-biodegraded light oils with high API gravities ($>37°$), low sulfur (<0.15 wt%), and whole-oil gas chromatograms that maximize about nC_8 or nC_9.

Terpanes and steranes

The terpane and sterane distributions for the six oil samples are remarkably similar and only a few selected ratios will be discussed (Table 2). GRANTHAM and WAKEFIELD (1988) observed that the C_{28}/C_{29} sterane ratios for crude oils generated from marine source rocks with no significant terrigenous organic matter input are less than 0.5 for Lower Paleozoic and older oils, 0.4–0.7 for Upper Paleozoic to Lower Jurassic oils, and greater than about

Table 2. Selected terpane and sterane data for the Algerian oil samples

	Terpanes				Steranes				
Well	Ts/ (Ts + Tm)	Tri/ (Tri + Hop)	C_{35}/C_{32} Hop	Tet/C_{26}	$C_{29}20S/$ (S + R)	$C_{29}\beta\beta/$ ($\beta\beta + \alpha\alpha$)	Total C_{28}	Total C_{29}	C_{28}/C_{29}
Assekaifaf	0.85	0.55	0.12	0.26	0.52	0.67	0.21	0.50	0.42
Dome	0.85	0.67	0.22	0.26	0.54	0.63	0.21	0.51	0.42
Hassi Messaoud	0.80	0.72	0.12	0.14	0.57	0.66	0.20	0.46	0.44
Oued Zenani	0.65	0.39	0.22	0.45	0.52	0.57	0.20	0.50	0.40
Zarzaitine	0.51	0.30	0.13	0.81	0.48	0.54	0.20	0.53	0.38
Zemlet	0.77	0.47	0.20	0.30	0.54	0.65	0.22	0.50	0.43

Ts/(Ts + Tm) = C_{27} 18α22,29,30-trisnorneohopane/(Ts + 17α22,29,30-trisnorhopane); Tri/(Tri + Hop) = 22 R + 22S epimers of 29,30-bisnorcyclohexaprenane and 30-norcyclohexaprenane/(Tri + C_{29} to C_{33} 17α-hopanes); C_{35}/C_{32}Hop = 22R + 22S epimers of the C_{35}/C_{32} 17α-hopanes; Tet/26 = C_{24} tetracyclic/C_{26} tricyclic terpanes; 20S/(20S + 20R) = C_{29} sterane 5α,14α,17α 20S/(20S + 20R); ββ/(ββ + αα) = C_{29} sterane (5α,14β,17β20S + 20 R)/(ββ + 5α,14α,17α20S + 20R); C_{28}/C_{29} = C_{28}/C_{29} [5α,14α,17α(H) 20S + 20R and 5α,14β,17β (H) 20S + 20R] steranes. Parameters are described in PETERS and MOLDOWAN (1993).

0.7 for Upper Jurassic to Miocene oils. The C_{28}/C_{29} sterane ratios for the six Algerian oil samples (0.38–0.44, Fig. 4) are consistent with Silurian or Devonian source rock. The Hassi Messaoud and Zemlet oil samples have slightly higher C_{28}/C_{29} sterane ratios than the other samples (0.42–0.44 versus 0.38–0.42, Table 2). However, because the trend of increasing C_{28}/C_{29} ratios with younger age is approximate, higher ratios for the Hassi Messaoud and Zemlet samples cannot be interpreted to indicate that they originated from younger source rock than the other four oil samples.

The Silurian Hassi Messaoud and Zemlet oil samples have lower C_{24} tetracyclic terpane ratios (0.14–0.30) than the Devonian Oued Zenani and Assekaifaf oil samples (0.45–0.81, Table 2). The Devonian Dome and Assekaifaf oil samples have C_{24} tetracyclic terpane ratios (0.26) that overlap the range of the two Silurian oil samples.

Sterane isomerization ratios indicate that the Assekaifaf, Dome, Hassi Messaoud, and Zemlet oil samples experienced thermal maturities near or past the peak of the oil window, while the Zarzaitine and Oued Zenani samples are less mature (Fig. 5). This interpretation is supported by terpane data, such as the Ts/(Ts + Tm) and tricyclics/(tricyclics + hopanes) ratios (Table 2), which are lower for the Zarzaitine and Oued Zenani samples than the other oil samples. Both of these terpane ratios depend partly on organic facies, but the data in the table suggest that maturity rather than source input is the main control on high tricyclic terpanes and Ts in these oil samples.

Aromatics

Conventional aromatic biomarker ratios for these oil samples are inadequate to differentiate Silurian from Devonian oils, either because they are not diagnostic or because of insensitivity related to low concentrations and co-elution (*e.g.* $C_{26}S/C_{28}S$ and $C_{27}R/C_{28}R$

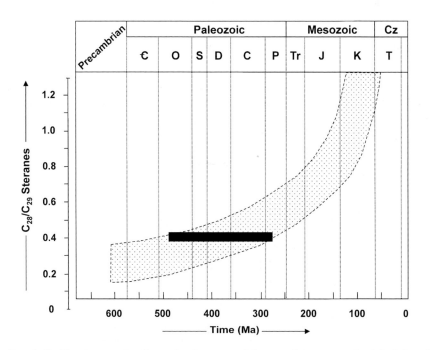

FIG. 4. C_{28}/C_{29} sterane ratios for crude oils generated from marine source rocks (stippled area) generally increase from the Paleozoic to the present due to increasing phytoplankton diversity (GRANTHAM and WAKEFIELD, 1988). The six Algerian oil samples have C_{28}/C_{29} sterane ratios in the range 0.38–0.44 (solid bar), consistent with Paleozoic source rock. Є, O, S, D, C, P, Tr, J, K, T, Cz—Cambrian, Ordovician, Silurian, Devonian, Carboniferous, Permian, Triassic, Jurassic, Cretaceous, Tertiary, and Cenozoic, respectively.

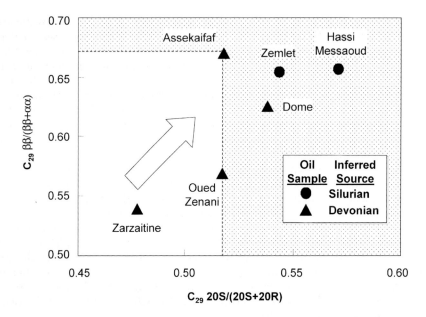

FIG. 5. Sterane isomerization ratios (Table 2) indicate that Oued Zenani and Zarzaitine oil samples are below the peak oil window, while the other oil samples have higher thermal maturity (diagonal arrow). Ratios are insensitive to further maturation in stippled areas (PETERS and MOLDOWAN, 1993).

triaromatic steroids, $\%C_{27}$ to C_{29} monoaromatic steroids). Aromatic maturity parameters, such as methylphenanthrene and dimethylnaphthalene indices, conflict with the sterane isomerization ratios and are not reported because they are generally unreliable for Paleozoic samples. However, several low-molecular-weight aromatic compound ratios (Table 3) may be useful to distinguish Silurian from Devonian oils, as discussed below.

Higher dibenzothiophene/2-methylnaphthalene in the Hassi Messaoud and Zemlet oil samples (DBT/2MeN = 0.07–0.12, Table 3) compared to the Devonian samples (0.02–0.06) suggests that the Silurian source rock was deposited under conditions of lower redox potential (Eh) and contained less iron in clay minerals that could react with sulfides to form pyrite. Excess sulfide in anoxic sediments with little available iron can result in sulfur incorporation into the kerogen, which is then incorporated into the generated oil

Table 3. Selected aromatic hydrocarbon data

| Well | Monoaromatic steroids | | Triaromatic steroids | | | | | |
	$\%C_{28}$	$\%C_{29}$	26/28S	27/28R	DBT/2MeN	MeDBT/MeP	P/Flu	Biph/Flu
Assekaifaf	43	35	0.31	0.38	0.03	0.072	6.9	2.9
Dome	41	34	0.27	0.42	0.03	0.087	10.1	3.9
Hassi M.	37	38	0.43	0.78	0.12	0.341	9.9	2.5
Oued Zenani	43	30	0.36	0.38	0.06	0.112	9.4	2.7
Zarzaitine	42	41	0.23	0.25	0.02	0.099	5.1	1.4
Zemlet	42	36	0.24	0.30	0.07	0.175	14.9	5.4

$\%C_{28} = C_{28}/(C_{27}$ to $C_{29})$ MA-steroids, including $5\alpha(20S + 20R)$ $5\beta(20S + 20R)$ and $10\beta \rightarrow 5\beta$ methyl-rearranged 20R, 20S and 24R, 24S isomers; 26/28S = C_{26}/C_{28} 20S triaromatic steroids; 27/28R = C_{27}/C_{28} 20R triaromatic steroids; DBT/2MeN = dibenzothiphene/2-methylnaphthalene; MeDBT/MeP = methyl-dibenzothiophene/methylphenanthrene; P/Flu = phenanthrene/fluorene; Biph/Flu = biphenyl/fluorene; monoaromatic and triaromatic steroid parameters are described in PETERS and MOLDOWAN (1993).

(PETERS and MOLDOWAN, 1993, p. 137). ILLICH et al. (1997) found a systematic regional decrease in the DBT/4-methylnaphthalene ratio for many Algerian oil samples, ranging from >0.12 at Hassi Messaoud to <0.06 at Zarzaitine, similar to our observations for DBT/2-MeN. They used chemometric analysis of biomarker and isotope data to conclude that their Hassi Messaoud and Zarzaitine oil samples originated from Silurian and Devonian source rocks, respectively, consistent with our interpretation. The ratio of sums of the methyldibenzothiophenes to methylphenanthrenes also separates the Silurian from Devonian oil samples (MeDBT/MeP = 0.18–0.34 versus 0.07–0.11, respectively, Table 3). These classes of compounds have similar partition coefficients between water and oil and thus, their ratio is not readily affected by water washing.

Lower Eh at the time of deposition for the Zemlet source rock is supported by a higher C_{35}/C_{32} hopane ratio for the Zemlet oil sample (0.20) compared to two of the Devonian samples (Assekaifaf and Zarzaitine = 0.12–0.13, Table 2). However, the C_{35}/C_{32} hopane ratio is also higher for Oued Zenani and Dome oil samples (0.22).

The sulfur content in the two Silurian oil samples (0.07–0.14, Table 1) is higher than most of the Devonian oil samples (0.04) except for the Zarzaitine sample (0.09), consistent with a lower Eh source rock. However, sulfur content is unsuitable to separate Silurian from Devonian oil samples because secondary processes readily alter this bulk parameter. For example, the comparatively high sulfur content of the Zarzaitine oil sample can be explained by its low thermal maturity compared to the other samples (Fig. 5). Sulfur content in oil decreases with increasing thermal maturity (ORR, 1974).

Stable carbon isotopes of saturated and aromatic hydrocarbons

Stable carbon isotope ratios for the saturated and aromatic hydrocarbon fractions of the Algerian oil samples support a genetic relationship between the Silurian Hassi Messaoud and Zemlet oil samples, which differ from the Devonian Assekaifaf, Dome, Zarzaitine, and Oued Zenani samples (Table 4). However, these isotopic differences are small (<1‰) and cannot be generally applied because of larger differences that might be imposed by secondary processes, such as thermal maturity. The calculated canonical variables (SOFER, 1984) for the six oil samples indicate marine source-rock organic matter (CV < 0.47), but lack systematic differences between the Silurian and Devonian samples.

Diamondoids

Diamondoid isomer ratios for the six oil samples are similar, but the Hassi Messaoud and Zemlet oil samples have lower ratios for the sum of adamantanes/sum of diamantanes (18.6–23.3 versus 26.4–36.1) and 1,3-/1,2-dimethyladamantane (0.73–0.90 versus 1.11–1.27)

Table 4. Stable carbon isotope ratios for saturated and aromatic hydrocarbons and the calculated canonical variable

Well	$\delta^{13}C_{sat}$ (‰)	$\delta^{13}C_{aro}$ (‰)	CV
Assekaifaf	− 29.3	− 28.2	− 0.13
Dome	− 29.5	− 28.5	− 0.29
Hassi Messaoud	− 29.6	− 28.9	− 0.92
Oued Zenani	− 29.0	− 28.1	− 0.66
Zarzaitine	− 29.4	− 28.2	− 0.13
Zemlet	− 29.8	− 28.9	− 0.41

Isotope data for saturated (sat) and aromatic (aro) hydrocarbon fractions are in parts per thousand (‰) relative to the PDB belemnite standard. Precision is about 0.05‰.

CV = canonical variable = $[-2.53\delta^{13}C_{sat} + 2.22\delta^{13}C_{aro}] - 11.65$; CV < 0.47 indicates marine organic matter input (SOFER, 1984).

Table 5. Selected diamondoid data

Well	Concentration (ppm)[a]									Ratio[b]			
	1	2	3	4	5	6	7	8	9	Ada/Dia	Sum Ada/Dia	1,3/1, 2-DiMeAda	1,3,5/1,3, 6-TriMeAda
Assekaifaf	4.2	18.8	15.9	10.3	12.4	12.6	13.9	6.0	9.1	6.9	26.4	1.14	0.65
Dome	6.0	22.1	17.4	12.1	12.8	12.2	14.0	5.5	8.6	16.3	36.1	1.24	0.63
Hassi M.	1.8	5.9	4.1	5.4	4.6	4.1	5.6	1.5	2.9	6.7	23.3	0.73	0.51
O. Zenani	4.0	15.2	11.3	9.3	9.1	8.4	10.0	4.0	6.0	8.6	27.4	1.12	0.66
Zarzaitine	4.5	16.3	12.5	9.2	9.4	9.4	9.8	4.2	6.7	16.6	32.9	1.27	0.63
Zemlet	2.0	7.4	5.5	6.0	5.3	5.1	6.1	2.0	3.7	5.9	18.6	0.90	0.53

[a] 1 = adamantane, 2 = 1-methyladamantane, 3 = 1,3-dimethyladamantane, 4 = 2-methyladamantane, 5 = 1,4-*cis*-dimethyladamantane, 6 = 1,4-*trans*-dimethyladamantane, 7 = 1,2-dimethyladamantane, 8 = 1,3,5-trimethyladamantane, 9 = 1,3,6-trimethyladamantane.
[b] Ada/Dia, adamantane/diamantane; Sum Ada/Dia, sum of adamantane and methylated adamantanes/sum of diamantane and methyldiamantanes; 1,3/1,2-DiMeAda, 1,3/1,2-dimethyladamantanes; 1,3,5/1,3,6-TriMeAda, 1,3,5-/1,3,6-trimethyladamantanes.

than Dome and the three Devonian samples (Table 5). A similar separation can be achieved using adamantane/diamantane (non-alkylated parent compounds; 5.9–6.7 versus 6.9–16.6) and 1,2,5-/1,3,6-trimethyladamantanes (0.51–0.53 versus 0.63–0.66; Table 5).

The Silurian-sourced Hassi Messaoud and Zemlet samples have lower diamondoid concentrations by a factor of at least 2–3 compared to the Devonian-sourced Assekaifaf, Dome, Oued Zenani, and Zarzaitine samples (Fig. 6), apparently due to low clay content in their source rocks. Biodegradation, thermal maturation, or source rock input can affect diamondoid concentrations in oil (DAHL et al., 1999). However, as discussed above, none of the oil samples show evidence for significant biodegradation. The Hassi Messaoud and Zemlet oil samples have slightly higher pristane/nC_{17} ratios than the other samples (0.49 versus 0.36–0.46), but these differences are probably due to source rock input rather than mild biodegradation or lower thermal maturity. Due to diamondoid stability, biodegradation or thermal maturation should yield higher rather than lower diamonodoid concentrations in oil samples. Furthermore, sterane isomerization ratios (Fig. 5) indicate that the Hassi

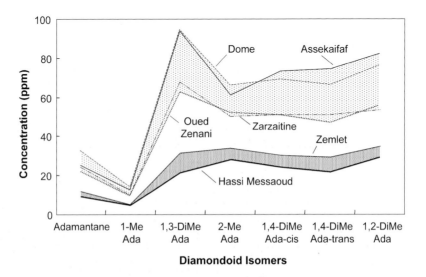

FIG. 6. Diamondoid isomer concentrations are systematically lower in the Silurian-sourced Hassi Messaoud and Zemlet oil samples compared to the Devonian samples, despite higher thermal maturity. The *x*-axis consists of compounds 1–7 in Table 5.

Messaoud and Zemlet oil samples have achieved thermal maturity near peak oil window. Thus, low diamondoids in the Hassi Messaoud and Zemlet oil samples are likely to be source related.

Formation of diamondoids requires acid catalysis by clay in the source rock (DAHL et al., 1999). Lower diamondoid concentrations in the Hassi Messaoud and Zemlet oil samples compared to the four Devonian samples, which have similar or lower thermal maturity, suggest lower clay content in the Silurian compared to the Devonian source rock. This is consistent with higher dibenzothiophene/2-methylnaphthalene ratios for the Hassi Messaoud and Zemlet oil samples, which also indicates lower Eh and possibly less clay in the source rock, as discussed above.

Stable carbon isotopes of light hydrocarbons

The isotopic composition and the pattern of isotopic compositions among certain n-alkanes and isoprenoids (Table 6) and gasoline-range branched and cyclic hydrocarbons (Table 7) measured by CSIA differ from the Silurian-sourced Hassi Messaoud and Zemlet oil samples compared to the four Devonian samples. Except at nC_7 and nC_9, the Zemlet oil sample has more negative stable carbon isotope ratios for the C_6-C_{19} n-alkanes than the other Algerian oil samples (Fig. 7). The isotopic compositions of n-alkanes in Hassi Messaoud oil sample more closely resemble the other samples than the Zemlet sample, except at nC_6 and possibly nC_7. Devonian source rock input may affect the heavier components in the Hassi Messaoud oil sample or their isotopic compositions may simply not be diagnostic of Silurian versus Devonian oil. The nC_6 in Hassi Messaoud and Zemlet oil samples is substantially depleted in ^{13}C (1.1–2.3‰) compared to the other samples.

Hassi Messaoud and Zemlet oil samples have more negative CSIA stable carbon isotope ratios for several gasoline-range cyclic hydrocarbons (Fig. 8), especially methylcyclopentane, cyclohexane, and methylcyclohexane; ≥ 0.6, 1.9, and 0.2‰, respectively) compared to the other Algerian oil samples. The pattern of isotopic ratios among these three compounds also systematically varies. Unlike the other oil samples, cyclohexane is more depleted in ^{13}C than methylcyclopentane and methylcyclohexane in the Hassi Messaoud and Zemlet oil samples.

Stable carbon isotopes of C_{15+} hydrocarbons

Although stable carbon isotope ratios measured by CSIA for most n-alkanes, pristane, and phytane in the Algerian oil samples partly overlap (Table 6), the isotopic difference between pristane and the nC_{17} alkane and between phytane and the nC_{18} alkane distinguish the Hassi Messaoud and Zemlet oil samples (Silurian) from the Devonian samples (Fig. 9). Particular caution must be applied when interpreting isotopic differences among C_{15+} hydrocarbons because co-elution of chromatographic peaks can occur, thus affecting the isotopic ratios.

Different assemblages of organisms in the depositional environment could contribute to the acyclic isoprenoids (e.g. pristane and phytane) and n-alkanes in the source rocks for these two oil families, including marine algae, archaebacteria, and land plants. For example, SCHOELL et al. (1992) showed that pristane ($-26.7‰$) and phytane ($-26.1‰$) in Miocene Monterey oil are isotopically enriched in ^{13}C compared to the $C_{16}-C_{22}$ n-alkanes ($-27.6 \pm 0.2‰$), apparently due to differences in source-rock organic matter input. They contend that pristane and phytane in the Monterey oil originated mainly from the phytol side chain of chlorophyll in marine phytoplankton and/or lipids in archaebacteria, while the n-alkanes originated from marine plankton lipids and land plant waxes.

Table 6. Compound-specific stable carbon isotope data for selected *n*-alkanes and isoprenoids

Well	nC_6	nC_7	nC_8	nC_9	nC_{10}	nC_{11}	nC_{12}	nC_{13}	nC_{14}	nC_{15}	nC_{16}	nC_{17}	Pr	nC_{18}	Ph	nC_{19}
Assekaifaf	−30.30	−29.09	−29.68	−30.05	−29.93	−29.56	−28.59	−28.85	−28.27	−29.26	−29.35	−29.61	−29.06	−29.17	−28.01	−29.22
Dome	−30.63	−30.58	−29.37	−30.05	−29.76	−29.42	−28.40	−28.58	−28.80	−30.73	−30.05	−32.11	−29.69	−31.58	NA	−30.90
Hassi M.	−32.63	−31.53	−29.66	−29.96	−29.70	−29.47	−28.39	−28.82	−29.16	−28.67	−29.15	−29.32	−29.78	−29.27	−29.25	−29.22
O. Zenani	−30.90	−30.92	−28.42	−29.14	−29.77	−29.31	−28.11	−28.41	−28.59	−29.22	−29.92	−30.50	−29.38	−31.07	−29.35	−30.46
Zarzaitine	−30.31	−29.30	−30.36	−30.57	−29.92	−30.18	−29.09	−29.55	−28.77	−29.33	−29.50	−29.95	−28.55	−29.62	−28.71	−29.37
Zemlet	−32.00	−30.66	−30.60	−30.45	−30.52	−30.32	−29.38	−29.65	−30.04	−30.89	−30.86	−30.95	−30.66	−31.40	−30.65	−30.76

Isotope data (δ^{13}C in ‰ relative to the PDB belemnite standard) for C_6–C_{14} *n*-alkanes are from whole oil analysis using PONA methylsilicone column. Precision of the measurement is ∼ 0.2‰ for each compound. Data for C_{15}–C_{19} *n*-alkanes and pristane and phytane are from DB-1 analysis of saturated hydrocarbon fractions using DB-1 column.

Table 7. Compound-specific stable carbon isotope data for gasoline-range branched and cyclic hydrocarbons

Well	MCyC5	CyC6	MCyC6	PrCyC6	4MC9	2MC9	iPC10	iPC11
Assekaifaf	− 23.25	− 24.34	− 26.18	− 27.09	− 27.74	− 27.07	− 26.34	− 26.60
Dome	− 24.22	− 25.18	− 26.75	− 26.52	− 27.72	− 27.09	− 26.36	− 26.08
Hassi Messaoud	− 24.86	− 28.86	− 27.21	− 27.38	− 28.73	− 28.04	− 26.11	− 27.17
Oued Zenani	− 23.74	− 24.43	− 26.22	− 25.90	− 27.66	− 25.79	− 25.33	− 25.32
Zarzaitine	− 24.17	− 25.60	− 26.66	− 27.20	− 27.98	− 27.68	− 26.63	− 27.40
Zemlet	− 25.52	− 27.50	− 29.96	− 27.34	− 28.38	− 27.79	− 26.80	− 27.40

Isotope data (δ^{13}C in ‰ relative to the PDB belemnite standard) are from whole oil analysis using PONA methylsilicone column. Precision of the measurement is ∼0.2‰ for each compound.

M: methyl; Cy: cyclo; C5, C6, C9, C10, C11: pentane, hexane, nonane, decane, and undecane, respectively; Pr: propyl; iP: isopropyl.

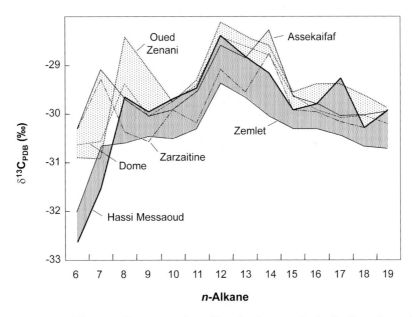

Fig. 7. Zemlet oil has generally more negative stable carbon isotope ratios for C_6–C_{19} n-alkanes than the other Algerian oil samples (Table 6). The largest isotopic difference between Hassi Messaoud and Zemlet (Silurian) oil samples and the Devonian samples occurs at nC_6.

CONCLUSIONS

Certain non-routine geochemical parameters distinguish crude oil that originated from presumed Silurian (Hassi Messaoud and Zemlet oil samples) and Devonian source rocks (Assekaifaf, Zarzaitine, Oued Zenani oil samples), but these interpretations are preliminary due to limited oil samples. The data suggest mainly Devonian source rock input to the Dome oil sample, which could have been a mixture based strictly on geologic considerations. Promising oil-to-oil correlation parameters for distinguishing Silurian from Devonian source rock input include diamondoids (*e.g.* concentrations, adamantanes/diamantanes, 1,3-/1,2-dimethylada-mantanes), light aromatic hydrocarbon ratios (*e.g.* dibenzothiophene/2-methylphenanthrenes), and stable carbon isotope ratios of gasoline-range hydrocarbons (*e.g.* nC_6 and nC_7 alkanes, methylcyclopentane, cyclohexane, and methylcyclohexane). The isotopic differences between pristane and the nC_{17} alkane and between phytane and the nC_{18} alkane distinguish Silurian oil from Devonian oil, apparently due to differences in assemblages of organisms that contributed organic matter to the source rock.

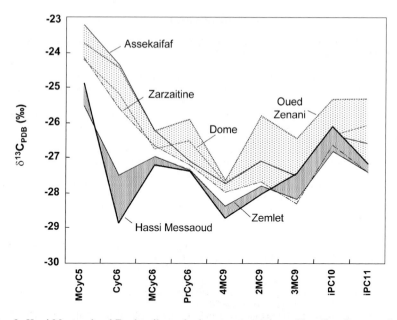

FIG. 8. Hassi Messaoud and Zemlet oil samples have more negative stable carbon isotope ratios for methylcyclopentane (MCyC5), cyclohexane (CyC6), methylcyclohexane (MCyC6), and several other light hydrocarbons compared to the Devonian oil samples (Table 7). The nC_6 and nC_7 alkanes show similar results (Fig. 7). Unlike the other samples, the Hassi Messaoud and Zemlet oil samples have cyclohexane that is isotopically more negative than methylcyclopentane and methylcyclohexane.

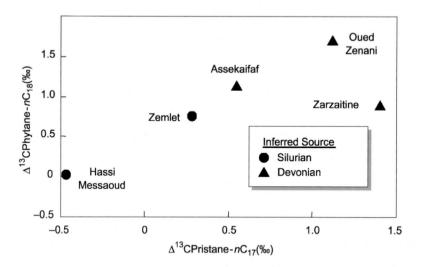

FIG. 9. Hassi Messaoud and Zemlet oil samples have smaller isotopic differences between pristane and nC_{17} and between phytane and nC_{18} than the Devonian samples. The Dome oil sample (Devonian) contained insufficient phytane for reliable measurement, but has pristane and nC_{17} that differ by 2.42‰ (Table 6).

Acknowledgements—The authors thank ExxonMobil Exploration Company for permission to publish these data. The analytical work was completed while the principal author was at ExxonMobil Upstream Research Company in Houston. We thank John Guthrie, Marit Johansen, John Okafor, Ngami Phan, and Marlene Schaps (ExxonMobil) for technical assistance. Ron Hill, Tim Klett, Les Magoon (USGS), and John Zumberge (Geomark Research, Inc.) provided helpful comments on a draft of the paper.

REFERENCES

BOOTE R. D., CLARK-LOWES D. D. and TRAUT M. W. (1998) Paleozoic petroleum systems of North Africa. In *Petroleum Geology of North Africa* (eds. D. S. MACGREGOR, R. T. J. MOODY and D. E. CLARK-LOWES), Geol. Soc. London Spec. Publ. **132**, pp. 7–68.

COCHRAN M. D. and PETERSEN L. E. (2001) Hydrocarbon exploration in the Berkine Basin, Grand Erg Oriental, Algeria. In *Petroleum Provinces of the Twenty-First Century* (eds. M. W. DOWNEY, J. C. THREET and W. A. MORGAN) *AAPG Mem.* **74**, 531–557.

COLE G. A., ABU-ALI M. A., AOUDEH S. M., CARRIGAN W. J., CHEN H. H., COLLING E. L., GWATHNEY W. J., AL-HAJII A. A., HALPERN H. I., JONES P. J., AL-SHARIDI S. H. and TOBEY M. H. (1994) Organic geochemistry of the Paleozoic petroleum system of Saudi Arabia. *Fuel* **8**, 1425–1442.

DAHL J. E., MOLDOWAN J. M., PETERS K. E., CLAYPOOL G. E., ROONEY M. A., MICHAEL G. E., MELLO M. R. and KOHNEN M. L. (1999) Diamondoid hydrocarbons as indicators of natural oil cracking. *Nature* **399**, 54–57.

DANIELS R. P. and EMME J. J. (1995) Petroleum system model, eastern Algeria, from source rock to accumulation: when, where and how? *Proceedings of the Seminar on Source Rocks and Hydrocarbon Habitat in Tunisia*, Enterprise Tunisienne D'Activités Peétrolières Mem. 9, pp. 101–124.

ECHIKH K. (1998) Geology and hydrocarbon occurrences in the Ghadames Basin, Algeria, Tunisia, Libya. In *Petroleum Geology of North Africa* (eds. D. S. MACGREGOR, R. T. J. MOODY and D. E. CLARK-LOWES), Geol. Soc. London Spec. Publ. **132**, pp. 109–129.

GRANTHAM P. J. and WAKEFIELD L. L. (1988) Variations in the sterane carbon number distributions of marine source rock derived crude oils through geological time. *Org. Geochem.* **12**, 61–73.

ILLICH H., ZUMBERGE J., SCHIEFELBEIN C. and BROWN S. (1997) Petroleum systems of the Ghadames basin and Illizi platform (abstract). *AAPG Bull.* **81**, 54 p.

JONES P. J. and STUMP T. E. (1999) Depositional and tectonic setting of the Lower Silurian hydrocarbon source rock facies, central Saudi Arabia. *AAPG Bull.* **83**, 314–332.

KLEMME H. D. and ULMISHEK G. F. (1991) Effective petroleum source rocks of the world; stratigraphic distribution and controlling depositional factors. *AAPG Bull.* **75**, 1809–1851.

KLETT T. R. (2000a) Total petroleum systems of the Illizi Province, Algeria and Libya—Tanezzuft-Illizi, U.S. Geol. Surv. Bull. 2202-A, 79 p.

KLETT T. R. (2000b) Total petroleum systems of the Grand Erg/Ahnet Province, Algeria and Morocco—the Tanezzuft-Timimoun, Tanezzuft-Ahnet, Tanezzuft-Sbaa, Tanezzuft-Mouydir, Tanezzuft-Benoud, and Tanezzuft-Béchar/Abadla, U.S. Geol. Surv. Bull. 2202-B, 144 p.

KLETT T. R. (2000c) Total petroleum systems of the Trias/Ghadames Province, Algeria, Tunisia, and Libya—the Tanezzuft-Oued Mya, Tanezzuft-Melrhir, and Tanezzuft-Ghadames, U.S. Geol. Surv. Bull. 2202-C, 118 p.

LÜNING S., CRAIG J., LOYDELL D. K., STORCH P. and FITCHES B. (2000) Lower Silurian 'hot shales' in North Africa and Arabia: regional distribution and depositional model. *Earth-Sci. Rev.* **49**, 121–200.

MACGREGOR D. S. (1998) Giant fields, petroleum systems and exploration maturity of Algeria. In *Petroleum Geology of North Africa* (eds. D. S. MACGREGOR, R. T. J. MOODY and D. E. CLARK-LOWES), Geol. Soc. London Spec. Publ. **132**, pp. 79–96.

MAGOON L. B. and DOW W. G. (1994) The petroleum system. In *The Petroleum System—From Source to Trap. AAPG Mem.* **60**, 3–24.

ORR W. L. (1974) Changes in sulfur content and isotopic ratios of sulfur during petroleum maturation. Study of Big Horn Basin Paleozoic oils. *AAPG Bull.* **58**, 2295–2318.

PETERS K. E. and MOLDOWAN J. M. (1993) *The Biomarker Guide. Interpreting Molecular Fossils in Petroleum and Ancient Sediments.* Prentice-Hall, Englewood Cliffs, NJ, 363 p.

SCHOELL M., MCCAFFREY M. A., FAGO F. J. and MOLDOWAN J. M. (1992) Carbon isotopic compositions of 28, 30-bisnorhopanes and other biological markers in a Monterey crude oil. *Geochim. Cosmochim. Acta* **56**, 1391–1399.

SOFER Z. (1984) Stable carbon isotope compositions of crude oils: application to source depositional environments and petroleum alteration. *AAPG Bull.* **68**, 31–49.

SONATRACH (1995) Geology of Algeria. Conference Sur L'Evaluation des Puits, Algerie, pp. I1–I93.

TISSOT B., ESPITALIÉ J., DEROO G., TEMPERE C. and JONATHAN D. (1984) Origin and migration of hydrocarbons in the eastern Sahara (Algeria). In *Petroleum Geochemistry and Basin Evaluation. AAPG Mem.* **35**, 315–324.

TYSON R. V. (1995) *Sedimentary Organic Matter: Organic Facies and Palynofacies.* Chapman & Hall, New York, 615 p.

ZUMBERGE J. E., MACKO S., ENGEL M., JOHANSSON F., SCHIEFELBEIN C. and BROWN S. (1996) Silurian shale origin for light oil, condensate, and gas in Algeria and the Middle East (abstract). *AAPG/SEPM Annu. Meet. Abstr.* **5**, 160 p.

Geochemical Investigations in Earth and Space Science: A Tribute to Isaac R. Kaplan
© The Geochemical Society, Publication No. 9, 2004
Editors: R.J. Hill, J. Leventhal, Z. Aizenshtat, M.J. Baedecker, G. Claypool,
R. Eganhouse, M. Goldhaber and K. Peters

C4-benzene and C4-naphthalene thermal maturity indicators for pyrolysates, oils and condensates[*]

RONALD J. HILL[1], SHAN-TAN LU[2], YONGCHUN TANG[3], MITCHELL HENRY[1] and
ISAAC R. KAPLAN[4]

[1]United States Geological Survey, Box 25046, MS 939, Denver Federal Center,
Denver, CO 80225, USA
[2]Zymax Forensics Inc., San Luis Obispo, CA 93401, USA
[3]Petroleum Energy and the Environment Research Center, California Institute of Technology,
Walnut, CA 91789, USA
[4]Institute of Geophysics and Planetary Physics, University of California,
Los Angeles, CA 90095, USA

Abstract—Determining the thermal maturity of light oils and condensates using chemical indicators can be difficult. Here we describe the use of C_4-benzenes (C_{10}) and C_4-naphthalenes (C_{14}) which are common constituents of oils and condensates as potential thermal maturity indicators. Using confined, dry pyrolysis of a saturate-rich Devonian oil from the Western Canada Sedimentary Basin, experiments were performed at 350–400°C, 650 bars and at time periods ranging from 3 to 33 days. The equivalent vitrinite reflectance ($\%R_0$) for the pyrolysis products (1.02–1.67%) was calculated from experimental conditions using Easy R_0 software (*Geochim. Cosmochim Acta* **53**, 1989, 2649–2657) to assess relative experimental thermal maturity. Ratios employing tetramethyl-, dimethylethyl-, and methylisopropylbenzene isomers and tetramethyl-naphthalene isomers correlate with calculated $\%R_0$ from pyrolysis. Excellent correlation of all C_4-benzene parameters and most C_4-naphthalene parameters was observed. The thermal maturity parameters defined by pyrolysis were applied to oils in the Fort Worth Basin, Texas, USA to determine applicability in natural systems. All C_4-benzene and C_4-naphthalene parameters showed positive correlation with the triaromatic steroid (TAS) maturity parameter commonly used to assess oil maturity. C_4-benzene and C_4-naphthalene ratios extend beyond the range of biomarker applicability, especially TAS, and are more abundant than biomarkers and less volatile than C_7 hydrocarbons.

INTRODUCTION

ALKYLBENZENES have been reported in solvent extracts and pyrolysates of recent sediments and ancient sediments, asphaltenes, coals (SINNINGHE DAMSTĚ *et al.*, 1988, 1993; VELLA and HOLZER, 1992; HARTGERS *et al.*, 1994; HOEFS *et al.*, 1995) and crude oils (SINNINGHE DAMSTĚ *et al.*, 1988, 1991; SCHENK *et al.*, 1997; WILKES *et al.*, 2000). Alkylnaphthalenes have been reported in ancient sediments (VANDENBROUKE, 1980; RADKE *et al.*, 1982a, 1986; ALEXANDER *et al.*, 1985; STRACHAN *et al.*, 1988), coals (RADKE *et al.*, 1982b, 1984, 1990a) and oils (VOLKMAN *et al.*, 1984; ALEXANDER *et al.*, 1983, 1984, 1985, 1993; PUTTMANN and VILLAR, 1987; RADKE *et al.*, 1990b, 1994; BASTOW *et al.*, 1998; VAN AARSSEN *et al.*, 1999). Alkyl aromatic hydrocarbons have been used as indicators of organic matter source and thermal maturity. The distribution and relative abundance of C_2–C_5 methylnaphthalene isomers (di-, tri -, tetra- and pentamethylnaphthalene) correlate with the thermal maturity in sediments and oils. RADKE *et al.* (1982a,b) found that changes in 1,5-dimethylnapthalene (DMN) relative to the 2,6- and 2,7-isomers correlate with increasing coal rank. ALEXANDER *et al.* (1984) showed that the relative abundance of 1,8-DMN decreased compared to other DMN isomers with

---

[*] Presented orally by the principal author at the Symposium to Honor Ian Kaplan (Ron Hill and Ken Peters, Organizers) Annual Geological Society of America Meeting, October 27-30, 2002, Denver, CO.

increasing thermal maturity in sediments. Subsequently, many DMN, trimethylnaphthalene (TMN), tetramethylnaphthalene (TeMN) and pentamethylnaphthalene (PMN) isomer ratios were shown to change with thermal maturity in sediments and oils (ALEXANDER *et al.*, 1985; VAN AARSSEN *et al.*, 1999). The relationship between alkylbenzene compound distribution and thermal maturity has not been investigated.

Determining the thermal maturity of light oils and condensates can be difficult. Biomarker concentrations in crude oils are low and biomarker maturity parameters have limited applicability at high levels of thermal maturity (PETERS and MOLDOWAN, 1993). Light hydrocarbons (C_6–C_7) are volatile, susceptible to biodegradation and maturity parameters derived from these compounds may be unreliable. In this chapter, we report on the correlation of C_4-benzene and C_4-naphthalene compounds with thermal maturity in oil cracking pyrolysis products of a Western Canada Sedimentary Basin (WCSB) oil. The use of C_4-benzene and C_4-naphthalene compound ratios as thermal maturity indicators in natural systems was evaluated using crude oils from the Fort Worth Basin, Texas, USA.

EXPERIMENTAL

Pyrolysis methods and starting material

A 35° API gravity Devonian oil from the WCSB in Alberta, Canada was topped at 35°C for 2.5 h under vacuum and the C_{9+} fraction was used as the starting material for the pyrolysis experiments The topped oil stable carbon isotope ratio is $-29.47‰$ relative to PDB. The composition of the topped oil is 64% saturates, 23% aromatics, 6.5% resins and 6.5% asphaltenes. The oil falls into the Cynthia shale oil family of EXPLORATION STAFF, CHEVRON STANDARD LIMITED (1979) and ALLAN and CREANEY (1991). Oil pyrolysis was performed isothermally in sealed gold tubes at temperatures ranging from 350°C to 400°C, time ranging from 3 to 33 days and pressure of 650 bars. Details of the experimental procedure were discussed previously (HILL *et al.*, 1994, 1996, 2003).

Fort Worth Basin oils

Fort Worth Basin oils and condensates were collected in Jack County and Wise County, Texas, USA with permission from Republic Energy (Fig. 1) The source of the oils and condensates is most likely the marine Mississippian Barnett Shale. API gravity for the oils ranges from 40 to 59° and sulfur contents are less than 0.5%.

Pyrolysis liquid product analysis

Following gas analysis, each gold tube was cut open with clippers and the residual oil was extracted three times by sonication in pentane to obtain the saturate and aromatic residual liquid fraction The pentane extract was decanted and passed through a 0.45 μm nylon filter to recover pyrobitumen. The dry ice/acetone bath used to trap the C_6–C_{12} fraction during gas analysis was removed and the trap rinsed thrice with pentane and then added to the gold tube pentane extract. Details of the gas analysis and solvent extraction procedure were discussed in HILL *et al.* (1996, 2003).

GC/MS analysis of C_4-benzenes and C_4-naphthalenes

The pentane extract from oil pyrolysis (1 μl) was injected directly into a Varian 3400 gas chromatograph (GC) connected to a Finnigan Model 4000 mass spectrometer. The 60 m DB-1 0.25 mm I.D. fused silica capillary column in the GC was programmed at 4°C/min, starting from 40°C (isothermal at 40°C for 5 min) to 310°C and held for 30 min at 310°C. Scanning was performed over a mass range of 50–550 amu (full scan). Mass spectral data were stored and processed with Finnigan INCOS 2300 data system.

The Fort Worth Basin crude oils and condensates were diluted 1:75 oil to methylene chloride and then 1 μl of diluted sample was injected directly into a Varian 3400 GC connected to a

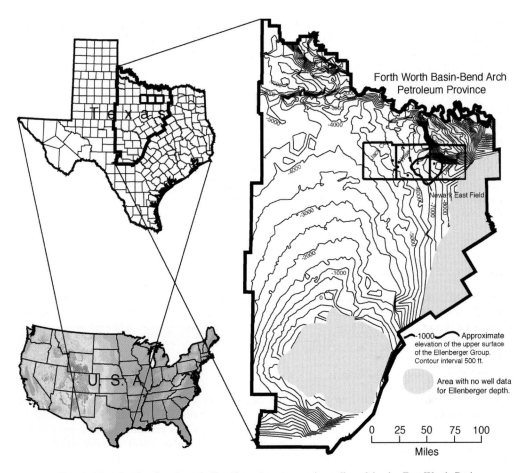

FIG. 1. Map showing location of oil and condensate samples collected in the Fort Worth Basin, Texas, USA.

Finnigan Model 4000 mass spectrometer and analyzed using the same conditions as those described for the oil pyrolysates.

GC/MS analysis of triaromatic steroids

The triaromatic steroids (TASs) in the Fort Worth Basin oils and condensates were analyzed using a Hewlett Packard 6890 GC connected to a JEOL GCMate double-focusing magnetic sector mass spectrometer scanning m/z 231.1174. The GC was equipped with a ZB-1701 60 m (l) X 0.31 mm I.D. fused silica column with 0.25 μ film thickness (bonded phase 14% cyanopropylphenyl, 86% dimethylpolysiloxane copolymer). The GC oven starting temperature was 50°C and was programmed to rise from 50°C to 150°C at 5°C/min and then from 150°C to 300°C at 3°C/min and hold for 9 min at 300°C.

RESULTS

The C$_4$-benzene and C$_4$-naphthalene compounds detected in the original WCBS oil, the oil pyrolysates and the Fort Worth Basin oils and condensates are listed in Table 1. The compounds listed in Table 1 are shown in the chromatograms in Fig. 2. The distributions of C$_4$-benzene and C$_4$-naphthalene compounds in the WCSB oil pyrolysates and the Fort Worth Basin oils

Table 1. C_4-benzene and C_4-naphthalene compounds detected by GC/MS in oil pyrolysates and Fort Worth Basin oils

	m/z 134		m/z 184
16	Sec-buytlbenzene	21	1-(1,1-Dimethylethyl)-naphthalene
17	1-Methyl-3-isopropylbenzene (MiPB)	22	1-Methyl-7-(1-methylethyl)-naphthalene
18	1-Methyl-4-isopropylbenzene	23	1,2-Diethylnapthalene
19	1-Methyl-2-isopropylbenzene	24	7-Ethyl-1,4-dimethyl-azulene
20	1,3-Diethylbenzene	25	1,3,5,7-Tetramethylnaphthalene (TeMN)
21	1-Methyl-3-propylbenzene	26	1,3,6,7-TeMN
22	Butylbenzene	27	1,2,4,6 + 1,2,4,7 + 1,4,6,7-TeMN
23	1,3-Dimethyl-5-ethylbenzene (DMEB)	28	1,2,5,7-TeMN
24	1,2-Diethylbenzene	29	2,3,6,7-TeMN
25	1-Methyl-2-propylbenzene	30	1,2,6,7-TeMN
26	1,4-Dimethyl-2-ethylbenzene	31	1,2,3,7-TeMN
27	1,3-Dimethyl-4-ethylbenzene	32	1,2,3,6-TeMN
28	1,2-Dimethyl-4-ethylbenzene	33	1,2,5,6 + 1,2,3,5-TeMN
29	1,3-Dimethyl-2-ethylbenzene		
30	1,2-Dimethyl-3-ethylbenzene		
31a	1,2,4,5-Tetramethylbenzene (TeMB)		
31	1,2,3,5-TeMB		
32	1,2,3,4-TeMB		

GC/MS, gas chromatography mass spectrometry.

and condensates are summarized in Tables 2 and 3, respectively. The C_4-benzene and C_4-naphthalene ratios used in this study are defined in Table 4.

Oil cracking pyrolysis experiments

C_4-benzenes were detected in the WCSB oil used in the oil cracking experiments and are most likely products of kerogen cracking during hydrocarbon generation (Table 1). HARTGERS *et al.* (1994) found that C_4-benzenes were ubiquitous products in flash pyrolysates of 47 kerogens and coals containing bacterial, algal and higher plant cellular debris. The authors postulated that flash pyrolysis results indicate the C_4-benzenes are covalently bonded in the kerogen and coal structures and are liberated by β and γ scission reactions during thermal decomposition of kerogen and coal. All the C_4-benzenes identified in the flash pyrolysates by HARTGERS *et al.* (1994) were also identified in the WCSB oil used in the oil cracking experiments.

Thermal maturity achieved during the oil cracking experiments is expressed as vitrinite reflectance ($\%R_0$), calculated using the software Easy R_0[1] (BURNHAM and SWEENEY, 1989) and the temperature and time conditions for each experiment. Calculated $\%R_0$ ranges from 1.02 to 1.67%. Figure 3 summarizes the changes in TeMB, DMEB and MiPB ratios that best correlate with calculated $\%R_0$ from 1.1 to 1.7. TeMB-1, the tetramethylbenzene ratio {1,2,4,5-TeMB/Total TeMB}, increases from 0.05 to 0.12; TeMB-2, the tetramethylbenzene ratio {1,2,3,5-TeMB/Total TeMB}, from 0.32 to 0.42; and TeMB-3, the tetramethylbenzene ratio {(1,2,3,5-TeMB + 1,2,4,5-TeMB)/Total TeMB}, from 0.38 to 0.55. The dimethylethylbenzene ratio DMEB-1 {1,4-DM2EB/(1,4-DM2EB + 1,3-DM2EB)} increases from 0.61 to 0.78; the dimethylethylbenzene ratio DMEB-2 {Total DMEB/(Total DMEB + Total TeMB)} from 0.24 to 0.52; and the methylisopropylbenzene ratio MiPB {1,4-MiPB/(1,4-MiPB + 1,3-MiPB)} from 0.77 to 0.90 with increase in calculated $\%R_0$.

Figure 4 summarizes changes in TeMN ratios that best correlate with increase in calculated $\%R_0$. TeMN-1 (1,3,6,7-TeMN/(1,3,6,7-TeMN + 1,2,5,6-TeMN + 1,2,3,5-TeMN)) was originally defined by VAN AARSSEN *et al.* (1999) from measurements on source rocks and crude oils and was found to increase as source rock and crude oil thermal maturity increased.

[1] Any use of trade, product, or firm names is for descriptive use only and does not imply endorsement by the US government.

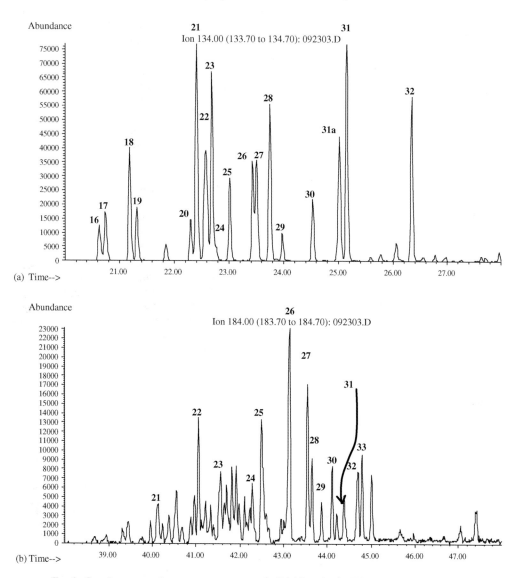

FIG. 2. Gas chromatography mass spectrometry m/z 134 (A) and m/z 184 (B) fragmentograms. Peak numbers correspond to peaks in Table 1.

TeMN-2 {1,3,6,7-TeMN/(1,3,6,7-TeMN + 1,2,5,7-TeMN)}, TeMN-3 {2,3,6,7-TeMN/ (2,3,6,7-TeMN + 1,2,3,7-TeMN)} and TeMN-4 (1,3,6,7-TeMN/Total TeMN) are based strictly on pyrolysis results from this study. In this study, TeMN-1 increases from 0.84 to 0.98 as calculated %R_0 increases. TeMN-2 increases from 0.77 to 0.94 and TeMN-3 increases from 0.45 to 0.70 although at maturities above ~ %R_0 = 1.6 there is some scatter in the data with the last two points showing a decrease in maturity. TeMN-4 increases from 0.28 to 0.38 as %R_0 increases to 1.3%. Above R_0 ~ 1.3%, TeMN-4 decreases from 0.38 at %R_0 ~ 1.3% to 0.15 at %R_0 ~ 1.67.

All the C$_4$-benzene and C$_4$-naphthalene ratios increase linearly with increase in calculated %R_0 except TeMN-4. Correlation of each C$_4$-benzene and C$_4$-naphthalene ratio with thermal maturity is strong with R^2 values exceeding 0.75 for all parameters except TeMN-3 (R^2 = 0.50) and TeMN-4 (non-linear). The TeMN-3 parameter shows increased scatter at higher maturities, contributing to the R^2 = 0.50.

Table 2. C₄-benzene and C₄-naphthalene distributions in oil cracking pyrolysates and compound ratios useful in thermal maturity determination

Pyrolysis temperature (°C)	Crude oil	350	350	350	350	360	360	360	360	360	370	380	390	400
Pyrolysis time (days)		3	4.33	12	20.7	3	6	12	24	33.5	3	3	3	3
Experimental $\%R_0$ (calculated)		1.02	1.08	1.3	1.41	1.13	1.29	1.44	1.59	1.67	1.25	1.37	1.5	1.65
C_4-benzenes														
Total tetramethyl (TeMB)[a]	58.6	66.44	67.51	57.00	54.91	62.07	57.58	49.01	42.10	37.27	53.55	51.21	37.18	30.41
1,2,4,5-TeMB[b]	0.29	0.08	0.09	0.07	0.11	0.07	0.07	0.08	0.13	0.11	0.06	0.05	0.14	0.16
1,2,3,5-TeMB[b]	0.17	0.35	0.32	0.38	0.35	0.32	0.31	0.41	0.41	0.44	0.38	0.40	0.38	0.42
1,2,3,4-TeMB[b]	0.54	0.57	0.60	0.55	0.54	0.61	0.62	0.51	0.46	0.45	0.56	0.55	0.48	0.42
Total dimethylethyl (DMEB)[a]	21.4	21.40	23.22	27.93	30.18	24.83	28.17	34.90	36.52	40.33	29.14	32.54	39.92	41.31
1,3-Dimethyl-5-ethylbenzene[c]	0.22	0.29	0.30	0.29	0.29	0.30	0.28	0.29	0.26	0.30	0.27	0.32	0.32	0.28
1,4-Dimethyl-2-ethylbenzene[c]	0.10	0.12	0.11	0.10	0.13	0.12	0.13	0.13	0.15	0.14	0.11	0.12	0.12	0.14
1,3-Dimethyl-4-ethylbenzene[c]	0.17	0.17	0.20	0.21	0.19	0.19	0.19	0.18	0.18	0.17	0.20	0.19	0.18	0.17
1,2-Dimethyl-4-ethylbenzene[c]	0.18	0.26	0.24	0.25	0.25	0.24	0.25	0.26	0.28	0.27	0.26	0.23	0.25	0.28
1,3-Dimethyl-2-ethylbenzene[c]	0.05	0.07	0.06	0.06	0.05	0.06	0.06	0.05	0.05	0.04	0.06	0.05	0.04	0.04
1,2-Dimethyl-3-ethylbenzene[c]	0.09	0.09	0.09	0.09	0.09	0.09	0.09	0.08	0.09	0.08	0.09	0.09	0.09	0.09
Total methylisopropyl (MiPB)[a]	11.1	8.37	9.62	10.10	9.87	9.16	10.05	11.49	11.90	14.71	11.37	12.44	15.46	16.86
1-Methyl-3-isopropylbenzene[d]	0.12	0.14	0.11	0.11	0.07	0.10	0.09	0.09	0.09	0.07	0.08	0.08	0.08	0.06
1-Methyl-4-isopropylbenzene[d]	0.45	0.45	0.45	0.47	0.49	0.45	0.45	0.52	0.52	0.53	0.45	0.47	0.54	0.58
1-Methyl-2-isopropylbenzene[d]	0.42	0.41	0.44	0.43	0.44	0.46	0.46	0.39	0.39	0.40	0.47	0.45	0.39	0.36
C_4-napthalenes														
Total tetramethylnaphthalenes (TeMN)[e]	78.54	76.01	74.39	67.55	63.59	69.54	66.27	59.77	56.68	53.39	64.39	60.56	53.59	50.67
1,3,5,7-TeMN[f]	0.07	0.07	0.11	0.22	0.30	0.02	0.20	0.32	0.58	0.70	0.23	0.26	0.56	0.74
1,3,6,7-TeMN[f]	0.21	0.28	0.27	0.35	0.38	0.35	0.35	0.34	0.21	0.16	0.32	0.30	0.26	0.16
1,2,4,6-TeMN + 1,2,4,7-TeMN + 1,4,6,7-TeMN[f]	0.23	0.19	0.19	0.14	0.10	0.19	0.12	0.09	0.03	0.03	0.11	0.14	0.04	0.02
1,2,5,7-TeMN[f]	0.09	0.10	0.09	0.05	0.03	0.08	0.06	0.04	0.03	0.02	0.07	0.06	0.03	0.02
2,3,6,7-TeMN[f]	0.03	0.05	0.05	0.07	0.06	0.06	0.06	0.08	0.02	0.04	0.06	0.07	0.04	0.03
1,2,6,7-TeMN[f]	0.17	0.08	0.10	0.06	0.04	0.11	0.06	0.04	0.09	0.01	0.08	0.06	0.04	0.03
1,2,3,7-TeMN[f]	0.07	0.08	0.06	0.03	0.03	0.07	0.04	0.04	0.02	0.03	0.05	0.05	0.03	0.01
1,2,3,6-TeMN[f]	0.08	0.08	0.08	0.05	0.06	0.07	0.06	0.04	0.03	0.01	0.04	0.03	0.00	0.01
1,2,5,6-TeMN + 1,2,3,5-TeMN[f]	0.06	0.07	0.04	0.03	0.01	0.06	0.04	0.03	0.01	0.01	0.04	0.04	0.01	0.00

[a] Percent of total C₄ benzenes measured.
[b] Fraction of total TeMB.
[c] Fraction of total DMEB.
[d] Fraction of total MiPB.
[e] Percent of total C₄ naphthalenes measured including azulene.
[f] Fraction of total TeMN.

Table 3. C₄-benzene and C₄-naphthalene distributions in Fort Worth Basin oils and compound ratios useful in thermal maturity determination

Well name	Craft TWB	Tarrant County Waterboard Unit 6	Mildred A. Durham	Della Christian WB	W.W.A. Murphy	Glenn George	Beaman	Collins	Caswell	Jerry North
Well number	2	4	14	3	4	2	2	3	1	1
C₄-benzenes										
Total tetramethylbenzene (TeMB)[a]	26.08	26.00	25.90	31.45	13.17	19.22	27.17	18.50	26.67	18.79
1,2,4,5-TeMB[b]	0.27	0.28	0.23	0.20	0.29	0.30	0.25	0.26	0.34	0.31
1,2,3,5-TeMB[b]	0.46	0.45	0.40	0.39	0.57	0.49	0.43	0.44	0.49	0.56
1,2,3,4-TeMB[b]	0.28	0.26	0.37	0.41	0.14	0.21	0.32	0.29	0.17	0.13
Total dimethylethyl (DMEB)[a]	34.01	35.09	33.02	32.26	34.57	33.33	34.21	34.34	37.76	40.73
1,3-Dimethyl-5-ethylbenzene[c]	0.29	0.29	0.28	0.26	0.40	0.33	0.30	0.35	0.41	0.51
1,4-Dimethyl-2-ethylbenzene[c]	0.17	0.17	0.17	0.16	0.16	0.15	0.16	0.16	0.14	0.13
1,3-Dimethyl-4-ethylbenzene[c]	0.17	0.16	0.17	0.18	0.15	0.16	0.16	0.15	0.14	0.14
1,2-Dimethyl-4-ethylbenzene[c]	0.24	0.25	0.24	0.25	0.17	0.23	0.25	0.22	0.24	0.16
1,3-Dimethyl-2-ethylbenzene[c]	0.04	0.04	0.04	0.05	0.05	0.04	0.04	0.04	0.01	0.02
1,2-Dimethyl-3-ethylbenzene[c]	0.10	0.09	0.10	0.10	0.08	0.08	0.10	0.08	0.05	0.04
Total methylisopropyl (MiPB)[a]	12.38	11.49	14.00	11.22	19.32	15.19	11.58	15.75	10.01	16.57
1-Methyl-3-isopropylbenzene[d]	0.20	0.20	0.21	0.22	0.20	0.28	0.23	0.19	0.10	0.12
1-Methyl-4-isopropylbenzene[d]	0.53	0.53	0.49	0.47	0.61	0.51	0.53	0.57	0.61	0.61
1-Methyl-2-isopropylbenzene[d]	0.27	0.27	0.29	0.31	0.20	0.22	0.25	0.24	0.28	0.28
C₄-naphthalenes										
Total tetramethylnaphthalenes (TeMN)[e]	73.53	72.65	75.22	75.90	72.65	72.83	74.54	76.17	72.22	67.01
1,3,5,7-TeMN[f]	0.16	0.13	0.13	0.11	0.12	0.15	0.15	0.13	0.11	0.12
1,3,6,7-TeMN[f]	0.31	0.31	0.24	0.25	0.29	0.29	0.25	0.26	0.40	0.29
1,2,4,6-TeMN + 1,2,4,7-TeMN + 1,4,6,7-TeMN[f]	0.18	0.20	0.21	0.22	0.22	0.18	0.18	0.21	0.16	0.21
1,2,5,7-TeMN[f]	0.08	0.09	0.09	0.12	0.09	0.08	0.10	0.10	0.07	0.08
2,3,6,7-TeMN[f]	0.06	0.07	0.05	0.05	0.05	0.06	0.05	0.04	0.08	0.05
1,2,6,7-TeMN[f]	0.07	0.07	0.09	0.08	0.09	0.07	0.09	0.08	0.06	0.10
1,2,3,7-TeMN[f]	0.03	0.03	0.04	0.04	0.06	0.04	0.03	0.03	0.03	0.05
1,2,3,6-TeMN[f]	0.05	0.05	0.06	0.05	0.06	0.06	0.05	0.06	0.05	0.05
1,2,5,6-TeMN + 1,2,3,5-TeMN[f]	0.06	0.05	0.09	0.08	0.03	0.07	0.10	0.09	0.06	0.06

[a] Percent of total C₄-benzenes measured.
[b] Fraction of total TeMB.
[c] Fraction of total DMEB.
[d] Fraction of total MiPB.
[e] Percent of total C₄-naphthalenes measured including azulene.
[f] Fraction of total TeMN.

Table 4. Maturity indicators based on C_4-benzene and C_4-naphthalene distributions

C_4-benzene and C_4-naphthalene ratios from pyrolysis

Experimental %R_0 (calculated)	Crude oil	1.02	1.08	1.3	1.41	1.13	1.29	1.44	1.59	1.67	1.25	1.37	1.5	1.65
TeMB-1	0.29	0.08	0.09	0.07	0.11	0.07	0.07	0.08	0.13	0.11	0.06	0.05	0.14	0.16
TeMB-2	0.17	0.35	0.32	0.38	0.35	0.32	0.31	0.41	0.41	0.44	0.38	0.40	0.38	0.42
TeMB-3	0.46	0.43	0.40	0.45	0.46	0.39	0.38	0.49	0.54	0.55	0.44	0.45	0.52	0.58
DMEB-1	0.65	0.63	0.62	0.65	0.72	0.66	0.69	0.74	0.76	0.77	0.64	0.70	0.73	0.77
DMEB-2	0.36	0.24	0.26	0.33	0.35	0.29	0.33	0.42	0.46	0.52	0.35	0.39	0.52	0.58
MiPB	0.79	0.77	0.80	0.81	0.87	0.82	0.83	0.85	0.86	0.88	0.85	0.85	0.88	0.91
TeMN-1	0.78	0.80	0.87	0.92	0.97	0.86	0.89	0.93	0.96	0.96	0.88	0.89	0.96	0.97
TeMN-2	0.69	0.74	0.75	0.88	0.94	0.81	0.87	0.91	0.90	0.91	0.83	0.84	0.89	0.91
TeMN-3	0.31	0.36	0.47	0.66	0.68	0.46	0.60	0.70	0.79	0.53	0.52	0.60	0.58	0.67
TeMN-4	0.21	0.28	0.27	0.35	0.38	0.35	0.35	0.34	0.21	0.16	0.32	0.30	0.26	0.16

C_4-benzene and C_4-naphthalene ratios from Fort Worth Basin oils

Well name	Craft TWB	Tarrant County Waterboard Unit 6	Mildred A. Durham	Della Christian WB	W.W.A. Murphy	Glenn George	Beaman	Collins	Caswell	Jerry North
Well number	2	4	14	3	4	2	2	3	1	1
API gravity	43	42	42	40	53	50	40	49	52	59
Triaromatic steroid ratio	0.59	0.57	0.47	0.42			0.48	0.51	0.67	
TeMB-1	0.27	0.28	0.23	0.20	0.29	0.30	0.25	0.26	0.34	0.31
TeMB-2	0.46	0.45	0.40	0.39	0.57	0.49	0.43	0.44	0.49	0.56
TeMB-3	0.72	0.74	0.63	0.59	0.86	0.79	0.68	0.71	0.83	0.87
DMEB-1	0.81	0.80	0.80	0.78	0.77	0.79	0.79	0.79	0.91	0.87
DMEB-2	0.57	0.57	0.56	0.51	0.72	0.63	0.56	0.65	0.59	0.68
MiPB	0.72	0.72	0.70	0.68	0.76	0.65	0.70	0.75	0.86	0.84
TeMN-1	0.84	0.86	0.73	0.77	0.92	0.80	0.71	0.75	0.88	0.82
TeMN-2	0.79	0.78	0.72	0.69	0.77	0.77	0.72	0.73	0.85	0.79
TeMN-3	0.64	0.67	0.57	0.55	0.46	0.58	0.59	0.56	0.75	0.50
TeMN-4	0.31	0.31	0.24	0.25	0.29	0.29	0.25	0.26	0.40	0.29

TeMB-1: 1,2,4,5-TeMB/Total TeMB; TeMB-2: 1,2,3,5-TeMB + 1,2,4,5-TeMB/Total TeMB; TeMB-3: (1,2,3,5-TeMB + 1,2,4,5-TeMB)/Total TeMB; DMEB-1: 1,4DM2EB/(1,4DM2EB + 1,3DM2EB); DMEB-2: Total DMEB/Total DMEB + TeMB; MiPB: 1M4iP/(1M4iP + 1M3iP); TeMN-1: 1,3,6,7-TeMN/(1,3,6,7-TeMN + 1,2,5,6-TeMN + 1,2,3,5-TeMN); TeMN-2: 1,3,6,7-TeMN/(1,3,6,7-TeMN + 1,2,5,7-TeMN); TeMN-3: 2,3,6,7-TeMN/(2,3,6,7-TeMN + 1,2,3,7-TeMN); TeMN-4: 1,2,6,7-TeMN/Total TeMN

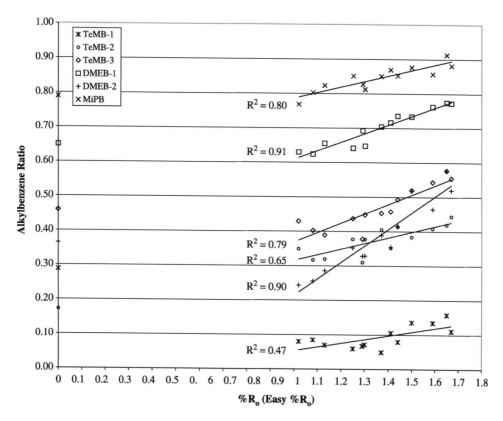

FIG. 3. Changes in C$_4$-benzene isomer ratios as a function of calculated %R$_0$ value for oil cracking pyrolysis products. Compound abbreviations are summarized in Table 1.

Fort Worth Basin oils

Thermal maturity for the Fort Worth Basin oils in this study was assessed using the TAS ratio (MACKENZIE *et al.*, 1981). In this study, the $(C_{21} + C_{22})/(C_{21} + C_{22} + C_{26} + C_{27} + C_{28})$ TAS ratio was used. The TAS ratio for the Fort Worth oils used in this study ranges from 0.42 to 0.68 with three oils being overmature and not having sufficient TAS content to determine a TAS value. The same C$_4$-benzene and C$_4$-naphthalene ratios were calculated for the Fort Worth Basin oils as were calculated for the oil cracking pyrolysates. Figure 5 summarizes the TeMB, DMEB and MiPB ratios plotted against the TAS maturity parameter. TeMB-1 ratio increases from 0.20 to 0.33, TeMB-2 from 0.39 to 0.49 and TeMB-3 from 0.60 to 0.73 as TAS increases. DMEB-1 increases from 0.77 to 0.87 as TAS increases. DMEB-2 increases from 0.55 to 0.60 as TAS increases from 0.42 to 0.68, although a DMEB-2 ratio of 0.65 is observed at TAS = 0.51. MiPB increases from 0.68 to 0.82 as TAS increases.

The 1,3,6,7-TeMN isomer is dominant in the unpyrolysed WCSB oil and all Fort Worth Basin oils. This observation is consistent with the results of VAN DUIN *et al.* (1997) that used molecular mechanics to show the 1,3,6,7-TeMN isomer is most stable. The use of TeMN distributions as an indicator of thermal maturity was originally established in crude oils from Australia (VAN AARSSEN *et al.*, 1999). In that study, TeMN-1 was shown to increase with increasing Ts/Tm in oils. The Fort Worth Basin oil results corroborate those of VAN AARSSEN *et al.* (1999) in that TeMN-1 correlates with TAS. In addition, TeMN-2, TeMN-3 and TeMN-4 also correlate with TAS for the Fort Worth Basin oils (Fig. 6). The magnitude of change in TeMN-1 and TeMN-2 is similar, although there is less data scatter in the TeMN-2 correlation. TeMN-1 increases from 0.72 to 0.88 and TeMN-2 increases

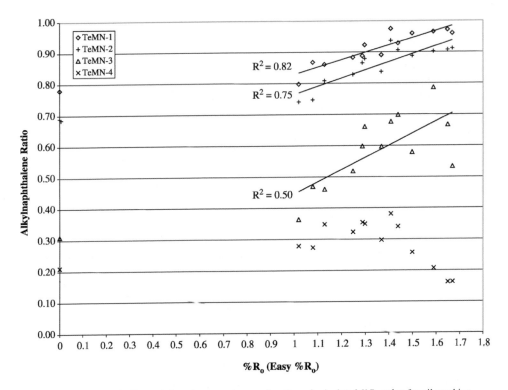

FIG. 4. Changes in C$_4$-napthalene isomer ratios as a function of calculated %R$_0$ value for oil cracking pyrolysis products. Compound abbreviations are summarized in Table 1.

from 0.69 to 0.85 as TAS increases. TeMN-3 and TeMN-4 have the advantage of showing a greater magnitude of change over the maturity range investigated. TeMN-3 ratio increases from 0.53 to 0.73 and TeMN-4 from 0.22 to 0.38 as TAS increases. The results show TeMN-2, TeMN-3 and TeMN-4 are equally effective thermal maturity indicators as TeMN-1. TeMN-2 appears to be more responsive to initial changes in maturity than the other TeMN ratios (Fig. 6).

All the C$_4$-benzene and C$_4$-naphthalene ratios increase linearly with increase in calculated TAS for Fort Worth Basin oils with the possible exception of MiPB. Correlation of each C$_4$-benzene and C$_4$-naphthalene ratio with TAS is good with R^2 values exceeding 0.68 for all parameters except MiPB ($R^2 = 0.16$). One datum point which could easily reflect an analytical error that falls off the otherwise linear trend, is the reason the MiPB correlation is low.

Comparison of C$_4$-benzene and C$_4$-naphthalene ratios from pyrolysates and Fort Worth Basin oils

The initial distribution of C$_4$-benzene and C$_4$-naphthalene compounds is different for the original WCSB oil and the Fort Worth Basin oils (Tables 2 and 3). The 1,2,3,5-tetramethylbenzene isomer is most abundant relative to the other two isomers in the Fort Worth Basin oils whereas the 1,2,3,4-tetramethylbenzene isomer is most abundant in WCSB oil. The 1,3-dimethyl-5-ethylbenzene and the 1,2-dimethyl-4-ethylbenzene isomers are most abundant and next most abundant in the WCSB and Fort Worth Basin oils when measured as a fraction of the total dimethylethylbenzenes. The 1-methyl-4-isopropylbenzene isomer is most abundant in the WCSB and Fort Worth Basin oils relative to the other two isomers. However, 1-methyl-2-isopropylbenzene, the next most abundant isomer in both oils is more abundant relatively in the WCSB oil than in the Fort Worth Basin oils. The 1,3,6,7-TeMN isomer is most

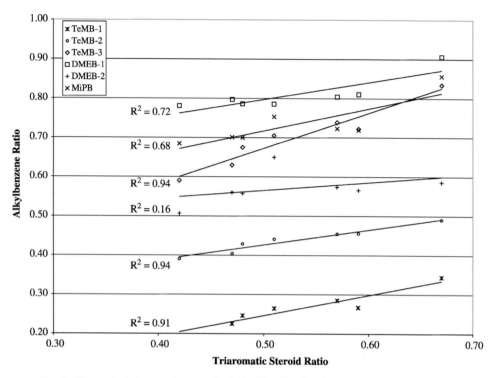

FIG. 5. Changes in C$_4$-benzene isomer ratios in Fort Worth Basin oils and condensates as a function of triaromatic steroid maturity parameter.

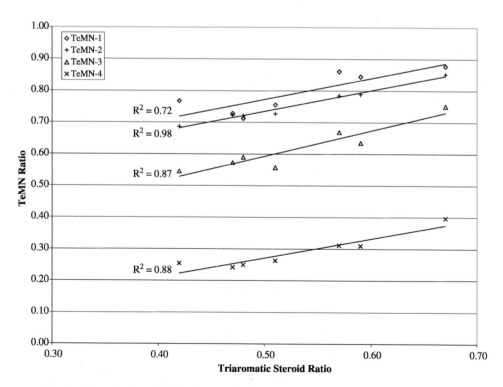

FIG. 6. Changes in C$_4$-naphthalene isomer ratios in Fort Worth Basin oils and condensates as a function of triaromatic steroid maturity parameter.

abundant in both the WCSB oil and the Fort Worth Basin oils, but the relative amounts of the other isomers differ. At high maturities, the 1,3,5,7-TeMN isomer becomes most abundant in the pyrolysates whereas 1,3,6,7-TeMN remains most abundant in the Fort Worth Basin oils.

DISCUSSION

The C_4-naphthalene and C_4-benzene data presented in this paper were collected as part of a larger study addressing the influence of pressure, temperature and time on hydrocarbon generation (HILL *et al.*, 1994, 1996, 2003). The goal of the study was to evaluate whether maturity parameters defined during oil pyrolysis could be extended to natural systems. Thus, the Devonian sourced oil from the WSCB was used to define C_4-naphthalene and C_4-benzene maturity parameters in the oil pyrolysis experiments and Mississippian oils from the Fort Worth Basin were used to evaluate C_4-naphthalene and C_4-benzene maturity parameters in natural systems. It was not the goal of this study to determine whether results from pyrolysis of a Devonian oil will compare exactly with oils from a Mississippian source. Due to facies differences between Devonian and Mississippian marine shales, the initial distribution of C_4-naphthalene and C_4-benzene isomers is different as discussed earlier. For this reason, we would also not expect maturity ratios from pyrolysis of a Devonian oil to exactly match maturity ratios from Mississippian Barnett Shale oils. However, if the maturity trends observed from oil pyrolysis results are valid, the C_4-naphthalene and C_4-benzene ratios should correlate with a well established thermal maturity parameter such as TAS in Mississippian Barnett Shale oils.

Calculating an equivalent $\%R_0$ value for pyrolysis experiments based on experimental conditions is a convenient way to compare the level of thermal stress achieved in experiments performed at various temperatures and times and eliminates uncertainty in biomarker maturity indicators that arise during pyrolysis. In this way, the level of thermal stress achieved for an experiment performed at 360°C and 12 days can be easily compared with an experiment performed at 400°C and 1 day, for instance. The positive correlation between various C_4-naphthalene and C_4-benzene ratios with increased thermal stress (calculated $\%R_0$) in the oil pyrolysis experiments motivated the evaluation of these ratios as maturity parameters in oils from the Fort Worth Basin. The Fort Worth Basin oils were analyzed as part of this study because the Barnett Shale is the only petroleum system in the basin and also because samples were readily available from various stratigraphic horizons. The TAS ratio is an accepted thermal maturity indicator for low API gravity oils (MACKENZIE *et al.*, 1981) and was used in this study to evaluate C_4-naphthalene and C_4-benzene as potential maturity indicators for high API gravity oils. Thus, correlation of C_4-naphthalene or C_4-benzene ratios with the TAS maturity ratio is viewed as a confirmation of the effectiveness of these parameters to estimate the thermal maturity of a light crude oil.

Biomarker parameters are particularly useful for evaluating the maturity of oils and early to mid-oil window source rocks, but are of limited use for high maturity condensates and source rocks. PETERS and MOLDOWAN (1993) provide an excellent summary of the utility of various terpane, sterane, aromatic steroid and porphyrin biomarker parameters and the applicable range of thermal maturity. Determining the thermal maturity of oils and condensates is difficult when the limits of biomarker maturity parameters are exceeded. Light hydrocarbons provide an alternative to biomarkers for evaluating thermal maturity of light oils and condensates. HUNT *et al.* (1980) found the ratio of quaternary to tertiary carbon centers in C_7 light hydrocarbons increases with increase in burial depth. SCHAEFER and LITTKE (1988) and SCHAEFER (1992) found a correlation between mean vitrinite reflectance and several C_7 light hydrocarbon ratios in the Toarcian Posidonia shale. THOMPSON (1983) found that several C_6 and C_7 light hydrocarbon ratios correlate with maximum subsurface temperature and vitrinite reflectance. MANGO (1987, 1990) suggested that temperature controls ring opening reactions of cyclopropane in the formation of C_7 isomers. Based on this concept, BEMENT *et al.* (1995) concluded that the 2,4-DMP/2,3-DMP (DMP: dimethylpentane) ratio was source independent and could be used to estimate the temperature of hydrocarbon generation. CHUNG *et al.* (1998) and WALTERS and HELLYER (1998) also concluded the 2,4-DMP/2,3-DMP ratio can be used as a thermal maturity indicator. There are disadvantages to use of light hydrocarbons as thermal maturity indicators. Light hydrocarbons are susceptible to secondary alteration effects such as evaporative loss and biodegradation, limiting the usefulness of these compounds in some situations.

Methylphenanthrene (RADKE et al., 1982a,b) and diamondoid (DAHL et al., 1999) compound distributions have been successfully applied as thermal maturity indicators and are less susceptible to secondary alteration effects. Methylphenanthrenes are less susceptible to volatile loss than the light hydrocarbons and are abundant in oils and source rocks. Methylphenanthrene distribution is most reliably used to evaluate thermal maturity in coals and Type III organic matter, but is of limited use for evaluating source rocks containing Type II organic matter. Diamondoid (saturated hydrocarbons with diamond structure) compound concentration and distribution can be used to assess the extent of oil cracking in natural systems (DAHL et al., 1999) and prove to be extremely useful in evaluating thermal maturity. C$_4$-benzenes and C$_4$-naphthalenes may prove to be a good compliment to diamondoids for assessment of thermal maturity.

The C$_4$-benzenes and C$_4$-naphthalenes are attractive as thermal maturity indicators in light oils and condensates due to their molecular weight (less volatile than C$_6$–C$_7$), boiling points $> C_7$ and abundance (more readily detected than biomarkers in light oils). They are also less biodegradable than C$_6$ and C$_7$ compounds. Compounds in the C$_{10}$–C$_{14}$ range are commonly present in light oils and condensates whereas aromatic steroids (C$_{21+}$), steranes (C$_{27+}$) and hopanes (C$_{32}$) are frequently absent in light oils and condensates or have equilibrated to such an extent that maturity parameters are no longer useful (PETERS and MOLDOWAN, 1993 and references within).

C$_4$-benzene and C$_4$-naphthalene thermal maturity indicators from pyrolysis experiments

Pyrolysis experiments provide a valuable starting point for evaluating and calibrating thermal maturity indicators for sediments, coals and oils. During pyrolysis, temperature, pressure and time are controlled such that changes in compound distributions can be attributed solely to thermal maturity. Compounds that increase in relative abundance with increased maturity are interpreted to be energetically stable compared to compounds that decrease. We targeted C$_4$-benzene and C$_4$-naphthalene isomers whose relative abundances show changes during pyrolysis for use as thermal maturity indicators.

The ratios TeMB-1, TeMB-2, TeMB-3, DMEB-1, DMEB-2 and MiPB correlate linearly with increasing calculated %R_0 in the oil cracking experiments (Fig. 3). The variation in these ratios with increase in thermal maturity demonstrates that there are differences in C$_4$-benzene isomer stability (Table 2). The 1,2,3,5-TeMN and 1,2,4,5-TeMN isomers appear to be more stable than the 1,2,3,4-TeMN isomer. Likewise, the 1,4-DM2EB and the 1-M4iPB isomers appear to be more stable than the 1,3-DM2EB and 1-M3iPB isomers, respectively. The differences in thermal stability may be induced by steric factors although confirmation from molecular mechanics calculations is required. Alternatively, the rates of generation and destruction of these compounds may vary depending on temperature with the compound ratios representing a balance between that generated from precursors and that destroyed by cracking.

Alkylnaphthalene maturity indicators are based on the greater stability of β-methyl over α-methyl substituted isomers (ALEXANDER et al., 1985; VAN AARSSEN et al., 1999). TeMN-1, TeMN-2 and TeMN-4 include 1,3,6,7-isomer whereas TeMN-3 includes the 2,3,6,7-isomer. The 1,3,6,7-isomer (αβββ) is dominantly β-methyl substituted and the 2,3,6,7-isomer is completely β-methyl substituted, making both isomers energetically favorable compared to other alkylnaphthalenes. The potential use of TeMNs as thermal maturity indicators was recognized by VAN DUIN et al. (1997). They used molecular mechanics thermodynamic calculations to predict the equilibrium composition of TeMNs and compared the calculated result to data obtained from experiments performed at 200°C. They found that the 1,3,6,7-TeMN isomer is the most abundant isomer followed by the 1,3,5,7- and 2,3,6,7-isomers under thermal stress of 200°C. The best agreement between experimental and calculated TeMN distributions occurred when entropy was included in the thermodynamic calculations. In our experiments, 1,3,6,7-TeMN was the most abundant isomer under thermal stress around 350–360°C followed by the 1,3,5,7-TeMN and 1,2,6,7-TeMN isomers, whereas 1,3,5,7-TeMN was most abundant at higher thermal stress around 380–400°C followed by 1,3,6,7-TeMN and 2,3,6,7-TeMN isomers. Results from our study are similar to those reported by van Duin *et al.* (1997); differences in experimental TeMN distribution results appear to be due to the higher temperatures (350–400°C) used in our study. Steric hindrance may play a role in 1,3,5,7-isomer stability at higher temperatures as this molecule has no adjacent methyl groups. Thermodynamic calculations might help to answer this question.

TeMN-4 also shows strong correlation with calculated %R_0 up to 1.5%, but decreases at higher calculated %R_0 values. The decrease may be due to lower stability of the 1,3,6,7-isomer at higher levels of thermal maturity or it may result from depletion of a TeMN-4 precursor at high maturity. There appears to be scatter in the TeMN-2 data at the highest %R_0 values which may represent formation–destruction processes of the 2,3,6,7-TeMN isomer. The TeMNs are particularly attractive as maturity indicators because they persist at high experimental temperatures.

C_4-benzene and C_4-naphthalene thermal maturity indicators in Fort Worth Basin oils

To assess applicability of the pyrolysis results to natural systems, C_4-benzenes were quantified and thermal maturity ratios calculated for oils and condensates from the Fort Worth Basin (Fig. 5). The Fort Worth Basin hydrocarbons in this study were generated from a single source, the Mississippian Barnett Shale. The only molecular maturity indicator available for use for these oils and condensates was TAS. The steranes and hopanes had reached equilibrium and even TAS was no longer useful for the highest maturity condensates (Table 2).

Prior to this study, the C_4-benzenes had not been evaluated as potential thermal maturity indicators. The C_4-benzenes in Table 4 were identified in the WCSB oil and all the Fort Worth Basin oils analyzed. The TeMB and MiPB ratios appear to be robust maturity indicators, increasing linearly with increase in TAS for the Fort Worth Basin hydrocarbons. This result corroborates observations from the oil cracking pyrolysis experiments. In general, the DEMB-2 ratio also increases with increase in thermal maturity, although one apparent outlier makes preliminary interpretation of the DMEB-2 data questionable.

C_4-benzene ratios provide a method to determine the relative maturity of light oils and condensates. The range of usefulness for C_4-benzene parameters appears to extend beyond the thermal maturity limits for all the biomarker parameters. Provided the C_4-benzene parameters can be calibrated to another maturity parameter that extends beyond TAS, such as vitrinite reflectance, the C_4-benzenes hold great potential as a thermal maturity indicator.

The TeMN compounds are common hydrocarbons in oils and occur in greater abundance than polycyclic biomarkers. The correlation of TeMN ratios with maturity also extends beyond the range of utility for the biomarkers, including TAS (Table 4). These data show that TeMNs represent another class of compounds that may be used to determine the thermal maturity of light oils and condensates provided TeMN ratios are calibrated to another maturity parameter such as vitrinite reflectance. The variation in TeMN-1, TeMN-2 and TeMN-3 ratios with thermal maturity in Fort Worth Basin oils mirror the pyrolysis results. Changes in these ratios are linear and provide confidence that pyrolysis results are applicable to natural systems. The trend in TeMN-4 ratio with increase in oil maturity in Fort Worth Basin oils is linear up to 1.3% R_0 (Fig. 4), but decreasing as maturity increases to 1.6% R_0 for pyrolysates. Peters *et al.* (1990) reported a similar reversal for sterane maturity parameters from hydrous pyrolysis of the phosphatic member of the Monterey Formation and the Phosphoria Retort shale. The siliceous member of the Monterey Formation did not show the reversal. The maturity parameter reversal was attributed to source rock mineralogy. The phosphatic member of the Monterey Formation and the Phosphoria Retort shale are enriched in carbonate and phosphate minerals whereas the siliceous member of the Monterey is relatively enriched in clay minerals. The TeMN-4 reversal in this study cannot be attributed to differences in mineralogy as the WCSB oil was cracked in the absence of minerals. Since TeMN-4 is the fraction of 1,3,6,7-TeMN measured relative to all TeMN isomers, the reversal is due to a reaction reducing the amount of 1,3,6,7-TeMN or to a reaction(s) forming other TeMN isomers in greater amounts relative to the 1,3,6,7-isomer. It is possible this reversal is an artifact of pyrolysis. In nature, the saturated/aromatic hydrocarbon ratio increases as thermal stress increases whereas in pyrolysis experiments the ratio decreases. The decrease in saturated/aromatic hydrocarbon ratio with increase in thermal stress during pyrolysis is a well-documented artifact of pyrolysis. BEHAR et al. (1999, 2002) demonstrated aromatic compounds are more stable at pyrolysis conditions relative to saturated hydrocarbons, whereas at geologic conditions, saturated hydrocarbons are more stable than aromatic hydrocarbons. It is possible 1,3,6,7-TeMN is less stable than one of the other TeMN isomers at pyrolysis conditions, resulting in the reversal. 1,3,5,7-TeMN is one candidate as it is the most abundant TeMN isomer at high levels of thermal stress during pyrolysis (Table 2). The difference between pyrolysis results and Fort Worth Basin oils requires further testing with additional oils and pyrolysates to determine whether the reversal is always observed.

Understanding the kinetics of TeMN isomer formation versus TeMN isomer degradation is important in understanding TeMN maturity parameters. The concept of the methyl pool was introduced by VAN AARSSEN et al. (1999) to help understand the distribution of methylated aromatic compounds in crude oils. Methylated naphthalene distributions are controlled by 1,2-methyl shifts (isomerisation) and transmethylation reactions (disproportionation) which they propose are mediated by clay catalysis. The methyl pool consists of methyl groups involved in these reactions and is available to every compound in a crude oil. In the context of this study, benzene compounds as well as naphthalene compounds may acquire methyl groups from or contribute methyl groups to the methyl pool. The 1,2-methyl shifts and transmethylation reactions will result in accumulation of the most stable isomers as less stable isomers interact with methyl groups in the methyl pool. This study has not addressed the relative importance of isomerisation versus disproportionation reactions in controlling aromatic compound distribution nor have we explored condensation and other reactions that may rearrange the molecular structure of aromatic compounds in pyrolysates and crude oils.

CONCLUSIONS

Determining the thermal maturity of light oils and condensates is difficult due to limitations of biomarker and light hydrocarbon parameters. C$_4$-benzenes (C$_{10}$) and C$_4$-naphthalenes (C$_{14}$) are abundant constituents in light oils and condensates, present in greater concentrations than biomarkers, less susceptible to biodegradation and less volatile than the C$_6$ and C$_7$ hydrocarbons which have previously been suggested as thermal maturity indicators. Confined, dry pyrolysis experiments performed at 350–400°C, 650 bars and at time periods ranging from 3 to 33 days demonstrate that ratios of C$_4$-benzenes and C$_4$-naphthalenes change systematically with increasing thermal stress. When measured in a group of oils and condensates from the Fort Worth Basin, C$_4$-benzene and C$_4$-naphthalene compound ratios correlate with the triaromatic steroid maturity parameter. C$_4$-benzene and C$_4$-naphthalene maturity parameters extend the range of maturity assessment beyond that possible using biomarkers and provide a complimentary method to C$_6$ and C$_7$ hydrocarbons for measuring light oil and condensate maturity. Of the C$_4$-benzene ratios investigated, TeMB ratios show the strongest correlation with TAS although DMEB-1 and MiPB also show good correlation. Of the C$_4$-naphthalene ratios investigated, TeMN-2 shows the strongest correlation with TAS although TeMN-1, TeMN-3 and TeMN-4 also show good correlations.

Acknowledgements—The authors thank Republic Energy for permission to sample oils from the Fort Worth Basin and publish the results. We also thank Dave King and Mike Pribil of the USGS for assistance in analysis of the Fort Worth Basin Oils. This paper was significantly improved by reviews from Mike Lewan and Ken Peters.

REFERENCES

ALEXANDER R., KAGI R. and SHEPPARD P. (1983) Relative abundance of dimethylnaphthalene isomers in crude oils. *J. Chromatogr.* **267**, 367–372.
ALEXANDER R., KAGI R. and SHEPPARD P. (1984) 1,8-dimethylnaphthalene as an indicator of petroleum maturity. *Nature* **308**, 442–443.

ALEXANDER R., KAGI R., ROWLAND S. J., SHEPPARD P. N. and CHIRILA T. V. (1985) The effects of thermal maturity on distributions of dimethylnaphthalenes and trimethylnaphthalenes in some ancient sediments and petroleums. *Geochim. Cosmochim. Acta* **49**, 385–395.

ALEXANDER R., BASTOW T. P., FISHER S. J. and KAGI R. I. (1993) Tetramethylnaphthalenes in crude oils. *J. Polycycl. Aromat. Comp.* **3**, 629–634.

ALLAN J. and CREANEY S. (1991) Oil families of the Western Canada basin. *Bull. Can. Petrol. Geol.* **39**, 107–122.

BASTOW T. P., ALEXANDER A., SOSROWIDJOJO I. B. and KAGI R. I. (1998) Pentamethylnaphthalenes and related compounds in sedimentary organic matter. *Org. Geochem.* **28**, 585–595.

BEHAR F., BUDZINSKI H., VANDENBROUKE M. and TANG Y. (1999) Methane generation from oil cracking: kinetics of 9-methylphenanthrene cracking and comparison with other pure compounds and oil fractions. *Energy Fuels* **13**, 471–481.

BEHAR F., LORANT F., BUDZINSKI H. and DESAVIS E. (2002) Thermal stability of alkylaromatics in natural systems: kinetics of thermal decomposition of dodecylbenzene. *Energy Fuels* **16**, 831–841.

BEMENT W. O., LEVEY R. A. and MANGO F. D. (1995) The temperature of oil generation as defined with C_7 chemistry maturity parameter (2,4-DMP/2,3-DMP). In *Organic Geochemistry: Developments and Applications to Energy, Climate, Environment and Human History* (eds. J. O. GRIMALT and C. DORRONSORO), pp. 505–507. Selected Papers from the 17th International Meeting on Organic Geochemistry, Donostia-San Sebastián, Spain.

BURNHAM A. K. and SWEENEY J. J. (1989) A chemical kinetic model of vitrinite maturation and reflectance. *Geochim. Cosmochim. Acta* **53**, 2649–2657.

CHUNG H. M., WALTERS C., BUCK S. and BRINHAM G. (1998) Mixed signals of the source and thermal maturity for petroleum accumulations from light hydrocarbons: an example of the Beryl field. *Org. Geochem.* **29**, 381–396.

DAHL J. E., MOLDOWAN J. M., PETERS K. E., CLAYPOOL G. E., ROONEY M. A., MICHAEL G. E., MELLO M. R. and KOHNEN M. L. (1999) Diamondoid hydrocarbons as indicators of natural oil cracking. *Nature* **399**, 54–57.

EXPLORATION STAFF, CHEVRON STANDARD LIMITED, (1979) The geology, geophysics and significance of the Nisku Reef discoveries, West Pembina area, Alberta Canada. *Bull. Can. Petrol. Geol.* **27**, 326–359.

HARTGERS W. A., SINNINGHE DAMSTË J. S. and DE LEEUW J. W. (1994) Geochemical significance of alkylbenzene distributions in flash pyrolysates of kerogens, coals and asphaltenes. *Geochim. Cosmochim. Acta* **58**, 1759–1775.

HOEFS M. J. L., van HEEMST J. D. H., GELIN F., KOOPMANS M. P., van KAAM-PETERS H. M. E., SCHOUTEN S., DE LEEUW J. W. and SINNINGHE DAMSTË J. S. (1995) Alternative biological sources for 1,2,3,4-tetramethylbenzene in flash pyrolysates of kerogen. *Org. Geochem.* **10**, 975–979.

HILL R. J., JENDEN P. D., TANG Y., TEERMAN S. C. and KAPLAN I. R. (1994) Influence of pressure on pyrolysis of coal. In *Vitrinite Reflectance as a Maturity Parameter* (eds. P. K. MUKHOPADHYAY and W. G. DOW). pp. 161–193. ACS Symposium Series 570.

HILL R. J., TANG Y., JENDEN P. D. and KAPLAN I. R. (1996) The influence of pressure on the thermal cracking of oil. *Energy Fuels* **10**, 873–882.

HILL R. J., TANG Y. and KAPLAN I. R. (2003) Insights into oil cracking based on laboratory experiments. *Org. Geochem.* **12**, 1651–1672.

HUNT J. M., HUC A. Y. and WHELAN J. K. (1980) Generation of light hydrocarbons in sedimentary rocks. *Nature* **288**, 688–690.

MACKENZIE A. S., HOFFMAN C. F. and MAXWELL J. R. (1981) Molecular parameters of maturation in the Toarcian shales, Paris Basin, France-III. Changes in aromatic steroid hydrocarbons. *Geochim. Cosmochim. Acta* **45**, 1345–1355.

MANGO F. D. (1987) An invariance in the isoheptanes of petroleum. *Science* **237**, 514–517.

MANGO F. D. (1990) The origin of light hydrocarbon in petroleum: a kinetic test of the steady state catalytic hypothesis. *Geochim. Cosmochim. Acta* **54**, 1315–1323.

PETERS K. E. and MOLDOWAN J. M. (1993) *The Biomarker Guide, Interpreting Molecular Fossils in Petroleum and Ancient Sediments*. Prentice Hall, Englewood Cliffs, NJ, 363 pp.

PETERS K. E., MOLDOWAN J. M. and SUNDARARAMAN P. (1990) Effects of hydrous pyrolysis on biomarker thermal maturity parameters: Monterey Phosphatic and Siliceous members. *Org. Geochem.* **15**, 249–265.

PUTTMANN W. and VILLAR H. (1987) Occurrence and geochemical significance of 1,2,5,6-tetramethylnaphthalene. *Geochim. Cosmochim. Acta* **51**, 3023–3029.

RADKE M., WELTE D. H. and WILLSCH H. (1982a) Geochemical study on a well in the Western Canada Basin: relation of the aromatic distribution pattern to maturity of organic matter. *Geochim. Cosmochim. Acta* **46**, 1–10.

RADKE M., WILLSCH H., LEYTHAEUSER D. and TEICHMÜLLER M. (1982b) Aromatic components of coal: relation of distribution pattern to rank. *Geochim. Cosmochim. Acta* **46**, 1833–1848.

RADKE M., LEYTHAEUSER D. and TEICHMÜLLER M. (1984) Relationship between rank and composition of aromatic hydrocarbons for coals of different origins. *Org. Geochem.* **6**, 423–430.

RADKE M., WELTE D. H. and WILLSCH H. (1986) Maturity parameters based on aromatic hydrocarbons: influence of the organic matter type. *Org. Geochem.* **10**, 51–63.

RADKE M., WILLSCH H. and TEICHMÜLLER M. (1990a) Generation and distribution of aromatic hydrocarbons in coals of low rank. *Org. Geochem.* **15**, 539–563.

RADKE M., GARRIGUES P. and WILLSCH H. (1990b) Methylated dicyclic and tricyclic aromatic hydrocarbons in crude oils from the Handil field, Indonesia. *Org. Geochem.* **15**, 17–34.

RADKE M., RULLKÖTTER J. and VRIEND S. P. (1994) Distribution of naphthalenes in crude oils from the Java Sea: source and maturation effects. *Geochim. Cosmochim. Acta* **58**, 3675–3689.

SCHAEFER R. G. (1992) Zur geochemie niedrigmolekularer Kohlenwasserstoffe im Posidonienschiefer der Hilsmulde. *Erdos Kohle—Erdgas Petrochem.* **45**, 73–78.

SCHAEFER R. G. and LITTKE R. (1988) Maturity-related compositional changes in the low-molecular-weight hydrocarbon fraction of Toarcian shales. *Org. Geochem.* **13**, 887–892.

SCHENK H. J., DI PRIMIO R. and HORSFIELD B. (1997) The conversion of oil into gas in petroleum reservoirs. Part 1: comparative kinetic investigation of gas generation from crude oils of lacustrine, marine and fluviodeltaic origin by programmed-temperature closed-system pyrolysis. *Org. Geochem.* **26**, 467–481.

SINNINGHE DAMSTÉ J. S., KOCK-VAN DALEN A. C. and DE LEEUW J. W. (1988) Identification of long-chain isoprenoid alkylbenzenes in sediments and crude oils. *Geochim. Cosmochim. Acta* **52**, 2671–2677.

SINNINGHE DAMSTÉ J. S., KOCK-VAN DALEN A. C., ALBRECHT P. A. and DE LEEUW J. W. (1991) Identification of long-chain 1,2-di-*n*-alkylbenzenes in Amposta crude oil from the Tarragona Basin (Spanish Mediterranean): implications for the origin and fate of alkylbenzenes. *Geochim. Cosmochim. Acta* **55**, 3677–3683.

SINNINGHE DAMSTÉ J. S., KEELY B. J., SUSANNAH E. B., BAAS M., MAXWELL J. R. and DE LEEUW J. W. (1993) Variations in abundances and distributions of isoprenoid chromans and long-chain alkylbenzenes in sediments of the Mulhouse Basin: a molecular sedimentary record of paleosalinity. *Org. Geochem.* **20**, 1201–1215.

STRACHAN M. G., ALEXANDER R. and KAGI R. I. (1988) Trimethylnaphthalenes in crude oils and sediments: effects of source and maturity. *Geochim. Cosmochim. Acta* **52**, 1255–1264.

THOMPSON K. F. M. (1983) Classification and thermal history of petroleum based on light hydrocarbons. *Geochim. Cosmochim. Acta* **47**, 303–316.

VAN AARSSEN B. G. K., BASTOW T. P., ALEXANDER A. and KAGI R. I. (1999) Distribution of methylated naphthalenes in crude oils: indicators of maturity, biodegradation and mixing. *Org. Geochem.* **30**, 1213–1227.

VAN DUIN A. C. T., BAAS J. M. A., VAN DE GRAAF B., DE LEEUW J. W., BASTOW T. P. and ALEXANDER R. (1997) Comparison of calculated equilibrium mixtures of alkylnaphthalenes and alkylphenanthrenes with experimental and sedimentary data; the importance of entropy calculations. *Org. Geochem.* **26**, 275–280.

VANDENBROUKE M. (1980) Structure of kerogens as seen by investigations on soluble extracts. In *Kerogen: Insoluble Organic Matter from Sedimentary Rocks* (ed. B. DURAND), pp. 415–443. Technip, Paris.

VELLA A. J. and HOLZER G. (1992) Distribution of isoprenoid hydrocarbons and alkylbenzenes in immature sediments: evidence for direct inheritance from bacterial/algal sources. *Org. Geochem.* **18**, 203–210.

VOLKMAN J. K., ALEXANDER R., KAGI R. I., ROWLAND S. J. and SHEPPARD P. N. (1984) Biodegradation of aromatic hydrocarbons in crude oils from the Barrow Sub-basin of Western Australia. *Org. Geochem.* **6**, 619–632.

WALTERS C. C. and HELLYER C. L. (1998) Multi-dimensional gas chromatography separation of C$_7$ hydrocarbons. *Org. Geochem.* **29**, 1033–1041.

WILKES H., BOREHAM C., HARMS G., ZENGLER K. and RABUS R. (2000) Anaerobic degradation and carbon isotopic fractionation of alkylbenzenes in crude oil by sulfate-reducing bacteria. *Org. Geochem.* **31**, 101–116.

Geochemical Investigations in Earth and Space Science: A Tribute to Isaac R. Kaplan
© The Geochemical Society, Publication No. 9, 2004
Editors: R.J. Hill, J. Leventhal, Z. Aizenshtat, M.J. Baedecker, G. Claypool,
R. Eganhouse, M. Goldhaber and K. Peters

Thermal alteration of Cretaceous black shale from the Eastern Atlantic. III: Laboratory simulations

BERND R. T. SIMONEIT[1], K. E. PETERS[2], B. G. ROHRBACK[3], S. BRENNER[4] and I. R. KAPLAN[5]

[1]Environmental and Petroleum Research Group, College of Oceanic and Atmospheric Sciences,
Oregon State University, Corvallis, OR 97331, USA
[2]US Geological Survey, MS 969, 345 Middlefield Road, Menlo Park, CA 94025, USA
[3]Infometrix, Inc., P.O. Box 1528, Woodinville, WA 98072, USA
[4]Department of Geography, Tel-Aviv University, Tel-Aviv, Israel
[5]Institute of Geophysics and Planetary Physics, University of California, Los Angeles, CA 90095, USA

Abstract—Laboratory thermal alteration (pyrolysis) experiments were carried out on composited Cretaceous black shale samples from DSDP Hole 368. Dried rock samples (low water–rock ratio) were heated in an inert atmosphere in the range of 250–500°C and the resultant bitumen and kerogen concentrates were characterized. Pyrobitumen forms initially (250°C), then major hydrocarbon generation occurs from 300–360°C and oxygenated products (*e.g.* alkanoic acids) have a maximum yield at 330–360°C with minor generation to 500°C. Dealkylation of the aromatic hydrocarbons to the parent polynuclear aromatic hydrocarbons occurs above 400°C. The kerogens become more aromatic with increasing pyrolysis temperature (atomic H/C decreases from 1.2 to 0.5 and vitrinite reflectance increases to 3%). The temperature ranges of pyrobitumen, bitumen, and PAH formation in dry pyrolysis experiments are of utility in correlating with field data on ore genesis.

INTRODUCTION

THE ORGANIC-RICH, black shales of Cretaceous age from Deep Sea Drilling Project (DSDP) Site 41-368 in the Atlantic Ocean have been studied extensively to assess the provenance and maturity of the organic matter (*e.g.* BOGOLYUBOVA and TIMOFEEV, 1978; KENDRICK *et al.*, 1977; PETERS *et al.*, 1978, 1983; SIMONEIT, 1977; SIMONEIT *et al.*, 1981). These shales were intruded by diabase sills during the early Miocene. Small quantities of light, extractable hydrocarbons were observed in cores onboard the drilling ship (LANCELOT *et al.*, 1975) and their presence was subsequently confirmed (BAKER *et al.*, 1977; DOW, 1977; SIMONEIT *et al.*, 1978). The geochemical analyses of rocks near these intrusions presented an excellent *in situ* experiment to monitor the thermal generation and expulsion of volatile organic components and the alteration of the residual organic matter (CONKRIGHT and SACKETT, 1992; DENNIS *et al.*, 1982; DOW, 1977; ISHIWATARI *et al.*, 1985; PETERS *et al.*, 1978, 1983; SIMONEIT *et al.*, 1978, 1981; VAN DE MEENT *et al.*, 1980). Similar studies for other shale sequences and coal beds that were intruded by dykes have also been reported (*e.g.* BISHOP and ABBOTT, 1994; CLAYTON and BOSTICK, 1986; FARRIMOND *et al.*, 1996; FREDERICKS *et al.*, 1985; GEORGE, 1992; GILBERT *et al.*, 1985; KHORASANI *et al.*, 1990; MEYERS and SIMONEIT, 1999; PERREGAARD and SCHIENER, 1979; RAYMOND and MURCHISON, 1989, 1992; SAXBY and STEPHENSON, 1987; UJIIÉ, 1986).

The purpose of this study, commenced in 1978, was to subject thermally unaltered rock samples from above and below the sills of DSDP Hole 368 to laboratory thermal alteration (pyrolysis) in order to simulate the natural intrusive event. Since that time petroleum generation and expulsion have been simulated in the laboratory mainly by hydrous pyrolysis (high water to source rock ratio) (*e.g.* LEWAN, 1993, 1997). "Dry" pyrolysis experiments (*i.e.* low water–rock ratio) are generally not applicable to address petroleum generation. However, the impetus of this paper is to better understand geothermal processes involving low water–rock ratios, including the use of organic tracers and organic matter interactions during ore formation (*e.g.* BLUMER, 1975; BROCKS *et al.*, 2003; CHEN *et al.*, 2003; GEISSMAN *et al.*, 1967; PARNELL *et al.*, 1993; PEABODY, 1993; WISE *et al.*, 1986). Two composited samples of shale from DSDP Site 41-368

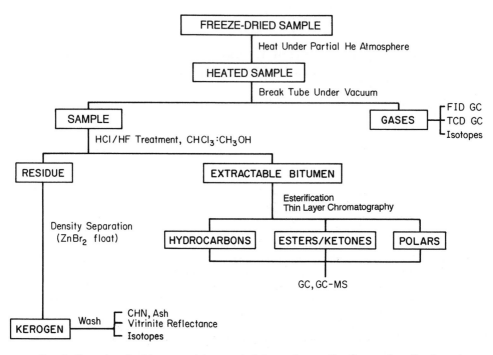

FIG. 1. Flow chart for bitumen and kerogen isolation and separation for samples after thermal alteration.

were subjected to dry pyrolysis (low water–rock ratio) over a temperature range of 250–500°C and the bitumens and kerogens from each step were characterized.

GEOLOGICAL BACKGROUND

Site 41-368 is located on the Cape Verde Rise (17°30.4'N, 21°21.2'W; water depth 3367 m) (LANCELOT *et al.*, 1975). The Turonian–Albian black shales were horizontally penetrated by two thin (~20 cm) and one thick (15 m) diabase sills (NATLAND, 1977). The age of the major intrusive unit was determined by the K/Ar method to be approximately 19 Ma (early Miocene) (DUNCAN and JACKSON, 1977). The Cretaceous section from Site 41-368 consists of black shale with variable organic carbon and carbonate contents (0.7–12% and 1–38%, respectively) due to lithologic differences (*e.g.* DENNIS *et al.*, 1982; SIMONEIT *et al.*, 1981). These values decrease to minima near the intrusions. The thermal stress from the intrusions affected the sediment to about 12 m above and 8 m below the major sill.

The samples were obtained from the working cores of the DSDP repository. The samples were freeze-dried, powdered in a glass mortar and pestle, composited, heated, and treated in the same manner for extraction, kerogen isolation, thin-layer chromatography (TLC) of the extracts, and analysis by gas chromatography (GC) and gas chromatography–mass spectrometry (GC–MS) as reported earlier (SIMONEIT *et al.*, 1981; Fig. 1).

EXPERIMENTAL METHODS

Heating experiments

The heating experiments were performed on a set of samples (10 g of composited rock each) sealed in Pyrex tubes after three successive evacuations and flushings with purified helium to remove all traces of air. The sealed tubes with sample and about 0.2–0.7 bar helium were then heated at the respective temperatures in the range of 200–500°C for 48 h. The warm-up time of

Table 1. Analytical results for the bitumen extracted from the unheated and heated rock samples

Sample	Total C_{org} (%)[a]	Total extract bitumen (μg/g)	Hydrocarbons Total (μg/g)	n-Alkanes CPI	n-Alkanes C_{max}[b]	Pr/Ph	Total esters, ketones, PAH (μg/g)	n-Alkanoic acids CPI[c]	n-Alkanoic acids C_{max}[b]
58/59-2/3/4/3 (composite)[d]									
1 Unheated	2.33 (2.70)	1930	680	1.41	**Ph, 31**	0.68	290	3.18 (2.65)	18, **30**
2 250°C	n.d.	2100	760	1.08	**Ph, 31**	0.84	400	2.39 (3.44)	16, **24**
3 300°C	n.d.	2710	1070	1.29	**15**, 29	1.8	490	1.72 (2.40)	**16**, 24
4 330°C	n.d.	2880	1220	1.06	**19**, 23	2.3	310	1.61 (1.40)	16
5 360°C	n.d.	5540	2830	0.91	16	–	770	1.57 (2.0)	10
6 400°C	n.d.	180	60	0.90	18	–	22	4.8	16
7 500°C	n.d.	75	18	0.85	16	–	4	13	16
63-3/4 (composite)[d]									
8 Unheated	7.31 (8.09)	7000	2600	1.66	**Ph, 17**	0.52	670	2.2–2.5	**18**, 24
9 250°C	n.d.	6800	2260	1.29	**Ph, 17**	0.9	700	2.01 (3.15)	**16**, 24
10 300°C	n.d.	8460	3400	1.18	**17**, Ph	0.95	810	1.82 (2.20)	**16**, 22
11 330°C	n.d.	18,900	3270	1.07	17	1.4	440	1.57 (1.30)	16
12 360°C	n.d.	6060	3800	1.05	15	2	200	1.45	16
13 400°C	n.d.	4350	2320	0.86	16	–	180	3.6	16
14 500°C	n.d.	1220	840	–	–	–	80	11	16

[a] Total carbon content given in parentheses.
[b] Dominant C_{max} is in bold.
[c] CPI for C_7–C_{20} and in parentheses for C_{20}–C_{36} if present.
[d] Samples were composited from cores 58 sections 2–4 and 59 section 3 from 921–948 mbsf (+9 to +35 m above the sill) and from core 63 sections 3, 4 and cc from 980.5 to 984.5 mbsf (−9 to −13 m below the sill) (for depth convention see SIMONEIT et al. (1981) and PETERS et al. (1983)).

the oven was about 15 min. The tubes were opened, contents removed for extraction and rinsed with extraction solvent which was combined with the associated sample extract.

Extraction and separation

After carbonate removal using 6N HCl the samples were extracted with chloroform and methanol (1:1) (Fig. 1; SIMONEIT *et al.*, 1981). Extract concentrates were separated by TLC on silica gel after esterification of the carboxylic acids (Fig. 1). Fractions corresponding to hydrocarbons, carboxylic acid esters, ketones and polar compounds were isolated for instrumental analysis (SIMONEIT *et al.*, 1981). The amounts of material in the hydrocarbon, ester, and ketone fractions from the TLC separations were determined by weighing replicate aliquots of a few microliters each on a Cahn microbalance and, after averaging, calculating the yields.

Kerogen concentration

Some of the extracted sediments were air-dried and checked for the presence of humic substances by leaching with 0.2N NaOH. Because the amounts of humic substances were negligible (SIMONEIT *et al.*, 1981), this step was not necessary prior to the isolation of kerogen. The samples were treated with a 1:1 mixture of 6N HCl and 60% HF (Baker analyzed reagent) in Teflon extraction vessels. Following brisk stirring with a Teflon stirring bar and settling, the clear supernatant was aspirated away. This process was repeated with increasing HF concentration, until 60% HF was added and no further reaction occurred. Samples were washed with double-distilled water and then mixed in a zinc bromide ($ZnBr_2$) solution of density 2.1 by ultrasonication for 20 min. Separation of the kerogen (float) by centrifugation at 2500 rpm for 20 min was followed by supernatant filtration and washing with doubly distilled water and then doubly distilled methanol through pre-washed glass fiber filters (55 mm Whatman Millipore GF/A). Finally, the kerogen was freeze-dried, extracted with chloroform, dried, and weighed.

Instrumental analysis of bitumen fractions

All GC analyses were carried out using a Hewlett Packard Model 5830A gas chromatograph equipped with a linear temperature programmer, flame ionization detector (FID), electronic integrator and fused silica capillary column (30 m × 0.25 mm) coated with DB-5 (J&W Inc). The temperature was programmed from 100 to 290°C at 4°C/min and held isothermal at 290°C for 30 min, with the helium carrier gas at a flow rate of 2.0 ml/min, injector temperature at 290°C, and the detector at 300°C. The concentrations of the resolved normal and isoprenoid alkanes, aromatic hydrocarbons, methyl *n*-alkanoates and *n*-alkanones were calculated using response factors derived from standards. In a few cases, corrections were made by comparison of peak heights.

The GC–MS analyses were carried out on a Finnigan Model 4021 quadrupole mass spectrometer interfaced directly to a Finnigan Model 9610 gas chromatograph equipped with a 30 m × 0.25 mm fused silica capillary column (J&W Inc.), which was coated with DB-5. The gas chromatograph for the GC–MS analyses was programmed from 40 to 290°C at 4°C/min, then held isothermal for 80 min, and the He carrier gas was at a flow rate of 2 ml/min. The MS ion source was at 200°C and operated at 70 eV ionization energy. The mass spectrometric data was acquired and processed using a Finnigan-Incos Model 2300 data system

Kerogen analyses

Elemental analyses (C, H, N and ash) were carried out by combustion at 900°C, using a procedure and apparatus similar to that described in STUMP and FRAZER (1973). Carbon dioxide, nitrogen and hydrogen were purified, measured volumetrically, and sealed in Pyrex tubes for isotopic analysis. The sulfur content was determined by Galbraith Laboratories, Knoxville, Tennessee. The high sulfur values may in part reflect inorganic sulfide decomposition during pyrolysis (Table 2).

Table 2. Analytical results for the kerogen concentrates from the unheated and heated samples

Sample	Temperature (°C)	C_{org} (%)	H (%)	N (%)	S (%)	Ash (%)	H/C	C/N	$\delta^{13}C$ (‰, PDB)[a]	Vitrinite reflectance (R_0 mode %)[a]	S1 (HC, mg/g rock)	S2 (HC, mg/g rock)	HI (mg HC/g C_{org})	S3 (CO_2, mg/g rock)	T_{max} (°C)
1[b]	Unheated	30.10	2.86	1.16	n.d.	n.d.	1.14	30	−24.82	0.40 ± 0.3	2.1	88	292	12	428
2	250	31.08	3.23	1.06	25.02	40.5	1.25	34	−24.55	0.27 ± 0.04	0	168	540	13	431
3	300	30.93	3.13	1.07	21.76	44.0	1.21	33	−24.53	0.34 ± 0.04	0	112	362	25	436
4	330	40.91	3.21	1.25	25.80	35.0	0.94	38	−24.24	0.39 ± 0.05	0	3	7	1.4	442
5	360	44.26	2.95	1.35	n.d.	n.d.	0.80	38	−24.25	0.77 ± 0.09	0.1	2	5	1	448
6	400	35.58	2.10	1.17	18.71	31.9	0.71	36	−23.96	2.06 ± 0.13	0.2	0.1	0.3	16	549
7	500	37.16	1.92	0.98	11.76	35.0	0.62	37	−23.84	3.24 ± 0.28	0	1	2.7	1	550
8[b]	Unheated	59.10	5.98	2.11	13.26	14.2	1.22	32	−27.94	0.48 ± 0.11	1.7	359	607	20	426
9	250	52.79	5.52	2.02	10.82	20.8	1.26	30	−27.81	0.23 ± 0.03	0	551	1044	22	428
10	300	53.27	5.31	2.11	10.86	26.1	1.20	29	−27.76	0.29 ± 0.05	0	304	570	27	435
11	330	62.77	5.81	2.31	11.26	12.2	1.11	32	−27.68	0.32 ± 0.06	0.2	27	43	2.6	438
12	360	63.60	4.76	2.69	11.59	11.9	0.90	28	−27.39	0.73 ± 0.11	0.1	8	12.6	2	448
13	400	60.55	3.13	2.79	8.13	15.6	0.62	26	−27.06	1.85 ± 0.19	0.1	0.1	0.2	15	549
14	500	57.96	2.36	2.14	4.70	18.6	0.49	31	−26.90	2.95 ± 0.25	0	0.1	0.2	13	549

n.d., not determined.
[a] $n = 2$ for $\delta^{13}C$; $n = 55–61$ particles for R_0.
[b] Unheated sediment composite, depth below seafloor: (1) 921–948 mbsf, (8) 980.5–984.5 mbsf.

Vitrinite reflectance (R_0) was measured on kerogen, which was embedded in epoxy briquettes and polished (BOSTICK and ALPERN, 1977; PETERS *et al.*, 1978). The measurement was carried out under oil immersion on approximately 100 vitrinite particles per sample.

Carbon isotope analyses were carried out by the method described by STUERMER *et al.* (1978), using Chicago PDB as the reference standard. The data are expressed in the conventional δ notation.

Programmed pyrolysis (Rock-Eval) analyses were performed on the kerogen concentrates. The operating parameters for the Rock-Eval II pyroanalyzer (Geocom, Inc.) were similar to those in CLEMENTZ (1979). Samples weighing up to 10 mg were pyrolyzed at 300°C for 4 min, followed by programmed pyrolysis at 25°C/min to 550°C in a helium atmosphere. Several measurements are obtained by this technique (ESPITALIÉ *et al.*, 1977). An FID senses any organic compound generated during pyrolysis. The first peak (S1) represents the milligrams of bitumen that can be thermally distilled from 1 g of sample. The second peak (S2) represents the milligrams of hydrocarbons generated by pyrolytic degradation of the kerogen in 1 g of sample. Carbon dioxide is formed from organic matter and trapped during temperature programming up to 390°C. It is released and analyzed by thermal conductivity detection as a third peak (S3) representing the milligrams of CO_2/g of sample. During pyrolysis the temperature is monitored by a thermocouple. The temperature for the maximum amount of S2 is called T_{max}.

RESULTS AND DISCUSSION

The extract yields of total bitumen and fractions from the unheated and heated rock samples are given in Table 1.

Aliphatic products

The aliphatic hydrocarbons have distributions characteristic of the full range from immature to fully mature to severely cracked over the temperature range of 250–500°C (Fig. 2). The *n*-alkanes of the unheated samples range from C_{14} to C_{36} with CPI values of 1.4–1.7, reflecting the mixed origin of the lipids from terrigenous plant waxes and marine autochthonous sources (SIMONEIT *et al.*, 1978, 1979; SIMONEIT, 1977, 1978, 2000). Rapid maturation and hydrocarbon generation from kerogen occurred from 300 to 360°C based on the high bitumen yields (Table 1). The alkanes were severely cracked to short-chain homologs with a smooth carbon number distribution (CPI = 0.9–1.1) at 400 and 500°C (condensate pattern at 500°C, Fig. 2g). The isoprenoid hydrocarbons range from C_{16} to C_{20}, dramatically decrease from 250 to 330°C, and are not detectable at higher temperatures. The ratios of Pr/Ph increase with increasing pyrolysis temperature (Table 1), consistent with maturation. This was also observed for the field samples altered by the sill intrusion (SIMONEIT *et al.*, 1978, 1981). The hopane and sterane biomarkers have relatively immature distributions in the unheated samples and do not show significant additional maturation with heating (Fig. 3). These biomarkers are destroyed at temperatures above 330°C (hopanes before steranes). Thus, biomarker maturation requires longer time and/or higher water–rock ratios (*e.g.* hydrothermal alteration, SIMONEIT, 2000, and references therein).

The *n*-alkanoic acids in the unheated samples range from C_7 to C_{36} with strong even carbon-number predominances (CPI = 2.2–3.2) and C_{max} at 18 and 24 or 30 (Fig. 4). Their relative concentrations decrease with increasing pyrolysis temperature, but do not go to zero at 500°C, and the CPI and C_{max} values decrease less severely (Table 1). This is interpreted to be due to losses by decarboxylation and cracking and continued generation of *n*-alkanoic acids (especially C_{16} and C_{18}) from the kerogen. The high CPI values for the samples heated to >400°C are due to the formation of mainly the C_{16} and C_{18} alkanoic acids. The hopanic acid biomarkers (analyzed as methyl esters) consist predominantly of the C_{31}–C_{34} 17β(H),21β(H)-22R-homohopanoic acids (C_{max} = 32) with lesser amounts of the 22S and 22R series of C_{31}–C_{33} 17α(H),21β(H)-homohopanoic acids (C_{max} = 32). The αβ series is present for C_{31}–C_{33} (C_{max} = 32) only at 250°C, and at 330°C no hopanoic acids are detectable.

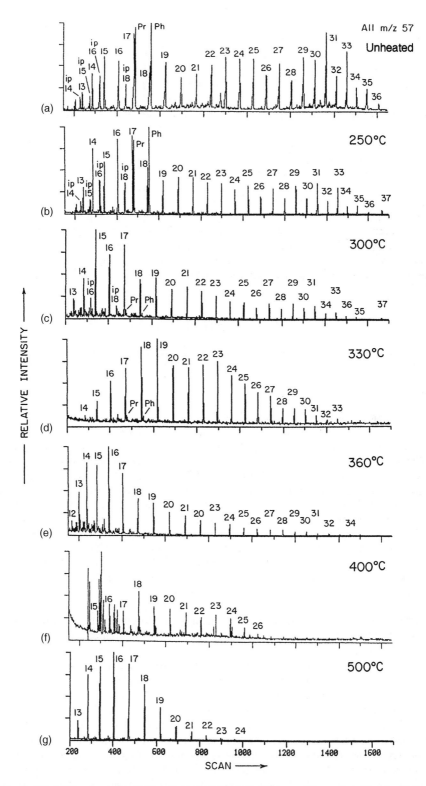

FIG. 2. GC–MS key ion traces for the alkanes (*m/z* 57) in the bitumen of the unaltered and thermally altered samples. Temperatures are shown; amb, source composite (all sample 1); numbers refer to carbon chain length; ip, isoprenoid; Pr, pristane; Ph, phytane.

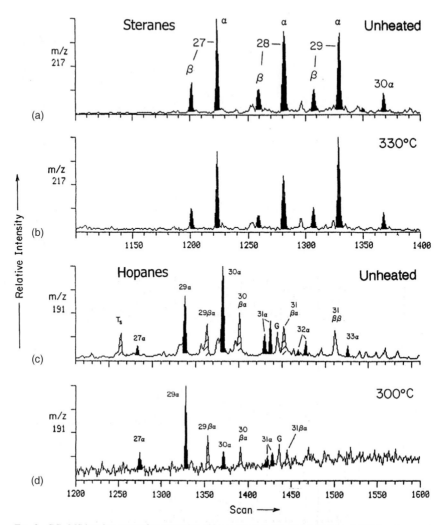

Fig. 3. GC–MS key ion traces for examples of biomarkers in the samples: (a) *m/z* 217 for steranes in the unheated sample 1; (b) *m/z* 217 for steranes in a 330°C pyrolysis sample 3; (c) *m/z* 191 for the triterpanes in the unheated sample 1; and (d) *m/z* 191 for the triterpanes in a 300°C pyrolysis sample 3 (for steranes β and α: 5β,14α,17α and 5α,14α,17α configurations, respectively; for triterpanes α: 17α (H),21β(H)-, βα: 17β(H),21α (H)-, and ββ: 17β(H),21β (H)-hopanes; G: gammacerane).

n-Alkan-2-ones and lesser amounts of *n*-alkan-3-ones occur in the unheated samples and range from C_{11} to C_{22}, with significant amounts of 6,10-dimethyltridecan-2-one and 6,10,14-trimethylpentadecan-2-one. The isoprenoid ketones and the alkan-3-ones are not detectable at 250°C and no ketones are present in the pyrolysis products at 300°C and above. Normal alkanols are not significant components in the unheated or heated samples.

Aromatic products

Alkylated aromatic hydrocarbons are dominant in the unheated samples and are altered to the parent polynuclear aromatic hydrocarbons (PAHs) with limited or no alkylation at higher pyrolysis temperatures. Similar results for aromatic hydrocarbons were reported for laboratory thermal alteration (150–410°C) of kerogen from recent marine sediments (ISHIWATARI and FUKUSHIMA, 1979). The pattern for the alkylnaphthalene series is shown in Fig. 5. Naphthalene is a trace component in the unheated samples, but becomes the major compound at 500°C.

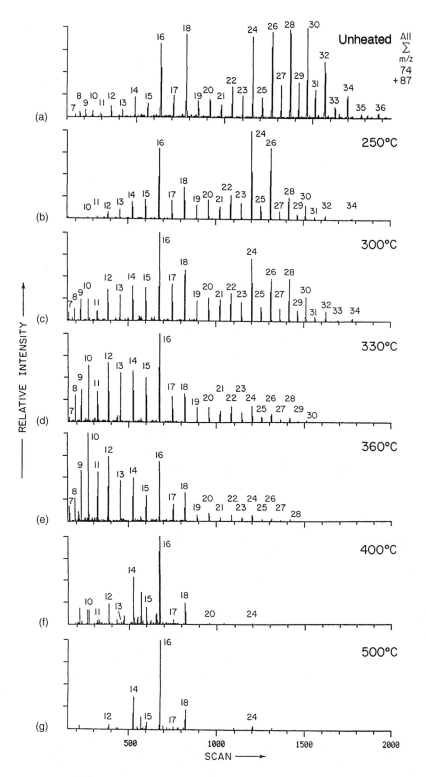

FIG. 4. GC–MS key ion traces for the alkanoic acids analyzed as methyl esters (sum of *m/z* 74 and 87). Pyrolysis temperatures are indicated; unheated, composited sample 1.

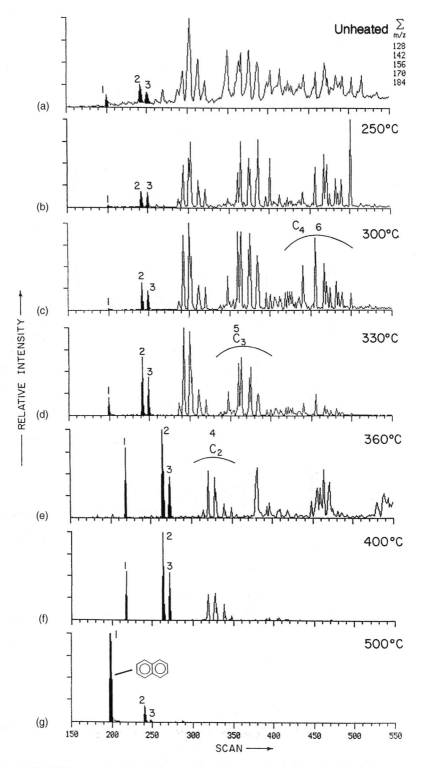

FIG. 5. GC–MS key ion plots for the alkylnaphthalene series in the unheated and pyrolyzed samples 1–7 (sum of *m/z* 128, 142, 156, 170 and 184; pyrolysis temperatures are indicated; 1, naphthalene; 2, 2-methylnaphthalene; 3, 1-methylnaphthalene; 4, dimethylnaphthalenes; 5, trimethylnaphthalenes; 6, tetramethylnaphthalenes and bibenzothiophene).

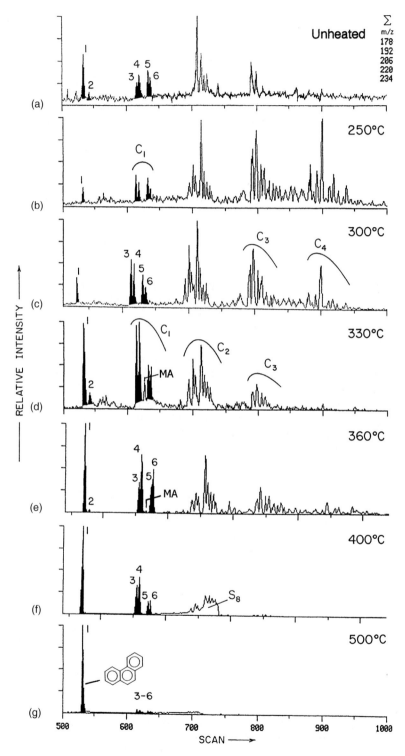

FIG. 6. GC–MS key ion plots for the alkylphenanthrene/alkylanthracene series in the unheated and pyrolyzed samples 1–7. Sum of m/z 178, 192, 206, 220 and 234; pyrolysis temperatures are indicated; 1, phenanthrene; 2, anthracene; 3, 3-methylphenanthrene; 4, 2-methylphenanthrene; MA, 2-methyl-anthracene; 5, 9-methylphenanthrene; 6, 1-methylphenanthrene; C_2, dimethylphenanthrenes; C_3, C_3-phenanthrenes; C_4, C_4-phenanthrenes; S_8, sulfur.

Concomitantly, the C_1–C_{4+} alkylation decreases progressively. The distribution patterns of the alkylphenanthrenes/anthracenes (P/A) are shown in Fig. 6. The unheated sample exhibits a P/A composition of intermediate maturity inherited from the *in situ* geothermal alteration, thus only additional maturation consisting of degradation and formation is observed for these samples. Further alkyl P/A are generated from the kerogen in the range of 250–330°C, which are then cracked to yield mainly phenanthrene at 500°C. Anthracene and methylanthracenes are present in the unheated sample, degrade from 250 to 300°C, are generated around 330°C, and again degrade above that temperature. Both alkylnaphthalene to naphthalene and alkylphenan-threne to phenanthrene ratios, commonly used as maturity indicators for sedimentary organic matter, are based on this dealkylation to the parent PAH with increasing temperature. Thus, degradation and generation of these compounds versus increasing temperature (*i.e.* maturity) must be taken into account with these ratios.

The higher molecular weight PAHs exhibit the same patterns as summarized in Fig. 7 for the m/z 202, 228 and 252 parent PAH compounds. The C_3 and higher alkylation dominance in the unheated sample inverts above about 360°C to the dominance of the $<C_3$ alkylation pattern and to the dominance of solely the parent PAHs at 500°C. PAH degradation to lower analogs (*i.e.* benzopyrene, $C_{20}H_{12}$, to pyrene, $C_{16}H_{10}$) is not a significant process. Thio-PAHs occur in the unheated samples and consist mainly of dibenzothiophene, three naphthobenzothiophene isomers and traces of five dinaphthothio-phene isomers (Fig. 8a). As the pyrolysis temperature increases, additional thio-PAH are generated and at 400°C and above they decrease in concentration and appear to revert to dibenzothiophene (Fig. 8).

The dealkylation of PAHs can be plotted for each temperature as in the example for the phenanthrene series (Fig. 9a), where at low temperatures the parent to C_3 increases with increasing molecular weight and the trend inverts above 360°C, until at 500°C phenanthrene is dominant with no C_2 or C_3 alkylation. Thus, dealkylation of alkyl PAHs to the parent PAHs proceeds rapidly above 400°C. This process can also be followed as a ratio of alkylated homologs to the parent PAH. This is illustrated in Fig. 9b with the ratio of the sum of the C_1-fluoranthenes + C_1-pyrenes over the sum of fluoranthene plus pyrene. There is a steady decrease above 300°C to essentially only parent PAHs at 500°C.

Kerogen alteration

The analytical results for the kerogen concentrates of the unheated and pyrolyzed rocks are given in Table 2. The pyrolysis did not affect the elemental composition (C, H, N, S) in any systematic manner, except the H content decreased with a concomitant increase in C_{org}. This is also reflected in the atomic H/C values, which decrease with increasing temperature, indicating aromatization. The atomic C/N shows no trend, suggesting that N is retained in the organic matter. The high sulfur content (Table 2) may be an artifact and thus no correlations (*e.g.* C/S) were attempted.

The stable carbon isotope data for the kerogen of the unheated and thermally altered samples cluster in two fields when correlated with the organic carbon content prior to demineralization (Fig. 10a). The organic carbon content of samples 1–7 above the sills ranges from 30.1 to 44.3% and the $\delta^{13}C$ values of the kerogen vary from -23.84 to $-24.82‰$, whereas the C_{org} for samples 8–14 below the sills ranges from 52.8 to 63.6% with $\delta^{13}C$ values for the kerogen from -26.9 to $-27.94‰$ (Table 2). The higher organic carbon contents of the samples below the sills may indicate that pyrobitumen is

---

Fɪɢ. 7. Selected GC–MS key ion plots for higher molecular weight alkyl PAH in the pyrolyzed samples (360–500°C): (a–c) sum of m/z 202, 216, 230, 244 for alkyl-fluoranthene (1) and pyrene (2) series; (d–f) sum of m/z 228, 242, 256, 270 for alkyl-benz[a]anthracene (1) and chrysene (2) series; and (g, h) sum of m/z 252, 266, 280, 294 for alkyl-benzo[b, j and k] fluoranthene (1) and benzo[e]pyrene (2) series.

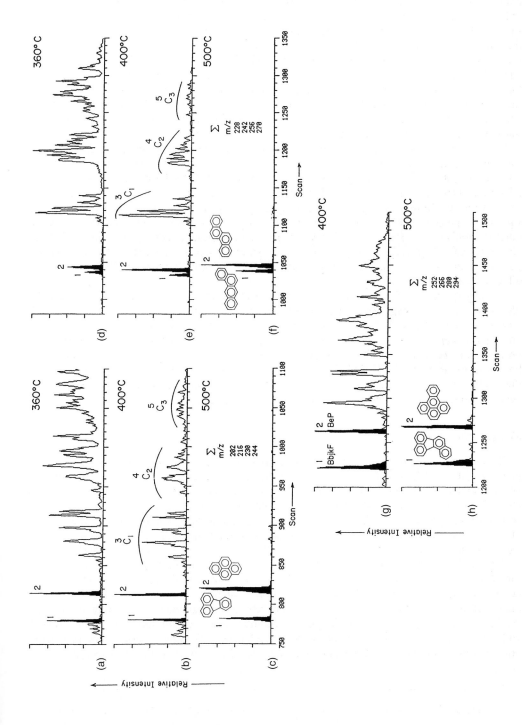

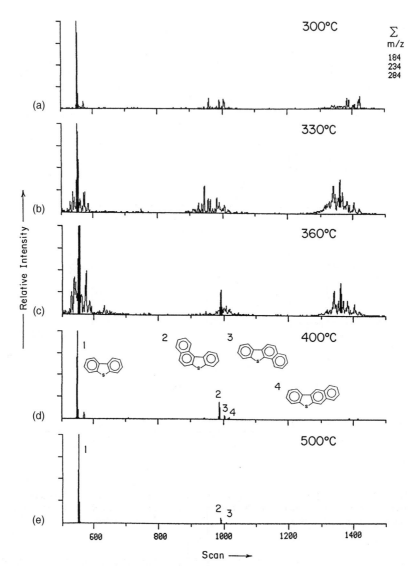

FIG. 8. GC–MS key ion plots for the thio-PAHs in the pyrolyzed samples 2–7. Sum of *m/z* 184, 234, 284; temperatures are indicated; 1, dibenzothiophene; 2, benz[b]naphtho[1,2-d]thiophene; 3, benzo[b]-naphtho[2,1-d]thiophene; 4, benzo[b]naphtho[2,3-d]thiophene; peak cluster at scans 1300–1420, dinaphthothiophenes.

more abundant in the kerogen due to limited bitumen migration and trapping during the intrusive event. The δ^{13}C values suggest a different admixture of terrigenous and marine organic matter in the source sediments above and below the sills. The samples above the sill are more enriched in terrigenous organic detritus, which is also supported by the bitumen data.

The effect of maturation on the stable carbon isotope compositions of the kerogen samples is plotted in Fig. 10b. The offset of about 3‰, due to the different source organic matter above versus below the sills, is evident. There is a progressive increase in δ^{13}C with increasing alteration temperature and the enrichment in ^{13}C is about 1‰. An enrichment of about 2‰ was reported for the *in situ* kerogen samples from DSDP Hole 368 (SIMONEIT *et al.*, 1981) and other investigators have also reported that carbon isotopic alteration of

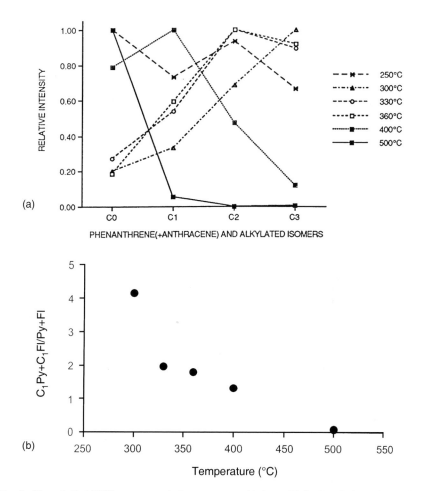

FIG. 9. Plots of alkyl PAHs versus pyrolysis temperature: (a) thermal behavior of phenanthrene plus anthracene and alkylated homologs and (b) ratio of methylpyrenes + methylfluoranthenes to pyrene + fluoranthene versus temperature.

sedimentary organic matter during diagenesis and catagenesis is minor (e.g. REDDING et al., 1980; YEH and EPSTEIN, 1981).

Vitrinite reflectance data (R_0) for the thermally altered samples (Fig. 11) follow the same trend above and below the sill. The R_0 values increase with increasing alteration temperature and plateau above 400°C. The R_0 values range from 0.23 to 3.24%, whereas the R_0 values of the in situ samples range from 0.25 to 4.81% (SIMONEIT et al., 1981; PETERS et al., 1978, 1983). The laboratory alteration temperatures and corresponding vitrinite reflectance values of the kerogens fit with the temperatures inferred from the in situ R_0 data (PETERS et al., 1978). The R_0 values of the unheated rocks combined for the composited samples (1 and 8) were estimated from the data for the in situ samples.

The programmed pyrolysis results from Rock-Eval analysis are given in Table 2. The unheated kerogens have minor S1 (free bitumen) peaks and major bitumen generation (S2) occurs for samples heated to 300°C. T_{max}, the temperature for maximum generation of S2, increases with increasing thermal alteration from 426 to 550°C (Note—the Rock-Eval oven had a temperature limit of 550°C). These values are in the same range as reported for the in situ samples (PETERS et al., 1983). A plot of the atomic H/C versus the hydrogen index (HI expressed as mg HC/g C_{org}) is shown in Fig. 12. The data reflect the thermal maturity of the kerogens and the organic matter transformations to bitumen as described by BASKIN (1997).

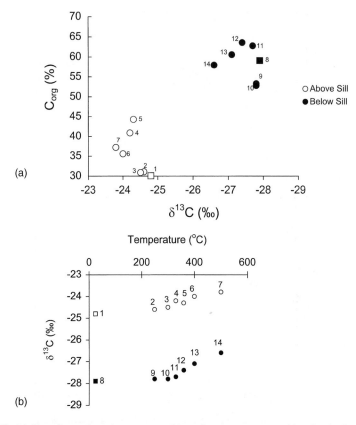

FIG. 10. (a) Plot of organic carbon versus stable carbon isotope composition for the kerogen concentrates of the composited unheated and pyrolyzed samples. (b) Plot of the stable carbon isotope composition versus pyrolysis temperature for the kerogen concentrates (sample numbers *cf.* Tables 1 and 2, samples 1 and 8 are the unheated composites).

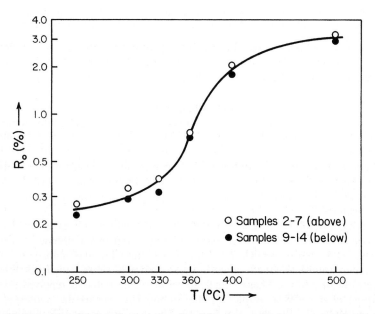

FIG. 11. Plot of the vitrinite reflectance data (log R_0) versus pyrolysis temperature for the kerogen concentrates.

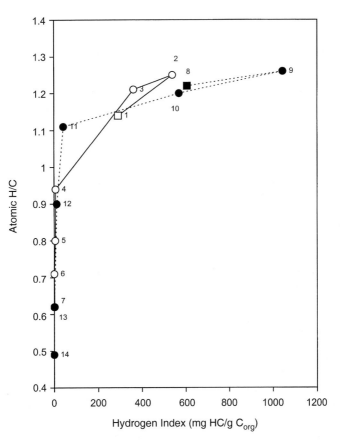

FIG. 12. Plot of atomic H/C versus hydrogen index for the two kerogen concentrate sets (square data points are starting composites 1 and 8; other sample numbers refer to kerogens heated from 250 to 500°C as given in Table 2).

The initial increases of HI and atomic H/C are due to pyrobitumen formation (samples 1, 2, 8, 9). The subsequent progressive decreases in atomic H/C and HI (samples 3–7, 10–13) follow the maturation trend (BASKIN, 1997).

CONCLUSIONS

The effects of laboratory thermal alteration experiments (low water–rock ratio) on the organic matter of composited Cretaceous black shale samples from DSDP Hole 368 are evident in the following parameters: (1) for the hydrocarbons—total yield, formation of PAH and thio-PAH compounds, and increase of Pr/Ph and (2) for kerogen—atomic H/C, HI, δ^{13}C, and vitrinite reflectance (*cf.* Tables 1 and 2).

The total hydrocarbon yield exhibits a maximum at 330–360°C and decreases above those temperatures due to severe cracking (gas formation). The Pr/Ph ratio increases with temperature to a maximum at 330°C and above that temperature the isoprenoid hydrocarbons are destroyed. The 17α-hopane biomarkers are still relatively immature in the unheated samples. They show no further maturation and are destroyed above 360°C. Alkanoic acids have a maximum yield at 330–360°C and are present over the complete temperature range. This indicates continual formation from kerogen and cracking of the products. PAHs increase in concentration with increasing temperature and the alkyl PAHs concomitantly decrease. At 500°C essentially only the parent PAHs remain in the extractable bitumen.

The kerogens become more aromatic in composition with increasing alteration temperature (*i.e.* hydrogen content decreases based on the atomic H/C). The unheated samples have atomic H/C ≈ 1.2, which progressively decreases to 0.5 at 500°C. Pyrobitumen forms initially (250°C) based on the atomic H/C to HI correlation. The $\delta^{13}C$ data reflect the effects of the thermal stress, but only as a 1‰ enrichment in ^{13}C over the temperature range. The thermally induced maturation is also evident in the vitrinite reflectance data, which exhibit a progressive increase to about 3% at 500°C, supporting the increase in aromaticity of the organic matter.

The major effect of "dry pyrolysis" of shales, *i.e.* thermal alteration under low water–rock ratio, is the initial formation of pyrobitumen (250°C), generation of bitumen (300–360°C) and formation of PAHs at high temperatures (>400°C), with concomitant loss of aliphatic components due to cracking to gas. Concurrently, the kerogen remains as a hydrogen-depleted, aromatic residue. The temperature ranges of pyrobitumen, bitumen and PAH formation, with the ratios of saturates/aromatics and PAH/alkyl PAH, are of utility for correlation of fluid temperatures and organic matter interactions during ore genesis (*e.g.* BROCKS *et al.*, 2003). Supergene ore formation models require percolation of oxygenated meteoric water (cold fluid) into weathering sediments, which results in no alteration of the *in situ* organic matter around the ore body. On the other hand, hypogene models involve hydrothermal fluid flow from deeper strata with concomitant alteration and migration of the organic matter in the aureole of the ore deposit (BROCKS *et al.*, 2003). Systematic changes in bitumen composition need to be determined around ore bodies in order to distinguish the supergene or hypogene processes. Ore deposits with associated bitumen, especially Carlin-type gold and sediment-hosted base metal, have approximate formation temperatures of 100–275°C (LEVENTHAL and GIORDANO, 2000). Thus, these compositional data may further refine the temperature windows and fluid interaction processes in ore genesis.

Acknowledgements—We acknowledge the National Science Foundation and the Deep Sea Drilling Project for making the core samples available. We thank Dr Monica A. Mazurek and Dr Vladimir O. Elias for technical assistance, Mr E. Ruth and Dr R. N. Leif for GC–MS data acquisition, and Prof. Zeev Aizenshtat, Dr Arie Nissenbaum and Dr Eli Tannenbaum for their review comments.

REFERENCES

BAKER E. W., HUANG W. Y., PALMER S. E. and RANKIN J. G. (1977) Mass and electron paramagnetic resonance spectrometric analyses of selected organic components of Cretaceous shales of marine origin. *Symposium on Analytical Chemistry of Tar Sands and Oil Shale.* Amer. Chem. Soc., New Orleans, March 20–25, 1977, 739–752.

BASKIN D. K. (1997) Atomic H/C ratio of kerogen as an estimate of thermal maturity and organic matter conversion. *Am. Assoc. Petrol. Geol. Bull.* **81**, 1437–1450.

BISHOP A. N. and ABBOTT G. D. (1994) Vitrinite reflectance and molecular geochemistry of Jurassic sediments: the influence of heating by Tertiary dykes (northwest Scotland). *Org. Geochem.* **22**, 165–177.

BLUMER M. (1975) Curtisite, idrialite and pendletonite, polycyclic aromatic hydrocarbon minerals. Their composition and origin. *Chem. Geol.* **16**, 245–256.

BOGOLYUBOVA L. I. and TIMOFEEV P. P. (1978) Composition of organic matter in "black shales" of the Cape Verde Basin (Eastern Atlantic) and their potential as petroleum source beds. *Lithol. Miner. Resour.* **13**, 519–530.

BOSTICK N. H. and ALPERN B. (1977) Principles of sampling, preparation and constituent selection of sedimentary organic matter. *J. Microsc.* **109**, 41–47.

BROCKS J. J., SUMMONS R. E., BUICK R. and LOGAN G. A. (2003) Origin and significance of aromatic hydrocarbons in giant ore deposits of the late Archean Hamersley Basin, Western Australia. *Org. Geochem.* **34**, 1161–1175.

CHEN J., WALTER M. R., LOGAN G. A., HINMAN M. C. and SUMMONS R. E. (2003) The Paleoproterozoic McArthur River (HYC) Pb/Zn/Ag deposit of northern Australia: organic geochemistry and ore genesis. *Earth Planet. Sci. Lett.* **210**, 467–479.

CLAYTON J. L. and BOSTICK N. H. (1986) Temperature effects on kerogen and on molecular and isotopic composition of organic matter in Pierre Shale near an igneous dyke. *Org. Geochem.* **10**, 135–143.

CLEMENTZ D. M. (1979) Effect of oil and bitumen saturation on source-rock pyrolysis. *Am. Assoc. Petrol. Geol. Bull.* **63**, 2227–2232.

CONKRIGHT M. E. and SACKETT W. M. (1992) Stable carbon isotope changes during the maturation of organic matter. In *Organic Matter: Productivity, Accumulation, and Preservation in Recent and Ancient Sediments* (eds. J. K. WHELAN and J. W. FARRINGTON), pp. 403–414. Columbia University Press, New York.

DENNIS L. W., MACIEL G. E., HATCHER P. G. and SIMONEIT B. R. T. (1982) ^{13}C nuclear magnetic resonance studies of kerogen from Cretaceous black shales thermally altered by basaltic intrusions and laboratory simulations. *Geochim. Cosmochim. Acta* **46**, 901–907.

Dow W. G. (1977) Contact metamorphism of kerogen in sediments from Leg 41 Cape Verde Rise and Basin. In *Initial Reports of the Deep Sea Drilling Project* (eds. Y. Lancelot, E. Seibold, W. E. Dean, L. F. Jansa, V. Eremeer, J. Gardner, P. Cepek, V. A. Krasheninnikov, U. Pflaumann, D. Johnson, J. G. Rankin and P. Trabant), Vol. 41, pp. 839–847. US Government Printing Office, Washington, DC.

Duncan R. A. and Jackson E. D. (1977) Geochronology of basaltic rocks recovered by DSDP Leg 41, Eastern Atlantic Ocean. In *Initial Reports of the Deep Sea Drilling Project* (eds. Y. Lancelot, E. Seibold, W. E. Dean, L. F. Jansa, V. Eremeer, J. Gardner, P. Cepek, V. A. Krasheninnikov, U. Pflaumann, D. Johnson, J. G. Rankin and P. Trabant), Vol. 41, pp. 1113–1118. US Government Printing Office, Washington, DC.

Espitalié J., Madec M. and Tissot B. (1977) Source rock characterization method for petroleum exploration. *Ninth Annual Offshore Technology Conference*, Houston, TX, OTC 2935, 439–448.

Farrimond P., Bevan J. C. and Bishop A. N. (1996) Hopanoid hydrocarbon maturation by an igneous intrusion. *Org. Geochem.* **25**, 149–164.

Fredericks P. M., Warbrooke P. and Wilson M. A. (1985) A study of the effect of igneous intrusions on the structure of an Australian high volatile bituminous coal. *Org. Geochem.* **8**, 329–340.

Geissman T. A., Sim K. Y. and Murdoch J. (1967) Organic minerals. Picene and chrysene as constituents of the mineral curtisite (idrialite). *Experientia* **23**, 793–794.

George S. C. (1992) Effect of igneous intrusion on the organic geochemistry of a siltstone and an oil shale horizon in the Midland Valley of Scotland. *Org. Geochem.* **18**, 705–723.

Gilbert T. D., Stephenson L. C. and Philp R. P. (1985) Effect of a dolerite intrusion on triterpane stereochemistry and kerogen in Rundle oil-shale, Australia. *Org. Geochem.* **8**, 163–169.

Ishiwatari R. and Fukushima K. (1979) Generation of unsaturated and aromatic hydrocarbons by thermal alteration of young kerogen. *Geochim. Cosmochim. Acta* **43**, 1343–1349.

Ishiwatari R., Morinaga S. and Simoneit B. R. T. (1985) Alkaline permanganate oxidation of kerogens from Cretaceous black shales thermally altered by diabase intrusions and laboratory simulations. *Geochim. Cosmochim. Acta* **49**, 1825–1835.

Kendrick J. W., Hood A. and Castaño J. R. (1977) Petroleum-generating potential of sediments from Leg 41, Deep Sea Drilling Project, In *Initial Reports of the Deep Sea Drilling Project* (eds. Y. Lancelot, E. Seibold, W. E. Dean, L. F. Jansa, V. Eremeer, J. Gardner, P. Cepek, V. A. Krasheninnikov, U. Pflaumann, D. Johnson, J. G. Rankin and P. Trabant), Vol. 41, pp. 817–819. US Government Printing Office, Washington, DC.

Khorasani G. K., Murchison D. G. and Raymond A. C. (1990) Molecular disordering in natural cokes approaching dyke and sill contact. *Fuel* **69**, 1037–1046.

Lancelot Y., Seibold E., Cepek P., Dean W. E., Eremeev V., Gardner J. V., Jansa L. F., Johnson D., Krasheninnikov V., Pflaumann U., Rankin J. G. and Trabant P. (1975) The eastern North Atlantic. *Geotimes* **20**, 18–21.

Leventhal J. S. and Giordano T. H. (2000) The nature and roles of organic matter associated with ores and ore-forming systems: an introduction. *Ore Genesis and Exploration: The Roles of Organic Matter*, 9, Rev. Econ. Geol., pp. 1–26.

Lewan M. D. (1993) Laboratory simulation of petroleum formation: hydrous pyrolysis. In *Organic Geochemistry, Principles and Applications* (eds. M. H. Engel and S. A. Macko), pp. 419–442. Plenum Press, New York.

Lewan M. D. (1997) Experiments on the role of water in petroleum formation. *Geochim. Cosmochim. Acta* **61**, 3691–3723.

Meyers P. A. and Simoneit B. R. T. (1999) Effects of extreme heating on the elemental and isotopic compositions of an Upper Cretaceous coal. *Org. Geochem.* **30**, 299–305.

Natland J. (1977) Composition of basaltic rocks recovered at Sites 367 and 368 Deep Sea Drilling Project, near the Cape Verde Islands. In *Initial Reports of the Deep Sea Drilling Project* (eds. Y. Lancelot, E. Seibold, W. E. Dean, L. F. Jansa, V. Eremeer, J. Gardner, P. Cepek, V. A. Krasheninnikov, U. Pflaumann, D. Johnson, J. G. Rankin and P. Trabant), Vol. 41, pp. 1107–1112. US Government Printing Office, Washington, DC.

Parnell J., Kcha H. and Landais P. (eds) (1993) *Bitumens in Ore Deposits*, 520 pp. Springer, Berlin.

Peabody C. E. (1993) The association of cinnabar and bitumen in mercury deposits of the California Coast Ranges. In *Bitumens in Ore Deposits* (eds. J. Parnell, H. Kucha and P. Landais), pp. 178–209. Springer, Berlin.

Perregaard J. and Schiener E. J. (1979) Thermal alteration of sedimentary organic matter by a basalt intrusive (Kimmeridgian Shales, Milne Land, East Greenland). *Chem. Geol.* **26**, 331–343.

Peters K. E., Simoneit B. R. T., Brenner S. and Kaplan I. R. (1978) Vitrinite reflectance–temperature determinations for intruded Cretaceous black shale in the Eastern Atlantic. In *Symposium in Geochemistry: Low Temperature Metamorphism of Kerogen and Clay Minerals* (ed. D. F. Oltz), pp. 53–58. SEPM Meeting, October 5, Los Angeles, 1978.

Peters K. E., Whelan J. K., Hunt J. M. and Tarafa M. E. (1983) Programmed pyrolysis of organic matter from thermally altered Cretaceous black shales. *Am. Assoc. Petrol. Geol. Bull.* **67**, 2137–2146.

Raymond A. C. and Murchison D. G. (1989) Organic maturation and its timing in a Carboniferous sequence in the central Midland Valley of Scotland: comparisons with northern England. *Fuel* **68**, 328–334.

Raymond A. C. and Murchison D. G. (1992) Effect of igneous activity on molecular-maturation indices in different types of organic matter. *Org. Geochem.* **18**, 725–735.

Redding C. E., Schoell M., Monin J. C. and Durand B. (1980) Hydrogen and carbon isotopic composition of coals and kerogens. In *Advances in Organic Geochemistry 1979* (eds. A. G. Douglas and J. R. Maxwell), pp. 711–723. Pergamon Press, New York.

Saxby J. D. and Stephenson L. C. (1987) Effect of an igneous intrusion on oil shale at Rundle (Australia). *Chem. Geol.* **63**, 1–16.

SIMONEIT B. R. T. (1977) Leg 41 sediment lipids, search for eolian organic matter in recent samples and examination of a black shale. In *Initial Reports of the Deep Sea Drilling Project* (eds. Y. LANCELOT, E. SEIBOLD, W. E. DEAN, L. F. JANSA, V. EREMEER, J. GARDNER, P. CEPEK, V. A. KRASHENINNIKOV, U. PFLAUMANN, D. JOHNSON, J. G. RANKIN and P. TRABANT), Vol. 41, pp. 855–858. US Government Printing Office, Washington, DC.

SIMONEIT B. R. T. (1978) The Organic Chemistry of Marine Sediments. In *Chemical Oceanography* (eds. J. P. RILEY and R. CHESTER), 2nd Ed., Vol. 7, pp. 233–311. Academic Press, New York, Chap. 39.

SIMONEIT B. R. T. (2000) Submarine and continental hydrothermal systems—a review of organic matter alteration and migration processes, and comparison with conventional sedimentary basins. In *Ore Genesis and Exploration: The Roles of Organic Matter* (eds. T. H. GIORDANO, R. M. KETTLER and S. A. WOOD). *Rev. Econ. Geol.* **9**, 193–215.

SIMONEIT B. R. T., BRENNER S., PETERS K. E. and KAPLAN I. R. (1978) Thermal alteration of Cretaceous black shale by basaltic intrusions in the Eastern Atlantic. *Nature* **273**, 501–504.

SIMONEIT B. R. T., MAZUREK M. A., BRENNER S., CRISP P. T. and KAPLAN I. R. (1979) Organic geochemistry of recent sediments from Guaymas Basin, Gulf of California. *Deep-Sea Res.* **26A**, 879–891.

SIMONEIT B. R. T., BRENNER S., PETERS K. E. and KAPLAN I. R. (1981) Thermal alteration of Cretaceous black shale by basaltic intrusions in the Eastern Atlantic. II. Effects on bitumen and kerogen. *Geochim. Cosmochim. Acta* **45**, 1581–1602.

STUERMER D. H., PETERS K. E. and KAPLAN I. R. (1978) Source indicators of humic substances and proto-kerogen. Stable isotope ratios, elemental compositions and electron spin resonance spectra. *Geochim. Cosmochim. Acta* **42**, 989–997.

STUMP R. K. and FRAZER J. W. (1973) Simultaneous determination of carbon, hydrogen and nitrogen in organic compounds. *Nucl. Sci. Abstr.* **28**, 7848 p.

UJIIÉ Y. (1986) Contact-metamorphic effect on parameters for kerogen maturation. *Org. Geochem.* **9**, 375–378.

VAN DE MEENT D., BROWN S. C., PHILP R. P. and SIMONEIT B. R. T. (1980) Pyrolysis–high resolution gas chromatography and pyrolysis–gas chromatography/mass spectrometry of kerogens and kerogen precursors. *Geochim. Cosmochim. Acta* **44**, 999–1013.

WISE S. A., CAMPBELL R. M., WEST W. R., LEE M. L. and BARTLE K. D. (1986) Characterization of polycyclic aromatic hydrocarbon minerals curtisite, idrialite and pendletonite using high-performance liquid chromatography, gas chromatography, mass spectrometry and nuclear magnetic resonance spectroscopy. *Chem. Geol.* **54**, 339–357.

YEH H. W. and EPSTEIN S. (1981) Hydrogen and carbon isotopes of petroleum and related organic matter. *Geochim. Cosmochim. Acta* **45**, 753–762.

Geochemical Investigations in Earth and Space Science: A Tribute to Isaac R. Kaplan
© The Geochemical Society, Publication No. 9, 2004
Editors: R.J. Hill, J. Leventhal, Z. Aizenshtat, M.J. Baedecker, G. Claypool,
R. Eganhouse, M. Goldhaber and K. Peters

Vitrinite alteration rate as a function of temperature, time, starting material, aqueous fluid pressure and oxygen fugacity—Laboratory corroboration of prior work

W.G. Ernst[1] and Rafael Ferreiro Mählmann[2]

[1]Department of Geological and Environmental Sciences, Geocorner, Building 320, Room 118,
Stanford University, Stanford, CA 94305-2115, USA
[2]Mineralogisch-Petrographisches Institut, Universität Basel, CH 4056 Basel, Switzerland

Abstract—Kinetic investigations were performed on disaggregated samples of lignite, mainly pure huminite fragments composed of angiosperm and gymnosperm xylite. We conducted experiments at 2.0 kbar aqueous fluid (= total) pressure, and at contrasting oxygen fugacity ranges defined by the hematite–magnetite (HM) and magnetite + quartz–fayalite (FMQ) buffers. Initial heating rates were 4–8°C/min; individual samples were held at 200, 250, 300 and 400°C for 5–204 days. After prolonged heating, mean vitrinite reflectance (VR) R_r values at these temperatures are 0.54, 0.74, 1.10 and 2.25%R_r, respectively. The experimentally determined alteration rate is defined by the expression %$R_r = K_0 t^{0.076}$, where K_0 is a function of temperature, and t is in days. The overall activation energy describing the kinetics of complex devolatilization reactions responsible for increased VR measured in our laboratory experiments is 21.8 ± 0.3 kJ mol^{-1}. Combined with earlier rate studies by Dalla Torre *et al.* in 1997, we conclude that the rate of vitrinite maturation apparently is unaffected by oxidation state and nature of the starting lignitic material (conifer or hardwood). Elevated total aqueous fluid pressure very slightly retards the rate of VR increase. Non-systematic trends are observed for the resinite–exudatinite–bituminite present in the run products. Our new experimental data confirm that VR is chiefly a function of temperature and time. In support of earlier field, theoretical and laboratory studies, the crucial variable that determines VR over geologic time intervals is host-rock temperature.

INTRODUCTION

Vitrinite is a diagenetic product of higher plant debris and is widely dispersed in clastic sedimentary rocks. Ligno-cellulose biopolymers that form during the peat and brown coal stages of burial are altered during further increases in temperature and lithostatic pressure, resulting in changes in the optical properties of the surviving, partly devolatilized organic matter— especially an increase in refractive index, vitrinite reflectance (VR) and anisotropy (Karweil, 1956; Suggate, 1959; Sharkey and McCartney, 1963). Huminite–vitrinite (*sensu stricto* ulminite B or telocollinite) macerals, the principal constituents of most coals, show continuous, temperature-sensitive, progressive evolution of optical properties (Stach *et al.*, 1982). VR measurement represents the most practical quantitative indicator of thermal history of the host rocks (Teichmüller, 1987a,b; Bustin, 1989). It is widely used to evaluate the maturation of hydrocarbon source rocks, the thermal history of sedimentary basins and to define temperatures of diagenesis and low-grade regional + contact metamorphism (*e.g.* Lopatin, 1971; Hunt, 1979; Waples, 1980; Tissot and Welte, 1984; Tissot *et al.*, 1987; Underwood *et al.*, 1999; Petmecky *et al.*, 1999; Ferreiro Mählmann, 2001).

Prior to the formation of a three-dimensional, ordered crystal structure (graphite), organic matter is thermodynamically metastable. Therefore, devolatilization and restructuring with concomitant increase in reflectivity of the remaining vitrinite must be a complex rate process depending on many variables (Franklin, 1951; Teichmüller and Teichmüller, 1954; Waples *et al.*, 1992; Suzuki *et al.*, 1993). As a simplifying approximation for sedimentary rocks, most investigators have disregarded influences of total pressure, aqueous fluid chemistry, total pressure, nature of the precursor organic material, mineralogic substrate and oxidation/ reduction potential. Instead they have employed VR as a direct proxy for temperature

(MCKENZIE, 1981; PRICE, 1983; ROKSANDIC, 1986; BARKER and GOLDSTEIN, 1991; BARKER and PAWLEWICZ, 1986, 1993; see also LAUGHLAND and UNDERWOOD, 1993). But because vitrinite is metastable, its reflectance is a function of reaction kinetics. Studies of natural occurrences have provided quantitative models for the relationship of VR to temperature (e.g. LAW, 1992; SCHEGG, 1994; EWBANK et al., 1995). More realistic empirical and theoretical models of the nature of organic material thermal maturation have considered time as well as temperature (HUCK and KARWEIL, 1955; LOPATIN, 1971; BOSTICK et al., 1979; WAPLES, 1980). The relationship was defined employing first-order Arrhenius reaction approaches by BURNHAM and SWEENEY (1989), SWEENEY and BURNHAM (1990), HUNT et al. (1991) and SUZUKI et al. (1993); see also WAPLES et al. (1992). Experimental studies quantifying the influence of variables such as the lithologic substrate, water content and laboratory control of ionic species + concentrations in aqueous solution, including pH, were undertaken by HUANG (1996), LEWAN (1997) and SEEWALD et al. (2000). Prior attempts to control oxygen fugacity by titration and by addition of hematite to experimental charges, increased the oxidation state of the run material but did not define fO_2 quantitatively.

DALLA TORRE et al. (1997) demonstrated in 2–7 day experiments between 0.5 and 20.0 kbar that high pressures modestly retard VR over the temperature range 200–350°C. All but two of their runs were conducted initially on the dry starting material. That is, H_2O was neither added to nor subtracted from the dry starting material, but volatiles were evolved during the course of their experiments. In this earlier kinetic study, sealed platinum capsules separated the charge from the H_2O pressure medium; but because of molecular dissociation of the pressure-medium water, H_2 was free to diffuse through the container and equilibrate with the run products. Oxygen fugacity was not controlled by DALLA TORRE et al. but was approximately maintained by the nickel-rich alloys employed in fabrication of the pressure vessels at fO_2, values near those defined by the Ni–NiO (intermediate fO_2, indicated as NNO) solid buffer assemblage (EUGSTER, 1957).

Several factors involved in the complex maturation process have yet to be assessed quantitatively. Thus, in the present laboratory study, we investigated the kinetics of vitrinite alteration under controlled conditions in order to re-evaluate the relative importance of temperature versus time over long-duration rate experiments. New kinetic data reported here extend results through comprehensive sets of runs over the temperature range 200–400°C at 2.0 kbar. We used the double-capsule solid buffer technique of Eugster, with oxygen fugacities defined by the hematite–magnetite (high fO_2, denoted as HM; individual runs indicated by H- prefix) and magnetite + quartz–fayalite (low fO_2, denoted as FMQ; individual runs indicated by F- prefix) solid buffers + H_2O. Experiments were run with H_2O pressure equal to total (= lithostatic) pressure. Laboratory temperatures and pressures somewhat greater than typical of petroliferous sedimentary basins were employed in order to achieve measurable rates of vitrinite maturation in reasonable time lengths. Thus results are also applicable to vitrinite-bearing diagenetic and low-grade metamorphic rocks. Most runs were carried out for much longer durations than the DALLA TORRE et al. experiments to obtain better control on the rate behavior (isothermal VR versus time curves). Moreover, we employed two different lignite samples as starting materials in order to determine if the initial nature of the organic matter would appreciably influence the kinetics of vitrinite alteration.

EXPERIMENTAL METHODS

Starting material A was the run material also used by DALLA TORRE et al. (1997). It is a clarite (huminite + resinite) gymnosperm lignite (conifer) from the Frimmersdorf seam a (Miocene), open cut Hambach, Lower Rhine Basin, Germany. Optical properties of the starting material, embedded in epoxy, polished and dried, include $R_r = 0.197 \pm 0.008\%$ (densinite = $0.300\%R_r$). Starting material B was a recently acquired specimen employed for most of the new experiments. It is a huminite (98% huminite) of an angiosperm lignite from the same area, open cut Hambach, Germany, but comes from Frimmersdorf seam b (hardwood xylite, angiosperm facies, Miocene). Optical properties of the telocollinite starting material, embedded in epoxy, polished and dried, include $R_r = 0.187 \pm 0.012\%$ (densinite = $0.303\%R_r$). Both samples

exhibit similar optical properties, but sample B shows a higher gelification grade. DALLA TORRE
et al. (1997) described additional details of experimental material A, the gymnosperm lignite.
Experimental material B, the angiosperm lignite, is a huminite with homogeneous curled-veined
layers detected under the binocular microscope and separated from layers with cellular
structures and tissue structures containing huminite, liptinite (resinite, cutinite, some sporinite
and sclerotinite) and inertinite (mostly semifusinite and micrinite); dispersed exudatinite–
bituminite (bitumen) was also present. Dry samples of both huminite materials were gently
disaggregated to coarse powder (crushed to 1.5 ± 1 mm) in an agate mortar and pestle prior to
loading into inner run capsules.

The inner run capsules, fabricated from seamless $Ag_{70}Pd_{30}$ tubing, were 20 mm in length,
with 3.0 mm outer diameter, and 0.12 mm wall thickness. Each run capsule was acetylene-torch
welded at one end. Approximately 10–15 mg of distilled water was introduced into the capsule
by hypodermic needle, then about 10 mg of granulated lignite was added. Finally, the open end
of the run capsule was crimped and the end folded over so as to effect a strong mechanical seal,
rather than being welded. This technique avoided potential thermal alteration of the charge
during welding, and allowed easier diffusion of hydrogen in order to effect oxygen equilibration
between the charge and the surrounding but spatially isolated buffer. The outer, oxygen buffer
capsules, fabricated from seamless Au tubing, were 32 mm in length, with 5.5 mm outer
diameter, and 0.25 mm wall thickness. Each buffer capsule was welded at one end, to which
30–40 mg of distilled water was added by hypodermic needle; the $Ag_{70}Pd_{30}$ run capsule was
then inserted into the Au buffer capsule and 150–175 mg of solid-buffer powder (HM or FMQ)
was packed around it. Lastly, the open end of the buffer capsule was crimped and the end folded
over to make a good mechanical seal. Under the temperature–pressure conditions of
the experiments and with aqueous fluid present in both capsules, hydrogen diffuses through
the silver–palladium alloy but not through the gold capsule. Due to the molecular dissociation
of H_2O, hydrogen diffusion and equilibration provides fO_2 control for the contents of the run
capsule as defined by the surrounding solid buffer (EUGSTER, 1957). Microscope examination at
the conclusion of each experiment assured that, at the temperatures maintained over the duration
of the runs (up to 6 months), all mineral species were retained in the solid buffers. This
demonstrated the integrity of the capsules and laboratory maintenance of the desired oxygen
fugacity, given the small proportion of the charge relative to the oxygen buffer.

Samples were placed in cone-in-cone "cold-seal" pressure vessels (KERRICK, 1987) connected
to a water pressure reservoir, a Heise bourdon tube gauge, and an air-driven hydraulic pump.
Prior to beginning this study, each autoclave and thermocouple assembly was temperature
calibrated against the melting point of NaCl at 1 atm. Thermal gradients in the pressure vessels
were plumbed in order to identify the hot spot for location of the run capsule. After
pressurization of an experimental sample to 2.0 kbar aqueous fluid pressure, the vessel was
placed in a nichrome-wound resistance furnace and externally heated to the desired temperature
employing an electronic proportional temperature regulator connected to a chromel–alumel
sensing thermocouple. The sample temperature was read from the emf (voltage) of a run
thermocouple inserted in an external well in the hot end of the autoclave. Heating and
cooling times for individual experiments were about 30–75 min (approximately 4–8°C/min);
these times are quite short compared to the duration of individual isothermal experiments
(5–204 days). Physical conditions were monitored daily, but in a few lengthy experiments,
only weekly. Temperatures and pressures reported are considered accurate to ±2°C and
±0.02 kbar, respectively.

At the conclusion of each run, the oxygen buffer was examined petrographically to ensure the
presence of all members of the solid buffer assemblage. In some experiments, a fetid odor was
detected on opening the run capsule. The vitrinite sample was removed—but not washed—and
dried at 105°C for about 10 min at atmospheric pressure.

OPTICAL ANALYSIS

Resin-mounted crushed run products were polished using 0.5 μm powder and felted
wool for the final polishing step. Mean random VR ($\%R_r$) was measured under orthoscopic

monochromatic light (546 nm) according to standard procedures (STACH *et al.*, 1982; DALLA TORRE *et al.*, 1997). One hundred telocollinite-measuring areas of 2 μm diameter and homogeneous reflectance were used to obtain a mean value. In cases of incipient bireflectance (anisotropy), maximum ($\%R_{max}$) and minimum reflectance ($\%R_{min}$) were determined under polarized light. Most samples yielded 100 measuring points on average, measured with a Prior (UK) point counter at a spacing of 0.5 mm. We attempted to measure at least 50 or more telocollinite values. Data involving fewer than 50 measuring points were not used for the rate study. Values with a standard deviation of >0.05% and a relative variation-coefficient of >0.005% were also not used in the kinetic analysis. The full data set describing optical properties of the run products are presented in Appendix A (including nine incompletely characterized samples).

MICROSCOPIC RESULTS

Gelified huminite macerals are present in the original starting materials. At low temperatures (200 and 250°C) and short run lengths (<100 days at 200°C and <20 days at 250°C), a volumetric increase of gelified particles was readily detected in both HM and FMQ buffered experiments. Probably part of the telocollinite reacted with moisture in the lignite sample, producing a higher VR of gelocollinite than observed in the non-gelified telocollinite. In some samples, areas with lower and others with higher grades of gelification were found, indicating a bimodal distribution of the measurement histogram and crushed particles with different VR values. Strong gelification is documented by an increase of VR versus lower values of structured telocollinite, becoming similar to corpocollinite and gelocollinite. For longer runs and higher temperatures, gelified and non-gelified particles show the same rank. Also at low temperatures and short run durations, some samples contain particles with a lighter rim and a darker core, or a dark core surrounded by a light mantle (possibly an oxidization effect), and at the margin a dark rim caused by high relief (a polishing problem). We conclude that the heating time may not be sufficient to establish stable conditions, and the maturation trend documents disequilibrium among the particles subjected to both oxygen fugacity ranges (samples H-2, H-20, F-20, H-24, F-24). In these cases, the maturity of huminite may be overestimated. Such samples with a relatively high VR do not fit with the proposed maturation trends. In the same experimental time and temperature range, another disturbing factor regarding the alteration is well demonstrated in our run products. In some samples, resinite is observed (0–4 vol.%), which at higher temperatures and heating times is altered to solid exudatinite–bituminite. At temperatures of 200, 250 and 300°C for short runs, some resinite particles initially show micro-pore formation, followed by granular disintegration (a coking effect) and also impregnation of the huminite by resinite (secondary bituminite or exudatinite), causing a strong darkening around the lumens (Plate 1A). Because original liptinite and experimentally slightly altered huminite consists of resinite filling the plant lumens (Plate 1C), we conclude that the resinite was injected into the lumen walls. In these samples, a general VR suppression is obvious, as indicated by poor fit with the visually estimated rate curves shown in Figs. 1–4. For these parts of the sample, Raman spectroscopy gives broad peaks typical for bituminite and resinite (research in progress by the second author). Thus, granular disintegration and resinite impregnation are responsible for the scattered maturation trends in the 200 and 250°C isothermal experiments. At 300°C for heating times >20 days, and at all 400°C experiments, lumens of the cellular tissue are empty, probably reflecting resinite and to some extent corpocollinite devolatilization.

With increasing temperature, the huminite exhibits more uniform reflectance, and the different huminite macerals cannot be distinguished; moreover, the huminite shows higher polishing relief at the rims. With increasing temperature and time, a micro-porous habit makes it nearly impossible to find easily measured homogeneous areas. Strong decomposition-produced material with a partly granular habit (Plate 1B) was difficult to measure because the grains are smaller than the measurement area of the photo-multiplier.

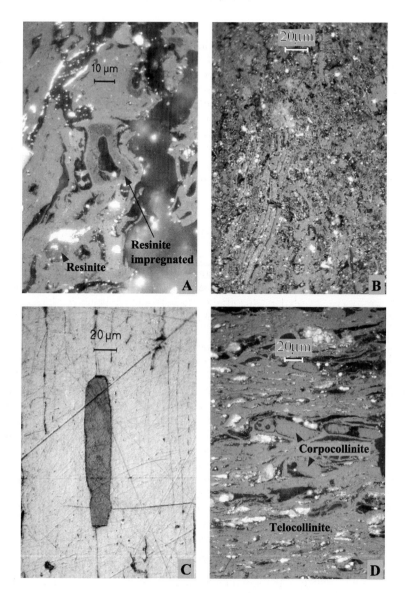

PLATE 1. (A) Liptinite impregnation of huminite (vitrinite) around mostly open lumens initially filled by resinite. Huminite impregnation caused dramatic lowering of the huminite reflectivity. (B) Granular decomposition of different huminite macerals at high experimental temperatures. (C) Resinite-filled lumen in telocollinite and gelocollinite. (D) Homogeneous mass of huminite around a corpocollinite. After long heating times and at high temperatures, all huminites exhibit the same organic matter reflectivity (vitrinite reflectance).

The HM experiments at 200°C for 22, 33, 55 and 71 days, and both the HM and FMQ VR data for the 173 day runs, were lowered by liptinite impregnation; the latter was caused by a dispersed volatilized, later condensed resinite, or by coal tar from hydrous pyrolysis, causing a strong bimodal reflectance distribution. In resinite-rich sample H-9, some slightly gelified and apparently non-impregnated huminite particles show a higher mean VR (fitting better with the rate curve presented in Fig. 1) of $0.53\%R_r$, $n = 21$. In short runs, rare resinite occurrences are well preserved (Plate 1C), and are only characterized by an increase of micro-pores with longer heating times. In runs > 33 days, resinite is partly disintegrated and has a granular habit. In HM samples H-24 and especially in H-20, strong oxidation (possibly due to rupture of the run

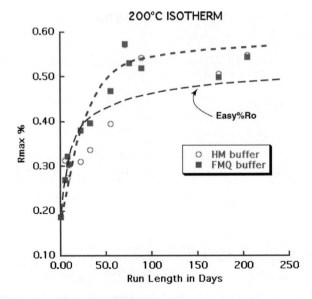

FIG. 1. Vitrinite reflectance measured on lignite B as a function of experimental run duration at 200°C, 2.0 kbar aqueous fluid pressure, with fO_2 defined by the hematite–magnetite and magnetite + quartz– fayalite solid buffers (data from Table 1 and Appendix A). The laboratory rate curve is visually estimated; the approximate quasi-threshold value for $\%R_r$ is 0.54. Also illustrated in the computed EASY$\%R_0$ model for vitrinite maturation at 200°C. EASY$\%R_0$ rate model curves agree with experimentally determined rates for the 250 and 300°C isotherms, but are low for the 200°C and slightly high for the 400°C runs (J. J. SWEENEY, personal communication, 01/10/03).

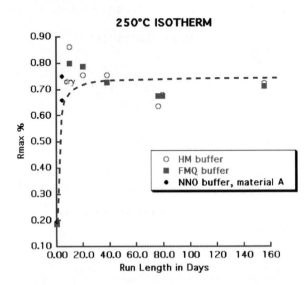

FIG. 2. Vitrinite reflectance measured on lignite B as a function of experimental run duration at 250°C, 2.0 kbar aqueous fluid pressure, with fO_2 defined by the hematite–magnetite and magnetite + quartz– fayalite solid buffers (data from Table 2 and Appendix A). The laboratory rate curve is visually estimated. The 1.0 kbar data of DALLA TORRE et al. (1997) for two runs on material A with fO_2 approximated by the nickel–nickel oxide buffer are also shown (initially dry experiment indicated). The approximate quasi- threshold value for $\%R_r$ is 0.74.

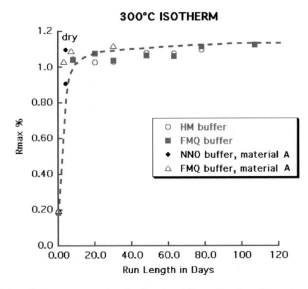

FIG. 3. Vitrinite reflectance measured on lignites A and B as a function of experimental run duration at 300°C, 2.0 kbar aqueous fluid pressure, with fO_2 defined by the hematite–magnetite and magnetite + quartz–fayalite solid buffers (data from Table 3 and Appendix A). The laboratory rate curve is visually estimated. The 1.0 kbar data of DALLA TORRE *et al.* (1997) for two runs on material A with fO_2 approximated by the nickel–nickel oxide buffer are also shown (initially dry experiment indicated). The approximate quasi-threshold value for $\%R_r$ is 1.10.

capsule?) is observed along pathways of probable reactive fluid expulsion, causing high VR values, with maxima of 0.72 and 0.85% R_r. Less altered and homogeneous parts exhibit a lower VR. Disequilibrium conditions and bitumen are responsible for the strong VR scatter; the reaction appears to be virtually of zero order.

Short 250°C runs show strong gelification; different huminite macerals were indistinguishable, in some samples from the very beginning of the reaction. Resinite shows degassing pores,

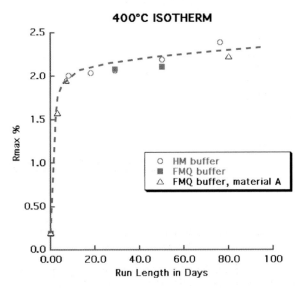

FIG. 4. Vitrinite reflectance measured on lignite samples A and B as a function of experimental run duration at 400°C, 2.0 kbar aqueous fluid pressure, with fO_2 defined by the hematite–magnetite and magnetite + quartz–fayalite solid buffers (data from Table 4 and Appendix A). The laboratory rate curve is visually estimated; the approximate quasi-threshold value for $\%R_r$ is 2.25.

very fine shrinkage fissures and some granular decomposition in all samples. In the longest runs, the granular decomposition of corpocollinite and vesicles in gelocollinite was also observed. In short runs (H-10, F-10, H-14 and F-14), less gelified areas have a characteristic VR value of about 0.75. Samples heated for 76 and 79 days are relatively rich in bituminite, and histograms are characterized by a bi- to trimodal distribution. In sample H-21, resinite-free grains have a mean value of $0.71\%R_r$ ($n = 14$), probably documenting the synthetic maturation. A relatively stable value apparently is reached after a few hours and does not change significantly over 155 days, but these findings are not well substantiated.

In short maturation runs <30 days at the 300°C isotherm, different huminite macerals are clearly recognized. Long heating times bring reflectance values into line, and huminite macerals are transformed into a homogeneous mass (the same is evident for all 400°C runs). Only corpocollinite is different, due to its characteristic rounded form (Plate 1D). In runs >80 days, micro-pores were formed; samples H-25 and F-25 are completely porous, and an accurate measurement therefore was impossible. The relief causes a lowering of VR by $0.30\%R_r$ (very few points show a value around $1.11\%R_r$). In samples F-25, H-25, F-13 and H-13, the first appearance of faint bireflectance indicates an incipient anisotropy of huminite–vitrinite (for F-13 and H-13, maximum $R_{max} = 1.15-1.22\%$).

All 400°C runs show vitrinite anisotropy. Bireflectance increases from 0.10 to $0.25\%R_{max}$.- Initially, anisotropy is undulatory, but becomes more homogeneous on longer experiments. Run products of samples F-29, F-28 and F-26 were completely decomposed to a fine submicroscopic granular mass, so accurate optical measurements were not possible. Complete decomposition also took place on run H-26, so the breakdown may not have been the result of low oxidation state; moreover, high rather than low fO_2 values would be expected to favor preferential decomposition of the organic matter. Coking is restricted to three FMQ samples, and may be an effect of different amounts of dissolved water in the organic matter resulting from accidental capsule leakage. Moreover, in the HM samples, less intensive localized granular decomposition was found in only a few grains. The more homogenous particles were used for the rate data presented in the figures.

RESULTS OF THE RATE STUDY

Results of isothermal kinetic experiments at 2.0 kbar aqueous fluid pressure, employing HM and FMQ oxygen buffers at 200, 250, 300 and 400°C, after initial heat up at about 4–8°C/min are presented in Tables 1–4, respectively. With increasing temperature, optical anisotropy increases (SHARKEY and MCCARTNEY, 1963), so we report the mean R_{max} values in Tables 1–4 (see also Appendix A). In addition to the new experiments on lignitic materials A and B, two pairs of initially dry starting material run at an external fluid pressure of 1 kbar, and unbuffered—but with fO_2 approximately defined by the NNO buffer—from the earlier research by DALLA TORRE et al. (1997) are also tabulated. The run data are illustrated in Figs. 1–4. Earlier, 2–7 day experiments of DALLA TORRE et al. (1997) coincide with the

Table 1. Kinetic study of vitrinite maturation, 2.0 kbar fluid pressure, 200°C

Material B, HM buffer			Material B, FMQ buffer		
Run no.	Length (days)	$\%R_{max}$	Run no.	Length (days)	$\%R_{max}$
H-1	204	0.55 ± 0.03	F-1	204	0.55 ± 0.02
H-2	89	0.54 ± 0.02	F-2	89	0.52 ± 0.02
H-3	9	0.30 ± 0.01	F-3	9	0.31 ± 0.02
H-4	55	0.40 ± 0.01	F-4	55	0.47 ± 0.02
H-5	22	0.31 ± 0.02	F-5	22	0.38 ± 0.01
H-6	8	0.27 ± 0.01	F-6	8	0.32 ± 0.03
H-7	33	0.34 ± 0.01	F-7	33	0.40 ± 0.02
H-8	5	0.31 ± 0.01	F-8	5	0.27 ± 0.01
H-9	173	0.51 ± 0.02	F-9	173	0.50 ± 0.03
H-24	71	0.57 ± 0.08	F-20	76	0.53 ± 0.02
			F-24	71	0.57 ± 0.01

Table 2. Kinetic study of vitrinite maturation, 2.0 kbar fluid pressure, 250°C

Run no.	Length (days)	$\%R_{max}$
Material B, HM buffer		
H-10	10	0.86 ± 0.02
H-14	20	0.75 ± 0.03
H-16	38	0.75 ± 0.03
H-18	79	0.68 ± 0.03
H-19	155	0.72 ± 0.04
H-21	76	0.64 ± 0.03
Material B, FMQ buffer		
F-10	10	0.80 ± 0.03
F-14	20	0.79 ± 0.03
F-16	38	0.73 ± 0.04
F-18	79	0.68 ± 0.03
F-19	155	0.71 ± 0.03
F-21	76	0.68 ± 0.04
Material A, NNO buffer[a]		
1.0 kbar	4	0.66
1.0 kbar, dry	4	0.75

[a] DALLA TORRE et al. (1997).

short-duration portions of the new rate curves. Inspection of the figures evokes the following conclusions:

(1) At constant temperature, rates of vitrinite alteration under contrasting oxidation/reduction conditions (HM and FMQ oxygen buffers) are indistinguishable within the limits of error for individual experiments. Provided buffer and charge equilibrated through reaction with H_2, oxidation state within this fO_2 range evidently does not play an important role in the kinetics of vitrinite maturation.

Table 3. Kinetic study of vitrinite maturation, 2.0 kbar fluid pressure, 300°C

Run no.	Length (days)	$\%R_{max}$
Material B, HM buffer		
H-11	78	1.10 ± 0.03
H-12	8	1.05 ± 0.05
H-13	107	1.13 ± 0.04
H-15	20	1.03 ± 0.04
H-17	48	1.08 ± 0.03
H-22	63	1.08 ± 0.04
H-23	30	1.03 ± 0.02
Material B, FMQ buffer		
F-11	78	1.12 ± 0.04
F-12	8	1.04 ± 0.02
F-13	107	1.13 ± 0.05
F-15	20	1.08 ± 0.01
F-17	48	1.07 ± 0.04
F-22	63	1.06 ± 0.03
F-23	30	1.04 ± 0.01
Material A, NNO buffer[a]		
1.0 kbar	4	0.91
1.0 kbar, dry	4	1.10
Material A, FMQ buffer		
R5	30	1.12 ± 0.03
R6	3	1.03 ± 0.03
R7	7	1.09 ± 0.02

[a] DALLA TORRE et al. (1997).

Table 4. Kinetic study of vitrinite maturation, 2.0 kbar fluid pressure, 400°C

Run no.	Length (days)	$\%R_{max}$
Material A, FMQ buffer		
R1	7	1.95 ± 0.07
R2	80	2.23 ± 0.07
R4	3	1.58 ± 0.07
Material B, FMQ buffer		
F-27	50	2.12 ± 0.07
F-30	29	2.09 ± 0.09
Material B, HM buffer		
H-26	76	2.39 ± 0.04
H-27	50	2.19 ± 0.08
H-28	18	2.04 ± 0.04
H-29	8	2.01 ± 0.09
H-30	29	2.07 ± 0.06

(2) Two different starting materials from the Frimmersdorf seam, Hambach, Germany A gymnosperm lignite and B angiosperm lignite, yielded the same kinetic results over all investigated experimental conditions. The contrasting nature of these specific lignitic materials did not result in detectably different rates of alteration.

(3) Within measurement uncertainties, both the pair of initially dry and the two wet experiments of DALLA TORRE et al. (1997) are compatible with our wet experiments. The relative abundance of H_2O per se imperceptibly retards the rate of vitrinite maturation.

(4) There is a faint suggestion that the slightly lower attendant pressure of the DALLA TORRE et al. (1997) 1.0 kbar experiments versus 2.0 kbar for the new data, resulted in a slightly enhanced rate of increased VR in the lower pressure runs (especially in the initially dry experiments). This finding supports the principal conclusion of DALLA TORRE et al. based on their experiments up to 20 kbar. HILL et al. (1994) reported similar conclusions.

(5) The increase of VR with reaction time at 400°C appears to be most systematic, whereas that at 250°C is the most irregular. As is commonly observed by laboratory experimentalists, metastability is more persistent at low temperatures, resulting in greater dispersion of the run data.

(6) Our main conclusion is that the new kinetic data appear to be quite compatible with the results of most prior experimental, theoretical and empirical VR rate studies, including the EASY$\%R_0$ rate model of SWEENEY and BURNHAM (1990). The Sweeney and Burnham rate model provides a good approximation to our laboratory alteration data, especially for the 250 and 300°C isotherms, whereas EASY$\%R_0$ calculated values of VR are slightly low for the 200°C experimental data and high for the 400°C runs (J. J. SWEENEY, personal communication, 01/10/03). The disparities between the model and the laboratory maturation rates are greatest for the 200°C isotherm, but as illustrated in Fig. 1, the computed and experimentally verified rate curves are topologically similar. Our long-duration laboratory results on vitrinite maturation demonstrate that, at constant conditions, a quasi-threshold R_r value is attained in a geologically insignificant time interval, and thereafter VR increases progressively more slowly with continued heating at constant temperature. These alteration values, determined in the laboratory over geologically insignificant time scales, are: 200°C, $\%R_r = 0.54$; 250°C, $\%R_r = 0.74$; 300°C, $\%R_r = 1.10$; and 400°C, $\%R_r = 2.25$. They are most appropriate for thermal alteration zone developed adjacent igneous dikes (N. H. BOSTICK, personal communication, 02/11/03).

KINETIC ANALYSIS

The new run data demonstrate that VR is a function chiefly of time and temperature, and to a much lesser extent, total pressure. VR increased the same amount for both studied lignite samples, and is virtually independent of fO_2. Of course, A and B lignites are similar in terms of untreated, initial VR and, although the fO_2 range investigated is appropriate for most diagenetic and low-grade metamorphic environments (FREY, 1987), we did not perform experiments under extremely oxidizing or reducing conditions. Our data reinforce earlier laboratory and field-based conclusions that VR provides a realistic measure of the temperature of host rocks subjected to long-term heating.

We analyzed our data employing an empirical rate equation

$$\%R_r = K_0 t^n \tag{1}$$

where K_0 is the rate constant, t the time in days and n an arbitrary constant that describes the kinetic behavior of vitrinite maturation (DALLA TORRE et al., 1997). Many different reactions are combined in this overall process including, at lower temperatures, the evolution of H_2O and CO_2, and at progressively higher temperatures, expulsion of methane and higher hydrocarbon gas species. The exponent n has no atomic or molecular significance in terms of stoichiometries of the reaction mechanism(s); it simply describes the observed kinetics of the alteration reaction(s). Rearranging this equation by taking the natural logarithm and plotting $\ln \%R_r$ versus $\ln t$ defines a straight line with a slope of n. All four isobars (Figs. 1–4) were employed in order to evaluate n, although the kinetic data are of varying qualities and thus differing reliabilities; 400°C runs are most reliable, those at 250°C least reliable. Values of n are as follows: 400°C, $n = 0.0715$; 300°C, $n = 0.0525$; 250°C, $n = 0.0180$; 200°C, $n = 0.1625$. The average of 0.076 is comparable to values derived for n on much shorter experiments by HUANG (1996) and DALLA TORRE et al. (1997), 0.078 and 0.0714, respectively. The rate constant was then calculated using Eq. (1), setting n to 0.076. Derived values for $\ln K_0$ using all experiments are presented in Table 5.

Table 5. $\ln K_0$ for HM, NNO and FMQ buffers, run materials A and B, $1000/T_K$

Material B, HM buffer		Material B, FMQ buffer		Material A, NNO buffer		Material A, FMQ buffer	
$1000/T_K$	$\ln K_0$	$1000/T_K$	$\ln K_0$	$1000/T_K$	$\ln K_0$	$1000/T_K$	$\ln K_0$
2.1142	−1.28386	2.1142	−1.43165	1.9120	−0.52088	1.7452	−0.14516
2.1142	−1.45267	2.1142	−1.29124	1.9120	−0.39304	1.7452	−0.05303
2.1142	−1.35772	2.1142	−1.35443	1.7452	−0.19967	1.7452	−0.05897
2.1142	−1.40288	2.1142	−1.19988	1.7452	−0.01005	1.4859	0.52056
2.1142	−1.35340	2.1142	−1.18955			1.4859	0.46837
2.1142	−1.23343	2.1142	−1.06385			1.4859	0.35876
2.1142	−0.95179	2.1142	−0.99699				
2.1142	−1.00384	2.1142	−1.01115				
2.1142	−1.07287	2.1142	−1.08680				
2.1142	−0.88258	2.1142	−0.96025				
1.9120	−0.32234	2.1142	−0.87909				
1.9120	−0.51003	1.9120	−0.44878				
1.9120	−0.55882	1.9120	−0.46725				
1.9120	−0.72627	1.9120	−0.59391				
1.9120	−0.78170	1.9120	−0.72362				
1.9120	−0.70626	1.9120	−0.72218				
1.7452	−0.20038	1.9120	−0.71417				
1.7452	−0.22799	1.7452	−0.15408				
1.7452	−0.21573	1.7452	−0.22099				
1.7452	−0.24128	1.7452	−0.22921				
1.7452	−0.23699	1.7452	−0.25388				
1.7452	−0.23891	1.7452	−0.23584				
1.7452	−0.11404	1.7452	−0.23041				
1.4859	0.45279	1.7452	−0.11804				
1.4859	0.49333	1.4859	0.47939				
1.4859	0.47509	1.4859	0.45299				
1.4859	0.48609						
1.4859	0.54477						

In order to express K_0 as a function of temperature, we employed an Arrhenius-type expression (BURNHAM and SWEENEY, 1989)

$$K_0 = k_1 \exp^{-E^*/RT} \tag{2}$$

where k_1 is a pre-exponential term, E^* the activation energy, R the universal gas constant (0.008314 kJ mol^{-1} K^{-1}), and T is in Kelvin. Raw values of K_0 from Table 5 were then plotted against $1000/T_K$. The slope of this line is $-E^*/R$. This formulation combines all the complex devolatilization reactions into a single pre-exponential term, and a single, integrated activation energy—clearly a major thermodynamic oversimplification, but one necessitated by the limited data set. Our experimental runs provide tight constraints at the 300°C and especially the 400°C isotherms, but lesser control at the 200 and 250°C isotherms. Some of the run products yielded anomalously high VR values on several short runs, particularly at lower temperatures; the fact

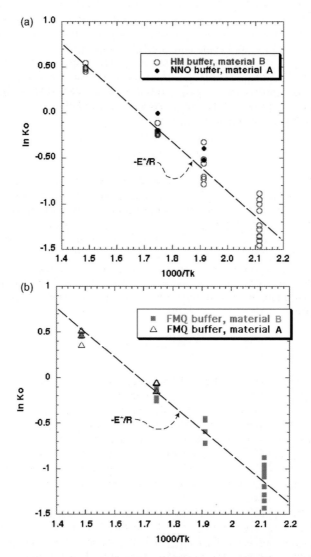

FIG. 5. Arrhenius plots of the natural log of the equilibrium constant, K_0, versus $1000/T_K$ for all experimental data (K_0 values from Table 5). The slope of the visually estimated curve is $-E^*/R$. All laboratory run data are plotted for: (a) controlled higher oxidation states, $E^* = 22.1$ kJ mol^{-1}; and (b) controlled lower oxidation states, $E^* = 21.6$ kJ mol^{-1}.

that they do not fit our ideal R_r versus time and K_0 versus $1000/T_K$ curves suggests a possible reason for ignoring these experiments (disequilibrium conditions in the sample, bituminite impregnation, oxidation by expelled fluids). However, we see no justifiable *a priori* reason to reject them. Plotting all the data for the higher oxidation states (HM and NNO) results in an activation energy of 22.1 kJ mol^{-1}, whereas a plot of all the data at the lower fO_2 range defined by the FMQ buffer yields an E^* of 21.6 kJ mol^{-1} (Fig. 5a and b, respectively). The two curves essentially coincide, but are illustrated separately in Fig. 5 to allow identification of possible differences in E^*. Activation energies such as evaluated here, although systematic and reproducible, are at best vaguely related to the complex chemical changes taking place in lignite during thermal maturation.

Similar but slightly higher activation energies for laboratory alteration of vitrinite have been reported by HUANG (1996) and DALLA TORRE *et al.* (1997); such values lie within the limits of error for experiments employing different experimental procedures and run lengths, impurities in the huminite starting material, and analytical uncertainties in measurement of VR. Using calorimetric data for the various vitrinite devolatilization reactions evolving H_2O, CO_2, CH_4 and higher hydrocarbons, BURNHAM and SWEENEY (1989) and SWEENEY and BURNHAM (1990) defined a series of first-order reactions ($n = 1.0$) yielding model E^* values approximately an order of magnitude greater than the composite activation energy measured in our laboratory study. However, these values are unrelated to ours due to the different rate expressions employed. More rigorous treatment of our run data seems unwarranted, but the kinetic and thermodynamic analyses presented here lend credence to the likelihood that we have measured a regular, systematic organic maturation process.

CONCLUSIONS

In spite of the fact that organic matter in general, and vitrinite in particular, are metastable in sedimentary and low-grade metamorphic environments, the crucial variable that influences the increase in VR over geologic time intervals is temperature. The peak temperature in host rocks probably outweighs prograde and retrograde temperature histories in terms of resultant VR. Such values should be compared with specific data in our isothermal experiments ($= T_{peak}$) for which heating and cooling times are negligible. The conclusion that temperature represents the dominating parameter in determining VR has been reached by numerous prior investigators, and has been practiced for decades in the exploration for oil and natural gas resources. For geologic durations, employing the simple formulation presented by BARKER (1988) for the peak burial temperature of vitrinite

$$T(°C) = 148 + 104[\ln(R_m)] \tag{3}$$

the derived VR at 200°C is calculated to be 1.65. Using our new experimental data (Tables 1 and 5) and rate equation (1), this VR value would be attained in slightly less than 11 m.y. Because of the exponential nature of the rate equation, the increase in VR with time is inappreciable at times longer than a few million years. Thus over the geologic time intervals implicit in the Barker formulation, the maturation value of lignite would closely approximate that given by Eq. (3). For instance, at 200°C, a VR value of 1.55—just 0.10 lower than that determined using Eq. (3)—would be obtained using our results for a heating time of less than 5 m.y., whereas a VR of 1.75—0.10 higher than the Eq. (3) value—would require the duration of such temperatures for about 24 m.y.

Application of rate equation (1) combined with the new kinetic data allows an independent assessment of the thermal regime attending the maturation of vitrinite in strata subjected to thermal conditions ranging from diagenetic to biotite zone of the greenschist facies, and in geologic environments where time is a significant variable. This is important because, for low-grade rocks, only fluid inclusion measurements and very sparse thermobarometric data currently are available to constrain the pressures and temperatures attending recrystallization. In such environments, neoblastic phases have equilibrated at temperatures below closure thresholds (cessation of diffusion) for most various radiometric systems, hence times of recrystallization may be determinable.

Total pressure has a very much smaller effect in reducing the elevation of reflectance at a given temperature, as concluded by DALLA TORRE et al. (1997). A major influence on VR reduction under low-temperature conditions in a closed experimental system is demonstrated by liptinite (alginite) impregnation of huminite, as also shown by HUANG (1996), and as reported from natural environments by WOLF (1978), HUTTON and COOK (1980) and PRICE and BARKER (1985). The latter authors generalized this relationship as indicative of increased preservation of vitrinite, leading to a H-rich analogue that exhibits lower reflectance compared with normal vitrinite subjected to the same thermal history. Our study demonstrates that the host-rock lithology, oxidation state and abundance of an aqueous pore fluid have an imperceptible influence on the kinetics of vitrinite alteration, adding confidence to the classical thermal maturation methods in wide use.

Acknowledgements—This research project grew out of an earlier study concerning the pressure effect on VR initiated by Michael Dalla Torre. We thank Monika Wolf, Ralf Littke, and Jochen Naeth for providing us with the xylite materials, as well as Brian Porter and Jun Liu for early laboratory help. We are especially grateful for the excellent quality of the resin-mounted polished microscope sections prepared by Willi Tschudin. The new laboratory-based kinetic study at Stanford University and the University of Basel was partially supported by grant 31963-AC2 from the Petroleum Research Fund of the American Chemical Society. A first-draft manuscript was reviewed and materially improved by J. J. Sweeney, N. H. Bostick and M. B. Underwood. Ch. E. Barker, R. J. Hill and an anonymous reviewer provided constructive criticism during journal review. We are grateful to these researchers and institutions for support and helpful feedback.

REFERENCES

BARKER C. E. (1988) Geothermics of petroleum systems: implications of the stabilization of kerogen thermal maturation after a geologically brief heating duration at peak temperature. In *Petroleum Systems of the United States* (ed. L. B. MAGOON), Vol. 1870, pp. 26–29. US Geological Survey Bulletin.

BARKER C. E. and GOLDSTEIN R. H. (1991) Fluid inclusion technique for determining maximum temperature in calcite and its comparison to the vitrinite reflectance geothermometer. *Geology* **18**, 1003–1006.

BARKER C. E. and PAWLEWICZ M. J. (1986) The correlation of vitrinite reflectance with maximum temperature in humic organic matter (eds. G. BUNTEBARTH and L. STEGENA), Vol. 5, pp. 79–228. Paleogeothermics, Lecture Notes in Earth Sciences, Springer, New York.

BARKER C. E. and PAWLEWICZ M. J. (1993) An empirical determination of the minimum number of measurements needed to estimate the mean random vitrinite reflectance of disseminated organic matter. *Org. Geochem.* **20**, 643–651.

BOSTICK N. H., CASHMAN S., MCCOLLOUGH T. H. and WADELL C. T. (1979) Gradients of vitrinite reflectance and present temperature in the Los Angeles and Ventura basins, California (ed. E. OLTZ), Symposium in Geochemistry: Low Temperature Metamorphism of Kerogen and Clay Minerals, pp. 65–96. SEPM Pacific Section, Los Angeles.

BURNHAM A. K. and SWEENEY J. J. (1989) A chemical kinetic model of vitrinite maturation and reflectance. *Geochim. Cosmochim. Acta* **53**, 2649–2657.

BUSTIN R. M. (1989) Diagenesis of kerogen (ed. I. E. HUTCHEON), Vol. 15, pp. 1–38. Short Course in Burial Diagenesis, Mineralogical Association Canada, Montreal.

DALLA TORRE M., FERREIRO MÄHLMANN R. and ERNST W. G. (1997) Experimental study on the pressure dependence of vitrinite maturation. *Geochim. Cosmochim. Acta* **61**, 2921–2928.

EUGSTER H. P. (1957) Heterogeneous reactions involving oxidation and reduction at high pressures and temperatures. *J. Chem. Phys.* **26**, 1760–1761.

EWBANK G., MANNING D. A. C. and ABBOTT G. D. (1995) The relationship between bitumens and mineralization in the south Pennine Orefield, central England. *J. Geol. Soc.* **152**, 751–765.

FERREIRO MÄHLMANN R. (2001) Correlation of very low grade data to calibrate a thermal maturity model in a nappe tectonic setting, a case study from the Alps. *Tectonophysics* **334**, 1–33.

FRANKLIN R. E. (1951) The structure of graphitic carbons. *Acta Crystallogr.* **4**, 253–261.

FREY, M. (ed) (1987) *Low Temperature Metamorphism*. Blacky, London, 351 pp.

HILL R. J., JENDEN P. D., TANG Y. C., TEERMAN S. C. and KAPLAN I. R. (1994) Influence of pressure on pyrolysis of coal. In *Vitrinite Reflectance as a Maturity Parameter: Applications and Limitations* (eds. P. K. MUKHOPADHYAY, and W. G. DOW), Vol. 570, pp. 161–193. American Chemical Society Symposium Series.

HUANG W. L. (1996) Experimental study of vitrinite maturation: effects of temperature, time, pressure, water, and hydrogen index. *Org. Chem.* **24**, 233–241.

HUCK G. and KARWEIL J. (1955) Physikalisch-chemische probleme der inkohlung. *Brennstoff-Chemie* **36/1**, 1–11.

HUNT J. M. (1979) *Petroleum Geochemistry and Geology*. Freeman, San Francisco, 339 pp.

HUNT J. M., LEWAN M. D. and HENNET R. J. C. (1991) Modeling oil generation with time-temperature index graphs based on the Arrhenius equation. *Am. Assoc. Petrol. Geol. Bull.* **75**, 795–807.

HUTTON A. C. and COOK A. C. (1980) Influence of alginite on the reflectance of vitrinite from Joadja, NSW, and some other coals and oil shales containing alginite. *Fuel* **59**, 711–714.

KARWEIL J. (1956) Die metamorphose der kohlen vom standpunkt der physikalischen chemie. *Zeit. Deutsch. Geol. Gesell.* **107**, 132–139.

KERRICK D. M. (1987) Cold-seal systems. In *Hydrothermal Experimental Techniques* (eds. G. C. ULMER and H. L. BARNES), pp. 293–323. Wiley, New York.

LAUGHLAND M. M. and UNDERWOOD M. B. (1993) *Vitrinite reflectance and estimates of paleotemperature within the Upper Shimanto Group, Muroto Peninsula, Shikoku, Japan*, Vol. 273, Geological Society of America Special Paper, pp. 25–43.

LAW B. E. (1992) Thermal maturity patterns of Cretaceous and Tertiary rocks, San Juan Basin, Colorado and New Mexico. *Geol. Soc. Am. Bull.* **104**, 192–207.

LEWAN M. D. (1997) Experiments on the role of water in petroleum formation. *Geochim. Cosmochim. Acta* **61**, 3691–3723.

LOPATIN N. V. (1971) Temperature and geologic time as factors in coalification. *Akad. Nauk SSSR Isv. Ser. Geol.* **3**, 95–106 (in Russian).

MCKENZIE D. T. (1981) The variation of temperature with time and hydrocarbon maturation in sedimentary basins formed by extension. *Earth Planet. Sci. Lett.* **55**, 87–98.

PETMECKY S., MEIER L., REISER H. and LITTKE R. (1999) High thermal maturity in the Lower Saxony Basin: intrusion or deep burial? *Tectonophysics* **304**, 317–344.

PRICE L. C. (1983) Geologic time as a parameter in organic metamorphism and vitrinite reflectance as an absolute paleogeothermometer. *J. Petrol. Geol.* **6**, 5–38.

PRICE L. C. and BARKER C. E. (1985) Suppression of vitrinite reflectance in amorphous rich kerogen—a major unrecognized problem. *J. Petrol. Geol.* **8**, 59–84.

ROKSANDIC M. M. (1986) Dynamic interpretation of organic–matter maturation and evolution of oil-generative window: discussion. *Am. Assoc. Petrol. Geol. Bull.* **70**, 1008–1010.

SCHEGG R. (1994) The coalification profile of the well Weggis (Subalpine Molasse, central Switzerland): implications for erosion estimates and the paleogeothermal regime in the external part of the Alps. *Bull. Swiss Assoc. Petrol. Geol. Engng.* **61/136**, 57–67.

SEEWALD J. S., EGLINTON L. B. and ONG Y. L. (2000) An experimental study of organic–inorganic interactions during vitrinite maturation. *Geochim. Cosmochim. Acta* **64**, 1577–1591.

SHARKEY A. G. Jr. and MCCARTNEY J. T. (1963) Physical properties of coal and its products. 2nd Supplement Volume (ed. M. A. ELLIOTT), Chemistry of Coal Utilization, pp. 167–169. Wiley, New York.

STACH F., MACKOWSKY M. T., TEICHMÜLLER M., TAYLOR G. H., CHANDRA D. and TEICHMÜILER R. (eds) (1982) *Textbook of Coal Petrology*, 3rd Ed. Bornträger, Stuttgart.

SUGGATE R. P. (1959) New Zealand coals, their geological setting and its influence on their properties. *N. Z. Dept. Sci. Ind. Res. Bull.* **13**, 113 pp.

SUZUKI N., MATSUBAYASHI H. and WAPLES D. W. (1993) A simpler kinetic model of vitrinite reflectance. *Am. Assoc. Petrol. Geol. Bull.* **77**, 1502–1508.

SWEENEY J. J. and BURNHAM A. K. (1990) Evaluation of a simple model of vitrinite reflectance based on chemical kinetics. *Am. Assoc. Petrol. Geol. Bull.* **74**, 1559–1570.

TEICHMÜLLER M (1987a) Recent advances in coalification studies and their application to geology. In *Coal and Coal-Bearing Strata* (ed. A. C. SCOTT), pp. 127–169. Blackwell, London.

TEICHMÜLLER M. (1987b) Organic material and very low-grade metamorphism. In *Low Temperature Metamorphism* (ed. M. FREY), pp. 114–161. Blacky, London.

TEICHMÜLLER M. and TEICHMÜLLER R. (1954) Die stoffliche und strukturelle metamorphose der kohle. *Geol. Rund.* **42**, 265–296.

TISSOT B. P. and WELTE D. H. (1984) *Petroleum Formation and Occurrence*. Springer, New York.

TISSOT B. P., PELET R. and UNGERER P. (1987) Thermal history of sedimentary basins, maturation indices, and kinetics of oil and gas generation. *Am. Assoc. Petrol. Geol. Bull.* **71**, 1445–1466.

UNDERWOOD M. B., SHELTON K. L., MCLAUGHLIN R. J., LAUGHLAND M. M. and SOLOMON R. M. (1999) Middle Miocene paleotemperature anomalies within the Franciscan Complex of northern California: thermo-tectonic responses near the Mendocino triple junction. *Geol. Soc. Am. Bull.* **111**, 1448–1467.

WAPLES D. W. (1980) Time and temperature in petroleum formation: application of Lopatin's method to petroleum exploration. *Am. Assoc. Petrol. Geol. Bull.* **64**, 916–926.

WAPLES D. W., KAMATA H. and SUIZU M. (1992) The art of maturity modeling. Part 1: finding a satisfactory geologic model. *Am. Assoc. Petrol. Geol. Bull.* **76**, 31–46.

WOLF M. (1978) Inkohlungsuntersuchungen im Hunsrück, Rheinisches Schiefergebirge. *Zeit. Deutsch. Geol. Gesell.* **129**, 217–227.

APPENDIX A

Table A1. Experimental run data on vitrinite maturation

Sample	T (°C)	t (days)	P (bars)	VR Re/R_{max}	Value (max)	Value (min)	s	v	n	Comments
Telocollinite	–	–	–	0.197	0.218	0.180	0.008	0.40	101	Gymnosperm lignite
Densinite	–	–	–	0.300	0.325	0.287	0.007	0.29	100	Gymnosperm lignite
R6 FMQ	300	3	2000	1.031	1.094	0.931	0.032	0.30	111	
R7 FMQ	300	7	2000	1.093	1.134	1.044	0.020	0.19	100	
R3 FMQ	300	7	2000	1.467	–	–	–	–	–	P dropped down to 500 bars during run
R5 FMQ	300	30	2000	1.129	1.179	0.910	0.027	0.27	100	Partly porous
R4 FMQ	400	3	2000	1.577	1.782	1.401	0.065	0.42	100	With bireflectance
R1 FMQ	400	7	2000	1.951	2.083	1.788	0.065	0.33	100	With bireflectance, granular decomposed
R2 FMQ	400	80	2000	2.228	2.375	2.129	0.067	0.45	46	
Telocollinite	–	–	–	0.187	0.205	0.168	0.008	0.41	100	Angiosperm lignite
Densinite	–	–	–	0.303	0.338	0.279	0.011	0.37	100	Angiosperm lignite
H-8 HM	200	5	300	0.313	0.347	0.284	0.012	0.39	100	
F-8 FMQ	200	5	300	0.270	0.303	0.238	0.013	0.49	100	Resinite impregnated (res. impr.)
H-3 HM	199	9	400	0.304	0.376	0.282	0.014	0.48	100	
F-3 FMQ	200	9	400	0.305	0.345	0.273	0.015	0.43	125	
H-6 HM	201	8	2000	0.274	0.317	0.236	0.014	0.47	120	Rich in liptinite, res. impr.
F-6 FMQ	202	8	2000	0.331	0.378	0.254	0.025	0.79	100	Bimodal histogram, different gelification
H-5 HM	201	22	1900	0.311	0.349	0.266	0.018	0.59	101	Some liptinite, res. impr.
F-5 FMQ	200	22	1900	0.381	0.429	0.350	0.014	0.34	119	Some liptinite
H-7 HM	199	33	2000	0.337	0.366	0.282	0.013	0.38	110	Different gelification grade, res. impr.
F-7 FMQ	201	33	2000	0.397	0.440	0.327	0.019	0.49	100	
H-4 HM	200	55	2000	0.395	0.427	0.360	0.013	0.33	100	Some liptinite
F-4 FMQ	201	55	2000	0.468	0.509	0.420	0.017	0.37	100	
H-24 HM	200	71	2000	0.572	0.587	0.552	0.008	0.14	98	(Oxidized parts 0.59–0.715%R_r)
F-24 FMQ	200	71	2000	0.574	0.625	0.530	0.011	0.19	100	10 Sporinite
H-20 HM	200	76	2000	0.625	0.658	0.591	0.014	0.23	100	Oxidation by a fluid $>0.85\%R_r$
F-20 FMQ	201	76	2000	0.532	0.573	0.486	0.015	0.27	110	Porous resinite
H-2 HM	200	89	2000	0.543	0.597	0.479	0.024	0.44	100	
F-2 FMQ	201	89	2000	0.519	0.557	0.463	0.015	0.29	100	
H-9 HM	199	173	2000	0.506	0.549	0.449	0.022	0.42	110	Partly resinite impregnated
F-9 FMQ	200	173	2000	0.499	0.571	0.445	0.026	0.52	100	Fissures, partly granular, many liptinite
H-1 HM	200	204	2000	0.549	0.630	0.429	0.034	0.59	110	Few resinite
F-1 FMQ	201	204	2000	0.545	0.577	0.480	0.020	0.37	100	Few resinite
H-10 HM	251	10	2000	0.863	0.912	0.810	0.021	0.24	99	Grains with different gelification grade
F-10 FMQ	252	10	2000	0.799	0.843	0.725	0.025	0.32	100	Grains with different gelification grade
H-14 HM	250	20	2000	0.754	0.816	0.694	0.032	0.44	96	Grains with different gelification grade

Table A1 (*continued*)

Sample	T (°C)	t (days)	P (bars)	VR Rt/Rmax	Value (max)	Value (min)	s	v	n	Comments
F-14 FMQ	251	20	2000	0.787	0.834	0.702	0.029	0.36	110	Grains with different gelification grade
H-16 HM	251	38	2000	0.754	0.826	0.686	0.025	0.34	100	
F-16 FMQ	251	38	2000	0.728	0.803	0.639	0.036	0.47	110	Some granular resinite
H-21 HM	249	76	2000	0.636	0.713	0.541	0.032	0.51	100	Strongly porous, partly granular
F-21 FMQ	250	76	2000	0.675	0.750	0.570	0.044	0.66	100	Strongly porous, partly granular
H-18 HM	250	79	2000	0.681	0.758	0.581	0.034	0.50	100	Strongly porous, partly granular
F-18 FMQ	250	79	2000	0.676	0.745	0.605	0.032	0.46	110	Strongly porous, partly granular
H-19 HM	250	155	2000	0.724	0.814	0.635	0.034	0.49	110	Bimodal histogram
F-19 FMQ	250	155	2000	0.714	0.786	0.651	0.027	0.37	110	
H-12 HM	300	8	2000	1.045	1.097	0.972	0.028	0.27	100	
F-12 FMQ	301	8	2000	1.041	1.084	0.994	0.021	0.21	100	
H-15 HM	300	20	2000	1.028	1.091	0.910	0.039	0.39	100	Some porous resinite
F-15 FMQ	300	20	2000	1.076	1.093	1.036	0.014	0.13	100	
H-23 HM	300	30	2000	1.038	1.069	1.000	0.014	0.13	100	Some porous resinite
F-23 FMQ	301	30	2000	1.031	1.061	0.990	0.017	0.17	100	Some porous resinite
H-17 HM	300	48	2000	1.082	1.163	0.999	0.035	0.31	110	
F-17 FMQ	302	48	2000	1.063	1.144	0.986	0.037	0.33	110	
H-22 HM	300	63	2000	1.076	1.160	0.933	0.040	0.37	100	
F-22 FMQ	301	63	2000	1.063	1.139	0.981	0.033	0.32	96	Corpocollinite partly decomposed
H-11 HM	300	78	2000	1.097	1.157	1.032	0.030	0.27	100	
F-11 FMQ	301	78	2000	1.116	1.227	1.009	0.043	0.39	100	
H-25 HM	299	84	2000	–	–	–	–	–	–	Strongly decomposed, granular
F-25 FMQ	300	84	2000	–	–	–	–	–	–	Strongly decomposed, granular
H-13 HM	300	107	2000	1.127	1.221	1.050	0.049	0.64	47	Slight bireflectance, partly granular
F-13 FMQ	300	107	2000	1.125	1.221	1.041	0.043	0.38	100	Slight bireflectance
H-29 HM	401	8	2000	**2.008**	2.200	1.817	0.086	0.43	100	Bireflectance, bimodal histogram
F-29 FMQ	402	8	2000	–	–	–	–	–	–	Strongly decomposed, granular
H-28 HM	400	18	2000	**2.041**	2.203	1.896	0.057	0.28	100	Bireflectance
F-28 FMQ	401	18	2000	–	–	–	–	–	–	Strongly decomposed, granular
H-30 HM	400	29	2000	**2.068**	2.237	1.935	0.058	0.28	100	Bireflectance
F-30 FMQ	401	29	2000	**2.086**	2.311	1.920	0.094	0.78	34	Partly decomposed, granular
H-27 HM	400	50	2000	**2.191**	2.412	2.060	0.075	0.34	100	Bireflectance
F-27 FMQ	400	50	2000	**2.116**	2.245	1.987	0.072	0.54	41	Partly decomposed, granular
H-26 HM	401	76	2000	**2.394**	2.453	2.284	0.038	0.16	100	Bireflectance
F-26 FMQ	402	76	2000	–	–	–	–	–	–	Strongly decomposed, granular

R_{max} in bold.

Geochemical Investigations in Earth and Space Science: A Tribute to Isaac R. Kaplan
© The Geochemical Society, Publication No. 9, 2004
Editors: R.J. Hill, J. Leventhal, Z. Aizenshtat, M.J. Baedecker, G. Claypool,
R. Eganhouse, M. Goldhaber and K. Peters

"…and the vale of Siddim was full of slime [= bitumen, asphalt?] pits" (Genesis, 14:10)

ARIE NISSENBAUM

Department of Environmental Sciences and Energy Research, Weizmann Institute of Science, Rehovot 76100, Israel

Abstract—The Dead Sea area has been associated with bitumen (= asphalt) for thousands of years. For this reason, it has commonly been taken for granted that pits of bitumen existed in the Dead Sea area, and into which the kings of Sodom and Gomorrah fell after losing a battle in the vale of Siddim in the Dead Sea region (Genesis, 14:10). However, physical evidence for the existence of such pits is practically non-existent. At times when the Dead Sea water level is low, as it is nowadays, large expanses of black mud covered with a carbonate crust are exposed along the coast of the lake. The black mud resembles asphalt in its shiny black color and sulfurous smell. It has been sometimes assumed that the mud contains asphalt, although this is not the case, and the color and smell are due to poorly crystallized iron sulfides. The solid looking carbonate veneer is quite frail and it is easy to sink through it into the underlying black mud. Thus, the biblical description may be of the kings of Sodom and Gomorrah fleeing through the mud flats when the lake level was low, and sinking into the black sulfurous mud.

INTRODUCTION

FEW BIBLICAL NARRATIVES have been the subject of such extensive scientific and para-scientific literature as the upheaval of Sodom and Gomorrah (Genesis 19:24–25). In a prelude to this momentous event, the Biblical narrative describes a raid by northern invader kings into the Dead Sea area during which they beat five kings of cities, including Sodom and Gomorrah, "in the Vale of Siddim which is the Dead Sea" (Genesis 14:3). After 12 years of conquest the defeated cities rebelled, but were beaten in the Vale of Siddim by Chedarlaomer and the other northern raider-kings. The Bible says "and the Vale of Siddim was full of slime pits: and the kings of Sodom and Gomorrah fled, and fell there" (Genesis 14:10, KJV). The term slime pits in the King James English version of the Bible corresponds to the Hebrew original *Be'erot Heimar* which means wells, or pits (*Be'erot*), of asphalt or bitumen (*Heimar*). The Arabic name for asphalt is also *Hummar* and the on-land asphalt mines in Hasbaya, Southern Lebanon were called "Biyar el Hummar" which means "asphalt wells or asphalt pits" (BURCKHARDT, 1823). The so-called Genesis Apocryphon, written in Aramaic and which is part of the Dead Sea Scrolls found in Qumran and epigraphically dated to the late 1st century BC/early 1st century AD, also says that the kings of Sodom and Gomorrah fell into "pits of…" but unfortunately the last word is missing (fragment 1QapGen, translated by L. Faulk and A. Scott). Josephus Flavius (1st century AD) writes "And when they were come over against Sodom, they pitched their camp at the vale called the Slime Pits, for at that time there were pits in that place; but now, upon the destruction of the city of Sodom, that vale became the Lake Asphaltites, as it is called" (FLAVIUS, 1959). Early translations of the Bible, such as the Greek Septuagint (*ca.* 2nd century BC) use the term *asfaltos* and the Aramaic Onkalos translation (2nd century AD) also refer to bitumen (in Aramaic *Himra*, which is equivalent to the Hebrew *Hemar)* pits. The Vulgate, the first Latin translation of the Bible by St Jerome, who was well versed in Hebrew, was completed in 405 AD, and uses the terminology "puteos multos bituminis" which was translated in the King James version as slime pits. The utilization of the term slime for bitumen was introduced in the 16th and 17th centuries by several authors, *e.g.* Tindale, Coverdale and Milton (see in the New Oxford Dictionary). For this reason it has been commonly accepted that the Kings of Sodom and Gomorrah fell into pits of bitumen, similar perhaps to the La Brea tar pits of Los Angeles or the Asphalt Lake of Trinidad. Indeed, the New American Standard Bible uses the term "tar pits" rather than "slime pits" and the New Revised Standard Version of the bible uses "bitumen pits".

The Dead Sea area has been closely associated with bitumen (or asphalt, both terms are used interchangeably in the literature) for thousands of years (NISSENBAUM, 1994b). Thus, it is not surprising that it had been taken for granted that indeed pits of bitumen did exist in the region. However, it has to be borne in mind that if the Sodom event took place in the 22nd century BC or thereabouts, then it describes an event, which occurred at least thousand years before the canonization of the Bible. Indeed, as Josephus Flavius comments, the pits existed many years before his time but were not personally known to him. Thus, it is possible that the allusion to asphalt may have been a "modern" application to an event which happened in the distant past (relative to the time of the writing up of the Sodom and Gomorrah narrative). It may be relevant that the Book of Jubilees, a non-canonical commentary on the Book of Genesis enhanced by stories and legends that was written in the late 2nd century BC, does not mention the asphalt pit and says instead that the kings fell through "wounds" (R. H. CHARLES translation, 1913).

The surface occurrences of asphalt in the Dead Sea basin are in two major forms: the first, as a cement of conglomerates that line dry river beds, such as in Nahal Heimar, or as small seepages and cavity fillings in Upper Cretaceous rocks as in Nahal Heimar and Massada (Fig. 1; RULLKOTTER et al., 1985; TANNENBAUM and AIZENSHTAT, 1985); the second, and most spectacular, is as large blocks of pure asphalt which can be sporadically found floating on the lake and which are carried to the shore by winds and currents (Figs. 1 and 2d; NISSENBAUM et al., 1980). The asphalt that is found today is usually quite hard, although some of the seepages are very viscous and can slowly flow when the ground temperature reaches above 40°C. In any case, it is difficult to reconcile those occurrences with that of asphalt pits into which a person, or animal, can sink.

Recently, another stimulating possibility to account for the biblical narrative was proposed by FRUMKIN and RAZ (2001). These authors suggested that the slime pits are sinkholes which were formed along the Dead Sea coast through the dissolution of sub-surface salt layer by groundwater that are flowing into the lake due to the decline in Dead Sea levels in 4000 years BP. This was suggested by an analogy to similar sinkholes, which are being developed in the Dead Sea coast since the late 1980s in a process that continues to this day. The sinkholes can be very large (up to 30 m in diameter and 20 m deep), and people and vehicles have been known to fall into them. Such sinkholes can certainly have an impact on any ground battle in the area, but it is difficult to explain the reference to slime (=bitumen) pits.

THE EVIDENCE FOR BITUMEN PITS

Scrutinizing the extensive amount of literature references which takes for granted that the Bible refers to bitumen or asphalt pits, one finds that practically the only physical evidence for the existence of asphalt pits in the Dead Sea basin is taken from a short description by the German traveler, H. Rothe, who traveled by foot along the eastern shore of the lake in 1874 (KERSTEN, 1879). Rothe saw in the coastal area of Ain el-Hummar (Fig. 1) at a distance of about 10 m from the shore, a shallow depression, 20 cm in diameter, filled with brown oily liquid, forming a 5 cm thick layer. As far as is known, this is only description of hydrocarbon seepage along the coast of the Dead Sea proper. Both location and description fits well with the occurrence of ozocerite in the same area, which occurs as small seepages and vein filling in the sandstones comprising the walls of the Rift Valley (NISSENBAUM and AIZENSHTAT, 1975). This ozocerite is probably related to the medium gravity oils which were found in the el-Hummar 2 well at a shallow depth of 327–427 m (HUGHES, 1996). However, other than the description by Rothe, no surface occurrence of hydrocarbons, which can be similar to asphalt pits, have been reported. Also, the geography of the area of the eastern coast of the lake, where the bitumen was observed, is characterized by the very steep cliffs of the Rift Valley, resulting in a very narrow coastal plain, which even when the Dead Sea lake level is low, as it is nowadays, cannot be an appropriate location for any valley war as described in the Bible.

Although fairly copious occurrences of asphalt are known in-land in the Nahal Heimar area (Fig. 1) in the western side of the lake, this deposit is couple of kilometers away from the Dead Sea coast and occurs mostly as gravel cement along the vertical walls of a dry river valley (NISSENBAUM and GOLDBERG, 1980). Thus, this deposit cannot be reconciled with the biblical

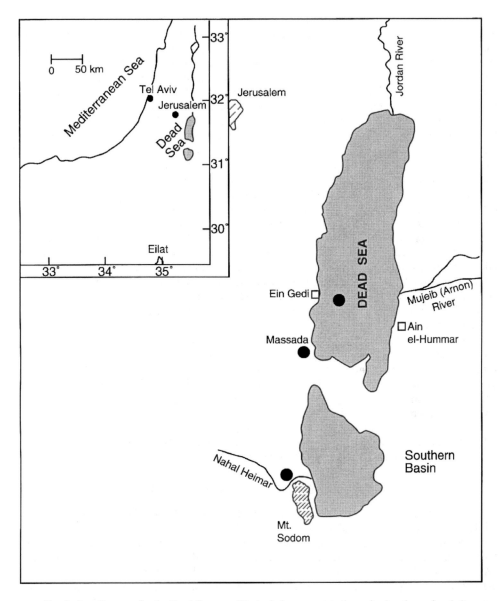

FIG. 1. Location map for the Dead Sea area. Black circles represents the major locations of asphalt occurrences.

description that emphasized the area was full of slime pits or with the Septuagint, the oldest translation of the Bible from the Hebrew original into Greek in the 3rd century BC, that says the valley "consists" of slime pits (LANCELOT BRENTON's translation, 1851).

Another possibility is that extensive asphalt seepages occurred in the southern basin of the Dead Sea which many people assume to be the location of the Vale of Siddim (FRUMKIN and ELITZUR, 2002). This basin was very shallow (only a few meters deep) in the 1960s and since then has dried out completely when the lake level dropped to −405 m, and was completely disconnected from the deep, northern basin (today about 320 m in depth). The southern basin is used now as an evaporation pond for the production of carnallite and the brines are pumped into it through a canal from the northern basin. However, the assumption that the asphalt pits occurred in this area is highly speculative since no asphalt seepages were observed recently in the exposed areas of the southern basin.

In the Middle Ages, two writers do mention asphalt pits in the Dead Sea area. Burchard of Mt Sion, a Dominican pilgrim to the Holy Land, who spent 10 years there between 1275 and 1285. Burchard is considered to be a careful observer of the geography, flora and fauna of the region. He wrote that there are many slime pits on the shore of the lake. There is always a pyramid built beside each pit, which he saw with his own eyes (BURCHARD OF MT SION, 1895–1896). It is not clear what the purpose of the pyramids were and neither is a description of the slime pits given. Another Dominican pilgrim to the holy land, Brother Felix Fabri who visited the holy land twice in 1480 and 1483, and who is considered a somewhat less reliable source, also writes that there are wells of bitumen along the Dead Sea shore. The bitumen is dug and sold to make an "exceedingly strong cement for walls" (FABRI, 1893). He also says near those wells, strong pyramids have been built. While the use of bitumen as a cement was widespread in ancient Mesopotamia, so far no description of medieval, or post-medieval, buildings in the holy land that used asphalt as a cement has been reported. The reference to pyramids might be an allusion to the heaps of salt that the inhabitants of the area extracted from small evaporation ponds along the shores and which were an important commercial resource of the region. The reference of Fabri to bitumen is puzzling since he visited the northwestern sector of the lake. However, most of the available information, and observations, indicate that today, and perhaps in the 20th century as well, asphalt is not known in this area. It might be that some such reports may actually refer to the so-called "Hajjar Musa", a thermally metamorphosed organic-rich carbonate of Senonian (Upper Cretaceous) age, blocks of which are common in this area (SPIRO et al., 1983). By visual inspection this very black and shiny, sulfur-rich rock can be confused with pure bitumen.

Thus, unless surface asphalt pits existed during biblical times, and they were large and common enough to influence the outcome of a battle, even if a minor one, there is no present day evidence for their existence and thus the indication that the Biblical slime pits are asphalt or bitumen pits is rather slender.

AN ALTERNATIVE PROPOSITION

The drastic lowering of water level in the Dead Sea since the 1930s and, in particular, during the last 20 years resulted in the exposure of large areas of mud flats along the rim of the lake. The mud flats are covered by a thin veneer of white aragonite and gypsum that is underlain by fine deep black muds that were formed underwater when the Dead Sea levels were higher (Fig. 2c). The deep black color of the mud, and its shiny appearance in fresh samples, strongly resembles the asphalt from the floating blocks of the Dead Sea (Fig. 2a and b). It had been sometimes claimed that the color (and therapeutic properties) of the black mud are due to the presence of asphalt. However, there is no organic geochemical evidence for the presence of asphalt in the black mud (NISSENBAUM et al., 2002; OLDENBURG et al., 2000a). The black color of the mud is due to the presence of poorly crystallized inorganic iron monosulfide in organic-poor sediment (NISSENBAUM and KAPLAN, 1976; OLDENBURG et al., 2000a,b). Also, the black mud reeks slightly of hydrogen sulfide which resembles somewhat the smell of the sulfur-rich asphalt.

The thin veneer of carbonates and gypsum on the mud flats can give the impression of a robust and strong cover, sufficient to support the weight of a human body. This, however, is quite illusionary and when walking over the crust, it is easy to sink into the black mud to a considerable depth (Fig. 1). Travelers to the Dead Sea area indeed reported such happenings. De Saulcy, who visited the Dead Sea in 1850–1851 describes, in a rather flowery language, how a sheik's horse sank into a mud near Mt Sodom (Fig. 1) "The hind legs of the charger sink at once. Horse and rider are starting to disappear. The rider without quitting its bridle tries to disengage his feet from the stirrup but sinks deeper in the attempt". However, the incident ended well when the horse "by effort almost superhuman" was dragged from his "tomb of sand". (DE SAULCY, 1854). However, such an incident was not confined to the southern extremity of the lake, and De Saulcy's horse managed to sink into the mud "up to his nostrils" in the coast of the north end of the lake (DE SAULCY, 1854). Similarly, Poole who visited the area near Mt Sodom in 1855 wrote "We passed a hole where a camel had fallen through the encrusted sand" (HILL, 1914). A somewhat less dramatic description was given by Browne who traveled to the northeastern coast of the Dead Sea and who wrote "We rode along the beach for several miles, and then stuck

FIG. 2. (a) A piece of asphalt from the floating blocks, Dead Sea. (b) A sediment core of black mud from the Dead Sea Coast. (c) A person sinking through the carbonate crust, Dead Sea coast. (d) Large blocks of asphalt strewn along the Dead Sea coast near Ein Gedi.

off into a morass, through which our horses plunged and staggered for some miles further" (BROWNE, 1853). Even at present, it is indeed possible for a human being to easily sink through the carbonate crust into the black sulfurous mud and care must be taken when walking over the crust (Fig. 2c).

Although the timing of the destruction of Sodom and Gomorrah and hence of the battle in the Vale of Siddim is unknown, several scholars dated it to around the 22nd century BC (RAST, 1982). During this period, the area had gone through a major climatic desiccation event resulting in a dramatic fall in the lake's water level (FRUMKIN et al., 1991; FRUMKIN and RAZ, 2001). This climatic change may, or may not, have been responsible for the ruin of Sodom and Gomorrah (NISSENBAUM, 1994a), but the fall in lake level would have produced large areas of exposed firm-looking mud flats of black mud, into which, as described above, man and animal can easily sink. The mud flats were covered by a veneer of aragonite and gypsum, which could have given an impression of fairly solid ground. However, this is misleading and the fragile crust can be easily broken through by anyone walking on top of it (Fig. 2c). Any attempt to move over the flats, as might have been done by the kings of Sodom and Gomorrah, when escaping the battlefield, probably in great panic, could have resulted in their sinking into the "quicksand". The similarity of the color, appearance and smell of the black mud to the famous Dead Sea asphalt might have resulted in the postulation that the fleeing kings sank into asphalt pits.

Acknowledgements—This article was written in tribute to Ian Kaplan, an exceptional mentor, a great teacher, an excellent scientist and above all a truly wonderful human being, a genuine "mentsch". Thanks are due to Dr A. Frumkin (Hebrew University, Jerusalem) and Dr Z. Ammar (Bar Ilan University) for reading and commenting on the manuscript and to Dr R. Garber (Chevron, Texas) for supplying one of the photographs. Dr T. Goldstein and Dr I. R. Kaplan contributed materially in improving the manuscript.

REFERENCES

BRENTON SIR LANCELOT C. L. (1851) *The Septuagint: Greek and English*. Samuel Bagster and Sons, Ltd, London.
BROWNE J. R. (1853) *Yusef: or Journey of the Frangi: a Crusade in the East*. Harper and Brothers, New York, 421 pp.
BURCHARD OF MT SION (1895–1896) *Itinerary*, Vol. 12, English translation by A. Stewart. Palestine Pilgrims Text Society, London, Reprinted 1971 by AMS Press, NY, 136 pp.
BURCKHARDT J. L. (1823) *Travels in Syria and the Holy Land*. John Murray, London, 720 pp.
CHARLES R. H. (1913) The Book of Jubilee. In *The Apocrypha and Pseudoepigrapha of the Old Testament* (Engl. trans. R. H. CHARLES). Clarendon Press, Oxford.
DE SAULCY F. (1854) *Narrative of a Journey Round the Dead Sea and in Bible Lands in 1850 and 1851*. Richard Bentley, London.
FABRI F. (1893) *The Book of Wandering of Felix Fabri*, Vol. 10, English translation by A. Stewart. Palestine Pilgrims Text Society, London, Reprinted 1971 by AMS Press, NY, 676 pp.
FLAVIUS J. (1959) *The Jewish Wars*, English translation by G. A. Williamson. Penguin Books, New York.
FRUMKIN A. and ELITZUR Y. (2002) Historic Dead Sea level fluctuations calibrated with geological and archaeological evidence. *Quat. Res.* **57**, 334–342.
FRUMKIN A. and RAZ E. (2001) Collapse and subsidence associated with salt karstification along the Dead Sea. *Carbonates Evaporites* **16**(2), 117–130.
FRUMKIN A., MAGARITZ M., CARMI I. and ZAK I. (1991) The Holocene climatic record of the salt caves of Mount Sedom, Israel. *Holocene* **1**, 5–15.
HILL J. G. (1914) The Dead Sea. *Palestine Expl. Fund Quart. St.*, pp. 23–29.
HUGHES W. R. (1996) Geochemistry of Ain El Hummar-2 crude oils, Jordan. *Geoarabia* **1**(1), 150 p.
KERSTEN O. (1879) Umwanderung des Todten Meeres im fruehjahr 1874. *Z. Deutsch. Palastina Vereins* **2**, 201–244.
NISSENBAUM A. (1994a) Sodom, Gomorrah and the lost cities of the plain—a climatic perspective. *Clim. Change* **26**, 435–446.
NISSENBAUM A. (1994b) Utilization of Dead Sea asphalt throughout history. *Rev. Chem. Engng.* **9**, 365–383.
NISSENBAUM A. and AIZENSHTAT Z. (1975) Geochemical studies on ozocerite from the Dead Sea. *Chem. Geol.* **16**, 121–127.
NISSENBAUM A. and GOLDBERG M. (1980) Asphalts, heavy oils, ozocerite and gases in the Dead Sea Basin. *Org. Geochem.* **2**, 167–180.
NISSENBAUM A and KAPLAN I. R. (1976) Sulfur and carbon isotopic evidence for biogeochemical processes in the Dead Sea. In *Environmental Biogeochemistry* (ed. J. O. NRIAGU), pp. 309–325. Ann Arbor Science Publishers, Ann Arbor, MI.
NISSENBAUM A., AIZENSHTAT Z. and GOLDBERG M. (1980) The floating asphalt blocks of the Dead Sea. In *Advances in Organic Geochemistry 1979* (eds. A. G. DOUGLAS and J. R. MAXWELL), pp. 157–161. Pergamon Press, London.
NISSENBAUM A., RULLKOTTER J. and YECHIELY Y. (2002) Are the curative properties of "black mud" from the Dead Sea due to the presence of bitumen(=asphalt) or other types of organic matter? *Environ. Geochem. Health* **24**, 327–335.
OLDENBURG T. B. P., RULLKOTTER J., BOTTCHER M. E. and NISSENBAUM A. (2000a) Molecular and isotopic characterization of organic matter from recent and sub-recent sediments from the Dead Sea. *Org. Geochem.* **31**, 251–265.
OLDENBURG T. B. P., RULLKOTTER J., BOTTCHER M. E. and NISSENBAUM A. (2000b) Addendum: molecular and isotopic characterization of organic matter from recent and sub-recent sediments from the Dead Sea. *Org. Geochem.* **31**, 773–774.
RAST W. E. (1982) Bab edh-Dhra and the origin of the Sodom saga. In *Archaeology and Biblical Interpretation: Essays in Memory of D. Glenn Rose* (eds. L. G. PERDUE, L. E. TOOMBS and G. L. JOHNSON), pp. 185–201. John Knox Press, Atlanta, GA.
RULLKOTTER J., SPIRO B. and NISSENBAUM A. (1985) Biological marker characteristics of oils and asphalts from carbonate rocks in a rapidly subsiding graben, Dead Sea, Israel. *Geochim. Cosmochim. Acta* **49**, 1357–1370.
SPIRO B., WELTE D. H., RULLKOTTER J. and SCHAEFER R. G. (1983) Asphalts, oils and bituminous rocks from the Dead Sea area—a geochemical correlation study. *Am. Assoc. Petrol. Geol. Bull.* **67**, 1163–1175.
TANNENBAUM E. and AIZENSHTAT Z. (1985) Formation of immature asphalt from organic-rich carbonate rocks—I: geochemical correlation. *Org. Geochem.* **8**, 181–192.

Geochemical Investigations in Earth and Space Science: A Tribute to Isaac R. Kaplan
© The Geochemical Society, Publication No. 9, 2004
Editors: R.J. Hill, J. Leventhal, Z. Aizenshtat, M.J. Baedecker, G. Claypool,
R. Eganhouse, M. Goldhaber and K. Peters

Geochemical and submicron-scale morphologic analyses of individual Precambrian microorganisms

J. William Schopf

Department of Earth and Space Sciences, Institute of Geophysics and Planetary Physics
(Center for the Study of Evolution and the Origin of Life), Molecular Biology Institute, and NASA
Astrobiology Institute (UCLA), University of California, 595 Charles Young Drive East, Box 951567,
Los Angeles, CA 90095-1567, USA

Abstract—In recent decades, the documented fossil record has been extended to some 3500 Ma (million years) ago. Hundreds of fossiliferous units have been discovered that contain thousands of microbial fossils, and the rules for accepting ancient microfossil-like objects as *bona fide* have come to be well established. Of these, criteria for establishing biogenicity have proven the most difficult to satisfy. Three new techniques have now been devised to help answer this need: (1) ion microprobe mass spectrometry has been used to measure the carbon isotopic compositions of individual Precambrian microfossils. (2) Laser-Raman imagery has been used to analyze the molecular structure of the carbonaceous matter comprising such cellular fossils and associated particulate organic matter. (3) Atomic force microscopy has been used to reveal the nanometer-scale structure of the kerogenous components of individual Precambrian microscopic fossils. These new techniques not only provide means for elucidation of the isotopic composition, molecular structure, and submicron-scale fine structure of individual microscopic fossils, but hold promise for understanding the geochemical maturation of ancient organic matter and clarifying the nature of minute fossil-like objects of putative but uncertain biogenicity, whether Precambrian or extraterrestrial.

STATEMENT OF THE PROBLEM

THE PRECAMBRIAN SEGMENT of geological time encompasses some 85% of the history of life, a record dominated by prokaryotic (bacterial, including cyanobacterial) microbes. Such microorganisms are minute, incompletely preserved in geological materials, and exhibit simple morphologies that can be mimicked by non-biologic mineralic microstructures; discrimination between true microbial fossils and microscopic pseudofossil look-alikes can therefore be difficult.

In the younger (Proterozoic, ~550 to 2500-Ma-old) Precambrian, differentiation of true fossils from pseudofossil mimics usually presents only minor problems—fossils are abundant, often well preserved, and not uncommonly so similar in morphology to modern microbial taxa that both their biogenicity and biological affinities can be readily established (MENDELSON and SCHOPF, 1992). But interpretation of the older (Archean, >2500 Ma (million years)) Precambrian fossil record is fraught with difficulties. Because of geological recycling, preserved rock units become increasingly rare with increasing geological age; few units dating from the Archean have survived to the present. Moreover, because of tectonism and related geologic processes, most Archean rocks have been metamorphosed, often severely, with older units generally being more altered than younger ones. Taken together, these two factors—the lack of survivability and the metamorphic alteration of the ancient rock record—play havoc with attempts to trace back the earliest records of life.

Nevertheless, at the current state of the science, only the known fossil record can provide clear-cut evidence of the antiquity of life on Earth, yielding a minimum age for life's existence and a basis for extrapolation as to when life may have actually emerged. The techniques used to investigate this problem have bearing on the broader issue of whether life once existed (or may now exist) elsewhere in our solar system, because the same questions will be asked and similar techniques applied in the search for evidence of past life in rock samples from other planetary bodies.

BONA FIDE EVIDENCE OF ANCIENT LIFE

The hunt for early records of life hinges critically, of course, on understanding what constitutes firm evidence, criteria for which have matured as studies of Precambrian paleobiology have come of age (SCHOPF and WALTER, 1983; SCHOPF, 1992a, 1993). Namely, to be acceptable, evidence of past life must meet the following five tests:

(1) *Provenance*: *Is the source of the rock sample that contains the putative biologic features established firmly?* In particular, the stratigraphic and geographic provenance should be known precisely, as demonstrated, *e.g.* by replicate sampling.

(2) *Age*: *Is the age of the rock sample known with appropriate precision?* For example, is the age tightly constrained by multiple geochronologic (radiometric) measurements and other geologic evidence?

(3) *Indigenousness*: *Are the putative biologic features indigenous to the rock?* Specifically, are the features embedded in the rock matrix rather than being recent surficial contaminants—a test that can be met for microscopic fossils by studies of petrographic thin sections and for chemical signatures by analyses *in situ* of particulate, immobile and insoluble, carbonaceous matter (kerogen).

(4) *Syngenicity*: *Are the features syngenetic with a primary mineral phase of the rock?* Or, on the contrary, are they of later origin, *e.g.* introduced into pores or fractures and lithified by secondary or later-generation minerals, a test that can usually be met by appropriately careful study of petrographic thin sections.

(5) *Biogenicity*: *Are the features assuredly biological?* This test, as discussed later, is almost always the most difficult to satisfy.

Evidence of biogenicity

Long recognized to be a critical problem in the hunt for early records of life (TYLER and BARGHOORN, 1963), questions regarding the biogenicity of putative Precambrian microscopic fossils—particularly those of especially great geologic age—persist to the present (BRASIER *et al.*, 2002; PASTERIS and WOPENKA, 2002; KAZMIERCZAK and KREMER, 2002; SCHOPF, in press). Yet in a general sense, establishment of the biogenicity of ancient fossil-like objects is relatively straightforward, as was shown decades ago when early workers in the field (BARGHOORN and TYLER, 1965; CLOUD, 1965; BARGHOORN and SCHOPF, 1965) first demonstrated that "Precambrian microfossils" are, indeed, true fossils. Namely, as was then shown, the biological origin of such microscopic objects can be established by demonstrating that they possess a suite of traits unique to life—a suite shared by fossils and living organisms but not by inanimate matter (a formulation, it may be noted, that is virtually identical to that promulgated in the early 1800s by Barron Georges Cuvier, a founder of paleontology, as he sought to establish that megascopic fossils were not merely "sports of nature"). Differing traits, of course, are used in differing situations, depending on the evidence available; and each of the traits so used is multifaceted, composed of a diverse set of factors and subfactors.

Of the various traits used to address the question of the biogenicity of minute fossil-like objects, three stand out as having proven particularly useful: the detailed (micron-scale) morphology of the objects in question; the carbon isotopic composition of associated kerogenous matter; and the molecular–structural makeup of fossil organic matter.

"Morphology", in this context, subsumes a great many variables—organismal shape (*e.g.* coccoid or filamentous); cell shape, size, and surface ornamentation; structure and thickness of an encompassing sheath, if present; and many others (SCHOPF, 1992b)—and in taxonomic studies, routinely includes quantitative analyses of intra-taxon morphological variability and population structure. Of the hundreds of Precambrian microbiotas thus analyzed (MENDELSON and SCHOPF, 1992), nearly half are preserved in microbially laminated megascopic stromatolites (SCHOPF, 1992b), their presence in which is yet another strong indicator of their biogenicity.

The second prime trait, the carbon isotopic composition of microfossil-associated organic matter, is supported by thousands of measurements in hundreds of Precambrian deposits

(STRAUSS and MOORE, 1992), studies that have traced isotopic signature of microbial photosynthesis to at least 3500 Ma ago (HAYES et al., 1983; STRAUSS et al., 1992b). Such analyses of bulk samples, however, yield average isotopic values of carbon derived from a mix of sources—primarily, microfossils of diverse types and states of preservation and syngenetically deposited sapropelic kerogen particles of multiple origins, in some rocks augmented by minor amounts of carbonaceous matter introduced in later-generation veinlets. As discussed below, by use of ion microprobe mass spectrometry such analyses have now been extended to the kerogenous materials to individual microscopic fossils (HOUSE et al., 2000; UENO et al., 2001), an analytical advance that ties isotopic composition directly to organismal morphology.

The third prime trait, the molecular–structural makeup of the organic matter comprising ancient fossils, has only recently begun to be investigated, most effectively by laser-Raman imagery (KUDRYAVTSEV et al., 2001; SCHOPF et al., 2002b) and atomic force microscopy (AFM) (KEMPE et al., 2002). Though new to the analysis of the kerogenous components of Precambrian organic-walled microfossils and associated sapropelic debris, such studies hold great promise. In particular, Raman imagery provides means by which to directly correlate optically discernable morphology with the molecular structure of individual carbonaceous microfossils, whereas studies by AFM permit visualization of the submicron-scale micromorphology of their preserved kerogenous constituents.

NEW ANALYTICAL TECHNIQUES

Ion microprobe mass spectrometry

Beginning with the pioneering studies of PARK and EPSTEIN (1963) and HOERING (1967), an impressive amount of data has been amassed on the carbon isotopic compositions of coexisting inorganic carbonate minerals and organic kerogens in Precambrian fossiliferous sediments (STRAUSS and MOORE, 1992). For the kerogenous components, most such data have been obtained on the total organic carbon fraction of selected rock specimens and on carbonaceous residues concentrated by acid maceration of large (kg-sized) rock samples, most notably as a result of two extensive studies carried out by the Precambrian Paleobiology Research Group (HAYES et al., 1983; STRAUSS et al., 1992a,b). Hundreds of such measurements have been reported for the kerogenous components of diverse fossil-bearing Precambrian shales and cherts (STRAUSS and MOORE, 1992), analyses that in comparison with values measured on coexisting carbonates yield a consistent signature of isotopic separation between the two carbon reservoirs of $25 \pm 10\%_o$—interpreted as evidencing biological photoautotrophic carbon fixation (HAYES et al., 1983; STRAUSS et al., 1992a,b)—that can be traced to at least as early as ~3500 Ma ago (Fig. 1). By their nature, however, such analyses can yield only an average value of the components measured, data providing strong evidence of the existence of photoautotrophic primary producers but no basis to either differentiate between the contributions of oxygenic and anoxygenic photosynthesizers or link carbon isotopic composition to particular microbial taxa of fossil communities.

This latter shortcoming, the inability to link isotopic composition to particular fossil microorganisms, has recently been remedied by the use of ion microprobe mass spectrometry to analyze the carbonaceous matter comprising the cell walls of individual Precambrian microscopic fossils, a technique developed and first applied to such studies by HOUSE et al. (2000). Used thus far to analyze the carbon isotopic compositions of microfossils in three Precambrian geologic units (Fig. 2)—the ~850-Ma-old Bitter Springs Formation of central Australia, the ~2100-Ma-old Gunflint Formation of southern Canada, and the ~ 3490-Ma-old Dresser Formation of Western Australia (HOUSE et al., 2000; UENO et al., 2001)—ion microprobe spectrometry requires that the specimens analyzed be exposed at the upper surfaces of polished petrographic thin sections. For relatively large (>20 μm diameter) specimens densely packed in richly fossiliferous deposits, this requirement can be met rather easily. But in sparsely fossiliferous assemblages, particularly those composed chiefly of narrow

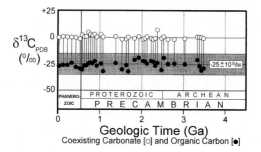

FIG. 1. Carbon isotopic values of coexisting carbonate and organic carbon measured in bulk samples of Phanerozoic and Precambrian sedimentary rocks, for the Precambrian represented by data from 100 fossiliferous cherts and shales shown as average values for groups of samples from 50-Ma-long intervals (STRAUSS and MOORE, 1992).

(~1 μm diameter) microbial filaments or similarly sized bacterial coccoids, identification of the fossils at the section surface by use of the low resolution optical microscopic apparatus typical of ion microprobes can be difficult. Moreover, because of instrument-inherent systematic drift of measurements obtained, attainment of the ±1‰ precision desired of such analyses can require repeated comparative measurements of the sample and an appropriate standard of known isotopic composition, making such analyses exceedingly time-consuming (HOUSE et al., 2000). Nevertheless, the efficacy of this new technique is firmly demonstrated by the consistency

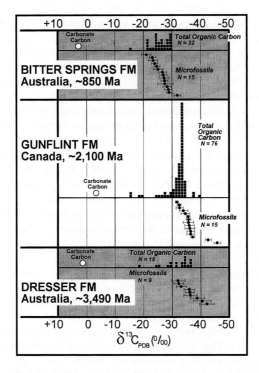

FIG. 2. Carbon isotopic values (with error bars) of individual Precambrian microfossils from single stratigraphic horizons of three geologic units measured by ion microprobe mass spectrometry compared with those of the carbonate and total organic carbon measured in bulk samples of the same geologic units. Values plotted for carbonate and total organic carbon are from STRAUSS and MOORE (1992); for microfossils from the Bitter Springs and Gunflint Formations, from HOUSE et al. (2000); and those for microfossils from the Dresser Formation, from UENO et al. (2001).

between the measurements acquired by ion microprobe spectrometry and those obtained by traditional bulk sample analyses (Fig. 2).

Laser-Raman imagery

Laser-Raman imagery, another technique new to paleobiology, can be used to analyze directly the molecular structure of fossil specimens as minute as 1 μm in size (KUDRYAVTSEV *et al.*, 2001; SCHOPF *et al.*, 2002b). Applied to cellularly preserved carbonaceous fossils, the high resolution virtual "chemical maps" produced by this technique show the spatial distribution of the kerogen comprising the specimens analyzed and its one-to-one correlation with optically discernable cellular morphology (Figs. 3 and 4). Unlike ion microprobe mass spectrometry, Raman imagery is a non-intrusive and non-destructive technique that can be used to analyze permineralized (petrified) fossils situated well within optically translucent petrographic thin sections (to depths >60 μm), regardless of whether the sections have been polished or unpolished and covered by a thin veneer of microscopy immersion oil (SCHOPF *et al.*, 2002b). Raman imagery is applicable also to analysis of such fossils exposed on the surface of cellulose acetate peels (prepared by the etch-and-peel technique; WALTON, 1928; JOY *et al.*, 1956) or freed from a rock matrix by palynological acid maceration (SCHOPF *et al.*, in press), and Raman point spectra have been acquired as well from Precambrian fossils preserved as flattened carbonaceous remnants in fine-grained clastic sediments (AROURI *et al.*, 2000) and particulate kerogens in acid-resistant residues (WOPENKA and PASTERIS, 1993). Thus, Raman spectroscopy can be used to analyze organic-walled specimens preserved by either of the two modes of fossilization most common in ancient sediments (SCHOPF, 1992b)—compression and permineralization—and is applicable to surface-exposed specimens as well as those present within thin sections or in macerations or peels, the three principal types of preparations used for paleontologic study of such materials. Moreover, because Raman spectroscopy is both non-destructive and non-intrusive, it can be applied even to specimens such as those archived in museum collections to protect them from physical or chemical alteration (SCHOPF *et al.*, 2002b).

In recent studies (SCHOPF, 2002; SCHOPF *et al.*, 2002a, in press), Raman imagery has been used to analyze the chemistry of carbonaceous microscopic fossils permineralized in 25 geologic units of essentially the same lithology (viz., fine-grained cherts), but of differing ages (ranging from Devonian to Archean) and notably different metamorphic histories that range from relatively "well preserved" to "poorly preserved" and include specimens that are among the oldest fossils known. The results demonstrate that the structure of the Raman spectra acquired varies systematically with the metamorphic grade of the fossil-bearing units sampled, the fidelity of preservation of the fossils studied, the geochemically altered color of the organic matter analyzed, and with both the H/C and N/C ratios measured in kerogens isolated from bulk samples of the fossil-bearing cherts. Deconvolution of the various spectra, facilitated by their comparison with those of multiple specimens from individual rock units and of fossils heated experimentally to simulate effects of low-grade metamorphism, has provided insight into the chemical makeup and catagenic alteration of these ancient kerogens. In particular, the Raman spectra establish that the kerogens are composed predominantly of disordered interlinked arrays of hydrogen-poor polycyclic aromatic hydrocarbons and document a progressive series of geochemical changes that accompany the processes of graphitization in geologically low-grade (greenschist facies and below) metamorphic terrains (SCHOPF *et al.*, in press).

Although no non-biologic mechanisms are known to be capable of organizing particulate organic matter (whether of biologic or non-biologic origin) into microscopic sports of nature that faithfully mimic populations of true cellular fossils, inorganic (mineralic) microscopic objects have repeatedly been misidentified as Precambrian microfossils (SCHOPF and WALTER, 1983). Thus, it has long been evident—dating from the first discovery of permineralized Precambrian microorganisms (TYLER and BARGHOORN, 1954) and recognition of the problems involved in distinguishing such fossils from mineralic pseudofossil look-alikes (*e.g.* TYLER and BARGHOORN, 1963)—that a need exists to directly correlate the chemical makeup of putative fossil-like objects with their morphology. Raman imagery has answered this need. The evidence

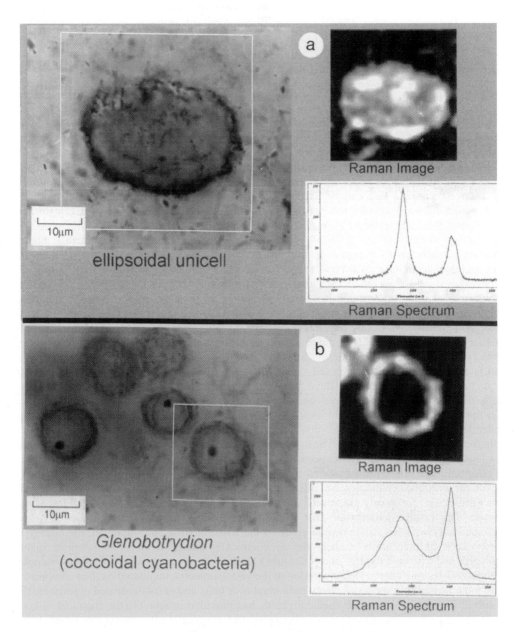

FIG. 3. Optical photomicrographs, Raman images, and Raman spectra of Precambrian permineralized microfossils analyzed in petrographic thin sections. White rectangles in the optical photomicrographs denote the areas from which Raman images were obtained. That these fossils are composed of disordered carbonaceous matter is established by the two prominent vibrational bands in the Raman spectra (at ~1350 and 1600 cm⁻¹) acquired from kerogen having the spatial distribution shown for each specimen in the accompanying Raman images, as measured by the techniques summarized by KUDRYAVTSEV et al. (2001) and SCHOPF et al. (2002a,b,c). (a) An unnamed ellipsoidal unicell (Auburn Dolomite, ~780-Ma-old, South Australia; SCHOPF, 1975, 1992c). (b) Glenobotrydion aenigmatis, a chroococcacean cyanobacterium (Bitter Springs Formation, ~850-Ma-old, central Australia; SCHOPF, 1968; SCHOPF and BLACIC, 1971).

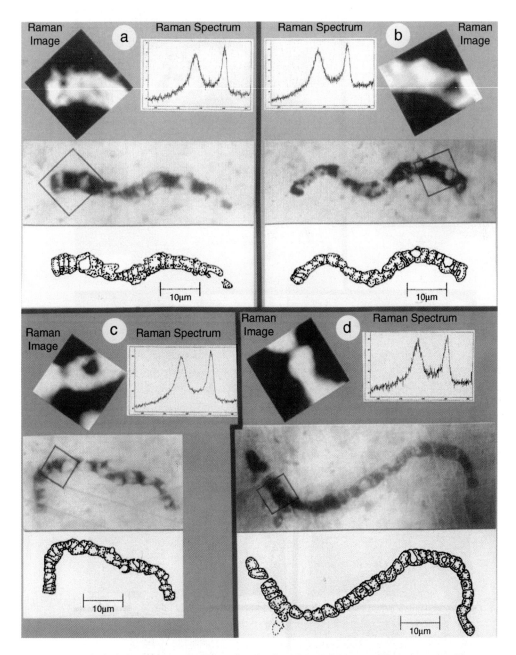

FIG. 4. Optical photomicrographs, interpretive drawings, Raman images, and Raman spectra of four permineralized specimens of *Primaevifilum amoenum*, a filamentous cyanobacterium-like prokaryote, in petrographic thin sections of the ~ 3465-Ma-old Apex chert of northwestern Western Australia (SCHOPF, 1992a, 1993). Black rectangles in the optical photomicrographs denote the areas from which Raman images were obtained. That these fossils are composed of disordered carbonaceous matter is established by the two prominent vibrational bands in the Raman spectra (at ~ 1350 and 1600 cm^{-1}), acquired from kerogen having the spatial distribution shown for each specimen in the accompanying Raman images, as measured by the techniques summarized by KUDRYAVTSEV *et al.* (2001) and SCHOPF *et al.* (2002a,b,c). (a) The Natural History Museum (NHM), London, specimen No V.63164[5]. (b) NHM specimen No V.63164[6]. (c) NHM specimen No V.63728[2]. (d) NHM specimen No 63166[1].

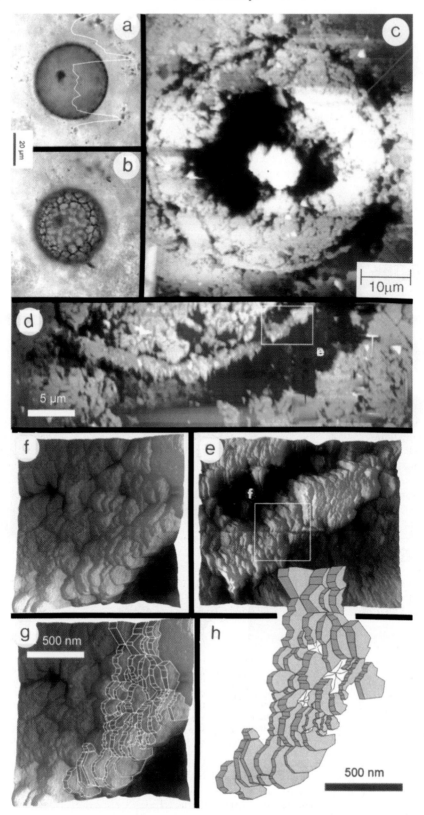

it can provide of the carbonaceous (kerogenous) molecular structure of the objects analyzed and the one-to-one spatial correlation of such organic matter with optically identifiable cellular morphology (Figs. 3 and 4) constitutes a firm and perhaps irrefutable indicator of biogenicity.

Atomic force microscopy

AFM, a third technique new to paleobiology, can be used to elucidate the nanometer-scale fine structure of materials analyzed, a technique of interest here because of the insight it can provide into the structural makeup of the kerogenous constituents of Precambrian microorganisms. Like ion microprobe spectrometry, AFM is a technique that permits analyses only of permineralized fossils situated at the upper surfaces of polished thin sections, in this instance exposed in bas-relief by gentle acid-etching. AFM is then used to obtain high-resolution virtual images that map the topography of the exposed cell walls and their component subunits. Combined with laser-Raman imagery, such studies have shown the kerogenous walls of Precambrian permineralized microfossils to be composed of stacked, platy, angular polyaromatic hydrocarbon subunits ~ 200 nm in size that are oriented perpendicular to the cell walls (Fig. 5; KEMPE et al., 2002). Not only can AFM provide insight into the submicron-scale morphology of such structures, but also in concert with Raman imagery this new technique holds promise for discrimination between true fossils and pseudofossil "look-alikes", a problem that has plagued Precambrian paleobiology since its inception (SCHOPF and WALTER, 1983). For example, by using AFM to determine the orientation of Raman-established aromatic subunits, true fossils, such as the one illustrated here in which the subunits are aligned parallel to the radii of the fossil (Fig. 5c–h), could be distinguished from such non-biologic particles as spheroidal carbon black in which the aromatic subunits are aligned parallel to the spheroid's circumference (OBERLIN et al., 1980).

MUTUALLY REINFORCING EVIDENCE

Applied to microfossil-like objects of uncertain biological origin, each of the three prime biogenic indicators—cellular morphology, carbon isotopic composition, and carbonaceous (kerogenous) molecular structure—provides strong evidence consistent with a biogenic interpretation. Nevertheless, none of these factors, if considered alone, is definitive: use of morphology, by itself, has led to numerous errors of interpretation (SCHOPF and WALTER, 1983; MENDELSON and SCHOPF, 1992); reliance on isotopic evidence, alone, has proven inconclusive (VAN ZUILEN et al., 2002; FEDO and WHITEHOUSE, 2002); and Raman spectra of geochemically mature organic matter can establish only its carbonaceous makeup, not its biological origin (SCHOPF et al., 2002c). Yet if the three factors are taken together, the strength of the interpretation markedly changes. Other than biology, no mechanism is known that can yield communities of fossil-like objects that exhibit cellular morphologies, produce "biological" isotopic compositions, and are themselves composed of carbonaceous (kerogenous) organic matter. Thus, rather than being weighed separately, the three biogenic indicators are best considered together as a suite of mutually reinforcing lines of evidence. It is in this regard that

---

FIG. 5. Optical photomicrographs (a, b), atomic force images (c–g), and computer-generated sketch (h) of parts of a permineralized spheroidal Proterozoic microfossil exposed at the upper surface of a petrographic thin section of stromatolitic chert from the ~ 650-Ma-old Chichkan Formation of southern Kazakhstan (KEMPE et al., 2002). (a) Specimen transected at its equatorial plane by the surface of the thin section, superimposed by a plot of an electron microprobe measurement showing that carbon is concentrated in and near its cell wall. (b) The lower half of the specimen, embedded in the chert matrix, showing the characteristic surface texture of this taxon of Precambrian microfossil. (c) Atomic force micrograph showing the exposed cell wall analyzed. (d) The portion of the wall imaged in (e–g) and illustrated in (h), a segment of the wall shown in the lowermost part of (c). (e) That portion of the wall indicated by the white rectangle in (d), shown here at higher magnification. (f) That portion of the wall indicated by the white rectangle in (e), shown here at higher magnification. (g) That portion of the wall shown in (f) with a superimposed computer-generated outline of its component parts. (h) A computer-generated sketch showing the ~200-nm-sized stacked angular platelets of kerogen that make up the permineralized cell wall.

the newly applied analytical techniques—ion microprobe mass spectrometry and Raman imagery combined with AFM—with their capability to link directly prime biogenic indicators in *individual* microscopic fossils, can be used most effectively.

EPILOGUE

In some quarters, it will be assumed that the impetus resulting in development of these novel techniques is a byproduct of the remarkable recent upsurge of interest in astrobiology, an exciting wide-ranging area of science almost universally regarded as "new". In actuality, however, astrobiology is new in name only, its focus on life elsewhere in the solar system stemming from NASA-supported exobiologic studies of the 1950s, and its concern with early life on Earth, from the emergence of Precambrian paleobiology a decade or so later. Indeed, this new field owes its sparkling success to groundwork laid years ago by pioneering interdisciplinary scientists who in their prescience saw that the combination of geology, chemistry, and biology, together, were far more likely to yield useful insight than any one of these disciplines alone. And in the vanguard of this seminal movement was none other than the honoree of the symposium for which this paper has been prepared—Issac R. Kaplan—a master of interdisciplinary science who deserves due credit as a pioneering practitioner of astrobiology long before it even existed!

Acknowledgements—For reviews of this manuscript, I thank J. Leventhal and C. Fedo, and for providing me the opportunity to contribute to this volume and for helpful suggestions regarding the research reported here, I thank, respectively, R. J. Hill and J. Shen-Miller—like myself, both appreciative admirers of I. R. Kaplan. Supported by NASA, through Grant NAG 5-12357 and the Astrobiology Institute.

REFERENCES

AROURI K. R., GREENWOOD P. F. and WALTER M. R. (2000) Biological affinities of Neoproterozoic acritarchs from Australia: microscopic and chemical characterisation. *Org. Geochem.* **31**, 75–89.

BARGHOORN E. S. and SCHOPF J. W. (1965) Microorganisms from the late Precambrian of central Australia. *Science* **150**, 337–339.

BARGHOORN E. S. and TYLER S. A. (1965) Microorganisms from the Gunflint chert. *Science* **147**, 563–577.

BRASIER M. D., GREEN O. R., JEPHCOAT A. P., KLEPPE A. K., VAN KRANENDONK M. J., LINDSAY J. F., STEELE A. and GRASSINEAU N. V. (2002) Questioning the evidence of Earth's oldest fossils. *Nature* **416**, 76–81.

CLOUD P. (1965) Significance of the Gunflint (Precambrian) microflora. *Science* **148**, 27–45.

FEDO C. and WHITEHOUSE M. J. (2002) Metasomatic origin of quartz-pyroxene rock, akilia, Greenland and implications for Earth's earliest biosphere. *Science* **296**, 1448–1452.

HAYES J. M., KAPLAN I. R. and WEDEKING K. W. (1983) Precambrian organic geochemistry, preservation of the record. In *Earth's Earliest Biosphere, Its Origin and Evolution* (ed. J. W. SCHOPF), pp. 93–134. Princeton University Press, Princeton, NJ.

HOERING T. C. (1967) The organic geochemistry of Precambrian rocks. In *Researches in Geochemistry.* (ed. P. H. ABELSON), Vol 2, pp. 87–111. Wiley, New York.

HOUSE C. H., SCHOPF J. W., MCKEEGAN K. D., COATH C. D., HARRISON T. M. and STETTER K. O. (2000) Carbon isotopic composition of individual Precambrian microfossils. *Geology* **28**, 707–710.

JOY J. W., WILLIS A. J. and LACEY W. S. (1956) A rapid cellulose peel technique in palaeobotany. *Ann. Bot. Nova Scotia* **20**, 635–637.

KAZMIERCZAK J. and KREMER B. (2002) Thermal alteration of Earth's oldest fossils. *Nature* **420**, 477–478.

KEMPE A., SCHOPF J. W., ALTERMANN W., KUDRYSVTSEV A. B. and HECKL W. M. (2002) Atomic force microscopy of Precambrian fossils. *Proc. Natl Acad. Sci. USA* **99**, 9117–9120.

KUDRYAVTSEV A. B., SCHOPF J. W., AGRESTI D. G. and WDOWIAK T. J. (2001) *In situ* laser-Raman imagery of Precambrian microscopic fossils. *Proc. Natl Acad. Sci. USA* **98**, 823–826.

MENDELSON C. V. and SCHOPF J. W. (1992) Proterozoic and selected Early Cambrian microfossils and microfossil-like objects. In *The Proterozoic Biosphere, a Multidisciplinary Study* (eds. J. W. SCHOPF and C. KLEIN), pp. 865–951. Cambridge University Press, Cambridge.

OBERLIN A., BOULMIER J. L. and VILLEY M. (1980) Electron microscopic study of kerogen mictrotexture. Selected criteria for determining the evolution path and evolution stage of kerogen. In *Kerogen, Insoluble Organic Matter from Sedimentary Rocks* (ed. B. DURAND), pp. 191–231. Éditions Technip, Paris.

PARK R. and EPSTEIN S. (1963) Carbon isotopic fractionation during photosynthesis. *Geochim. Cosmochim. Acta* **21**, 110–115.

PASTERIS J. D. and WOPENKA B. (2002) Images of the Earth's earliest fossils? *Nature* **420**, 476–477.

SCHOPF J. W. (1968) Microflora of the Bitter Springs Formation, late Precambrian, central Australia. *J. Paleontol.* **42**, 651–688.

SCHOPF J. W. (1975) Precambrian paleobiology: problems and perspectives. *Annu. Rev. Earth Planet. Sci.* **3**, 213–249.

SCHOPF J. W. (1992a) Paleobiology of the Archean. In *The Proterozoic Biosphere, a Multidisciplinary Study* (eds. J. W. SCHOPF and C. KLEIN), pp. 25–39. Cambridge University Press, Cambridge.

SCHOPF J. W. (1992b) Proterozoic prokaryotes: affinities, geologic distribution, and evolutionary trends. In *The Proterozoic Biosphere, a Multidisciplinary Study* (eds. J. W. SCHOPF and C. KLEIN), pp. 195–218. Cambridge University Press, Cambridge.

SCHOPF J. W. (1992c) Evolution of the Proterozoic biosphere: benchmarks, tempo, and mode. In *The Proterozoic Biosphere, a Multidisciplinary Study* (eds. J. W. SCHOPF and C. KLEIN), pp. 583–600. Cambridge University Press, Cambridge.

SCHOPF J. W. (1993) Microfossils of the Early Archean Apex chert: new evidence of the antiquity of life. *Science* **260**, 640–646.

SCHOPF J. W. (2002) *Geochemistry and Submicron-Scale Structure of Individual Microfossils*, Abstracts, Annual Meeting. Geological Society of America, Denver, CO, p. 152.

SCHOPF J. W. Earth's earliest biosphere: status of the hunt. In *Tempos and Events in Precambrian Time* (ed. W. Altermann). Elsevier, New York, (in press).

SCHOPF J. W. and BLACIC J. M. (1971) New microorganisms from the Bitter Springs Formation (late Precambrian) of the north-central Amadeus Basin, Australia. *J. Paleontol.* **45**, 925–960.

SCHOPF J. W. and WALTER M. R. (1983) Archean microfossils: new evidence of ancient microbes. In *Earth's Earliest Biosphere, Its Origin and Evolution* (ed. J. W. SCHOPF), pp. 214–239. Princeton University Press, Princeton, NJ.

SCHOPF J. W., KUDRYAVTSEV A. B., AGRESTI D. G., CZAJA A. D. and WDOWIAK T. J. (2002a) Laser-Raman chemistry and organic metamorphism of Precambrian microscopic fossils. *Eos. Trans. Am. Geophys. Union* **83**(19), S372–S373 (Abstract).

SCHOPF J. W., KUDRYAVTSEV A. B., AGRESTI D. G., WDOWIAK T. J. and CZAJA A. D. (2002b) Laser-Raman imagery of Earth's earliest fossils. *Nature* **416**, 73–76.

SCHOPF J. W., KUDRYAVTSEV A. B., AGRESTI D. G., WDOWIAK T. J. and CZAJA A. D. (2002c) Images of the Earth's earliest fossils—reply. *Nature* **420**, 477 p.

SCHOPF J.W., KUDRYAVTSEV A. B., AGRESTI D. G., CZAJA A. D. and WDOWIAK T. J. Laser-Raman chemistry and organic metamorphism of Precambrian microscopic fossils. *Precambrian Res*, (in press).

STRAUSS H. and MOORE T. B. (1992) Abundances and isotopic compositions of carbon and sulfur species in whole rock and kerogen samples. In *The Proterozoic Biosphere, a Multidisciplinary Study* (eds. J. W. SCHOPF and C. KLEIN), pp. 709–798. Cambridge University Press, Cambridge.

STRAUSS H., DES MARAIS D. J., HAYES J. M., LAMBERT I. B. and SUMMONS R. E. (1992a) Procedures of whole rock and kerogen analysis. In *The Proterozoic Biosphere, a Multidisciplinary Study* (eds. J. W. SCHOPF and C. KLEIN), pp. 699–708. Cambridge University Press, Cambridge.

STRAUSS H., DES MARAIS D. J., HAYES J. M. and SUMMONS R. E. (1992b) Concentrations of organic carbon and maturities, and elemental compositions of kerogens. In *The Proterozoic Biosphere, a Multidisciplinary Study* (eds. J. W. SCHOPF and C. KLEIN), pp. 95–99. Cambridge University Press, Cambridge.

TYLER S. A. and BARGHOORN E. S. (1954) Occurrence of structurally preserved plants in pre-Cambrian rocks of the Canadian shield. *Science* **119**, 606–608.

TYLER S. A. and BARGHOORN E. S. (1963) Ambient pyrite grains in Precambrian cherts. *Am. J. Sci.* **261**, 424–432.

UENO Y., ISOZAKI Y., YURIMOTO H. and MARUYAMA S. (2001) Carbon isotopic signatures of individual Archean microfossils(?) from Western Australia. *Int. Geol. Rev.* **40**, 196–212.

VAN ZUILEN M. A., LEPLAND A. and ARRHENIUS G. (2002) Reassessing the evidence for the earliest traces of life. *Nature* **418**, 627–630.

WALTON J. (1928) A method of preparing sections of fossil plants contained in coal balls or in other types of petrifaction. *Nature* **122**, 571 p.

WOPENKA B. and PASTERIS J. D. (1993) Structural characterization of kerogens to granulite-facies graphite: applicability of Raman microprobe spectroscopy. *Am. Mineral.* **78**, 533–557.

Geochemical Investigations in Earth and Space Science: A Tribute to Isaac R. Kaplan
© The Geochemical Society, Publication No. 9, 2004
Editors: R.J. Hill, J. Leventhal, Z. Aizenshtat, M.J. Baedecker, G. Claypool,
R. Eganhouse, M. Goldhaber and K. Peters

Origin and migration of methane in gas hydrate-bearing sediments in the Nankai Trough

AMANE WASEDA[1] and TAKASHI UCHIDA[2]

[1]JAPEX Research Center, Japan Petroleum Exploration Co., Ltd, 1-2-1 Hamada, Mihama-ku,
Chiba 261-0025, Japan
[2]Technology Research Center, Japan National Oil Corporation, Mihama-ku, Chiba 261-0025, Japan

Abstract—Carbon and hydrogen isotope compositions of methane and hydrocarbon compositions in gas hydrate-bearing shallow sediments in the Nankai Trough show that the methane is generated by microbial reduction of CO_2. The $\delta^{13}C$ values of CH_4 range from -96 to $-63‰$ in the upper 300 m sediments. Both $\delta^{13}C$ values of CH_4 and CO_2 become more positive with increasing depth. The preferential depletion of $^{12}CO_2$, progressive decrease in microbial activity with depth and upward gas migration through the sediments column explain the $\delta^{13}C$ depth profiles. In deeper horizons, the origins of gases change from microbial to thermogenic at around 1500 mbsf (meters below seafloor). Gases shallower than 1500 mbsf have lower $\delta^{13}C$ values of CH_4 (lower than $-59‰$), while gases deeper than 1500 mbsf have higher $\delta^{13}C$ values of CH_4 (-48 to $-35‰$), typical for gases generated by thermal decomposition of organic matter. The measured total organic carbon (TOC) in the Nankai Trough is around 0.5%, which is considered too low for *in situ* formation of gas hydrate. Consequently, some gas migration and accumulation processes are required for the concentrated formation of the gas hydrates (up to 80% in pore space) in the Nankai Trough. This process may be related to the geological setting of the Nankai Trough, where fluid flow containing methane is active through thrust systems within Nankai accretionary prism sediments. There is, however, no indication of thermogenic gases in shallow sediment including the hydrate-bearing intervals, suggesting that the fluid migration is rather local and restricted to the shallow sediments.

INTRODUCTION

NATURAL GAS HYDRATES in sediments are widely distributed in offshore continental margins and in polar regions. The amount of methane contained in gas hydrates is currently estimated to be between 10^{15} and 10^{17} m^3 (KVENVOLDEN, 2000), considered an important natural resource in the future. Natural gas hydrates are also believed to be a factor in global climate changes (HAQ, 2000) and submarine geologic hazards (PAULL *et al.*, 2000b).

In 1995, Japan's Ministry of Economy, Trade and Industry (METI), formerly Ministry of International Trade and Industry (MITI), launched a project to explore for marine gas hydrate accumulations around Japan. From late 1999 to early 2000, an exploratory hole "MITI Nankai Trough" was drilled on the landward side of the eastern Nankai Trough, offshore Japan (Fig. 1) by Japan National Oil Corporation (JNOC) along with Japan Petroleum Exploration Co., Ltd (JAPEX) as the well operator. The water depth at the drill site was 945 m and the sub-bottom depth of the hole was 2355 m. The seismic bottom simulating reflector (BSR) is present at around 295 mbsf. There were two exploration objectives: one was a gas hydrate survey in shallow Quaternary sediments and the other was conventional oil and gas exploration in deeper Tertiary sediments. In addition to the main hole, seven short holes (two site survey, two pilot and three post-survey holes) were also drilled for the gas hydrate survey around the main hole. In this paper, we will clarify the origins of methane in gas hydrates found in the MITI Nankai Trough Well and discuss gas migration and hydrate formation in the sediments.

SAMPLES AND ANALYTICAL METHODS

Gas and sediment samples derived from the main hole and a post-survey hole were analyzed. Gas samples were collected using the headspace gas method. Cuttings were placed in a can,

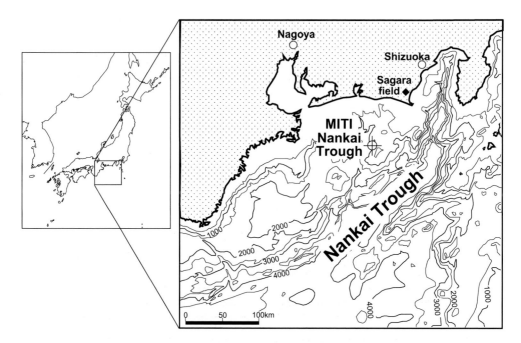

FIG. 1. Locality map of the MITI Nankai Trough Well. Contour lines: water depth (m).

previously prepared with a septa-covered entry port. Distilled water was added to the can leaving about 200 cm^3 headspace; then, the can was firmly sealed. In the laboratory, the cans were shaken with ultrasonic wave to partition sediment gas into the headspace, and a portion of the resulting headspace gas was analyzed. Gas samples were also collected from the pressure–temperature-core sampler (PTCS) used in gas hydrate-bearing intervals. Data from the site survey holes and part of the data from the main hole have been reported elsewhere (WASEDA et al., 1998; WASEDA and UCHIDA, 2002).

The molecular compositions of hydrocarbon gases were determined on a Shimadzu GC-7A gas chromatograph. For isotopic analyses methane was separated from the other gas components by gas chromatography and subsequently combusted to CO_2 and H_2O over CuO at 850°C, using a vacuum preparation line (SCHOELL, 1980). The H_2O was then reduced to H_2 by reaction with zinc in sealed glass tubes at 480°C (VENNEMANN and O'NEIL, 1993). The stable carbon and hydrogen isotope compositions of methane were measured using a VG Isotech Sira Series II mass spectrometer. Isotope ratios are reported in the usual δ-notation relative to the PDB (Pee Dee Belemnite) standard for carbon and SMOW (Standard Mean Ocean Water) standard for hydrogen. The reproducibility of isotope values is ±0.15‰ for $\delta^{13}C$ and ±3‰ for δD.

TOC measurements were performed on sediment samples. Wet sediment samples were dried for 16 h at 60°C and then crushed and acidified with 6N HCl to remove carbonate. The TOC contents of carbonate-free residue were determined using a Yanaco MT-3 CHN Corder. The TOC contents of the whole rock were then calculated with correction for carbonate removal.

For vitrinite reflectance (R_0) measurements, coarsely crushed samples were acidified with hydrochloric and hydrofluoric acid to remove carbonate and silicate. The solution was then centrifuged with heavy liquid to separate kerogen. The kerogen was embedded in a resin plug and polished to a flat shiny surface. Measurements of the percentage of incident light reflected from vitrinite particles under oil immersion with a Carl Zeiss MPM-03 microspectrophotometry system were made.

GAS HYDRATE OCCURRENCE

In addition to the visible gas hydrates recovered in cores, several lines of indirect evidence for gas hydrate occurrences were derived. Downhole geophysical logging shows that the gas

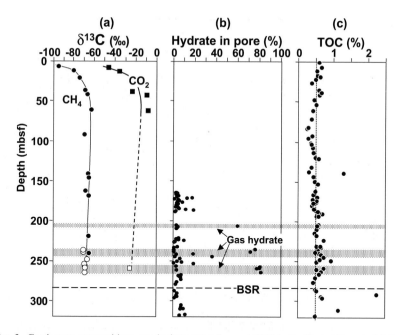

Fig. 2. Gas isotope compositions, gas hydrate saturation and TOC in the Nankai Trough. (a) Depth-trends of $\delta^{13}C$ of CH_4 and CO_2. Open symbols, PTCS samples; solid symbols, headspace gas samples. Some of the data are from WASEDA et al. (1998) and WASEDA and UCHIDA (2002). (b) Depth-trend of gas hydrate saturation estimated from Cl^- anomalies. Data are from UCHIDA et al. (2004). (c) Depth-trend of TOC. Some of the data are from WASEDA et al. (1998).

hydrates are present in three discrete intervals (1–10 m thickness) around 207 mbsf, between 235 and 245 mbsf and between 255 and 265 mbsf. Gas hydrate saturation in pore space is estimated to range from 40 to 80% in the three intervals, based on the resistivity logs. Anomalies of low chloride contents of interstitial water are used as a proxy for the hydrate saturation, since Cl^- is excluded during hydrate formation. The chloride anomalies also show the three gas hydrate layers, and gas hydrate saturation in pore space in the layers is estimated from Cl^- dilution to be as high as 80% (Fig. 2), which is consistent with the log data. Moreover, core temperature depressions were also observed in each hydrate layer. The sediments are composed of sandy and muddy beds. The gas hydrate was observed as intergranular pore fillings in the sandy beds.

ORIGINS OF METHANE IN THE NANKAI TROUGH GAS HYDRATE

Molecular and isotopic compositions of the gas samples recovered from gas hydrate-bearing intervals of the MITI Nankai Trough Well indicate that the origin of the gases is microbial reduction of CO_2. The $C_1/(C_2 + C_3)$ ratios are greater than 4000 and their $\delta^{13}C$ values of CH_4 range from -71 to $-66‰$. The compositions of the Nankai Trough gas hydrates are similar to the Blake Ridge gas hydrates (Fig. 3), indicating that the origin of methane in both localities is microbial.

There are two main pathways for microbial methanogenesis: CO_2 reduction and acetate fermentation. These two processes can be distinguished from their hydrogen isotope compositions of methane. Fermentation-derived methane is characterized by its depletion in deuterium, resulting in very negative δD values less than $-250‰$ (SCHOELL, 1988). δD values of CH_4 range from -193 to $-189‰$, similar to the Blake Ridge gas hydrate, indicating that the methanogenesis pathway is CO_2 reduction (Fig. 4). CO_2 reduction is generally the major methanogenic pathway in marine environment (RICE and CLAYPOOL, 1981; WHITICAR et al., 1986).

The $\delta^{13}C$ parallel depth-trends of CH_4 and CO_2 in shallow sediments also suggest that the origin of gas is microbial CO_2 reduction. $\delta^{13}C$ data from the main and post-survey holes

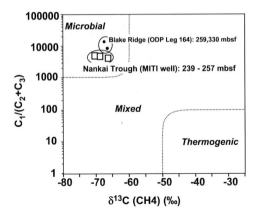

FIG. 3. Interpretive plot of molecular ratios of hydrocarbons vs. $\delta^{13}C$ of methane in gas hydrate from the Nankai Trough and the Blake Ridge (MATSUMOTO *et al.*, 2000). Genetic classification is modified from BERNARD (1978).

(Table 1) are shown in Fig. 2, together with the data from the site survey holes (WASEDA *et al.*, 1998). The $\delta^{13}C$ values of CH_4 increases with depth parallel to the depth-trend of $\delta^{13}C$ of co-existing CO_2 in the upper 60 m sediments in the Nankai Trough (Fig. 2). Similar parallel trends have been observed in other gas hydrate-bearing sediments, such as the Blake Ridge (Fig. 5a; PAULL *et al.*, 2000a), offshore Guatemala (JEFFREY *et al.*, 1985) and offshore Peru (KVENVOLDEN and KASTNER, 1990) and also in many other gas hydrate-free sediments, such as Cariaco Trench (CLAYPOOL and KAPLAN, 1974).

These parallel $\delta^{13}C$ changes of the CH_4 and CO_2 with depth in the uppermost 100 or 200 m sediments are consistent with gradual ^{13}C-depletion of substrate CO_2 (Fig. 6; CLAYPOOL and KAPLAN, 1974). However, the curvature of the trend line for the Rayleigh fractionation model supposing a constant conversion rate (Fig. 6) is different from the observed data (Figs. 2 and 5). This indicates that the conversion rate, namely microbial activity, progressively decreases with depth. Though the change in activity of methanogenic archaea with depth in marine sediments is not clear, it is reported that the total number of bacteria decreases logarithmically with depth (GETLIFF *et al.*, 1992; PARKES *et al.*, 1994). Another difference between the closed-system Rayleigh model and the observed data is found in the deeper sediments. Below 100 or 200 mbsf,

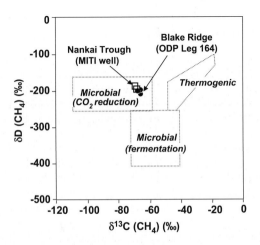

FIG. 4. Interpretive plot of hydrogen and $\delta^{13}C$ of methane in gas hydrate from the Nankai Trough and the Blake Ridge (MATSUMOTO *et al.*, 2000). Genetic classification is modified from SCHOELL (1988).

Table 1. Gas molecular composition and $\delta^{13}C$ of methane in pressure–temperature-core sampler (PTCS) from the post-survey hole-2 and headspace gas samples from the main hole of the MITI Nankai Trough Well

Depth (mbsf)	Sample type	O_2 (%)	N_2 (%)	C_1 (ppm)	C_2 (ppm)	C_3 (ppm)	i-C_4 (ppm)	n-C_4 (ppm)	CO_2[a] (ppm)	$\frac{C_2 - C_4}{C_1 - C_4}$ (%)	$\delta^{13}C$ (C_1) (‰)
Gas in PTCS from the post-survey hole-2											
254	PTCS	0.12	1.84	980,355	0	0	0	0	0	0.00	− 69.6
266	PTCS	0.34	5.07	944,343	0	0	0	0	1572	0.00	− 69.1
Headspace gas from the main hole											
168.20	Core	9.31	89.74	9035	0	0	0	0	470	0.00	− 65.3
755	Cuttings	1.31	95.12	35,136	0	0	0	0	510	0.00	− 67.4
855	Cuttings	1.65	90.22	81,247	0	0	0	0	0	0.00	− 65.9
955	Cuttings	1.28	90.75	79,763	0	0	0	0	0	0.00	− 65.8
1055	Cuttings	1.28	96.67	5102	0	0	0	0	15,489	0.00	n.d.
1155	Cuttings	1.36	93.22	54,161	0	0	0	0	0	0.00	− 62.1
1255	Cuttings	1.35	95.90	25,924	0	0	0	0	1547	0.00	− 60.5
1263.95	Core	2.33	93.28	42,091	0	0	0	0	1858	0.00	− 63.1
1355	Cuttings	1.92	94.74	33,398	0	0	0	0	0	0.00	− 62.7
1455	Cuttings	1.51	95.66	27,266	0	0	0	0	1028	0.00	− 59.0
1555	Cuttings	1.24	84.28	67,948	1599	0	0	0	75,213	2.30	− 47.2
1655	Cuttings	1.28	89.90	85,753	2511	0	0	0	0	2.84	− 40.8
1755	Cuttings	1.29	92.34	61,313	2346	0	0	0	0	3.69	− 37.1
1855	Cuttings	1.34	94.90	34,404	1465	0	0	0	1676	4.09	− 35.9
1955	Cuttings	1.42	95.93	24,497	2013	0	0	0	0	7.59	− 37.3
2055	Cuttings	1.34	92.80	53,633	4358	653	0	0	0	8.54	− 38.9
2103.05	Core	1.94	90.47	53,517	4903	923	0	0	16,548	9.82	− 36.4
2155	Cuttings	1.29	90.88	69,056	6882	1591	821	0	0	11.86	− 41.4
2255	Cuttings	1.32	90.99	65,254	9006	1825	807	0	0	15.14	n.d.
2355	Cuttings	2.50	95.80	12,310	3482	1202	0	0	0	27.57	− 41.2

[a] Erratic high CO_2 data might be due to microbial activities in the containers after sampling.
n.d.: not determined.

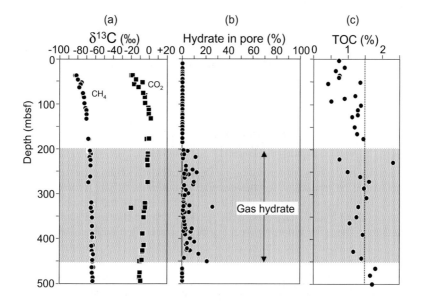

FIG. 5. Gas isotope compositions, gas hydrate saturation and TOC in Site 977, ODP Leg 164, Blake Ridge. (a) Depth-trends of $\delta^{13}C$ of CH_4 and CO_2. Data are from PAULL et al. (2000a,b). (b) Depth-trend of gas hydrate saturation estimated from Cl^- anomalies. Modified from SHIPBOARD SCIENTIFIC PARTY (1996). (c) Depth-trend of TOC. Data are from SHIPBOARD SCIENTIFIC PARTY (1996).

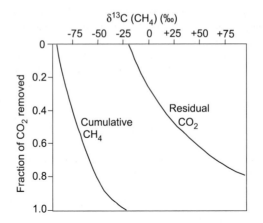

FIG. 6. Carbon isotope fractionation in closed-system single-stage Rayleigh processes. Modified from CLAYPOOL and KAPLAN (1974).

the $\delta^{13}C$ values of CH_4 and CO_2 no longer increase in the Nankai Trough and the Blake Ridge sediments (Figs. 2 and 5). PAULL et al. (2000a) explained this phenomenon, based on a model simulation of CH_4 and CO_2 profiles of the Blake Ridge gases, supposing upward gas migration through the sediment column in addition to the progressive decrease in microbial activity with depth.

Consequently, the molecular and isotope data are consistent with CO_2 reduction being the dominant methanogenic process in marine sediments. The isotope depth profiles are due to progressive decrease in microbial activity with depth and upward gas migration through the sediment column.

GAS HYDRATE SATURATION AND FLUID FLOW

As shown above, the molecular and isotope composition of the methane in gas hydrates of the Nankai Trough is almost identical to that of the Blake Ridge, and the isotope profiles in the shallow sediments including the hydrate-bearing sediments of the two areas are also similar. However, the distribution and pore saturation of the gas hydrate in the sediments are different between the two areas. In the Nankai Trough, gas hydrates exist as thin layers (1–10 m) in sands with high pore saturation, up to 80% (Fig. 2), while in the Blake Ridge, they are distributed over a thick zone (250 m thickness) of muds with low pore saturation, 5–6% on average (Fig. 5). In the following section, we discuss the *in situ* production, migration and accumulation of methane as gas hydrates in the two areas taking into account the TOC levels in the sediments and the geological settings.

Since the source of carbon for microbial methanogenesis is organic matter in sediments, the TOC content would be important for the generative-potential of gas hydrates (KVENVOLDEN, 1993; WASEDA, 1998). Let us assume that all the excess amount of methane beyond the solubility in pore water can form hydrate and ignore the porosity change due to the conversion of organic matter to methane. Then, the volume percent of pore space filled with gas hydrate *in situ* at the base of gas hydrate stability zone can be calculated. WASEDA (1998) summarized the results of the calculated hydrate volume percent filled in pore space by *in situ* microbial methane production as a function of TOC under various marine conditions. Water depth, sediment porosity and microbially utilizable organic carbon in TOC are the major factors influencing the amount of hydrate. Among these factors, the utilizable organic carbon (UOC) percent has the strongest influence on the calculated results. CLAYTON (1992) estimated that the amount of carbon utilized in the methanogenesis zone is about 10% of the original TOC based on a mass balance calculation of atomic C, H and O in organic matter.

In the Blake Ridge, gas hydrate formation by *in situ* microbial methanogenesis can roughly explain the average amount of the hydrates in the sediments. The average volume percent of the hydrate filled in pore space of the sediments is estimated to be about 5–6% by the pore water chloride content anomalies between 200 and 450 mbsf (MATSUMOTO *et al.*, 1996). If the hydrate-filled pore space by *in situ* microbial methanogenesis is 5 vol% of the total pore space, about 2 wt% TOC is required in the case of water depth of 3000 m and UOC of 10% (WASEDA, 1998). The TOC values in the gas hydrate-bearing sediments between 200 and 450 mbsf range from 0.8 to 2.3 wt%, and the average is 1.4 wt% (Fig. 5; SHIPBOARD SCIENTIFIC PARTY, 1996). Therefore, the calculated TOC value is somewhat higher but comparable to the measured values in the hydrate-bearing sediments.

On the other hand, in the Nankai Trough, the TOC contents in the sediment are too low to explain the highly saturated hydrate layers. The TOC data from the site survey holes (WASEDA *et al.*, 1998) and the main hole (Table 2) are shown together in Fig. 2. The TOC contents in the Nankai Trough are around 0.5 wt%. In the above calculation, gas hydrate is hardly formed by *in situ* microbial methane production with 0.5% TOC. However, gas hydrate concentrated layers were found at the MITI well and the saturation in pore is estimated to be as high as 80 vol%. Consequently, some migration and accumulation processes are likely required for the concentrated formation of the gas hydrates in the Nankai Trough. Gas hydrate samples were also recovered from other marine sediments such as offshore Oregon, Mexico, Guatemala and Peru. These gas hydrate-bearing sediments have higher TOC values (1.5–3.0% on average) than the Nankai Trough sediments (WASEDA, 1998).

Occurrences of gas hydrates, average TOC, geological settings and fluid migration patterns in the Nankai Trough and the Blake Ridge are summarized in Table 3. The difference in the gas hydrate distribution in the sediments between the Nankai Trough and the Blake Ridge may be related to the difference in geological settings. The Nankai Trough is located on an active margin with accretionary prism, where fluid flows are active through thrust systems in the accretionary prism sediments. In such geological settings, fluids containing methane migrate selectively in sand layers with higher porosities and permeabilities and excess methane beyond the solubility in pore water could be trapped in pore space as hydrate under $P–T$ conditions of hydrate stability, provided that the flow rate is not too high. Moreover, coarse-grained sand layers with lower capillary pressures would have concentrated methane by diffusion from nearby mud sediment layers with higher capillary pressures. As shown above, the Nankai Trough sediments are mainly composed of sand and mud layers of turbidite origin and the gas hydrates are concentrated only in the sand layers. On the other hand, the Blake Ridge is located on a passive margin, where the fluid flows are rather slow and pervasive, and the sediments are muddy. In such geological settings, gas hydrate may be uniformly formed according to the TOC contents in the nearby sediments.

THERMOGENIC GAS IN DEEP SEDIMENTS

In the deeper horizons of the MITI Nankai Trough Well, the origins of gases change from microbial to thermogenic at around 1500 mbsf (Fig. 7). Gases shallower than 1500 mbsf show lower $\delta^{13}C$ values of CH_4 (lower than $-59‰$), while gases deeper than 1500 mbsf show higher $\delta^{13}C$ values of CH_4 (-48 to $-35‰$), typical for gases generated by thermal decomposition of organic matter. The sediments above 2100 mbsf have R_0 values less than 0.5%, suggesting that the organic matter is immature for petroleum generation. The sediments below 2100 mbsf have R_0 values higher than 0.5%, indicating that the organic matter in the sediments is within the range of thermogenic hydrocarbon generation. Therefore, the thermogenic gases existing between 1500 and 2100 mbsf are interpreted to have migrated from the thermally mature zone below 2100 mbsf.

Although the gases shallower than 1500 mbsf are mainly microbial in origin, the $\delta^{13}C$ values of CH_4 in gas samples between 1000 and 1500 mbsf are higher than those in the gas hydrate. This may indicate that the fluid migration is rather local and restricted within the upper 1000 m of sediments utmost. The $\delta^{13}C$ values of CH_4 in headspace gases between 755 and 955 mbsf

Table 2. Total organic carbon (TOC) content and vitrinite reflectance (R_0) of cuttings and cores collected from the main hole of the MITI Nankai Trough Well

Depth (mbsf)	Sample type	TOC (%)	R_0 (%)
166.65	Core	0.47	0.24
171.00	Core	0.52	
176.00	Core	0.55	
180.15	Core	0.42	
185.50	Core	0.49	
189.30	Core	0.69	
206.75	Core	0.59	
210.40	Core	0.54	
223.20	Core	0.72	
233.40	Core	0.65	
236.00	Core	0.59	
240.30	Core	0.63	
242.50	Core	0.66	
245.10	Core	0.75	
248.75	Core	0.95	
251.00	Core	0.54	
258.60	Core	0.73	0.26
260.85	Core	0.55	
262.45	Core	0.67	
265.60	Core	0.61	
270.60	Core	0.51	
277.00	Core	0.46	
285.00	Core	0.48	
288.20	Core	0.54	
291.00	Core	2.27	
295.00	Core	0.66	
297.00	Core	0.71	
311.90	Core	1.14	
318.25	Core	0.49	
755	Cuttings	0.55	0.26
855	Cuttings	0.69	0.24
955	Cuttings	0.67	0.25
1055	Cuttings	0.67	0.29
1078.0	SWC	0.88	
1155	Cuttings	0.59	0.28
1174.5	SWC	0.58	
1255	Cuttings	0.59	0.27
1256.00	Core	0.25	0.29
1261.85	Core	0.42	0.33
1263.70	Core	0.40	0.31
1286.0	SWC	1.07	
1326.0	SWC	0.91	
1355	Cuttings	0.34	0.24
1455	Cuttings	0.83	0.33
1508.0	SWC	0.53	
1552.0	SWC	0.94	
1555	Cuttings	1.04	0.31
1647.0	SWC	0.44	
1655	Cuttings	0.90	0.23
1755	Cuttings	0.99	0.28
1772.0	SWC	0.68	
1791.0	SWC	0.56	
1855	Cuttings	0.74	0.36
1924.0	SWC	0.43	
1955	Cuttings	0.71	0.31
2033.0	SWC	0.91	
2055	Cuttings	0.33	0.35
2103.00	Core	0.90	0.59
2106.00	Core	0.60	0.65
2155	Cuttings	0.79	0.63
2238.5	SWC	0.95	
2255	Cuttings	0.83	0.53
2355	Cuttings	0.67	0.58

Blanks: not determined. SWC: sidewall core.

Table 3. Occurrences of gas hydrates, origin of gas, average TOC, geological settings and fluid migration patterns in the Nankai Trough and the Blake Ridge

Area	Gas hydrate layer		Lithology	Origin of gas	Average TOC (*in situ* methane production)	Geological setting	Fluid flow
	Thickness	Saturation					
Nankai Trough MITI well	Thin (1–10 m) three layers (200–270 mbsf)	High (max. 80% in pore)	Sand and mud (hydrate in sand)	Microbial CO$_2$ reduction	0.5% (too low to explain the highly saturated hydrate layers)	Active margin (accretionary prism)	Active flow to permeable sand layer
Blake Ridge ODP Leg 164	Thick (250 m) zone (200–450 mbsf)	Low (av. 5–6%, max. 25% in pore)	Mud	Microbial CO$_2$ reduction	1.5% (roughly enough to explain the average hydrate saturation)	Passive margin	Slow pervasive flow

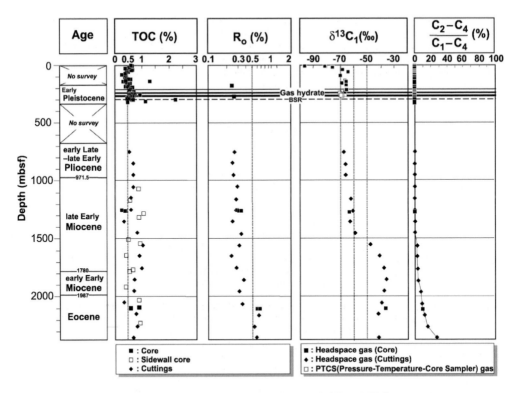

FIG. 7. Geochemical log of the MITI Nankai Trough Well.

range from −67.4 to −65.8‰, similar to the $\delta^{13}C$ values of CH_4 in gas hydrate-bearing intervals between 207 and 265 mbsf, while the $\delta^{13}C$ values of CH_4 in headspace gases between 1055 and 1455 mbsf range from −63.1 to −59.0‰, slightly higher than the shallower gases, suggesting some contribution of thermogenic methane from deeper sediments.

Although no indication of thermogenic gas was found in the gas hydrate at the well location, it is possible that thermogenic gases could migrate to the shallow sediments and form gas hydrate that seal gas beneath the gas hydrate in the Nankai Trough area. The recently depleted Sagara oil field is located 70 km northeast of the MITI Nankai Trough Well (Fig. 1) and produced a total of 4560 kl oil from shallow reservoirs between 150 and 250 m (INOMA, 1982), providing evidence that long migration of thermogenic hydrocarbon is possible in this area. The possibility of thermogenic hydrocarbon generation is also supported by the existence of the mature Eocene–Miocene petroleum source rocks, which have been confirmed in the MITI well and other wells in this area.

CONCLUSIONS

Molecular and isotopic compositions indicate that the methane in gas hydrate is generated by microbial reduction of CO_2 in the sediments of the Nankai Trough. Progressive decrease in microbial activity with depth and some upward gas migration through the sediment column are inferred from the isotope depth profiles. The *in situ* generative-potential of microbial methane is relatively low in the Nankai Trough due to low TOC. Consequently, upward migration of microbial methane and selective accumulation into permeable sands may account for the high concentration of methane hydrates in discrete sand layers. Since the Nankai Trough is located on an active margin with accretionary prism, fluid migration should be active. However, no indication of thermogenic gases in the hydrate-bearing intervals suggests that the fluid migration is rather local and restricted within the upper 1000 m of sediments at the well location.

Acknowledgements—Researchers participated in the gas hydrate study of the MITI Nankai Trough Well are gratefully acknowledged for their collaboration. The authors are indebted to Ryo Matsumoto of Tokyo University for his valuable comments and advice. The assistance of Hirotsugu Iwano and Yoshiteru Kajiwara in TOC and gas analyses is appreciated. We are grateful to Japan National Oil Corporation for authorizing the publication of the data from the MITI Nankai Trough Well. Reviews by George Claypool and Thomas Lorenson greatly improved the manuscript.

REFERENCES

BERNARD B. B. (1978) Light hydrocarbons in marine sediments. Ph.D. Dissertation, Texas A&M University, College Station, TX.

CLAYPOOL G. E. and KAPLAN I. R. (1974) The origin and distribution of methane in marine sediments. In *Natural Gases in Marine Sediments* (ed. I. R. KAPLAN), pp. 99–139. Plenum Publishing, New York.

CLAYTON C. (1992) Source volumetrics of biogenic gas generation. In *Bacterial Gas* (ed. R. VIALLY), pp. 191–204. Technip, Paris.

GETLIFF J. M., FRY J. C., CRAGG B. A. and PARKES R. J. (1992) The potential for bacteria growth in deep sediments layers of the Japan Sea, Hole 798B—Leg 128. In *Proc. ODP, Sci. Results 127/128 (Pt. 1)* (eds. K. A. PISCIOTTO, J. C. INGLE JR., M. T. VON BREYMANN, J. BARRON *et al.*), pp. 755–760. Ocean Drilling Program, College Station, TX.

HAQ B. U. (2000) Climate impact of natural gas hydrate. In *Natural Gas Hydrate in Oceanic and Permafrost Environments* (ed. M. D. MAX), pp. 137–148. Kluwer Academic Publishers, Dordrecht.

INOMA A. (1982) Sagara and Kakegawa areas. *Japanese Oil and Natural Gas Resources.* Japanese Association of Natural Gas Mining, pp. 246–276 (in Japanese).

JEFFREY A. W. A., PFLAUM R. C., McDONALD T. J., BROOKS J. M. and KVENVOLDEN K. A. (1985) Isotopic analysis of core gases at Sites 565–570, Deep Sea Drilling Project Leg 84. In *Init. Repts. DSDP 84* (eds. R. VON HUENE, J. AUBOUIN *et al.*), pp. 719–726. US Government Printing Office, Washington, DC.

KVENVOLDEN K. A. (1993) Gas hydrates—geological perspective and global change. *Rev. Geophys.* **31**, 173–187.

KVENVOLDEN K. A. (2000) Gas hydrate and humans. *Ann. NY Acad. Sci.* **912**, 17–22.

KVENVOLDEN K. A. and KASTNER M. (1990) Gas hydrates of the Peruvian outer continental margin. In *Proc. ODP, Sci. Results 112* (eds. E. SUESS, R. VON HUENE *et al.*), pp. 517–526. Ocean Drilling Program, College Station, TX.

MATSUMOTO R., WATANABE Y., SATOH M., OKADA H., HIROKI Y., KAWASAKI M., ODP LEG 164 SHIPBOARD SCIENTIFIC PARTY, (1996) Distribution and occurrence of marine gas hydrates—preliminary results of ODP Leg 164: Blake Ridge Drilling. *J. Geol. Soc. Jpn* **102**, 932–944 (in Japanese with English abstract).

MATSUMOTO R., UCHIDA T., WASEDA A., UCHIDA T., TAKEYA S., HIRANO T., YAMADA K., MAEDA Y. and OKUI T. (2000) Occurrence, structure and composition of natural gas hydrates recovered from the Blake Ridge, Northwest Atlantic. In *Proc. ODP, Sci. Results 164* (eds. C. K. PAULL, R. MATSUMOTO, P. J. WALLACE and W. P. DILLON), pp. 13–28. Ocean Drilling Program, College Station, TX.

PARKES R. J., CRAGG B. A., BALE S. J., GETLIFF J. M., GOODMAN K., ROCHELLE P. A., FRY J. C., WEIGHTMAN A. J. and HARVEY S. M. (1994) Deep bacterial biosphere in Pacific Ocean sediments. *Nature* **371**, 410–413.

PAULL C. K., LORENSON T. D., BOROWSKI W. S., USSLER W. III and RODRIGUEZ N. M. (2000a) Isotopic composition of CH_4, CO_2 species, and sedimentary organic matter within samples from the Blake Ridge: gas source implications. In *Proc. ODP, Sci. Results 164* (eds. C. K. PAULL, R. MATSUMOTO, P. J. WALLACE and W. P. DILLON), pp. 67–78. Ocean Drilling Program, College Station, TX.

PAULL C. K., USSLER W. III and DILLON W. P. (2000b) Potential role of gas hydrate decomposition in generating submarine slope failures. In *Natural Gas Hydrate in Oceanic and Permafrost Environments* (ed. M. D. MAX). pp. 149–156. Kluwer Academic Publishers, Dordrecht.

RICE D. D. and CLAYPOOL G. E. (1981) Generation, accumulation and resource potential of biogenic gas. *Am. Assoc. Pet. Geol. Bull.* **65**, 5–25.

SCHOELL M. (1980) The hydrogen and carbon isotopic composition of methane from natural gases of various origins. *Geochim. Cosmochim. Acta* **44**, 649–661.

SCHOELL M. (1988) Multiple origins of methane in the earth. *Chem. Geol.* **71**, 1–10.

SHIPBOARD SCIENTIFIC PARTY (1996) Site 997. In *Proc. ODP. Init. Repts. [CD-ROM] 164* (eds. C. K. PAULL, R. MATSUMOTO, P. J. WALLACE *et al.*), pp. 77–334. Ocean Drilling Program, College Station, TX.

UCHIDA T., LU H., TOMARU T., DALLIMORE S. R. and NANKAI TROUGH SCIENTIFIC PARTY (2004) Subsurface occurrence of natural gas hydrate in the Nankai Trough area: implication for gas hydrate concentration. *Resource Geol.* **54**, 35–44.

VENNEMANN T. W. and O'NEIL J. R. (1993) A simple and inexpensive method of hydrogen isotope and water analyses of minerals and rocks based on zinc reagent. *Chem. Geol.* **103**, 227–234.

WASEDA A. (1998) Organic carbon content, bacterial methanogenesis, and accumulation processes of gas hydrates in marine sediments. *Geochem. J.* **32**, 143–157.

WASEDA A. and UCHIDA T. (2002) Origin of methane in gas hydrates from the Mackenzie Delta and Nankai Trough. *Proceedings of the Fourth International Conference on Gas Hydrates, Yokohama*, pp. 169–174.

WASEDA A., BABA K., YAGI M., MATSUMOTO R., LU H., HIROKI Y. and FUJII T. (1998) Geochemical characteristics of gases and sediments in the Nankai Trough, offshore Tokai. *Proceedings of the International Symposium on Methane Hydrates: Resources in the Near Future?, Chiba*, pp. 55–59.

WHITICAR M. J., FABER E. and SCHOELL M. (1986) Biogenic methane formation in marine and freshwater environments: CO_2 reduction vs. acetate fermentation—isotope evidence. *Geochim. Cosmochim. Acta* **50**, 693–709.

Geochemical Investigations in Earth and Space Science: A Tribute to Isaac R. Kaplan
© The Geochemical Society, Publication No. 9, 2004
Editors: R.J. Hill, J. Leventhal, Z. Aizenshtat, M.J. Baedecker, G. Claypool,
R. Eganhouse, M. Goldhaber and K. Peters

Seasonal methane emissions by diffusion and ebullition from oligohaline marsh environments in coastal Louisiana

JOEL S. LEVENTHAL[1,2] and GLENN R. GUNTENSPERGEN[3]

[1]Diversified Geochemistry, 8944 W. Warren Dr. Lakewood, CO 80227, USA
[2]U.S. Geological Survey, Federal Center MS 973, Denver, CO 80225, USA
[3]U.S. Geological Survey, Patuxent Wildlife Research Center, Laurel, MD 20708, USA

Abstract—Methane is an important atmospheric greenhouse gas that is emitted from many natural and anthropogenic sources. In order to evaluate the global methane budget, precise data are needed from the diverse sources including coastal wetlands. Over 100 time-series determinations of methane emissions from an oligohaline wetland (brackish marsh) in coastal Louisiana show large variability during five seasonal sampling periods. Emission by both diffusion and ebullition (bubbles) was measured, however, neither of these emission modes were strongly dependent on either water depth or temperature (except in winter). Methane emission to static collectors placed over plants (*Scirpus olneyi* and *Spartina patens*) was not significantly different from shallow open water or mud. However, considerable heterogeneity in methane emissions and processes occurs even at a single site. Thus, establishing a reasonable estimate of the overall methane emission for a particular marsh environment and season requires multiple measurements at several sites. The average emissions for April, May, July, and September ranged from 31 to 54 mg/m^2/h (744–1296 mg/m^2/day). This can be separated into emissions from diffusion ranging from 8.3 to 20 mg/m^2/h (18–50% of total) and emissions due to ebullition of 20–44 mg/m^2/h (50–82%). January emissions were much lower, amounting to 0.2 mg/m^2/h (6 mg/m^2/day), mainly by diffusion with only one episode of ebullition. Extrapolating these data to annual emissions gives total annual methane emissions of 203 g/m^2/yr (61 g/m^2/yr by diffusion and 142 g/m^2/yr by ebullition).

INTRODUCTION

Global change and methane from wetlands

INCREASED GREENHOUSE gas concentrations in the atmosphere are resulting in warmer atmosphere and surface ocean temperatures (IPCC, 2001; WIGLEY and RAPER, 1992). This warming will lead to rising sea level by both thermal expansion and melting of polar ice. Rising sea level will cause changes in coastal wetlands due to drowning of salt marshes and increased salinization of brackish, intermediate, and freshwater coastal marshes. Salinization will also result in a net decrease in freshwater marsh areal extent (PARK *et al.*, 1989) and a net decrease in methane emissions from coastal wetlands worldwide (TITUS, 1990) because methane emission from marsh environments is inversely related to salinity (DELAUNE *et al.*, 1983; BARTLETT *et al.*, 1987). The emission gradient generated by changing salinity supports the negative feedback between sea level rise and decreased coastal marsh methane emissions that has been proposed (LEVENTHAL, 1991) and results from the presence of alternative electron acceptors, such as sulfate.

Methane emissions from wetlands are an important part of the methane budget (CICERONE and OREMLAND, 1988; KHALIL and RASMUSSEN, 1990) and were even more important contributors to the atmospheric methane content in the past (CHAPPELLEZ *et al.*, 1990). WHITING and CHANTON (1993) have proposed that methane emission is controlled by primary production because they found that about 3% of net primary production is emitted as methane in a wide range of environments from subtropical to arctic. Trends in methane emission are important because methane is a greenhouse gas and is probably responsible for as much as 25% of the increased greenhouse warming during the 1980s (DICKENSON and CICERONE, 1986; LASHOF, 1989; LASHOF and AHUJA, 1990). However, the ranges of methane emissions observed from various wetlands vary by more than three orders of magnitude depending on the type of wetland and season (MATTHEWS and FUNG, 1987; ASELMANN and CRUTZEN, 1989, Table 2; BARTLETT and HARRISS, 1992, Tables 1 and 3) and landscape heterogeneity (DISE, 1993; DISE *et al.*, 1993; BUBIER *et al.*, 1995; DUCHEMIN *et al.*, 1995; PRIEME, 1994).

As a result of these uncertainties, the actual contribution of wetlands to the atmospheric methane budget is even more poorly known than the factor-of-two that is usually quoted (CICERONE and OREMLAND, 1988). Another important consideration is that the areal extent of various types of wetlands worldwide is only known to a factor of two (ASELMANN and CRUTZEN, 1989; FUNG *et al.*, 1991; CHAPMAN, 1977). BARTLETT *et al.* (1989) have discussed the large variations of methane fluxes measured from adjacent but diverse wetlands (Everglades) and the related problems of estimating a global wetland flux. Thus, the uncertainty in methane emissions from wetlands may vary by a factor of two or three.

Coastal marsh methane

Only a few sets of measurements of methane emissions and methane porewater depth profiles from coastal transitional oligohaline marshes have been made (DELAUNE *et al.*, 1983; BARTLETT *et al.*, 1987). DELAUNE *et al.* (1983) were the first to measure methane emissions along a salinity gradient (Barataria basin) in coastal Louisiana. DELAUNE *et al.* data from three differing salinity regimes (saline, brackish, and fresh marsh) showed an increase in methane emission as water salinity ranged from 14 ppt (parts per thousand) salinity to fresh (0.2 ppt). Their results spanned 14 months and represented two or three measurements (replicates, from the same collector chamber) every other month. They give mean values for each site, but variations at a given site and at a given time are not presented. However, the average bimonthly values presented show large variations (more than 50%) for the summer months (May to September). BARTLETT *et al.* (1987) present emissions data for Virginia coastal marshes that ranged in salinity from 24 ppt to fresh. Their data for two sites (at each salinity regime) per month show variations of 20–50% for the summer months during highest emissions. However, neither of these groups discusses the reasons for the high variability of the summer paired monthly values or mentions how much, if any, of the emissions are due to ebullition (bubbles). After our work was completed, ALFORD *et al.* (1997) continued the work of the Louisiana State University group measuring monthly methane fluxes from three wetland habitats over a 3-year period again using permanent base units. Their results included new observations that they recognized as ebullition and considerable seasonal and spatial variation as before.

Ebullition

Bubble ebullition is the one-way transport of gases from organic-rich sediments to the water column and then to the atmosphere. It represents one of the major mechanisms for gas exchange in freshwater and coastal marine ecosystems (CHANTON and WHITING, 1995). BARTLETT *et al.* (1990) and DEVOL *et al.* (1990) both report that ebullition is the major mechanism of methane emission to the troposphere from the Amazon River floodplain. Work by WILSON *et al.* (1989) and HAPPELL and CHANTON (1993) reports that ebullition is 19 and 45%, respectively, of the annual methane flux for fresh water marshes and that it is more important during the warm summer months of maximum daily emissions. More recently, KELLER and STALLARD (1994) report on bubbling from shallow lakes in Panama and show that ebullition events are triggered by wind. However, the published data (up to 1996) for coastal marshes in Louisiana do not give information on the amount or importance of ebullition. WEYHENMEYER (1999) reported on methane emission from beaver ponds in a forest from Ontario, Canada where 65% of the total flux was due to ebullition.

Present study

Work reported here was part of the U.S. Geological Survey efforts in the U.S. National Global Change Research Program (COMMITTEE ON EARTH AND ENVIRONMENTAL SCIENCES, 1994).

This project evaluates the varying emissions of methane along a salinity gradient in Louisiana wetlands along the modern Gulf Coast of North America. These wetlands and their methane emissions are expected to be severely impacted by sea level rise in the next century (GORNITZ, 1991). This study area is particularly relevant because it has already been affected by apparent sea level rise. These sediments are subsiding due to lack of sediment input from rivers that have been diverted and dammed, compaction and organic matter oxidation of Holocene sediments, and withdrawal of oil and gas.

Our original aim was to determine more precise and detailed methane emissions data from this wetland type and to measure the various controls (water depth, vegetation, etc.) on the quantity of emissions. For logistical reasons we sampled only four times (July and September 1991 and January and May 1992) before Hurricane Andrew in August 1992 and once after that (April 1994). Hurricane Andrew was an extratropical cyclone with peak winds of 225 km/h. It made landfall along the Louisiana coast 20–40 km from our study sites (GUNTENSPERGEN et al., 1995; GUNTENSPERGEN and VAIRIN, 1996). In this report, we show the characteristics of methane emissions from four to eight collector locations at 2–5 sites in the oligohaline (intermediate to brackish) environment of the Terrebonne Basin, Gulf Coast, Louisiana. We show the characteristics of the methane emissions from individual ebullition events and the diffusion measurements during 5-month (seasonal) sampling periods. We also describe the physical and biological setting from a variety of locations that are all within this oligohaline marsh. All of these details can be combined to show the heterogeneity encountered in the natural environment and that the seasonal methane emissions we document are greater than those reported by DeLAUNE et al. (1983). Our results (LEVENTHAL, 1995) stress the importance of multiple measurements of methane fluxes and distinguishing the contributions of diffusion and ebullition to arrive at a meaningful flux for a given site for a particular time (day/week/month). The results also suggest the need for many more measurements from the important, but diverse, wetland methane sources in order to understand the ranges in emission, and the need to better understand some of the major controls on emissions from these sources. We believe that the global methane budgets and atmospheric models for methane will have considerable uncertainty until this detailed work is done.

Study sites

Our sites are located in Terrebonne Parish, Louisiana, about 100 km southwest of New Orleans, near Lake DeCade (Latitude 29°22′30″N, Longitude 98°57′30″W; Fig. 1) (LEVENTHAL et al., 1993). This region is part of the Terrebonne Basin, which is an ancient deposition center for the Mississippi River (PENLAND et al., 1990). All of this area is less than 2 m above mean sea level (Lake Mechant and Lake Penchant USGS Quadrangle maps, 7.5 min series, 1:24,000). This area is as undisturbed as can be found in the Gulf Coast, it is more than 15 km from any road and, in general, any boats in the area use natural channels or man-made canals that are more than 500 m away. These oligohaline marshes adjacent to Jug Lake and Otter Bayou are typical of approximately 2500 km² of Gulf Coast marsh (O'NEIL, 1949; REYER et al., 1988; CHABRECK and LINSCOMBE, 1978; PENLAND et al., 1990).

At these sites, the National Biological Service established a marsh management and research project to study the effects of impoundment on the stability of marsh ecosystems (GUNTENSPERGEN et al., 1993; PENDELTON et al., 1992). Methane measurements were made from sites established by placement of raised boardwalks in the spring of 1991. Under most conditions, the boardwalks were between 6 and 24 cm above the water surface. All measurements were made at these sites before any of the water-holding, -control, and -diversion structures were emplaced during Spring, 1992 (FOOTE et al., 1993), except for the May 1992 data that were made only at control sites and the April 1994 measurements made after Hurricane Andrew (August 1992).

The wetlands at this site are classified as brackish marshes, based on the presence of indicator plants (PENFOUND and HATHAWAY, 1938; O'NEIL, 1949; WICKER, 1980; GOSSELINK, 1984) and are mostly dominated by *Spartina patens*, with subordinate amounts of *Eleocharis* spp., *Vigna lutea*, *Baopa monneiri*, *Aster tenufolius*, and *Scirpus olneyi* (GUNTENSPERGEN et al., 1995).

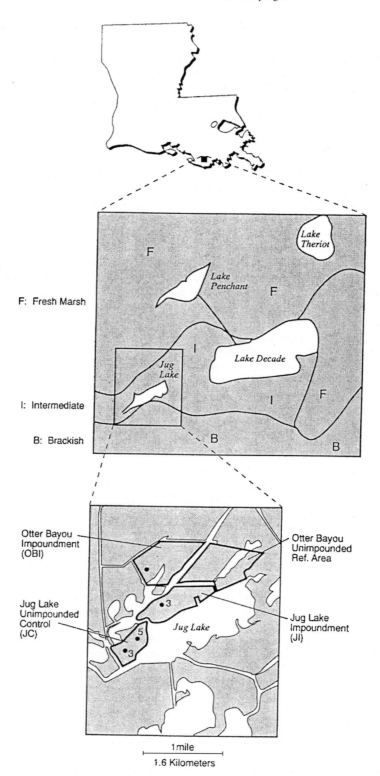

F: Fresh Marsh

I: Intermediate

B: Brackish

Otter Bayou
Impoundment
(OBI)

Otter Bayou
Unimpounded
Ref. Area

Jug Lake
Unimpounded
Control
(JC)

Jug Lake
Impoundment
(JI)

1 mile
1.6 Kilometers

Water hyacinth (a floating macrophyte) was also present at some sites. However, in spite of these brackish to intermediate vegetation indicators, surface water at the time of these measurements, and throughout most of the year has less than 0.8 ppt salinity and interstitial water 0.8–3 ppt (from 1 to 50 cm depth) during our 1991 and 1992 sampling (HARGIS et al., 1993; JACKSON et al., 1995). Thus, based on salinity, these marshes would be classified as oligohaline (COWARDIN et al., 1979) or intermediate (GOSSELINK, 1984; for review see HACKNEY et al., 1996). The marsh study areas vary from 20 to 50% open water (i.e. 50–80% vegetated) in part because of their deteriorating conditions resulting from regional subsidence due to lack of sediment accretion keeping pace with apparent sea level rise to maintain surface elevation. Pipeline canals are another impact in coastal Louisiana, that account for about 10% of the marsh area and along with the natural channels also about 10% of the area (REED and ROZAS, 1995) that allow more rapid water movement along a salinity gradient. During this century, Louisiana marshes have been disappearing at an alarming rate, although the rate has slowed during the past decade (BRITSCH and KEMP, 1990). However, rates of land loss in the Terrebonne basin continue to be high. Yearly loss rates were somewhat higher in 1978–1990 (26.4 km^2/yr) than in 1956–1978 (24.1 km^2/yr) (BARRAS et al., 1994).

Water height (depth) conditions in the marsh are minimally influenced by the Gulf tides, accounting for less than 15 cm increase. The main controls on water depth are the frontal passages bringing fresh water from the north, southerly winds and offshore tropical storms pushing saline Gulf water into the marshes and local/regional rainfall (ROBERTS and VAN HEERDEN, 1992). Some of this rainfall comes from afternoon thunderstorms that can be local and quite violent in terms of wind and amount of precipitation. Local thunderstorms occurred in this area during the weeks of the July 1991 and May 1992 sampling. For each sampling series, the season, salinity, and temperature variables were relatively constant during our measurements and therefore cannot be responsible for the differences in methane emission during a single week of measurements (July 1991, September 91, and May 92 sediment temperatures were 29–30°C; in April 1994 sediment temperature was 25.5°C). January 1992 sediment temperature was much cooler (11°C).

We made measurements at seven locations that represent the range of natural variability in this environment. The sites represented 7 of the 20 boardwalks put in place for the marsh management study (GUNTENSPERGEN et al., 1993). At each boardwalk we used four collectors at 4–8 locations to quantify methane emissions. Some measurements were made by placing collectors over live vegetation, some over open water above a mat, and some above non-vegetated sediment (mud). The year 1991 was very wet (the wettest year on record) and by mid-July the rainfall was 100% of normal for the full year. Thus, water depths were somewhat greater and the water less saline than in other years. For example, 1992 (also a wet year, 20% above normal) showed considerably higher surface salinity values of 1–2 ppt in July 1992 than in 1991 (HARGIS et al., 1993).

FIELD AND MEASUREMENT METHODS

Field procedures

Approach to the boardwalks was made by airboat at a speed as slow as possible and with as little environmental disturbance as possible. Despite this caution, bubbles (that we now know are mainly methane) were sometimes observed from nearby sediments and we could smell H$_2$S immediately after we stopped the airboat at the boardwalk. The airboat approach path was always to the same end of the boardwalk, whereas the methane measurements were made from

FIG. 1. Location of oligohaline (brackish-intermediate) marshes in Terrebonne Parish, Louisiana: top: map of Louisiana, small square is area shown as blow-up in center; center: marsh area around Lake Decade (90°45′W, 29°40′N) showing F—fresh, I—intermediate, and B—brackish marsh in 1988 (from CHABRECK and LINSCOMBE, 1992), outlined square (Jug Lake) is enlarged at bottom; bottom: location of marsh measurements (filled circles) adjacent to Jug Lake and Otter Bayou. Note scale: 1 mile = 1.6 km (from USGS Lake Penchant Quad).

the middle to the far end of the boardwalks. To minimize any possible disturbance, only one person (JL) was on the boardwalk at the time of the measurements.

Methane collectors and measurement methods are described elsewhere (LEVENTHAL, 1992; CLAYTON et al., 1993). Briefly, a portable flame ionization detector and hydrogen carrier gas with an internal pump (SIP-1000, Summitt Interests, Lyons, CO) for sampling the collectors and providing real-time results at regular intervals are used to measure methane accumulated in the static collectors. The detection range of the instrument is from 0.2 to 10,000 ppm CH_4, but methane accumulated in the collectors from the marsh was generally in the range of 10–200 ppm. The instrument was calibrated with methane standards of 10, 50 and 100 ppm (Scott Specialty Gases, Longmont, CO).

Methane collectors and measurement

The collectors are 11-liter plastic buckets approximately 30 cm in diameter with a 6 mm O.D. plastic tube that is 1–2 m long for sampling the accumulated gas. The buckets float on the surface of the water (ZIMMERMAN, 1979) or positioned over small plants or in open water areas and also directly adjacent to plants that were too large (over 30 cm in diameter or height) to fit under the collector. At each site, four buckets were carefully positioned (with floats) on the water surface in order that the underlying sediment is not disturbed. Our portable static chamber method contrasts with the often used method of a permanently placed (i.e. one location) collar that penetrates the sediments and is left in place, but requires physical connection and sealing of the above-water collector to the base before measurement.

During our experiments, gas usually accumulates for 30–50 min with sampling/measurement of CH_4 at 6–10 min intervals. Using these real-time measurements, it is possible to observe ebullition events or air leakage (loss of seal of the collector) or other disturbance during collection. Thus, any special or unusual events were noted (and sometimes followed-up with replicate or modified measurements) or spurious readings were discarded and/or collectors resealed or repositioned.

Porewater measurements

Samples of porewater (and contained gas) were taken at the sites just after the CH_4 emission measurements were completed. Samples were collected using a "sipper", which is a 3 mm O.D., 100 cm long rigid plastic tube that is inserted into the sediment. An attached syringe (10 or 20 ml) with a valve allows collection of samples of porewater (and gas) from various depths (5–100 cm). The samples were usually analyzed within 6 h (but sometimes refrigerated overnight and analyzed within 24 h). For samples without at least 20% gas phase in the syringe, air (for head space) was added just prior to analysis at room temperature. The syringes were individually shaken for 60 s to transfer the methane to the gas phase just prior to analysis. A small septum was fitted on the valve on the syringe. A 0.5 cc gas-tight syringe (Precision Sampling Corp., Baton Rouge, LA) was used to remove 0.2 cc of gas, which was then injected into the 0.7 m long × 3 mm O.D. Teflon column packed with Porapak Q using H_2 carrier gas and detected with the flame ionization detector (for more details see LEVENTHAL, 1992).

Calculations of emission rates

Emission rates were calculated from the slopes of the lines on the methane vs. time plots or the final accumulated methane quantity. For diffusion, which is relatively constant for an individual 30–45 min collection, the slope can be multiplied by a constant factor to get emissions per minute (or per hour, day, month or year). Thus, instrument signal increases over a given time (ppm methane per minute) and can be multiplied by a factor (14.7, see LEVENTHAL, 1992) to convert to mg/m^2/day for each time interval (or for the entire 30–45 min experiment) for which there is a set of measurements, based on the area and volume of the collection chamber.

Abrupt increases in collected methane due to ebullition are treated as occurring during the time between measurements (usually 6–10 min), during which the diffusion is usually masked. Methane due to ebullition is the difference between the total methane accumulated (at the end of the 45 min experiment) minus the rate calculated from the linear (diffusion) part of the total emission. In the collections where ebullition did not occur or occurred during the first or last timed interval, the data were tested with linear-least-squares regressions of time vs. methane accumulated. The data show correlation coefficients of >0.96 (15 sets), or 0.83–0.91 (4 sets) (for example for the July data). Some collections with shorter time (fewer points) or ebullition in the middle collection interval or very large ebullition events and/or multiple ebullition events could not be tested statistically.

RESULTS

Methane emissions

Typical results of methane accumulation in the collectors at individual locations are shown in Fig. 2. In Fig. 2, collectors "m" and "o" show linear increases in methane accumulation, whereas collector "p" shows an abrupt increase between 30 and 37 min due to ebullition. Figure 3 shows results for eight locations of the JI-3 boardwalk site. If the ebullition event occurred during the time between setting the collector in place and the first measurement it has the number 1 at the side of the bar (Fig. 3). This first-interval ebullition accounts for 13 of 27 ebullition events during the July 1991, measurements, far more than expected for random ebullition events. In addition, five of these first-interval events were for collectors placed over emergent live plants. Emissions due to ebullition (shaded areas on the bars, Fig. 3) account for 40–80% of the total methane emissions but differ greatly between sites and even within each site (Fig. 4a–e).

Porewater methane

Figure 5a–d shows methane in porewater at various depths near one (or several) of the collectors used for emissions. The porewater measurements were always made after the emissions measurements were completed to avoid disturbance of the sediment or sediment–water interface. In July 1991 (Fig. 5a), below 20 cm all the sites show water that is near saturation with methane (1 mM at this water temperature). Below 25 cm, the methane is supersaturated in the samples from the two profiles in the mud at site JC-3. Similar data confirm saturation or supersaturation at at least one depth for all the other seasons. The methane profiles for May 92 show the most methane. Interestingly, the profiles with the most methane are those with the lowest measured diffusion and ebullition emissions for collectors at location OB-4, whereas higher emissions are associated with lower porewater methane for JC-3 and JC-5.

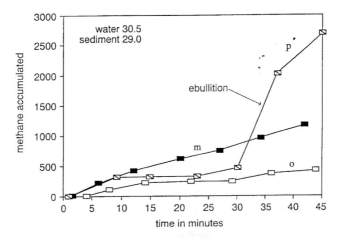

FIG. 2. Boardwalk JC-3 (July 1991). Results of simultaneous measurements at three nearby locations; collector m over dead spartina, water depth 20–25 cm; collector o—open water, 5–15 cm depth; collector p—open water, depth 12–20 cm. The precision of the measurement is about the size of data point.

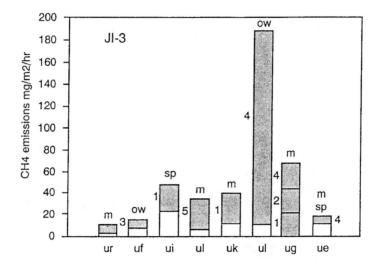

FIG. 3. Histograms of methane emissions from site JI-3 (July 1991) boardwalk (8 collectors/locations). Solid bars show methane emission by ebullition, open bar is diffusion. Symbols above bar: ow—open water; m—over vegetation mat; sp—collector over *S. patens*; scir—*S. olneyi*. The interval during which the ebullition occurred is indicated by the number to the left of the bar, 1 (for the first interval), 2, 3, etc. for subsequent intervals.

Methane ebullition and emission variability

Figures 2–4 show many collectors that have large ebullition events. Some occur during the first 5 min of methane accumulation. While it is possible that this may have been caused by the initial placement of the collectors on the surface of the water (between 4 and 8 cm deep; depth determined after the methane measurements were completed), there are many examples of ebullition that were not associated with the initial placement of the collectors. During the

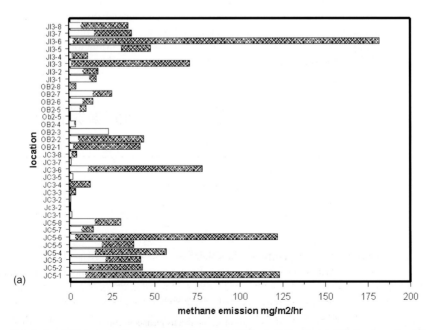

FIG. 4. (a–e) Histograms of methane emission by ebullition (solid bars) and diffusion (open bars) for the five seasonal sets of data. Note varying scale for emissions values. (a) July 1991.

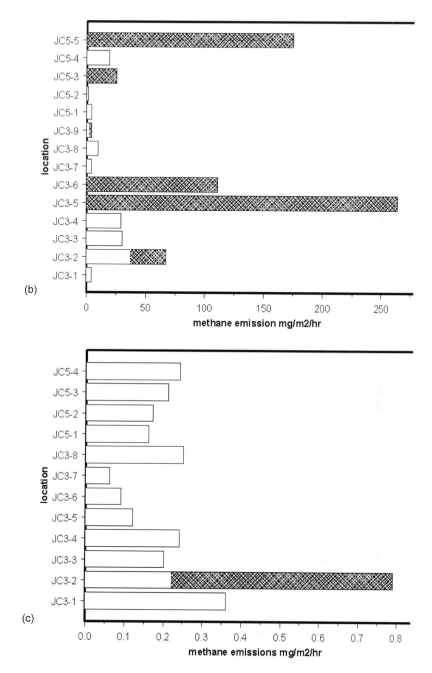

FIG. 4 (*continued*) (b) September 1991, (c) January 1992.

collection and measurement of methane, bubbles were sometimes observed to be emitted spontaneously from the water surface, in one case associated with a documented methane ebullition event. Furthermore, if the water–sediment system is supersaturated with methane (*i.e.* a bubble has already formed in the porewater and is poised for release) then ebullition could be activated by wind, barometric pressure decrease, burrowing or other animal activity, and represents natural methane emission to the atmosphere. This is true even if the placement of the chamber triggered the release somewhat prematurely. The transport of methane through live

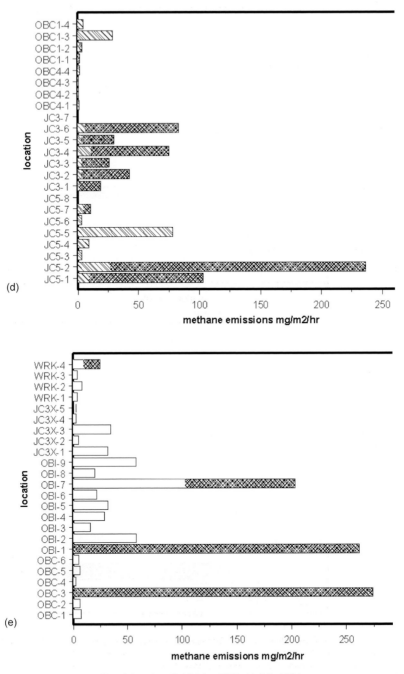

FIG. 4 (*continued*) (d) May 1992, (e) July 1994.

S. patens and *S. olneyi* was tested by the placement of collectors over plants, but there is no systematic increase (excess) in those emissions compared to collectors that were not over plants.

At all sites, one or more of the collectors showed some methane emission by ebullition (except in January 1992). Thus, ebullition is not only interesting but is also important because it is often the dominant mode of emission during those months when the marsh sediments are supersaturated. In contrast to the study sites of MARTENS and BERNER (1974), the ebullition

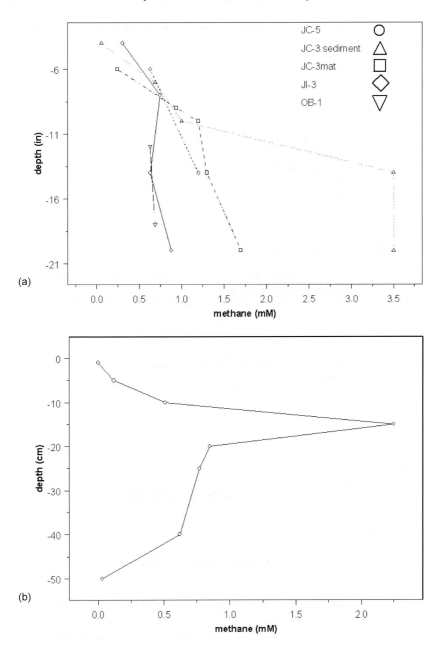

Fig. 5. (a) July 1991. Five depth profiles for methane porewater (and gas) samples. Note that water saturation is 1.06 mM. Thus, two of the JC-3 sediment samples are quite supersaturated and have methane in the gas phase; three deeper samples from JC-3 mat and one from OBI-1 are slightly supersaturated. (b) September 1991 at site JC-3.

here is probably not driven by tidal interaction or forcing because of the distance from the open Gulf and lack of direct connection and a characteristic microtidal event would result in less than 15 cm change in water height.

The relative proportion of methane emission by ebullition varies from 0 to 99%. This methane ebullition and its heterogeneity was unexpected because ebullition had not been reported from brackish marshes in Louisiana (DELAUNE *et al.*, 1983) or Virginia (BARTLETT *et al.*, 1987). Methane ebullition was first reported by MARTENS and BERNER (1974) for tidally influenced

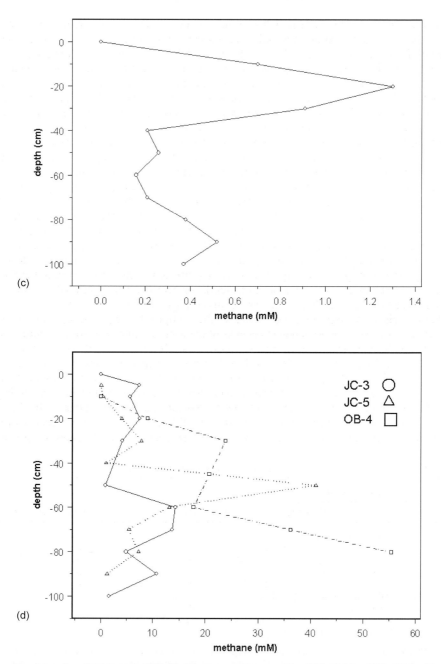

Fig. 5 (*continued*) (c) January 1992. Methane is not supersaturated at 11C. (d) May 1992. OB-4 with highest porewater methane has lowest emissions (see Table 1 and text).

areas and also more recently in freshwater areas of the Amazon (BARTLETT *et al.*, 1990; DEVOL *et al.*, 1990; KELLER and STALLARD, 1994). In July 1991, there were a total of 10 experiments (time series measurements) where collectors were over or adjacent to emergent vegetation. During initial placement of the collector it is possible that the collector touched the emergent plant causing it to move in the water or its below water parts to move the sediment and that this disturbance was the cause of the observed ebullition. However, great effort was made not to disturb the plant, thus these methane bubbles probably were preformed and ready for release and would have been released during the next gust of wind or disturbance by an aquatic animal. Four collector locations showed a second ebullition event and one location showed a third ebullition event. In summary, for July 1991, 14 of 31 ebullition events were during the first timed collection interval. Similar results were noted for May 1992, with 7 of 12 being first interval events and September 1991, with four of eight being first events and for April 1994 with three of four first events. Tables 1 and 2 separate the results into diffusion only, diffusion + ebullition, and diffusion + ebullition but with the first interval ebullition being omitted (indicated by " * " Table 1). In some cases, a second set of measurements (made at five of the locations in July 1991) at the same location during the next hour showed similar emission mode (diffusion and ebullition) and rate at the particular location. Thus, it appears that the methane emission is not temporally variable for a given site. However, when the second measurement was made at a nearby (2–5 m distant) location it often gave different emission results that indicate that the methane emission is spatially variable. These differences in rate of methane emission and mode observed from collectors on the same boardwalk but approximately 5 m apart illustrate the large range of methane emissions that occurs on even a very fine geographic scale. Thus, the use of many locations (collectors) allows a better estimate of the overall methane emission from the marsh than measurements from only one or two fixed location collectors as has been done in the past. Plotting these overall (diffusion + ebullition) results for the five seasonal mean values on a log scale tends to mask some of the differences but still shows the large variation of emissions due to the complex controls on emissions (Fig. 6).

Methane emissions from ebullition events in these oligohaline marshes are quite variable. This variation is especially pronounced because of the irregular (but common) ebullition events that mask the diffusion emission for the next 5–10 min. The size of bubbles needed to account for the observed ebullition is calculated to be 0.1–5 mm in diameter or multiple bubbles of this size or smaller. For example, in July 1991, 60% of the collectors showed one or two ebullition events that were responsible for between 50 and 95% of the accumulated methane during the 45 min to 1 h collection time for those chambers. Similar results were obtained for May and September (but not for January). We believe that ebullition is very important because of these timed collection results and because we have observed spontaneous ebullition events during

Table 1. Methane emissions by location and site

Date site	dif	dif + ebu	dif + ebu[a]	n	ebu (%)	ebu (%)[a]	# n w/ebu	# ebu events	# 1st ebu[a]
CH$_4$ emissions (mg/m^2/h)									
July 1991 (30°C, sediment salinity n.d.)									
JC5	12	43	14	8	72	14	4	4	3
JC3	2	12	4	8	83	50	5	8	2
JI3	10	54	42	8	81	75	8	10	3
OBI1	10	16	11	8	37	10	7	9	6
September 1991 (30°C, sediment salinity 0.5–3‰)									
JC5	5.2	46	24	5	89	79	2	3	2
JC3	13	58	32	9	78	59	4	6	3
January 1992 (11°C, sediment salinity 1.3–3.1‰)									
JC5	0.2	0.26	0.2	8	25	0	1	1	1
JC3	0.2	0.2	0.2	4	0	0	0	0	0
May 1992 (29°C, sediment salinity 0.8–3.3‰)									
JC5	12	63	21	7	81	43	3	3	3
JC3	4.5	40	12	7	89	62	7	8	6
OBC4	0.6	0.8	0.8	4	25	25	1	1	1
OBC1	9.1	9.8	9.8	4	7	7	2	2	1
April 1994 (25°C, sediment salinity n.d.)									
OBC	4.3	50	4.3	6	16	0	1	1	1
OBI	49	78	49	9	11	0	1	1	0
JC3x	19	19	19	4	0	0	0	0	0
Wrack	6.3	7.5	7.4	4	10	10	1	2	1

[a] 1st ebullition event excluded.
dif, diffusion; ebu, ebullition.

Table 2. Jug Lake sites summary

Date	CH$_4$ emissions (mg/m^2/h)			n collect	ebu (%)	ebu (%)[a]	n sites w/ebu	# ebu event	CH$_4$ emissions (mg/m^2/day)	
	dif	dif + ebu	dif + ebu[a]						dif + ebu	dif + ebu[a]
July 91	8.4	31	18	32	73	53	24	31	744	432
September 91	10	54	28	14	82	64	6	9	1296	672
January 92	0.2	0.24	0.2	12	20	1	1	1	6	5
May 92	8.3	31	11	21	73	25	12	14	744	264
April 1994	20	39	20	23	50	3	3	4	936	480

[a] 1st ebullition excluded.
dif, diffusion; ebu, ebullition.

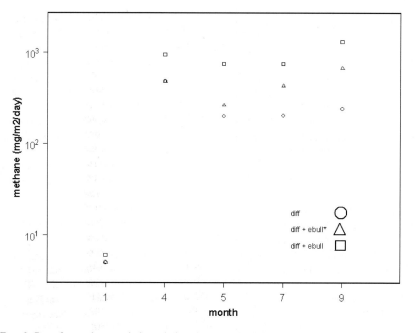

FIG. 6. Data for methane emissions during the year. ○ diffusion only; □ diffusion + ebullition; △ diffusion + ebullition (first interval omitted).

collection or induced events due to gentle disturbances—probing with a stick, passage of the air boat or even due to crustaceans crawling on the surface.

DISCUSSION

Comparison to earlier methane emissions work

Methane emissions that we measured from these sites are comparable to the brackish site in Louisiana reported by DeLaune et al. (1983) for summer months. The results from Alford et al. (1997) range from 13 to 3900 mg/m²/day from January and September, respectively, are about a factor of 2–3 higher than our results for January and September. Alford et al. (1997) report an average value of 912 mg/m²/day (S.D. = 923) or 333 g/m²/yr which is about 50% higher than our results. This difference is probably not significant considering that we were in different marshes, our seasonal vs. their monthly sampling, our multiple (4–8) non-fixed, floating collectors vs. their two collared permanent collectors. Both sets of results show the variability associated with methane emissions from these marsh environments. Typically, water depth, temperature, salinity, and vegetation type have been suggested as factors affecting methane emissions. Bartlett et al. (1989) show that even during the mid-winter low emissions from wetlands, there were large variations in fluxes from what appeared to be very similar wetlands. In contrast to earlier reports, there was no obvious control due to water depth for their results. Bartlett et al. (1989) conclude that there is no substitute for multiple measurements in the field because none of the other measured parameters seem to show a clear relationship or control to methane emission. Our results are consistent with this suggestion and show that no single one of these factors appears to have the major influence on the emissions in a particular wetland. This is particularly true for the ebullition component that is responsible for about one-half of the overall emissions.

Dise (1993) and Dise et al. (1993) discussed the large variability in methane emission rates observed from Minnesota peat. Their work explains the disparate ranges reported earlier by several groups working on northern peatlands. DISE (1993) shows that the two orders-of-magnitude range observed also varies with temperature during the year.

WILSON et al. (1989) first pointed out the importance of ebullition in the methane emissions budget for a freshwater swamp in Virginia. Ten sites showed large variations of more than three orders of magnitude during the year. Ebullition events were observed in 19% of their measurements but accounted for 34% of the total methane flux. Peak emissions occurred at different times in different wetlands and show the "importance of smaller scale environmental parameters controlling methane flux". In a study at a north Florida swamp, HAPPELL and CHANTON (1993) report that ebullition accounts for 45% of the total methane emissions but during the flooded summer months ebullition made up as much as 80% of the methane flux. These authors also show that variation in methane fluxes between two nearby sites sometimes varies by a factor of 2 during the times of high emissions. WEYHENMEYER (1999) reported on methane emission from beaver ponds in a forest from Ontario, Canada where 65% of the total flux was due to ebullition, which was variable in space and time.

Other workers have also noted large differences in methane emissions from closely spaced collectors in a local area: KLINGER et al. (1994) reported large variations using 4–6 collectors at five sites from Hudson Bay lowland: intrasite variation was generally from $+100\%$ to -100% of the mean for the site but the standard deviation of the individual collectors was generally only ± 5–30% (excluding the lowest emission rates). This was partly explained by the depth to the water table (and sealing/leaking of the chamber with the ground surface or water table). MOORE et al. (1994) report within site spatial variability of methane emissions to be 50–100% (coefficient of variation) between visually uniform sites.

Methane in porewater

The methane content of the porewater profiles shows increasing concentration from the surface to 20–30 cm depth and then remains nearly constant. For a surface water temperature and sediment water temperature of 30°C the saturation of methane occurs at approximately 1 mM, which is exceeded by many of the porewaters at depths below 25 cm. Presumably this supersaturation is required for ebullition to be significant. The decrease in methane toward shallower water is probably due to consumption in the oxygenated zone and escape by diffusion and ebullition from the porewater to the atmosphere. It is interesting to note that at site JC-3 (July 91), the sediment below 25 cm has more than twice as much methane as does the sediment below a S. patens mat (Fig. 5a). Because the mat is not completely floating all the time, we suggest that when the mat is close to or contacts the sediment it disturbs the sediment sufficiently to cause methane release. Another possibility is that the plant stem and/or roots connection of the mat to the sediment causes disturbance and release of the trapped methane.

The location with the lowest methane emissions, JC-3, is also one with the most methane in the porewater below about 20 cm depth. The second lowest methane emissions are from OBI-1, and this location also has porewater methane just at saturation at the deepest sample. Therefore, the two locations with the largest emissions (by diffusion and ebullition) are JC-5 and JI-3 and these two sites have porewater that is not methane supersaturated below 25 cm depth. It should be noted, however, that ebullition events were observed at both JC-5 and JI-3, so that there must be some gas phase methane in the sediments even though it is not recorded in our incomplete depth profiles (that were even less complete for JI-3, only two samples). For the overall marsh this emission vs. degree of supersaturation at depth may compensate for the relative methane emissions and relative amount of methane stored in the sediments and to make it less variable for the total methane (methane emissions plus methane stored in porewaters).

Annual methane emissions

Using the data of Tables 1 and 2 and Fig. 6, we can estimate the annual methane emissions for the Terrebonne Parish, Atchafalaya Basin oligohaline marshes in Gulf Coast, Louisiana. We have data for four groups of measurements (April through September) that do not show

Table 3. Comparison of oligohaline methane emissions

Site state	Salinity (‰)	CH$_4$ emissions (mg/m^2/day) Average	CH$_4$ emissions	
			Average (April–September), mg/m^2/day	Annual (g/m^2)
Virginia[a]	1.4–8	49	114	18
Louisiana Barataria[b]	1	266	450	97
Terrebonne[c]	0.5–3			
Diffusion only		200	288	73
Total		564	930	206

[a] BARTLETT et al. (1987).
[b] DeLAUNE et al. (1983).
[c] This work.

a summer peak. Therefore, to estimate the emissions for the 6 months (April through September) we use the daily mean (930 ± 225 mg/m^2/day) times 182 days to get 170 g/m^2. We assume that March and October have one-half the daily emissions of the April through September months (giving 28 g/m^2), that November has one-fourth (7.4 g/m^2). For December and February, we assume the same emissions as measured for January, but the total for these 3 months is only 0.5 g/m^2. This gives us 206 g methane/m^2/yr considering both diffusion and ebullition. (If we recalculate the sum *without* including the first interval ebullition event the annual methane emission becomes 102 g/m^2.)

To estimate the annual methane emissions by diffusion (only) we use the above data and the measured factor 0.31 for the proportion of diffusion to get 62 g/m^2/yr. This value (Table 3) is lower than the Barataria data (only diffusion) of DeLAUNE et al. (1983). (An alternate calculation uses 0.31 for April–September, 0.5 for March and October, 0.75 for February and November, and 1.0 for January and December with no ebullition to give 73 g/m^2/yr that is closer to the DeLaune et al. diffusion result.) (see Table 3). We believe that the ebullition we have measured is a natural event and these data give new higher total methane emissions from this marsh type than previously reported (DeLAUNE et al., 1983). Table 3 summarizes the methane values reported for coastal brackish-oligohaline wetlands in Louisiana and Virginia.

CONCLUSIONS

Ebullition has been documented for Gulf Coast marshes and has been shown to be very important part of the overall methane emissions from oligohaline-brackish marshes in coastal Louisiana. The total annual emissions (ebullition and diffusion) estimated from five seasonal measurements are 205 g/m^2/yr. If all of the ebullition events are included, the annual methane emissions are much higher than those previously reported. Approximately one-half of the methane is emitted by ebullition during the first 6 min of accumulation/collection, but because the collectors are floated on the surface of the water, we do not believe this emission is an artifact or due to disturbance during placement of the collectors. We believe this first ebullition event is the release of preformed bubbles that have migrated to the sediment–water interface and are "poised" for release at the first slight change in pressure or water movement. The variation in methane emission during the year is not too different during April to October but is much lower during mid-winter (January), due to lower ambient (sediment) temperatures. The variation of methane emission during a single day over a spatial area of 10 m by 2 m is more than 100% and is partly related to water depth and to vegetation cover, but no large flux through the two major plants present was noted. For this reason multiple (at least four) measurements and collectors need to be made at a single site to get an average methane emission rate.

Acknowledgements—One of us (JL) first learned about methane in sediments while at Ian Kaplan's lab at UCLA in 1968 (what an exciting time to be at the Kaplan lab) for a summer project to set up an operational system to measure gases in porewaters for Leg 1 of DSDP. This project was a prelude to JL's later scientific work including this project reported here and in 1999 as a shipboard organic geochemist on ODP Leg 184 using the newest generation of porewater gas analysis system, which was, in fact, rather similar to the original system.—Research reported here was part of the USGS Geologic Division Global Change and Climate History Program. The project was initiated when one of us (GG) was at U.S. Fish and Wildlife Service in Louisiana; support there was also by the National Biological Service, National Wetlands Research Center, Robert E. Stewart, Jr., Director. Our thanks go to National Biological Service—National Wetlands Research Center personnel: Dr Lee Foote for logistics, Kathy Reynolds for airboat transportation. Dr Denise Reed for arranging for laboratory facilities at Louisiana Universities Marine Center LUMCON (Dr Michael Dagg, Interim Director). We thank U.S.G.S. reviewers Maggie Hinkle and Ted McConnaughey for initial review, Ken Peters and several anonymous reviewers for their suggestions. We especially appreciate Dr Kaplan's review and comments.

REFERENCES

ALFORD D. P., DELAUNE R. D. and LINDAU C. W. (1997) Methane flux from Mississippi River Deltaic plain wetlands. *Biogeochemistry* **37**, 227–236.

ASELMANN I. and CRUTZEN P. (1989) Global distribution of natural freshwater wetlands and rice paddies: their net primary productivity, seasonality and possible methane emissions. *J. Atmos. Chem.* **8**, 307–358.

BARRAS J. A., BOURGEOIS P. E. and HANDLEY L. R. (1994) *Land loss in coastal Louisiana 1956–1990. National Biological Service*, National Wetlands Research Center, Open File Report 94-01, 4 p, 10 color plates.

BARTLETT K. B. and HARRISS R. C. (1992) Review and assessment of methane emissions from wetlands. In *Atmospheric Methane: Sources, Sinks and Role in Global Change* (eds. M. KHALIL and M. SHEARER), Shearer proceedings of NATO workshop, October 1991, pp. 261–320. Pergamon Press, New York.

BARTLETT K., BARTLETT D., HARRISS R. and SEBACHER D. (1987) Methane emissions along a salt marsh salinity gradient. *Biogeochemistry* **4**, 183–202.

BARTLETT D., BARTLETT K., HARTMAN J., HARRISS R., SEBACHER D., PELLETIER-TRAVIS R., DOW D. and BRANNON D. (1989) Methane emissions from the Florida Everglades: patterns of variability in a regional wetland ecosystem. *Global Biogeochem. Cycles* **3**, 363–374.

BARTLETT K., CRILL P., BONSAI J., RICHEY J. and HARRISS R. (1990) Wet season methane flux from the Amazon River floodplains. *J. Geophys. Res.* **95**, 16733–16788.

BLAKE D. R. and ROWLAND F. S. (1988) Continuing worldwide increase in tropospheric methane, 1978 to 1987. *Science* **239**, 1124–1131.

BRITSCH L. D. and KEMP E. B. (1990) *Land loss rates: Mississippi River deltaic plain*, Tech. Rpt., GL-90-2. U.S. Army Corp of Engineers, New Orleans, LA, 25 p.

BUBIER J. L., MOORE T., BELLISARIO L., COMER N. and CRILL P. (1995) Ecological controls on methane emissions from a northern peatland complex in the zone of discontinuous permafronst, Manitoba, Canada. *Global Biogeochem. Cycles* **9**, 455–470.

CHABRECK R. H. and LINSCOMBE G. (1978) *Vegetative Type Map of the Louisiana Coastal Marshes 1978.* Louisiana Department of Wildlife and Fisheries, New Orleans.

CHABRECK R. H. and LINSCOMBE G. (1992) *Louisiana Coastal Marsh Vegetation Type Map 1988, Terrebonne Bay Sheet.* Louisiana Department of Wildlife and Fisheries, Baton Rouge.

CHANTON J. P. and WHITING G. J. (1995) Trace gas exchange in freshwater and coastal marine environments: ebullition and transport by plants. In *Biogenic Trace Gases: Measuring Emissions from Soil and Water* (eds. P. A. MATSON and R. C. HARRISS), pp. 98–125. Blackwell Science, Oxford.

CHAPMAN V. J. (ed.) (1977) *Ecosystems of the World 1, Wet Coastal Ecosystems*, pp. 428. Elsevier, Amsterdam.

CHAPPELLEZ J., BARNOLA J., RAYNAUD D., KOROTKEVICH Y. and LORIUS C. (1990) Ice-core record of atmospheric methane over the past 160,000 years. *Nature* **345**, 127–131.

CICERONE R. and OREMLAND R. (1988) Biogeochemical aspects of atmospheric methane. *Global Biogeochem. Cycles* **2**, 299–327.

CLAYTON J., LEVENTHAL J., RICE D., PASHIN J., MOSHER B. and CZEPIEL P. (1993) Atmospheric methane flux from coals in Alabama. In *Energy Gas* (ed. D. HOWELL), US Geol. Surv. Prof. Paper 1570, pp. 471–492.

COMMITTEE ON EARTH and ENVIRONMENTAL SCIENCES, (1994) *Our Changing Planet, The FY 1994 U.S. Global Change Research Program.* National Science Foundation, Washington, DC, 84 p.

COWARDIN L. M., CARTER V., GOLET F. and LaROE E. (1979) Classification of wetlands and deepwater habitats in the U.S., U.S. Fish Wildlife Serv. FWS/OBS-79/31, 103 pp.

DELAUNE R., SMITH C. and PATRICK W. (1983) Methane release from Gulfcoast wetlands. *Tellus* **35B**, 8–15.

DEVOL A. H., RICHEY J. E., FORSBERG B. R. and MARTINELLI L. A. (1990) Seasonal dynamics in methane emissions form the Amazon Floodplain. *J. Geophys. Res.* **95**, 16417–16426.

DICKENSON R. E. and CICERONE R. J. (1986) Future global warming from atmospheric trace gases. *Nature* **319**, 109–115.

DISE N. (1993) Methane emissions from Minnesota peatlands: spatial and seasonal variability. *Global Biogeochem. Cycles* **7**, 123–142.

DISE N., GORHAM E. and VERRY E. S. (1993) Environmental factors controlling methane emissions from peatlands in northern Minnesota. *J. Geophys. Res.* **98**, 10583–10594.

DUCHEMIN E., LUCOTTE M., CANUEL R. and CHAMBERLAND A. (1995) Production of greenhouses gases CH_4 and CO_2 by hydroelectric reservoirs in the boreal region. *Global Biogeochem. Cycles* **9**, 529–540.

FOOTE A. L., BURKETT V. and WILLIAMS S.J. (1993) Natural resource problem solving: an interdisciplinary approach in coastal Louisiana. In *Coastal Zone '93, Proceedings of 8th Symposium on Coastal and Ocean management*, New Orleans, pp. 258–266.

FUNG I., JOHN J., LERNER J., MATTHEWS E., PRATHER M., STEELE L. and FRASER P. (1991) Three-dimensional model synthesis of the global methane cycle. *J. Geophys. Res.* **96**, 13033–13065.

GORNITZ V. (1991) Global coastal hazards from future sea level rise. *Paleogeogr. Paleoclimatol. Paleoecol. (Global and Planetary Change Sect.)* **89**, 379–398.

GOSSELINK J. G. (1984) The ecology of delta marshes of coastal Louisiana: a community profile. U.S. Fish and Wildlife FWS/OBS-84/09, 133 p.

GUNTENSPERGEN G. R. and VAIRIN B. A. (1996) Willful winds: Hurricane Andrew and Louisiana's coast. Louisiana Sea Grant College Program and U.S. Dept. Interior, National Biological Service, Lafayette, Louisiana, 16 p.

GUNTENSPERGEN G. R., FOOTE A. L. and PENDLETON E. C. (1993) Experiment to determine value of impoundments and hydrologic manipulation in restoration of Louisiana coastal wetlands. *Restoration and Management Notes* **11**, 64–65.

GUNTENSPERGEN G. R., CAHOON D. R., GRACE J., STEYER G. D., FOURNET S., TOWSON M. A. and FOOTE A. L. (1995) Disturbance and recovery of the Louisiana coastal marsh landscape from the impacts of Hurricane Andrew. *J. Coastal Res.* **21**, 324–339.

HACKNEY C. T., BRADY S., STEMMY L., BORIS M., DENNIS C., HANCOCK T., O'BRYON M., TILTON C. and BARBEE E. (1996) Does intertidal vegetation indicate specific soil and hydrologic conditions. *Wetlands* **16**, 89–94.

HAPPELL J. and CHANTON J. (1993) Carbon remineralization in a north Florida swamp forest: effects of water level on the pathways and rates of soil organic matter decomposition. *Global Biogeochem. Cycles* **7**, 475–490.

HARGIS T., RAFFERTY P., LYNCH J., BILLOCK A. and TWILLEY R. (1993) Chemical analysis of pore water in marsh ecosystems subjected to alterations in hydrology. In *Wetlands: Proceedings of 13th Annual Soc. Wetland Scientists meeting* (ed. M. LANDIN), pp. 852–856. SWS, Utica MS.

IPCC, INTERGOVERNMENTAL PANEL ON CLIMATE CHANGE, (2001) *Working Group I, Scientific Assessment of Climate Change*. Blackwell, Cambridge University Press, UK, 364 p.

JACKSON L. L., BALISTRIERI L., SMITH K. S., WALTON-DAY K., KIRSCHEMANN D., BRIGGS P., FEY D. and SUTLEY S. (1995) Geochemistry of sediments from coastal marshes of Louisiana. U.S. Geol. Survey Open-File Report, 95-251, 146 p.

KELLER M. and STALLARD R. F. (1994) Methane emission by bubbling from Gatun Lake, Panama. *J. Geophys. Res.* **99**, 8703–8719.

KHALIL M. and RASMUSSEN R. (1990) Constraints on the global sources of methane and analysis of recent budgets. *Tellus* **42B**, 229–236.

KLINGER L. F., ZIMMERMAN P. R., GREENBERG J. P., HEIDT L. E. and GUENTHER A. B. (1994) Carbon trace gas fluxes along a successional gradient in the Hudson Bay lowland. *J. Geophys. Res.* **99D**, 1469–1494.

LASHOF D. (1989) The dynamic greenhouse: feedback processes that may influence future concentrations of atmospheric trace gases and climatic change. *Climatic Change* **14**, 213–231.

LASHOF D. and AHUJA A. (1990) Relative contributions of greenhouse gas emissions to global warming. *Nature* **344**, 529–531.

LEVENTHAL J. S. (1991) Global change: negative feedback between sea level rise and methane production from estuarine wetlands. *EOS Trans. Am. Geophys. Union* **72**, 109 p.

LEVENTHAL J. (1992) Modern methane measurements in marshes. U.S. Geol. Survey Open-File Report 92-445, 24 p.

LEVENTHAL J. (1995) Methane budget and global change in coastal wetlands. USGS McKelvey Forum on Energy and the Environment, Washington DC. *U.S. Geol. Surv. Circ.* **1108**, 81–82.

LEVENTHAL J., JACKSON L., BALISTRIERI L., SMITH K. S., WALTON-DAY K., FOOTE A. L. and GUNTENSPERGEN G. (1993) Biogeochemistry along an estuarine salinity gradient: a proxy for sea level rise due to global change. In *Wetlands: Proceedings of 13th Annual Soc. Wetland Scientists* (ed. M. LANDIN), pp. 95–100. SWS, Utica, MS.

MARTENS C. and BERNER R. (1974) Methane production in the interstitial waters of sulfate 1 depleted marine sediments. *Science* **185**, 1167–1169.

MATTHEWS E. and FUNG I. (1987) Methane emission from natural wetlands: global distribution, area, and environmental characteristics of sources. *Global Biogeochem. Cycles* **1**, 61–86.

MOORE T. R., HEYES A. and ROULET N. T. (1994) Methane emissions from wetlands, southern Hudson Bay lowland. *J.Geophys. Res.* **99D**, 1455–1467.

O'NEIL T. (1949) The muskrat in the coastal Louisiana marshes (and everything else you wanted to know about the marshes). Louisiana Wildlife and Fisheries Commission, New Orleans, 134 p + vegetation map (color).

PARK R. A., TREHAN M., MAUSEL P. and HOWE R. (1989) Coastal wetlands in the twentyfirst century: profound alteration due to rising sea level. In *Wetlands: Successes* (ed. D. FISK), pp. 71–80. Am. Water Res. Assoc.

PENDELTON E. C., FOOTE A. L. and GUNTENSPERGEN G. R. (1992) Managing Louisiana marshes—an experimental approach. In *Coastal Wetlands* (eds. H. S. BOLTON and O. T. MAGOON), pp. 235–246. Am. Soc. Civil Engineers, New York.

PENFOUND W. T. and HATHAWAY E. S. (1938) Plant communities in the marshland of southeastern Louisiana. *Ecol. Monogr.* **8**, 1–56.

PENLAND S., ROBERTS H., SALLENGER A., CAHOON D., DAVIS D. and GROAT C. (1990) Coastal land loss in Louisiana. *Trans. Gulf Coast Assoc. Geol. Soc.* **XL**, 685198 p.

PRIEME A. (1994) Production and emission of methane in a brackish and freshwater wetland. *Soil Biol. Biochem.* **26**, 7–18.

REED D. J. and ROZAS L. P. (1995) An evaluation of the potential for infilling existing pipeline canals in Louisiana coastal marshes. *Wetlands* **15**, 149–158.

REYER A. J., HOLLAND C., FIELD D. W., CASSELIS J. and ALEXANDER C. (1988) National Coastal Wetlands Inventory: The distribution and areal extent of coastal wetlands in estuaries of the Gulf of Mexico, NOAA Report, 18 pp.

ROBERTS H. and VAN HEERDEN I. (1992) Atchafalaya-Wax Lake delta complex: The new Mississippi River delta lobe. Coastal studies Institute Industrial Assoc. Research Program, Report No. 1, Baton Rouge.

TITUS J. (1990) Greenhouse effect and coastal wetland policy. *Environ. Manag.* **15**, 39–58.

WEYHENMEYER C. E. (1999) Methane emissions from beaver ponds: rates patterns and transport mechanisms. *Global Biogeochem. Cycles* **13**, 1079–1090.

WHITING G. J. and CHANTON J. P. (1993) Primary production control of methane emission from wetlands. *Nature* **364**, 794–795.

WICKER K. M. (1980) Mississippi deltaic plain region ecological characterization: a habitat mapping study. Office of Biological Services, U.S. Fish and Wildlife Service, FWS/OBS-79/07, 45 p + appendices.

WIGLEY T. and RAPER S. (1992) Implications for climate and sea level of revised IPCC emissions scenarios. *Nature* **357**, 293–300.

WILSON J., CRILL P., BARTLETT K., SEBACHER D., HARRISS R. and SASS R. (1989) Seasonal variation of methane emissions from a temperate swamp. *Biogeochemistry* **8**, 55–71.

ZIMMERMAN P. R. (1979) Testing of hydrocarbon emissions from vegetation and development of methodology. EPA, Office of Air Quality, EPA 450/4-79-004, 71 pp.

Geochemical Investigations in Earth and Space Science: A Tribute to Isaac R. Kaplan
© The Geochemical Society, Publication No. 9, 2004
Editors: R.J. Hill, J. Leventhal, Z. Aizenshtat, M.J. Baedecker, G. Claypool,
R. Eganhouse, M. Goldhaber and K. Peters

Organic geochemistry of lipids in marine sediments in the Canary Basin: Implications for origin and accumulation of organic matter

STEN LINDBLOM[1] and ULF JÄRNBERG[2]

[1]Department of Geology and Geochemistry, Stockholm University, S-10591 Stockholm, Sweden
[2]Institute of Applied Environmental Research, Stockholm University, S-10691 Stockholm, Sweden

Abstract—The Canary basin consists of the continental margin off the coast of Northwest Africa, the deep basin, the Madeira Abyssal Plain (MAP) and the volcanic Canary Islands.

The continental margin has an area of upwelling marine productivity with special characteristics: (a) a wide shelf area, (b) a wide irregular zone of upwelling, and (c) proximity to a large dust source—the Saharan desert. The Canary Islands divide the continental slope into a northern area and a southern area of primary productivity and accumulation of organic matter (OM). Accumulation of marine OM in the northern area is augmented by input of Saharan dust containing adsorbed terrigenous OM.

n-Alkane distributions show a pronounced maximum around C_{29} and C_{31} with a strong odd-to-even predominance typical of terrigenous OM. There are less pronounced maxima near $C_{15}-C_{17}$, indicating a recent marine origin for the OM.

Fatty acids show carbon number distributions similar to the n-alkanes, with a higher content and substantially higher C_{16}/C_{26} ratios in near-surface sediments of the slope samples, which decrease with depth in the sediment.

The above data from the slope sediments can be compared to the most recent organic turbidite "a" on the MAP. Turbidite "a" was emplaced about 1000 years ago, incorporating slope derived material representing a 200,000-year time period.

TOC and biomarker signatures of sediments on the slope are comparable to turbidite "a". Biomarker analyses of deeper sediments on the MAP show similar data continuing back to early Miocene.

Comparison of recent seafloor surface biomarkers with deeper samples indicates a varying contribution of terrigenous OM and a strong degradation of marine OM components in favor of leaving the long-chained compounds that increase a terrigenous signature of the sediments. The marine component is higher than analysis of deeper biomarkers tends to indicate.

Polyaromatic hydrocarbons (PAH) are low in concentration, and no human pollution effects were detected. Perylene data suggest active diagenesis down the sediment core.

INTRODUCTION

THE CONTINENTAL MARGIN off the coast of Northwest Africa is an area of upwelling and marine productivity with special characteristics: (a) a wide shelf area, (b) a wide zone of upwelling, and (c) proximity to a large dust source—the Saharan desert. The Canary Islands divide a northern area of primary productivity and accumulation of organic matter (OM) and a southern area. Accumulation of marine OM is supplemented by input of Saharan dust containing adsorbed terrigenous OM.

Upwelling conditions prevailed along the African coastline for a long time and was studied during several recent ODP legs (SCHMINCKE *et al.*, 1995; RUDDIMAN *et al.*, 1988; WEFER *et al.*, 1998). Several studies focused on the continental margin along the Northwestern African coast (MÜLLER *et al.*, 1983; FÜTTERER, 1983; TEN HAVEN *et al.*, 1989). Although much sedimentary and micropaleontological data have been gathered to show an active continental margin with respect to ocean currents and paleoproductivity, very little biomarker analysis has been done.

In this study we look at the origin and alteration of OM and the heterogeneous composition of OM on the slope as a precursor to turbidites on the MAP. We do this by studying different biomarkers (n-alkanes, fatty acids, alcohols), their variation with depth (reflecting age), and their aerial distribution. We attempt to show that the long and on-going turbidite emplacement process is reflected in variations in biomarkers of slope sediments and MAP turbidites.

THE CANARY BASIN

The Canary Basin contains a varied geological setting along the continental shelf and slope off the Northwest African coast, the Canary Islands, and the Madeira Abyssal Plain (MAP) (Fig. 1). The MAP is situated in the deepest part (about 5400 m) of the Canary Basin about 700 km west of the Canary Islands (JARVIS and HIGGS, 1987).

The Eastern Boundary current sweeps down along the Northwest African coast. This causes an upwelling zone extending 50 km from the coast. Inhomogeneities in current patterns and coastal and island irregularities cause filaments of upwelling waters to stretch out into the Atlantic (FREUDENTHAL *et al.*, 2001). The accumulation of organic sediments occurs on the continental slope, where upwelling facilitates high productivity and depletion of oxygen during oxidation of settling OM. Where the oxygen minimum layer intersects the slope, organic sediments tend to be preserved.

Studies of sedimentation on the MAP revealed large sequences of turbidites. Turbidites are sediments transported by turbidity currents that are rapidly accumulated sedimentary units (WEAVER and KUIJPERS, 1983; WEAVER *et al.*, 1998b). There are three different types of

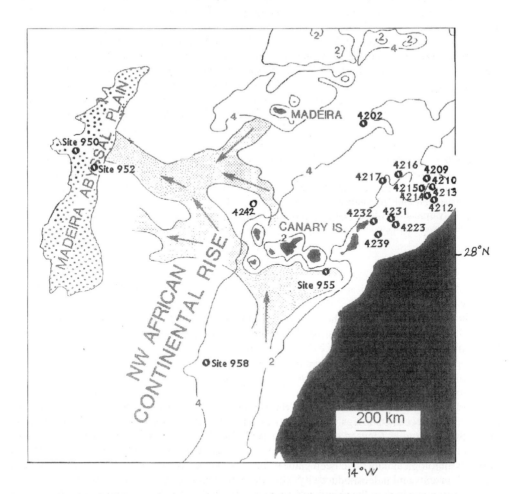

FIG. 1. Map of the Canary Basin with ODP Leg 157 Sites 950 and 952, Leg 159T Site 958 and Meteor cruise 37/1 Sites GeoB 4202 etc. Bathymetric lines with 4 and 2 denote water depths in thousands of meters. Solid arrows depict pathways for turbidites.

turbidites found in the MAP: volcanic, calcareous, and organic. Volcanic turbidites may originate from the oceanic islands of the Canaries and Madeira (DE LANGE, 1998; JARVIS and HIGGS, 1987). They contain considerably more volcanogenic and other igneous material than other turbidites on the MAP. Calcareous turbidites probably originate from the Great Meteor-Cruiser Seamount chain to the west of the MAP (WEAVER and ROTHWELL, 1987). They are characterized by very high carbonate contents (more than 75% $CaCO_3$). Organic turbidites, which are the main focus of this work, probably originate from a highly productive area on the northwest African continental margin (DE LANGE et al., 1987; JARVIS and KUIJPERS, 1987; WEAVER and KUIJPERS, 1983). They were defined as "organic-rich" by workers on the MAP sediments when they contain more than 0.3% TOC and are green in color. This is not organic-rich by most standards and we keep to the term "organic" in this study.

TURBIDITE EMPLACEMENT

Changes in sea level at the beginning and end of glacial periods have caused instability and mass wasting on the continental slope and on the flanks of the Canary Islands (WEAVER et al., 1998a,b). WEAVER et al. (1992, 1995) developed a model for basin development and filling in the Canary Basin. Individual mass wasting events can be traced from the slope to the abyssal plain. These studies suggest that all major mass wasting in the area produces turbidity currents sufficiently large to reach and cover the abyssal plain. Turbidite emplacement was continuous from Early Miocene until now, but sediment flux varied greatly. The major input occurred from 17 to 13 Ma and from 6.5 Ma until the present (WEAVER et al., 1998a,b).

Studies of possible pathways have revealed "canyons" leading from the African Coast and the Canary Islands (Fig. 1; MASSON, 1994). Based on these expectations, we approached core sampling in profiles along the continental shelf off Morocco during Meteor cruise 37/1. In practice, we had difficulties sampling the points nearest to the coast due to uneven bottom conditions. However, GeoB 4223 gave an 8 m core at 775 m water depth with dominant olive green sediments. The organic carbon content in this core was lower than the turbidite samples, but consistently between 0.5 and 0.8 wt% TOC (TIM FREUDENTAHL, personal communication, 1998). This is taken to represent the original type of sediments acting as precursor for the organic-rich turbidites.

The emplacement process produced repeated units of turbidites spread over the whole area of the abyssal plain with well-mixed material in each individual unit. Carbonate dissolves very rapidly and the carbonate content may in part be dissolved during transport in turbidity currents, thus causing a relative increase in TOC.

SAMPLING

During ODP Leg 157, sampling of organic turbidites was done immediately after splitting the cores as they came on deck, after an equilibration period of few hours. The samples were immediately stored frozen. During the Meteor cruise 37/1, similar sampling procedures were maintained and all samples were kept frozen until processed for analysis. Multicores, box cores, and gravity cores were taken (WEFER et al., 1997).

METHODS

The analytical procedure for extraction and class separation was adapted from VENKATESAN et al. (1987) and consisted of repeated ultrasonic solvent extractions using hexane/acetone followed by saponification of the free fatty acids in KOH/methanol. The organic extract was then divided into three fractions containing aliphatics (n-alkanes), PAH and n-alcohols, and sterols using open column chromatography on 10% water-deactivated silica. Free fatty acids and n-alcohols were derivatized by BSTFA prior to analysis on GC–MS. All determinations were performed using a Hewlett Packard 5890 series II gas chromatograph equipped with a split/splitless injector and an electronic pressure control. The column was a 30 m × 0.25 mm ID coated with a 0.25 μm, 5% phenyl-methylsilicone phase (J&W DB-5MS) at a head pressure of 13 Psi. The detector was a JEOL Automass II low resolution mass spectrometer operating in the electron ionization (EI) mode with an electron energy of 70 eV. For n-alkanes and PAH analysis, selected ion monitoring (SIM) was used, while fatty acids and n-alcohols were analyzed with the MS operating in scan mode. Quantification was performed using reference mixtures covering the range in carbon numbers as well as deuterium-labeled recovery standards. Blank samples were run in parallel with the samples through the entire analytical procedure. Estimated error expressed as standard deviation was 15%. Total organic carbon (TOC) of slope sediments was obtained using a CHN Leco 900 analyzer; overall accuracy for C and N measurements was $\pm 3\%$ and subtracting carbonate carbon measured on a coulometer, Coulometrics 5011, with an accuracy of $\pm 3\%$. ODP TOC and C/N data were measured on the ship.

RESULTS

The continental slope off the coast of Morrocco

The type of core, location, water depth, and recovery for the slope samples are given in Table 1. Table 2 shows location, TOC and calcium carbonate, nitrogen, C/N ratios, and approximate ages.

Analyses compared seafloor samples from successive locations through the sediments on the slope areas. Table 3 gives total n-alkanes, fatty acids, alkanols, and PAH for the samples from the continental slope. Also included are ratios of CPI, C_{17}/C_{31} (alkanes), C_{16}/C_{26} (fatty acids), phenantrene to anthracene ratio (PHN/ANT), and fluoranthene to pyrene ratio (FLO/PYR). Total values of retene and perylene are given separately. Profile data are represented as ratios C_{17}/C_{31} and C_{16}/C_{26} for n-alkanes and fatty acids, respectively.

n-Alkane distributions show a pronounced maximum around C_{29} and C_{31}, with a strong odd over even predominance (carbon preference index, CPI = $5-10$, $C_{23}-C_{34}$). A second much less pronounced maximum around $C_{15}-C_{17}$, indicative of a recent marine origin for the OM, occurs in only a few samples. In the aliphatic hydrocarbon fraction, no unresolved complex mixture (UCM), typical of petrogenic sources, was observed for any of the samples. Isoprenoid alkanes, such as pristane and phytane, generally appeared at lower concentrations than their corresponding n-alkanes (Fig. 2a). The pristane/phytane ratio was below one for all samples.

In the surface samples, the free fatty acid fraction was dominated by straight-chain saturated fatty acids with lauric (12:0), palmitic (16:0) and stearic (18:0) acid being the most abundant (Fig. 2b). Most of the samples from deeper core layers showed a different fatty acid profile, with a second pronounced maximum around hexacosanoic (26:0) acid.

Table 1. Location of cores on the continental slope

Core (GeoB)	Type	Latitude (North)	Longitude (West)	Water depth (m)	Recovery (cm)
4202-1	Box core	32°28.6	13°39.8	4289	39
4217-5	Gravity core	30°25.2	12°54.7	2506	716
4216-1	Gravity core	30°37.8	12°23.8	2324	1117
4209-1	Box core	30°21.4	11°05.0	2150	33
4215-1	Multicore	30°02.2	11°33.2	2106	33
4215-2	Gravity core	30°02.2	11°33.1	2105	766
4210-2	Multicore	30°18.0	10°58.8	1959	16
4214.1	Gravity core	29°46.9	11°11.9	1791	952
4213.1	Multicore	29°41.8	11°04.7	1547	22
4212.3	Multicore	29°36.2	10°57.0	1256	32
4212-2	Gravity core	29°36.3	10°57.0	1258	857
4231-2	Multicore	29°05.3	12°33.3	1207	31
4232-1	Multicore	29°01.3	13°23.2	1161	13
4239-1	Multicore	28°29.7	13°10.8	884	12
4232-2	Multicore	29°01.0	12°28.0	775	779

Gravity cores, box cores and multicores used in this study from Meteor cruise 37/1.

Table 2. Core samples and parameters: TOC, water depth at core locations, carbonate content C/N ratios and approximate age estimates using sediment rates of HENDERIKS *et al.* (2001)

Core number	Section cm below seafloor	Water depth (m)	TOC (%)	C/N	%CaCO₃	Age (year)
GeoB 4202	18–21	4284	0.14	3	48	1000
GeoB 4209	18–21	2169	0.31	4	51	2800
GeoB 4210	0–2	1959	0.42	4	45	–
GeoB 4210	10–12	1959	0.07	1	52	2200
GeoB 4212	0–2	1256	0.36	4	56	–
GeoB 4212	10–12	1256	0.06	1	60	2200
GeoB 4212	20–22	1256	<0.05	–	62	2900
GeoB 4212	30–32	1256	0.07	1	58	3700
GeoB 4212	219–221	1256	0.34	5	59	17,000
GeoB 4213	0–2	1549	0.37	4	57	–
GeoB 4213	10–12	1549	<0.05	–	59	2200
GeoB 4213	20–22	1549	<0.05	–	60	2900
GeoB 4214	59–64	1791	0.51	5	50	5900
GeoB 4214	309–313	1791	0.30	4	40	22,000
GeoB 4215	0–2	2105	0.35	5	68	–
GeoB 4215	10–12	2105	0.34	8	60	2200
GeoB 4215	20–22	2105	0.24	5	61	2900
GeoB 4215	28–30	2105	<0.05	–	60	3600
GeoB 4215	48–52	2105	0.61	6	46	5100
GeoB 4216	195–198	2324	0.60	–	48	35,200
GeoB 4217	372–376	2506	0.05	1	66	72,000
GeoB 4223	333–337	775	0.67	–	–	24,000
GeoB 4223	513–517	775	0.85	–	–	53,200
GeoB 4231	0–2	1207	0.37	4	52	–
GeoB 4231	10–12	1207	0.24	3	53	2200
GeoB 4231	20–22	1207	0.18	3	53	2900
GeoB 4232	0–2	1160	0.47	6	50	–
GeoB 4232	10–12	1160	0.08	1	62	2200
GeoB 4239	0–2	884	0.55	5	47	–
GeoB 4239	10–12	884	0.41	4	50	2200
ODP 157 950A	13H03, 20–24	5438	1.70	16	40	
ODP 157 950A	18XCC, 19–23	5438	0.58	12	45	
ODP 157 952A	36X04, 108–112	5432	1.00	15	12	
ODP 157 955A	13H01, 52–55	2854	0.40	8	44	
ODP 157 955A	42X02, 75–78	2854	0.40	–	20	
ODP 157 958A	3H03, 84–87	3728	1.98	–	–	

The C_{16}/C_{26} *n*-fatty acid ratio decreased with both increasing water column depth and increasing depth below sea floor (mbsf), emphasizing the change toward a terrigenous profile with time.

Mono- and di-saturated fatty acids were present in a 1:1 ratio to the saturated moiety in the surface sediment samples, while the deeper layers were almost depleted in these fatty acids. Similarly, various sterols were detected in appreciable amounts in the surface sediment samples, but not in the deeper layers.

The C_{10}–C_{30} *n*-alcohol profile in the surface sediment samples (Fig. 2c) is analogous to the fatty acids, and maximized around C_{15}–C_{20} and shows an almost identical profile among all the seafloor surface sediment samples, regardless of water depth.

The deep atlantic madeira abyssal plain

Turbidite samples from Sites 950, 951 and 952 (ODP Leg 157) on the MAP have between 0.75 and 1.68 wt% TOC and C/N ratios near 10. Rock Eval pyrolysis data indicate a fair source rock and immature sediments (SCHMINCKE *et al.*, 1995). The overall organic geochemical shipboard data indicate a predominantly marine origin of the OM (SCHMINCKE *et al.*, 1995), but including a terrigenous component.

Table 3. Summary of organic parameters

Sample section	n-Alkanes (µg/g)	Fatty acids (µg/g)	PAH (ng/g)	CPI alkanes	C_{17}/C_{31} alkanes	C_{16}/C_{26} fatty ac.	PHN/ANT	FLO/PYR	Retene	Perylene
4202-1, 18–21	5.38	1.88	21	1.3	0.060	3	235,829	0.227	<0.02	2.738
4209-1, 18–21	0.07	1.10	2.29	4.1	0.237	67	8496	0.211	<0.02	<0.05
4210-2, 1–2	0.50	5.24	2.11	9.3	0.025	52	14,395	0.060	<0.02	<0.05
4210-2, 10–12	0.92	1.12	5	9.9	0.010	23	45,283	0.396	0.060	<0.05
4212-3, 0–2	0.39	2.74	2.38	8.9	0.035	50	17,447	0.071	<0.02	<0.05
4212-3, 10–12	0.58	0.62	6	7.7	0.022	13	20,404	0.281	<0.02	<0.05
4212-3, 20–22	0.70	0.97	3	9.0	0.050	5	32,644	0.444	<0.02	<0.05
4212-3, 30–32	0.40	1.07	4	8.4	0.030	21	40,663	0.529	<0.02	0.067
4212-1, 219–223	5.05	1.73	19	1.4	0.035	2	213,563	0.333	0.298	<0.05
4213-1, 0–2	0.54	4.27	0.92	7.5	0.030	35	4732	0.041	<0.02	<0.05
4213-1, 10–12	0.84	1.36	9	7.8	0.037	13	300,000	0.312	0.080	<0.05
4213-1, 20–22	0.67	1.10	6	6.6	0.027	18	70,973	1.061	0.075	<0.05
4214-1, 59–64	0.77	2.89	8	9.3	0.035	1	36,270	1.271	<0.02	<0.05
4214-1, 309–313	1.08	2.19	6	10.6	0.034	2	86,072	1.066	0.040	<0.05
4215-1, 0–2	0.53	7.00	3.13	4.5	0.068	34	8722	0.048	<0.02	<0.05
4215-1, 10–12	2.18	1.65	3	2.7	0.014	18	55,726	1.462	<0.02	<0.05
4215-1, 20–22	2.01	1.34	9	2.7	0.039	14	127,183	0.489	<0.02	<0.05
4215-1, 28–30	2.70	1.23	5	1.7	0.011	7	93,186	0.509	0.040	<0.05
4215-1, 48–52	0.81	1.95	10	5.1	0.057	5	12,626	0.760	0.040	0.280
4216-1, 195–198	2.57	3.40	9	2.7	0.032	3	76,450	0.354	0.310	0.122
4217-6, 372–376	4.59	1.42	7	1.4	0.056	2	22,546	0.209	0.100	<0.05
4223-2, 333–337	0.54	2.75	5	6.2	0.048	1	72,040	0.540	<0.02	<0.05
4223-2, 513–517	1.42	2.47	6	3.6	0.014	1	11,147	0.356	0.040	0.108
4231-1, 0–2	0.18	10.50	2.52	4.9	0.111	265	6397	0.070	<0.02	<0.05
4231-1, 10–12	0.77	0.87	4	4.4	0.029	5	15,667	0.253	<0.02	<0.05
4231-1, 20–22	0.29	7.22	0.43	5.6	0.050	115	—	—	<0.02	<0.05
4232-1, 0–2	0.13	1.56	1.95	3.9	0.136	35	5759	0.040	<0.02	<0.05
4232-1, 10–12	0.07	0.76	1.13	5.4	0.155	25	3261	0.295	<0.02	<0.05
4239-1, 0–2	0.09	2.78	0.97	2.3	0.178	38	6977	0.173	<0.02	<0.05
4239-1, 10–12	0.63	0.82	4	10.2	0.026	4	55,937	0.273	<0.02	<0.05
157 950A 13H 03, 20–24	6.55	4.20	22	2.6	0.030	0.17	243,937	0.575	0.140	6.865
157 950A 18X CC,19–23	2.58	3.76	11	6.5	0.020	1	—	—	0.450	<0.05
157 952A 36X 04,108–112	1.26	4.46	20	5.8	0.027	0.20	111,098	0.207	0.030	0.626
157 955A 13H 01, 52–55	1.82	2.57	4	3.1	0.020	1	60,630	0.435	0.040	0.083
157 958A 3H 3, 84–87	3.42	3.60	11	7.7	0.009	0.42	4273	0.666	0.065	0.275
157 955A 42X02, 75–78	1.75	1.99	3	6.8	0.038	1	35,211	0.362	<0.02	0.237

Analysis results of n-alkanes, fatty acids, PAH. Also given are CPI (Carbon Preference Index) of alkanes, the ratios C_{17}/C_{31} (indicating short-chain to long-chain alkanes), and C_{16}/C_{26} (indicating short-chain to long-chain fatty acids). Included are phenantrene to anthracene ratios (PHN/ANT) and of fluoranthene to pyrene (FLO/PYR) with the selected compounds retene and perylene added in separate columns.

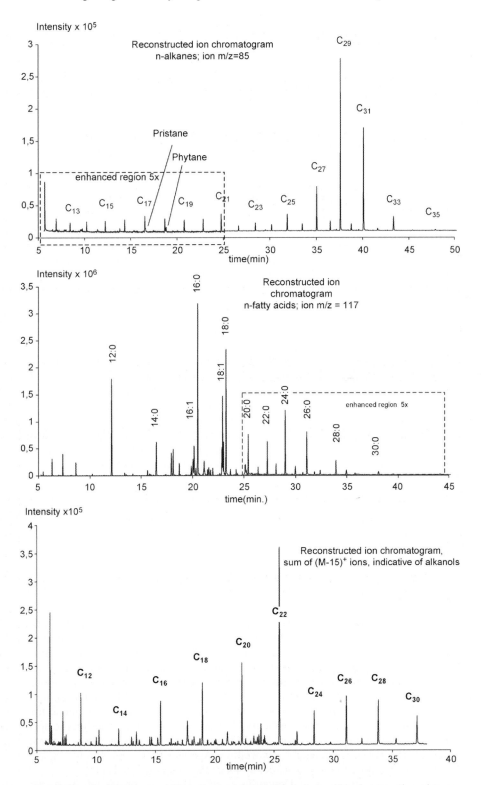

FIG. 2. Sample chromatograms illustrating peak-height distributions within the respective substance class: (a) *n*-alkane and isoprenoids (GeoB4231-10-12), (b) *n*-fatty acids (GeoB 4215-0-2) and (c) *n*-alkanols (GeoB 4215-0-2).

Table 4. Location of ODP cores

Core	Section	Latitude (North)	Longitude (West)	Water depth (m)	Recovery (m)
ODP 157 950A	13H03, 20–24	31°9.011	25°36.004	5437	339
ODP 157 950A	18XCC, 19–23	31°9.011	25°36.004	5437	339
ODP 157 952A	36X04, 108–112	30°47.449	24°30.574	5431	415
ODP 157 955A	13H01, 52–55	27°19.548	15°13.848	2854	599
ODP 157 955A	42X02, 75–78	27°19.548	15°13.848	2854	599
ODP 157 958A	3H03, 84–87			3789	133

Separate data on location of cores from ODP Legs 157 (Sites 950 to 955) and 159T (Site 958).

Clay mineral content may possibly affect the alteration process and influence the distribution of organic compounds. It is especially sensitive when using pyrolysis (PETERS, 1986). The abyssal plain samples have smectite as the main clay mineral (MARTINE GERARD, personal communication). The Shipboard Rock Eval pyrolysis gave the following parameters for MAP samples with TOC > 1.3 wt%: T_{max} from 400 to 430°C and PI average around 0.2; S_2 range 2.15–4.73 mg C/g rock with HI 20–252 mg HC/g TOC.

There is a prominent color change following oxidation at the top of the emplaced turbidites. Oxidation changes the color of OM from dark olive green to light turquoise.

DE LANGE (1998) studied this oxic alteration on the MAP and found a uniform near 80% decrease in TOC. None of this oxic alteration was seen in the slope samples. Here we have only used unoxidized samples from the MAP.

With this MAP material in mind, we want to do some comparative calculations further on regarding turbidite emplacement. Therefore, data from turbidite samples are also given in Tables 3 and 4. Vertical profiles of fatty acids are given in Fig. 3. The decrease in shorter chain compounds appears to be linear on a logarithmic scale. Earlier work on the abyssal plain has given lower TOC on near-surface samples of similar age (JARVIS and HIGGS, 1987). Pyrolysis studies of some of the younger turbidites showed the OM to be mainly of marine origin, with little or no terrigenous material (DELEEUW et al., 1982). Together with the presence of siliceous microplankton, this indicates that the sediments formed below areas of high surface productivity.

The Canary Islands and their southern approaches

We have also included a location on the southeastern flank of Gran Canaria Island (ODP Site 955) and a location further south of the Canary Islands at ODP Site 958 from the lower continental slope (Tables 3 and 4).

DISCUSSION

Accumulation and age considerations on the continental slope

Holocene (from 10 ka to the present) sedimentation rates were calculated (HENDERIKS et al., 2002) taking into account (a) sediment loss at the top of gravity cores and (b) the occurrence of turbidites. Sediment loss was estimated by comparing multicores for porosity, TOC, total carbonate, and fine fraction stable oxygen isotope profiles. Mean sedimentation rates (in cm/ka) were calculated for each time slice, assuming constant rates between tie-points of correlation and isotopic events with the SPECEMAP time scale (HENDERIKS et al., 2002). Such calibrations for Site GeoB 4223 indicate higher sediment accumulation near the coast (under upwelling conditions). Similar calculations for Site GeoB 4242 gave sedimentation rates for the open ocean sites. Between these environments, Site GeoB 4216 has intermediate rates, where productivity and accumulation may be influenced by upwelling. Ages for the samples in the present study were calculated according to these considerations and are included in Table 2.

If we adopt the age divisions of Henderiks et al. (2002), we obtain the following major events. Holocene: between present and 7000 years; Last Glacial Maximum (LGM), oxygen isotope stage 2.2: 16,000–24,000 years; Last Interglacial Climax (LIC), oxygen isotope

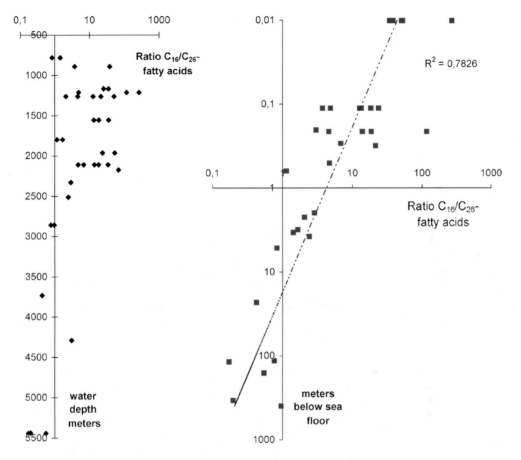

FIG. 3. Vertical profiles of fatty acids in surface and subbottom sediments. The decrease in shorter chain compounds appear in a linear trait on a logarithmic scale. This is taken to point out a major diagenetic component in the distribution of fatty acids in the analysed samples.

stage 5.5: 122,000 years; end of Penultimate Glaciation (ePG), oxygen isotope event 6.2: 135,000 years. All calculations by HENDERIKS et al. (2002) show that Site GeoB 4223 has the highest or very high values of TOC (proxy for OM). Site GeoB 4240 sediments reach further back than Site GeoB 4223, but otherwise show similar characteristics.

Data from several locations in this study reach from the surface down 20–30 cm into the sediment. These samples are all within the near-coast upwelling zone and reach back 3600–5000 years, after the end of the last glaciation.

Lipid variations below the seafloor

The occurrence of short-chain (*i.e.* 15–20 carbon atoms) alkanes, alkanols, and fatty acids is generally attributed to a contribution from marine algae, whereas longer chain molecules (*i.e.* 25–30 carbon atoms) originate mainly from terrigenous plant waxes (KILLOPS and KILLOPS, 1993). The presence or absence of unsaturated compounds can be indicative of the degree of preservation of the OM (KILLOPS and KILLOPS, 1993).

Polycyclic aromatic compounds (PAH) are a complex group with compounds that can be indicative of petrogenic sources, anthropogenic combustion, forest fires, or biogenic transformation processes. Perylene has been suggested as a biogenic marker in particular from diatomaceous rich sediments (GASKELL et al., 1975; KILLOPS and KILLOPS, 1993).

Holocene variations of lipids–Data from four core locations give similar variations in *n*-alkane and fatty acid content from the seafloor downward. TOC is fairly constant around 0.30–0.35 wt% in many surface samples (0–2 cm). Immediately below this, TOC decreases to less than 0.05 in many samples.

If we take Site GeoB 4223 as a model of OM accumulation, we notice that

(a) TOC varies between 0.18 and 1.07 wt% over the 7.68 m long core.

(b) TOC shows fluctuations and maxima at 103, 143–168, 353–383, 463–488, and 512–523 cm.

(c) Near the surface, 0–33 cm below the seafloor, TOC is fairly constant—only dropping from 0.45 to 0.37 wt% during the first 3 cm.

We compared the cores where we have analysis from 0, 10, 20, 30 cm.

These cores are GeoB 4212 and 4231. In addition, cores GeoB 4213 and 4215 have data from 0 to 10 cm. In most of these cores, TOC drops from 0.30–0.40 wt% to below 0.10 wt% from 0 cm and downward below seafloor surface. In GeoB 4231 and 4239 the decrease in TOC is much less. We believe that only at GeoB 4223, 4231, and 4239 do we find true oxygen minimum, and possibly more intense upwelling conditions (*e.g.* at GeoB 4223, GeoB 4239, GeoB 4231, GeoB 4231, where TOC stays fairly constant through greater depths). *n*-Alkanes, on the other hand, increase at 10 cm followed by a decrease further down (Fig. 3), as evidenced in four locations. Fatty acids have higher values at the seafloor surface, with a sharp decrease further down. This could indicate a mainly marine origin of OM, which then undergoes alteration at depth to evolve into a more terrigenous signature (SANTOS *et al.*, 1994; WAKEHAM, 1995).

Lipid variations–Three samples, GeoB4212, 219–223 cm (TOC = 0.34 wt%), GeoB 4214, 309–313 cm (TOC = 0.30 wt%), and GeoB 4223, 333–337 cm (TOC = 0.67 wt%) represent LGM and are approximately 17,000, 22,000 and 24,000 years old, respectively. Total *n*-alkanes decrease within this time span and fatty acids increase, although C_{17}/C_{31} (for *n*-alkanes) and C_{16}/C_{26} (for fatty acids) stay about the same.

The aerial distribution of *n*-alkanes and fatty acids is difficult to interpret due to varying water depth, currents, and oxygen distribution. The vertical profiles show slight variations between locations, which probably reflect variations in productivity. The windblown supply of OM is dominated by terrigenous compounds, *e.g.* C_{20}–C_{31}, and would give a more even distribution between all samples in the area.

Lipid variation in deeper ODP samples–TOC varies from 0.6–1.70 wt% in three samples from different depths below the seafloor on the MAP. There is no systematic correlation between TOC and different types of lipids.

Compared to the slope sediments, these samples have very low C_{16}/C_{26}- (fatty acids) and C_{17}/C_{31}- (*n*-alkanes) values. This is consistent with diagenetic breakdown of lower molecular weight compounds compared to higher molecular weight compounds in a longer time perspective (SANTOS *et al.*, 1994; BERTRAND *et al.*, 1996). The samples from the MAP turbidites are a mixture of different age sediments that were originally accumulated on the slope.

Preservation of organic matter

The accumulation of marine organic carbon in sediments off Northwest Africa has probably been caused mainly by high surface-water productivity. High productivity may result from fluvial nutrient supply and/or coastal upwelling of nutrient-rich waters. Earlier work south of Cape Blanc indicated a higher fluvial nutrient supply during humid interglacial periods (GASKELL *et al.*, 1975; STEIN *et al.*, 1989). Increased upwelling during the arid glacial periods may have had similar effects on surface-water productivity off northwest Africa (STEIN *et al.*, 1989). This means that the supply of marine organic carbon to the sediments could have reached similar levels during both glacials and interglacials.

Many other upwelling areas worldwide have steeper bathymetric gradients than our study area, resulting in a narrower zone of oxygen depletion. This can cause higher concentrations of organic carbon, *e.g.* up to 7–8 wt%, outside Peru (McCAFFREY *et al.*, 1991).

In the present chapter, there is a predominance of longer-chain *n*-alkanes, indicating a terrigenous origin. This conflicts with other evidence that points toward a marine origin for the OM (SCHMINCKE *et al.*, 1995; WEAVER *et al.*, 1998a,b). However, there is little terrigenous supply along the Morroccan coast. No major river occurs and the Saharan dust is the most obvious source of terrigenous OM. On the other hand, we have an upwelling zone along the Morroccan coast with a potential high marine OM productivity. Again TOC is not high and only Site GeoB 4223 shows a consistent level of TOC several meters below seafloor. Even here TOC only reaches 1.07 wt% at its highest.

Our observations merit the following comments:

n-Alkanes.   Our data illustrate that *n*-alkane distributions alone cannot be used as a tool to elucidate the origin and fate of OM. If we consider *n*-alkane distributions alone, an overestimation of the terrigenous influence may result.

Fatty acids.   Selective reduction in short-chain fatty acids (C < 20) and input of long-chain fatty acids (C > 20) in deeper layers suggest an alteration with time. The 12, 16, and 18 fatty acids as detected in the seafloor surface samples, would reflect original composition of OM and indicate a marine origin of the slope sediments.

PAH.   Perylene data suggest a biogenic contribution during diagenesis, which proceeds with increasing sediment depth (VENKATESAN, 1988). The presence of retene, in minute quantities in deep sediment layers, suggests minor terrigenous influence from coniferous plants, and may provide some hints as to the deforestation of the Sahara, which many claim to have taken place during the last 10,000 years.

MORRIS (1987) reported phytoplankton debris as the major component of OM in the MAP h-turbidite. This is found at 9.0–9.4 mbsf and is approximately 190–247 k.y. old. COWIE *et al.* (1995) found that 80% of the original OM had been removed in the oxidized upper part of the f-turbidite. The C/N ratio changed from 11 to 5. They concluded that degradation was non-selective regarding individual OM compounds. PRAHL *et al.* (1997) determined that TOC degradation was compound selective: higher plant wax derived *n*-alkanes were enriched relative to TOC and some marine biomarkers and plant wax fatty acids were lost. The relative presence of the terrigenous component was proposed to have increased from 15% of the TOC in the unoxidized samples to 30–60% of TOC in the oxidized samples. HOEFS *et al.* (1998) show that the original composition of OM is very homogenous, even on a molecular level, with identical internal distributions of compound classes in pyrolysates from different turbidite samples.

Marine or Terrigenous origin of OM

The possible sources of OM in the Canary Basin area are (1) marine organisms formed during upwelling, (2) windblown material from the African continent, (3) material brought by river discharge, and (4) drifting marine extraneous matter.

Productivity and accumulation at the NW African continental margin has been high both south and north of the Canary Islands, recently in summer/winter relations (STEIN, 1991). Upwelling and primary production is still prominent north of the Canary Islands (HENDERIKS *et al.*, 2001).

Terrigenous material is either fluvial or aeolian. River supply in recent times has been limited, but windblown material has been indicated in several studies of the Canary area (SARNTHEIN *et al.*, 1982). Studies of aerosol populations show that organic compounds may be adsorbed on windblown particles, as well as within pieces of vegetation (GAGOSIAN *et al.*, 1981, 1987). Terrigenous matter exposed to the atmosphere for a long time should be more prone to alteration than river transported material due to more effective oxygen exposure.

Windblown material from the Sahara desert has been increasing to the south in recent times. A clockwise wind pattern brings material northward along the coast at 1000 m altitude blowing the opposite way with regard to the prevailing tradewinds at lower altitudes (SARNTHEIN *et al.*, 1982). The sample from Site GeoB 4223 represents an area of high TOC accumulation, where olive green pelagic sediments have elevated TOC between 0.3 and 0.8 wt%.

The C_{16}/C_{26} ratio (and C_{12}/C_{26} ratio) is high for fatty acids in the surface samples and decreases further down in the sediment. The initial high values of total fatty acids at the seafloor surface, mainly high C_{16}, represent lower molecular weight fatty acids which are characteristic of marine OM. C_{16} fatty acid is also more easily degraded than the higher molecular weight compounds, which originate from terrigenous sources.

With diagenetic alteration, the characteristic marine compounds (*e.g.* C_{10}–C_{17} compounds) are degraded more than the higher molecular weight compounds that serve as protective waxes. With time we observe a terrigenous signature, although *n*-alkanes and fatty acids were originally more marine in character. But in the seafloor surface samples, we can see whether there is a distinct marine component in that region.

The proportions of the differently sourced OM in our samples are not yet clear. It has been shown by VENKATESAN *et al.* (1987) that on the slope off the coast of New York, marine OM increased compared to that on the shelf, where terrigenous material dominates. We propose that a terrigenous signature in our results in part is due to selective preservation of C_{21+} compounds, and not solely caused by dominantly terrigenous input.

Calculation of slope-to-Abyssal-Plain emplacement process

The aquisition of slope samples in the upwelling area north of Gran Canaria makes it possible to apply the mass calculation of WEAVER and THOMSON (1993) and have OM data both from the source area on the slope and the discharge area on the Plain. Our study of the slope sediments would be comparable to the most recent organic-rich turbidite "a" on the MAP. Turbidite "a" was emplaced about 1000 years ago, incorporating slope deposited material from a 200,000-year time period (WEAVER and ROTHWELL, 1987).

The volume calculation is based on the occurrence of dated nannofossil species together with appropriate oxygen isotope measurements. This gives the time span of original slope sediment accumulation. The limited occurrence of mixed nannofossil species along the transport path precludes major erosion during movement of the turbidity current (WEAVER and THOMSON, 1993). The "a" turbidite is around 0.90 m thick (WEAVER and ROTHWELL, 1987). The central area of MAP, covered by the three Leg 157 boreholes, is roughly 140×140 km^2 = 19,600 km^2. The volume of "a" turbidite in this area is $0.0009 \times 19,600$ km^3, which is almost 18 km^3.

On the slope, a slab today, taking a sediment slice with a TOC around 0.5 wt%, as exemplified by GeoB4223, and thickness around 10 m will require an area of 18 km^3 divided by 0.01 km or $18/0.01 = 1800$ km^2.

The age calculation from the spread of nannofossil species encountered in turbidite "a" together with an assumed sedimentation rate of 8 cm/1000 years results in a necessary depth of 15 m below seafloor to be eroded in the source area. The erosion is visualized as the dislocation of a slab with vertical sides and a square area with a 42 km side. If we disregard oxidation of the upper part, turbidite "a" has a TOC of 0.5 wt% over the 50 cm section (COLLEY and THOMSON, 1985; JARVIS and HIGGS, 1987). The "a" turbidite covers an approximate area of 1000 km by 150 km on the MAP. WEAVER and THOMSON (1993) assume a sedimentation rate of 8 cm/1000 years. HENDERIKS *et al.* (2002) calculated a sedimentation rate varying from 8 to 11 cm/1000 years in recent times in the near-coast zone, assuming a constant rate from now back to 150,000 years. Thus, the calculation of WEAVER and THOMSON (1993) still provides a reasonable estimate in the light of the recent slope data. Data from the most recent turbidite, "a", on the MAP agrees well with the data from the probable source— the slope off the coast of Morocco, even when using a maximum thickness of 0.90 m for turbidite "a".

The deeper turbidite samples have higher TOC, higher C/N, and slightly higher proportions of higher molecular weight *n*-alkanes and fatty acids. Lower C_{16}/C_{26} ratios and C_{17}/C_{31} ratios may imply somewhat higher terrigenous OM contributions. Countercurrents at intermediate water depths trek northwards compared to the main south-heading Canary Current. These counter currents could contribute terrigenous material from the Senegal River in times of increased fluvial runoff to the slope regions south of the Canary Islands, the southern source area (GASKELL et al., 1975).

PAH and contamination

This study examines surface and near-surface samples, which require investigation of whether our data have been influenced by human activity. Pollution could give signals that interfere with indigenous biomarkers (KILLOPS and KILLOPS, 1993). Parent compounds at typical pre-industrial concentrations dominate polycyclic aromatic compounds in the samples. Alkylated species occur at levels similar to their respective parent PAH, excluding any direct petrogenic source to the OM. The high phenantrene to anthracene ratio (PHN/ANT in Table 4) is indicative of selective degradation of anthracene, presumably during long-range air transport but also with increasing age of the OM, since the deeper sediment layers in particular have high PHN/ANT. The low ratio of fluoranthene to pyrene (FLO/PYR < 1), which is in reverse to what is found in recent sediments close to urban sources, excludes any recent influence of anthropogenically derived PAH. Traces of markers, such as retene, occur in only a few samples, but were mostly below the detection limit. Some of the deep sea samples also have anomalously high levels of perylene, which could have been formed through diagenesis (WAKEHAM, 1996; VENKATESAN, 1988). Occasional higher values of retene could point to increased earlier forestation in the Sahara area.

Human contamination might emanate from towns on the islands or ships passing the area. Near-coast contamination has been studied in many parts of the world. No such evidence has been found in the present study. The low concentrations of PAH and the absence of an UCM means that the sampling localities, even at the sea floor, have not been subject to human input of contaminating organic compounds.

CONCLUSIONS

Upwelling and high productivity of marine organisms resulted in the accumulation of organic sediments in an oxygen-free zone along the continental slope on the Northwest African margin. Preservation of the OM varies through time and space. Surface sediments indicate a uniform deposition of around 0.35 wt% organic carbon over the entire region along the slope. OM is mainly of marine origin, as shown by fatty acid distributions with a maximum $C_{14}-C_{18}$ compositions together with analogous alkanol distributions. These compounds decrease with depth in the sediment at the same time as the marine $C_{10}-C_{20}$ signature is changed to a terrigenous $C_{21}-C_{32}$ signature.

n-Alkane distributions are lower at the surface, but increase with depth in the sediment. Freshwater diatoms and quartz in the sediments support input by aeolian dust (WEFER et al., 1997). Windblown material has probably also brought terrigenous OM. The relative proportions of marine and terrigenous OM are not clear, but most of the OM appears to be of marine origin.

The continental slope is a source area for organic turbidites on the MAP. The most recently emplaced "a" turbidite probably originated north of 28°N from the investigated area in this study. TOC and biomarker contents from slope sediments, off the coast of Morrocco, have characteristics similar to the "a" turbidite. The deeper turbidites on MAP have higher organic carbon contents and slightly different characteristics, including a higher proportion of terrigenous OM. Polycyclic aromatic hydrocarbons (PAH) show no evidence of anthropogenic contamination in our samples. Sporadic higher values of retene could point to increased earlier forestation in the Sahara area.

Acknowledgements—We thank the crews and Shipboard scientists of JOIDES Resolution during ODP Leg 157 and RV Meteor, cruise 37/1 for an excellent time on board. We acknowledge Anna Winberg, Paul Frogner, Klara Hajnal and Sofi Kruckenberg for valuable technical assistance.—We also thank participants in the CANIGO project for a stimulating research climate and in particular Gregorio Parilla for doing the impossible very efficiently. Indira Venkatesan is thanked for inter-laboratory comparisons. Finally, we thank reviewers Russ Kaufman and Ken Peters for their useful comments. This is a contribution to the MAST program CANIGO (Subproject 3, "Particle Flux and Oceanography in the Eastern Boundary Current system", EC contract No. MAS-CT9-0060.

REFERENCES

BERTRAND P., SHIMMIELD G., MARTINEZ P., GROUSSET F., JORISSEN F., PATERNE M., PUJOL C., BOULOUBASSI I., BUAT MENARD P., PEYPOUQUET J.-P., BEAUFORT L., SICRE M.-A., LALLIER-VERGES E., FOSTER J. M., TERNOIS Y. AND THE OTHER PARTICIPANTS OF THE SEDORQUA PROGRAM (1996) The glacial ocean productivity hypothesis: the importance of regional, temporal and spatial studies. *Mar. Geol.* **130**, 1–9.

COLLEY S. and THOMSON J. (1985) Recurrent uranium relocations in distal turbidites emplaced in pelagic conditions. *Geochim. Cosmochim. Acta* **49**, 2339–2348.

COWIE G. L., HEDGES J. I., PRAHL F. G. and DE LANGE G. J. (1995) Elemental and major biochemical changes across an oxidation front in a relict turbidite: an oxygen effect. *Geochim. Cosmochim. Acta* **59**, 33–46.

DE LANGE G. J. (1998) Oxic vs. Anoxic diagenetic alteration of turbiditic sediments in the Madeira Abyssal Plain, Eastern North Atlantic. *Proceedings of ODP, Scientific Results*, Vol. 157(ed. P. P. E. WEAVER, H.-U. SCHMINCKE, J. V. FIRTH, W. DUFFIELD,), Ocean Drilling Program, College Station, TX, pp. 573–580.

DE LANGE G. J., JARVIS I. and KUIJPERS A. (1987) Geochemical characteristics and provenance of late Quaternary sediments from the Madeira Abyssal Plain, North Atlantic. *Geology and Geochemistry of Abyssal Plains*, Vol. 431(ed. P. P. E. WEAVER, J. THOMSON), Geological Society Special Publication, London, pp. 71–86.

DELEEUW J. W., VIETS T. C. and SCHENK P. A. (1982) *Een Vergelijkend Onderzoek van 8 Sedimentmonsters uit het Angola-Basin met Curie-Punt Pyrolyse Masspectrometrie*, Rep. TH-Delft-FOM Institute for Atomic and Molecular Physics, Amsterdam, pp. 1–6.

FREUDENTHAL T., MEGGERS H., MORENO A., HENDERIKS J., KUHLMANN H. and WEFER G. (2001) Upwelling intensity and filament activity off Morrocco during the last 250,000 years. *Deep-Sea Res.* **49**, 3655–3674.

FÜTTERER D. K. (1983) The modern upwelling record off Northwest Africa. In *Coastal Upwelling. Its sediment record. Part B: Sedimentary records of ancient coastal upwelling* (eds. J. THIEDE and E. SUESS), NATO Conference Series, Series IV: Marine Science 10b, pp. 105–121. Plenum Press, New York.

GAGOSIAN R. B., PELTZER E. T. and ZAFIRIOU O. C. (1981) Atmospheric transport of continentally derived lipids to the tropical North Pacific. *Nature* **291**, 312–314.

GAGOSIAN R. B., PELTZER E. T. and MERRILL J. T. (1987) Long-range transport of terrestrially derived lipids in aerosols from the south Pacific. *Nature* **325**, 800–803.

GASKELL S. J., MORRIS R. J., EGLINTON G. and CALVERT S. E. (1975) The geochemistry of a recent marine sediment off nortwest Africa. An assessment of source of input and early diagenesis. *Deep-Sea Res.* **22**, 777–879.

HENDERIKS J., FREUDENTHAL T., MEGGERS H., NAVE S., ABRANTES F., BOLLMANN J. and THIERSTIEN H. R. (2002) Glacial-interglacial variability of particle accumulation in the Canary Basin: A time-slice approach. *Deep-Sea Res. II* **49**, 3675–3705.

HOEFS M. J. L., SINNINGHE DAMSTÉ J. S., DE LANGE G. J. and DE LEEUW J. W. (1998) Changes in kerogen composition across an oxidation front in Madeira Abyssal Plain turbidites as revealed by pyrolysis GC-MS. *Proceedings of ODP, Scientific Results*, Vol. 157(ed. P. P. E. WEAVER, H.-U. SCHMINCKE, J. V. FIRTH and W. DUFFIELD), Ocean Drilling Program, College Station, TX, pp. 591–607.

JARVIS I. and HIGGS N. (1987) Trace-element mobility during early diagenesis in distal turbidites: late Quaternary of the Madeira Abyssal Plain, N. Atlantic. *Geology and Geochemistry of Abyssal Plains*, 31(ed. P. P. E. WEAVER, J. THOMSON,), Geological Society Special Publication London, pp. 179–214.

KILLOPS S. D. and KILLOPS V. J. (1993) *An introduction to organic geochemistry*. Longman Scientific & Technical, Longman Group, UK, 265 pp.

MCCAFFREY M. A., FARRINGTON J. W. and REPETA D. J. (1991) The organic geochemistry of Peru margin surface sediments: II. Paleoenvironmental implications of hydrocarbon and alcohol profiles. *Geochim. Cosmochim. Acta* **55**, 483–498.

MASSON D. G. (1994) Late Quaternary turbidity current pathways to the Madeira Abyssal Plain and some constraints on turbidity current mechanisms. *Basin Res.* **6**, 17–33.

MORRIS R. J. (1987) Turbidite flows as a source of organic matter in deep water marine deposits: evidence from Quaternary sediments on the Madeira Abyssal Plain. *Mem. Soc. Geol. Fr.* **151**, 43–53.

MÜLLER P. J., ERLENKEUSER H. and VON GRAFENSTEIN R. (1983) Glacial-interglacial cycles in oceanic productivity inferred from organic carbon contents in Eastern North Atlantic sediment cores. In *Coastal Upwelling. Its sediment record. Part B: Sedimentary records of ancient coastal upwelling* (eds. J. THIEDE and E. SUESS), NATO Conference Series, Series IV: Marine Science 10b, pp. 365–398. Plenum Press, New York.

PETERS K. E. (1986) Guidelines for evaluating petroleum source rock using programmed pyrolysis. *AAPG Bull.* **70**, 318–329.

PRAHL F. G., DE LANGE G. J., SCHOLTEN S. and COWIE G. I. (1997) A case for postdepositional aerobic degradation of terrestrial organic matter in turbidite deposits from the Madeira Abyssal Plain. *Org. Geochem.* **27**, 141–152.

RUDDIMAN W., SARNTHEIN M., BALDAUF J. and SHIPBOARD SCIENTIFIC PARTY (1988) *Proceedings. ODP, Initial Reports (Pt. A)*, Vol. 108, Ocean Drilling Program, College Station, TX.

SANTOS V., BILLETT D. S. M., RICE A. L. and WOLFF G. A. (1994) Organic matter in deep-sea sediments in the northeast Atlantic Ocean. I-Lipids. *Deep-Sea Res.* **41**, 787–819.

SARNTHEIN M., THIEDE J., PFLAUMANN U., ERLENKEUSER H., FÜTTERER D., KOOPMANN B., LANGE H. and SEIBOLD E. (1982) Atmospheric and oceanic circulation patterns off Northwest Africa during the past 25 million years. In *Geology of the Northwest African Continental Margin* (eds. U. VON RAD, K. HINZ, M. SARNTHEIN and E. SEIBOLD), pp. 545–604. Springer, Berlin.

SCHMINCKE H.-U., WEAVER P. P. E., FIRTH J. V. and SHIPBOARD SCIENTIFIC PARTY (1995) *Proceedings of ODP, Initial Reports*, Vol. 157, Ocean Drilling Program, College Station, TX.

STEIN R. (1991) *Accumulation of Organic Carbon in Marine Sediments*, Lecture Notes in Earth Sciences, 34. Springer, Berlin-Heidelberg, 217 pp.

STEIN R., TEN HAVEN H. L., LITTKE R., RULLKÖTTER J. and WELTE D. H. (1989) Accumulation of marine and terrigenous organic carbon at up-welling Site 658 and non-up-welling Sites 657 and 659: implications for the reconstruction of paleoenvironments in the eastern subtropical Atlantic through late Cenozoic times. *Proceedings of ODP, Scientific Results* (eds. W. RUDDIMAN, M. SARNTHEIN, *et al.*,), Vol. 108, pp. 361–385. Ocean Drilling Program, College Station, TX.

TEN HAVEN H. L., RULLKÖTTER J. and STEIN R. (1989) Preliminary analysis of extractable lipids in sediments from the eastern North Atlantic (Leg 108): comparison of a coastal upwelling area (Site 658) with a nonupwelling area Site 659). *Proceedings of ODP, Scientific Results*, (eds. W. RUDDIMAN, M. SARNTHEIN, *et al.*,) Vol. 108, pp. 351–359. Ocean Drilling Program, College Station, TX.

VENKATESAN M. I. (1988) Occurrence and possible sources of perylene in marine sediments: a review. *Mar. Chem.* **25**, 1–27.

VENKATESAN M. I., RUTH E., STEINBERG S. and KAPLAN I. R. (1987) Organic geochemistry of sediments from the continental margin off southern New England, USA-Part II. Lipids. *Mar. Chem.* **21**, 267–299.

WAKEHAM S. G. (1995) Lipid biomarkers for heterotrophic alteration of suspended particulate organic matter in oxygenated and anoxic water columns of the ocean. *Deep-Sea Res.* **42**, 1749–1771.

WAKEHAM S. G. (1996) Aliphatic and polycyclic aromatic hydrocarbons in Black Sea sediments. *Mar. Chem.* **53**, 187–205.

WEAVER P. P. E. and KUIJPERS A. (1983) Climatic control of turbidite deposition on the Madeira Abyssal Plain. *Nature* **306**, 360–363.

WEAVER P. P. E. and ROTHWELL R. G. (1987) Sedimentation on the Madeira Abyssal Plain. *Geology and Geochemistry of Abyssal Plains* (eds. P. P. E. WEAVER and J. THOMSON), Vol. 31, pp. 71–86. Geological Society Special Publication, London.

WEAVER P. P. E. and THOMSON J. (1993) Calculating erosion by deep-sea turbidity currents during initiation and flow. *Nature* **364**, 136–138.

WEAVER P. P. E., ROTHWELL R. G., EBBING J., GUNN D. E. and HUNTER P. M. (1992) Correlation, frequency of emplacement and source directions of megaturbidites on the Madeira Abyssal Plain. *Mar. Geol.* **109**, 1–20.

WEAVER P. P. E., MASSON D. G., GUNN D. E., KIDD R. B., ROTHWELL R. G. and MADDISON D. A. (1995) Sediment mass-wasting in the Canary Basin. In *Atlas of Deep Water Environments: Architectural Style in Turbidite systems* (eds. K. T. PICKERING, R. N. HISCOTT, N. H. KENYON, F. RICCI LUCHI and R. D. A. SMITH), pp. 287–296. Chapman & Hall, London.

WEAVER P. P. E., SCHMINCKE H.-U., FIRTH J. V. and DUFFIELD W. (eds) (1998a). *Proceedings of ODP, Scientific Results*, Vol. 157, Ocean Drilling Program, College Station, TX.

WEAVER P. P. E., JARVIS I., LEBREIRO S. M., ALIBÉS B., BARAZA J., HOWE R. and ROTHWELL R. G. (1998b) Neogene turbidite sequence on the Madeira Abyssal Plain: Basin filling and diagenesis in the deep ocean. In *Proceedings of ODP, Scientific Results* (eds. P. P. E. WEAVER, H.-U. SCHMINCKE, J. V. FIRTH and W. DUFFIELD), Vol. 157, pp. 619–634. Ocean Drilling Program, College Station, TX.

WEFER G., ABRANTES F. and CRUISE PARTICIPANTS (1997) Report and preliminary results of METEOR Cruise M37/1, Lisbon-Las Palmas, 04.12.1996-23.12.1996. Berichte, Fachbereich Geowissenschaften, Universität Bremen No. 90, 79 pp.

WEFER G., BERGER W. H., RICHTER C. and SHIPBOARD SCIENTIFIC PARTY (1998) *Proceedings of ODP, Initial Reports*, Vol. 175. Available from World Wide Web (online): E-mail: http://www-odp.tamu.edu/publications/175_IR/175TOC. HTM.

Geochemical Investigations in Earth and Space Science: A Tribute to Isaac R. Kaplan
© The Geochemical Society, Publication No. 9, 2004
Editors: R.J. Hill, J. Leventhal, Z. Aizenshtat, M.J. Baedecker, G. Claypool,
R. Eganhouse, M. Goldhaber and K. Peters

Formation of Eastern Mediterranean sapropels—What can be learnt from Baltic Sea sapropels?

ROLF O. HALLBERG

Department of Geology and Geochemistry, Stockholm University, S-106 91, Stockholm, Sweden

Abstract—The 1947–1948 Swedish Deep Sea Expedition with M/S Albatross revealed a remarkable Late Quaternary development of the Eastern Mediterranean Sea (Meddelanden från Oceanografiska Institutet i Göteborg, 21 (1952) 1–38). Several organic rich layers (sapropels S1–S11) with black sulfides indicated long periods of stagnation when the preservation of organic matter increased and hydrogen sulfide was accumulated in the bottom water column. In the literature usually two hypotheses are favored to explain these stagnant periods.

(1) High primary production, which during degradation increases oxygen utilization and, hence, the preservation of organic matter settling to the ocean bottom.
(2) A highly stratified water column with bottom water stagnation without oxygen renewal, which would increase organic carbon preservation.

A model is presented on the fate of organic carbon and shows that the most plausible explanation is a combination of these two hypotheses, which is the present situation in the Baltic Sea, where organic rich layers are accumulated together with iron sulfides as black minute laminas in a sapropel type of sediments. The present environment in the Eastern Mediterranean will not produce a sapropel even with anoxic bottom water. A primary production in the order of 3–4 times more than the present is required. The Baltic Sea experienced a stagnation period at around the time of the latest sapropel, S1, and is presently suffering from intermittent stagnation of the bottom water and can thus serve to explain the mechanism behind sapropel formation of the Eastern Mediterranean (E.M.). The development mirrors the general sedimentological and geochemical sequence of a semi-enclosed sea affected by water stratification due to salinity gradients. This paper presents similarities between the development of redox facies of the latest Eastern Mediterranean sapropel (S1) and the redox transition in the Baltic Sea during and after a turnover from fresh to brackish water. E.M. trace elements (Mn and Zn) and stable sulfur isotope data are used as paleogeochemical proxies. They form the base for a discussion on redox chemistry and possible diagenesis during formation.

INTRODUCTION

SEMI-ENCLOSED SEAS are found at margins of ocean bodies. They contain relatively large bodies of water that to varying degrees are restricted in circulation and in exchange with the ocean bodies. The restricted circulation of the enclosed sea often creates a vertical separation into two water bodies because of differences in density/salinity. A characteristic chemistry differentiates the two bodies with the lower usually being depleted in oxygen due to reduced ventilation, sometimes to such an extent that hydrogen sulfide is built up in the bottom water like in the present days Black Sea (*e.g.* DEUSER, 1974; SOROKIN, 1983) and Baltic Sea (FONSELIUS, 1969; MATTHÄUS, 1979; STERNBECK and SOHLENIUS, 1997; SOHLENIUS *et al.*, 2001). This development is mirrored in the sediments that become laminated with dark gray to black layers alternating with light gray to white. The former containing appreciable amounts of organic material and sulfides while the latter are rich in carbonates. Carbon is mainly buried in either organic or inorganic form dependent on varying biogeochemistry of the water body.

When the Albatross Expedition collected sediment cores from the Eastern Mediterranean (E.M.) they found several dark layers with high organic carbon concentrations (in extreme cases close to 20% by dry weight) indicating periods of so-called stagnant water bodies depleted in oxygen and therefore enhancing organic preservation. Such unconsolidated sedimentary layers rich in organic carbon are usually referred to as sapropels. A sapropel is defined (KIDD *et al.*, 1978) as a discrete sediment layer more than 1 cm thick and containing more than 2% by dry weight of organic carbon. They are distinguished from peat in being rich in fatty and waxy substances and poor in cellulose material. The Albatross Expedition identified 11 sapropels in the E.M. cores. Later expeditions have revealed additional sapropels. Deposition of the last sapropel, S1, has been dated to approximately 9.5–6 ka BP. The entire time span of the sapropel forming conditions for the 12 latest sapropels in the E.M. is about 450 ka.

Two prerequisites have been suggested for the formation of sapropels.

(i) stagnant bottom water body with reducing conditions,
(ii) high primary production.

The previous will preserve organic matter to a higher extent than an oxygenic environment and the latter will produce high amounts of organic carbon and thus increase the net export of organic carbon to the sediments. The driving force for these events is the water circulation pattern, which in the E.M. can be explained with a dry/wet oscillation in the Mediterranean climate (ARIZTEGUI *et al.*, 2000).

The present circulation of the Mediterranean Sea is a surface inflow from the Atlantic Ocean through the straight of Gibraltar. Due to evaporation, surface water salinity increases towards the east where it becomes dense enough to sink and form intermediate water usually referred to as Mediterranean Intermediate Water (MIW). By diffusion and advection, MIW is mixed with the more saline bottom water to produce Eastern Mediterranean Deep Water (EMDW). This mixing process thereby supplies EMDW with oxygen. Formation of the present EMDW is compensated by a continuous exchange, overflowing the sills of Sicily and further over the sills of Gibraltar back to the Atlantic Ocean (ROHLING, 1994). This type of circulation in an enclosed sea where the oxygenated ocean water in the top part is circulated to the bottom of the enclosed sea and compensated by a subsurface outflow of more saline water back to the ocean is called an anti-estuarine circulation or because of its typical example, Mediterranean circulation. The oligotrophic situation of the present Mediterranean Sea has probably lasted for the last 6000 years since the end of the youngest sapropel, S1. The present circulation does not give rise to the formation of sapropels. Two main prerequisites are missing namely anoxic conditions in EMDW and a high primary production.

To explain the sapropel formation in the E.M., a reversal in circulation has been proposed (*e.g.* STANLEY *et al.*, 1975; CALVERT, 1983; SARMIENTO *et al.*, 1988; THUNELL and WILLIAMS, 1989; HOWELL and THUNELL, 1992). They favor an estuarine circulation where river input and precipitation exceeds evaporation. A reversal in circulation over the sills of Sicily and Gibraltar cannot be verified (VERGNAUD-GRAZZINI *et al.*, 1989; MYERS and HAINES, 1998). However, the same authors state that the buoyancy of the fresh surface water will create a low-salinity cap and hence decrease or even halt the oxygen advection to EMDW. The low salinity cap hypothesis is supported by the decrease of $\delta^{18}O$ values found in most sapropels. This decrease in $\delta^{18}O$ has frequently been used to infer a top layer of water with low salinity, thus supporting the water column stratification hypothesis (*e.g.* CITA *et al.*, 1977; THUNELL and WILLIAMS, 1983; TANG and SCOTT, 1993). $\delta^{18}O$ values become more negative to the north and south over the continents in comparison with ocean water. The ratio of fresh and marine water inflow to the Mediterranean is mirrored in the foraminiferal $\delta^{18}O$ values (TANG and SCOTT, 1993). The prevailing hypothesis for the increasing fresh water inflow to the Mediterranean during sapropel formation is the periodic changes of precipitation over North Africa (ROSSIGNOL-STRICK, 1985). This is regulated by the monsoonal intensity, resulting in high correlation between the insolation monsoon index and the geological record of East Mediterranean sapropels (ROSSIGNOL-STRICK *et al.*, 1982; ROSSIGNOL-STRICK, 1985). The monsoon was the driving force for a more humid climate over northern Africa, increased river run-off and formation of the low salinity cap. Also the monsoon is the driving force to produce upwelling conditions of the E.M., as upwelling is usually wind driven. Sapropel formation thus appears to reflect special climatic conditions at these latitudes. These climatic changes can in turn be correlated with the Milankovitch precession cycles (CITA *et al.*, 1991).

As a result of the increased fresh water inflow, estuarine circulation is developed that is characterized by a strong halocline and restricted circulation of the bottom water. It has been proposed that melt water influx from the ice-sheet over northern Europe and Asia may have influenced the formation of the sapropels to a large extent (OLAUSSON, 1961). This may only be the case for the older sapropels S6 and S8, which contain cold water foraminiferal assemblages in contrast to other sapropels that have warm foraminiferal assemblages (CITA *et al.*, 1977). Warm foraminiferal assemblages might be expected from increased inflow from the Nile River due to high monsoonal summer precipitation over Ethiopia. Possible melt water influx from the ice-sheet during the formation of S6 and S8 would make the low salinity cap even more effective. BARD *et al.* (2002) is using $\delta^{18}O$ values from a Tyrrhenian stalagmite as evidence that S6 was formed during humid conditions with increased rainfall over the entire Mediterranean Sea rather than just restricted to the river Nile. Another support for increased flux of continental water and their accompanying sedimentary weathering product during sapropel formation is the increase in general average grain size, which is true for all sapropels. The prerequisites for the formation of sapropels in the Mediterranean are most likely a combination of water stratification and a significant increase of primary production resulting in high deposition of organic matter and preservation under anoxic conditions (see review by ROHLING, 1994).

Estuarine circulation is dominant in the Baltic Sea at present. Fresh water inflow mainly from rivers in the northern area is compensated by a surface outflow through the Danish straights to the North Sea. A subsurface inflow of marine water from the south creates a saline gradient from the south to the north. The northern part of the Baltic Sea exhibits almost fresh water conditions. The high river run-off and low evaporation results in a low-salinity cap and a stratification of the water body. The present halocline is situated between 50 and 70 m, which is in the range of the average water depth (56 m). The saline bottom water is mixed with the overlaying fresh water by advection, making the Baltic Sea the largest brackish sea in the world. To some extent this mixing weakens the stratification of the water column but even so the present bottom water salinity is double that of the surface water. The intrusions of the comparatively more dense salt-water from the North Sea deliver oxygen to the lower water body. These intrusions depend to a large extent on strong westerly winds and low pressure over the Baltic Sea in combination with high pressure over the North Sea. Due to these meteorological factors the temporal and spatial distribution of oxygenic conditions in the basins of the Baltic Sea exhibit a considerable variation. Oxygen is rapidly consumed by biota and reducing components like hydrogen sulfide and the anoxic condition of the bottom water is the dominating situation resulting in a sapropel formation. These anoxic conditions result in release of nutrients (mainly phosphorous) that are distributed to the surface waters by upwelling along the coasts thereby stimulating the primary production. The Baltic Sea is presently on the wedge of being classified as eutrophic. The latest sapropel formation in the Baltic Sea started about 1.5 ka later than sapropel S1 in the Mediterranean Sea. The Baltic sapropel was a consequence of a transition from a fresh to a brackish water environment, which in turn was due to an opening of a connection with the North Sea via the Danish sills. The causes behind a water-stratified body in the Baltic are thus different from those in the Mediterranean but the consequences for the preservation of organic matter are the same. The situation in the Baltic Sea is such that sediments of sapropel type are still forming and the mechanism for their formation can be studied in the recent environment. The Baltic Sea may thus serve as a model to study the prerequisites for the E.M. sapropel formation though much more shallow than the Mediterranean Sea and with lower salinity in the deep water. Despite these differences, the mechanisms for flux and preservation of organic carbon during sapropel formation are, however, the same.

MATERIAL AND METHODS

Sampling

A core from the Eastern Mediterranean, NE of Crete (Latitude: 35°55′, Longitude: 26°03′) at a water depth of 1220 m was collected on R/V Aegaeo in 1991. Because of a crack in the sediment core, almost 1 cm of sample from around 35 to 36 cm from the top of the core is "missing".

Core MIPC 17 contains a representative 7 cm thick section, 38–46 cm (including the crack), of sapropel S1. This sapropel is formed approximately between 6 and 9.5 ^{14}C ky BP. One date was determined on the sediment at a depth of 36 cm and estimated to 8.3 ± 0.1 ^{14}C ky BP. The sampling was made with a piston corer (i.d. 8 cm). Subsampling of 0.5 cm thick slices was made with a scalpel in the laboratory.

Total organic carbon (TOC) and inorganic carbon (IC)

Analyses of inorganic and organic carbon were made on dry sediment samples by combustion on a Shimadzu TOC-5000 total organic carbon analyzer equipped with a solid sample module SSM-5000A.

Trace metals

Trace metal analyses were made on freeze-dried samples after leaching with 7 M HNO_3 at 125°C for 30 min. Leached samples were centrifuged for 10 min at 13,800g and decanted to achieve a clear solution. Analyses were made with a Varian SpectrAA 220. International geochemical standard Mag-1 was used for calibration.

Sulfate and sulfide

A first extraction of sulfate was made from sediments with hot 6 M HCl (15 ml/1.5 g sediment) during 2 h under continuous stirring. The residue was filtered (0.45 m) and washed with distilled water. The solution was treated with $BaCl_2$ to precipitate $BaSO_4$ and sulfate was determined by weight.

A second parallel extraction was performed on 1.5 g sediment with 0.04 M $NaHCO_3$ to verify that only sulfate was extracted (VAN STEMPVOORT et al., 1990). The residue from this extraction was further extracted with cold 3 M HCl. Sulfate in these solutions was converted to $BaSO_4$ by precipitation with $BaCl_2$.

The difference between the results of the two extraction methods was within 11% or based on sulfur isotope mass balances, 1‰. The data from the second extraction was always higher. The data from the first extraction has been used in this paper.

Sulfide (mainly pyrite) analyses were made on the residual sediment from the first extraction method by oxidation to sulfate with a combination of 14 M HNO_3 and Br_2 (KROUSE and TABATABAI, 1986). The sulfate was converted to $BaSO_4$ by precipitation with $BaCl_2$.

Sulfur isotopes

Approximately 0.5 mg of $BaSO_4$ was mixed with equal amounts of V_2O_5 and reacted in an online Elemental Analyzer to a ConFlo SIRMS (Finigan Delta +). The sulfur isotope ratio is defined as the deviation in ‰ of the ratio $^{34}S/^{32}S$ between a sample and a standard, expressed in the conventional $\delta^{34}S$ notation relative to Canyon Diablo Troilite (CDT). The accuracy of the measurements based on standard measurements was better than ± 0.3‰.

RESULTS

TOC and IC

The concentrations of TOC and IC are depicted in Fig. 1. An increase of TOC up to 3% is observed between 43 and 41 cm followed by a short decrease before TOC stabilizes around 3%. The top part of the sapropel, 35–33 cm, exhibits a decrease in TOC to values below 1%. This may be explained by a true decrease in the amount of deposited organic carbon and/or oxidation of organic matter after deposition. Post-depositional oxidation has been reported for several sapropels and in some cases all organic carbon have been oxidized (e.g. THOMSON et al., 1995; JUNG et al., 1997). As the colors of the core are black and gray they do not indicate any substantial oxidation. Mn concentrations only show a slight increase at the top of the

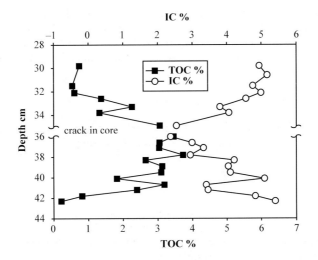

FIG. 1. Dry weight of total organic carbon (TOC) and inorganic carbon (IC).

sapropel (Fig. 2). A strong increase is used as evidence that a substantial oxidation has taken place (THOMSON *et al.*, 1999). The IC data inversely mirror the TOC data, which is expected as carbon is buried in either of the two forms.

Trace metals as proxies of sapropel formation

Two metals (Zn and Mn) have been chosen as proxies of sapropel formation because they show distinct concentration features during the sapropel formation (Fig. 2). Zn in contrast to Mn is not a redox sensitive element and therefore its concentrations merely reflect the overall mobilization and accumulation of trace metals. Zn on the other hand exhibits almost 20% higher concentrations in the sapropel compared to the ambient sediment above and below the sapropel. A ratio, 10·Zn/Mn, mirrors the sapropel formation to a very high degree (Fig. 3). It increases very sharply from 0.6 at a sediment depth close to 42 cm to 1.1 at the beginning of the sapropel at a sediment depth of 39 cm. This initial stage is usually referred to as a protosapropel. The metal ratio then stays between 1 and 1.1

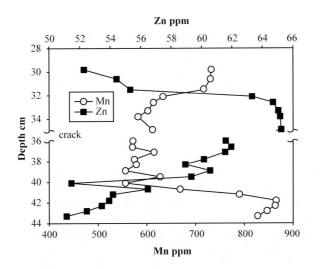

FIG. 2. Dry weight of manganese and zinc.

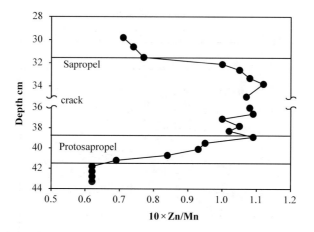

Fig. 3. Boundaries of protosapropel and sapropel are indicated in accordance with color differences of the sediment core. The depicted graph of the ratio $10 \cdot Zn/Mn$ connects well to these boundaries.

during the "true" sapropel formation and thereafter it shows a sharp decrease back to a value close to 0.6. This graph is thus a better indicator of the boundaries of the sapropel S1 formation at this site than TOC/IC. Mn data accordingly show significantly higher values below and above the sapropel with almost 30% lower values in the sapropel compared to the sediments below its formation. Other scientists have made similar observations (*e.g.* SUTHERLAND *et al.*, 1984) and stated that enrichment of Mn is a reliable indicator of bottom water oxygenation (CALVERT and PEDERSEN, 1993). Mn is a redox sensitive element and is precipitated as carbonate in the Baltic Sea during oxic bottom water conditions. STERNBECK and SOHLENIUS (1997) describe the mechanism for this. During anoxic conditions high concentrations of Mn^{2+} are built up in the pore water of the sediments and will further release to the bottom water column. When this is oxidized, Mn^{4+} is precipitated as Ca-rich rhodochrocite and oxide. In the Baltic Proper the release of Mn from the reduced sediments results in a precipitation and enrichment of Mn in the oxidized sediments along the borders of the deep anoxic basins (*e.g.* the discussion in HALLBERG (1973)).

Sulfate, sulfide and sulfur isotopes

The sulfate data depicted in Fig. 4 shows a sudden increase from 0.95 to 1.8 mg/g at a depth of about 39 cm. The δ^{34}S-sulfate also exhibits a sudden change at the same depth but to 9.5‰, which is half of the original 17.5‰ (Fig. 5). A small decrease from 19.8 to 17.5‰ can also be observed at a depth of 42 cm during the initial protosapropel formation.

DISCUSSION

The development of anoxia in the bottom water

The MIPC 17 sapropel S1 shows a distinct black color between sediment depths 32 and 39 cm (including a crack). The thickness of the MIPC 17 sapropel S1 is 6 cm (excluding the crack), which is a common value reported for S1. It also has 3 cm sediment of the characteristic gray color immediately below the sapropel. The sediment layer on top of the sapropel was also gray in contrast to the ochre color frequently observed (DE LANGE *et al.*, 1989; THOMSON *et al.*, 1995, 1999). The depositional sequence of sapropels with a characteristic gradual build-up from beige to gray oxidized ooze and further into increasingly darker layers peaking in black is most probably a result of increasing EMDW oxygen depletion. The transition of foraminiferal assemblages indicates that the development of the anoxic conditions was not a sudden event as

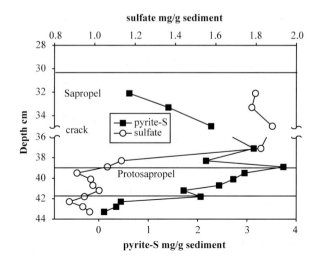

FIG. 4. Boundaries of the sapropel are indicated as in Fig. 3. Sulfate and pyrite–sulfide concentrations are given as content of sulfur.

may be assumed from the usually sharp lower boundary of the black sapropel, but on the contrary a gradual depletion of oxygen concentrations of deep water (VISMARA-SCHILLING, 1986). High primary production will result in increased depletion of oxygen during the degradation of the organic matter. Changes in the benthic fauna indicate in the case of several sapropels that reduced deep water ventilation had started an appreciable time before the formation of the true sapropel (*i.e.* CITA and PODENZANI, 1980; DE LANGE *et al.*, 1989; JORRISEN *et al.*, 1993). The fossil ostracode record further supports decreased oxygen content of the bottom water before the sapropel began to form (VAN HARTEN, 1987).

$\delta^{34}S$ *of pyrite*—The $\delta^{34}S$ of pyrite and sulfate can give information on the type of environment for the sulfide formation. The $\delta^{34}S$ of pyrite in MIPC 17 below the sediment depth of 42 cm decreases sharply from -36 to $-46\%o$ (Fig. 5). At the beginning of the sapropel formation it returns to $-37\%o$. Such negative values have been attributed to a very low sulfate reduction rate, which in turn will give rise to higher fractionation of the sulfur isotopes (*e.g.* HARRISON and THODE, 1958; KAPLAN and RITTENBERG, 1962; NAKAI and JENSEN, 1964).

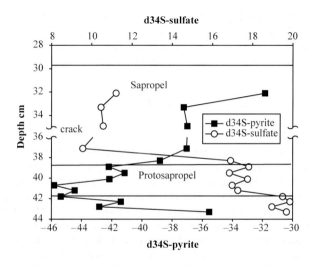

FIG. 5. Graphs showing stable isotope data of dry sediment sulfate and sulfide.

It has also been proposed that pyrite was formed by downward diffusion of hydrogen sulfide due to limited amount of iron in the deposited sediment (PASSIER et al., 1996). During the protosapropel formation the bottom water was probably still ventilated to some extent and the formation of HS⁻ was absent from the bottom water or restricted to a thin blanket of water at the sediment–water interface.

PASSIER et al. (1996) made mass-balance calculations that indicate the formation of HS⁻ in an open system at, or close to, the sediment–water interface during the sapropel formation. In the Baltic Sea, at the transition between fresh and brackish water, δ^{34}S-pyrite experiences a similar rapid excursion to more negative values. It starts at a value around zero and decreases to $-30‰$ and thereafter it increases to $-20‰$ (STERNBECK and SOHLENIUS, 1997). The sapropel environments of E.M. and the Baltic Sea exhibit a rapid decrease at the beginning of the oxygen depletion followed by less negative values when hydrogen sulfide is becoming more prominent in the bottom water.

Experiments with benthic chambers in the Baltic Sea revealed δ^{34}S of sulfides in the bottom water to be very close to zero (HALLBERG, 1985). The negative values and general decrease in δ^{34}S-pyrite with sediment depth during the protosapropel formation do not give evidence for a diffusion of HS⁻ from bottom water at the sediment–water interface if this HS⁻ had δ^{34}S values close to zero. Instead the δ^{34}S values would be expected to be much more positive. The δ^{34}S of monosulfides in the Baltic Sea sediments at the transition from well ventilated to stagnant bottom water environment is also very close to zero. It then stabilizes between 0 and $-5‰$ for thousands of years (STERNBECK and SOHLENIUS, 1997). A δ^{34}S of monosulfides close to zero would thus be in agreement with the Baltic Sea data. The hypothesis by PASSIER et al. (1996) that pyrite was formed by downward diffusion of hydrogen sulfide cannot be verified by this study. It is thus more likely that pyrite in the protosapropel as well as in the sapropel was formed in the sediment either directly from interstitial sulfate reduction at low rate or from transformation of metastable greigite formed in the sediment.

δ^{34}S of monosulfides—The δ^{34}S of the monosulfides could not be measured due to too small samples but can be deduced from the sulfate data (Fig. 4). The observed doubling of the sulfate concentrations at a depth of about 39 cm has no natural explanation from the site of deposition. It can be interpreted as a fifty/fifty mixture of seawater sulfate and sulfate from an oxidation of the sedimentary iron sulfides. This may be the result of an oxidation of the sediment due to a diffusion of oxygen down the core since the time of deposition and/or an oxidation after sampling. The core does not exhibit any sign of depositional oxidation like an ochre layer at the top of the sapropel. However, the sediment core was not protected from exposure to air during sampling, storage and analysis. The average δ^{34}S of the sulfate data of the deepest samples is $+19.5‰$, which can be considered to be representative of the actual seawater. As a consequence an estimated δ^{34}S value of the oxidized sulfide in a fifty/fifty mixture with seawater sulfate would have been $+0.25‰$ to give rise to the observed δ^{34}S value of sulfate of about $+10‰$ in the sapropel. We can exclude pyrite sulfur from oxidation because the δ^{34}S of the pyrite is far too negative and varies between -37 and $-46‰$ in the region where sulfate was contributed by oxidation of iron sulfides (Fig. 5). Also the amount of pyrite–sulfide increases steadily during the protosapropel formation while the sulfate concentrations remain rather constant (Fig. 4). The dramatic increase in sulfate concentrations is not balanced by a proportional decrease in the sulfide data. When the sulfide data decrease the sulfate data remain constant.

However, monosulfides of iron are known to be more susceptible to oxidation than pyrite (ALLER and RUDE, 1988; FOSSING and JØRGENSEN, 1990; MORSE, 1991). If the δ^{34}S of mono-sulfides were close to zero in agreement with the discussion above, an oxidation of monosulfides would accordingly be a plausible explanation for the doubling of sulfate values at a depth of 39 cm as well as the excursion of δ^{34}S-sulfate to less positive values at the same depth. The smaller decrease in δ^{34}S-sulfate at 42 cm (Fig. 5) compared to the decrease at the beginning of the true sapropel may then be explained with a smaller amount of monosulfides that could be oxidized in the protosapropel layers. Usually the monosulfides give the sediment a black color and as the layers below 39 cm are gray, they may contain lower amounts of monosulfides compared to the black layers between 39 and 32 cm, which can be used as an indication of a progressive anoxic environment. During the sapropel formation the EMDW was probably reduced with sulfate reduction occurring in the water column as well as in the sediments.

In continuous cultures of sulfate-reducing bacteria the biosynthesis of iron monosulfides was preferentially formed at pH close to 8 (HALLBERG, 1972), which is the common pH of water in the marine environment. In contrast, greigite (Fe_3S_4) and pyrite were preferentially formed at a pH close to 6, which also is in agreement with thermodynamic data (BERNER, 1964). In reduced Baltic Sea sediments the pH is slightly lower than 7. *In situ* formation of sulfides in the sediments would thus favor the formation of pyrite.

Accumulation of organic carbon during anoxic conditions

The mass flow of organic carbon from primary production to the final preservation of organic carbon in the bottom sediments is summarized in a conceptual model (Fig. 6). This is the basis for modeling of the fate of organic carbon in the E.M. A hypothetical vertical water/sediment column with a horizontal area of 1 m^2 was chosen and for simplicity it does not take into account turbidity inputs from shallow water areas into deeper basins. A total carbon budget model for the entire E.M. was not chosen for the reason that the volume of water and bathometry has changed considerably in the E.M. during the time span of sapropel formations. The sea level has increased 120 m during the last 17,000 years (FAIRBANKS, 1989), which has influenced the water volume of E.M. significantly. The Sicily sill depth has become shallower due to tectonic events during the same period (BENSON, 1972).

As a first approach a model was used to test whether anoxic deep water can be the sole explanation to the high concentrations of organic matter in the sapropel layers. Data from the literature has been combined to make an algorithm using Model Maker ver. 3 software. The model uses the present primary productivity of 40 mg $\pm$ 10 mg C m^{-2} day^{-1}. This is an average of the 12 g C m^{-2} a^{-1} given by BETHOUX (1989) and 16.7 g C m^{-2} a^{-1} by KROM *et al.* (1992). The net export of carbon from the euphotic zone is set to 25 $\pm$ 5% of the primary production. This is a compromise between the 20% given by SUESS (1980) for ocean waters and the 29% estimated for the Baltic Sea (JONSSON and CARMAN, 1994). This net export from the euphotic zone is oxidized to a large extent by the deep water as long as oxic conditions are prevailing.

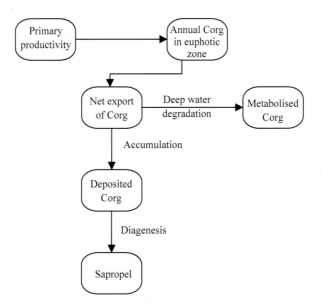

FIG. 6. Conceptual model of fate of carbon from primary productivity in the surficial waters to the final conservation as "sapropel".

In the model only $15 \pm 5\%$ of this net export of organic matter is finally deposited in the bottom sediments during oxic conditions. PASSIER et al. (1999), calculated a burial efficiency of 16% for sapropel S1, which falls well within the lower part of the range of 13–48% reported by CANFIELD (1989) for low sedimentation sites in euxinic environments. During anoxic conditions of the deep water the degradation is set to 15%, leaving 85% of the net export of carbon from the euphotic zone to be deposited in the bottom sediments. This is based on the arguments that nitrate, ferric iron and sulfate are the main electron acceptors of bacteria during anoxic conditions and from the Baltic Sea we know that sulfate-reducing bacteria play the major role. In the Baltic basin sediments, 10–20% of the deposited organic carbon is degraded by sulfate reduction (BÅGANDER, 1977). KAPLAN et al. (1963) have estimated that 16% of organic matter is used for sulfate reduction in the surface sediments of Santa Barbara Basin. Diagenetic processes further degrade the deposited organic matter. JONSSON and CARMAN (1994) have estimated the surface sediment content of organic carbon to be 2.5 times larger than the background values of Baltic deep basin cores. Thus, 40% of deposited organic matter in the sediment is finally preserved after diagenesis. A simulation over a period of 250 years with a time step of $\frac{1}{2}$ year and with increasing anoxia from year 50 is depicted in Fig. 7. The decrease in EMDW oxygen is similar to the scenario in the Baltic Sea during last century. In the E.M., however, the time period for the oxygen decrease may have been more than 1000 years. When oxygen is depleted the degradation of carbon in the deep water is set to 15%. The final "sapropel" compartment in the E.M. at present primary productivity conditions will during stagnant conditions result in an annual carbon content of $0.3 \mathrm{~g} \mathrm{~m}^{-2} \mathrm{a}^{-1}$. The confidence interval allows this value to reach $0.4 \mathrm{~g} \mathrm{~m}^{-2} \mathrm{a}^{-1}$. This is in very good agreement with observed concentrations of 0.1–0.4% in top part of the MIPC 17 core.

Several papers report an increased amount of clastic material during the formation of a sapropel. This indicates an increased river run-off and thus an increased sedimentation rate during sapropel formation. MURAT and GOT (1987) report sedimentation rates between 1.5 and

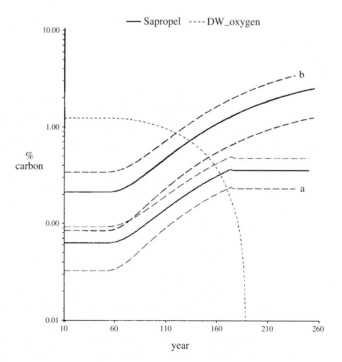

FIG. 7. Depicted graphs of the model data described in the text. Confidence intervals are depicted with long broken lines. (a) Sapropel formation with an input equal to the present primary productivity in E.M. of $40 \pm 10 \mathrm{~mg} \mathrm{~m}^{-2} \mathrm{day}^{-1}$. (b) Sapropel formation with an input of $120 \pm 50 \mathrm{~mg} \mathrm{~m}^{-2} \mathrm{day}^{-1}$.

3 cm/1000 year. It is thus probable that the sedimentation rate is closer to the upper limit given by Murat and Got. A number given in many publications (*e.g.* MANGINI and DOMINIK, 1979) is 2 cm/1000 year, which can also be calculated from data in ROSSIGNOL-STRICK *et al.* (1982). This is the sedimentation rate used in this model and agrees with the MIPC 17 site of 6 cm/3 ka. This would result in an annual sedimentation of 20 cm^3 m^{-2}. The dry bulk density of the sediment core at the MIPC 17 site is 1.3 g cm^{-3}, which is equal to the value given for several cores by NIJENHUIS *et al.* (2001). If we use this number the dry weight of the amount of annual sediment is estimated to be 26 g. To be classified as a sapropel with 2% of organic carbon this sediment should thus have an annual preservation of 0.5 g m^{-2}, which is only slightly higher than the 0.3 g m^{-2} estimated in the model. CALVERT (1983) concluded that a sapropel containing more than a few percent carbon could not be produced at the present production rate in the Mediterranean Sea.

Stagnant deep water in combination with the present E.M. primary production would thus barely give rise to a sapropel and definitely not to a sapropel with 10% or more of organic carbon, which is the case in some of the older sapropels. To explain these high numbers we need an increased primary production, which in turn requires a higher input of phosphorous to the surface waters. The limiting factor for primary production in the Eastern Mediterranean Sea is phosphorous (KROM *et al.*, 1991; SLOMP *et al.*, 2002). Part of this phosphorous will be contributed by the fresh water input, which is necessary to strengthen the density stratification of the water column. However, even if the model does not predict a sapropel, the present primary production with a stagnant EMDW could be the trigging factor for a sapropel with more than 2% organic carbon. This is because a release of nutrients from the bottom sediments will take place during such conditions. *In situ* investigations in the Baltic Sea show that phosphate is liberated from sediments during anoxic conditions (HALLBERG *et al.*, 1972). This phosphate is brought to the surface waters by upwelling. The basins of the Baltic Sea are shallower than those of the Mediterranean Sea but even so upwelling can take place during prevailing wind stress, which brings surface water in one direction compensated by deep water going in the opposite direction. In a stratified sea the pycnocline is sloped during such conditions and thus causes a shoaling of deep water (MYERS and HAINES, 1998). A decrease in the excess evaporation of E.M. would give rise to a concomitant change in the thermal balance, which will invoke a shoaling of the pycnocline (ROHLING, 1991). In order to stimulate the primary productivity in E.M. the phosphate must be brought to the photic zone, which probably meant a maximum water depth of this zone of 150 m. The fossil phytoplanktons found in the sapropels indicate a dominance of species capable of photosynthesis in deeper waters than "normal" phytoplankton (CASTRADORI, 1993). Abundance variations of *F. profunda*, the only species of calcareous nanoplankton which inhabits the lower part of the photic zone, suggest that sapropel formation was triggered by increased primary production, either confined to a deep chlorophyll maximum (DCM) or extended to a thicker layer of the photic zone (CASTRADORI, 1993). The same result can be achieved with an expanded hydrogen sulfide zone in the water column, as described by REPETA (1993). Repeta observed in the Black Sea that occasionally the hydrogen sulfide of the deeper water is shoaling and can be present in the euphotic zone. This also shifts the dominant phytoplankton populations in the water column to noncarbonate-bearing species and result in the formation of intense dark bands in the sediments (REPETA, 1993).

Increased primary productivity in combination with anoxic conditions

The first to challenge the hypothesis of bottom water stagnation as a sole explanation to sapropel formation was CALVERT (1983). According to Calvert the formation of the sapropels in E.M. cannot be explained simply by the preservation of deposited organic material under anoxic conditions at the prevailing rates of sedimentation and accumulation of organic carbon. Instead the periods of sapropel formation must coincide with periods of greatly increased marine primary productivity. There are several evidences from nitrogen isotope ratios for enhanced primary productivity during formation of E.M. sapropels (CALVERT *et al.*, 1992; STRUCK *et al.*, 2001). MORRIS *et al.* (1984) presented the occurrence of sapropel deposits from the Guinea Basin, South Atlantic. They argue that this basin most likely has not been subjected to stagnant conditions and that the organic rich deposits instead are a direct result of increased productivity in the overlaying water column. SUTHERLAND *et al.* (1984) state that periods of sapropel formation in E.M. coincided with periods of greatly increased marine primary production brought about by reversal in circulation in the Mediterranean during periods of high run-off. They base their explanation on *e.g.* the significantly higher Ba levels in the sapropel layers. VAN OS *et al.* (1991) also state that the "high Ba" argument could be used as an indication of increased primary production. They base their conclusion on millimeter scale geochemical investigations of several sapropels but also called for careful consideration before using the distribution of Ba in sediments as a strict measure of primary production. CITA (1991) reported a sharp increase in Ba concentrations that coincide with a disappearance of oxygen at the pycnocline in the water column of the present Tyro Basin in the Mediterranean, thus indicating that high Ba concentrations do not exclude anoxic conditions. WEHAUSEN and BRUMSACK (1999) have estimated the primary production (new production) to be $26–62 \, \mathrm{g \, C \, m^{-2} \, a^{-1}}$ during sapropel formation. These estimated values correspond to an input of primary production of $120 \, \mathrm{mg} \pm 50 \, \mathrm{mg \, C \, m^{-2} \, day^{-1}}$, which was used in the same model as above. The corresponding graph is depicted in Fig. 7 and shows that the estimated primary production by Wehausen and Brumsack agrees very well with a sapropel formation. According to the model, the final amount of carbon preserved in the sediment is $1–3 \, \mathrm{g \, C \, m^{-2} \, a^{-1}}$. The minimum amount required is $0.5 \, \mathrm{g \, C \, m^{-2} \, a^{-1}}$ (see calculation above). In this model the degradation of organic carbon in the DW was allowed to reach zero. The present primary production of the Baltic Sea is $160 \, \mathrm{g \, m^{-2} \, a^{-1}}$ and results in a carbon content of about 3% (JONSSON and CARMAN, 1994). The sedimentation rate in the basins of the Baltic Sea is about $1 \, \mathrm{mm \, a^{-1}}$, which is comparatively high and would correspond to an annual deposition of $30 \, \mathrm{g \, C \, m^{-2} \, a^{-1}}$.

CONCLUSIONS

The sapropel environments of E.M. and the Baltic Sea exhibit several similarities in the beginning of their depositional sequence like a rapid decrease in the $\delta^{34}S$ values of pyrite at the beginning of the oxygen depletion followed by less negative values when hydrogen sulfide is becoming more prominent in the bottom water. This is interpreted as an indication of a progressive anoxic environment.

$\delta^{34}S$ of monosulfides are close to zero at the beginning of the sapropel formation in the Baltic Sea. Indirect estimates of $\delta^{34}S$ of monosulfides in the E.M. sapropel are in agreement with the Baltic Sea data and can be used as a plausible explanation for the doubling of sulfate values at the beginning of the sapropel formation as well as the excursion to less positive $\delta^{34}S$-sulfate values during the same sedimentary sequence, also taking into consideration that monosulfides are more susceptible to oxidation than pyrite.

The different type of iron sulfides can be used to identify whether the sulfide was formed in the free water as well as in the sediment based on experimental data that monosulfides are preferentially formed at pH 8, which is the common value in the free water column.

A modeling of the fate of carbon from primary productivity to the final preservation in the sediment revealed a clear view that the today's E.M. environment will not produce a sapropel even if the bottom water was turned anoxic. The most important factor is most likely the intensity of the primary production, which must be in the order of 3–4 times higher than the present in combination with anoxic bottom water. This is also in agreement with the present situation in the Baltic Sea.

The metal ratio $10 \cdot Zn/Mn$ is a useful proxy for identification of the sapropel formation boundaries including the initial protosapropel stage in the studied core.

Acknowledgements—I thank Dr Christos Anagnostos and Aristomenis Karageorgis of the National Centre for Marine Research, Greece for providing the MIPC 17 core.

REFERENCES

ALLER R. C. and RUDE P. D. (1988) Complete oxidation of solid phase sulfides by manganese and bacteria in anoxic marine sediments. *Geochim. Cosmochim. Acta* **52**, 751–765.

ARIZTEGUI D., ASIOLI A., LOWE J. J., TRINCARDI F., VIGLIOTTI L., TAMBURINI F., CHONDROGIANNI C., ACCORSI C. A., BANDINI MAZZANTI M., MERCURI A. M., VAN DER KAARS S., MCKENZIE J. A. and OLDFIELD F. (2000) Paleoclimate and the formation of sapropel S1: inferences from Late Qarternary lacustrine and marine sequences in the central Mediterranean region. *Paleogeogr. Paleoclimatol. Paleoecol.* **158**, 215–240.

BÅGANDER L. E. (1977) In situ studies of bacterial sulfate reduction at the sediment–water interface. *Ambio Spec. Rep.* **5**, 147–155.

BARD E., DELAYGUE G., ROSTEK F., ANTONIOLI F., SILENZI S. and SCHRAG D. P. (2002) Hydrological conditions over the western Mediterranean basin during the deposition of the cold Sapropel 6 (ca. 175 kyr BP). *Earth Planet. Sci. Lett.* **202**, 481–494.

BENSON R. H. (1972) Ostracodes as indicators of threshold depth in the Mediterranean during the Pliocene. In *The Mediterranean Sea: A Natural Sedimentation Laboratory* (ed. D. J. STANLEY), pp. 63–73. Dowden, Hutchinson & Ross, Inc., Stroudsburg, PA.

BERNER R. A. (1964) Stability fields of iron minerals in anaerobic marine sediments. *J. Geol.* **72**, 826–834.

BETHOUX J. P. (1989) Oxygen consumption, new production, vertical advection and environmental evolution in the Mediterranean Sea. *Deep-Sea Res.* **36**, 769–781.

CALVERT S. E. (1983) Geochemistry of Pleistocene sapropels and associated sediments from the eastern Mediterranean. *Oceanol. Acta* **6**, 255–267.

CALVERT S. E. and PEDERSEN T. F. (1993) Geochemistry of recent oxic and anoxic marine sediments: implications for the geological record. *Mar. Geol.* **113**, 67–88.

CALVERT S. E., NIELSEN B. and FONTUGNE M. R. (1992) Evidence from nitrogen isotope ratios for enhanced productivity during formation of eastern Mediterranean sapropels. *Nature* **359**, 223–225.

CANFIELD D. E. (1989) Sulfate reduction and oxic respiration in marine sediments: implications for organic carbon preservation in euxinic environments. *Deep-Sea Res.* **36**, 121–138.

CASTRADORI D. (1993) Calcareous nanofossils and the origin of eastern Mediterranean sapropels. *Paleoceanography* **8**, 459–471.

CITA M. B. (1991) Anoxic basins of the eastern Mediterranean: an overview. *Paleoceanography* **6**, 133–141.

CITA M. B. and PODENZANI M. (1980) Destructive effects of oxygen starvation and ash falls on benthic life; a pilot study. *Quart. Res.* **13**, 230–241.

CITA M. B., VERGNAUD GRAZZINI C., CHAMLEY R. C., CIARANFI N. and D'ONOFRIO S. (1977) Paleoclimatic record of a long deep sea core from the eastern Mediterranean. *Quart. Res.* **8**, 205–235.

CITA M. B., PARISI E. and BAXIU M. (1991) Climatically modulated anoxic episodes and productivity in the Middle Pleistocene (Crotonian) of the eastern Mediterranean. *Riv. It. Paleont. Strat.* **97**, 677–693.

DE LANGE G. J., MIDDELBURG J. J. and PRUYSERS P. A. (1989) Middle and Late Quaternary depositional sequences and cycles in the eastern Mediterranean. *Sedimentology* **36**, 151–158.

DEUSER W. G. (1974) Evolution of anoxic conditions in Black Sea during Holocene. *Am. Assoc. Petrol. Geol. Bull.* **20**, 133–136.

FAIRBANKS R. G. (1989) A 17,000-year glacio-eustatic sea level record: influence of glacial melting rates on the Younger Dryas event and deep-ocean circulation. *Nature* **342**, 637–642.

FONSELIUS S. H. (1969) Hydrography of the Baltic deep basins. Fishery Board of Sweden, Series Hydrography, Rep. No. 23, 97 pp.

FOSSING H. and JØRGENSEN B. B. (1990) Oxidation and reduction of radiolabeled inorganic sulfur compounds in an estuarine sediment, Kysing Fjord, Denmark. *Geochim. Cosmochim. Acta* **54**, 2731–2742.

HALLBERG R. O. (1972) Iron and zinc sulfides formed in a continuous culture of sulfate-reducing bacteria. *Neues Jahrb. Mineral. Monatsh.* **11**, 481–500.

HALLBERG R. O. (1973) The microbiological C–N–S cycles in sediments and their effect on the ecology of the sediment–water interface. *Oikos Suppl.* **15**, 51–62.

HALLBERG R. O. (1985) Computer simulation of sulfur isotope fractionation in a closed sulfuretum. *Geomicrobiol. J.* **4**, 131–152.

HALLBERG R. O., BÅGANDER L. E., ENGVALL A.-G. and SCHIPPEL F. A. (1972) Method for studying geochemistry of sediment–water interface. *Ambio* **1**, 71–73.

HARRISON A. G. and THODE H. G. (1958) Mechanism of the bacterial reduction of sulphate from isotope fractionation studies. *Trans. Faraday Soc.* **54**, 84–92.

HOWELL M. W. and THUNELL R. C. (1992) Organic carbon accumulation in Bannock Basin: evaluating the role productivity in the formation of eastern Mediterranean sapropels. *Mar. Geol.* **103**, 461–471.

JONSSON P. and CARMAN R. (1994) Changes in deposition of organic matter and nutrients in the Baltic Sea during the twentieth century. *Mar. Pollut. Bull.* **28**, 417–426.

JORRISEN F. J., ASIOLI A., BORSETTI A. M., CAPOTONDI L., DE VISSER J. P., HILGEN F. J., ROHLING E. J., VAN DER BORG K., VERGNAUD GRAZZINI C. and ZACHARIASSE W. J. (1993) Late Quaternary central Mediterranean biochronology. *Mar. Micropaleontol.* **21**, 169–189.

JUNG M., ILMBERGER J., MANGINI A. and EMEIS K.-C. (1997) Why some Mediterranean sapropels survived burn-down (and others did not). *Mar. Geol.* **141**, 51–60.

KAPLAN I. R. and RITTENBERG S. C. (1962) The microbial fractionation of sulfur isotopes. In *Biogeochemistry of Sulfur Isotopes* (ed. M. L. JENSEN), pp. 80–93. Yale University Press, New Haven.

KAPLAN I. R., EMERY K. O. and RITTENBERG S. C. (1963) The distribution and isotopic abundance of sulfur in recent marine sediments of southern California. *Geochim. Cosmochim. Acta* **27**, 297–331.

KIDD R. B., CITA M. B. and RYAN W. B. F. (1978) Stratigraphy of eastern Mediterranean sapropel sequences recovered during DSDP Leg 42A and their paleoenvironmental significance. *Initial Rep. DSDP* **42A**, 421–443.

KROM M. D., BRENNER S., ISRAILOV L. and KRUMGALZ B. (1991) Dissolved nutrients, preformed nutrients and calculated elemental ratios in the south-east Mediterranean Sea. *Oceanol. Acta* **14**, 189–194.

KROM M. D., BRENNER S., KRESS N., NEORI A. and GORDON L. I. (1992) Nutrient dynamics and new production in a warm-core eddy from the Eastern Mediterranean Sea. *Deep-Sea Res.* **39**, 467–480.

KROUSE H. R. and TABATABAI M. A. (1986) Stable sulfur isotopes. In *Sulfur in Agriculture* (ed. M. A. TABATABAI), pp. 160–205. Am. Soc. Agron.–Soil Sci. Soc. Am., Madison, WI.

KULLENBERG B. (1952) On the salinity of the water contained in marine sediments. *Meddelanden från Oceanografiska Institutet i Göteborg* **21**, 1–38.

MANGINI A. and DOMINIK J. (1979) Late Quaternary sapropel on the Mediterranean Ridge; U-budget and evidence for low sedimentation rates. *Sedim. Geol.* **23**, 113–125.

MATTHÄUS VON W. (1979) Langzeitvariationen von Temperatur, Salzgehalt und Sauerstoffgehalt im Tiefenwasser der zentralen Ostsee. *Beitr. Meeresk.* **42**, 41–93.

MORRIS R. J., MCCARTNEY M. J. and WEAVER P. P. E. (1984) Sapropelic deposits in a sediment from the Guinea Basin, South Atlantic. *Nature* **309**, 611–614.

MORSE J. W. (1991) Oxidation kinetics of sedimentary pyrite in seawater. *Geochim. Cosmochim. Acta* **55**, 3665–3667.

MURAT A. and GOT H. (1987) Middle and late Quaternary depositional sequences and cycles in the eastern Mediterranean. *Sedimentology* **34**, 885–899.

MYERS P. G. and HAINES K. (1998) Modeling the paleocirculation of the Mediterranean: the last glacial maximum and the Holocene with emphasis on the formation of sapropel S_1. *Paleoceanography* **13**, 586–606.

NAKAI N. and JENSEN M. L. (1964) The kinetic isotope effect in the bacterial reduction and oxidation of sulfur. *Geochim. Cosmochim. Acta* **28**, 1893–1912.

NIJENHUIS I. A., BECKER J. and DE LANGE G. J. (2001) Geochemistry of coeval marine sediments in Mediterranean ODP cores and land section: implications for sapropel formation models. *Paleogeogr. Paleoclimatol. Paleoecol.* **165**, 97–112.

OLAUSSON E. (1961) Studies in deep-sea cores. *Rep. Swed. Deep-Sea Exped. 1947–1948* **8**, 337–391.

PASSIER H. F., MIDDELBURG J. J., van Os B. J. H. and DE LANGE G. J. (1996) Diagenetic pyritisation under eastern Mediterranean sapropels caused by downward sulphide diffusion. *Geochim. Cosmochim. Acta* **5**, 751–763.

PASSIER H. F., BÖTTCHER M. E. and DE LANGE G. J. (1999) Sulphur enrichment in organic matter of Eastern Mediterranean sapropels: a study of sulphur isotope partitioning. *Aquat. Geochem.* **5**, 99–118.

REPETA D. J. (1993) A high resolution historical record of Holocene anoxygenic primary production in the Black Sea. *Geochim. Cosmochim. Acta* **57**, 4337–4342.

ROHLING E. J. (1991) A simple two-layered model for shoaling of the eastern Mediterranean pycnocline due to glacio-eustatic sea level lowering. *Paleoceanography* **6**, 537–541.

ROHLING E. J. (1994) Review and new aspects concerning the formation of eastern Mediterranean sapropels. *Mar. Geol.* **122**, 1–28.

ROSSIGNOL-STRICK M. (1985) Mediterranean Quartenary sapropels, an immediate response of the African Monsoon to variation of insolation. *Paleogeogr. Paleoclimatol. Paleoecol.* **49**, 237–263.

ROSSIGNOL-STRICK M., NESTEROFF W., OLIVE P. and VERGNAUD-GRAZZINI C. (1982) After the deluge: Mediterranean stagnation and sapropel formation. *Nature* **295**, 105–110.

SARMIENTO J. L., TIMOTHY H. and TOGGWEILER J. R. (1988) Mediterranean nutrient balance and episodes of anoxia. *Global Biogeochem. Cycles* **2**, 427–444.

SLOMP C. P., THOMSON J. and DE LANGE G. J. (2002) Enhanced generation of phosphorus during formation of the most recent eastern Mediterranean sapropel (S1). *Geochim. Cosmochim. Acta* **66**, 1171–1184.

SOHLENIUS G., EMEIS K.-C., ANDRÉN E., ANDRÉN T. and KOHLY A. (2001) Development of anoxia during Holocene fresh–brackish water transition in the Baltic Sea. *Mar. Geol.* **177**, 221–242.

SOROKIN Yu. I. (1983) The Black Sea. In *Ecosystems of the World 16, Estuaries and Enclosed Seas* (ed. B. H. KETCHUM), pp. 253–292. Elsevier, Amsterdam.

STANLEY D. J., MALDONADO A. and STUCKENRATH R. (1975) Strait of Sicily depositional rates and patterns, and possible reversal of currents in the late Quaternary. *Paleogeogr. Paleoclimatol. Paleoecol.* **18**, 279–291.

STERNBECK J. and SOHLENIUS G. (1997) Authigenic sulfide and carbonate mineral formation in Holocene sediments of the Baltic Sea. *Chem. Geol.* **135**, 55–73.

STRUCK U., EMEIS K.-C., VOSS M., KROM M. D. and RAU G. H. (2001) Biological productivity during sapropel S5 formation in the eastern Mediterranean sea: evidence from stable isotopes of nitrogen and carbon. *Geochim. Cosmochim. Acta* **65**, 3249–3266.

SUESS E. (1980) Particulate organic carbon flux in the oceans; surface productivity and oxygen utilization. *Nature* **288**, 260–263.

SUTHERLAND H. E., CALVERT S. E. and MORRIS R. J. (1984) Geochemical studies of the recent sapropel and associated sediment from the Hellenic outer ridge, Eastern Mediterranean Sea. I: Mineralogy and chemical composition. *Mar. Geol.* **56**, 79–92.

TANG C. M. and SCOTT L. D. (1993) Seasonal salinity changes during Mediterranean sapropel deposition 9000 years B.P.: evidence from isotopic analyses of individual planktonic foraminifera. *Paleoceanography* **8**, 473–493.

THOMSON J., HIGGS N. C., WILSON T. R. S., CROUDACE I. W., DE LANGE G. J. and van Santvoort P. J. M. (1995) Redistribution and geochemical behaviour of redox-sensitive elements around the most recent eastern Mediterranean sapropel. *Geochim. Cosmochim. Acta* **59**, 3487–3501.

THOMSON J., MERCONE D., DE LANGE G. J. and VAN SANTVOORT P. J. M. (1999) Review of recent advances in the interpretation of eastern Mediterranean sapropel S1 from geochemical evidence. *Mar. Geol.* **153**, 77–89.

THUNELL R. C. and WILLIAMS D. F. (1983) Paleotemperature and paleosalinity history of the eastern Mediterranean during the late Quaternary. *Paleogeogr. Paleoclimatol. Paleoecol.* **44**, 23–39.

THUNELL R. C. and WILLIAMS D. F. (1989) Glacial–Holocene salinity changes in the Mediterranean Sea: hydrographic and depositional effects. *Nature* **338**, 493–496.

VAN HARTEN D. (1987) Ostracodes and the early Holocene anoxic event in the eastern Mediterranean—evidence and implications. *Mar. Geol.* **75**, 263–269.

VAN OS B. J. H., MIDDELBURG J. J. and DE LANGE G. J. (1991) Possible diagenetic mobilization of barium in sapropelic sediment from the eastern Mediterranean. *Mar. Geol.* **100**, 125–136.

VAN STEMPVOORT D. R., REARDON E. J. and FRITZ P. (1990) Fractionation of S and oxygen isotopes in sulfate by soil sorption. *Geochim. Cosmochim. Acta* **54**, 2817–2826.

VERGNAUD-GRAZZINI C., CARALP M., FAUGÈRES J.-C., GONTHIER E., GROUSSET F., PUJOL C. and SALIÈGE J.-F. (1989) Mediterranean outflow through the Strait of Gibraltar since 18000 years BP. *Oceanol. Acta* **12**, 305–324.

VISMARA-SCHILLING A. (1986) Foraminiferi bentonici profondi associati a eventi anossici del Pleistocene medio e superioer del Mdeiterraneo Orientale. *Riv. It. Paleont. Strat.* **92**, 103–147.

WEHAUSEN R. and BRUMSACK H.-J. (1999) Cyclic variations in the chemical composition of eastern Mediterranean Pliocene sediments: a key for understanding sapropel formation. *Mar. Geol.* **153**, 161–176.

Geochemical Investigations in Earth and Space Science: A Tribute to Isaac R. Kaplan
© The Geochemical Society, Publication No. 9, 2004
Editors: R.J. Hill, J. Leventhal, Z. Aizenshtat, M.J. Baedecker, G. Claypool,
R. Eganhouse, M. Goldhaber and K. Peters

Carbon–sulfur–iron relationships in the rapidly accumulating marine sediments off southwestern Taiwan

SHUH-JI KAO[1], SHIH-CHIEH HSU[1], CHORNG-SHERN HORNG[1] and KON-KEE LIU[2,3,*]

[1]Institute of Earth Sciences, Academia Sinica, Taipei, Taiwan, ROC
[2]Institute of Hydrological Sciences, National Central University, Jungli, Taoyuan 320, Taiwan, ROC
[3]National Center for Ocean Research, Taipei, Taiwan, ROC

Abstract—The continental slope off southwestern Taiwan is characterized by rapid sedimentation as a result of high denudation rates in southern Taiwan. Analyses of total organic carbon (TOC), total sulfur (TS), total iron (Fe_T) and cold acid extractable iron (Fe_A) in sediment samples from piston and box cores show unusually low sulfur contents (0.01–0.12%) in sediments against a background of moderate organic carbon contents (0.5–1.3%) and rich reactive iron (1.2–2.0%). Consequently, the degrees of sulfidation (less than 0.1) are very low for iron. Porewater sulfate concentration dropped rapidly in the top 30 cm of the piston core but decreased very gradually below. Below the top layer of relatively rapid sulfate depletion, only 5 mM of porewater sulfate was further consumed as the depth reaching 3 m below the sediment surface. The low content of reduced sulfur and low porewater sulfate depletion in the sediment column suggest very slow net sulfate reduction, which may result from competition for organic matter degradation or re-oxidation of reduced sulfur during microbial reduction of Fe(III) or Mn(IV) oxides. This may explain the exceptionally low S/C ratios ($\sim 1/30$) and low degrees of sulfidation (mostly below 0.06) found in sediments off southwestern Taiwan. In light of the high fraction of fossil carbon in river-borne sediments observed in northern Taiwan, the highly refractory terrigenous material, which probably constitutes a major fraction of sedimentary organic matter in the study area, may also limit sulfate reduction.

INTRODUCTION

THE DEPOSITION of organic carbon and reduced sulfur, mostly as pyrite, in sediments beneath oxygenated seawater often produces a chemical signature, which is a rather consistent S/C ratio of 0.36 (BERNER, 1982). The ratio has been widely used to indicate a normal marine environment in modern and ancient settings (BERNER, 1970, 1982; LEVENTHAL, 1983; BERNER and RAISWELL, 1984; RAISWELL and BERNER, 1985; DEAN and ARTHUR, 1989). The relative abundance of buried organic carbon, porewater sulfate and available iron for iron sulfidation affects sulfur and carbon cycling at the Earth's surface (RAISWELL and BERNER, 1986; HENRICHS and REEBURGH, 1987; CANFIELD, 1993; THAMDRUP and CANFIELD, 1996), which in turn controls the level of O_2 in the atmosphere over geological time (BERNER and RAISWELL, 1983; BERNER and CANFIELD, 1989). The constancy in S/C ratios is used as a basic constraint in modeling the evolution of atmospheric oxygen (BERNER, 1989, 1991). However, the S/C ratio of pyrite vs. organic carbon burial in the shelf sediments north of Taiwan is as low as 0.09 (LIN et al., 2000).

During pyrite formation, the initial and intermediate products include iron monosulfide, mackinawite and greigite (GOLDHABER and KAPLAN, 1974; MORSE et al., 1987), which are later transformed to pyrite by further reactions with intermediate sulfur species, such as elemental sulfur or polysulfides (GOLDHABER and KAPLAN, 1974; LUTHER, 1991; SCHOONEN and BARNES, 1991a,b; LYONS, 1997; BENNING et al., 2000) or directly with H_2S (HOWARTH, 1979; HURTGEN et al., 1999). However, greigite (Fe_3S_4) and pyrrhotite (Fe_7S_8)

Present address: Institute of Hydrological Sciences, National Central University, Jungli, Taoyuan 320, Taiwan, ROC.

(SWEENEY and KAPLAN, 1973; BERNER, 1984; WILKIN and BARNES, 1997), which are considered metastable and not expected to survive over long periods of geological time, had been found extensively in the thick (>3000 m) Plio-Pleistocene marine sequences of southwestern Taiwan (HORNG *et al.*, 1992a,b, 1998; TORII *et al.*, 1996; JIANG *et al.*, 2001). The depositional environment and geochemical characteristics therein should illuminate the major factors affecting the phase transformations of these authigenic iron sulfide minerals.

Oceanic islands, which make up only 3% of Earth's land area, contribute >40% of global riverine sediment flux (MILLIMAN and SYVITSKI, 1992; MILLIMAN *et al.*, 1999) and 17–35% of the particulate organic carbon entering the world's oceans (LYONS *et al.*, 2002). The importance of high-standing islands in contributing metal fluxes from land to the ocean has also been highlighted recently (CAREY *et al.*, 2002). High sediment fluxes coupled with narrow shelves around these islands result in high sediment delivery to the deep seafloor, which may serve as important geochemical sinks on a global scale (NITTROUER *et al.*, 1995). Mounting evidence points to unique geochemical pathways for carbon, sulfur and metals around these islands. However, the geochemical characteristics and diagenetic processes in such sedimentary environments are poorly known.

In this paper, we present data for total organic carbon (TOC), sulfur and iron contents in sediment samples from piston and box cores taken from the shelf and slope off southwestern Taiwan. We also present data of acid soluble iron, sulfidic iron and porewater sulfate in selected samples. The goal of this study is to illuminate the unusual inter-relationships among carbon, sulfur and iron in the rapidly accumulating sediments off southwestern Taiwan, which may serve as a typical example of sedimentation environments around high-standing islands with rapid denudation rates.

GEOLOGICAL SETTING

The oblique collision between the Luzon Arc and Chinese continental margin from the Late Miocene to the present has resulted in the formation of Taiwan Island (SUPPE, 1981; TENG, 1990). The arc–continent collision in the Taiwan orogen resulted in a foreland basin to the west and a mountain belt, namely the Central Range (Fig. 1a) to the east (COVEY, 1984). The modern depositional environment we studied is the shelf to slope area off southwestern Taiwan (Fig. 1a). Huge amount of sediments is discharged from rivers in southern Taiwan, including the Tsengwen, Erhjen and Kaoping Rivers. A narrow shelf, a broad slope extending to depths greater than 3000 m and three submarine canyon systems characterize the study area. The narrow Kaoping Shelf of about 10 km width extends along the southwestern coast from the southern tip of Taiwan to mid-way between mouths of the Kaoping and Erhjen Rivers at the southern entrance of the Taiwan Strait (Fig. 1a). Three submarine canyon systems (Fig. 1a) are the multi-headed Penghu Canyon (PH) extending southward from the Taiwan Strait, the Kaoping Canyon (KP) extending southwestward from the Kaoping River mouth and the Fangliao Canyon (FL) extending offshore from the axis of the Chaochou Fault on land (Fig. 1a). These canyons serve as the major conduits for seaward transport of sediments.

Sediment accumulation rates on the shelf and slope off southwestern Taiwan range from 0.1 to 1.44 g cm^{-2} yr^{-1} (stations marked by crosses in Fig. 1b), which were determined by TSAI and CHUNG (1989), LEE *et al.* (1993) and CHEN and LEU (1984) using ^{210}Pb, ^{10}Be and magnetic inclination methods, respectively. These rates are similar to or even higher than typical values (0.02–0.92 g cm^{-2} yr^{-1}) reported for the Washington continental shelf and canyon systems (CARPENTER *et al.*, 1982). The highest sediment accumulation rate was observed at a depth of ~300 m on the upper slope. Sedimentation decreases with increasing water depth and increasing distance from the source region (LEE *et al.*, 1993).

MATERIALS AND METHODS

Sampling of marine sediments (Fig. 1a) was conducted on Cruises 346 and 405 of R/V Ocean Researcher-I. Station locations and measurements conducted are listed in Table 1. There were 30 sampling stations over a wide range of depths from 30 to 2809 m. Box cores were

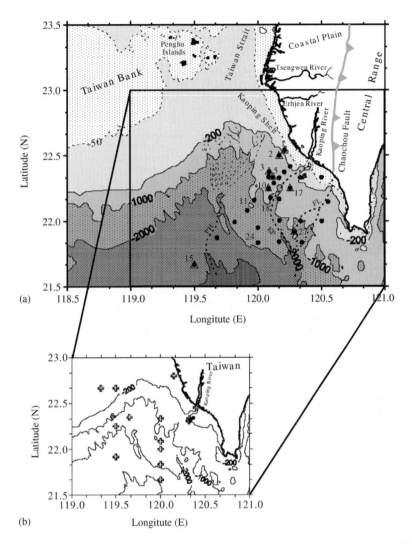

Fig. 1. (a) Study area and sampling locations (●). Triangles: down-core sampling locations. Three submarine canyons are shown in bold dashed curves and marked by PH, KP and FL, respectively, for Penghu, Kaoping and Fangliao Canyons. Isobaths are in meters. The axis of the Chaochou Fault on land is shown. (b) The station locations for reported sediment accumulation rates (see text).

taken at 29 stations. A 3 m long piston core with core liner of 6 cm inner diameter was collected at Station 17 with water depth of 670 m; once on deck it was stored vertically in a walk-in refrigerator on board the research vessel. Subcores from the box cores were obtained with plastic liners of 5.3 cm inner diameter and 60 cm length. All subcores were immediately sealed with plastic caps and frozen in a freezer. In the laboratory, surficial sediments (~2 cm in thickness) were scraped from the core top with a stainless steel spatula. Down-core samples, at a depth increment of about 2–3 cm, were sliced from subcores of eight box cores obtained at stations 1, 2, 3, 5, 10, 15, 17 and 27 (marked by triangles in Fig. 1a). The sediment samples were partitioned for three types of analyses: (1) analysis of porewater sulfate; (2) extraction and analysis of sulfidic iron from wet sediments; (3) determinations of TOC, total sulfur (TS), total iron and acid soluble iron in dried sediments.

Porewater samples were obtained from subsamples of the piston core by centrifugation within 72 h from coring. Porewater sulfate was determined by the turbidity method (APHA, 1981).

Table 1. Station locations, types of samples analyzed and measurements performed. All samples were collected by the box corer except the one collected at Station 17

Station	Latitude	Longitude	Depth (m)	Samples analyzed	Measurements[a]
1	22.541	120.210	133	Down-core	BA, seq. extract.
2	22.500	120.167	299	Down-core	BA, seq. extract.
3	22.376	120.087	690	Down-core	BA, seq. extract.
4	22.337	120.085	731	Surficial	BA
5	22.333	120.117	680	Down-core	BA
6	22.417	120.252	205	Surficial	BA
7	22.374	120.207	370	Surficial	BA
8	22.330	120.167	620	Surficial	BA
9	22.289	120.124	750	Surficial	BA
10	22.248	120.084	860	Down-core	BA
11	22.160	119.971	940	Surficial	BA, mineral extract.
12	22.082	119.916	1561	Surficial	BA
13	21.991	119.817	1508	Surficial	BA
14	21.872	119.678	2183	Surficial	BA
15	21.667	119.499	2809	Down-core	BA
16	22.332	120.333	244	Surficial	BA
17	22.250	120.253	670	Down-core (piston core)	BA, mineral extract., porewater SO_4^{2-}
18	22.180	120.099	1507	Surficial	BA, mineral extract.
19	22.167	120.167	734	Surficial	BA
20	22.333	120.501	30	Surficial	BA
21	22.167	120.335	587	Surficial	BA
22	22.003	120.170	1234	Surficial	BA
23	21.949	119.999	1207	Surficial	BA
24	21.833	120.000	1570	Surficial	BA, mineral extract.
25	22.148	120.550	633	Surficial	BA
26	22.000	120.334	812	Surficial	BA
27	21.917	120.283	1050	Down-core	BA
28	21.840	120.165	1713	Surficial	BA
29	21.999	120.503	286	Surficial	BA
30	21.836	120.345	1287	Surficial	BA

[a] "BA" indicates the bulk analyses of TOC, TS, Fe_T and Fe_A. "Mineral extract." represents magnetic mineral extraction. "Seq. extract." represents sequential extraction (see text for details).

Concentration of the resulting $BaSO_4$ colloid suspension was measured with a Shimadzu UV-150 spectrophotometer.

For the bulk analyses, freshly thawed sediments were washed with deionized distilled water to remove porewater salt and then freeze-dried. Removal of salt was critical for the accurate determination of sediment-bound sulfur because of the very low sulfur contents. This is illustrated by the comparison of TS determined for washed and non-washed sediments presented in Appendix A and Fig. A1. The dried sediments were later ground to powder by hand with mortar and pestle for analyses.

TS and TOC were analyzed with a HORIBA EMIA model CS500 analyzer equipped with a resistance furnace and an ND-IR detector. For TS, about 0.2 g of sample, mixed with 1 g of granular tin, was combusted at 1350°C in the analyzer. For TOC analysis, about 0.1 g of sediment was fumed with HCl to remove carbonates. The decarbonated samples were combusted at 1350°C in the analyzer. The precision for TS and TOC analyses is better than ± 5%.

For the determination of total iron concentration (Fe_T), 25 mg of powdered samples were digested in a mixture of 5 ml HF, 10 ml HNO_3 and 0.5 ml $HClO_4$ at approximately 200°C on a hot plate (KOKOT *et al.*, 1992). For the determination of reactive iron, different methods, which differ in the acids, concentrations, time and solid/solution ratios used for extraction, have been applied by various authors. Among these methods, extractions with dithionite or HCl solutions are the most widely used. In order to compare results with observations on sediments off northern Taiwan (HUANG and LIN, 1995), we have chosen cold 1N HCl (16 h) extractable iron

(Fe$_A$) as a proxy of the reactive iron fraction (LIN and MORSE, 1991). Some previous studies reported good agreement between extractions using sodium dithionite or 1N HCl solutions (HUERTA-DIAZ and MORSE, 1990; LEVENTHAL and TAYLOR, 1990). On the other hand, several recent studies have shown that the HCl-soluble fraction often includes iron silicates that require prolonged exposure ($\geq 10^2$ yr) to react with dissolved sulfide (CANFIELD et al., 1992; RAISWELL et al., 1994; RAISWELL and CANFIELD, 1996; LYONS, 1997). In addition, 1N cold HCl may leach an unknown portion of acid-volatile sulfide (AVS) (CORNWELL and MORSE, 1987). This operationally defined fraction may represent the maximum reactive iron in the sediments, which is the sum of the highly and the poorly reactive iron (RAISWELL et al., 1994; LYONS, 1997; RAISWELL and CANFIELD, 1998). However, goethite and hematite are only slightly soluble in cold 1N HCl solution but much more soluble in the dithionite solution (RAISWELL et al., 1994); in this respect the cold 1N HCl is less potent than the dithionite solution.

The sulfidic iron (Fe$_S$) was determined for samples from box cores collected at Stations 1, 2 and 3. Freshly thawed subcores were subsampled carefully and quickly to minimize oxidation of reduced Fe resulting from exposure to ambient oxygen. The samples were treated with a two-stage sequential extraction, which was modified from a six-stage sequential extraction proposed by CHESTER et al. (1988). The extract of the first stage includes metals bound with manganese oxides and crystalline iron oxides and more labile associations such as the exchangeable, carbonate and easily reducible fractions. The second stage extraction is aimed at pyritic iron and other metals bound with sulfide. About 1 g of wet sample was put into 50 ml centrifuge tubes and reacted with 25 ml of a reagent mixture of 0.25 M hydroxylamine hydrochloride and 25% acetic acid for 6 h at 75°C. The sample was centrifuged for 15 min at 14,000g, and the supernatant solution was kept for analysis. This fraction may represent highly reactive iron, which differs from Fe$_A$ (see "Results" and "Discussion" sections). It is noted that AVS may be leached at this stage, but the leaching efficiency is not known. The residue was subsequently reacted with the mixture of 8 ml of 30% hydrogen peroxide and 5 ml 0.01N nitric acid for 5 h at 85°C. Ten milliliters of ammonium acetate solution (pH 2) were then added with continuous stirring overnight. The supernatant was separated by the same centrifugation procedure and saved for later analyses. The fraction of the second stage is operationally defined as sulfidic iron, but it also includes some of the iron bound with organic matter. Other transition metals, including Cd, Pb, Zn, Co and Ni, were also measured to assess the relative importance of other metals in sulfidation.

Iron concentrations in extracts were measured in triplicates with a Hitachi-Z8100 atomic absorption spectrophotometer equipped with a Zeeman correction system. The flame atomizer was used for extracts from total digestions and acid extractions; the flameless graphite furnace was used for extracts of sulfidic iron. The contents of iron from sequential extractions were corrected for water contents (but not for salt contents) in sediments in order to get concentrations on a dry weight basis. Accuracy and precision for Fe analysis were checked by replicate extraction analysis ($n = 5$) of standard reference material BCSS-1, which is issued by the National Research Council, Canada and has a certified iron content of 3.287 $\pm$ 0.098%; our analytical value was 3.266 $\pm$ 0.056%, indicating good accuracy of our analyses. The relative precision for iron determination in this study is better than ± 5%.

RESULTS

Surficial sediments

TOC contents in surficial sediments off southwestern Taiwan range from 0.35 to 1.25%. Sediments relatively enriched in TOC (>0.9%; Fig. 2a) were deposited primarily in the area between the two submarine canyons (KP and FL). A plume-like patch of relatively high TOC contents extends southward from the Kaoping River mouth. This distribution pattern suggests the Kaoping River to be the main source of organic-rich sediments and implies that the terrestrial organics may have significant contribution to the sedimentary organic in the

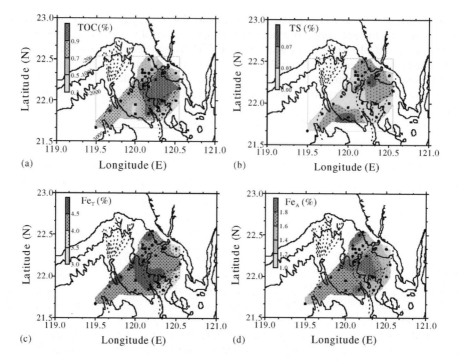

Fig. 2. Distribution of (a) total organic carbon (TOC, %), (b) total sulfur (TS, %), (c) total iron (Fe_T, %) and (d) acid-extractable iron (Fe_A, %) in surficial sediments. Isobaths and submarine canyons are also shown as in Fig. 1.

study area. Outside this patch, one high TOC value (>0.9%) was found at a water depth of 2180 m in the southwestern part of the study area. This station is located in the middle reach of another submarine canyon (PH), which originates from the Penghu Channel at the southern end of the Taiwan Strait (Fig. 2a). The relationship between TOC distribution and submarine canyons reveals that the canyons are major conduits for the transport of organic-rich, fine-grained sediments.

TS contents in surficial sediments range from 0.01 to 0.13%, with most values falling below 0.07%. This range is much lower than the TS contents (0.1–1.8%) commonly found in non-euxinic modern fine-grained marine sediments (data compiled by BERNER and RAISWELL (1983)) and the range (0.27–0.40%) observed in ancient shales (HOLSER and KAPLAN, 1966). The consistently low TS content in surficial sediments (<0.03%) has also been observed in shelf and slope sediments off northern Taiwan (HUANG and LIN, 1995; LIN *et al.*, 2002). The spatial distribution of TS (Fig. 2b) shows an area of higher values (>0.03%) near the KP River mouth, which overlaps with the high TOC patch. However, the overall distribution pattern differs considerably from that of TOC. The decoupled distribution pattern may be attributed to the relative supply of terrestrial (*i.e.* refractory) with respect to fresh marine (*i.e.* labile) organics. The higher TS values appear in shallower region suggesting higher delivery of labile organics than that in the deeper areas. It is noted that the highest TS value (0.13%) occurs at a water depth of 1570 m (Station 24 in Fig. 1a), where the TOC content is not exceptionally high. It has been recently discovered that gas hydrate may be abundant in the middle and lower slope between the KP and PH Canyons (CHI *et al.*, 1998). It is worth of further investigation whether the local high TS is associated with vents of methane gas released from gas hydrate deposits. However, this TS value is still very low compared to other marine environments, suggesting limited sulfate reduction through anaerobic methane oxidation even if it had occurred.

Fe_T contents in surficial sediments range from 3.1 to 5.4%, with most values lying above 4.0%. This range agrees with the reported range (0.8–7.6%) of total iron in shales

(DAVIS *et al.*, 1988) but falls on the higher end of the Fe_T range (1.8–4.5%) observed in most continental margin sediments (RAISWELL and CANFIELD, 1998). Higher Fe_T values (>4.5%) occur mostly at water depths between 700 and 1500 m in the middle reach of the KP submarine canyon system (Fig. 2c).

Contents of Fe_A in surficial sediments range from 0.97 to 2.40%, with most values falling between 1.2 and 2.0% (Fig. 2d). These concentrations are 50–120% higher than those found in the shelf and slope sediments in the Gulf of Mexico (LIN and MORSE, 1991). The patch of higher Fe_A occurs at shallower depths than that of higher Fe_T.

Down-core properties

The depth profile of the TOC content (Fig. 3) in the piston core shows the highest TOC content (1.07%) at the surface and an abrupt decrease to 0.6% at 7 cm. Between 7 and 50 cm, the TOC content fluctuates between 0.47 and 0.75%. Below 50 cm, TOC content shows little change. The down-core profile of sulfate in porewater (Fig. 3) shows a typical depletion trend with depth. The sulfate depletion rate is very rapid in the upper 40 cm, with a mean gradient of -0.18 mM cm^{-1}, whereas the depletion rate becomes much lower (-0.02 mM cm^{-1}) below 50 cm. TS contents measured in this piston core (Fig. 3) range from 0.02 to 0.10% with no apparent down-core trend. These values are also significantly lower than the mean value (0.22%) of modern fine-grained sediments (BERNER, 1982).

Figure 4 shows the down-core variations in TOC, TS, Fe_T and Fe_A in seven box cores. The data from the piston cores are also plotted as a reference. Also plotted are values observed in surficial sediments from the same region (see Table 1 and Fig. 2 for their localities). No common trend is found in any of these down-core distributions and the ranges of down-core variation match those of the surficial sediments. The lack of systematic diagenetic trends for chemical constituents in the upper sediment column is common in areas of rapid sedimentation near Taiwan. Similar features have also been found in many sediment cores from the southern East China Sea north of Taiwan (HUANG and LIN, 1995; LIN *et al.*, 2000, 2002). This relationship may reflect the highly dynamic depositional environment that characterizes the high-standing

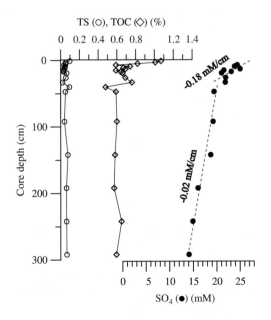

FIG. 3. The depth profiles of TOC (%), TS (%) and porewater sulfate concentration (mM) of the piston core. The mean gradients of sulfate depletion in the upper and lower sediment column are shown by numerals and illustrated by dashed lines.

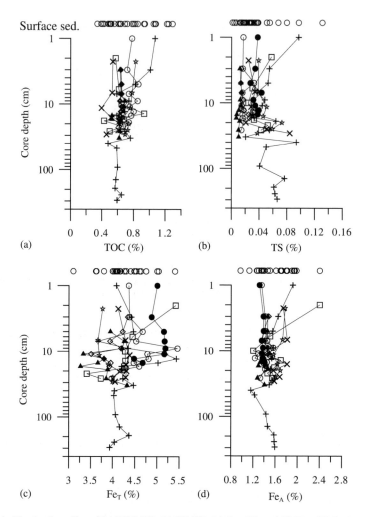

FIG. 4. The depth profiles of (a) TOC (%), (b) TS (%), (c) Fe_T (%) and (d) Fe_A (%) in the seven box cores. The depth profiles of the same properties in the piston core represented by plus signs (+) are also shown for reference. Similarly those of surficial sediments from the same area are also plotted at the top of each panel.

island. High seismic activity (WANG and SHIN, 1998) may further disturb the sedimentary sequences along the slopes.

Reactive iron

The fraction of reactive iron in total iron in the sediments may be indicated by the relationship between Fe_A and Fe_T (Fig. 5). For all surficial or down-core samples, Fe_A contents fall within a narrow band of 1/4 to 1/2 of Fe_T. The fractions of cold acid soluble iron overlaps with, but are mostly higher than, the reported average (28 ± 11%) of the total reactive iron for continental margin sediments by RAISWELL and CANFIELD (1998), who defined the total reactive iron as the sum of the hot HCl extractable iron and pyrite iron. Because some of the original reactive iron must have been converted to sulfidic iron during diagenesis, we have employed the sequential extraction to assess the sulfidic iron.

Table 2 lists contents (%) of iron from the two-stage sequential extraction of samples from three box cores. The contents of extractable iron from the first stage (Fe_{1st}) range from 0.84 to

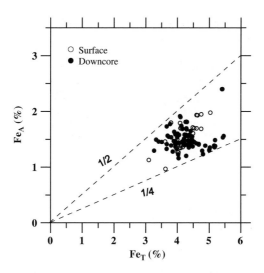

FIG. 5. Relationship between Fe_T and Fe_A in surficial sediments (○) and down-core sediments (●). Dashed lines for Fe_A/Fe_T ratios of 1/2 and 1/4 are shown as references.

1.47% with a mean value of $1.14 \pm 0.17\%$. The weight percentages of iron in the sulfidic fraction range from 0.03 to 0.106% with an average of $0.065 \pm 0.015\%$, which are much lower than values of Fe_{1st}. The concentrations of other transition metals (Cd, Pb, Zn, Co and Ni) are much lower than those of sulfidic iron by a factor of 100. In other words, Fe is the predominant metal in the sulfide fraction.

Table 2. Abundances (%) of iron in the stage-1 extract (Fe_{1st}) and the sulfidic fraction (Fe_S) obtained by sequential extraction (see text). Also listed are total sulfur (TS) and cold acid soluble iron (Fe_A) contents in salt-free sediments from three box cores and the atomic ratios of TS/Fe_S

Station	Depth (cm)	Fe_{1st} (%)	Fe_S (%)	TS (%)	TS/Fe_S (atom/atom)	Fe_A (%)
1	0–4.5	1.16	0.073	0.025	0.59	1.74
	4.5–9	1.15	0.07	0.032	0.80	1.82
	9–13.5	0.99	0.057	0.029	0.89	1.43
	13.5–18	1.14	0.075	0.030	0.70	1.81
	18–22.5	0.99	0.069	0.020	0.51	1.56
	22.5–27	1.26	0.077	0.053	1.20	1.68
	27–31.5	1.07	0.077	0.084	1.90	1.58
2	0–4	1.42	0.03	0.057	3.29	2.40
	4–8	1.47	0.055	0.031	0.99	1.51
	8–12	1.06	0.049	0.025	0.89	1.20
	12–16	1.12	0.051	0.032	1.10	1.70
	16–20	1.05	0.05	0.032	1.10	1.58
	20–24	0.94	0.106	0.054	0.89	1.46
	24–28	0.84	0.067	0.042	1.10	1.55
3	0–4.5	0.89	0.068	0.039	0.99	1.77
	4.5–9	1.11	0.063	0.040	1.10	1.73
	9–13.5	1.24	0.077	0.053	1.20	1.57
	13.5–18	1.16	0.064	0.044	1.20	1.64
	18–22.5	1.35	0.06	0.072	2.09	1.69
	22.5–27	1.16	0.068	0.058	1.50	1.60
	27–32.5	1.40	0.055	0.038	1.20	1.54
Average		1.14	0.065	0.042	1.20	1.65
S.D.		0.17	0.015	0.016	0.61	0.22

The sum of iron contents from the first and the second stage extractions has a mean value of $1.21 \pm 0.17\%$, which is slightly higher than the mean value of the total reactive iron $(0.82 \pm 0.31\%)$ in continental margin sediments (RAISWELL and CANFIELD, 1998). The similarity suggests that it is reasonable to assume that iron removed in the first stage represents the highly reactive iron. However, in most cases, sulfidic iron or pyritic iron represents a major fraction of the total reactive iron, whereas, in our case, sulfidic iron stands for but a very small fraction, indicating the excess of highly reactive iron. Compared to Fe_A contents of the same samples (Table 2), the Fe_{1st} contents are all lower, falling mostly within 60 and 80% of the former. This implies that 60–80% of Fe_A in our samples are probably highly reactive iron.

Table 2 also lists the TS contents and the atomic ratio (TS/Fe_S) between TS and the operationally defined sulfidic iron. The ratio ranges from 0.51 to 3.29 with a mean of 1.20 ± 0.61 ($n = 21$), which corresponds to a ratio of 0.69 ± 0.35 by weight. All ratios except one are less than or close to 2, which represents the pyrite stoichiometry. The average is close to the stoichiometry of greigite (Fe_3S_4) and pyrrhotite (Fe_7S_8), which have S/Fe ratios of 1.33 and 1.1, respectively. In fact, greigite and pyrrhotite have been identified in magnetic mineral extracts from these sediments by X-ray diffraction (C. S. HORNG, unpublished data). It is noted that some of the AVS may be removed during the first stage extraction because of the relatively low pH of the solution. Hence the Fe_S obtained in the second stage may be an underestimate of sulfidic iron. However, such errors are probably limited because the total AVS may represent only a small fraction of TS. The analyses of samples from the continental slope off northeastern Taiwan yield a mean AVS fraction of about 15% of Fe_T in the top 1 m of sediments (LIN *et al.*, 2002). In light of the very low TS content, the AVS-bound iron should have very limited contribution to the acid extractable iron.

DISCUSSION

The supplies of organic matter, iron and sulfate during sediment burial are the three major factors controlling the formation of iron sulfide minerals (BERNER, 1984). In the following we discuss what may have caused the unusually low sulfur contents in the sediments that we observed. First, we discuss the availability of reactive iron and then the retention of reduced sulfur.

Degree of sulfidation

The degree of pyritization has been defined to quantify the extent to which reactive iron has been fixed as pyrite (BERNER, 1970; RAISWELL *et al.*, 1988). In our case, we use the degree of sulfidation (DOS) instead (LEVENTHAL and TAYLOR, 1990). Following the definition by BOESEN and POSTMA (1988), we define the DOS as follows:

$$\text{DOS} = \text{sulfidic iron/total reactive iron}, \tag{1}$$

where the sulfidic iron includes pyrite iron and iron in monosulfides and other intermediate iron sulfides, and the "total reactive iron" is the sum of sulfidic iron and the non-sulfidic reactive iron. As mentioned earlier, the sulfidic iron obtained from the sequential extraction may have experienced possible loss of AVS in the first stage of extraction. The reactive iron may be represented either by that extracted in the first stage (Fe_{1st}) or by Fe_A. Both Fe_{1st} and Fe_A may contain small portion of AVS, other metastable sulfides, and their oxidation products that are soluble in 1N HCl (CORNWELL and MORSE, 1987); therefore, they may not be strictly non-sulfidic. But the errors are probably small.

We first illustrate the DOS for the three box cores that have been analyzed for sulfidic iron by sequential extraction (Fig. 6a). The reactive iron defined by different extraction procedures yields slightly different values of the DOS; nevertheless, the DOS is a useful index of the diagenetic condition for the formation of iron sulfide minerals. Both cases are plotted in Fig. 6a. The former definition ("total reactive iron" = $Fe_S + Fe_{1st}$) leads to DOS values in the range of 2–10% with most falling within 4–8%; the latter definition ("total reactive iron" = $Fe_S + Fe_A$) makes the DOS values shift toward lower values with most falling within 3–5%.

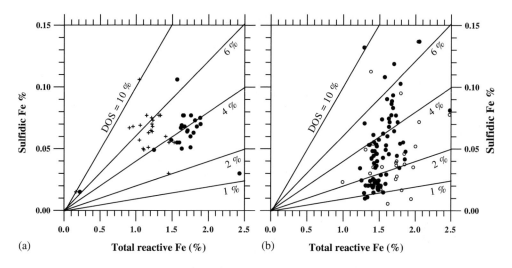

FIG. 6. (a) Scatter plot of sulfidic iron (Fe_S) vs. total reactive iron for the three box cores, which have been analyzed by sequential extraction (see text). The total reactive iron is defined as the sum of sulfidic iron and reactive iron that is represented by either Fe_{1st} or Fe_A; the former representation is marked by the plus signs ($+$) and the latter by the open circles ($\bigcirc$). Lines for degrees of iron sulfidation (DOS) from 1 to 10% are shown as references. (b) The same as (a) except that the sulfidic iron contents are calculated from TS for surficial sediments ($\bigcirc$) and down-core sediments ($\bullet$). The reactive iron is represented by Fe_A. The values of DOS in down-core sediments are somewhat higher.

HUANG and LIN (1995) reported similarly low values (mostly less than 5%) of DOP based on Fe_A in shelf sediments off northern Taiwan. When they used oxalate extractable iron to represent reactive iron, they still obtained DOP no more than 16%.

In order to assess the full range of DOS in all samples, we use the mean TS/Fe_S ratio, which is 0.69 (w/w), obtained from the three box cores to calculate the sulfidic iron contents in other samples as follows:

$$Fe_S = TS/0.69. \tag{2}$$

Similarly, we pool together Fe_S and Fe_A to represent the total reactive iron. The scatter plot of estimated Fe_S vs. "total reactive iron" illustrates the DOS of surficial and down-core samples (Fig. 6b). The contents of "total reactive iron" are in the range of 1.2–2.0%. The sulfidic iron ranges from 0.01 to 0.14%. The DOS values scatter over a more expanded range than those found for the three box cores but still all fall below 10%. It is noteworthy that all but three surficial samples have DOS less than 4%, whereas about 1/3 of the down-core sediments have DOS values higher than 4%. This is expected as a result of progressive sulfidation with depth, but the increase is limited. The DOS may be overestimated because of the fixed weight ratio (0.69) of TS/Fe_S used for the calculation of Fe_S as the ratio may increase with depth. If the maximum ratio of 1.15, namely for pyrite, is used for the calculation, the DOS values would be even smaller.

The range of DOS found in our study area is extremely low relative to that of most marine settings, including continental margins, deep seas and dysaerobic environments as defined by RAISWELL and CANFIELD (1998). The low DOS values imply that there is an excess of reactive iron available for sulfidation in the sediments of our study sites.

Retention of reduced sulfur

Apart from iron, the availability of sulfate can also limit the production of iron sulfides in sediments, once the rate of sulfate consumption exceeds the diffusive supply, especially in environments of rapid sediment accumulation. However, the measured sulfate concentrations

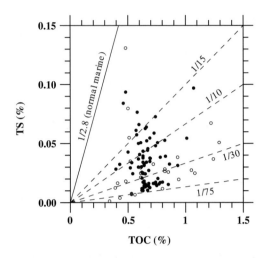

FIG. 7. Scatter plot of TS vs. TOC in surficial sediments (○) and down-core sediments (●). Lines for S/C ratios from 1/75 to 1/2.8 (normal marine ratio) are shown as references. The S/C ratios are somewhat higher in the down-core sediments, though a few surficial samples also show S/C ratios approaching the normal value.

in porewater from the piston core never dropped below 14 mM over the entire length of 3 m. In other words, 50% or more of the seawater sulfate (~ 28 mM) is available for sulfate reduction. Although porewater sulfate was not measured in other cores, the much higher concentrations of TS in non-washed samples relative to salt-free samples (Fig. A1) also indicate the availability of sulfate in these samples.

The scatter diagram of TS vs. TOC illustrates the S/C ratios of our samples (Fig. 7). Although the study area is a "normal marine" environment (*i.e.* clastic sediments overlain by oxic waters of typical oceanic salinity), most of the S/C ratios in the sediments are considerably lower than the average ratio of 1/2.8 obtained for reduced sulfur and organic carbon in sediments beneath oxygenated seawater (BERNER, 1982). By contrast, the S/C ratios in our samples are mostly lower than 1/10, typically occurring in fresh water environments. Similarly, low S/C ratios have been observed in Amazon inner shelf muds (ALLER *et al.*, 1986; ALLER and BLAIR, 1996). The authors attribute the low ratios to the oxidation power of iron oxides and reworking of sediments. Unlike the Amazon case, our samples are mostly from the slope and submarine canyons rather than the shelf and, therefore, deserve further discussion.

MORSE and BERNER (1995) discussed the factors determining S/C ratio in normal marine sediments under non-limiting conditions for iron and sulfate, which match conditions in our environment. The molar ratio between organic carbon and pyrite sulfur buried in sediment can be represented as

$$S/C = 0.5 f_{CS} f_B / (1/f_M - 1), \tag{3}$$

where f_M is the fraction of metabolized organic carbon in the TOC initially deposited, f_{CS} is the relative proportion of metabolized organic carbon decomposed via sulfate reduction and f_B is the fraction of reduced sulfur that is buried instead of being re-oxidized.

MORSE and BERNER (1995) used the relationship between sediment accumulation rate and percentage of organic carbon burial (*i.e.* burial efficiency) reported by CANFIELD (1993) to infer the fraction of metabolized organic matter, *i.e.* f_M, under different conditions. According to this relationship, f_M should range from 0.2 to 0.8 in our study area, where the sediment accumulation ranges between 0.1 and 1.4 g cm^{-2} yr^{-1}. According to studies of river-borne organic matter in the northeastern Taiwan (KAO and LIU, 1996, 1997), up to 75% of the TOC in river discharged sediments is fossil carbon, which is probably rather refractory and not readily available to sulfate-reducing bacteria. The fraction of metabolized organic carbon may be considerably less than those predicted from the relationship of CANFIELD (1993). Therefore, we adopt the lower limit (0.2) for f_M, yielding the following relationship:

$$S/C = f_{CS} f_B / 8. \tag{4}$$

Apparently, the product of f_{CS} and f_B must be small to create low S/C ratios. Assuming an S/C ratio of 1/30, which is the mean of TS/TOC of our samples, one can get a value of 0.24 for $f_{CS} f_B$. This value is considerably lower than that (>2) predicted for normal marine sediments at similar f_M (MORSE and BERNER, 1995). There are two possibilities that may result in this low value. The first possibility, opting for a normal value of f_{CS}, requires a very low value of f_B, which is caused probably by a high rate of sulfide re-oxidation; the other requires a low value of f_{CS} under a normal burial efficiency (f_B). We discuss the two possibilities below.

LIN et al. (2000, 2002) suggested a normal or even high sulfate reduction rate with high percentages (95–98% for the shelf and 13–66% for the slope) of sulfide re-oxidation in the surface layer of sediments north of Taiwan to explain the low sulfur burial rate and low S/C ratios. Similar situation may have occurred in shelf and slope sediments off southwestern Taiwan.

However, how the re-oxidation occurs is not clear. The profile of porewater sulfate from the piston core shows a very low sulfate depletion rate in the sediment column deeper than 40 cm, suggesting slow net sulfate reduction. Therefore, it is unlikely that the dissolved sulfide diffuses upward and gets oxidized near the surface. Instead the dissolved sulfide may be oxidized to sulfate by iron(III) or manganese(IV) oxides as suggested by ALLER et al. (1986) for the Amazon shelf. However, ALLER et al. (1986) have noted the lack of direct evidence for such reactions. Even if the reactions are allowed, they must compete for the dissolved Fe(III) or Mn(IV) against other reactions that also use the same ions (see discussion below). Yet another alternative is re-oxidation during reworking of the sediments as what happens in the Amazon Shelf (ALLER and BLAIR, 1996). Moderate reworking of sediments at a few localities has been observed for the shelf and canyon sediments in the same study area, but traces of benthic fauna activities are evident in the general area (YU et al., 1993). This is quite different from that observed in the inner Amazon shelf, where reworking has wiped out all traces of bioturbation (ALLER and BLAIR, 1996). Therefore, reworking does not appear to be an important process accountable for the low TS in our case.

The other possibility requires a low value of f_{CS}, which may be attributed to oxidation of organic matter by iron(III) or manganese(IV) oxides (ALLER et al., 1986). Compared with other marine environments in the world, the TOC contents (0.4–1.3%) are rather low in view of the rapidly accumulating sediments (STEIN, 1990). The low TOC contents are attributable to mineral dilution resulting from high denudation rate (KAO and LIU, 1996). The materials contributing to the TOC dilution effect are mainly from iron-bearing argillite formations rather than from carbonates. This is evidenced by the low $CaCO_3$ contents (average $< 5\%$, $n = 108$, unpublished data) in the sediments. Recent studies have suggested that such reactions are more important than previously thought (HELDER, 1989; CANFIELD, 1993; THAMDRUP and CANFIELD, 1996). The high percentage of reactive iron may have enhanced such reactions. Moreover, the higher iron reactivity may suppress sulfate reduction under substrate-limited conditions (LOVLEY and PHILLIPS, 1987), namely the low reactivity and low concentration of organic matter in our samples.

More studies are needed to examine the relative importance of different pathways proposed above in order to better understand the cycling of carbon, sulfur and iron in the unusual marine environments surrounding high-standing islands. In view of the high fraction of fossil carbon in river-discharged sediments in Taiwan (KAO and LIU, 1996, 1997), exceptionally low reactivity of organic matter, namely very low f_M, may also contribute to the low S/C ratios in our case. Geochemical characterization and quantification of the reactivity of the organic matter are needed to verify this hypothesis.

CONCLUSIONS

C–S–Fe relationships and relevant parameters in rapidly deposited sediments from southwestern Taiwan reveal occurrences of low S/C ratios relative to normal marine sediments and low degree of iron sulfidation. Abundant porewater sulfate and reactive iron in sediments could not have limited sulfidation. Low degree of sulfate depletion in the upper 3 m of slope sediments implies low net sulfate reduction rates below the surface layer. The high fraction of fossil carbon in terrigenous sediments observed elsewhere in Taiwan suggests the possibility of low reactivity of terrigenous organic matter, which may result in the low production of reduced sulfur. On the other hand, the very fast sedimentation rates may protect metabolizable organics of marine origin from oxidation at the sediment surface and favor their burial in the sediments. If this is the case, sulfate reduction should not be limited by the lack of metabolizable organic matter. Instead the low S/C ratios may stem from re-oxidation of reduced sulfur or consumption of organic matter during reduction of iron(III) oxides. The rich supply of reactive iron precludes the possibility of diffusive loss of H_2S by re-oxidation near the surface of the sediment column. Therefore, the very high contents (1–2.5%) of reactive iron may induce an active microbial utilization of iron(III) or manganese(IV) oxides as electron acceptors, which may re-oxidize reduced sulfur or out-compete sulfate for organic carbon, producing the observed low S/C ratios. Further studies are warranted to test our hypotheses by quantifying the importance of Fe(III) and Mn(IV) reduction and characterizing the reactivity of sedimentary organic matter.

Acknowledgements—We thank Juo-Hou Lee previously at National Taiwan University for assistance with sampling and chemical analyses. This work benefits from constructive comments and rich suggestions from T. W. Lyons and R. Raiswell. This study was supported by the National Science Council of the Republic of China under grants NSC 92-2116-M-001-003 and NSC 91-2611-M-002-005-OP3. This is IESAS contribution no. IESAS-910, NCU-IHS contribution no. 9 and NCOR contribution no. 80.

REFERENCES

ALLER R. C. and BLAIR N. E. (1996) Sulfur diagenesis and burial on the Amazon shelf: major control by physical sedimentation processes. *Geo-Mar. Lett.* **16**, 3–10.

ALLER R. C., MACKIN J. E. and COX R. T. (1986) Diagenesis of Fe and S in Amazon inner shelf muds: apparent dominance of Fe reduction and implications for the genesis of ironstones. *Cont. Shelf Res.* **6**, 263–289.

APHA (1981) *Standard Methods for the Examination of Water and Wastewater*, 15th Ed. American Public Health Association, Washington, DC.

BENNING G. L., WILKIN R. T. and BARNES H. L. (2000) Reaction pathways in the Fe–S system below 100°C. *Chem. Geol.* **167**, 25–51.

BERNER R. A. (1970) Sedimentary pyrite formation. *Am. J. Sci.* **268**, 1–23.

BERNER R. A. (1982) Burial of organic carbon and pyrite sulfur in the modern ocean: its geochemical and environmental significance. *Am. J. Sci.* **282**, 451–473.

BERNER R. A. (1984) Sedimentary pyrite formation: an update. *Geochim. Cosmochim. Acta* **48**, 605–615.

BERNER R. A. (1989) A biogeochemical cycles of carbon and sulfur and their effect on atmospheric oxygen over Phanerozoic time. *Paleogeogr. Paleoclimatol. Paleoecol.* **75**, 97–122.

BERNER R. A. (1991) A model for atmospheric CO_2 over Phanerozoic time. *Am. J. Sci.* **291**, 339–376.

BERNER R. A. and CANFIELD D. (1989) A new model for atmospheric oxygen over Phanerozoic time. *Am. J. Sci.* **289**, 333–361.

BERNER R. A. and RAISWELL R. (1983) Burial of organic carbon and pyrite sulfur in sediments over Phanerozoic time: a new theory. *Geochim. Cosmochim. Acta* **47**, 855–862.

BERNER R. A. and RAISWELL R. (1984) C/S method for distinguishing freshwater from marine sedimentary rocks. *Geology* **12**, 365–368.

BOESEN C. and POSTMA D. (1988) Pyrite formation in anoxic environments of the Baltic. *Am. J. Sci.* **288**, 575–603.

CANFIELD D. E. (1993) Organic matter oxidation in marine sediments. In *Interaction of C, N, P and S Biogeochemical Cycles* (eds. R. WOLLAST, F. T. MACKENZIE and L. CHOU), pp. 333–363. Springer, Berlin.

CANFIELD D. E., RAISWELL R. and BOTTRELL S. (1992) The reactivity of sedimentary iron minerals toward sulfide. *Am. J. Sci.* **292**, 659–683.

CAREY A. E., NEZAT C. A., LYONS W. B., KAO S.-J., HICKS D. M. and OWEN J. S. (2002) Trace metal fluxes to the ocean: the importance of high-standing oceanic islands. *Geophys. Res. Lett.* **29**(23), 2099 (doi: 10.1029/2002GL015690).

CARPENTER R., PETERSON M. L. and BENNETT J. T. (1982) Pb-derived sediment accumulation and mixing rates for the Washington continental slope. *Mar. Geol.* **48**, 135–164.

CHEN M.-P. (1983) Physical properties and depositional environments of continental slope sediments, southern Taiwan Strait. *Acta Oceanogr. Taiwanica* **14**, 42–63.

CHEN M.-P. and LEU T.-C. (1984) Paleomagnetic inclination variations in Taiwan strait after Holocene transgression. *Acta Oceanogr. Taiwanica* **15**, 53–70.

CHESTER R., THOMAS A., LIN F.-J., BASAHAM A. S. and JACINTO G. (1988) The solid state speciation of copper in surface water particulates and oceanic sediments. *Mar. Chem.* **24**, 261–292.

CHI W.-C., REED D.-L., LIU C.-S. and LUNDBERG N. (1998) Distribution of the bottom-simulating reflector in the offshore Taiwan collision zone. *Terrest. Atmos. Ocean. Sci.* **9**, 779–794.

CORNWELL J. C. and MORSE J. W. (1987) The characterization of iron sulfide minerals in anoxic marine sediments. *Mar. Chem.* **22**, 193–206.

COVEY M. C. (1984) Sedimentary and tectonic evolution of the western Taiwan foredeep. Ph.D. Thesis, Princeton University.

DAVIS H. R., BYERS C. W. and DEAN W. E. (1988) Pyrite formation in the lower Cretaceous Mowry shale: effect of organic matter type and reactive iron content. *Am. J. Sci.* **288**, 873–890.

DEAN W. E. and ARTHUR M. A. (1989) Iron–sulfur–carbon relationships in organic-carbon-rich sequences: I. Cretaceous Western Interior Seaway. *Am. J. Sci.* **289**, 708–743.

GOLDHABER M. B. and KAPLAN I. R. (1974) The sulfur cycle. In *The Sea.* (ed. E. D. GOLDBERG), Vol. 5, pp. 569–656. Wiley, New York.

HELDER W. (1989) Early diagenesis and sediment–water exchange in the Savu Basin (Eastern Indonesia). *Neth. J. Sea Res.* **24**, 555–572.

HENRICHS S. M. and REEBURGH W. S. (1987) Anaerobic mineralization of marine sediment organic matter: rates and the role of anaerobic processes in the oceanic carbon economy. *Geomicrobial. J.* **5**, 191–237.

HOLSER W. T. and KAPLAN I. R. (1966) Isotope geochemistry of sedimentary sulfates. *Chem. Geol.* **1**, 93–135.

HORNG C.-S., CHEN J.-C. and LEE T.-Q. (1992a) Variations in magnetic minerals from two Plio-Pleistocene marine-deposited sections, southwestern Taiwan. *J. Geol. Soc. China* **35**, 323–335.

HORNG C.-S., LAJ C., LEE T.-Q. and CHEN J.-C. (1992b) Magnetic characteristics of sedimentary rocks from the Tsengwen Hsi and Erhjen Hsi sections in southwestern Taiwan. *Terrest. Atmos. Ocean. Sci.* **3**, 519–532.

HORNG C. S., TORII M., SHEA K. S. and KAO S. J. (1998) Inconsistent magnetic polarities between greigite- and pyrrhotite/magnetite-bearing marine sediments from the Tsailiao-chi section, southwestern Taiwan. *Earth Planet. Sci. Lett.* **164**, 467–481.

HOWARTH R. W. (1979) Pyrite: its rapid formation in a salt marsh and its importance in ecosystem metabolism. *Science* **203**, 49–51.

HUANG K.-M. and LIN S. (1995) The carbon–sulfide–iron relationship and sulfate reduction rate in the East China Sea continental shelf sediments. *Geochem. J.* **29**, 301–315.

HUERTA-DIAZ M. A. and MORSE J. W. (1990) A quantitative method for determination of trace metal concentrations in sedimentary pyrite. *Mar. Chem.* **29**, 119–144.

HURTGEN M. T., LYONS T. W., INGALL E. D. and CRUSE A. M. (1999) Anomalous enrichments of iron monosulfide in euxinic marine sediments and the role of H_2S in iron sulfide transformations: examples from Effingham Inlet, Orca Basin and the Black Sea. *Am. J. Sci.* **299**, 556–588.

JIANG W.-T., HORNG C.-S., ROBERTS A. P. and PEACOR D. R. (2001) Contradictory magnetic polarities in sediments and variable timing of neoformation of authigenic greigite. *Earth Planet. Sci. Lett.* **193**, 1–12.

KAO S.-J. and LIU K.-K. (1996) Particulate organic carbon export from a subtropical mountainous river (Lanyang-Hsi) in Taiwan. *Limnol. Oceanogr.* **41**, 1749–1757.

KAO S.-J. and LIU K.-K. (1997) Fluxes of dissolved and nonfossil particulate organic carbon from an Oceania small river (Lanyang-Hsi) in Taiwan. *Biogeochemistry* **39**, 255–269.

KOKOT S., KING G., LELLER H. R. and MASSART D. L. (1992) Application of chemometrics for the selection of microwave digestion procedures. *Anal. Chim. Acta* **268**, 81–94.

LEE T., YOU C.-F. and LIU T.-K. (1993) Model-dependent ^{10}Be sedimentation rates for the Taiwan Strait and their tectonic significance. *Geology* **21**, 423–426.

LEVENTHAL J. S. (1983) An interpretation of carbon and sulfur relationships in Black Sea sediments as indicators of environments of deposition. *Geochim. Cosmochim. Acta* **47**, 133–137.

LEVENTHAL J. S. and TAYLOR C. (1990) Comparison of methods to determine the degree of pyritisation. *Geochim. Cosmochim. Acta* **54**, 2621–2625.

LIN S. and MORSE J. W. (1991) Sulfate reduction and iron sulfide mineral formation in the Gulf of Mexico anoxic sediments. *Am. J. Sci.* **291**, 55–89.

LIN S., HUANG K.-M. and CHEN S.-K. (2000) Organic carbon deposition and its control on iron sulfide formation of the southern East China Sea continental shelf sediments. *Cont. Shelf Res.* **20**, 619–635.

LIN S., HUANG K.-M. and CHEN S.-K. (2002) Sulfate reduction and iron sulfide mineral formation in the southern East China Sea continental slope sediment. *Deep Sea Res. I* **49**, 1837–1852.

LOVLEY D. R. and PHILLIPS E. J. P. (1987) Competitive mechanisms for inhibition of sulfate reduction and methane production in the zone of ferric iron reduction in sediments. *Appl. Environ. Microbiol.* **53**, 2636–2641.

LUTHER G. W. III (1991) Pyrite synthesis via polysulphide compounds. *Geochim. Cosmochim. Acta* **55**, 2839–2849.

LYONS T. W. (1997) Sulfur isotopic trends and pathways of iron sulfide formation in upper Holocene sediments of the anoxic Black Sea. *Geochim. Cosmochim. Acta* **61**, 3367–3382.

LYONS W. B., NEZAT C. A., CAREY A. E. and HICKS D. M. (2002) Organic carbon fluxes to the ocean from high-standing islands. *Geology* **30**(5), 443–446.

MILLIMAN J. D. and SYVITSKI J. P. M. (1992) Geomorphic/tectonic control of sediment discharge to the ocean: the importance of small mountainous rivers. *J. Geol.* **91**, 1–21.

MILLIMAN J. D., FARNSWORTH K. L. and ALBERTIN C. S. (1999) Flux and fate of fluvial sediments leaving large islands in the East Indies. *J. Sea Res.* **41**, 97–107.

MORSE J. W. and BERNER R. A. (1995) What determines sedimentary C/S ratio? *Geochim. Cosmochim. Acta* **59**, 1073–1077.

MORSE J. W., MILLERO F. J., CORNWELL J. C. and RICKARD D. (1987) The chemistry of the hydrogen sulfide and iron sulfide system in natural waters. *Earth Sci. Rev.* **24**, 1–42.

NITTROUER C. A., BRUNSKILL G. J. and FIGUEIREDO A. G. (1995) Importance of tropical coastal environments. *Geo-Mar. Lett.* **15**, 121–126.

RAISWELL R. and BERNER R. A. (1985) Pyrite formation in euxinic and semi-euxinic sediments. *Am. J. Sci.* **285**, 710–724.

RAISWELL R. and BERNER R. A. (1986) Pyrite and organic-matter in Phanerozoic normal marine shales. *Geochim. Cosmochim. Acta* **50**, 1967–1976.

RAISWELL R. and CANFIELD D. E. (1996) Rates of reaction between silicate iron and dissolved sulfide in Peru Margin sediments. *Geochim. Cosmochim. Acta* **60**, 2777–2787.

RAISWELL R. and CANFIELD D. E. (1998) Sources of iron for pyrite formation in marine sediments. *Am. J. Sci.* **298**, 219–245.

RAISWELL R., BUCKLEY F., BERNER R. A. and ANDERSON T. F. (1988) Degree of pyritization of iron as a paleoenvironmental indicator. *J. Sedim. Petrol.* **58**, 812–819.

RAISWELL R., CANFIELD D. E. and BERNER R. A. (1994) A comparison of iron extraction methods for the determination of degree of pyritisation and the recognition of iron-limited pyrite formation. *Chem. Geol.* **111**, 101–110.

SCHOONEN M. A. A. and BARNES H. L. (1991a) Reactions forming pyrite and marcasite from solution: I. Nucleation of FeS$_2$ below 100°C. *Geochim. Cosmochim. Acta* **55**, 1495–1504.

SCHOONEN M. A. A. and BARNES H. L. (1991b) Reactions forming pyrite and marcasite from solution: II. Via FeS precursors below 100°C. *Geochim. Cosmochim. Acta* **55**, 1505–1514.

STEIN R. (1990) Organic carbon content/sedimentation rate relationship and its paleoenvironmental significance for marine sediments. *Geo-Mar. Lett.* **10**, 37–44.

SUPPE J. (1981) Mechanics of mountain building and metamorphism in Taiwan. *Mem. Geol. Soc. China* **4**, 67–89.

SWEENEY R. E. and KAPLAN I. R. (1973) Pyrite framboid formation: laboratory synthesis and marine sediments. *Econ. Geol.* **5**, 618–634.

TENG L.-S. (1990) Geotectonic evolution of late Cenozoic arc–continent collision in Taiwan. *Tectonophysics* **183**, 57–76.

THAMDRUP B. and CANFIELD D. E. (1996) Pathways of carbon oxidation in continental margin sediments off central Chile. *Limnol. Oceanogr.* **41**, 1629–1650.

TORII M., FUKUMA K., HORNG C.-S. and LEE T.-Q. (1996) Magnetic discrimination of pyrrhotite- and greigite-bearing sediment samples. *Geophys. Res. Lett.* **23**, 1813–1816.

TSAI S.-W. and CHUNG Y. (1989) Pb-210 in the sediments of Taiwan Strait. *Acta Oceanogr. Taiwanica* **22**, 1–13.

WANG C.-Y. and SHIN T.-C. (1998) Illustrating 100 years of Taiwan seismicity. *Terrest. Atmos. Ocean. Sci.* **9**, 589–614.

WILKIN R. T. and BARNES H. L. (1997) Pyrite formation in an anoxic estuarine basin. *Am. J. Sci.* **297**, 620–650.

YU H.-S., AUSTER P. J. and COOPER R. A. (1993) Surface geology and biology at the head of Kaoping Canyon off southwestern Taiwan. *Terrest. Atmos. Ocean. Sci.* **4**, 441–455.

APPENDIX A: EFFECT OF WASHING ON TOTAL SULFUR CONTENTS

In order to demonstrate the necessity of washing the sediment samples for the determination of sediment-bound sulfur in sulfur-deficient sediments, we performed a comparison of TS measurements for salt-free (washed) samples and non-washed samples. The results are shown in Fig. A1. Compared to salt-free samples, all non-washed samples except one show significant enrichment of sulfur. As expected, surficial sediments show higher degrees of enrichment than down-core samples as a result of higher porosity and higher sulfate concentration in the former. The amounts of enrichment ranges from 0.05 to 0.25% for surficial samples and 0 to 0.12% for down-core samples.

We use the following calculation to illustrate the effect of porewater sulfate on the TS content in non-washed samples. The sulfur content contributed by porewater sulfate may be expressed as follows:

$$S\,(\%) = \phi[SO_4^{2-}] \times 32\,(\text{g mol}^{-1})/(\rho_s(1 - \phi) + \rho_w \phi S) \times 100\%,$$

where ϕ is the porosity, ρ_s is the specific gravity of sediments, ρ_w is the density of porewater and S is the salinity of porewater. The specific gravity of solid sediments was in the range of 2.65–2.78 (CHEN, 1983) in the study area. We assume the porosity of 0.55–0.85, the salinity of 35‰ and density of porewater of 1.03 g ml^{-1}, then we get the sulfur content of 0.04–0.18% for porewater sulfate concentration of 28 mM. This range roughly matches those obtained for the surficial sediments. Because the TS contents are very low in the samples, the porewater sulfate may constitute a major fraction of sulfur in non-washed sediments. Hence, it is critical to remove salt by washing for the determination of sediment-bound sulfur for our samples, which are depleted in sulfur.

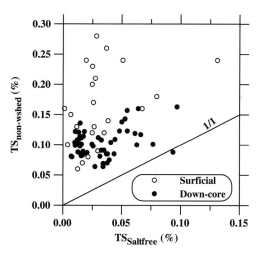

FIG. A1. Scatter plot of TS values in non-washed samples vs. those in salt-free (washed) samples: surficial sediments (○) and down-core sediments (●). Line of 1/1 is shown as a reference. The significant enrichment of TS in non-washed samples indicates the necessity of removing salt for determination of sediment-bound sulfur in sulfur poor sediments.

Author Index

Subject Index